Student Solutions Manual

for

Kaufmann and Schwitters's

Elementary and Intermediate Algebra

A Combined Approach

Fourth Edition

Karen L. Schwitters
Seminole Community College

Laurel Fischer

Australia • Canada • Mexico • Singapore • Spain • United Kingdom • United States

Cover image: Rowan Moore

Printed in the United States of America
1 2 3 4 5 6 7 09 08 07 06 05

Printer: Thomson/West

0-534-49064-6

For more information about our products, contact us at:
Thomson Learning Academic Resource Center
1-800-423-0563

For permission to use material from this text or product, submit a request online at **http://www.thomsonrights.com.**
Any additional questions about permissions can be submitted by email to **thomsonrights@thomson.com.**

Thomson Higher Education
10 Davis Drive
Belmont, CA 94002-3098
USA

Asia (including India)
Thomson Learning
5 Shenton Way
#01-01 UIC Building
Singapore 068808

Australia/New Zealand
Thomson Learning Australia
102 Dodds Street
Southbank, Victoria 3006
Australia

Canada
Thomson Nelson
1120 Birchmount Road
Toronto, Ontario M1K 5G4
Canada

UK/Europe/Middle East/Africa
Thomson Learning
High Holborn House
50–51 Bedford Row
London WC1R 4LR
United Kingdom

Latin America
Thomson Learning
Seneca, 53
Colonia Polanco
11560 Mexico
D.F. Mexico

Spain (including Portugal)
Thomson Paraninfo
Calle Magallanes, 25
28015 Madrid, Spain

Table of Contents

Chapter 10: Exponents and Radicals

Chapter 11: Quadratic Equations and Inequalities

Chapter 12: Coordinate Geometry: Lines, Parabolas, Circles, Ellipses, and Hyperbolas

Chapter 13: Functions

Chapter 1 Some Basic Concepts of Arithmetic and Algebra

PROBLEM SET 1.1 Numerical and Algebraic Expressions

1. $9 + 14 - 7$
$23 - 7$
16

3. $7(14 - 9)$
$7(5)$
35

5. $16 + 5 \cdot 7$
$16 + 35$
51

7. $4(12 + 9) - 3(8 - 4)$
$4(21) - 3(4)$
$84 - 12$
72

9. $4(7) + 6(9)$
$28 + 54$
82

11. $6 \cdot 7 + 5 \cdot 8 - 3 \cdot 9$
$42 + 40 - 27$
$82 - 27$
55

13. $(6 + 9)(8 - 4)$
$(15)(4)$
60

15. $6 + 4[3(9 - 4)]$
$6 + 4[3(5)]$
$6 + 4(15)$
$6 + 60$
66

17. $16 \div 8 \cdot 4 + 36 \div 4 \cdot 2$
$2 \cdot 4 + 36 \div 4 \cdot 2$
$8 + 36 \div 4 \cdot 2$
$8 + 9 \cdot 2$
$8 + 18$
26

19. $\frac{8 + 12}{4} - \frac{9 + 15}{8}$
$\frac{20}{4} - \frac{24}{8}$
$5 - 3$
2

21. $56 - [3(9 - 6)]$
$56 - [3(3)]$
$56 - (9)$
$56 - 9$
47

23. $7 \cdot 4 \cdot 2 \div 8 + 14$
$28 \cdot 2 \div 8 + 14$
$56 \div 8 + 14$
$7 + 14$
21

25. $32 \div 8 \cdot 2 + 24 \div 6 - 1$
$4 \cdot 2 + 4 - 1$
$8 + 4 - 1$
$12 - 1$
11

27. $4 \cdot 9 \div 12 + 18 \div 2 + 3$
$36 \div 12 + 18 \div 2 + 3$
$3 + 18 \div 2 + 3$
$3 + 9 + 3$
$12 + 3$
15

29. $\frac{6(8 - 3)}{3} + \frac{12(7 - 4)}{9}$
$\frac{6(5)}{3} + \frac{12(3)}{9}$
$\frac{30}{3} + \frac{36}{9}$
$10 + 4$
14

Problem Set 1.1

31. $83 - \frac{4(12-7)}{5}$

$83 - \frac{4(5)}{5}$

$83 - \frac{20}{5}$

$83 - 4$

79

33. $\frac{4 \cdot 6 + 5 \cdot 3}{7 + 2 \cdot 3} + \frac{7 \cdot 9 + 6 \cdot 5}{3 \cdot 5 + 8 \cdot 2}$

$\frac{24 + 15}{7 + 6} + \frac{63 + 30}{15 + 16}$

$\frac{39}{13} + \frac{93}{31}$

$3 + 3$

6

35. $7x + 4y$ for $x = 6$, $y = 8$

$7(6) + 4(8)$

$42 + 32$

74

37. $16a - 9b$ for $a = 3$, $b = 4$

$16(3) - 9(4)$

$48 - 36$

12

39. $4x + 7y + 3xy$ for $x = 4$, $y = 9$

$4(4) + 7(9) + 3(4)(9)$

$16 + 63 + 108$

$79 + 108$

187

41. $14xz + 6xy - 4yz$ for $x = 8$, $y = 5$, $z = 7$

$14(8)(7) + 6(8)(5) - 4(5)(7)$

$784 + 240 - 140$

$1024 - 140$

884

43. $\frac{54}{n} + \frac{n}{3}$ for $n = 9$

$\frac{54}{9} + \frac{9}{3}$

$6 + 3$

9

45. $\frac{y + 16}{6} + \frac{50 - y}{3}$ for $y = 8$

$\frac{8 + 16}{6} + \frac{50 - 8}{3}$

$\frac{24}{6} + \frac{42}{3}$

$4 + 14$

18

47. $(x + y)(x - y)$ for $x = 8$, $y = 3$

$(8 + 3)(8 - 3)$

$(11)(5)$

55

49. $(5x - 2y)(3x + 4y)$ for $x = 3$, $y = 6$

$[5(3) - 2(6)][3(3) + 4(6)]$

$[15 - 12][9 + 24]$

$(3)(33)$

99

51. $6 + 3[2(x + 4)]$ for $x = 7$

$6 + 3[2(7 + 4)]$

$6 + 3[2(11)]$

$6 + 3(22)$

$6 + 66$

72

53. $81 - 2[5(n + 4)]$ for $n = 3$

$81 - 2[5(3 + 4)]$

$81 - 2[5(7)]$

$81 - 2[35]$

$81 - 70$

11

55. $\frac{bh}{2}$, for $b = 8$, $h = 12$

$\frac{8(12)}{2}$

$\frac{96}{2}$

48

57. $\frac{bh}{2}$, for $b = 7$, $h = 6$

$\frac{7(6)}{2}$

$\frac{42}{2}$

21

59. $\frac{bh}{2}$, for $b = 16$, $h = 5$

$\frac{16(5)}{2}$

$\frac{80}{2}$

40

61. $\frac{h(b_1 + b_2)}{2}$, for $h = 17$, $b_1 = 14$, $b_2 = 6$

$\frac{17(14+6)}{2}$

$\frac{17(20)}{2}$

$\frac{340}{2}$

170

63. $\frac{h(b_1 + b_2)}{2}$, for $h = 8$, $b_1 = 17$, $b_2 = 24$

$\frac{8(17+24)}{2}$

$\frac{8(41)}{2}$

$\frac{328}{2}$

164

65. $\frac{h(b_1 + b_2)}{2}$, for $h = 18$, $b_1 = 6$, $b_2 = 11$

$\frac{18(6+11)}{2}$

$\frac{18(17)}{2}$

$\frac{306}{2}$

153

Further Investigations

71. $36 + 12 \div 3 + 3 + 6 \cdot 2$
becomes $36 + 12 \div (3 + 3) + 6 \cdot 2$
in order to equal 50.

73. $36 + 12 \div 3 + 3 + 6 \cdot 2$
becomes $36 + (12 \div 3 + 3) + 6 \cdot 2$
in order to equal 55.

PROBLEM SET 1.2 Prime and Composite Numbers

1. Since $8(7) = 56$, "8 divides 56" is a true statement.

3. Since $6(9) = 54$, "6 does not divide 54" is a false statement.

5. Since $8(12) = 96$, "96 is a multiple of 8" is a true statement.

7. Since there is no whole number, k, such that $4(k) = 54$, "54 is not a multiple of 4" is a true statement.

9. Since $4(36) = 144$, "144 is divisible by 4" is a true statement.

11. Since there is no whole number, k, such that $3(k) = 173$, "173 is divisible by 3" is a false statement.

13. Since $11(13) = 143$, "11 is a factor of 143" is a true statement.

15. Since there is no whole number, k, such that $9(k) = 119$, "9 is a factor of 119" is a false statement.

17. Since $3(19) = 57$ and 3 is a prime number, "3 is a prime factor of 57" is a true statement.

19. Since $2(2) = 4$, 4 is not a prime number, therefore "4 is a prime factor of 48" is a false statement.

21. Since 53 is only divisible by itself and 1, it is a prime number.

23. Since 59 is only divisible by itself and 1, it is a prime number.

25. Since $91 = 7(13)$, it is a composite number.

27. Since 89 is only divisible by itself and 1, it is a prime number.

29. Since $111 = 3(37)$, it is a composite number.

31. $26 = 2 \cdot 13$

33. $36 = 4 \cdot 9 = 2 \cdot 2 \cdot 3 \cdot 3$

35. $49 = 7 \cdot 7$

37. $56 = 8 \cdot 7 = 2 \cdot 4 \cdot 7 = 2 \cdot 2 \cdot 2 \cdot 7$

39. $120 = 3 \cdot 40 = 3 \cdot 4 \cdot 10 =$
$3 \cdot 2 \cdot 2 \cdot 2 \cdot 5 = 2 \cdot 2 \cdot 2 \cdot 3 \cdot 5$

41. $135 = 5 \cdot 27 = 5 \cdot 3 \cdot 9 =$
$5 \cdot 3 \cdot 3 \cdot 3 = 3 \cdot 3 \cdot 3 \cdot 5$

43. $12 = 2 \cdot 2 \cdot 3$
$16 = 2 \cdot 2 \cdot 2 \cdot 2$
The greatest common factor is $2 \cdot 2 = 4$.

45. $56 = 2 \cdot 2 \cdot 2 \cdot 7$
$64 = 2 \cdot 2 \cdot 2 \cdot 2 \cdot 2 \cdot 2$
The greatest common factor is $2 \cdot 2 \cdot 2 = 8$.

47. $63 = 3 \cdot 3 \cdot 7$
$81 = 3 \cdot 3 \cdot 3 \cdot 3$
The greatest common factor is $3 \cdot 3 = 9$.

49. $84 = 2 \cdot 2 \cdot 3 \cdot 7$
$96 = 2 \cdot 2 \cdot 2 \cdot 2 \cdot 2 \cdot 3$
The greatest common factor is $2 \cdot 2 \cdot 3 = 12$.

51. $36 = 2 \cdot 2 \cdot 3 \cdot 3$
$72 = 2 \cdot 2 \cdot 2 \cdot 3 \cdot 3$
$90 = 2 \cdot 3 \cdot 3 \cdot 5$
The greatest common factor is $2 \cdot 3 \cdot 3 = 18$.

53. $48 = 2 \cdot 2 \cdot 2 \cdot 2 \cdot 3$
$60 = 2 \cdot 2 \cdot 3 \cdot 5$
$84 = 2 \cdot 2 \cdot 3 \cdot 7$
The greatest common factor is $2 \cdot 2 \cdot 3 = 12$.

55. $6 = 2 \cdot 3$
$8 = 2 \cdot 2 \cdot 2$
The least common multiple is $2 \cdot 2 \cdot 2 \cdot 3 = 24$.

57. $12 = 2 \cdot 2 \cdot 3$
$16 = 2 \cdot 2 \cdot 2 \cdot 2$
The least common multiple is $2 \cdot 2 \cdot 2 \cdot 2 \cdot 3 = 48$.

59. $28 = 2 \cdot 2 \cdot 7$
$35 = 5 \cdot 7$
The least common multiple is $2 \cdot 2 \cdot 5 \cdot 7 = 140$.

61. $49 = 7 \cdot 7$
$56 = 2 \cdot 2 \cdot 2 \cdot 7$
The least common multiple is $2 \cdot 2 \cdot 2 \cdot 7 \cdot 7 = 392$.

63. $8 = 2 \cdot 2 \cdot 2$
$12 = 2 \cdot 2 \cdot 3$
$28 = 2 \cdot 2 \cdot 7$
The least common multiple is $2 \cdot 2 \cdot 2 \cdot 3 \cdot 7 = 168$.

65. $9 = 3 \cdot 3$
$15 = 3 \cdot 5$
$18 = 2 \cdot 3 \cdot 3$
The least common multiple is $2 \cdot 3 \cdot 3 \cdot 5 = 90$.

69. Since all even numbers are divisible by 2, 2 is the only even number that is divisible only by itself and 1.

71. The smallest whole number, greater than 1, that would produce a remainder of 1 when divided by 2,3,4,5,6 would be the least common multiple of those numbers plus 1. The prime factors are: $2 = 2$, $3 = 3$, $4 = 2 \cdot 2$, $5 = 5$, $6 = 2 \cdot 3$. The least common multiple is $2 \cdot 2 \cdot 3 \cdot 5 = 60$. The least common multiple plus 1 is 61.

73. The greatest common factor of x and y if x and y are nonzero whole numbers and y is a multiple of x is x, because x is a factor of x and y and x would be the largest whole number that would divide x.

75. The least common multiple of x and y if the greatest common factor of x and y is 1 would be xy, because all prime factors of x and y would need to be included in the least common multiple.

77. $76 = 2 \cdot 38 = 2 \cdot 2 \cdot 19$
Rule for 2 used twice.

79. $123 = 3 \cdot 41$ Rule for 3

81. $115 = 5 \cdot 23$ Rule for 5

83. $441 = 9 \cdot 49 = 3 \cdot 3 \cdot 7 \cdot 7$
Rule for 9
Rule for 3

85. $153 = 9 \cdot 17 = 3 \cdot 3 \cdot 17$
Rule for 9
Rule for 3

PROBLEM SET 1.3 Integers: Addition and Subtraction

1. $5 + (-3) = 2$

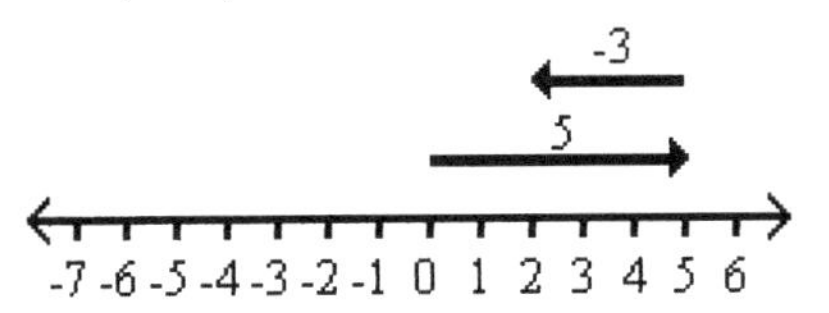

3. $-6 + 2 = -4$

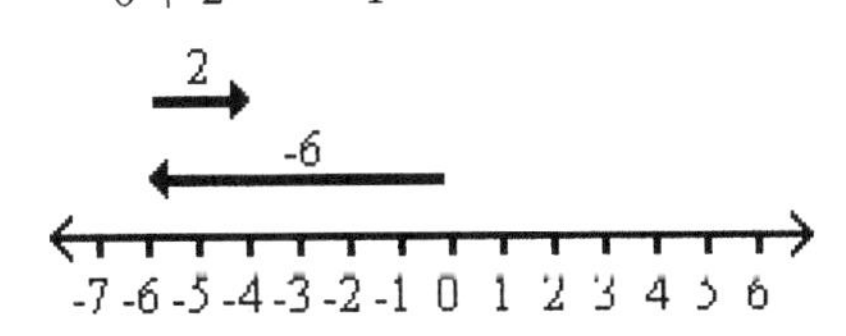

5. $-3 + (-4) = -7$

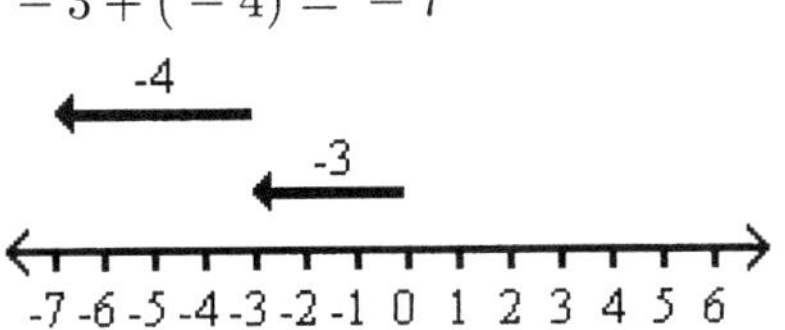

7. $8 + (-2) = 6$

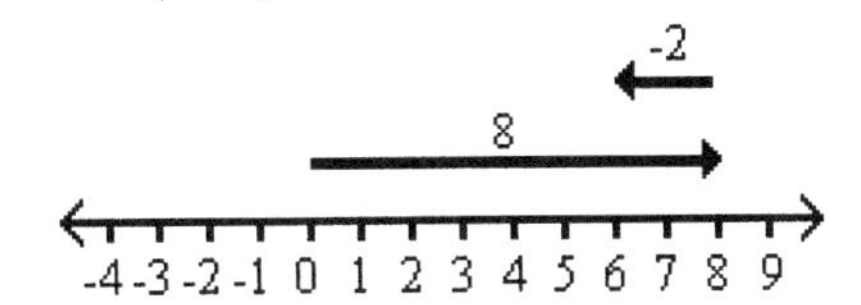

9. $5 + (-11) = -6$

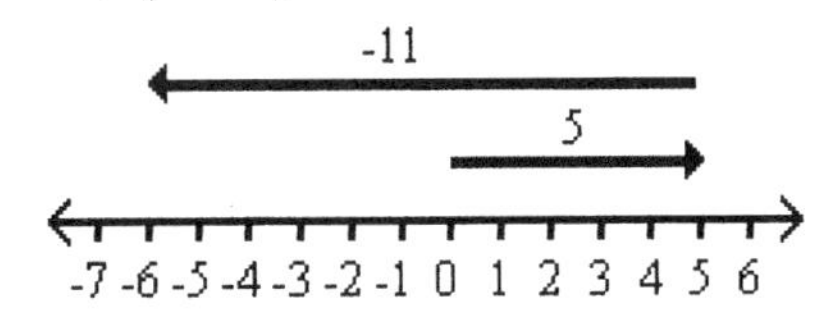

11. $17 + (-9) = |17| - |-9| = 17 - 9 = 8$

13. $8 + (-19) = -(|-19| - |8|) =$
$-(19 - 8) = -11$

15. $-7 + (-8) = -(|-7| + |-8|) =$
$-(7 + 8) = -15$

17. $-15 + 8 = -(|-15| - |8|) =$
$-(15 - 8) = -7$

19. $-13 + (-18) = -(|-13| + |-18|) =$
$-(13 + 18) = -31$

21. $-27 + 8 = -(|-27| - |8|) =$
$-(27 - 8) = -19$

23. $32 + (-23) = +(|32| - |-23|) =$
$+(32 - 23) = 9$

25. $-25 + (-36) = -(|-25| + |-36|) =$
$-(25 + 36) = -61$

Problem Set 1.3

27. $54 + (-72) = -(|-72| - |54|) =$
$-(72 - 54) = -18$

29. $-34 + (-58) =$
$-(|-34| + |-58|) =$
$-(34 + 58) = -92$

31. $3 - 8 = 3 + (-8) = -5$

33. $-4 - 9 = -4 + (-9) = -13$

35. $5 - (-7) = 5 + 7 = 12$

37. $-6 - (-12) = -6 + 12 = 6$

39. $-11 - (-10) = -11 + 10 = -1$

41. $-18 - 27 = -18 + (-27) = -45$

43. $34 - 63 = 34 + (-63) = -29$

45. $45 - 18 = 45 + (-18) = 27$

47. $-21 - 44 = -21 + (-44) = -65$

49. $-53 - (-24) = -53 + 24 = -29$

51. $6 - 8 - 9 = 6 + (-8) + (-9) = -11$

53. $-4 - (-6) + 5 - 8 =$
$-4 + 6 + 5 + (-8) = -1$

55. $5 + 7 - 8 - 12 =$
$5 + 7 + (-8) + (-12) = -8$

57. $-6 - 4 - (-2) + (-5) =$
$-6 + (-4) + 2 + (-5) = -13$

59. $-6 - 5 - 9 - 8 - 7 =$
$-6 + (-5) + (-9) + (-8) + (-7) = -35$

61. $7 - 12 + 14 - 15 - 9 =$
$7 + (-12) + 14 + (-15) + (-9) = -15$

63. $-11 - (-14) + (-17) - 18 =$
$-11 + 14 + (-17) + (-18) = -32$

65. $16 - 21 + (-15) - (-22) =$
$16 + (-21) + (-15) + 22 = 2$

67. $\begin{array}{r} 5 \\ \underline{-9} \\ -4 \end{array}$ This problem in horizontal format is $5 + (-9) = -4$

69. $\begin{array}{r} -13 \\ \underline{-18} \\ -31 \end{array}$ This problem in horizontal format is $(-13) + (-18) = -31$

71. $\begin{array}{r} -18 \\ \underline{9} \\ -9 \end{array}$ This problem in horizontal format is $(-18) + 9 = -9$

73. $\begin{array}{r} -21 \\ \underline{39} \\ 18 \end{array}$ This problem in horizontal format is $(-21) + 39 = 18$

75. $\begin{array}{r} 27 \\ \underline{-19} \\ 8 \end{array}$ This problem in horizontal format is $27 + (-19) = 8$

77. $\begin{array}{r} -53 \\ \underline{24} \\ -29 \end{array}$ This problem in horizontal format is $(-53) + 24 = -29$

79. $\begin{array}{r} 5 \\ \underline{12} \\ -7 \end{array}$ Change the sign of the bottom number and add $\begin{array}{r} 5 \\ \underline{-12} \\ -7 \end{array}$

81. $\begin{array}{r} 6 \\ \underline{-9} \\ 15 \end{array}$ Change the sign of the bottom number and add $\begin{array}{r} 6 \\ \underline{9} \\ 15 \end{array}$

83. $\begin{array}{r} -7 \\ \underline{-8} \\ 1 \end{array}$ Change the sign of the bottom number and add $\begin{array}{r} -7 \\ \underline{8} \\ 1 \end{array}$

85. $\begin{array}{r} 17 \\ \underline{-19} \\ 36 \end{array}$ Change the sign of the bottom number and add $\begin{array}{r} 17 \\ \underline{19} \\ 36 \end{array}$

87. $\begin{array}{r} -23 \\ \underline{16} \\ -39 \end{array}$ Change the sign of the bottom number and add $\begin{array}{r} -23 \\ \underline{-16} \\ -39 \end{array}$

89. $\begin{array}{r} -12 \\ \underline{12} \\ -24 \end{array}$ Change the sign of the bottom number and add $\begin{array}{r} -12 \\ \underline{-12} \\ -24 \end{array}$

91. $x - y$ for $x = -6,\ y = -13$
$(-6) - (-13)$
$(-6) + 13$
7

93. $-x + y - z$ for $x = 3,\ y = -4,\ z = -6$
$-(3) + (-4) - (-6) =$
$-3 + (-4) + 6 = -1$

95. $-x - y - z$ for $x = -2,\ y = 3,\ z = -11$
$-(-2) - (3) - (-11) =$
$2 + (-3) + 11 = 10$

97. $-x + y + z$ for $x = -11,\ y = 7,\ z = -9$
$-(-11) + (7) + (-9) =$
$11 + 7 + (-9) = 9$

99. $x - y - z$ for $x = -15,\ y = 12,\ z = -10$
$(-15) - (12) - (-10) =$
$(-15) + (-12) + 10 = -17$

101. $4 + (-7) = -3$ (3 yards behind original scrimmage line)

103. $-4 + (-6) = -10$ (10 yards behind original scrimmage line)

105. $-5 + 2 = -3$ (3 yards behind original scrimmage line)

107. $-4 + 15 = 11$ (11 yards ahead of original scrimmage line)

109. $-12 + 17 = 5$ (5 yards ahead of original scrimmage line)

111. $60 + (-125) = -65$
(overall expense of $65)

113. $-55 + (-45) = -100$
(overall expense of $100)

115. $-70 + 45 = -25$
(overall expense of $25)

117. $-120 + 250 = 130$
(overall income of $130)

119. $145 + (-65) = 80$
(overall income of $80)

121. $-17 + 14 = -3$ (temperature at noon is -3°F)

123. $(+3) + (-2) + (-3) + (-5) = -7$
(7 under par for tournament)

125. $(-2)+(+1)+(+3)+(+1)+(-2) = +1$
($1 gain for the five-day period)

PROBLEM SET 1.4 Integers: Multiplication and Division

1. $5(-6) = -(|5| \cdot |-6|) =$
$-(5 \cdot 6) = -30$

3. $\dfrac{-27}{3} = -\left(\dfrac{|-27|}{|3|}\right) =$
$-\left(\dfrac{27}{3}\right) = -9$

5. $\dfrac{-42}{-6} = \dfrac{|-42|}{|-6|} = \dfrac{42}{6} = 7$

7. $(-7)(8) = -(|-7| \cdot |8|) =$
$-(7 \cdot 8) = -(56) = -56$

9. $(-5)(-12) = |-5| \cdot |-12|$
$5 \cdot 12 = 60$

11. $\dfrac{96}{-8} = -\left(\dfrac{|96|}{|-8|}\right) =$
$-\left(\dfrac{96}{8}\right) = -12$

Problem Set 1.4

13. $14(-9) = -(|14| \cdot |-9|) =$
$-(14 \cdot 9) = -126$

15. $(-11)(-14) = |-11| \cdot |-14| =$
$11 \cdot 14 = 154$

17. $\frac{135}{-15} = -\left(\frac{|135|}{|-15|}\right) =$
$-\left(\frac{135}{15}\right) = -9$

19. $\frac{-121}{-11} = \frac{|-121|}{|-11|} = \frac{121}{11} = 11$

21. $(-15)(-15) = |-15| \cdot |-15| =$
$15 \cdot 15 = 225$

23. $\frac{112}{-8} = -\left(\frac{|112|}{|-8|}\right) =$
$-\left(\frac{112}{8}\right) = -14$

25. $\frac{0}{-8} = 0$ because $(-8)(0) = 0$

27. $\frac{-138}{-6} = \frac{|-138|}{|-6|} = \frac{138}{6} = 23$

29. $\frac{76}{-4} = -\left(\frac{|76|}{|-4|}\right) =$
$-\left(\frac{76}{4}\right) = -(19) = -19$

31. $(-6)(-15) = |-6| \cdot |-15| =$
$(6)(15) = 90$

33. $(-56) \div (-4) = |-56| \div |-4| =$
$56 \div 4 = 14$

35. $(-19) \div 0$ is undefined because no number times zero produces -19.

37. $(-72) \div 18 = -(|-72| \div |18|) =$
$-(72 \div 18) = -4$

39. $(-36)(27) = -(|-36| \cdot |27|) =$
$-(36 \cdot 27) = -972$

41. $3(-4) + 5(-7) =$
$-12 + (-35) = -47$

43. $7(-2) - 4(-8) = -14 - (-32) =$
$-14 + 32 = 18$

45. $(-3)(-8) + (-9)(-5) =$
$24 + 45 = 69$

47. $5(-6) - 4(-7) + 3(2) =$
$-30 - (-28) + 6 = -30+28+6 = 4$

49. $\frac{13 + (-25)}{-3} = \frac{-12}{-3} = 4$

51. $\frac{12 - 48}{6} = \frac{-36}{6} = -6$

53. $\frac{-7(10) + 6(-9)}{-4} = \frac{-70 + (-54)}{-4} =$
$\frac{-124}{-4} = 31$

55. $\frac{4(-7) - 8(-9)}{11} = \frac{-28 - (-72)}{11} =$
$\frac{-28 + 72}{11} = \frac{44}{11} = 4$

57. $-2(3) - 3(-4) + 4(-5) - 6(-7) =$
$-6 - (-12) + (-20) - (-42) =$
$-6 + 12 + (-20) + 42 = 28$

59. $-1(-6) - 4 + 6(-2) - 7(-3) - 18 =$
$6 - 4 + (-12) - (-21) - 18 =$
$6 - 4 + (-12) + 21 - 18 = -7$

61. $7x + 5y$ for $x = -5, y = 9$
$7(-5) + 5(9) =$
$-35 + 45 = 10$

63. $9a - 2b$ for $a = -5, b = 7$
$9(-5) - 2(7) = -45 - 14 =$
$-45 + (-14) = -59$

65. $-6x - 7y$, for $x = -4, y = -6$
$-6(-4) - 7(-6) =$
$24 + 42 = 66$

67. $\dfrac{5x-3y}{-6}$ for $x=-6,\ y=4$

$\dfrac{5(-6)-3(4)}{-6}=\dfrac{-30-12}{-6}=$

$\dfrac{-30+(-12)}{-6}=\dfrac{-42}{-6}=7$

69. $3(2a-5b)$ for $a=-1,\ b=-5$

$3[2(-1)-5(-5)]=$

$3[-2-(-25)]=$

$3[-2+25]=3(23)=69$

71. $-2x+6y-xy$ for $x=7,\ y=-7$

$-2(7)+6(-7)-(7)(-7)=$

$-14+(-42)-(-49)=$

$-56+49=-7$

73. $-4ab-b$ for $a=2,\ b=-14$

$-4(2)(-14)-(-14)=$

$-8(-14)+14=$

$112+14=126$

75. $(ab+c)(b-c)$ for $a=-2,\ b=-3,\ c=4$

$[(-2)(-3)+(4)][(-3)-(4)]=$

$(6+4)(-7)=10(-7)=-70$

77. $\mathrm{F}=59$: $\dfrac{5(\mathrm{F}-32)}{9}=\dfrac{5[(59)-32]}{9}=$

$\dfrac{5(27)}{9}=\dfrac{135}{9}=15$

79. $\mathrm{F}=14$: $\dfrac{5(\mathrm{F}-32)}{9}=\dfrac{5[(14)-32]}{9}=$

$\dfrac{5(-18)}{9}=\dfrac{-90}{9}=-10$

81. $\mathrm{F}=-13$: $\dfrac{5(\mathrm{F}-32)}{9}=\dfrac{5[(-13)-32]}{9}=$

$\dfrac{5(-45)}{9}=\dfrac{-225}{9}=-25$

83. $\mathrm{C}=25$: $\dfrac{9\mathrm{C}}{5}+32=$

$\dfrac{9(25)}{5}+32=\dfrac{225}{5}+32=45+32=77$

85. $\mathrm{C}=40$: $\dfrac{9\mathrm{C}}{5}+32=$

$\dfrac{9(40)}{5}+32=\dfrac{360}{5}+32=72+32=104$

87. $\mathrm{C}=-10$: $\dfrac{9\mathrm{C}}{5}+32=$

$\dfrac{9(-10)}{5}+32=\dfrac{-90}{5}+32=$

$-18+32=14$

89. $\begin{pmatrix}\text{Value of}\\ \text{800 shares}\end{pmatrix}=\begin{pmatrix}\text{800 times the price}\\ \text{of shares at closing}\end{pmatrix}$:

$$\begin{aligned}\text{Value}&=800\left[\text{price}+\left(\tfrac{\text{1 day's}}{\text{increase}}\right)+\left(\tfrac{\text{4 day's}}{\text{decrease}}\right)\right]\\&=800[19+2+4(-1)]\\&=800[19+2+(-4)]\\&=800(17)\\&=\$13,600\end{aligned}$$

91. $\begin{pmatrix}\text{Temperature}\\ \text{at 10 pm}\end{pmatrix}=$

$\begin{pmatrix}\text{Starting}\\ \text{temp}\end{pmatrix}+\begin{pmatrix}\text{Hours temp}\\ \text{dropped}\end{pmatrix}\begin{pmatrix}\text{Degree change}\\ \text{per hour}\end{pmatrix}=$

$5+4(-3)=5+(-12)=-7°\mathrm{F}.$

PROBLEM SET 1.5 Use of Properties

1. $3(7+8)=3(7)+3(8)$:
Distributive property

3. $-2+(5+7)=(-2+5)+7$:
Associative property of addition

5. $143(-7)=-7(143)$:
Commutative property of multiplication

7. $-119+119=0$:
Additive inverse property

9. $-56+0=-56$:
Identity property of addition

11. $[5(-8)]4=5[-8(4)]$:
Associative property of multiplication

Problem Set 1.5

13. $(-18+56)+18=$
$[56+(-18)]+18=$
$56+(-18+18)=$
$56+0=56$

15. $36-48-22+41=$
$-12-22+41=$
$-34+41=7$

17. $(25)(-18)(-4)=$
$(25)(-4)(-18)=$
$(-100)(-18)=1,800$

19. $(4)(-16)(-9)(-25)=$
$(4)(-25)(-16)(-9)=$
$(-100)(144)=-14{,}400$

21. $37(-42-58)=37(-100)=-3{,}700$

23. $59(36)+59(64)=59(36+64)=$
$59(100)=5{,}900$

25. $15(-14)+16(-8)=$
$-210-128=-338$

27. $17+(-18)-19-14+13-17=$
$[17+(-17)]+(-18)-19-14+13=$
$0+(-37)+(-14+13)=$
$-37-1=-38$

29. $-21+22-23+27+21-19=$
$(-21+22)+(-23+27)+(21-19)=$
$1+4+2=7$

31. $9x-14x=(9-14)x=-5x$

33. $4m+m-8m=$
$(4+1-8)m=-3m$

35. $-9y+5y-7y=$
$(-9+5-7)y=-11y$

37. $4x-3y-7x+y=$
$4x-7x-3y+y=$
$(4-7)x+(-3+1)y=$
$-3x-2y$

39. $-7a-7b-9a+3b=$
$-7a-9a-7b+3b=$
$(-7-9)a+(-7+3)b=$
$-16a-4b$

41. $6xy-x-13xy+4x=$
$6xy-13xy-x+4x=$
$-7xy+3x$

43. $5x-4+7x-2x+9=$
$5x+7x-2x-4+9=$
$10x+5$

45. $-2xy+12+8xy-16=$
$-2xy+8xy+12-16=$
$6xy-4$

47. $-2a+3b-7b-b+5a-9a=$
$-2a+5a-9a+3b-7b-b=$
$-6a-5b$

49. $13ab+2a-7a-9ab+ab-6a=$
$13ab-9ab+ab+2a-7a-6a=$
$5ab-11a$

51. $3(x+2)+5(x+6)=$
$3(x)+3(2)+5(x)+5(6)=$
$3x+6+5x+30=$
$3x+5x+6+30=$
$8x+36$

53. $5(x-4)+6(x+8)=$
$5x-20+6x+48=11x+28$

55. $9(x+4)-(x-8)=$
$9(x+4)-1(x-8)=$
$9x+36-x+8=$
$8x+44$

57. $3(a-1)-2(a-6)+4(a+5)=$
$3a-3-2a+12+4a+20=$
$5a+29$

59. $-2(m+3)-3(m-1)+8(m+4)=$
$-2m-6-3m+3+8m+32=$
$3m+29$

61. $(y+3)-1(y-2)-1(y+6)-7(y-1)=$
$y+3-y+2-y-6-7y+7=$
$-8y+6$

63. $3x+5y+4x-2y=7x+3y=$
$7(-2)+3(3)$ for $x=-2$ and $y=3$
$-14+9=-5$

65. $5(x-2)+8(x+6)=$
$5x-10+8x+48=$
$13x+38$ for $x=-6$
$13(-6)+38=-78+38=-40$

67. $8(x+4)-10(x-3)=$
$8x+32-10x+30=$
$-2x+62$ for $x=-5$
$-2(-5)+62=$
$10+62=72$

69. $(x-6)-(x+12)=$
$x-6-x-12=-18$
for $x=-3$

71. $2(x+y)-3(x-y)=$
$2x+2y-3x+3y=$
$-x+5y$ for $x=-2,\ y=7$
$-(-2)+5(7)=2+35=37$

73. $2xy+6+7xy-8=$
$9xy-2$ for $x=2,\ y=-4$
$9(2)(-4)-2=-72-2=-74$

75. $5x-9xy+3x+2xy=$
$8x-7xy$ for $x=12$ and $y=-1$
$8(12)-7(12)(-1)=96+84=180$

77. $(a-b)-(a+b)=$
$a-b-a-b=$
$-2b$ for $a=19,\ b=-17$
$-2(-17)=34$

79. $-3x+7x+4x-2x-x=$
$5x$ for $x=-13$
$5(-13)=-65$

CHAPTER 1 | Review Problem Set

1. $7+(-10)=-(|-10|-|7|)=$
$-(10-7)=-(3)=-3$

2. $(-12)+(-13)=-(|-12|+|-13|)=$
$-(12+13)=-(25)=-25$

3. $8-13=8+(-13)=-5$

4. $-6-9=-6+(-9)=-15$

5. $-12-(-11)=-12+11=-1$

6. $-17-(-19)=-17+19=2$

7. $(13)(-12)=-(|13|\cdot|-12|)=$
$-(13\cdot 12)=-(156)=-156$

8. $(-14)(-18)=|-14|\cdot|-18|=$
$14\cdot 18=252$

9. $(-72)\div(-12)=|-72|\div|-12|=$
$72\div 12=6$

10. $(117)\div(-9)=-(|117|\div|-9|)=$
$-(117\div 9)=-(13)=-13$

11. 73, prime number

12. Because $87=3(29)$, it is a composite number.

13. Because $63=9(7)$, it is a composite number.

14. Because $81=9(9)$, it is a composite number.

15. Because $91=7(13)$, it is a composite number.

16. $24=4\cdot 6=2\cdot 2\cdot 2\cdot 3$

17. $63=9\cdot 7=3\cdot 3\cdot 7$

18. $57=3\cdot 19$

19. $64 = 16 \cdot 4 = 4 \cdot 4 \cdot 4 = 2 \cdot 2 \cdot 2 \cdot 2 \cdot 2 \cdot 2$

20. $84 = 4 \cdot 21 = 2 \cdot 2 \cdot 3 \cdot 7$

21. $36 = 2 \cdot 2 \cdot 3 \cdot 3$
$54 = 2 \cdot 3 \cdot 3 \cdot 3$
The greatest common factor is $2 \cdot 3 \cdot 3 = 18$.

22. $48 = 2 \cdot 2 \cdot 2 \cdot 2 \cdot 3$
$60 = 2 \cdot 2 \cdot 3 \cdot 5$
$84 = 2 \cdot 2 \cdot 3 \cdot 7$
The greatest common factor is $2 \cdot 2 \cdot 3 = 12$.

23. $18 = 2 \cdot 3 \cdot 3$
$20 = 2 \cdot 2 \cdot 5$
The least common multiple is $2 \cdot 2 \cdot 3 \cdot 3 \cdot 5 = 180$.

24. $15 = 3 \cdot 5$
$27 = 3 \cdot 3 \cdot 3$
$35 = 5 \cdot 7$
The least common multiple is $3 \cdot 3 \cdot 3 \cdot 5 \cdot 7 = 945$.

25. $(19 + 56) + (-9) = 75 + (-9) = 66$

26. $43 - 62 + 12 = 43 + (-62) + 12 =$
$-19 + 12 = -7$

27. $8 + (-9) + (-16) + (-14) + 17 + 12 =$
$37 + (-39) = -2$

28. $19 - 23 - 14 + 21 + 14 - 13 = 4$

29. $3(-4) - 6 = -12 - 6 = -18$

30. $(-5)(-4) - 8 = 20 - 8 = 12$

31. $5(-2) + 6(-4) = -10 - 24 = -34$

32. $(-6)(8) + (-7)(-3) =$
$-48 + 21 = -27$

33. $(-6)(3) - (-4)(-5) =$
$-18 - 20 = -38$

34. $(-7)(9) - (6)(5) = -63 - 30 = -93$

35. $\dfrac{4(-7) - 3(-2)}{-11} = \dfrac{-28 - (-6)}{-11} =$
$\dfrac{-28 + 6}{-11} = \dfrac{-22}{-11} = 2$

36. $\dfrac{(-4)(9) + (5)(-3)}{1 - 18} = \dfrac{-36 + (-15)}{-17} =$
$\dfrac{-51}{-17} = 3$

37. $3 - 2[4(-3 - 1)] =$
$3 - 2[4(-3 + (-1))] =$
$3 - 2[4(-4)] = 3 - 2(-16) =$
$3 - (-32) = 3 + 32 = 35$

38. $-6 - [3(-4 - 7)] =$
$-6 - [3(-4 + (-7))] =$
$-6 - [3(-11)] = -6 - (-33) =$
$-6 + 33 = 27$

39. $125 - (-50) = 175$
The difference between record high and record low temperatures is 175°F.

40. $|20,320 - (-282)| = |20,320 + 282| =$
$|20,602| = 20,602$
The absolute value of the difference in elevation is 20,602 feet.

41. $2(6) - 4 + 3(8) - 1 =$
$12 - 4 + 24 - 1 = 31$

42. $3278 + 175(4) - 50 - 189 - 160 - 20 - 115 =$
$3278 + 700 - 534 = 3444$
Her balance is \$3,444.

43. $12x + 3x - 7x = (12 + 3 - 7)x = 8x$

44. $9y + 3 - 14y - 12 = 9y - 14y + 3 - 12 =$
$(9 - 14)y + (3 - 12) = -5y + (-9) =$
$-5y - 9$

45. $8x + 5y - 13x - y = 8x - 13x + 5y - y =$
$(8 - 13)x + (5 - 1)y = -5x + 4y$

46. $9a + 11b + 4a - 17b = 9a + 4a + 11b - 17b =$
$(9 + 4)a + (11 - 17)b = 13a + (-6b) =$
$13a - 6b$

47. $3ab - 4ab - 2a = -ab - 2a$

48. $5xy - 9xy + xy - y = -3xy - y$

49. $3(x+6)+7(x+8)=$
$3x+18+7x+56=10x+74$

50. $5(x-4)-3(x-9)=$
$5x-20-3x+27=2x+7$

51. $-3(x-2)-4(x+6)=$
$-3x+6-4x-24=-7x-18$

52. $-2x-3(x-4)+2x=$
$-2x-3x+12+2x=-3x+12$

53. $2(a-1)-a-3(a-2)=$
$2a-2-a-3a+6=-2a+4$

54. $-(a-1)+3(a-2)-4a+1=$
$-a+1+3a-6-4a+1=-2a-4$

55. $5x+8y$ for $x=-7$ and $y=-3$
$5(-7)+8(-3)=-35-24=-59$

56. $7x-9y$ for $x=-3$ and $y=4$
$7(-3)-9(4)=-21-36=-57$

57. $\dfrac{-5x-2y}{-2x-7}$ for $x=6$ and $y=4$
$\dfrac{-5(6)-2(4)}{-2(6)-7}=\dfrac{-30-8}{-12-7}=\dfrac{-38}{-19}=2$

58. $\dfrac{-3x+4y}{3x}$ for $x=-4,\ y=-6$
$\dfrac{-3(-4)+4(-6)}{3(-4)}=$
$\dfrac{12+(-24)}{-12}=\dfrac{-12}{-12}=1$

59. $-2a+\dfrac{a-b}{a-2}$ for $a=-5,\ b=9$
$-2(-5)+\dfrac{(-5)-(9)}{(-5)-2}=$
$10+\dfrac{(-5)+(-9)}{(-5)+(-2)}=$
$10+\dfrac{-14}{-7}=10+2=12$

60. $\dfrac{2a+b}{b+6}-3b$ for a $=3,\ b=-4$
$\dfrac{2(3)+(-4)}{(-4)+6}-3(-4)=$
$\dfrac{6+(-4)}{2}-(-12)=$
$\dfrac{2}{2}+12=1+12=13$

61. $5a+6b-7a-2b=$
$-2a+4b$ for $a=-1,\ b=5$
$-2(-1)+4(5)=2+20=22$

62. $3x+7y-5x+y=$
$-2x+8y$ for $x=-4$ and $y=3$
$-2(-4)+8(3)=8+24=32$

63. $2xy+6+5xy-8=$
$7xy-2$ for $x=-1,\ y=1$
$7(-1)(1)-2=(-7)(1)-2=-9$

64. $7(x+6)-9(x+1)=$
$7x+42-9x-9=$
$-2x+33$ for $x=-2$
$-2(-2)+33=4+33=37$

65. $-3(x-4)-2(x+8)=$
$-3x+12-2x-16=$
$-5x-4$ for $x=7$
$-5(7)-4=-35-4=-39$

66. $2(x-1)-(x+2)+3(x-4)=$
$2x-2-x-2+3x-12=$
$4x-16$ for $x=-4$
$4(-4)-16=-16-16=-32$

67. $(a-b)-(a+b)-b=$
$a-b-a-b-b=-3b$
for $a=-1,\ b=-3$
$-3(-3)=9$

68. $2ab-3(a-b)+b+a=$
$2ab-3a+3b+b+a=$
$2ab-2a+4b$ for $a=2,\ b=-5$
$2(2)(-5)-2(2)+4(-5)=$
$4(-5)-4-20=-20-4-20=-44$

CHAPTER 1 Test

1. $6 + (-7) - 4 + 12$
$-1 - 4 + 12 =$
$-5 + 12 = 7$

2. $7 + 4(9) + 2 =$
$7 + 36 + 2 = 45$

3. $-4(2 - 8) + 14 =$
$-4(-6) + 14 =$
$24 + 14 = 38$

4. $5(-7) - (-3)(8) =$
$-35 - (-24) =$
$-35 + 24 = -11$

5. $8 \div (-4) + (-6)(9) - 2 =$
$-2 + (-6)(9) - 2 =$
$-2 + (-54) - 2 = -58$

6. $(-8)(-7) + (-6) - (9)(12) =$
$56 + (-6) - 108 = -58$

7. $\dfrac{6(-4) - (-8)(-5)}{-16} = \dfrac{-24 - 40}{-16} =$
$\dfrac{-64}{-16} = 4$

8. $-14 + 23 - 17 - 19 + 26 =$
$-14 - 17 - 19 + 23 + 26 =$
$-50 + 49 = -1$

9. $(-14)(4) \div 4 + (-6) =$
$-56 \div 4 + (-6) =$
$-14 + (-6) = -20$

10. $6(-9) - (-8) - (-7)(4) + 11 =$
$-54 + 8 - (-28) + 11 =$
$-54 + 8 + 28 + 11 = -7$

11. $7 - 13 = -6$
By 6:00 am the temperature will be $-6°$F.

12. $7x - 9y$ for $x = -4,\ y = -6$
$7(-4) - 9(-6) =$
$-28 - (-54) =$
$-28 + 54 = 26$

13. $-4a - 6b$ for $a = -9,\ b = 12$
$-4(-9) - 6(12) =$
$36 - 72 = -36$

14. $3xy - 8y + 5x$ for $x = 7,\ y = -2$
$3(7)(-2) - 8(-2) + 5(7) =$
$21(-2) - (-16) + 35 =$
$-42 + 16 + 35 = 9$

15. $5(x - 4) - 6(x + 7) =$
$5x - 20 - 6x - 42 =$
$-x - 62$ for $x = -5$
$-(-5) - 62 = 5 - 62 = -57$

16. $3x - 2y - 4x - x + 7y =$
$-2x + 5y$ for $x = 6,\ y = -7$
$-2(6) + 5(-7) = -12 + (-35) = -47$

17. $\dfrac{-x - y}{y - x}$ for $x = -9,\ y = -6$
$\dfrac{-(-9) - (-6)}{(-6) - (-9)} = \dfrac{9 + 6}{-6 + 9} =$
$\dfrac{15}{3} = 5$

18. 79, prime number

19. $360 = 8 \cdot 45 = 8 \cdot 9 \cdot 5$
$= 2 \cdot 2 \cdot 2 \cdot 3 \cdot 3 \cdot 5$

20. $36 = 2 \cdot 2 \cdot 3 \cdot 3$
$60 = 2 \cdot 2 \cdot 3 \cdot 5$
$84 = 2 \cdot 2 \cdot 3 \cdot 7$
The greatest common factor is $2 \cdot 2 \cdot 3 = 12$.

21. $9 = 3 \cdot 3$
$24 = 2 \cdot 2 \cdot 2 \cdot 3$
The least common multiple is $2 \cdot 2 \cdot 2 \cdot 3 \cdot 3 = 72$.

22. $[-3+(-4)]+(-6) = -3+[(-4)+(-6)]$:
Associative Property of Addition

23. $8(25+37) = 8(25)+8(37)$:
Distributive Property

24. $-7x+9y-y+x-2y-7x =$
$-7x+x-7x+9y-y-2y =$
$-13x+6y$

25. $-2(x-4)-5(x+7)-6(x-1) =$
$-2x+8-5x-35-6x+6 =$
$-2x-5x-6x+8-35+6 =$
$-13x-21$

Chapter 2 The Real Numbers

PROBLEM SET 2.1 Rational Numbers: Multiplication and Division

1. $\frac{8}{12}=\frac{4\cdot 2}{4\cdot 3}=\frac{2}{3}$

3. $\frac{16}{24}=\frac{8\cdot 2}{8\cdot 3}=\frac{2}{3}$

5. $\frac{15}{9}=\frac{3\cdot 5}{3\cdot 3}=\frac{5}{3}$

7. $\frac{-8}{48}=-\frac{2\cdot 2\cdot 2}{2\cdot 2\cdot 2\cdot 3}=-\frac{1}{6}$

9. $\frac{27}{-36}=-\frac{9\cdot 3}{9\cdot 4}=-\frac{3}{4}$

11. $\frac{-54}{-56}=\frac{2\cdot 27}{2\cdot 28}=\frac{27}{28}$

13. $\frac{24x}{44x}=\frac{4\cdot 6\cdot x}{4\cdot 11\cdot x}=\frac{6}{11}$

15. $\frac{9x}{21y}=\frac{3\cdot 3\cdot x}{3\cdot 7\cdot y}=\frac{3x}{7y}$

17. $\frac{14xy}{35y}=\frac{2\cdot 7\cdot x\cdot y}{5\cdot 7\cdot y}=\frac{2x}{5}$

19. $\frac{-20ab}{52bc}=-\frac{4\cdot 5\cdot a\cdot b}{4\cdot 13\cdot b\cdot c}=-\frac{5a}{13c}$

21. $\frac{-56yz}{-49xy}=\frac{7\cdot 8\cdot y\cdot z}{7\cdot 7\cdot x\cdot y}=\frac{8z}{7x}$

23. $\frac{65abc}{91ac}=\frac{5\cdot 13\cdot a\cdot b\cdot c}{7\cdot 13\cdot a\cdot c}=\frac{5b}{7}$

25. $\frac{3}{4}\cdot\frac{5}{7}=\frac{3\cdot 5}{4\cdot 7}=\frac{15}{28}$

27. $\frac{2}{7}\div\frac{3}{5}=\frac{2}{7}\cdot\frac{5}{3}=\frac{10}{21}$

29. $\frac{3}{8}\cdot\frac{12}{15}=\frac{\overset{1}{\cancel{3}}\cdot\overset{3}{\cancel{12}}}{\underset{2}{\cancel{8}}\cdot\underset{5}{\cancel{15}}}=\frac{3}{10}$

31. $\frac{-6}{13}\cdot\frac{26}{9}=-\frac{\overset{2}{\cancel{6}}\cdot\overset{2}{\cancel{26}}}{\underset{1}{\cancel{13}}\cdot\underset{3}{\cancel{9}}}=-\frac{4}{3}$

33. $\frac{7}{9}\div\frac{5}{9}=\frac{7}{9}\cdot\frac{9}{5}=\frac{7\cdot\overset{1}{\cancel{9}}}{\underset{1}{\cancel{9}}\cdot 5}=\frac{7}{5}$

35. $\frac{1}{4}\div\frac{-5}{6}=-\frac{1}{4}\cdot\frac{6}{5}=-\frac{1\cdot\overset{3}{\cancel{6}}}{\underset{2}{\cancel{4}}\cdot 5}=-\frac{3}{10}$

37. $\left(-\frac{8}{10}\right)\left(-\frac{10}{32}\right)=\frac{\overset{1}{\cancel{8}}\cdot\overset{1}{\cancel{10}}}{\underset{1}{\cancel{10}}\cdot\underset{4}{\cancel{32}}}=\frac{1}{4}$

39. $-9\div\frac{1}{3}=-\frac{9}{1}\cdot\frac{3}{1}=-27$

41. $\frac{5x}{9y}\cdot\frac{7y}{3x}=\frac{5\cdot 7\cdot\cancel{x}\cdot\cancel{y}}{9\cdot 3\cdot\cancel{x}\cdot\cancel{y}}=\frac{35}{27}$

43. $\frac{6a}{14b}\cdot\frac{16b}{18a}=\frac{\overset{1}{\cancel{6}}\cdot\overset{8}{\cancel{16}}\cdot\cancel{a}\cdot\cancel{b}}{\underset{7}{\cancel{14}}\cdot\underset{3}{\cancel{18}}\cdot\cancel{a}\cdot\cancel{b}}=\frac{8}{21}$

45. $\frac{10x}{-9y}\cdot\frac{15}{20x}=-\frac{\overset{1}{\cancel{10}}\cdot\overset{5}{\cancel{15}}\cdot\cancel{x}}{\underset{3}{\cancel{9}}\cdot\underset{2}{\cancel{20}}\cdot\cancel{x}\cdot y}=-\frac{5}{6y}$

47. $ab\cdot\frac{2}{b}=\frac{a\cdot\cancel{b}\cdot 2}{\cancel{b}}=2a$

49. $\left(-\frac{7x}{12y}\right)\left(-\frac{24y}{35x}\right)=\frac{\overset{1}{\cancel{7}}\cdot\overset{2}{\cancel{24}}\cdot\cancel{x}\cdot\cancel{y}}{\underset{1}{\cancel{12}}\cdot\underset{5}{\cancel{35}}\cdot\cancel{x}\cdot\cancel{y}}=\frac{2}{5}$

51. $\frac{3}{x} \div \frac{6}{y} = \frac{\overset{1}{\cancel{3}}}{x} \cdot \frac{y}{\underset{2}{\cancel{6}}} = \frac{y}{2x}$

53. $\frac{5x}{9y} \div \frac{13x}{36y} = \frac{5x}{9y} \cdot \frac{36y}{13x} =$

$\frac{5 \cdot \overset{4}{\cancel{36}} \cdot \cancel{x} \cdot \cancel{y}}{\underset{1}{\cancel{9}} \cdot 13 \cdot \cancel{x} \cdot \cancel{y}} = \frac{20}{13}$

55. $\frac{-7}{x} \div \frac{9}{x} = \frac{-7}{\cancel{x}} \cdot \frac{\cancel{x}}{9} = -\frac{7}{9}$

57. $\frac{-4}{n} \div \frac{-18}{n} = \frac{-\overset{2}{\cancel{4}}}{\cancel{n}} \cdot \frac{\cancel{n}}{-\underset{9}{\cancel{18}}} = \frac{2}{9}$

59. $\frac{3}{4} \cdot \frac{8}{9} \cdot \frac{12}{20} = \frac{\overset{1}{\cancel{3}} \cdot \overset{2}{\cancel{8}} \cdot \overset{3}{\cancel{12}}}{\underset{1}{\cancel{4}} \cdot \underset{3}{\cancel{9}} \cdot \underset{5}{\cancel{20}}} = \frac{6}{15} = \frac{2}{5}$

61. $\left(-\frac{3}{8}\right)\left(\frac{13}{14}\right)\left(-\frac{12}{9}\right) =$

$\frac{\overset{1}{\cancel{3}} \cdot 13 \cdot \overset{3}{\cancel{12}}}{\underset{2}{\cancel{8}} \cdot 14 \cdot \underset{3}{\cancel{9}}} = \frac{13 \cdot \overset{1}{\cancel{3}}}{14 \cdot \underset{2}{\cancel{6}}} = \frac{13}{28}$

63. $\left(\frac{3x}{4y}\right)\left(\frac{8}{9x}\right)\left(\frac{12y}{5}\right) -$

$\frac{\overset{1}{\cancel{3}} \cdot \overset{2}{\cancel{8}} \cdot 12 \cdot \cancel{x} \cdot \cancel{y}}{\underset{1}{\cancel{4}} \cdot \underset{3}{\cancel{9}} \cdot 5 \cdot \cancel{x} \cdot \cancel{y}} = \frac{2 \cdot \overset{4}{\cancel{12}}}{\underset{1}{\cancel{3}} \cdot 5} = \frac{8}{5}$

65. $\left(-\frac{2}{3}\right)\left(\frac{3}{4}\right) \div \frac{1}{8} = -\frac{\overset{1}{\cancel{2}} \cdot \overset{1}{\cancel{3}}}{\underset{1}{\cancel{3}} \cdot \underset{2}{\cancel{4}}} \div \frac{1}{8} =$

$-\frac{1}{2} \div \frac{1}{8} = -\frac{1}{2} \cdot 8 = -4$

67. $\frac{5}{7} \div \left(-\frac{5}{6}\right)\left(-\frac{6}{7}\right) =$

$\frac{5}{7} \cdot \left(-\frac{6}{5}\right)\left(-\frac{6}{7}\right) =$

$\frac{\overset{1}{\cancel{5}} \cdot 6 \cdot 6}{7 \cdot \underset{1}{\cancel{5}} \cdot 7} = \frac{36}{49}$

69. $\left(-\frac{6}{7}\right) \div \left(\frac{5}{7}\right)\left(-\frac{5}{6}\right) =$

$\left(-\frac{6}{7}\right) \cdot \left(\frac{7}{5}\right)\left(-\frac{5}{6}\right) =$

$\frac{\overset{1}{\cancel{6}} \cdot \overset{1}{\cancel{7}} \cdot \overset{1}{\cancel{5}}}{\underset{1}{\cancel{7}} \cdot \underset{1}{\cancel{5}} \cdot \underset{1}{\cancel{6}}} = \frac{1}{1} = 1$

71. $\left(\frac{4}{9}\right)\left(-\frac{9}{8}\right) \div \left(-\frac{3}{4}\right) =$

$\left(\frac{4}{9}\right)\left(-\frac{9}{8}\right) \cdot \left(-\frac{4}{3}\right) =$

$\frac{\overset{1}{\cancel{4}} \cdot \overset{1}{\cancel{9}} \cdot 4}{\underset{1}{\cancel{9}} \cdot \underset{2}{\cancel{8}} \cdot 3} = \frac{\overset{2}{\cancel{4}}}{\underset{1}{\cancel{2}} \cdot 3} = \frac{2}{3}$

73. $\left(\frac{5}{2}\right)\left(\frac{2}{3}\right) \div \left(-\frac{1}{4}\right) \div (-3) =$

$\left(\frac{5}{2}\right)\left(\frac{2}{3}\right) \cdot \left(-\frac{4}{1}\right) \cdot \left(-\frac{1}{3}\right) =$

$\frac{5 \cdot \overset{1}{\cancel{2}} \cdot 4 \cdot 1}{\underset{1}{\cancel{2}} \cdot 3 \cdot 1 \cdot 3} = \frac{5 \cdot 4}{3 \cdot 3} = \frac{20}{9}$

75. $\begin{pmatrix}\text{portion of}\\\text{accounts in}\\\text{Maria's dept.}\end{pmatrix}\cdot\begin{pmatrix}\text{portion of}\\\text{accounts}\\\text{Maria has}\end{pmatrix}=\begin{pmatrix}\text{Maria's portions}\\\text{of all the agency's}\\\text{accounts}\end{pmatrix}$

$$\left(\frac{3}{4}\right)\cdot\left(\frac{1}{3}\right)=\begin{pmatrix}\text{Maria's portions}\\\text{of all the agency's}\\\text{accounts}\end{pmatrix}$$

$$\frac{3}{12}=\frac{1}{4}$$

Maria is responsible for $\frac{1}{4}$ of all the agency's accounts.

77. $\begin{pmatrix}\text{cups of}\\\text{sugar}\end{pmatrix}\cdot\begin{pmatrix}\text{number of}\\\text{cakes}\end{pmatrix}=\begin{pmatrix}\text{amount of}\\\text{sugar}\end{pmatrix}$

$$\left(\frac{3}{4}\right)\cdot(3)=\begin{pmatrix}\text{amount of}\\\text{sugar}\end{pmatrix}$$

$$\frac{3\cdot 3}{4\cdot 1}=\frac{9}{4}=2\frac{1}{4}$$

The amount of sugar needed is $2\frac{1}{4}$ cups of sugar.

79. $\begin{pmatrix}\text{amount of milk in}\\\text{original recipe}\end{pmatrix}\cdot\left(\frac{1}{2}\right)=\begin{pmatrix}\text{amount of milk in}\\\text{half of recipe}\end{pmatrix}$

$$\left(3\frac{1}{2}\right)\cdot\left(\frac{1}{2}\right)=\begin{pmatrix}\text{amount of milk in}\\\text{half of recipe}\end{pmatrix}$$

$$\left(\frac{7}{2}\right)\cdot\left(\frac{1}{2}\right)=\frac{7\cdot 1}{2\cdot 2}=\frac{7}{4}=1\frac{3}{4}$$

She should use $1\frac{3}{4}$ cups of milk.

81. $\begin{pmatrix}\text{material}\\\text{to make}\\\text{one drape}\end{pmatrix}\cdot\begin{pmatrix}\text{number}\\\text{of}\\\text{drapes}\end{pmatrix}=\begin{pmatrix}\text{material}\\\text{to make}\\\text{5 drapes}\end{pmatrix}$

$$\left(3\frac{1}{4}\right)\cdot(5)=\begin{pmatrix}\text{material to make}\\\text{5 drapes}\end{pmatrix}$$

$$\left(\frac{13}{4}\right)\cdot\left(\frac{5}{1}\right)=\frac{13\cdot 5}{4\cdot 1}=\frac{65}{4}=16\frac{1}{4}$$

The amount of material needed to make drapes for 5 windows is $16\frac{1}{4}$ yards.

85. **a.** 8 **b.** 12 **c.** 40
d. 42 **e.** 5 **f.** 7

87. **a.** $\frac{99}{117} = \frac{9 \cdot 11}{9 \cdot 13} = \frac{11}{13}$

b. $\frac{175}{225} = \frac{5 \cdot 35}{5 \cdot 45} = \frac{35}{45} = \frac{5 \cdot 7}{5 \cdot 9} = \frac{7}{9}$

c. $\frac{-111}{123} = -\frac{3 \cdot 37}{3 \cdot 41} = -\frac{37}{41}$

d. $\frac{-234}{270} = -\frac{9 \cdot 26}{9 \cdot 30} =$

$-\frac{26}{30} = -\frac{2 \cdot 13}{2 \cdot 15} = -\frac{13}{15}$

e. $\frac{270}{495} = \frac{9 \cdot 30}{9 \cdot 55} = \frac{30}{55} = \frac{5 \cdot 6}{5 \cdot 11} = \frac{6}{11}$

f. $\frac{324}{459} = \frac{9 \cdot 36}{9 \cdot 51} = \frac{36}{51} =$

$\frac{3 \cdot 12}{3 \cdot 17} = \frac{12}{17}$

g. $\frac{91}{143} = \frac{7 \cdot 13}{11 \cdot 13} = \frac{7}{11}$

h. $\frac{187}{221} = \frac{11 \cdot 17}{13 \cdot 17} = \frac{11}{13}$

PROBLEM SET 2.2 Rational Numbers: Addition and Subtraction

1. $\frac{2}{7} + \frac{3}{7} = \frac{2+3}{7} = \frac{5}{7}$

3. $\frac{7}{9} - \frac{2}{9} = \frac{7-2}{9} = \frac{5}{9}$

5. $\frac{3}{4} + \frac{9}{4} = \frac{3+9}{4} = \frac{12}{4} = 3$

7. $\frac{11}{12} - \frac{3}{12} = \frac{11-3}{12} = \frac{8}{12} = \frac{2}{3}$

9. $\frac{1}{8} - \frac{5}{8} = \frac{1-5}{8} = \frac{-4}{8} = -\frac{1}{2}$

11. $\frac{5}{24} + \frac{11}{24} = \frac{5+11}{24} = \frac{16}{24} = \frac{2}{3}$

13. $\frac{8}{x} + \frac{7}{x} = \frac{8+7}{x} = \frac{15}{x}$

15. $\frac{5}{3y} + \frac{1}{3y} = \frac{5+1}{3y} = \frac{6}{3y} = \frac{2}{y}$

17. $\frac{1}{3} + \frac{1}{5} = \left(\frac{1 \cdot 5}{3 \cdot 5}\right) + \left(\frac{1 \cdot 3}{5 \cdot 3}\right) =$

$\frac{5}{15} + \frac{3}{15} = \frac{8}{15}$

19. $\frac{15}{16} - \frac{3}{8} = \frac{15}{16} - \left(\frac{3 \cdot 2}{8 \cdot 2}\right) =$

$\frac{15}{16} - \frac{6}{16} = \frac{9}{16}$

21. $\frac{7}{10} + \frac{8}{15} = \left(\frac{7 \cdot 3}{10 \cdot 3}\right) + \left(\frac{8 \cdot 2}{15 \cdot 2}\right) =$

$\frac{21}{30} + \frac{16}{30} = \frac{37}{30}$

23. $\frac{11}{24} + \frac{5}{32} = \left(\frac{11 \cdot 4}{24 \cdot 4}\right) + \left(\frac{5 \cdot 3}{32 \cdot 3}\right) =$

$\frac{44}{96} + \frac{15}{96} = \frac{59}{96}$

25. $\frac{5}{18} - \frac{13}{24} = \left(\frac{5 \cdot 4}{18 \cdot 4}\right) - \left(\frac{13 \cdot 3}{24 \cdot 3}\right) =$

$\frac{20}{72} - \frac{39}{72} = -\frac{19}{72}$

27. $\frac{5}{8} - \frac{2}{3} = \left(\frac{5 \cdot 3}{8 \cdot 3}\right) - \left(\frac{2 \cdot 8}{3 \cdot 8}\right) =$

$\frac{15}{24} - \frac{16}{24} = -\frac{1}{24}$

Problem Set 2.2

29. $-\frac{2}{13}-\frac{7}{39}=-\left(\frac{2\cdot 3}{13\cdot 3}\right)-\frac{7}{39}=$
$-\frac{6}{39}-\frac{7}{39}=-\frac{13}{39}=-\frac{1}{3}$

31. $-\frac{3}{14}+\frac{1}{21}=-\left(\frac{3\cdot 3}{14\cdot 3}\right)+\left(\frac{1\cdot 2}{21\cdot 2}\right)=$
$-\frac{9}{42}+\frac{2}{42}=-\frac{7}{42}=-\frac{1}{6}$

33. $-4-\frac{3}{7}=-\left(\frac{4}{1}\cdot\frac{7}{7}\right)-\frac{3}{7}=$
$-\frac{28}{7}-\frac{3}{7}=-\frac{31}{7}$

35. $\frac{3}{4}-6=\frac{3}{4}-\left(\frac{6}{1}\cdot\frac{4}{4}\right)=$
$\frac{3}{4}-\frac{24}{4}=-\frac{21}{4}$

37. $\frac{3}{x}+\frac{4}{y}=\left(\frac{3\cdot y}{x\cdot y}\right)+\left(\frac{4\cdot x}{y\cdot x}\right)=$
$\frac{3y}{xy}+\frac{4x}{xy}=\frac{3y+4x}{xy}$

39. $\frac{7}{a}-\frac{2}{b}=\left(\frac{7\cdot b}{a\cdot b}\right)-\left(\frac{2\cdot a}{b\cdot a}\right)=$
$\frac{7b}{ab}-\frac{2a}{ab}=\frac{7b-2a}{ab}$

41. $\frac{2}{x}+\frac{7}{2x}=\left(\frac{2\cdot 2}{x\cdot 2}\right)+\frac{7}{2x}=$
$\frac{4}{2x}+\frac{7}{2x}=\frac{11}{2x}$

43. $\frac{10}{3x}-\frac{2}{x}=\frac{10}{3x}-\left(\frac{2\cdot 3}{x\cdot 3}\right)=$
$\frac{10}{3x}-\frac{6}{3x}=\frac{4}{3x}$

45. $\frac{1}{x}-\frac{7}{5x}=\left(\frac{1\cdot 5}{x\cdot 5}\right)-\frac{7}{5x}=$
$\frac{5}{5x}-\frac{7}{5x}=-\frac{2}{5x}$

47. $\frac{3}{2y}+\frac{5}{3y}=\left(\frac{3\cdot 3}{2y\cdot 3}\right)+\left(\frac{5\cdot 2}{3y\cdot 2}\right)=$
$\frac{9}{6y}+\frac{10}{6y}=\frac{19}{6y}$

49. $\frac{5}{12y}-\frac{3}{8y}=\left(\frac{5\cdot 2}{12y\cdot 2}\right)-\left(\frac{3\cdot 3}{8y\cdot 3}\right)=$
$\frac{10}{24y}-\frac{9}{24y}=\frac{1}{24y}$

51. $\frac{1}{6n}-\frac{7}{8n}=\left(\frac{1\cdot 4}{6n\cdot 4}\right)-\left(\frac{7\cdot 3}{8n\cdot 3}\right)=$
$\frac{4}{24n}-\frac{21}{24n}=-\frac{17}{24n}$

53. $\frac{5}{3x}+\frac{7}{3y}=\left(\frac{5\cdot y}{3x\cdot y}\right)+\left(\frac{7\cdot x}{3y\cdot x}\right)=$
$\frac{5y}{3xy}+\frac{7x}{3xy}=\frac{5y+7x}{3xy}$

55. $\frac{8}{5x}+\frac{3}{4y}=\left(\frac{8\cdot 4y}{5x\cdot 4y}\right)+\left(\frac{3\cdot 5x}{4y\cdot 5x}\right)=$
$\frac{32y}{20xy}+\frac{15x}{20xy}=\frac{32y+15x}{20xy}$

57. $\frac{7}{4x}-\frac{5}{9y}=\left(\frac{7\cdot 9y}{4x\cdot 9y}\right)-\left(\frac{5\cdot 4x}{9y\cdot 4x}\right)=$
$\frac{63y}{36xy}-\frac{20x}{36xy}=\frac{63y-20x}{36xy}$

59. $-\frac{3}{2x}-\frac{5}{4y}=-\left(\frac{3\cdot 2y}{2x\cdot 2y}\right)-\left(\frac{5\cdot x}{4y\cdot x}\right)=$
$-\frac{6y}{4xy}-\frac{5x}{4xy}=\frac{-6y-5x}{4xy}$

61. $3+\frac{2}{x}=\left(\frac{3\cdot x}{1\cdot x}\right)+\frac{2}{x}=$
$\frac{3x}{x}+\frac{2}{x}=\frac{3x+2}{x}$

63. $2-\frac{3}{2x}=\left(\frac{2}{1}\cdot\frac{2x}{2x}\right)-\frac{3}{2x}=$
$\frac{4x}{2x}-\frac{3}{2x}=\frac{4x-3}{2x}$

65. $\frac{1}{4}-\frac{3}{8}+\frac{5}{12}-\frac{1}{24}=$

$\frac{6}{24}-\frac{9}{24}+\frac{10}{24}-\frac{1}{24}=\frac{6}{24}=\frac{1}{4}$

67. $\frac{5}{6}+\frac{2}{3}\cdot\frac{3}{4}-\frac{1}{4}\cdot\frac{2}{5}=$

$\frac{5}{6}+\frac{1}{2}-\frac{1}{10}=$

$\frac{25}{30}+\frac{15}{30}-\frac{3}{30}=\frac{37}{30}$

69. $\frac{3}{4}\cdot\frac{6}{9}-\frac{5}{6}\cdot\frac{8}{10}+\frac{2}{3}\cdot\frac{6}{8}=$

$\frac{1}{2}-\frac{2}{3}+\frac{1}{2}=$

$\frac{3}{6}-\frac{4}{6}+\frac{3}{6}=\frac{2}{6}=\frac{1}{3}$

71. $4-\frac{2}{3}\cdot\frac{3}{5}-6=4-\frac{2}{5}-6$

$=-2-\frac{2}{5}=-\left(\frac{2}{1}\cdot\frac{5}{5}\right)-\frac{2}{5}$

$=-\frac{10}{5}-\frac{2}{5}=-\frac{12}{5}$

73. $\frac{4}{5}-\frac{10}{12}-\frac{5}{6}\div\frac{14}{8}+\frac{10}{21}=$

$\frac{4}{5}-\frac{10}{12}-\frac{5}{6}\cdot\frac{8}{14}+\frac{10}{21}=$

$\frac{4}{5}-\frac{10}{12}-\frac{10}{21}+\frac{10}{21}=\frac{4}{5}-\frac{10}{12}=$

$\left(\frac{4}{5}\cdot\frac{12}{12}\right)-\left(\frac{10}{12}\cdot\frac{5}{5}\right)=$

$\frac{48}{60}-\frac{50}{60}=-\frac{2}{60}=-\frac{1}{30}$

75. $24\left(\frac{3}{4}-\frac{1}{6}\right)=$

$24\left(\frac{3}{4}\right)-24\left(\frac{1}{6}\right)=18-4=14$

77. $64\left(\frac{3}{16}+\frac{5}{8}-\frac{1}{4}+\frac{1}{2}\right)=$

$64\left(\frac{3}{16}\right)+64\left(\frac{5}{8}\right)-64\left(\frac{1}{4}\right)+64\left(\frac{1}{2}\right)=$

$12+40-16+32=68$

79. $\frac{7}{13}\left(\frac{2}{3}-\frac{1}{6}\right)=\frac{7}{13}\left(\frac{2}{3}\right)-\frac{7}{13}\left(\frac{1}{6}\right)=$

$\frac{14}{39}-\frac{7}{78}=\left(\frac{14\cdot 2}{39\cdot 2}\right)-\frac{7}{78}=$

$\frac{28}{78}-\frac{7}{78}=\frac{21}{78}=\frac{7}{26}$

81. $\frac{1}{3}x+\frac{2}{5}x=\left(\frac{1}{3}+\frac{2}{5}\right)x=$

$\left(\frac{5}{15}+\frac{6}{15}\right)x=\frac{11}{15}x$

83. $\frac{1}{3}a-\frac{1}{8}a=\left(\frac{1}{3}-\frac{1}{8}\right)a=$

$\left(\frac{8}{24}-\frac{3}{24}\right)a=\frac{5}{24}a$

85. $\frac{1}{2}x+\frac{2}{3}x+\frac{1}{6}x=\left(\frac{1}{2}+\frac{2}{3}+\frac{1}{6}\right)x=$

$\left(\frac{3}{6}+\frac{4}{6}+\frac{1}{6}\right)x=\frac{8}{6}x=\frac{4}{3}x$

87. $\frac{3}{5}n-\frac{1}{4}n+\frac{3}{10}n=\left(\frac{3}{5}-\frac{1}{4}+\frac{3}{10}\right)n=$

$\left(\frac{12}{20}-\frac{5}{20}+\frac{6}{20}\right)n=\frac{13}{20}n$

89. $n+\frac{4}{3}n-\frac{1}{9}n=\left(1+\frac{4}{3}-\frac{1}{9}\right)n=$

$\left(\frac{9}{9}+\frac{12}{9}-\frac{1}{9}\right)n=\frac{20}{9}n$

Problem Set 2.2

91. $-n - \frac{7}{9}n - \frac{5}{12}n = \left(-1 - \frac{7}{9} - \frac{5}{12}\right)n =$

$\left(-\frac{36}{36} - \frac{28}{36} - \frac{15}{36}\right)n = -\frac{79}{36}n$

93. $\frac{3}{7}x + \frac{1}{4}y + \frac{1}{2}x + \frac{7}{8}y =$

$\frac{3}{7}x + \frac{1}{2}x + \frac{1}{4}y + \frac{7}{8}y =$

$\left(\frac{3}{7} + \frac{1}{2}\right)x + \left(\frac{1}{4} + \frac{7}{8}\right)y =$

$\left(\frac{6}{14} + \frac{7}{14}\right)x + \left(\frac{2}{8} + \frac{7}{8}\right)y =$

$\frac{13}{14}x + \frac{9}{8}y$

95. $\frac{2}{9}x + \frac{5}{12}y - \frac{7}{15}x - \frac{13}{15}y =$

$\frac{2}{9}x - \frac{7}{15}x + \frac{5}{12}y - \frac{13}{15}y =$

$\left(\frac{2}{9} - \frac{7}{15}\right)x + \left(\frac{5}{12} - \frac{13}{15}\right)y =$

$\left(\frac{10}{45} - \frac{21}{45}\right)x + \left(\frac{25}{60} - \frac{52}{60}\right)y =$

$-\frac{11}{45}x - \frac{27}{60}y = -\frac{11}{45}x - \frac{9}{20}y$

97. $\left(\begin{matrix}12'' \\ \text{round}\end{matrix}\right) + \left(\begin{matrix}18'' \\ \text{square}\end{matrix}\right) + \left(\begin{matrix}12''\text{x}16'' \\ \text{rectangle}\end{matrix}\right) = \left(\begin{matrix}\text{total} \\ \text{fabric}\end{matrix}\right)$

$\left(\frac{1}{2}\text{yd}\right) + \left(\frac{3}{4}\text{yd}\right) + \left(\frac{7}{8}\text{yd}\right) =$

$\frac{1}{2} + \frac{3}{4} + \frac{7}{8} =$

$\frac{4}{8} + \frac{6}{8} + \frac{7}{8} =$

$\frac{17}{8} = 2\frac{1}{8}$ yards

99. $\left(\begin{matrix}\text{Amount walk} \\ \text{shortened}\end{matrix}\right) = \left(\begin{matrix}\text{Daily} \\ \text{walk}\end{matrix}\right) - \left(\begin{matrix}\text{Amount} \\ \text{walked}\end{matrix}\right)$

$= 2\frac{1}{2} - \frac{3}{4}$

$= \left(2 + \frac{1}{2}\right) - \frac{3}{4}$

$= \frac{8}{4} + \frac{2}{4} - \frac{3}{4} = \frac{7}{4}$

$= 1\frac{3}{4}$ miles

101. Perimeter is the sum of the lengths of the three sides of the triangle.

$P = a + b + c$

$= 14\frac{1}{2} + 12\frac{1}{3} + 9\frac{5}{6}$

$= 14 + \frac{1}{2} + 12 + \frac{1}{3} + 9 + \frac{5}{6}$

$= 14 + 12 + 9 + \frac{1}{2} + \frac{1}{3} + \frac{5}{6}$

$= 35 + \frac{3}{6} + \frac{2}{6} + \frac{5}{6}$

$= 35 + \frac{10}{6} = 35 + 1\frac{4}{6} = 36\frac{2}{3}$

The plot of ground needs $36\frac{2}{3}$ yards of fencing.

PROBLEM SET 2.3 Real Numbers and Algebraic Expressions

1. -2 is real, rational, integer, and negative.

3. $\sqrt{5}$ is real, irrational, and positive.

5. 0.16 is real, rational, noninteger and positive.

7. $-\frac{8}{7}$ is real, rational, noninteger and negative.

9. $0.37 + 0.25 = 0.62$

11. $2.93 - 1.48 = 1.45$

13. $(-4.7) + 1.4 = -3.3$

15. $-3.8 + 11.3 = 7.5$

17. $6.6 - (-1.2) = 6.6 + 1.2 = 7.8$

19. $-11.5 - (-10.6) = -11.5 + 10.6 = -0.9$

21. $(0.4)(2.9) = 1.16$

23. $(-0.8)(0.34) = -0.272$

25. $(9)(-2.7) = -24.3$

27. $(-0.7)(-64) = 44.8$

29. $1.56 \div 1.3 = 1.2$

31. $5.92 \div (-0.8) = -7.4$

33. $-0.266 \div (-0.7) = 0.38$

35. $16.5 - 18.7 + 9.4 =$
$16.5 + 9.4 - 18.7 =$
$25.9 - 18.7 = 7.2$

37. $0.34 - 0.21 - 0.74 + 0.19 =$
$0.34 + 0.19 - 0.21 - 0.74 =$
$0.53 - 0.95 = -0.42$

39. $0.76(0.2 + 0.8) = 0.76(1.0) = 0.76$

41. $0.6(4.1) + 0.7(3.2) = 2.46 + 2.24 = 4.7$

43. $7(0.6) + 0.9 - 3(0.4) + 0.4 =$
$4.2 + 0.9 - 1.2 + 0.4 =$
$4.2 + 0.9 + 0.4 - 1.2 =$
$5.5 - 1.2 = 4.3$

45. $(0.96) \div (-0.8) + 6(-1.4) - 5.2 =$
$-1.2 + 6(-1.4) - 5.2 =$
$-1.2 + (-8.4) - 5.2 = -14.8$

47. $5(2.3) - 1.2 - 7.36 \div 0.8 + 0.2 =$
$11.5 - 1.2 - 7.36 \div 0.8 + 0.2 =$
$11.5 - 1.2 - 9.2 + 0.2 =$
$11.5 + 0.2 - 1.2 - 9.2 =$
$11.7 - 10.4 = 1.3$

49. $x - 0.4x - 1.8x =$
$(1 - 0.4 - 1.8)x = -1.2x$

51. $5.4n - 0.8n - 1.6n =$
$(5.4 - 0.8 - 1.6)n = 3n$

53. $-3t + 4.2t - 0.9t + 0.2t =$
$(-3 + 4.2 - 0.9 + 0.2)t = 0.5t$

55. $3.6x - 7.4y - 9.4x + 10.2y =$
$3.6x - 9.4x - 7.4y + 102.y =$
$(3.6 - 9.4)x + (-7.4 + 10.2)y =$
$-5.8x + 2.8y$

57. $0.3(x - 4) + 0.4(x + 6) - 0.6x =$
$0.3x - 1.2 + 0.4x + 2.4 - 0.6x =$
$0.3x + 0.4x - 0.6x - 1.2 + 2.4 =$
$(0.3 + 0.4 - 0.6)x + (-1.2 + 2.4) =$
$0.1x + 1.2$

59. $6(x - 1.1) - 5(x - 2.3) - 4(x + 1.8) =$
$6x - 6.6 - 5x + 11.5 - 4x - 7.2 =$
$6x - 5x - 4x - 6.6 + 11.5 - 7.2 =$
$(6 - 5 - 4)x + (-6.6 + 11.5 - 7.2) =$
$-3x - 2.3$

61. $x + 2y + 3z$ for $x = \frac{3}{4}$, $y = \frac{1}{3}$, $z = -\frac{1}{6}$
$\left(\frac{3}{4}\right) + 2\left(\frac{1}{3}\right) + 3\left(-\frac{1}{6}\right) =$
$\frac{3}{4} + \frac{2}{3} - \frac{1}{2} = \frac{11}{12}$

63. $\frac{3}{5}y - \frac{2}{3}y - \frac{7}{15}y =$
$\left(\frac{3}{5} - \frac{2}{3} - \frac{7}{15}\right)y =$
$-\frac{8}{15}y$ for $y = -\frac{5}{2}$
$-\frac{8}{15}\left(-\frac{5}{2}\right) = \frac{4}{3}$

65. $-x-2y+4z$
for $x=1.7,\ y=-2.3,\ z=3.6$
$-(1.7)-2(-2.3)+4(3.6)=$
$-1.7+4.6+14.4=17.3$

67. $5x-7y$ for $x=-7.8,\ y=8.4$
$5(-7.8)-7(8.4)=$
$-39-58.8=-97.8$

69. $0.7x+0.6y$ for $x=-2,\ y=6$
$0.7(-2)+0.6(6)=$
$-1.4+3.6=2.2$

71. $1.2x+2.3x-1.4x-7.6x=$
$(1.2+2.3-1.4-7.6)x=$
$-5.5x$ for $x=-2.5$
$-5.5(-2.5)=13.75$

73. $-3a-1+7a-2=$
$-3a+7a-1-2=$
$4a-3$ for $a=0.9$
$4(0.9)-3=3.6-3=0.6$

75. $\begin{pmatrix}\text{value of}\\\text{stock}\end{pmatrix}=\begin{pmatrix}\text{number of}\\\text{shares}\end{pmatrix}\cdot\begin{pmatrix}\text{price per}\\\text{share}\end{pmatrix}$

$\begin{pmatrix}\text{value of}\\\text{400 shares}\end{pmatrix}=(400)\cdot(14.35)=5740$

$\begin{pmatrix}\text{value of}\\\text{250 shares}\end{pmatrix}=(250)\cdot(16.68)=4170$

$\begin{matrix}\text{value of}\\\text{650 shares}\end{matrix}=5740+4170=9,910$

Total value of stock is $9,910.

77. $\begin{pmatrix}\text{length of}\\\text{each piece}\end{pmatrix}=\begin{pmatrix}\text{total}\\\text{length}\end{pmatrix}\div\begin{pmatrix}\text{number of}\\\text{pieces}\end{pmatrix}$
$=76.4\div 4=19.1$

The length of each piece is 19.1 cm.

79. Perimeter $=4\cdot\begin{pmatrix}\text{length of}\\\text{each side}\end{pmatrix}$

$\frac{\text{Perimeter}}{4}=$ length of each side

$\frac{18.8}{4}=4.7$

The length of each side of the square is 4.7 cm.

81. $\begin{matrix}\text{cost of}\\\text{apples}\end{matrix}=\begin{pmatrix}\text{number of}\\\text{pounds}\end{pmatrix}\cdot\begin{pmatrix}\text{price per}\\\text{pound}\end{pmatrix}$

$\begin{matrix}\text{price of}\\\text{gala apples}\end{matrix}=(2)(1.79)=3.58$

$\begin{matrix}\text{price of}\\\text{fuji apples}\end{matrix}=(3)(0.99)=2.97$

Total cost of apples $=3.58+2.97=\$6.55$.

89. **(a)** Denominator has only factors of 2.
(b) Denominator has a factor of 3.
(c) $\frac{7}{8},\ \frac{11}{16},\ \frac{13}{32},\ \frac{17}{40},\ \frac{9}{20},\ \frac{3}{64}$

PROBLEM SET 2.4 Exponents

1. $2^6=2\cdot2\cdot2\cdot2\cdot2\cdot2=64$

3. $3^4=3\cdot3\cdot3\cdot3=81$

5. $(-2)^3=(-2)(-2)(-2)=-8$

7. $-3^2=-(3\cdot3)=-9$

9. $(-4)^2=(-4)(-4)=16$

11. $\left(\frac{2}{3}\right)^4=\left(\frac{2}{3}\right)\left(\frac{2}{3}\right)\left(\frac{2}{3}\right)\left(\frac{2}{3}\right)=\frac{16}{81}$

13. $-\left(\frac{1}{2}\right)^3=-\left(\frac{1}{2}\right)\left(\frac{1}{2}\right)\left(\frac{1}{2}\right)=-\frac{1}{8}$

15. $\left(-\frac{3}{2}\right)^2=\left(-\frac{3}{2}\right)\left(-\frac{3}{2}\right)=\frac{9}{4}$

17. $(0.3)^3=(0.3)(0.3)(0.3)=0.027$

19. $-(1.2)^2=-(1.2)(1.2)=-1.44$

21. $3^2 + 2^3 - 4^3 = 9 + 8 - 64 = -47$

23. $(-2)^3 - 2^4 - 3^2 =$
$-8 - 16 - 9 = -33$

25. $5(2)^2 - 4(2) - 1 = 5(4) - 8 - 1 =$
$20 - 8 - 1 = 11$

27. $-2(3)^3 - 3(3)^2 + 4(3) - 6 =$
$-2(27) - 3(9) + 12 - 6 =$
$-54 - 27 + 12 - 6 = -75$

29. $-7^2 - 6^2 + 5^2 =$
$-49 - 36 + 25 = -60$

31. $-3(-4)^2 - 2(-3)^3 + (-5)^2 =$
$-3(16) - 2(-27) + 25 =$
$-48 + 54 + 25 = 31$

33. $\frac{-3(2)^4}{12} + \frac{5(-3)^3}{15} =$
$\frac{-3(16)}{12} + \frac{5(-27)}{15} =$
$\frac{-48}{12} + \frac{-135}{15} = -4 - 9 = -13$

35. $9 \cdot x \cdot x = 9x^2$

37. $3 \cdot 4 \cdot x \cdot y \cdot y = 12xy^2$

39. $-2 \cdot 9 \cdot x \cdot x \cdot x \cdot x \cdot y = -18x^4y$

41. $(5x)(3y) = 5 \cdot 3 \cdot x \cdot y = 15xy$

43. $(6x^2)(2x^2) = 6 \cdot 2 \cdot x \cdot x \cdot x \cdot x = 12x^4$

45. $(-4a^2)(-2a^3) =$
$(-4)(-2) \cdot a \cdot a \cdot a \cdot a \cdot a = 8a^5$

47. $3x^2 - 7x^2 - 4x^2 =$
$(3 - 7 - 4)x^2 = -8x^2$

49. $-12y^3 + 17y^3 - y^3 =$
$(-12 + 17 - 1)y^3 = 4y^3$

51. $7x^2 - 2y^2 - 9x^2 + 8y^2 =$
$(7 - 9)x^2 + (-2 + 8)y^2 =$
$-2x^2 + 6y^2$

53. $\frac{2}{3}n^2 - \frac{1}{4}n^2 - \frac{3}{5}n^2 =$
$\left(\frac{2}{3} - \frac{1}{4} - \frac{3}{5}\right)n^2 =$
$\left(\frac{40}{60} - \frac{15}{60} - \frac{36}{60}\right)n^2 =$
$-\frac{11}{60}n^2$

55. $5x^2 - 8x - 7x^2 + 2x =$
$(5 - 7)x^2 + (-8 + 2)x =$
$-2x^2 - 6x$

57. $x^2 - 2x - 4 + 6x^2 - x + 12 =$
$x^2 + 6x^2 - 2x - x - 4 + 12 =$
$7x^2 - 3x + 8$

59. $\frac{9xy}{15x} = \frac{\overset{3}{\cancel{9}} \cdot \cancel{x} \cdot y}{\underset{5}{\cancel{15}} \cdot \cancel{x}} = \frac{3y}{5}$

61. $\frac{22xy^2}{6xy^3} = \frac{\overset{11}{\cancel{22}} \cdot \cancel{x} \cdot \cancel{y} \cdot \cancel{y}}{\underset{3}{\cancel{6}} \cdot \cancel{x} \cdot \cancel{y} \cdot \cancel{y} \cdot y} = \frac{11}{3y}$

63. $\frac{7a^2b^3}{17a^3b} = \frac{7 \cdot \cancel{a} \cdot \cancel{a} \cdot \cancel{b} \cdot b \cdot b}{17 \cdot \cancel{a} \cdot \cancel{a} \cdot a \cdot \cancel{b}} = \frac{7b^2}{17a}$

65. $\frac{-24abc^2}{32bc} = \frac{\overset{3}{\cancel{24}} \cdot a \cdot \cancel{b} \cdot \cancel{c} \cdot c}{\underset{4}{\cancel{32}} \cdot \cancel{b} \cdot \cancel{c}} = -\frac{3ac}{4}$

67. $\frac{-5x^4y^3}{-20x^2y} = \frac{\cancel{5} \cdot \cancel{x} \cdot \cancel{x} \cdot x \cdot x \cdot \cancel{y} \cdot y \cdot y}{4 \cdot \cancel{5} \cdot \cancel{x} \cdot \cancel{x} \cdot \cancel{y}} =$
$\frac{x^2y^2}{4}$

69. $\left(\frac{7x^2}{9y}\right)\left(\frac{12y}{21x}\right) = \frac{\overset{1}{\cancel{7}} \cdot \overset{4}{\cancel{12}} \cdot \cancel{x} \cdot x \cdot \cancel{y}}{\underset{3}{\cancel{9}} \cdot \underset{3}{\cancel{21}} \cdot \cancel{x} \cdot \cancel{y}} = \frac{4x}{9}$

Problem Set 2.4

71. $\left(\frac{5c}{a^2b^2}\right) \div \left(\frac{12c}{ab}\right) = \frac{5c}{a^2b^2} \cdot \frac{ab}{12c} =$

$\frac{5 \cdot \cancel{a} \cdot \cancel{b} \cdot \cancel{c}}{12 \cdot \cancel{a} \cdot a \cdot \cancel{b} \cdot b \cdot \cancel{c}} = \frac{5}{12ab}$

73. $\frac{6}{x} + \frac{5}{y^2} = \frac{6 \cdot y^2}{x \cdot y^2} + \frac{5 \cdot x}{y^2 \cdot x} =$

$\frac{6y^2}{xy^2} + \frac{5x}{xy^2} = \frac{6y^2 + 5x}{xy^2}$

75. $\frac{5}{x^4} - \frac{7}{x^2} = \frac{5}{x^4} - \frac{7x^2}{x^2} \cdot \frac{x^2}{x^2} =$

$\frac{5}{x^4} - \frac{7x^2}{x^4} = \frac{5 - 7x^2}{x^4}$

77. $\frac{3}{2x^3} + \frac{6}{x} = \frac{3}{2x^3} + \frac{6}{x} \cdot \frac{2x^2}{2x^2} =$

$\frac{3}{2x^3} + \frac{12x^2}{2x^3} = \frac{3 + 12x^2}{2x^3}$

79. $\frac{-5}{4x^2} + \frac{7}{3x^2} = \frac{-5 \cdot 3}{4x^2 \cdot 3} + \frac{7 \cdot 4}{3x^2 \cdot 4} =$

$\frac{-15}{12x^2} + \frac{28}{12x^2} = \frac{13}{12x^2}$

81. $\frac{11}{a^2} - \frac{14}{b^2} = \frac{11 \cdot b^2}{a^2 \cdot b^2} - \frac{14 \cdot a^2}{b^2 \cdot a^2} =$

$\frac{11b^2}{a^2b^2} - \frac{14a^2}{a^2b^2} = \frac{11b^2 - 14a^2}{a^2b^2}$

83. $\frac{1}{2x^3} - \frac{4}{3x^2} = \frac{1 \cdot 3}{2x^3 \cdot 3} - \frac{4 \cdot 2x}{3x^2 \cdot 2x} =$

$\frac{3}{6x^3} - \frac{8x}{6x^3} = \frac{3 - 8x}{6x^3}$

85. $\frac{3}{x} - \frac{4}{y} - \frac{5}{xy} = \frac{3 \cdot y}{x \cdot y} - \frac{4 \cdot x}{y \cdot x} - \frac{5}{xy} =$

$\frac{3y}{xy} - \frac{4x}{xy} - \frac{5}{xy} = \frac{3y - 4x - 5}{xy}$

87. $4x^2 + 7y^2$ for $x = -2, y = -3$

$4(-2)^2 + 7(-3)^2 =$

$4(4) + 7(9) =$

$16 + 63 = 79$

89. $3x^2 - y^2$ for $x = \frac{1}{2}, y = -\frac{1}{3}$

$3\left(\frac{1}{2}\right)^2 - \left(-\frac{1}{3}\right)^2 =$

$3\left(\frac{1}{4}\right) - \left(\frac{1}{9}\right) =$

$\frac{3}{4} - \frac{1}{9} = \frac{3 \cdot 9}{4 \cdot 9} - \frac{1 \cdot 4}{9 \cdot 4} =$

$\frac{27}{36} - \frac{4}{36} = \frac{23}{36}$

91. $x^2 - 2xy + y^2$ for $x = -\frac{1}{2}, y = 2$

$\left(-\frac{1}{2}\right)^2 - 2\left(-\frac{1}{2}\right)(2) + (2)^2 =$

$\left(\frac{1}{4}\right) + 2 + 4 = \frac{1}{4} + \frac{2 \cdot 4}{1 \cdot 4} + \frac{4 \cdot 4}{1 \cdot 4} =$

$\frac{1}{4} + \frac{8}{4} + \frac{16}{4} = \frac{25}{4}$

93. $-x^2$ for $x = -8$

$-x^2 = -(-8)^2 =$

$-(-8)(-8) = -(64) = -64$

95. $-x^2 - y^2$ for $x = -3, y = -4$

$-(-3)^2 - (-4)^2 =$

$-9 - 16 = -25$

97. $-a^2 - 3b^3$ for $a = -6, b = -1$

$-(-6)^2 - 3(-1)^3 =$

$-36 - 3(-1) =$

$-36 + 3 = -33$

99. $y^2 - 3xy$ for $x = 0.4, y = -0.3$

$(-0.3)^2 - 3(0.4)(-0.3) =$

$0.09 - 3(-0.12) =$

$0.09 + 0.36 = 0.45$

PROBLEM SET 2.5 Translating from English to Algebra

For problems, 1-12, the answers will vary.

1. The difference of a and b
3. One-third of the product of B and h
5. Two times the quantity, l plus w
7. The quotient of A divided by w
9. The quantity, a plus b, divided by 2
11. Two more than three times y
13. $l + w$
15. ab
17. $\frac{d}{t}$
19. lwh
21. $y - x$
23. $xy + 2$
25. $7 - y^2$
27. $\frac{x - y}{4}$
29. $10 - x$
31. $10(n + 2)$
33. $xy - 7$
35. $xy - 12$

For problems 37-68, it may help to do a specific example before trying to formulate the general expression.

37. Suppose that the sum of two numbers is 35 and one of the numbers is 14. Then we subtract to find the other number, $35 - 14$. Thus, if one of the numbers is n, then the other number is $35 - n$.
39. Since the smaller number plus the difference must equal the larger number, the other number must be $n + 45$.
41. In ten years, Janet's age will be increased by 10, so it is represented as $y + 10$.
43. Twice Debra's age is $2x$, so Debra's mother is 3 years less than $2x$ or $2x - 3$.
45. Three dimes and five quarters is $10(3) + 25(2) = 80$ cents. Thus d dimes and q quarters is $10d + 25q$ cents.
47. If a car travels 200 miles in 5 hours, then it would be traveling at a rate of $\frac{200 \text{ miles}}{5 \text{ hours}} = 40$ miles per hour. Therefore, the rate of the car is $\frac{d}{t}$.
49. If 5 pounds of candy cost \$15, then to find the price per pound we divide 15 by 5. Thus, if p pounds cost d dollars, $\frac{d}{p}$ represents the price per pound.
51. To find a monthly salary, the annual salary is divided by 12, so Larry's monthly salary is $\frac{d}{12}$.
53. If 6 is the whole number, then the next larger whole number would be $6 + 1 = 7$. Thus, if n is the whole number, the next larger whole number is $n + 1$.
55. Suppose 7 is the odd number, then the next larger odd number would be $7 + 2 = 9$. Thus, if n is the odd number, the next larger odd number would be $n + 2$.
57. If Willie is y years old, then twice Willie's age is $2y$. Since his father is 2 years less than twice Willie's age, the father's age is $2y - 2$. The sum of their ages is $y + (2y - 2) = 3y - 2$.

Problem Set 2.5

59. To convert yards to inches, the yards must be multiplied by 36. To convert feet to inches, the number of feet must be multiplied by 12. If the perimeter of a rectangle is 5 yards and 2 feet, then the perimeter in inches is $36(5) + 12(2) =$ 204 inches. Thus, the perimeter of a rectangle that is y yards and f feet is $36y + 12f$ inches.

61. To change feet to yards, we divide the number of feet by 3. Therefore, f feet equals $\frac{f}{3}$ yards.

63. The width of a rectangle is w feet and the length is three times the width, or $3w$ feet. The perimeter of the rectangle is the sum of the lengths of the four sides. There are 2 widths and 2 lengths in a rectangle so the perimeter is 2 times the width plus 2 times the length or $2(w) + 2(3w) = 8w$ feet.

3w = length

w   w = width

3w

65. See problem 63. The length is l inches and the width is $\left(\frac{l}{2} - 2\right)$ inches. The perimeter is $2l + 2(\frac{l}{2} - 2) = 3l - 4$ inches.

l = length

$\frac{l}{2}$ - 2 = width

67. The first side of a triangle is f feet long. The second side is 2 feet longer, or $f + 2$ feet long. The third side is twice as long as the second side, or $2(f + 2)$. The perimeter is the sum of the lengths of the sides, or $f + (f+2) + 2(f+2) = (4f + 6)$feet. To convert to inches, multiply the number of feet by 12. The perimeter of the triangle in inches is $12(4f + 6) = 48f + 72$ inches.

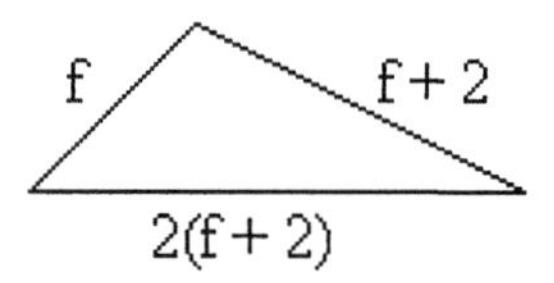

69. Let $l =$ length and $w =$ width.
Length is twice the width: $l = 2w$
Area of Rectangle $=$ (length)(width)
$A = lw$
$A = (2w)(w) = 2w^2$
The area of the rectangle is $2w^2$ square yards.

71. Let $s =$ length of side in yards, so $3s =$ length of side in feet.
Area of Square $=$ (side)^2
$A = s^2$
$A = (3s)^2$
The area of the square in feet is $9s^2$ square feet.

CHAPTER 2 Review Problem Set

1. $2^6 = 2 \cdot 2 \cdot 2 \cdot 2 \cdot 2 \cdot 2 = 64$

2. $(-3)^3 = (-3)(-3)(-3) = -27$

3. $-4^2 = -(4 \cdot 4) = -(16) = -16$

4. $5^3 = 5 \cdot 5 \cdot 5 = 125$

5. $\left(\frac{3}{4}\right)^2 = \left(\frac{3}{4}\right)\left(\frac{3}{4}\right) = \frac{9}{16}$

6. $-\left(\frac{1}{2}\right)^2 = -\left(\frac{1}{2}\right)\left(\frac{1}{2}\right) = -\frac{1}{4}$

7. $\left(\frac{1}{2}+\frac{2}{3}\right)^2=\left(\frac{3}{6}+\frac{4}{6}\right)^2=$
$\left(\frac{7}{6}\right)^2=\left(\frac{7}{6}\right)\left(\frac{7}{6}\right)=\frac{49}{36}$

8. $(0.6)^3=(0.6)(0.6)(0.6)=0.216$

9. $(0.12)^2=(0.12)(0.12)=0.0144$

10. $(0.06)^2=(0.06)(0.06)=0.0036$

11. $\left(-\frac{2}{3}\right)^3=$
$\left(-\frac{2}{3}\right)\left(-\frac{2}{3}\right)\left(-\frac{2}{3}\right)=-\frac{8}{27}$

12. $\left(-\frac{1}{2}\right)^4=$
$\left(-\frac{1}{2}\right)\left(-\frac{1}{2}\right)\left(-\frac{1}{2}\right)\left(-\frac{1}{2}\right)=\frac{1}{16}$

13. $\left(\frac{1}{4}-\frac{1}{2}\right)^3=\left(\frac{1}{4}-\frac{2}{4}\right)^3=$
$\left(-\frac{1}{4}\right)\left(-\frac{1}{4}\right)\left(-\frac{1}{4}\right)=-\frac{1}{64}$

14. $(0.5)^2=(0.5)(0.5)=0.25$

15. $\left(\frac{1}{2}+\frac{1}{3}-\frac{1}{6}\right)^2=\left(\frac{3}{6}+\frac{2}{6}-\frac{1}{6}\right)^2=$
$\left(\frac{4}{6}\right)^2=\left(\frac{2}{3}\right)^2=\left(\frac{2}{3}\right)\left(\frac{2}{3}\right)=\frac{4}{9}$

16. $\frac{3}{8}+\frac{5}{12}=\frac{3}{8}\cdot\frac{3}{3}+\frac{5}{12}\cdot\frac{2}{2}=$
$\frac{9}{24}+\frac{10}{24}=\frac{19}{24}$

17. $\frac{9}{14}-\frac{3}{35}=\frac{9}{14}\cdot\frac{5}{5}-\frac{3}{35}\cdot\frac{2}{2}=$
$\frac{45}{70}-\frac{6}{70}=\frac{39}{70}$

18. $\frac{2}{3}+\frac{-3}{5}=\frac{2}{3}\cdot\frac{5}{5}+\frac{-3}{5}\cdot\frac{3}{3}=$
$\frac{10}{15}+\frac{-9}{15}=\frac{1}{15}$

19. $\frac{7}{x}+\frac{9}{2y}=\frac{7\cdot 2y}{x\cdot 2y}+\frac{9\cdot x}{2y\cdot x}=$
$\frac{14y}{2xy}+\frac{9x}{2xy}=\frac{14y+9x}{2xy}$

20. $\frac{5}{xy}-\frac{8}{x^2}=\frac{5\cdot x}{xy\cdot x}-\frac{8\cdot y}{x^2\cdot y}=$
$\frac{5x}{x^2y}-\frac{8y}{x^2y}=\frac{5x-8y}{x^2y}$

21. $\left(\frac{7y}{8x}\right)\left(\frac{14x}{35}\right)=\frac{\overset{1}{\cancel{7}}y}{\underset{4}{\cancel{8}}\cancel{x}}\cdot\frac{\overset{7}{\cancel{14}}\cancel{x}}{\underset{5}{\cancel{35}}}=\frac{7y}{20}$

22. $\left(\frac{6xy}{9y^2}\right)\div\left(\frac{15y}{18x^2}\right)=\frac{6xy}{9y^2}\cdot\frac{18x^2}{15y}=$
$\frac{108x^3y}{135y^3}=\frac{4x^3}{5y^2}$

23. $\left(\frac{-3x}{12y}\right)\left(\frac{8y}{-7x}\right)=\frac{-3x}{12y}\cdot\frac{8y}{-7x}=$
$\frac{-24xy}{-84xy}=\frac{2}{7}$

24. $\left(-\frac{4y}{3x}\right)\left(-\frac{3x}{4y}\right)=-\frac{4y}{3x}\cdot-\frac{3x}{4y}=$
$\frac{12xy}{12xy}=1$

25. $\left(\frac{6n}{7}\right)\left(\frac{9n}{8}\right)=\frac{6n}{7}\cdot\frac{9n}{8}=$
$\frac{54n^2}{56}=\frac{27n^2}{28}$

26. $\frac{1}{6}+\frac{2}{3}\cdot\frac{3}{4}-\frac{5}{6}\div\frac{8}{6}=$
$\frac{1}{6}+\frac{2}{4}-\frac{5}{6}\div\frac{8}{6}=$
$\frac{1}{6}+\frac{1}{2}-\frac{5}{6}\cdot\frac{6}{8}=$

$$\frac{1}{6}+\frac{1}{2}-\frac{5}{8}=\frac{4}{24}+\frac{12}{24}-\frac{15}{24}=\frac{1}{24}$$

27. $\frac{3}{4}\cdot\frac{1}{2}-\frac{4}{3}\cdot\frac{3}{2}=$
$\frac{3}{8}-\frac{4}{2}=\frac{3}{8}-\frac{16}{8}=-\frac{13}{8}$

28. $\frac{7}{9}\cdot\frac{3}{5}+\frac{7}{9}\cdot\frac{2}{5}=\frac{7}{9}\left(\frac{3}{5}+\frac{2}{5}\right)=$
$\frac{7}{9}\left(\frac{5}{5}\right)=\frac{7}{9}\cdot 1=\frac{7}{9}$

29. $\frac{4}{5}\div\frac{1}{5}\cdot\frac{2}{3}-\frac{1}{4}=\left(\frac{4}{5}\cdot\frac{5}{1}\right)\cdot\frac{2}{3}-\frac{1}{4}=$
$\frac{4}{1}\cdot\frac{2}{3}-\frac{1}{4}=\frac{8}{3}-\frac{1}{4}=\frac{8\cdot 4}{3\cdot 4}-\frac{1\cdot 3}{4\cdot 3}=$
$\frac{32}{12}-\frac{3}{12}=\frac{29}{12}$

30. $\frac{2}{3}\cdot\frac{1}{4}\div\frac{1}{2}+\frac{2}{3}\cdot\frac{1}{4}=$
$\frac{2}{12}\div\frac{1}{2}+\frac{2}{12}=$
$\left(\frac{2}{12}\cdot\frac{2}{1}\right)+\frac{2}{12}=$
$\frac{4}{12}+\frac{2}{12}=\frac{6}{12}=\frac{1}{2}$

31. $0.48+0.72-0.35-0.18=$
$1.20-0.53=0.67$

32. $0.81+(0.6)(0.4)-(0.7)(0.8)=$
$0.81+0.24-0.56=$
$1.05-0.56=0.49$

33. $1.28\div 0.8-0.81\div 0.9+1.7=$
$1.6-0.81\div 0.9+1.7=$
$1.6-0.9+1.7=2.4$

34. $(0.3)^2+(0.4)^2-(0.6)^2=$
$0.09+0.16-0.36=-0.11$

35. $(1.76)(0.8)+(1.76)(0.2)=$
$1.76(0.8+0.2)=1.76(1.0)=1.76$

36. $(2^2-2-2^3)^2=(4-2-8)^2=$
$(-6)^2=36$

37. $1.92(0.9+0.1)=1.92(1.0)=1.92$

38. $\frac{3}{8}x^2-\frac{2}{5}y^2-\frac{2}{7}x^2+\frac{3}{4}y^2=$
$\frac{3}{8}x^2-\frac{2}{7}x^2-\frac{2}{5}y^2+\frac{3}{4}y^2=$
$\left(\frac{3}{8}-\frac{2}{7}\right)x^2+\left(-\frac{2}{5}+\frac{3}{4}\right)y^2=$
$\frac{5}{56}x^2+\frac{7}{20}y^2$

39. $0.24ab+0.73bc-0.82ab-0.37bc=$
$0.24ab-0.82ab+0.73bc-0.37bc=$
$(0.24-0.82)ab+(0.73-0.37)bc=$
$-0.58ab+0.36bc$

40. $\frac{1}{2}x+\frac{3}{4}x-\frac{5}{6}x+\frac{1}{24}x=$
$\frac{12}{24}x+\frac{18}{24}x-\frac{20}{24}x+\frac{1}{24}x=$
$\frac{31}{24}x-\frac{20}{24}x=\frac{11}{24}x$

41. $1.4a-1.9b+0.8a+3.6b=2.2a+1.7b$

42. $\frac{2}{5}n+\frac{1}{3}n-\frac{5}{6}n=$
$\frac{12}{30}n+\frac{10}{30}n-\frac{25}{30}n=$
$\frac{22}{30}n-\frac{25}{30}n=-\frac{3}{30}n=-\frac{1}{10}n$

43. $n-\frac{3}{4}n+2n-\frac{1}{5}n=$
$\left(1-\frac{3}{4}+2-\frac{1}{5}\right)n=$
$\left(\frac{20}{20}-\frac{15}{20}+\frac{40}{20}-\frac{4}{20}\right)n=\frac{41}{20}n$

44. $\frac{1}{4}x - \frac{2}{5}y$ for $x = \frac{2}{3},\ y = -\frac{5}{7}$

$\frac{1}{4}\left(\frac{2}{3}\right) - \frac{2}{5}\left(-\frac{5}{7}\right) =$

$\frac{1}{6} + \frac{2}{7} = \frac{7}{42} + \frac{12}{42} = \frac{19}{42}$

45. $a^3 + b^2$ for $a = -\frac{1}{2},\ b = \frac{1}{3}$

$\left(-\frac{1}{2}\right)\left(-\frac{1}{2}\right)\left(-\frac{1}{2}\right) + \left(\frac{1}{3}\right)\left(\frac{1}{3}\right) =$

$-\frac{1}{8} + \frac{1}{9} = -\frac{9}{72} + \frac{8}{72} = -\frac{1}{72}$

46. $2x^2 - 3y^2$ for $x = 0.6,\ y = 0.7$

$2(0.6)^2 - 3(0.7)^2 =$

$2(0.36) - 3(0.49) =$

$0.72 - 1.47 = -0.75$

47. $0.7w + 0.9z$ for $w = 0.4,\ z = -0.7$

$0.7(0.4) + 0.9(-0.7) =$

$0.28 - 0.63 = -0.35$

48. $\frac{3}{5}x - \frac{1}{3}x + \frac{7}{15}x - \frac{2}{3}x =$

$\left(\frac{3}{5} - \frac{1}{3} + \frac{7}{15} - \frac{2}{3}\right)x =$

$\left(\frac{9}{15} - \frac{5}{15} + \frac{7}{15} - \frac{10}{15}\right)x =$

$\frac{1}{15}x$ for $x = \frac{15}{17}$

$\frac{1}{15}\left(\frac{15}{17}\right) = \frac{1}{17}$

49. $\frac{1}{3}n + \frac{2}{7}n - n =$

$\left(\frac{1}{3} + \frac{2}{7} - 1\right)n =$

$\left(\frac{7}{21} + \frac{6}{21} - \frac{21}{21}\right)n =$

$-\frac{8}{21}n$ for $n = 21$

$-\frac{8}{21}\left(21\right) = -8$

50. If one number is n,
the other number is $72 - n$.

51. If Joan has 3 pennies and 4 dimes,
she has $3 + 10(4) = 43$ cents.
Thus, if she has p pennies and d dimes,
she has $p + 10d$ cents.

52. 1 hour = 60 minutes

$\frac{x \text{ words}}{1 \text{ hour}} \cdot \frac{1 \text{ hour}}{60 \text{ minutes}} = \frac{x}{60}$ words/minute

53. Twice Harry's age is $2y$. If his brother is
3 years less than twice Harry's age,
his brother is $2y - 3$ years old.

54. Larry chose n. Cindy chose 3 more than
5 times Larry's number or $5n + 3$.

55. To convert yards to inches, the yards must be multiplied by 36. To convert feet to inches, the feet must be multiplied by 12. If the file cabinet is y yards and f feet tall, then it would be $36y + 12f$ inches tall.

56. 100 cm $= 1$m

m meters $\cdot$ 100cm/m $= (100m)$cm

57. If Corinne has 3 nickels, 4 dimes,
and 2 quarters, she has
$5(3) + 10(4) + 25(2) = 105$ cents.
Thus, if she has n nickels, d dimes, and
q quarters she has $5n + 10d + 25q$ cents.

58. $n - 5$

59. $5 - n$

60. $10(x - 2)$

61. $10x - 2$

62. $x - 3$

63. $\frac{d}{r}$

64. $x^2 + 9$

65. $(x + 9)^2$

66. $x^3 + y^3$

67. $xy - 4$

CHAPTER 2 | Test

1 **a.** $(-3)^4 = (-3)(-3)(-3)(-3) = 81$

b. $-2^6 = -(2)(2)(2)(2)(2)(2) = -64$

c. $(0.2)^3 = (0.2)(0.2)(0.2) = 0.008$

2. $\dfrac{42}{54} = \dfrac{6 \cdot 7}{6 \cdot 9} = \dfrac{7}{9}$

3. $\dfrac{18xy^2}{32y} = \dfrac{2 \cdot 9 \cdot x \cdot y \cdot y}{2 \cdot 16 \cdot y} = \dfrac{9xy}{16}$

4. $5.7 - 3.8 + 4.6 - 9.1 =$
$5.7 + 4.6 - 3.8 - 9.1 =$
$10.3 - 12.9 = -2.6$

5. $0.2(0.4) - 0.6(0.9) + 0.5(7) =$
$0.08 - 0.54 + 3.5 = 3.04$

6. $-0.4^2 + 0.3^2 - 0.7^2 =$
$-0.16 + 0.09 - 0.49 = -0.56$

7. $\left(\dfrac{1}{3} - \dfrac{1}{4} + \dfrac{1}{6}\right)^4 = \left(\dfrac{4}{12} - \dfrac{3}{12} + \dfrac{2}{12}\right)^4 =$
$\left(\dfrac{3}{12}\right)^4 = \left(\dfrac{1}{4}\right)^4 =$
$\left(\dfrac{1}{4}\right)\left(\dfrac{1}{4}\right)\left(\dfrac{1}{4}\right)\left(\dfrac{1}{4}\right) = \dfrac{1}{256}$

8. $\dfrac{5}{12} \div \dfrac{15}{8} = \dfrac{\overset{1}{\cancel{5}}}{\underset{3}{\cancel{12}}} \cdot \dfrac{\overset{2}{\cancel{8}}}{\underset{3}{\cancel{15}}} = \dfrac{2}{9}$

9. $-\dfrac{2}{3} - \dfrac{1}{2}\left(\dfrac{3}{4}\right) + \dfrac{5}{6} =$
$-\dfrac{2}{3} - \dfrac{3}{8} + \dfrac{5}{6} =$
$-\dfrac{16}{24} - \dfrac{9}{24} + \dfrac{20}{24} = -\dfrac{5}{24}$

10. $3\left(\dfrac{2}{5}\right) - 4\left(\dfrac{5}{6}\right) + 6\left(\dfrac{7}{8}\right) =$
$\dfrac{6}{5} - \dfrac{10}{3} + \dfrac{21}{4} =$
$\dfrac{72}{60} - \dfrac{200}{60} + \dfrac{315}{60} = \dfrac{187}{60}$

11. $4\left(\dfrac{1}{2}\right)^3 - 3\left(\dfrac{2}{3}\right)^2 + 9\left(\dfrac{1}{4}\right)^2 =$
$4\left(\dfrac{1}{8}\right) - 3\left(\dfrac{4}{9}\right) + 9\left(\dfrac{1}{16}\right) =$
$\dfrac{1}{2} - \dfrac{4}{3} + \dfrac{9}{16} =$
$\dfrac{24}{48} - \dfrac{64}{48} + \dfrac{27}{48} = -\dfrac{13}{48}$

12. $\dfrac{8x}{15y} \cdot \dfrac{9y^2}{6x} = \dfrac{\overset{4}{\cancel{72}}\ xy^2}{\underset{5}{\cancel{90}}\ xy} = \dfrac{4y}{5}$

13. $\dfrac{6xy}{9} \div \dfrac{y}{3x} = \dfrac{6xy}{9} \cdot \dfrac{3x}{y} =$
$\dfrac{18x^2y}{9y} = 2x^2$

14. $\dfrac{4}{x} - \dfrac{5}{y^2} = \left(\dfrac{4}{x} \cdot \dfrac{y^2}{y^2}\right) - \left(\dfrac{5}{y^2} \cdot \dfrac{x}{x}\right) =$
$\dfrac{4y^2}{xy^2} - \dfrac{5x}{xy^2} = \dfrac{4y^2 - 5x}{xy^2}$

15. $\dfrac{3}{2x} + \dfrac{7}{6x} = \left(\dfrac{3}{2x} \cdot \dfrac{3}{3}\right) + \dfrac{7}{6x} =$
$\dfrac{9}{6x} + \dfrac{7}{6x} = \dfrac{16}{6x} = \dfrac{8}{3x}$

16. $\dfrac{5}{3y} + \dfrac{9}{7y^2} = \left(\dfrac{5}{3y} \cdot \dfrac{7y}{7y}\right) + \left(\dfrac{9}{7y^2} \cdot \dfrac{3}{3}\right) =$
$\dfrac{35y}{21y^2} + \dfrac{27}{21y^2} = \dfrac{35y + 27}{21y^2}$

17. $\left(\frac{15a^2b}{12a}\right)\left(\frac{8ab}{9b}\right)=\frac{120a^3b^2}{108ab}=\frac{10a^2b}{9}$

18. $3x-2xy-4x+7xy=$
$3x-4x-2xy+7xy=$
$-x+5xy$

19. $-2a^2+3b^2-5b^2-a^2=$
$-2a^2-a^2+3b^2-5b^2=$
$-3a^2-2b^2$

20. x^2-xy+y^2 for $x=\frac{1}{2}$, $y=-\frac{2}{3}$
$\left(\frac{1}{2}\right)^2-\left(\frac{1}{2}\right)\left(-\frac{2}{3}\right)+\left(-\frac{2}{3}\right)^2=$
$\frac{1}{4}-\left(-\frac{1}{3}\right)+\frac{4}{9}=$
$\frac{9}{36}+\frac{12}{36}+\frac{16}{36}=\frac{37}{36}$

21. $0.2x-0.3y-xy$ for $x=0.4$, $y=0.8$
$0.2(0.4)-0.3(0.8)-(0.4)(0.8)=$
$0.08-0.24-0.32=-0.48$

22. $\frac{3}{4}x-\frac{2}{3}y$ for $x=-\frac{1}{2}$, $y=\frac{3}{5}$
$\frac{3}{4}\left(-\frac{1}{2}\right)-\frac{2}{3}\left(\frac{3}{5}\right)=$
$-\frac{3}{8}-\frac{2}{5}=-\frac{3}{8}\cdot\frac{5}{5}-\frac{2}{5}\cdot\frac{8}{8}=$
$-\frac{15}{40}-\frac{16}{40}=-\frac{31}{40}$

23. $3x-2y+xy$ for $x=0.5$, $y=-0.9$
$3(0.5)-2(-0.9)+(0.5)(-0.9)=$
$1.5-(-1.8)+(-0.45)=$
$1.5+1.8+(-0.45)=2.85$

24. If David has 3 nickels, 4 dimes, and 2 quarters, he has $5(3)+10(4)+25(2)=105$ cents. So, if he has n nickels, d dimes, and q quarters, then he has $5n+10d+25q$ cents.

25. Hal chose n. Sheila chose 3 less than 4 times Hal's number or $4n-3$.

CHAPTERS 1-2 Cumulative Review

1. $16-18-14+21-14+19=$
$56-46=10$

2. $7(-6)-8(-6)+4(-9)=$
$-42-(-48)+(-36)=$
$-42+48+(-36)=-30$

3. $6-[3-(10-12)]=$
$6-[3-(-2)]=$
$6-[3+2]=$
$6-(5)=$
$6-5=1$

4. $-9-2[4-(-10+6)]-1=$
$-9-2[4-(-4)]-1=$
$-9-2(8)-1=$
$-9-16-1=-26$

5. $\frac{-7(-4)-5(-6)}{-2}=$
$\frac{28+30}{-2}=\frac{58}{-2}=-29$

6. $\frac{5(-3)+(-4)(6)-3(4)}{-3}=$
$\frac{-15+(-24)-12}{-3}=\frac{-51}{-3}=17$

7. $\frac{3}{4}+\frac{1}{3}\div\frac{4}{3}-\frac{1}{2}=$
$\frac{3}{4}+\frac{1}{3}\cdot\frac{3}{4}-\frac{1}{2}=$
$\frac{3}{4}+\frac{1}{4}-\frac{1}{2}=$
$\frac{4}{4}-\frac{1}{2}=$
$1-\frac{1}{2}=\frac{1}{2}$

8. $\left(\frac{2}{3}\right)\left(-\frac{3}{4}\right)-\left(\frac{5}{6}\right)\left(\frac{4}{5}\right)=$

$-\frac{1}{2}-\frac{2}{3}=-\frac{3}{6}-\frac{4}{6}=-\frac{7}{6}$

9. $\left(\frac{1}{2}-\frac{2}{3}\right)^2=\left(\frac{3}{6}-\frac{4}{6}\right)^2=$

$\left(-\frac{1}{6}\right)^2=\left(-\frac{1}{6}\right)\cdot\left(-\frac{1}{6}\right)=\frac{1}{36}$

10. $-4^3=-(4^3)=-(4\cdot4\cdot4)=$
$-(64)=-64$

11. $\frac{0.0046}{0.000023}=\left(\frac{0.0046}{0.000023}\right)\left(\frac{10000}{10000}\right)=$

$\frac{46}{0.23}=200$

12. $(0.2)^2-(0.3)^3+(0.4)^2=$
$0.04-0.027+0.16=0.173$

13. $3xy-2x-4y$ for $x=-6$ and $y=7$
$3(-6)(7)-2(-6)-4(7)=$
$-126+12-28=$
$-154+12=-142$

14. $-4x^2y-2xy^2+xy$ for $x=-2,\ y=-4$
$-4(-2)^2(-4)-2(-2)(-4)^2+(-2)(-4)=$
$-4(4)(-4)-2(-2)(16)+8=$
$64+64+8=136$

15. $\frac{5x-2y}{3x}$ for $x=\frac{1}{2},\ y=-\frac{1}{3}$

$\frac{5\left(\frac{1}{2}\right)-2\left(-\frac{1}{3}\right)}{3\left(\frac{1}{2}\right)}=$

$\frac{\frac{5}{2}-\left(-\frac{2}{3}\right)}{\frac{3}{2}}=\frac{\frac{5}{2}+\frac{2}{3}}{\frac{3}{2}}=$

$\frac{\frac{15}{6}+\frac{4}{6}}{\frac{3}{2}}=\frac{\frac{19}{6}}{\frac{3}{2}}=$

$\frac{19}{6}\div\frac{3}{2}=\frac{19}{6}\cdot\frac{2}{3}=\frac{19}{9}$

16. $0.2x-0.3y+2xy$ for $x=0.1,\ y=0.3$
$0.2(0.1)-0.3(0.3)+2(0.1)(0.3)=$
$0.02-0.09+0.06=-0.01$

17. $-7x+4y+6x-9y+x-y=$
$-7x+6x+x+4y-9y-y=$
$-6y$ for $x=0.2,\ y=0.4$
$-6(0.4)=-2.4$

18. $\frac{2}{3}x-\frac{3}{5}y+\frac{3}{4}x-\frac{1}{2}y=$

$\frac{2}{3}x+\frac{3}{4}x-\frac{3}{5}y-\frac{1}{2}y=$

$\left(\frac{2}{3}+\frac{3}{4}\right)x+\left(-\frac{3}{5}-\frac{1}{2}\right)y=$

$\frac{17}{12}x+\left(-\frac{11}{10}\right)y$ for $x=\frac{6}{5},\ y=-\frac{1}{4}$

$\frac{17}{12}\left(\frac{6}{5}\right)+\left(-\frac{11}{10}\right)\left(-\frac{1}{4}\right)=$

$\frac{17}{10}+\frac{11}{40}=\left(\frac{17}{10}\cdot\frac{4}{4}\right)+\frac{11}{40}=$

$\frac{68}{40}+\frac{11}{40}=\frac{79}{40}$

19. $\frac{2}{n}-\frac{3}{n^2}$ for $n=2$

$\frac{2}{2}-\frac{3}{(2)^2}=1-\frac{3}{4}=\frac{4}{4}-\frac{3}{4}=\frac{1}{4}$

20. $-ab+\frac{1}{5}a-\frac{2}{3}b$ for $a=-2,\ b=\frac{3}{4}$

$-(-2)\left(\frac{3}{4}\right)+\frac{1}{5}(-2)-\left(\frac{2}{3}\right)\left(\frac{3}{4}\right)=$

$\frac{3}{2}+\left(-\frac{2}{5}\right)-\frac{1}{2}=$

$\frac{15}{10}+\left(-\frac{4}{10}\right)-\frac{5}{10}=\frac{6}{10}=\frac{3}{5}$

21. $54 = 2 \cdot 3 \cdot 3 \cdot 3$

22. $78 = 2 \cdot 3 \cdot 13$

23. $91 = 7 \cdot 13$

24. $153 = 3 \cdot 3 \cdot 17$

25. $\left.\begin{array}{l} 42 = 2 \cdot 3 \cdot 7 \\ 70 = 2 \cdot 5 \cdot 7 \end{array}\right\}$
The greatest common factor is $2 \cdot 7 = 14$.

26. $\left.\begin{array}{l} 63 = 3 \cdot 3 \cdot 7 \\ 81 = 3 \cdot 3 \cdot 3 \cdot 3 \end{array}\right\}$
The greatest common factor is $3 \cdot 3 = 9$.

27. $\left.\begin{array}{l} 28 = 2 \cdot 2 \cdot 7 \\ 36 = 2 \cdot 2 \cdot 3 \cdot 3 \\ 52 = 2 \cdot 2 \cdot 13 \end{array}\right\}$
The greatest common factor is $2 \cdot 2 = 4$.

28. $\left.\begin{array}{l} 48 = 2 \cdot 2 \cdot 2 \cdot 2 \cdot 3 \\ 66 = 2 \cdot 3 \cdot 11 \\ 78 = 2 \cdot 3 \cdot 13 \end{array}\right\}$
The greatest common factor is $2 \cdot 3 = 6$.

29. $\left.\begin{array}{l} 20 = 2 \cdot 2 \cdot 5 \\ 28 = 2 \cdot 2 \cdot 7 \end{array}\right\}$
The least common multiple is $2 \cdot 2 \cdot 5 \cdot 7 = 140$.

30. $\left.\begin{array}{l} 40 = 2 \cdot 2 \cdot 2 \cdot 5 \\ 100 = 2 \cdot 2 \cdot 5 \cdot 5 \end{array}\right\}$
The least common multiple is $2 \cdot 2 \cdot 2 \cdot 5 \cdot 5 = 200$.

31. $\left.\begin{array}{l} 12 = 2 \cdot 2 \cdot 3 \\ 18 = 2 \cdot 3 \cdot 3 \\ 27 = 3 \cdot 3 \cdot 3 \end{array}\right\}$
The least common multiple is $2 \cdot 2 \cdot 3 \cdot 3 \cdot 3 = 108$.

32. $\left.\begin{array}{l} 16 = 2 \cdot 2 \cdot 2 \cdot 2 \\ 20 = 2 \cdot 2 \cdot 5 \\ 80 = 2 \cdot 2 \cdot 2 \cdot 2 \cdot 5 \end{array}\right\}$
The least common multiple is $2 \cdot 2 \cdot 2 \cdot 2 \cdot 5 = 80$.

33.
$$\frac{2}{3}x - \frac{1}{4}y - \frac{3}{4}x - \frac{2}{3}y =$$
$$\frac{2}{3}x - \frac{3}{4}x - \frac{1}{4}y - \frac{2}{3}y =$$
$$\frac{8}{12}x - \frac{9}{12}x - \frac{3}{12}y - \frac{8}{12}y =$$
$$-\frac{1}{12}x - \frac{11}{12}y$$

34.
$$-n - \frac{1}{2}n + \frac{3}{5}n + \frac{5}{6}n =$$
$$\left(-1 - \frac{1}{2} + \frac{3}{5} + \frac{5}{6}\right)n =$$
$$\left(-\frac{30}{30} - \frac{15}{30} + \frac{18}{30} + \frac{25}{30}\right)n =$$
$$-\frac{2}{30}n = -\frac{1}{15}n$$

35.
$$3.2a - 1.4b - 6.2a + 3.3b =$$
$$3.2a - 6.2a - 1.4b + 3.3b =$$
$$-3a + 1.9b$$

36.
$$-(n-1) + 2(n-2) - 3(n-3) =$$
$$-n + 1 + 2n - 4 - 3n + 9 =$$
$$-2n + 6$$

37.
$$-x + 4(x-1) - 3(x+2) - (x+5) =$$
$$-x + 4x - 4 - 3x - 6 - x - 5 =$$
$$-x + 4x - 3x - x - 4 - 6 - 5 =$$
$$-x - 15$$

38.
$$2a - 5(a+3) - 2(a-1) - 4a =$$
$$2a - 5a - 15 - 2a + 2 - 4a =$$
$$2a - 5a - 2a - 4a - 15 + 2 =$$
$$-9a - 13$$

39.
$$\frac{5}{12} - \frac{3}{16} = \frac{5}{12} \cdot \frac{4}{4} - \frac{3}{16} \cdot \frac{3}{3} =$$
$$\frac{20}{48} - \frac{9}{48} = \frac{11}{48}$$

Chapters 1-2 Cumulative Review

40. $\frac{3}{4} - \frac{5}{6} - \frac{7}{9} =$

$\frac{3}{4} \cdot \frac{9}{9} - \frac{5}{6} \cdot \frac{6}{6} - \frac{7}{9} \cdot \frac{4}{4} =$

$\frac{27}{36} - \frac{30}{36} - \frac{28}{36} = -\frac{31}{36}$

41. $\frac{5}{xy} - \frac{2}{x} + \frac{3}{y} =$

$\frac{5}{xy} - \frac{2}{x} \cdot \frac{y}{y} + \frac{3}{y} \cdot \frac{x}{x} =$

$\frac{5}{xy} - \frac{2y}{xy} + \frac{3x}{xy} =$

$\frac{5 - 2y + 3x}{xy}$

42. $-\frac{7}{x^2} + \frac{9}{xy} = \left(-\frac{7}{x^2} \cdot \frac{y}{y}\right) + \left(\frac{9}{xy} \cdot \frac{x}{x}\right) =$

$-\frac{7y}{x^2y} + \frac{9x}{x^2y} = \frac{-7y + 9x}{x^2y}$

43. $\left(\frac{7x}{9y}\right)\left(\frac{12y}{14}\right) = \frac{\overset{1}{\cancel{7}}x}{\underset{3}{\cancel{9}}y} \cdot \frac{\overset{4}{\cancel{12}}\,y}{\underset{2}{\cancel{14}}} =$

$\frac{4xy}{6y} = \frac{2x}{3}$

44. $\left(-\frac{5a}{7b^2}\right)\left(-\frac{8ab}{15}\right) = -\frac{\overset{1}{\cancel{5}}a}{7b^2} \cdot -\frac{8ab}{\underset{3}{\cancel{15}}} =$

$\frac{8a^2b}{21b^2} = \frac{8a^2}{21b}$

45. $\left(\frac{6x^2y}{11}\right) \div \left(\frac{9y^2}{22}\right) =$

$\frac{\overset{2}{\cancel{6}}x^2y}{\underset{1}{\cancel{11}}} \cdot \frac{\overset{2}{\cancel{22}}}{\underset{3}{\cancel{9}}y^2} =$

$\frac{4x^2y}{3y^2} = \frac{4x^2}{3y}$

46. $\left(\frac{-9a}{8b}\right) \div \left(\frac{12a}{18b}\right) =$

$-\frac{\overset{3}{\cancel{9}}a}{\underset{4}{\cancel{8}}b} \cdot \frac{\overset{9}{\cancel{18}}\,b}{\underset{4}{\cancel{12}}\,a} = -\frac{27ab}{16ab} = -\frac{27}{16}$

47. p pennies and n nickels and d dimes is $p + 5n + 10d$ cents.

48. $4n - 5$

49. y yards and f feet and i inches is $36y + 12f + i$ inches.

50. Perimeter of a Rectangle $=$ the sum of the lengths of the sides

Perimeter in meters $= x + x + y + y$

$= (2x + 2y)$ meters

1 meter $=$ 100 centimeters

To change to centimeters multiply meters by 100, so

$(2x + 2y)$meters $= (2x + 2y) \cdot 100$ cm $=$

$= (200x + 200y)$cm

Perimeter in centimeters is $(200x + 200y)$cm.

Chapter 3 Equations, Inequalities and Problem Solving

PROBLEM SET 3.1 Solving First-Degree Equations

1. $x + 9 = 17$
$x + 9 - 9 = 17 - 9$
Subtract 9 from both sides.
$x = 8$
The solution set is $\{8\}$.

3. $x + 11 = 5$
$x + 11 - 11 = 5 - 11$
Subtract 11 from both sides.
$x = -6$
The solution set is $\{-6\}$.

5. $-7 = x + 2$
$-7 - 2 = x + 2 - 2$
Subtract 2 from both sides.
$-9 = x$
The solution set is $\{-9\}$.

7. $8 = n + 14$
$8 - 14 = n + 14 - 14$
Subtract 14 from both sides.
$-6 = n$
The solution set is $\{-6\}$.

9. $21 + y = 34$
$21 + y - 21 = 34 - 21$
Subtract 21 from both sides.
$y = 13$
The solution set is $\{13\}$.

11. $x - 17 = 31$
$x - 17 + 17 = 31 + 17$
Add 17 to both sides.
$x = 48$
The solution set is $\{48\}$.

13. $14 = x - 9$
$14 + 9 = x - 9 + 9$
Add 9 to both sides.
$23 = x$
The solution set is $\{23\}$.

15. $-26 = n - 19$
$-26 + 19 = n - 19 + 19$
Add 19 to both sides.
$-7 = n$
The solution set is $\{-7\}$.

17. $y - \frac{2}{3} = \frac{3}{4}$
$y - \frac{2}{3} + \frac{2}{3} = \frac{3}{4} + \frac{2}{3}$
Add $\frac{2}{3}$ to both sides.
$y = \frac{9}{12} + \frac{8}{12} = \frac{17}{12}$
The solution set is $\left\{\frac{17}{12}\right\}$.

19. $x + \frac{3}{5} = \frac{1}{3}$
$x + \frac{3}{5} - \frac{3}{5} = \frac{1}{3} - \frac{3}{5}$
Subtract $\frac{3}{5}$ from both sides.
$x = \frac{5}{15} - \frac{9}{15} = -\frac{4}{15}$
The solution set is $\left\{-\frac{4}{15}\right\}$.

21. $b + 0.19 = 0.46$
$b + 0.19 - 0.19 = 0.46 - 0.19$
Subtract 0.19 from both sides.
$b = 0.27$
The solution set is $\{0.27\}$.

23. $n - 1.7 = -5.2$
$n - 1.7 + 1.7 = -5.2 + 1.7$
Add 1.7 to both sides.
$n = -3.5$
The solution set is $\{-3.5\}$.

25. $15 - x = 32$
$15 - x - 15 = 32 - 15$
Subtract 15 from both sides.
$-x = 17$
$-1(-x) = -1(17)$
Multiply both sides by -1.
$x = -17$
The solution set is $\{-17\}$.

27. $-14 - n = 21$
$-14 - n + 14 = 21 + 14$
Add 14 to both sides.
$-n = 35$
$-1(-n) = -1(35)$
Multiply both sides by -1.
$n = -35$
The solution set is $\{-35\}$.

29. $7x = -56$
$\frac{7x}{7} = \frac{-56}{7}$
Divide both sides by 7.
$x = -8$
The solution set is $\{-8\}$.

31. $-6x = 102$
$\frac{-6x}{-6} = \frac{102}{-6}$
Divide both sides by -6.
$x = -17$
The solution set is $\{-17\}$.

33. $5x = 37$
$\frac{5x}{5} = \frac{37}{5}$
Divide both sides by 5.
$x = \frac{37}{5}$
The solution set is $\left\{\frac{37}{5}\right\}$.

35. $-18 = 6n$
$\frac{-18}{6} = \frac{6n}{6}$
Divide both sides by 6.
$-3 = n$
The solution set is $\{-3\}$.

37. $-26 = -4n$
$\frac{-26}{-4} = \frac{-4n}{-4}$
Divide both sides by -4.
$\frac{13}{2} = n$
The solution set is $\left\{\frac{13}{2}\right\}$.

39. $\frac{t}{9} = 16$
$9\left(\frac{t}{9}\right) = 9(16)$
Multiply both sides by 9.
$t = 144$
The solution set is $\{144\}$.

41. $\frac{n}{-8} = -3$
$-8\left(\frac{n}{-8}\right) = -8(-3)$
Multiply both sides by -8.
$n = 24$
The solution set is $\{24\}$.

43. $-x = 15$
$-1(-x) = -1(15)$
Multiply both sides by -1.
$x = -15$
The solution set is $\{-15\}$.

45. $\frac{3}{4}x = 18$

$\frac{4}{3}\left(\frac{3}{4}x\right) = \frac{4}{3}(18)$

Multiply both sides by $\frac{4}{3}$.

$x = 24$

The solution set is $\{24\}$.

47. $-\frac{2}{5}n = 14$

$-\frac{5}{2}\left(-\frac{2}{5}n\right) = -\frac{5}{2}(14)$

Multiply both sides by $-\frac{5}{2}$.

$n = -35$

The solution set is $\{-35\}$.

49. $\frac{2}{3}n = \frac{1}{5}$

$\frac{3}{2}\left(\frac{2}{3}n\right) = \frac{3}{2}\left(\frac{1}{5}\right)$

Multiply both sides by $\frac{3}{2}$.

$n = \frac{3}{10}$

The solution set is $\left\{\frac{3}{10}\right\}$.

51. $\frac{5}{6}n = -\frac{3}{4}$

$\frac{6}{5}\left(\frac{5}{6}n\right) = \frac{6}{5}\left(-\frac{3}{4}\right)$

Multiply both sides by $\frac{6}{5}$.

$n = -\frac{9}{10}$

The solution set is $\left\{-\frac{9}{10}\right\}$.

53. $\frac{3x}{10} = \frac{3}{20}$

$\frac{10}{3}\left(\frac{3x}{10}\right) = \frac{10}{3}\left(\frac{3}{20}\right)$

Multiply both sides by $\frac{10}{3}$.

$x = \frac{1}{2}$

The solution set is $\left\{\frac{1}{2}\right\}$.

55. $\frac{-y}{2} = \frac{1}{6}$

$-2\left(\frac{-y}{2}\right) = -2\left(\frac{1}{6}\right)$

Multiply both sides by -2.

$y = -\frac{1}{3}$

The solution set is $\left\{-\frac{1}{3}\right\}$.

57. $-\frac{4}{3}x = -\frac{9}{8}$

$-\frac{3}{4}\left(-\frac{4}{3}x\right) = -\frac{3}{4}\left(-\frac{9}{8}\right)$

Multiply both sides by $-\frac{3}{4}$.

$x = \frac{27}{32}$

The solution set is $\left\{\frac{27}{32}\right\}$.

59. $-\frac{5}{12} = \frac{7}{6}x$

$\frac{6}{7}\left(-\frac{5}{12}\right) = \frac{6}{7}\left(\frac{7}{6}x\right)$

Multiply both sides by $\frac{6}{7}$.

$-\frac{5}{14} = x$

The solution set is $\left\{-\frac{5}{14}\right\}$.

61. $-\frac{5}{7}x = 1$

$-\frac{7}{5}\left(-\frac{5}{7}x\right) = -\frac{7}{5}(1)$

Multiply both sides by $-\frac{7}{5}$.

$x = -\frac{7}{5}$

The solution set is $\left\{-\frac{7}{5}\right\}$.

63. $-4n = \frac{1}{3}$

$-\frac{1}{4}(-4n) = -\frac{1}{4}\left(\frac{1}{3}\right)$

Multiply both sides by $-\frac{1}{4}$.

$n = -\frac{1}{12}$

The solution set is $\left\{-\frac{1}{12}\right\}$.

65. $-8n = \frac{6}{5}$

$-\frac{1}{8}(-8n) = -\frac{1}{8}\left(\frac{6}{5}\right)$

Multiply both sides by $-\frac{1}{8}$.

$n = -\frac{3}{20}$

The solution set is $\left\{-\frac{3}{20}\right\}$.

67. $1.2x = 0.36$

$\frac{1.2x}{1.2} = \frac{0.36}{1.2}$

Divide both sides by 1.2.

$x = 0.3$

The solution set is $\{0.3\}$.

69. $30.6 = 3.4n$

$\frac{30.6}{3.4} = \frac{3.4n}{3.4}$

Divide both sides by 3.4.

$9 = n$

The solution set is $\{9\}$.

71. $-3.4x = 17$

$\frac{-3.4x}{-3.4} = \frac{17}{-3.4}$

Divide both sides by -3.4.

$x = -5$

The solution set is $\{-5\}$.

PROBLEM SET 3.2 Equations and Problem Solving

1. $2x + 5 = 13$

$2x + 5 - 5 = 13 - 5$

Subtract 5 from both sides.

$2x = 8$

$\frac{2x}{2} = \frac{8}{2}$

Divide both sides by 2.

$x = 4$

The solution set is $\{4\}$.

3. $5x + 2 = 32$

$5x + 2 - 2 = 32 - 2$

Subtract 2 from both sides.

$5x = 30$

$\frac{5x}{5} = \frac{30}{5}$

Divide both sides by 5.

$x = 6$

The solution set is $\{6\}$.

5.
$$3x - 1 = 23$$
$$3x - 1 + 1 = 23 + 1$$
Add 1 to both sides.
$$3x = 24$$
$$\frac{3x}{3} = \frac{24}{3}$$
Divide both sides by 3.
$$x = 8$$
The solution set is $\{8\}$.

7.
$$4n - 3 = 41$$
$$4n - 3 + 3 = 41 + 3$$
Add 3 to both sides.
$$4n = 44$$
$$\frac{4n}{4} = \frac{44}{4}$$
Divide both sides by 4.
$$x = 11$$
The solution set is $\{11\}$.

9.
$$6y - 1 = 16$$
$$6y - 1 + 1 = 16 + 1$$
Add 1 to both sides.
$$6y = 17$$
$$\frac{6y}{6} = \frac{17}{6}$$
Divide both sides by 6.
$$x = \frac{17}{6}$$
The solution set is $\left\{\frac{17}{6}\right\}$.

11.
$$2x + 3 = 22$$
$$2x + 3 - 3 = 22 - 3$$
Subtract 3 from both sides.
$$2x = 19$$
$$\frac{2x}{2} = \frac{19}{2}$$
Divide both sides by 2.
$$x = \frac{19}{2}$$
The solution set is $\left\{\frac{19}{2}\right\}$.

13.
$$10 = 3t - 8$$
$$10 + 8 = 3t - 8 + 8$$
Add 8 to both sides.
$$18 = 3t$$
$$\frac{18}{3} = \frac{3t}{3}$$
Divide both sides by 3.
$$6 = t$$
The solution set is $\{6\}$.

15.
$$5x + 14 = 9$$
$$5x + 14 - 14 = 9 - 14$$
Subtract 14 from both sides.
$$5x = -5$$
$$\frac{5x}{5} = \frac{-5}{5}$$
Divide both sides by 5.
$$x = -1$$
The solution set is $\{-1\}$.

17.
$$18 - n = 23$$
$$18 - n - 18 = 23 - 18$$
Subtract 18 from both sides.
$$-n = 5$$
$$-1(-n) = -1(5)$$
Multiply both sides by -1.
$$n = -5$$
The solution set is $\{-5\}$.

19.
$$-3x + 2 = 20$$
$$-3x + 2 - 2 = 20 - 2$$
Subtract 2 from both sides.
$$-3x = 18$$
$$\frac{-3x}{-3} = \frac{18}{-3}$$
Divide both sides by -3.
$$x = -6$$
The solution set is $\{-6\}$.

Problem Set 3.2

21.
$$7+4x=29$$
$$7+4x-7=29-7$$
Subtract 7 from both sides.
$$4x=22$$
$$\frac{4x}{4}=\frac{22}{4}$$
Divide both sides by 4.
$$x=\frac{11}{2}$$
The solution set is $\left\{\frac{11}{2}\right\}$.

23.
$$16=-2-9a$$
$$16+2=-2-9a+2$$
Add 2 to both sides.
$$18=-9a$$
$$\frac{18}{-9}=\frac{-9a}{-9}$$
Divide both sides by -9.
$$-2=a$$
The solution set is $\{-2\}$.

25.
$$-7x+3=-7$$
$$-7x+3-3=-7-3$$
Subtract 3 from both sides.
$$-7x=-10$$
$$\frac{-7x}{-7}=\frac{-10}{-7}$$
Divide both sides by -7.
$$x=\frac{10}{7}$$
The solution set is $\left\{\frac{10}{7}\right\}$.

27.
$$17-2x=-19$$
$$17-2x-17=-19-17$$
Subtract 17 from both sides.
$$-2x=-36$$
$$\frac{-2x}{-2}=\frac{-36}{-2}$$
Divide both sides by -2.
$$x=18$$
The solution set is $\{18\}$.

29.
$$-16-4x=9$$
$$-16-4x+16=9+16$$
Add 16 to both sides.
$$-4x=25$$
$$\frac{-4x}{-4}=\frac{25}{-4}$$
Divide both sides by -4.
$$x=-\frac{25}{4}$$
The solution set is $\left\{-\frac{25}{4}\right\}$.

31.
$$-12t+4=88$$
$$-12t+4-4=88-4$$
Subtract 4 from both sides.
$$-12t=84$$
$$\frac{-12t}{-12}=\frac{84}{-12}$$
Divide both sides by -12.
$$t=-7$$
The solution set is $\{-7\}$.

33.
$$14y+15=-33$$
$$14y+15-15=-33-15$$
Subtract 15 from both sides.
$$14y=-48$$
$$\frac{14y}{14}=\frac{-48}{14}$$
Divide both sides by 14.
$$y=-\frac{24}{7}$$
The solution set is $\left\{-\frac{24}{7}\right\}$.

35.
$$32-16n=-8$$
$$32-16n-32=-8-32$$
Subtract 32 from both sides.
$$-16n=-40$$
$$\frac{-16n}{-16}=\frac{-40}{-16}$$
Divide both sides by -16.
$$n=\frac{5}{2}$$
The solution set is $\left\{\frac{5}{2}\right\}$.

37. $17x - 41 = -37$

$17x - 41 + 41 = -37 + 41$

Add 41 to both sides.

$17x = 4$

$\frac{17x}{17} = \frac{4}{17}$

Divide both sides by 17.

$x = \frac{4}{17}$

The solution set is $\left\{\frac{4}{17}\right\}$.

39. $29 = -7 - 15x$

$29 + 7 = -7 - 15x + 7$

Add 7 to both sides.

$36 = -15x$

$\frac{36}{-15} = \frac{-15x}{-15}$

Divide both sides by -15.

$-\frac{12}{5} = x$

The solution set is $\left\{-\frac{12}{5}\right\}$.

41. Let n represent the number.

$n + 12 = 21$

$n = 9$

The number is 9.

43. Let n represent the number.

$n - 9 = 13$

$n = 22$

The number is 22.

45. Let c represent the cost of the other item.

$c + 25 = 43$

$c = 18$

The cost of the other item is $18.

47. Let x represent Nora's age now; therefore, $x + 6$ represents her age 6 years from now.

$x + 6 = 41$

$x = 35$

Nora is presently 35 years old.

49. Let h represent his hourly rate. Then the product of the number of hours times the hourly rate equals total amount earned.

$6h = 43.50$

$h = \frac{43.50}{6} = 7.25$

His hourly rate was $7.25.

51. Let x represent the number.

Then 3 times the number is $3x$.

$3x + 6 = 24$

$3x = 18$

$x = 6$

The number is 6.

53. Let n represent the number.

Then 3 times the number is $3n$.

$19 = 3n + 4$

$15 = 3n$

$5 = n$

The number is 5.

55. Let n represent the number.

Then 5 times the number is $5n$.

$49 = 5n - 6$

$55 = 5n$

$11 = n$

The number is 11.

57. Let n represent the number.

Then 6 times the number is $6n$.

$6n - 1 = 47$

$6n = 48$

$n = 8$

The number is 8.

59. Let x represent the number.

$27 - 8x = 3$

$-8x = -24$

$x = 3$

The number is 3.

61. Let c represent the cost of the ring.
$550 = 2c - 50$
$600 = 2c$
$300 = c$
The cost of the ring was \$300.

63. Let w represent the width of the floor.
$18 = 5w - 2$
$20 = 5w$
$4 = w$
The width of the floor is 4 meters.

65. Let c represent speakers of the English language.
$874,000,000 = 3c - 149,000,000$
$1,023,000,000 = 3c$
$341,000,000 = c$
In 2001 there were $341,000,000$ speakers of the English language.

67. Let h represent the hours of labor.
$454 = 379 + 60h$
$75 = 60h$
$1.25 = h$
There were 1.25 hours of labor in the repair bill.

PROBLEM SET 3.3 More on Solving Equations and Problem Solving

1.
$$\begin{aligned} 2x + 7 + 3x &= 32 \\ 5x + 7 &= 32 \\ 5x + 7 - 7 &= 32 - 7 \\ 5x &= 25 \\ \frac{5x}{5} &= \frac{25}{5} \\ x &= 5 \end{aligned}$$
The solution set is $\{5\}$.

3.
$$\begin{aligned} 7x - 4 - 3x &= -36 \\ 4x - 4 &= -36 \\ 4x - 4 + 4 &= -36 + 4 \\ 4x &= -32 \\ \frac{4x}{4} &= \frac{-32}{4} \\ x &= -8 \end{aligned}$$
The solution set is $\{-8\}$.

5.
$$\begin{aligned} 3y - 1 + 2y - 3 &= 4 \\ 5y - 4 &= 4 \\ 5y - 4 + 4 &= 4 + 4 \\ 5y &= 8 \\ \frac{5y}{5} &= \frac{8}{5} \\ y &= \frac{8}{5} \end{aligned}$$
The solution set is $\left\{\frac{8}{5}\right\}$.

7.
$$\begin{aligned} 5n - 2 - 8n &= 31 \\ -3n - 2 &= 31 \\ -3n - 2 + 2 &= 31 + 2 \\ -3n &= 33 \\ \frac{-3n}{-3} &= \frac{33}{-3} \\ n &= -11 \end{aligned}$$
The solution set is $\{-11\}$.

9.
$$\begin{aligned} -2n + 1 - 3n + n - 4 &= 7 \\ -4n - 3 &= 7 \\ -4n - 3 + 3 &= 7 + 3 \\ -4n &= 10 \\ \frac{-4n}{-4} &= \frac{10}{-4} \\ n &= -\frac{5}{2} \end{aligned}$$
The solution set is $\left\{-\frac{5}{2}\right\}$.

11.
$$\begin{aligned} 3x + 4 &= 2x - 5 \\ 3x + 4 - 2x &= 2x - 5 - 2x \\ x + 4 - 4 &= -5 - 4 \\ x &= -9 \end{aligned}$$
The solution set is $\{-9\}$.

13.
$$\begin{aligned} 5x - 7 &= 6x - 9 \\ 5x - 7 - 6x &= 6x - 9 - 6x \\ -x - 7 + 7 &= -9 + 7 \\ -x &= -2 \\ x &= 2 \end{aligned}$$
The solution set is $\{2\}$.

15.
$$\begin{aligned} 6x + 1 &= 3x - 8 \\ 6x + 1 - 3x &= 3x - 8 - 3x \\ 3x + 1 - 1 &= -8 - 1 \\ 3x &= -9 \\ \frac{3x}{3} &= \frac{-9}{3} \\ x &= -3 \end{aligned}$$
The solution set is $\{-3\}$.

17.
$$\begin{aligned} 7y - 3 &= 5y + 10 \\ 7y - 3 - 5y &= 5y + 10 - 5y \\ 2y - 3 + 3 &= 10 + 3 \\ 2y &= 13 \\ \frac{2y}{2} &= \frac{13}{2} \\ y &= \frac{13}{2} \end{aligned}$$
The solution set is $\left\{\frac{13}{2}\right\}$.

19.
$$\begin{aligned} 8n - 2 &= 11n - 7 \\ 8n - 2 - 8n &= 11n - 7 - 8n \\ -2 + 7 &= 3n - 7 + 7 \\ 5 &= 3n \\ \frac{5}{3} &= \frac{3n}{3} \\ \frac{5}{3} &= n \end{aligned}$$
The solution set is $\left\{\frac{5}{3}\right\}$.

21.
$$\begin{aligned} -2x - 7 &= -3x + 10 \\ -2x - 7 + 3x &= -3x + 10 + 3x \\ x - 7 + 7 &= 10 + 7 \\ x &= 17 \end{aligned}$$
The solution set is $\{17\}$.

23.
$$\begin{aligned} -3x + 5 &= -5x - 8 \\ -3x + 5 + 5x &= -5x - 8 + 5x \\ 2x + 5 - 5 &= -8 - 5 \\ 2x &= -13 \\ \frac{2x}{2} &= \frac{-13}{2} \\ x &= -\frac{13}{2} \end{aligned}$$
The solution set is $\left\{-\frac{13}{2}\right\}$.

25.
$$\begin{aligned} -7 - 6x &= 9 - 9x \\ -7 - 6x + 9x &= 9 - 9x + 9x \\ -7 + 3x &= 9 \\ -7 + 3x + 7 &= 9 + 7 \\ 3x &= 16 \\ \frac{3x}{3} &= \frac{16}{3} \\ x &= \frac{16}{3} \end{aligned}$$
The solution set is $\left\{\frac{16}{3}\right\}$.

27.
$$\begin{aligned} 2x - 1 - x &= 3x - 5 \\ x - 1 &= 3x - 5 \\ x - 1 - x &= 3x - 5 - x \\ -1 + 5 &= 2x - 5 + 5 \\ 4 &= 2x \\ \frac{4}{2} &= \frac{2x}{2} \\ 2 &= x \end{aligned}$$
The solution set is $\{2\}$.

29.
$$\begin{aligned} 5n - 4 - n &= -3n - 6 + n \\ 4n - 4 &= -2n - 6 \\ 4n - 4 + 2n &= -2n - 6 + 2n \\ 6n - 4 + 4 &= -6 + 4 \\ 6n &= -2 \\ \frac{6n}{6} &= \frac{-2}{6} = -\frac{1}{3} \end{aligned}$$
The solution set is $\left\{-\frac{1}{3}\right\}$.

31. $-7-2n-6n=7n-5n+12$
$-7-8n=2n+12$
$-7-8n-2n=2n+12-2n$
$-7-10n+7=12+7$
$-10n=19$
$\frac{-10n}{-10}=\frac{19}{-10}=-\frac{19}{10}$
The solution set is $\left\{-\frac{19}{10}\right\}$.

33. Let n represent the number. Then $4n$ represents four times the number.
$n+4n=85$
$5n=85$
$n=17$
The number is 17.

35. Let n represent the first odd number, then $n+2$ represents the next odd number.
$n+(n+2)=72$
$2n+2=72$
$2n=70$
$n=35$
The two consecutive odd numbers are 35 and 37.

37. Let n, $n+2$, and $n+4$ represent the three consecutive even numbers.
$n+(n+2)+(n+4)=114$
$3n+6=114$
$3n=108$
$n=36$
The numbers are 36, 38, and 40.

39. Let n represent the number.
$3n+2=7n-4$
$-4n+2=-4$
$-4n=-6$
$n=\frac{-6}{-4}=\frac{3}{2}$
The number is $\frac{3}{2}$.

41. Let n represent the number.
$n+5n=3n-18$
$6n=3n-18$
$3n=-18$
$n=-6$
The number is -6.

43. Let a represent the first angle. Then the other angle is represented by $2a-6$. Since the angles are complementary, the sum of their measures is 90°.
$a+(2a-6)=90$
$3a-6=90$
$3a=96$
$a=32$
The measures of the angles are 32° and $2(32)-6=58°$.

45. Let a represent the smaller angle. Then $3a-20$ represents the larger angle. Since they are supplementary angles, the sum of their measures is 180°.
$a+(3a-20)=180$
$4a-20=180$
$4a=200$
$a=50$
The measures of the angles are 50°, and $3(50)-20=130°$.

47. Let a represent the smaller of the other two angles; then $a+10$ represents the larger angle. The sum of the measures of the three angles of a triangle is 180°.
$40+a+a+10=180$
$50+2a=180$
$2a=130$
$a=65$
The other two angles of the triangle would be 65° and 75°.

49. Let x represent the price paid for the vase.
$69 = 2x - 15$
$84 = 2x$
$42 = x$
Marci paid \$42 for the antique vase.

51. Let h represent Bob's normal hourly rate.
Bob worked 8 hours at the higher rate.
$40h + 8(2h) = 504$
$40h + 16h = 504$
$56h = 504$
$h = 9$
Bob's normal hourly rate is \$9.00 per hour.

53. Let m represent the number of men, then $3m$ represents the number of women.
$m + 3m = 600$
$4m = 600$
$m = 150$
Therefore, 150 men and $3(150) = 450$ women attended the concert.

55. Let x represent the cost of a used textbook.
Then $15 + 2x$ is the cost of the biology text.
$129 = (15 + 2x) + x$
$129 = 15 + 3x$
$114 = 3x$
$x = 38$
The used text is \$38, so the biology text costs $15 + 2(38) = 15 + 76 = \$91$.

57. Let x represent the amount of tip for the busboy, then $25 + 3x$ represents the tip for the server and $50 + x$ represents the tip for the bartender.
$x + (25 + 3x) + (50 + x) = 275$
$5x + 75 = 275$
$5x = 200$
$x = 40$
The server will receive $25 + 3(40) = \$145$.

60a. $7x - 3 = 4x - 3$
$7x - 3 - 4x = 4x - 3 - 4x$
$3x - 3 + 3 = -3 + 3$
$3x = 0$
$x = 0$
The solution set is $\{0\}$.

b. $-x - 4 + 3x = 2x - 7$
$2x - 4 = 2x - 7$
$2x - 4 - 2x = 2x - 7 - 2x$
$-4 = -7$ (contradiction)
The solution set is $\emptyset$.

c. $-3x + 9 - 2x = -5x + 9$
$-5x + 9 = -5x + 9$
$-5x + 9 + 5x = -5x + 9 + 5x$
$9 = 9$ (identity)
The solution set is $\{$all real numbers$\}$.

d. $5x - 3 = 6x - 7 - x$
$5x - 3 = 5x - 7$
$5x - 3 - 5x = 5x - 7 - 5x$
$-3 = -7$ (contradiction)
The solution set is $\emptyset$.

e. $7x + 4 = -x + 4 + 8x$
$7x + 4 = 7x + 4$
$7x + 4 - 7x = 7x + 4 - 7x$
$4 = 4$ (identity)
The solution set is $\{$all real numbers$\}$.

f. $3x - 2 - 5x = 7x - 2 - 5x$
$-2x - 2 = 2x - 2$
$-2x - 2 - 2x = 2x - 2 - 2x$
$-4x - 2 = -2$
$-4x = 0$
$x = 0$
The solution set is $\{0\}$.

g. $-6x - 8 = 6x + 4$
$-6x - 8 - 6x = 6x + 4 - 6x$
$-12x - 8 = 4$
$-12x = 12$
$x = -1$
The solution set is $\{-1\}$.

h.
$$\begin{aligned} -8x+9 &= -8x+5 \\ -8x+9+8x &= -8x+5+8x \\ 9 &= 5 \text{ (contradiction)} \end{aligned}$$
The solution set is $\emptyset$.

PROBLEM SET 3.4 Equations Involving Parentheses and Fractional Forms

1.
$$\begin{aligned} 7(x+2) &= 21 \\ 7x+14 &= 21 && \text{Apply distributive property.} \\ 7x &= 7 && \text{Subtract 14 from both sides.} \\ x &= 1 && \text{Divide both sides by 7.} \end{aligned}$$
The solution set is $\{1\}$.

3.
$$\begin{aligned} 5(x-3) &= 35 \\ 5x-15 &= 35 && \text{Apply distributive property.} \\ 5x &= 50 && \text{Add 15 to both sides.} \\ x &= 10 && \text{Divide both sides by 5.} \end{aligned}$$
The solution set is $\{10\}$.

5.
$$\begin{aligned} -3(x+5) &= 12 \\ -3x-15 &= 12 && \text{Apply distributive property.} \\ -3x &= 27 && \text{Add 15 to both sides.} \\ x &= -9 && \text{Divide both sides by } -3. \end{aligned}$$
The solution set is $\{-9\}$.

7.
$$\begin{aligned} 4(n-6) &= 5 \\ 4n-24 &= 5 && \text{Apply distributive property.} \\ 4n &= 29 && \text{Add 24 to both sides.} \\ n &= \frac{29}{4} && \text{Divide both sides by 4.} \end{aligned}$$
The solution set is $\left\{\frac{29}{4}\right\}$.

9.
$$\begin{aligned} 6(n+7) &= 8 \\ 6n+42 &= 8 && \text{Apply distributive property.} \\ 6n &= -34 && \text{Subtract 42 from both sides.} \\ n &= \frac{-34}{6} && \text{Divide both sides by 6.} \\ n &= -\frac{17}{3} && \text{Reduce.} \end{aligned}$$
The solution set is $\left\{-\frac{17}{3}\right\}$.

11.
$$\begin{aligned} -10 &= -5(t-8) \\ -10 &= -5t+40 && \text{Apply distributive property.} \\ -50 &= -5t && \text{Subtract 40 from both sides.} \\ 10 &= t && \text{Divide both sides by } -5. \end{aligned}$$
The solution set is $\{10\}$.

13.
$$\begin{aligned} 5(x-4) &= 4(x+6) \\ 5x-20 &= 4x+24 && \text{Apply distributive property.} \\ x-20 &= 24 && \text{Subtract } 4x \text{ from both sides.} \\ x &= 44 && \text{Add 20 to both sides.} \end{aligned}$$
The solution set is $\{44\}$.

We will discontinue giving reasons for each step but we will continue to show enough of the work so that you can follow the steps. If a new technique is introduced, then we will indicate some of the reasons again.

15.
$$\begin{aligned} 8(x+1) &= 9(x-2) \\ 8x+8 &= 9x-18 \\ -x+8 &= -18 \\ -x &= -26 \\ x &= 26 \end{aligned}$$
The solution set is $\{26\}$.

17.
$$\begin{aligned} 8(t+5) &= 4(2t+10) \\ 8t+40 &= 8t+40 \\ 40 &= 40 \end{aligned}$$
The result is a true statement.
The solution set is {all reals}.

19.
$$\begin{aligned} 2(6t+1) &= 4(3t-1) \\ 12t+2 &= 12t-4 \\ 2 &= -4 \end{aligned}$$
The result is a false statement.
The solution set is the empty set, $\emptyset$.

21. $$\begin{aligned} -2(x-6) &= -(x-9) \\ -2x+12 &= -x+9 \\ -x+12 &= 9 \\ -x &= -3 \\ x &= 3 \end{aligned}$$
The solution set is $\{3\}$.

23. $$\begin{aligned} -3(t-4)-2(t+4) &= 9 \\ -3t+12-2t-8 &= 9 \\ -5t+4 &= 9 \\ -5t &= 5 \\ t &= -1 \end{aligned}$$
The solution set is $\{-1\}$.

25. $$\begin{aligned} 3(n-10)-5(n+12) &= -86 \\ 3n-30-5n-60 &= -86 \\ -2n-90 &= -86 \\ -2n &= 4 \\ n &= -2 \end{aligned}$$
The solution set is $\{-2\}$.

27. $$\begin{aligned} 3(x+1)+4(2x-1) &= 5(2x+3) \\ 3x+3+8x-4 &= 10x+15 \\ 11x-1 &= 10x+15 \\ x-1 &= 15 \\ x &= 16 \end{aligned}$$
The solution set is $\{16\}$.

29. $$\begin{aligned} -(x+2)+2(x-3) &= -2(x-7) \\ -x-2+2x-6 &= -2x+14 \\ x-8 &= -2x+14 \\ 3x-8 &= 14 \\ 3x &= 22 \\ x &= \frac{22}{3} \end{aligned}$$
The solution set is $\left\{\frac{22}{3}\right\}$.

31. $$\begin{aligned} 5(2x-1)-(2x+4) &= 4(2x+3)-21 \\ 10x-5-2x-4 &= 8x+12-21 \\ 8x-9 &= 8x-9 \\ -9 &= -9 \end{aligned}$$
The result is a true statement.
The solution set is $\{\text{all reals}\}$.

33. $$\begin{aligned} -(a-1)-(3a-2) &= 6+2(a-1) \\ -a+1-3a+2 &= 6+2a-2 \\ -4a+3 &= 2a+4 \\ -6a+3 &= 4 \\ -6a &= 1 \\ a &= -\frac{1}{6} \end{aligned}$$
The solution set is $\left\{-\frac{1}{6}\right\}$.

35. $$\begin{aligned} 3(x-1)+2(x-3) &= -4(x-2)+10(x+4) \\ 3x-3+2x-6 &= -4x+8+10x+40 \\ 5x-9 &= 6x+48 \\ -x-9 &= 48 \\ -x &= 57 \\ x &= -57 \end{aligned}$$
The solution set is $\{-57\}$.

37. $$\begin{aligned} 3-7(x-1) &= 9-6(2x+1) \\ 3-7x+7 &= 9-12x-6 \\ -7x+10 &= -12x+3 \\ 5x+10 &= 3 \\ 5x &= -7 \\ x &= -\frac{7}{5} \end{aligned}$$
The solution set is $\left\{-\frac{7}{5}\right\}$.

For Problems 39 – 60, we begin each solution by multiplying both sides of the given equation by the least common denominator of all of the denominators in the equation. This has the effect of "clearing the equation of all fractions."

39. $$\begin{aligned} \frac{3}{4}x-\frac{2}{3} &= \frac{5}{6} \\ 12\left(\frac{3}{4}x-\frac{2}{3}\right) &= 12\left(\frac{5}{6}\right) \\ 9x-8 &= 10 \\ 9x &= 18 \\ x &= 2 \end{aligned}$$
The solution set is $\{2\}$.

Problem Set 3.4

41.
$$\begin{aligned}\frac{5}{6}x+\frac{1}{4}&=-\frac{9}{4}\\12\left(\frac{5}{6}x+\frac{1}{4}\right)&=12\left(-\frac{9}{4}\right)\\10x+3&=-27\\10x&=-30\\x&=-3\end{aligned}$$
The solution set is $\{-3\}$.

43.
$$\begin{aligned}\frac{1}{2}x-\frac{3}{5}&=\frac{3}{4}\\20\left(\frac{1}{2}x-\frac{3}{5}\right)&=20\left(\frac{3}{4}\right)\\10x-12&=15\\10x&=27\\x&=\frac{27}{10}\end{aligned}$$
The solution set is $\left\{\frac{27}{10}\right\}$.

45.
$$\begin{aligned}\frac{n}{3}+\frac{5n}{6}&=\frac{1}{8}\\24\left(\frac{n}{3}+\frac{5n}{6}\right)&=24\left(\frac{1}{8}\right)\\8n+20n&=3\\28n&=3\\n&=\frac{3}{28}\end{aligned}$$
The solution set is $\left\{\frac{3}{28}\right\}$.

47.
$$\begin{aligned}\frac{5y}{6}-\frac{3}{5}&=\frac{2y}{3}\\30\left(\frac{5y}{6}-\frac{3}{5}\right)&=30\left(\frac{2y}{3}\right)\\25y-18&=20y\\5y-18&=0\\5y&=18\\y&=\frac{18}{5}\end{aligned}$$
The solution set is $\left\{\frac{18}{5}\right\}$.

49.
$$\begin{aligned}\frac{h}{6}+\frac{h}{8}&=1\\24\left(\frac{h}{6}+\frac{h}{8}\right)&=24(1)\\4h+3h&=24\\7h&=24\\h&=\frac{24}{7}\end{aligned}$$
The solution set is $\left\{\frac{24}{7}\right\}$.

51.
$$\begin{aligned}\frac{x+2}{3}+\frac{x+3}{4}&=\frac{13}{3}\\12\left(\frac{x+2}{3}+\frac{x+3}{4}\right)&=12\left(\frac{13}{3}\right)\\4(x+2)+3(x+3)&=4(13)\\4x+8+3x+9&=52\\7x+17&=52\\7x&=35\\x&=5\end{aligned}$$
The solution set is $\{5\}$.

53.
$$\begin{aligned}\frac{x-1}{5}-\frac{x+4}{6}&=-\frac{13}{15}\\30\left(\frac{x-1}{5}-\frac{x+4}{6}\right)&=30\left(-\frac{13}{15}\right)\\6(x-1)-5(x+4)&=2(-13)\\6x-6-5x-20&=-26\\x-26&=-26\\x&=0\end{aligned}$$
The solution set is $\{0\}$.

55.
$$\begin{aligned}\frac{x+8}{2}-\frac{x+10}{7}&=\frac{3}{4}\\28\left(\frac{x+8}{2}-\frac{x+10}{7}\right)&=28\left(\frac{3}{4}\right)\\14(x+8)-4(x+10)&=7(3)\\14x+112-4x-40&=21\\10x+72&=21\\10x&=-51\\x&=-\frac{51}{10}\end{aligned}$$
The solution set is $\left\{-\frac{51}{10}\right\}$.

57.
$$\frac{x-2}{8} - 1 = \frac{x+1}{4}$$
$$8\left(\frac{x-2}{8} - 1\right) = 8\left(\frac{x+1}{4}\right)$$
$$1(x-2) - 8 = 2(x+1)$$
$$x - 2 - 8 = 2x + 2$$
$$x - 10 = 2x + 2$$
$$-10 = x + 2$$
$$-12 = x$$
The solution set is $\{-12\}$.

59.
$$\frac{x+1}{4} = \frac{x-3}{6} + 2$$
$$12\left(\frac{x+1}{4}\right) = 12\left(\frac{x-3}{6} + 2\right)$$
$$3(x+1) = 2(x-3) + 24$$
$$3x + 3 = 2x - 6 + 24$$
$$3x + 3 = 2x + 18$$
$$x + 3 = 18$$
$$x = 15$$
The solution set is $\{15\}$.

61. Let n and $n+1$ represent the consecutive whole numbers.
$$n + 4(n+1) = 39$$
$$n + 4n + 4 = 39$$
$$5n = 35$$
$$n = 7$$
The numbers are 7 and 8.

63. Let n, $n+1$, and $n+2$ represent the consecutive whole numbers.
$$2(n + n + 1) = 3(n+2) + 10$$
$$2n + 2n + 2 = 3n + 6 + 10$$
$$4n + 2 = 3n + 16$$
$$n = 14$$
The numbers are 14, 15, and 16.

65. Let n represent the smaller number, then $17 - n$ represents the larger number.
$$2n = (17 - n) + 1$$
$$2n = 17 - n + 1$$
$$2n = 18 - n$$
$$3n = 18$$
$$n = 6$$
The numbers are 6 and 11.

67. Let n represent the number.
$$\frac{1}{3}n + 20 = \frac{3}{4}n$$
$$12\left(\frac{1}{3}n + 20\right) = 12\left(\frac{3}{4}n\right)$$
$$4n + 240 = 9n$$
$$240 = 5n$$
$$48 = n$$
The number is 48.

69. Let n represent the larger number, then $n - 6$ represents the smaller number.
$$\frac{1}{2}n = 5 + \frac{1}{3}(n-6)$$
$$6\left(\frac{1}{2}n\right) = 6\left[5 + \frac{1}{3}(n-6)\right]$$
$$3n = 6(5) + 6\left[\frac{1}{3}(n-6)\right]$$
$$3n = 30 + 2(n-6)$$
$$3n = 30 + 2n - 12$$
$$3n = 2n + 18$$
$$n = 18$$
The numbers are 18 and 12.

71. Let x represent the shorter piece of board, then $20 - x$ represents the longer piece.
$$4x = 3(20 - x) - 4$$
$$4x = 60 - 3x - 4$$
$$4x = 56 - 3x$$
$$7x = 56$$
$$x = 8$$
The length of the pieces are 8 feet and 12 feet.

73. Let x represent the number of nickels, then $35 - x$ represents the number of quarters.

$$0.05x + 0.25(35 - x) = 5.75$$
$$0.05x + 8.75 - 0.25x = 5.75$$
$$-0.20x + 8.75 = 5.75$$
$$-0.20x = -3$$
$$x = \frac{-3}{-0.20}$$
$$x = 15$$

There are 15 nickels and $35 - 15 = 20$ quarters.

75. Let n represent the number of nickels, $2n$ the number of dimes, and $2n + 10$ the number of quarters.

$$n + 2n + 2n + 10 = 210$$
$$5n + 10 = 210$$
$$5n = 200$$
$$n = 40$$

Max has 40 nickels, $2(40) = 80$ dimes, and $2(40) + 10 = 90$ quarters.

77. Let d represent the number of dimes and $18 - d$ the number of quarters. The value in cents of the dimes is $10d$ and $25(18 - d)$ is the value in cents of the quarters.

$$10d + 25(18 - d) = 330$$
$$10d + 450 - 25d = 330$$
$$-15d + 450 = 330$$
$$-15d = -120$$
$$d = 8$$

Maida has 8 dimes and $18 - 8 = 10$ quarters.

79. Let x represent the number of crabs, then $3x$ represents the number of fish and $x + 2$ represents the number of plants.

$$x + 3x + x + 2 = 22$$
$$5x + 2 = 22$$
$$5x = 20$$
$$x = 4$$

There are 4 crabs, $3(4) = 12$ fish, and $4 + 2 = 6$ plants.

81. Let a represent the measure of the angle, then $180 - a$ represents its supplement and $90 - a$ represents its complement.

$$180 - a = 2(90 - a) + 30$$
$$180 - a = 180 - 2a + 30$$
$$180 - a = -2a + 210$$
$$a = 30$$

The angle has a measure of 30°.

83. Let c represent the measure of angle C. Then $\frac{1}{5}c - 2$ represents angle A and $\frac{1}{2}c - 5$ represents angle B. The sum of the measures of all angles in a triangle is 180°.

$$c + \frac{1}{5}c - 2 + \frac{1}{2}c - 5 = 180$$
$$10\left[c + \frac{1}{5}c - 2 + \frac{1}{2}c - 5\right] = 10(180)$$
$$10c + 2c - 20 + 5c - 50 = 1800$$
$$17c - 70 = 1800$$
$$17c = 1870$$
$$c = 110$$

Angle A measures $\frac{1}{5}(110) - 2 = 20°$, angle B measures $\frac{1}{2}(110) - 5 = 50°$, and angle C measures 110°.

85. Let a represent the measure of the angle, then $180 - a$ represents its supplement and $90 - a$ represents its complement.

$$180 - a = 3(90 - a) - 10$$
$$180 - a = 270 - 3a - 10$$
$$180 - a = -3a + 260$$
$$2a = 80$$
$$a = 40$$

The angle has a measure of 40°.

92 a.

$$-2(x - 1) = -2x + 2$$
$$-2x + 2 = -2x + 2$$
$$2 = 2 \text{ (identity)}$$

The solution set is {all real numbers}.

b. $3(x+4) = 3x - 4$
$$3x + 12 = 3x - 4$$
$$12 = -4 \text{ (contradiction)}$$
The solution set is $\emptyset$.

d. $$\frac{x-3}{3} + 4 = 3$$
$$3\left(\frac{x-3}{3}\right) + 3(4) = 3(3)$$
$$1(x-3) + 12 = 9$$
$$x + 9 = 9$$
$$x = 0$$
The solution set is $\{0\}$.

e. $$\frac{x+2}{3} + 1 = \frac{x-2}{3}$$
$$3\left(\frac{x+2}{3} + 1\right) = 3\left(\frac{x-2}{3}\right)$$
$$x + 2 + 3 = x - 2$$
$$x + 5 = x - 2$$
$$5 = -2 \text{ (contradiction)}$$
The solution set is $\emptyset$.

f. $$\frac{x-1}{5} - 2 = \frac{x-11}{5}$$
$$5\left(\frac{x-1}{5} - 2\right) = 5\left(\frac{x-11}{5}\right)$$
$$x - 1 - 10 = x - 11$$
$$x - 11 = x - 11$$
$$-11 = -11 \text{ (identity)}$$
The solution set is {all real numbers}.

i. $7(x-1) + 4(x-2) = 15(x-1)$
$$7x - 7 + 4x - 8 = 15x - 15$$
$$11x - 15 = 15x - 15$$
$$-4x = 0$$
$$x = 0$$
The solution set is $\{0\}$.

93. Let n represent the first integer, $n+1$ the second integer, and $n+2$ the third integer.
$$n + n + 2 = 2(n+1)$$
$$2n + 2 = 2n + 2$$
$$2 = 2 \text{ (identity)}$$
Any three consecutive integers would satisfy the conditions given.

PROBLEM SET 3.5 Inequalities

1. The left side simplifies to
$2(3) - 4(5) = 6 - 20 = -14$
and the right side to
$5(3) - 2(-1) + 4 = 15 + 2 + 4 = 21.$
Since $-14 < 21$, the given inequality is true.

3. The left side simplifies to
$$\frac{2}{3} - \frac{3}{4} + \frac{1}{6} = \frac{8}{12} - \frac{9}{12} + \frac{2}{12} = \frac{1}{12}$$
and the right side to
$$\frac{1}{5} + \frac{3}{4} - \frac{7}{10} = \frac{4}{20} + \frac{15}{20} - \frac{14}{20} = \frac{5}{20} = \frac{1}{4}.$$
Since $\frac{1}{12} < \frac{1}{4}$, the given inequality is false.

5. The left side simplifies to
$\left(-\frac{1}{2}\right)\left(\frac{4}{9}\right) = -\frac{2}{9}$ and the right side to $\left(\frac{3}{5}\right)\left(-\frac{1}{3}\right) = -\frac{1}{5}$.
Since $-\frac{2}{9} < -\frac{1}{5}$, the given inequality is false.

7. The left side simplifies to
$$\frac{3}{4} + \frac{2}{3} \div \frac{1}{5} = \frac{3}{4} + \frac{2}{3} \cdot \frac{5}{1} = \frac{3}{4} + \frac{10}{3}$$
$$= \frac{9}{12} + \frac{40}{12} = \frac{49}{12}$$ and the right side to
$$\frac{2}{3} + \frac{1}{2} \div \frac{3}{4} = \frac{2}{3} + \frac{1}{2} \cdot \frac{4}{3} = \frac{2}{3} + \frac{2}{3} = \frac{4}{3}.$$
Since $\frac{49}{12} > \frac{4}{3}$, the given inequality is true.

9. The left side simplifies to
$0.16 + 0.34 = 0.50$ and the right side to
$0.23 + 0.17 = 0.40$. Since $0.50 > 0.40$, the given inequality is true.

Problem Set 3.5

11. $\{x|x > -2\}$ or $(-2, \infty)$

-7 -6 -5 -4 -3 -2 -1 0 1 2 3 4 5 6

13. $\{x|x \leq 3\}$ or $(-\infty, 3]$

15. $\{x|x > 2\}$ or $(2, \infty)$

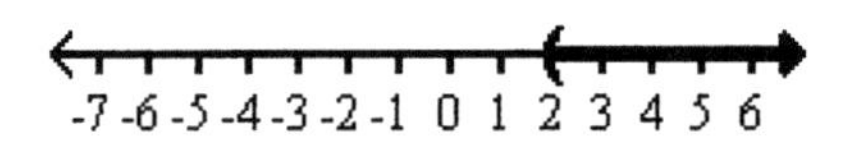

17. $\{x|x \leq -2\}$ or $(-\infty, -2]$

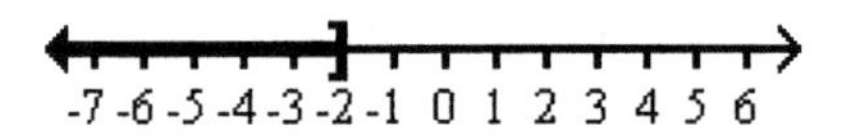

19. $\{x|x < -1\}$ or $(-\infty, -1)$

-7 -6 -5 -4 -3 -2 -1 0 1 2 3 4 5 6

21. $\{x|x < 2\}$ or $(-\infty, 2)$

-7 -6 -5 -4 -3 -2 -1 0 1 2 3 4 5 6

23. $x + 6 < -14$

Subtract 6 from both sides.

$x + 6 - 6 < -14 - 6$

$x < -20$

The solution set is

$\{x|x < -20\}$ or $(-\infty, -20)$.

25. $x - 4 \geq -13$

Add 4 to both sides.

$x - 4 + 4 \geq -13 + 4$

$x \geq -9$

The solution set is

$\{x|x \geq -9\}$ or $[-9, \infty)$.

27. $4x > 36$

Divide both sides by 4.

$$\frac{4x}{4} > \frac{36}{4}$$

$x > 9$

The solution set is

$\{x|x > 9\}$ or $(9, \infty)$.

29. $6x < 20$

Divide both sides by 6.

$$\frac{6x}{6} < \frac{20}{6}$$

$$x < \frac{10}{3}$$

The solution set is

$\left\{x|x < \frac{10}{3}\right\}$ or $\left(-\infty, \frac{10}{3}\right)$.

31. $-5x > 40$

Divide both sides by -5,

which reverses the inequality.

$$\frac{-5x}{-5} < \frac{40}{-5}$$

$x < -8$

The solution set is

$\{x|x < -8\}$ or $(-\infty, -8)$.

33. $-7n \leq -56$

Divide both sides by -7,

which reverses the inequality.

$$\frac{-7n}{-7} \geq \frac{-56}{-7}$$

$n \geq 8$

The solution set is

$\{n|n \geq 8\}$ or $[8, \infty)$.

35. $48 > -14n$

Divide both sides by -14, which reverses the inequality.

$\frac{48}{-14} < \frac{-14n}{-14}$

$-\frac{24}{7} < n$

$-\frac{24}{7} < n$ means $n > -\frac{24}{7}$.

$n > -\frac{24}{7}$

The solution set is

$\left\{n|n > -\frac{24}{7}\right\}$ or $\left(-\frac{24}{7}, \infty\right)$.

37. $16 < 9 + n$

Subtract 9 from both sides.

$16 - 9 < 9 + n - 9$

$7 < n$

$7 < n$ means $n > 7$.

$n > 7$

The solution set is

$\{n|n > 7\}$ or $(7, \infty)$.

39. $3x + 2 > 17$

Subtract 2 from both sides.

$3x + 2 - 2 > 17 - 2$

$3x > 15$

Divide both sides by 3.

$\frac{3x}{3} > \frac{15}{3}$

$x > 5$

The solution set is

$\{x|x > 5\}$ or $(5, \infty)$.

41. $4x - 3 \leq 21$

Add 3 to both sides.

$4x - 3 + 3 \leq 21 + 3$

$4x \leq 24$

Divide both sides by 4.

$\frac{4x}{4} \leq \frac{24}{4}$

$x \leq 6$

The solution set is

$\{x|x \leq 6\}$ or $(-\infty, 6]$.

43. $-2x - 1 \geq 41$

Add 1 to both sides.

$-2x - 1 + 1 \geq 41 + 1$

$-2x \geq 42$

Divide both sides by -2, which reverses the inequality.

$\frac{-2x}{-2} \leq \frac{42}{-2}$

$x \leq -21$

The solution set is

$\{x|x \leq -21\}$ or $(-\infty, -21]$.

45. $6x + 2 < 18$

Subtract 2 from both sides.

$6x + 2 - 2 < 18 - 2$

$6x < 16$

Divide both sides by 6.

$\frac{6x}{6} < \frac{16}{6}$

$x < \frac{8}{3}$

The solution set is

$\left\{x|x < \frac{8}{3}\right\}$ or $\left(-\infty, \frac{8}{3}\right)$.

47. $3 > 4x - 2$

Add 2 to both sides.

$3 + 2 > 4x - 2 + 2$

$5 > 4x$

Divide both sides by 4.

$\frac{5}{4} > \frac{4x}{4}$

$\frac{5}{4} > x$

$\frac{5}{4} > x$ means $x < \frac{5}{4}$.

$x < \frac{5}{4}$

The solution set is

$\left\{x|x < \frac{5}{4}\right\}$ or $\left(-\infty, \frac{5}{4}\right)$.

Problem Set 3.5

49. $-2 < -3x + 1$

Subtract 1 from both sides.

$-2 - 1 < -3x + 1 - 1$

$-3 < -3x$

Divide both sides by -3, which reverses the inequality.

$\frac{-3}{-3} > \frac{-3x}{-3}$

$1 > x$

$1 > x$ means $x < 1$.

$x < 1$

The solution set is $\{x|x < 1\}$ or $(-\infty, 1)$.

51. $-38 \geq -9t - 2$

Add 2 to both sides.

$-38 + 2 \geq -9t - 2 + 2$

$-36 \geq -9t$

Divide both sides by -9, which reverses the inequality.

$\frac{-36}{-9} \leq \frac{-9t}{-9}$

$4 \leq t$

$4 \leq t$ means $t \geq 4$.

$t \geq 4$

The solution set is $\{t|t \geq 4\}$ or $[4, \infty)$.

53. $5x - 4 - 3x > 24$

Combine similar terms on the left side.

$2x - 4 > 24$

Add 4 to both sides.

$2x - 4 + 4 > 24 + 4$

$2x > 28$

Divide both sides by 2.

$\frac{2x}{2} > \frac{28}{2}$

$x > 14$

The solution set is $\{x|x > 14\}$ or $(14, \infty)$.

55. $4x + 2 - 6x < -1$

Combine similar terms.

$-2x + 2 < -1$

Subtract 2 from both sides.

$-2x + 2 - 2 < -1 - 2$

$-2x < -3$

Divide both sides by -2, which reverses the inequality.

$\frac{-2x}{-2} > \frac{-3}{-2}$

$x > \frac{3}{2}$

The solution set is $\left\{x|x > \frac{3}{2}\right\}$ or $\left(\frac{3}{2}, \infty\right)$.

57. $-5 \geq 3t - 4 - 7t$

Combine similar terms.

$-5 \geq -4t - 4$

Add 4 to both sides.

$-5 + 4 \geq -4t - 4 + 4$

$-1 \geq -4t$

Divide both sides by -4, which reverses the inequality.

$\frac{-1}{-4} \leq \frac{-4t}{-4}$

$\frac{1}{4} \leq t$

$\frac{1}{4} \leq t$ means $t \geq \frac{1}{4}$

$t \geq \frac{1}{4}$

The solution set is $\left\{t|t \geq \frac{1}{4}\right\}$ or $\left[\frac{1}{4}, \infty\right)$.

59. $-x-4-3x>5$

Combine similar terms.

$$-4x-4>5$$

Add 4 to both sides.

$$-4x-4+4>5+4$$
$$-4x>9$$

Divide both sides by -4, which reverses the inequality.

$$\frac{-4x}{-4}<\frac{9}{-4}$$
$$x<-\frac{9}{4}$$

The solution set is

$\left\{x|x<-\frac{9}{4}\right\}$ or $\left(-\infty,-\frac{9}{4}\right)$.

65. $x-4<x+6$

$-4<6$ (identity)

The solution set is $\{x|x$ is any real number$\}$.

67. $5x+2>5x+7$

$2>7$ (contradiction)

The solution set is $\emptyset$.

69. $-2x+7+2x>1$

$7>1$ (identity)

The solution set is $\{x|x$ is any real number$\}$.

71. $-7\geq 5x-2-5x$

$-7\geq -2$ (contradiction)

The solution set is $\emptyset$.

PROBLEM SET 3.6 Inequalities, Compound Inequalities, and Problem Solving

1. $3x+4>x+8$

Subtract x from both sides.

$$3x+4-x>x+8-x$$
$$2x+4>8$$

Subtract 4 from both sides.

$$2x+4-4>8-4$$
$$2x>4$$

Divide both sides by 2.

$$\frac{2x}{2}>\frac{4}{2}$$
$$x>2$$

The solution set is $\{x|x>2\}$ or $(2,\infty)$.

3. $7x-2<3x-6$

Subtract $3x$ from both sides.

$$7x-2-3x<3x-6-3x$$
$$4x-2<-6$$

Add 2 to both sides.

$$4x-2+2<-6+2$$
$$4x<-4$$

Divide both sides by 4.

$$\frac{4x}{4}<\frac{-4}{4}$$
$$x<-1$$

The solution set is $\{x|x<-1\}$ or $(-\infty,-1)$.

5. $6x+7>3x-3$

Subtract $3x$ from both sides.

$$6x+7-3x>3x-3-3x$$
$$3x+7>-3$$

Subtract 7 from both sides.

$$3x+7-7>-3-7$$
$$3x>-10$$

Divide both sides by 3.

$$\frac{3x}{3}>\frac{-10}{3}$$
$$x>-\frac{10}{3}$$

The solution set is

$\left\{x|x>-\frac{10}{3}\right\}$ or $\left(-\frac{10}{3},\infty\right)$.

Problem Set 3.6

7. $5n - 2 \le 6n + 9$

Subtract $6n$ from both sides.

$5n - 2 - 6n \le 6n + 9 - 6n$

$-n - 2 \le 9$

Add 2 to both sides.

$-n - 2 + 2 \le 9 + 2$

$-n \le 11$

Multiply by -1, which reverses the inequality.

$n \ge -11$

The solution set is

$\{n|n \ge -11\}$ or $[-11, \infty)$.

9. $2t + 9 \ge 4t - 13$

Subtract $4t$ from both sides.

$2t + 9 - 4t \ge 4t - 13 - 4t$

$-2t + 9 \ge -13$

Subtract 9 from both sides.

$-2t + 9 - 9 \ge -13 - 9$

$-2t \ge -22$

Divide both sides by -2, which reverses the inequality.

$$\frac{-2t}{-2} \le \frac{-22}{-2}$$

$t \le 11$

The solution set is

$\{t|t \le 11\}$ or $(-\infty, 11]$.

11. $-3x - 4 < 2x + 7$

Subtract $2x$ from both sides.

$-3x - 4 - 2x < 2x + 7 - 2x$

$-5x - 4 < 7$

Add 4 to both sides.

$-5x - 4 + 4 < 7 + 4$

$-5x < 11$

Divide both sides by -5, which reverses the inequality.

$$\frac{-5x}{-5} > \frac{11}{-5}$$

$$x > -\frac{11}{5}$$

The solution set is

$\left\{x|x > -\frac{11}{5}\right\}$ or $\left(-\frac{11}{5}, \infty\right)$.

13. $-4x + 6 > -2x + 1$

Add $2x$ to both sides.

$-4x + 6 + 2x > -2x + 1 + 2x$

$-2x + 6 > 1$

Subtract 6 from both sides.

$-2x + 6 - 6 > 1 - 6$

$-2x > -5$

Divide both sides by -2, which reverses the inequality.

$$\frac{-2x}{-2} < \frac{-5}{-2}$$

$$x < \frac{5}{2}$$

The solution set is

$\left\{x|x < \frac{5}{2}\right\}$ or $\left(-\infty, \frac{5}{2}\right)$.

15. $5(x - 2) \le 30$

Apply distributive property.

$5x - 10 \le 30$

Add 10 to both sides.

$5x - 10 + 10 \le 30 + 10$

$5x \le 40$

Divide both sides by 5.

$$\frac{5x}{5} \le \frac{40}{5}$$

$x \le 8$

The solution set is

$\{x|x \le 8\}$ or $(-\infty, 8]$.

17. $2(n + 3) > 9$

Apply distributive property.

$2n + 6 > 9$

Subtract 6 from both sides.

$2n + 6 - 6 > 9 - 6$

$2n > 3$

Divide both sides by 2.

$$\frac{2n}{2} > \frac{3}{2}$$

$$n > \frac{3}{2}$$

The solution set is

$\left\{n|n > \frac{3}{2}\right\}$ or $\left(\frac{3}{2}, \infty\right)$.

19. $-3(y-1) < 12$
Apply distributive property.
$-3y+3 < 12$
Subtract 3 from to both sides.
$-3y+3-3 < 12-3$
$-3y < 9$
Divide both sides by -3,
which reverses the inequality.
$$\frac{-3y}{-3} > \frac{9}{-3}$$
$y > -3$
The solution set is
$\{y|y > -3\}$ or $(-3, \infty)$.

21. $-2(x+6) > -17$
Apply distributive property.
$-2x-12 > -17$
Add 12 to both sides.
$-2x-12+12 > -17+12$
$-2x > -5$
Divide both sides by -2,
which reverses the inequality.
$$\frac{-2x}{-2} < \frac{-5}{-2}$$
$$x < \frac{5}{2}$$
The solution set is
$\left\{x|x < \frac{5}{2}\right\}$ or $\left(-\infty, \frac{5}{2}\right)$.

23. $3(x-2) < 2(x+1)$
Apply distributive property.
$3x-6 < 2x+2$
Subtract $2x$ from both sides.
$3x-6-2x < 2x+2-2x$
$x-6 < 2$
Add 6 to both sides.
$x-6+6 < 2+6$
$x < 8$
The solution set is
$\{x|x < 8\}$ or $(-\infty, 8)$.

25. $4(x+3) > 6(x-5)$
Apply distributive property.
$4x+12 > 6x-30$
Subtract $6x$ from both sides.
$4x+12-6x > 6x-30-6x$
$-2x+12 > -30$
Subtract 12 from both sides.
$-2x+12-12 > -30-12$
$-2x > -42$
Divide both sides by -2,
which reverses the inequality.
$$\frac{-2x}{-2} < \frac{-42}{-2}$$
$x < 21$
The solution set is
$\{x|x < 21\}$ or $(-\infty, 21)$.

27. $3(x-4)+2(x+3) < 24$
Apply distributive property.
$3x-12+2x+6 < 24$
Combine similar terms.
$5x-6 < 24$
Add 6 to both sides.
$5x-6+6 < 24+6$
$5x < 30$
Divide both sides by 5.
$$\frac{5x}{5} < \frac{30}{5}$$
$x < 6$
The solution set is
$\{x|x < 6\}$ or $(-\infty, 6)$.

Problem Set 3.6

29. $5(n+1)-3(n-1) > -9$

Apply distributive property.

$$5n+5-3n+3 > -9$$

Combine similar terms.

$$2n+8 > -9$$

Subtract 8 from both sides.

$$2n+8-8 > -9-8$$
$$2n > -17$$

Divide both sides by 2.

$$\frac{2n}{2} > \frac{-17}{2}$$
$$n > -\frac{17}{2}$$

The solution set is

$\left\{n \middle| n > -\frac{17}{2}\right\}$ or $\left(-\frac{17}{2}, \infty\right)$.

31. $\frac{1}{2}n - \frac{2}{3}n \geq -7$

Multiply both sides by 6.

$$6\left(\frac{1}{2}n - \frac{2}{3}n\right) \geq 6(-7)$$
$$3n - 4n \geq -42$$
$$-n \geq -42$$

Multiply both sides by -1, which reverses the inequality.

$$n \leq 42$$

The solution set is

$\{n|n \leq 42\}$ or $(-\infty, 42]$.

33. $\frac{3}{4}n - \frac{5}{6}n < \frac{3}{8}$

Multiply both sides by 24.

$$24\left(\frac{3}{4}n - \frac{5}{6}n\right) < 24\left(\frac{3}{8}\right)$$
$$18n - 20n < 9$$

Combine similar terms.

$$-2n < 9$$

Divide both sides by -2, which reverses the inequality.

$$\frac{-2n}{-2} > \frac{9}{-2}$$
$$n > -\frac{9}{2}$$

The solution set is

$\left\{n \middle| n > -\frac{9}{2}\right\}$ or $\left(-\frac{9}{2}, \infty\right)$.

35. $\frac{3x}{5} - \frac{2}{3} > \frac{x}{10}$

Multiply both sides by 30.

$$30\left(\frac{3x}{5} - \frac{2}{3}\right) > 30\left(\frac{x}{10}\right)$$
$$18x - 20 > 3x$$
$$18x - 20 + 20 > 3x + 20$$
$$18x > 3x + 20$$
$$15x > 20$$

Divide both sides by 15.

$$\frac{15x}{15} > \frac{20}{15}$$
$$x > \frac{4}{3}$$

The solution set is

$\left\{x \middle| x > \frac{4}{3}\right\}$ or $\left(\frac{4}{3}, \infty\right)$.

37. $n \geq 3.4 + 0.15n$

Subtract $0.15n$ from both sides.

$$n - 0.15n \geq 3.4$$
$$0.85n \geq 3.4$$
$$n \geq 4$$

The solution set is

$\{n|n \geq 4\}$ or $[4, \infty)$.

39. $$\begin{aligned} 0.09t + 0.1(t + 200) &> 77 \\ 0.09t + 0.1t + 20 &> 77 \\ 0.19t + 20 &> 77 \\ 0.19t &> 57 \\ t &> 300 \end{aligned}$$
The solution set is
$\{t|t > 300\}$ or $(300, \infty)$.

41. $$\begin{aligned} 0.06x + 0.08(250 - x) &\geq 19 \\ 0.06x + 20 - 0.08x &\geq 19 \\ -0.02x + 20 &\geq 19 \\ -0.02x &\geq -1 \\ x &\leq 50 \end{aligned}$$
The solution set is
$\{x|x \leq 50\}$ or $(-\infty, 50]$.

43. $$\frac{x-1}{2} + \frac{x+3}{5} > \frac{1}{10}$$
Multiply both sides by 10.
$$\begin{aligned} 10\left(\frac{x-1}{2} + \frac{x+3}{5}\right) &> 10\left(\frac{1}{10}\right) \\ 5(x-1) + 2(x+3) &> 1(1) \\ 5x - 5 + 2x + 6 &> 1 \\ 7x + 1 &> 1 \\ 7x &> 0 \\ x &> 0 \end{aligned}$$
The solution set is
$\{x|x > 0\}$ or $(0, \infty)$.

45. $$\frac{x+2}{6} - \frac{x+1}{5} < -2$$
Multiply both sides by 30.
$$\begin{aligned} 30\left(\frac{x+2}{6} - \frac{x+1}{5}\right) &< 30(-2) \\ 5(x+2) - 6(x+1) &< -60 \\ 5x + 10 - 6x - 6 &< -60 \\ -x + 4 &< -60 \\ -x &< -64 \end{aligned}$$
Multiply both sides by -1,
which reverses the inequality.
$$x > 64$$
The solution set is
$\{x|x > 64\}$ or $(64, \infty)$.

47. $$\frac{n+3}{3} + \frac{n-7}{2} > 3$$
Multiply both sides by 6.
$$\begin{aligned} 6\left(\frac{n+3}{3} + \frac{n-7}{2}\right) &> 6(3) \\ 2(n+3) + 3(n-7) &> 18 \\ 2n + 6 + 3n - 21 &> 18 \\ 5n - 15 &> 18 \\ 5n &> 33 \\ n &> \frac{33}{5} \end{aligned}$$
The solution set is
$\left\{n|n > \frac{33}{5}\right\}$ or $\left(\frac{33}{5}, \infty\right)$.

49. $$\frac{x-3}{7} - \frac{x-2}{4} \leq \frac{9}{14}$$
Multiply both sides by 28.
$$\begin{aligned} 28\left(\frac{x-3}{7} - \frac{x-2}{4}\right) &\leq 28\left(\frac{9}{14}\right) \\ 4(x-3) - 7(x-2) &\leq 2(9) \\ 4x - 12 - 7x + 14 &\leq 18 \\ -3x + 2 &\leq 18 \\ -3x &\leq 16 \\ x &\geq -\frac{16}{3} \end{aligned}$$
The solution set is
$\left\{x|x \geq -\frac{16}{3}\right\}$ or $\left[-\frac{16}{3}, \infty\right)$.

51. Since it is an "and" statement, we need to satisfy both inequalities at the same time. Thus, all numbers between -1 and 2, but not including -1 and 2, are solutions.

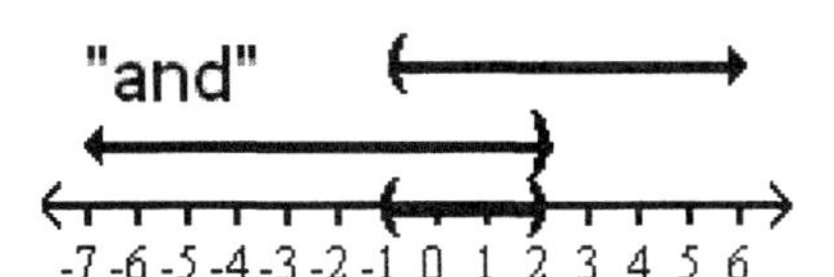

Problem Set 3.6

53. Since it is an "or" statement, the solution set consists of all numbers less than (but not including) -2 along with all numbers greater than (but not including) 1.

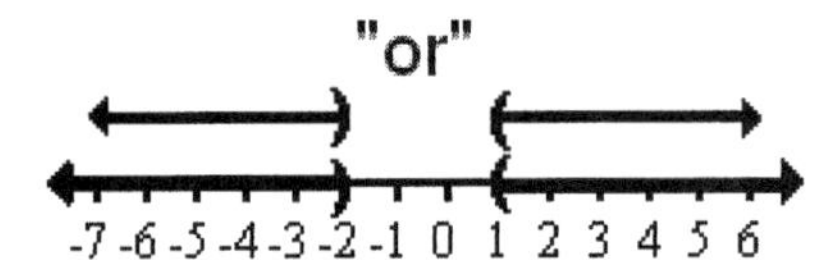

55. Since it is an "and" statement, we need to satisfy both inequalities at the same time. Thus, all numbers between -2 and 2, including 2 (but not including -2), are solutions.

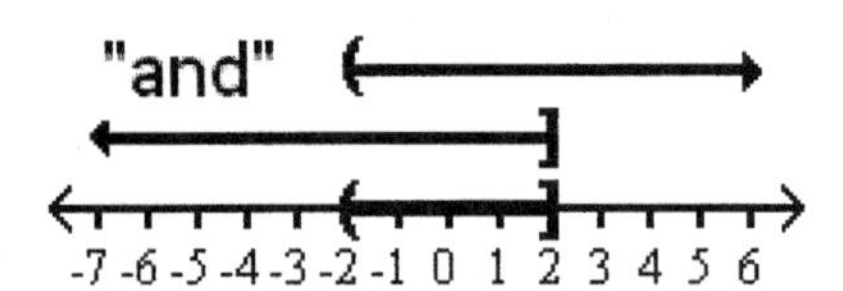

57. Since it is an "and" statement we are looking for all numbers that satisfy both inequalities at the same time. Thus, any number greater than 2 will work.

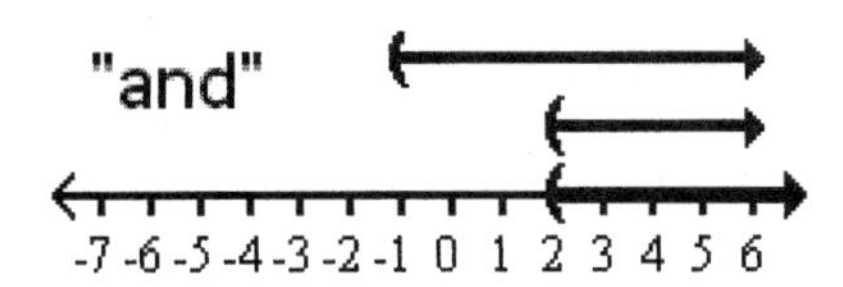

59. Since it is an "or" statement we are looking for all numbers that satisfy either inequality. Thus, any number greater than -4 will work.

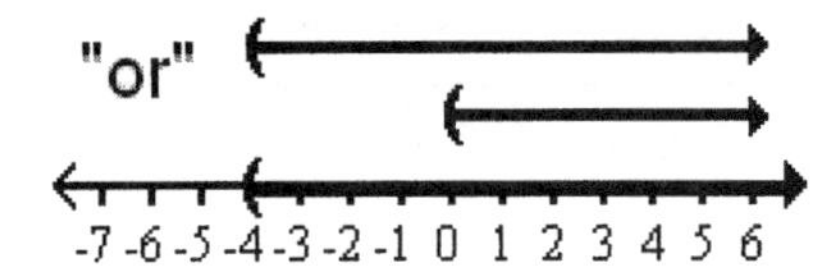

61. Since it is an "and" statement we are looking for all numbers that satisfy both inequalities at the same time. There are no numbers that are both greater than 3 and less than -1. So the solution set is $\emptyset$.

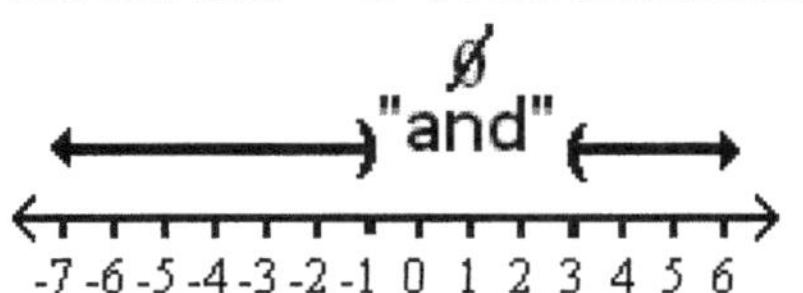

63. Since it is an "or" statement, the solution set consists of all numbers less than or equal to 0 along with all numbers greater than or equal to 2.

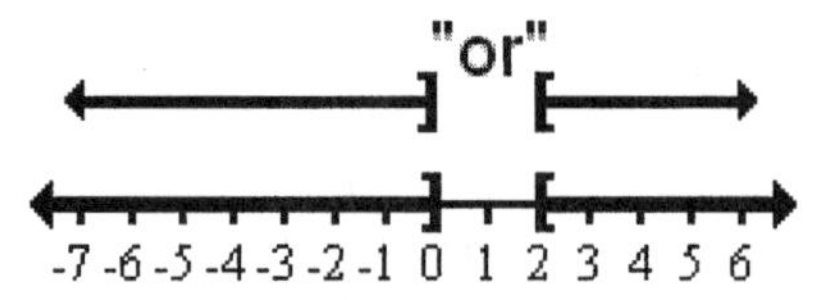

65. Since it is an "or" statement, we want all numbers greater than -4 along with all numbers less than 3. Thus, the solution set is the entire set of real numbers.

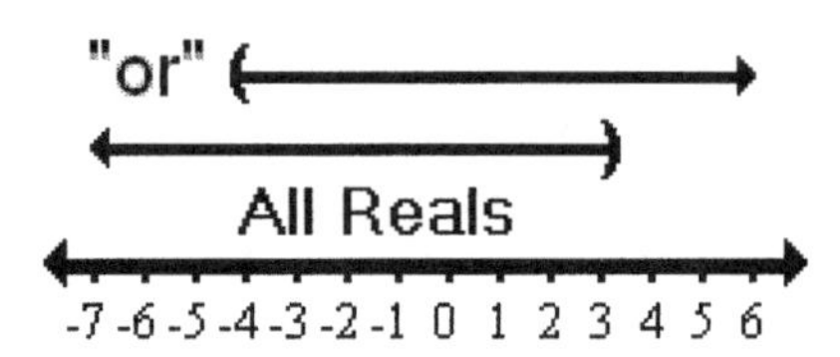

67. Let n represent the number.

$3n + 5 > 26$

$3n > 21$

$n > 7$

The numbers must be greater than 7.

69. Let w represent the width of the rectangle. Also remember that "length plus width equals one-half of the perimeter" of a rectangle.

$w + 20 \leq 35$

$w \leq 15$

Thus, 15 inches is the largest possible value for the width.

71. Let x be her score in the last game.

$$\frac{132 + 160 + x}{3} \geq 150$$
$$292 + x \geq 450$$
$$x \geq 158$$

She must bowl 158 or better in her last game.

73. Let x be his average on the last two exams.

$$\frac{96 + 90 + 94 + 2x}{5} > 92$$
$$280 + 2x > 460$$
$$2x > 180$$
$$x > 90$$

He must have an average better than 90 on the last two exams.

75. Let r represent the number of sales.

$$48r > 4000 + 32r$$
$$16r > 4000$$
$$r > 250$$

The business must have more than 250 sales.

77. Let x be his score on the final round.

$$\frac{82 + 84 + 78 + 79 + x}{5} \leq 80$$
$$323 + x \leq 400$$
$$x \leq 77$$

He must shoot 77 or less.

CHAPTER 3 Review Problem Set

1.
$$9x - 2 = -29$$
$$9x - 2 + 2 = -29 + 2$$
$$9x = -27$$
$$x = -3$$

The solution set is $\{-3\}$.

2.
$$-3 = -4y + 1$$
$$-4 = -4y$$
$$1 = y$$

The solution set is $\{1\}$.

3.
$$7 - 4x = 10$$
$$7 - 4x - 7 = 10 - 7$$
$$-4x = 3$$
$$x = -\frac{3}{4}$$

The solution set is $\left\{-\frac{3}{4}\right\}$.

4.
$$6y - 5 = 4y + 13$$
$$6y - 5 - 4y = 4y + 13 - 4y$$
$$2y - 5 = 13$$
$$2y = 18$$
$$y = 9$$

The solution set is $\{9\}$.

5.
$$4n - 3 = 7n + 9$$
$$4n - 3 - 7n = 7n + 9 - 7n$$
$$-3n - 3 = 9$$
$$-3n - 3 + 3 = 9 + 3$$
$$-3n = 12$$
$$n = -4$$

The solution set is $\{-4\}$.

6.
$$7(y - 4) = 4(y + 3)$$
$$7y - 28 = 4y + 12$$
$$3y - 28 = 12$$
$$3y = 40$$
$$y = \frac{40}{3}$$

The solution set is $\left\{\frac{40}{3}\right\}$.

7.
$$\begin{aligned} 2(x+1)+5(x-3) &= 11(x-2) \\ 2x+2+5x-15 &= 11x-22 \\ 7x-13 &= 11x-22 \\ 7x-13-11x &= 11x-22-11x \\ -4x-13 &= -22 \\ -4x-13+13 &= -22+13 \\ -4x &= -9 \\ x &= \frac{9}{4} \end{aligned}$$
The solution set is $\left\{\frac{9}{4}\right\}$.

8.
$$\begin{aligned} -3(x+6) &= 5x-3 \\ -3x-18 &= 5x-3 \\ -8x-18 &= -3 \\ -8x &= 15 \\ x &= -\frac{15}{8} \end{aligned}$$
The solution set is $\left\{-\frac{15}{8}\right\}$.

9.
$$\begin{aligned} \frac{2}{5}n-\frac{1}{2}n &= \frac{7}{10} \\ 10\left(\frac{2}{5}n-\frac{1}{2}n\right) &= 10\left(\frac{7}{10}\right) \\ 4n-5n &= 7 \\ -n &= 7 \\ n &= -7 \end{aligned}$$
The solution set is $\{-7\}$.

10.
$$\begin{aligned} \frac{3n}{4}+\frac{5n}{7} &= \frac{1}{14} \\ 28\left(\frac{3n}{4}+\frac{5n}{7}\right) &= 28\left(\frac{1}{14}\right) \\ 21n+20n &= 2 \\ 41n &= 2 \\ n &= \frac{2}{41} \end{aligned}$$
The solution set is $\left\{\frac{2}{41}\right\}$.

11.
$$\begin{aligned} \frac{x-3}{6}+\frac{x+5}{8} &= \frac{11}{12} \\ 24\left(\frac{x-3}{6}+\frac{x+5}{8}\right) &= 24\left(\frac{11}{12}\right) \\ 4(x-3)+3(x+5) &= 2(11) \\ 4x-12+3x+15 &= 22 \\ 7x+3 &= 22 \\ 7x &= 19 \\ x &= \frac{19}{7} \end{aligned}$$
The solution set is $\left\{\frac{19}{7}\right\}$.

12.
$$\begin{aligned} \frac{n}{2}-\frac{n-1}{4} &= \frac{3}{8} \\ 8\left(\frac{n}{2}-\frac{n-1}{4}\right) &= 8\left(\frac{3}{8}\right) \\ 4n-2(n-1) &= 3 \\ 4n-2n+2 &= 3 \\ 2n+2 &= 3 \\ 2n &= 1 \\ n &= \frac{1}{2} \end{aligned}$$
The solution set is $\left\{\frac{1}{2}\right\}$.

13.
$$\begin{aligned} -2(x-4) &= -3(x+8) \\ -2x+8 &= -3x-24 \\ -2x+8+3x &= -3x-24+3x \\ x+8 &= -24 \\ x &= -32 \end{aligned}$$
The solution set is $\{-32\}$.

14.
$$\begin{aligned} 3x-4x-2 &= 7x-14-9x \\ -x-2 &= -2x-14 \\ -x-2+2x &= -2x-14+2x \\ x-2 &= -14 \\ x &= -12 \end{aligned}$$
The solution set is $\{-12\}$.

15.
$$\begin{aligned} 5(n-1)-4(n+2) &= -3(n-1)+3n+5 \\ 5n-5-4n-8 &= -3n+3+3n+5 \\ n-13 &= 8 \\ n &= 21 \end{aligned}$$
The solution set is $\{21\}$.

16. $\dfrac{x-3}{9}=\dfrac{x+4}{8}$

$8(x-3)=9(x+4)$
$8x-24=9x+36$
$-x=60$
$x=-60$

The solution set is $\{-60\}$.

17. $\dfrac{x-1}{-3}=\dfrac{x+2}{-4}$

$-4(x-1)=-3(x+2)$
$-4x+4=-3x-6$
$-x+4=-6$
$-x=-10$
$x=10$

The solution set is $\{10\}$.

18. $-(t-3)-(2t+1)=3(t+5)-2(t+1)$

$-t+3-2t-1=3t+15-2t-2$
$-3t+2=t+13$
$-4t=11$
$t=-\dfrac{11}{4}$

The solution set is $\left\{-\dfrac{11}{4}\right\}$.

19. $\dfrac{2x-1}{3}=\dfrac{3x+2}{2}$

$2(2x-1)=3(3x+2)$
$4x-2=9x+6$
$-5x-2=6$
$-5x=8$
$x=-\dfrac{8}{5}$

The solution set is $\left\{-\dfrac{8}{5}\right\}$.

20. $3(2t-4)+2(3t+1)=-2(4t+3)-(t-1)$

$6t-12+6t+2=-8t-6-t+1$
$12t-10=-9t-5$
$21t-10=-5$
$21t=5$
$t=\dfrac{5}{21}$

The solution set is $\left\{\dfrac{5}{21}\right\}$.

21. $3x-2>10$

$3x>12$
$x>4$

The solution set is $\{x|x>4\}$ or $(4,\infty)$.

22. $-2x-5<3$

$-2x<8$
$x>-4$

The solution set is $\{x|x>-4\}$ or $(-4,\infty)$.

23. $2x-9\geq x+4$

$x-9\geq 4$
$x\geq 13$

The solution set is $\{x|x\geq 13\}$ or $[13,\infty)$.

24. $3x+1\leq 5x-10$

$-2x+1\leq -10$
$-2x\leq -11$
$x\geq \dfrac{11}{2}$

The solution set is $\left\{x|x\geq\dfrac{11}{2}\right\}$ or $\left[\dfrac{11}{2},\infty\right)$.

25. $6(x-3)>4(x+13)$

$6x-18>4x+52$
$2x-18>52$
$2x>70$
$x>35$

The solution set is $\{x|x>35\}$ or $(35,\infty)$.

26. $2(x+3)+3(x-6)<14$

$2x+6+3x-18<14$
$5x-12<14$
$5x<26$
$x<\dfrac{26}{5}$

The solution set is $\left\{x|x<\dfrac{26}{5}\right\}$ or $\left(-\infty,\dfrac{26}{5}\right)$.

27. $\frac{2n}{5} - \frac{n}{4} < \frac{3}{10}$

$20\left(\frac{2n}{5} - \frac{n}{4}\right) < 20\left(\frac{3}{10}\right)$

$8n - 5n < 6$

$3n < 6$

$n < 2$

The solution set is

$\{n|n < 2\}$ or $(-\infty, 2)$.

28. $\frac{n+4}{5} + \frac{n-3}{6} > \frac{7}{15}$

$30\left(\frac{n+4}{5} + \frac{n-3}{6}\right) > 30\left(\frac{7}{15}\right)$

$6(n+4) + 5(n-3) > 2(7)$

$6n + 24 + 5n - 15 > 14$

$11n + 9 > 14$

$11n > 5$

$n > \frac{5}{11}$

The solution set is

$\left\{n|n > \frac{5}{11}\right\}$ or $\left(\frac{5}{11}, \infty\right)$.

29. $-16 < 8 + 2y - 3y$

$-16 < 8 - y$

$-24 < -y$

$24 > y$

$y < 24$

The solution set is

$\{y|y < 24\}$ or $(-\infty, 24)$.

30. $-24 > 5x - 4 - 7x$

$-24 > -2x - 4$

$-20 > -2x$

$10 < x$

$x > 10$

The solution set is $\{x|x > 10\}$ or $(10, \infty)$.

31. $-3(n-4) > 5(n+2) + 3n$

$-3n + 12 > 5n + 10 + 3n$

$-3n + 12 > 8n + 10$

$-11n + 12 > 10$

$-11n > -2$

$n < \frac{2}{11}$

The solution set is

$\left\{n|n < \frac{2}{11}\right\}$ or $\left(-\infty, \frac{2}{11}\right)$.

32. $-4(n-2) - (n-1) < -4(n+6)$

$-4n + 8 - n + 1 < -4n - 24$

$-5n + 9 < -4n - 24$

$-n + 9 < -24$

$-n < -33$

$n > 33$

The solution set is $\{n|n > 33\}$ or $(33, \infty)$.

33. $\frac{3}{4}n - 6 \leq \frac{2}{3}n + 4$

$12\left(\frac{3}{4}n - 6\right) \leq 12\left(\frac{2}{3}n + 4\right)$

$9n - 72 \leq 8n + 48$

$n - 72 \leq 48$

$n \leq 120$

The solution set is $\{n|n \leq 120\}$ or $(-\infty, 120]$.

34. $\frac{1}{2}n - \frac{1}{3}n - 4 \geq \frac{3}{5}n + 2$

$30\left(\frac{1}{2}n - \frac{1}{3}n - 4\right) \geq 30\left(\frac{3}{5}n + 2\right)$

$15n - 10n - 120 \geq 18n + 60$

$5n - 120 \geq 18n + 60$

$-13n \geq 180$

$n \leq -\frac{180}{13}$

The solution set is

$\left\{n|n \leq -\frac{180}{13}\right\}$ or $\left(-\infty, -\frac{180}{13}\right]$.

35. $-12 > -4(x-1)+2$
$-12 > -4x+4+2$
$-12 > -4x+6$
$-18 > -4x$
$\frac{18}{4} < x$
$x > \frac{9}{2}$
The solution set is $\left\{x|x > \frac{9}{2}\right\}$ or $\left(\frac{9}{2}, \infty\right)$.

36. $36 < -3(x+2)-1$
$36 < -3x-6-1$
$36 < -3x-7$
$43 < -3x$
$-\frac{43}{3} > x$
$x < -\frac{43}{3}$
The solution set is
$\left\{x|x < -\frac{43}{3}\right\}$ or $\left(-\infty, -\frac{43}{3}\right)$.

37. Since it is an "and" statement, we need to satisfy both inequalities at the same time. Thus, all numbers between -3 and 2, but not including -3 and 2, are solutions.

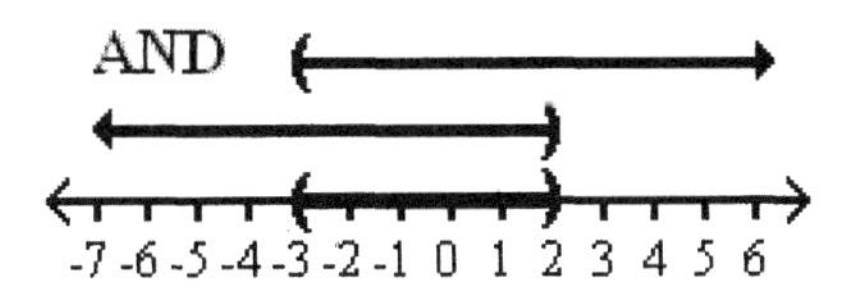

38. Since it is an "or" statement, the solution set consists of all numbers less than (but not including) -1 along with all numbers greater than (but not including) 4.

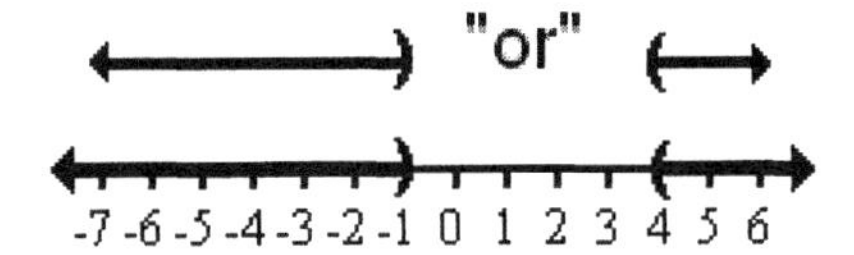

39. Since it is an "or" statement, we want all numbers less than 2 along with all numbers greater than 0. Thus, the solution set is the entire set of real numbers.

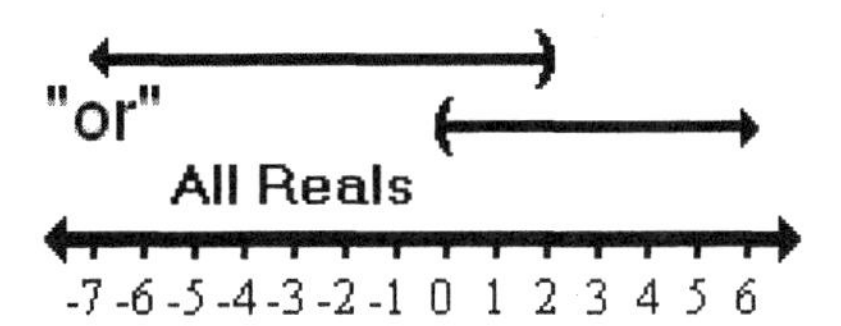

40. Since it is an "and" statement, we are looking for all numbers that satisfy both inequalities at the same time. Thus, any number greater than 1 will work.

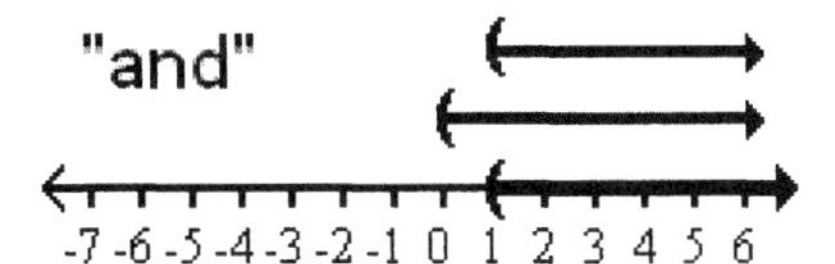

41. Let n represent the number.
$\frac{3}{4}n = 18$
$4\left(\frac{3}{4}n\right) = 4(18)$
$3n = 72$
$n = 24$
The number is 24.

42. Let n represent the number.
$19 = 3n-2$
$21 = 3n$
$7 = n$
The number is 7.

43. Let n represent the larger number, then $n-12$ represents the smaller number.
$n-21 = 12$
$n = 33$
The larger number is 33.

44. Let n represent the number.
$9n - 1 = 7n + 15$
$2n - 1 = 15$
$2n = 16$
$n = 8$
The number is 8.

45. Let x represent the score on the fifth exam.
$\frac{83 + 89 + 78 + 86 + x}{5} \geq 85$
$336 + x \geq 425$
$x \geq 89$
She must have a score of 89 or better.

46. Let n represent the smaller number, then $40 - n$ represents the larger number.
$6n = 4(40 - n)$
$6n = 160 - 4n$
$10n = 160$
$n = 16$
The smaller number is 16 and the larger number is 24.

47. Let n represent the number.
$\frac{2}{3}n - 2 = \frac{1}{2}n + 1$
$6\left(\frac{2}{3}n - 2\right) = 6\left(\frac{1}{2}n + 1\right)$
$4n - 12 = 3n + 6$
$n - 12 = 6$
$n = 18$
The number is 18.

48. Let x represent the score on the fourth exam. The total score from the first three exams was $3(84) = 252$.
$\frac{252 + x}{4} \geq 85$
$252 + x \geq 340$
$x \geq 88$
She must have a score of 88 or better.

49. Let x represent the number of nickels, then $30 - x$ represents the number of dimes. The value in cents of the nickels is $5x$ and $10(30 - x)$ is the value in cents of the dimes.
$5x + 10(30 - x) = 260$
$5x + 300 - 10x = 260$
$-5x + 300 = 260$
$-5x = -40$
$x = 8$
There would be 8 nickels and $30 - (8) = 22$ dimes.

50. Let n represent the number of nickels, $3n + 1$ the number of dimes, and $2(3n + 1)$ the number of quarters. The value in cents of the nickels is $5n$, $10(3n + 1)$ is the value in cents of the dimes, and $25(6n + 2)$ is the value in cents of the quarters.
$5n + 10(3n + 1) + 25(6n + 2) = 1540$
$5n + 30n + 10 + 150n + 50 = 1540$
$185n + 60 = 1540$
$185n = 1480$
$n = 8$
There would be 8 nickels, $3(8) + 1 = 25$ dimes, and $2(25) = 50$ quarters.

51. Let a represent the measure of the angle, then $180 - a$ represents its supplement and $90 - a$ represents its complement.
$180 - a = 3(90 - a) + 14$
$180 - a = 270 - 3a + 14$
$180 - a = 284 - 3a$
$180 + 2a = 284$
$2a = 104$
$a = 52$
The measure of the angle is 52°.

52. Let m represent the number of miles that she drove.

$$3(25) + 0.20m = 215$$
$$75 + 0.20m = 215$$
$$0.20m = 140$$
$$20m = 14,000$$
$$m = 700$$

She drove 700 miles.

CHAPTER 3 | Test

1.
$$7x - 3 = 11$$
$$7x = 14$$
$$x = 2$$
The solution set is $\{2\}$.

2.
$$-7 = -3x + 2$$
$$-9 = -3x$$
$$3 = x$$
The solution set is $\{3\}$.

3.
$$4n + 3 = 2n - 15$$
$$2n + 3 = -15$$
$$2n = -18$$
$$n = -9$$
The solution set is $\{-9\}$.

4.
$$3n - 5 = 8n + 20$$
$$-5n - 5 = 20$$
$$-5n = 25$$
$$n = -5$$
The solution set is $\{-5\}$.

5.
$$4(x - 2) = 5(x + 9)$$
$$4x - 8 = 5x + 45$$
$$-x - 8 = 45$$
$$-x = 53$$
$$x = -53$$
The solution set is $\{-53\}$.

6.
$$9(x + 4) = 6(x - 3)$$
$$9x + 36 = 6x - 18$$
$$3x = -54$$
$$x = -18$$
The solution set is $\{-18\}$.

7.
$$5(y - 2) + 2(y + 1) = 3(y - 6)$$
$$5y - 10 + 2y + 2 = 3y - 18$$
$$7y - 8 = 3y - 18$$
$$4y - 8 = -18$$
$$4y = -10$$
$$y = -\frac{10}{4} = -\frac{5}{2}$$
The solution set is $\left\{-\frac{5}{2}\right\}$.

8.
$$\frac{3}{5}x - \frac{2}{3} = \frac{1}{2}$$
$$30\left(\frac{3}{5}x - \frac{2}{3}\right) = 30\left(\frac{1}{2}\right)$$
$$18x - 20 = 15$$
$$18x = 35$$
$$x = \frac{35}{18}$$
The solution set is $\left\{\frac{35}{18}\right\}$.

9.
$$\frac{x - 2}{4} = \frac{x + 3}{6}$$
$$12\left(\frac{x - 2}{4}\right) = 12\left(\frac{x + 3}{6}\right)$$
$$3(x - 2) = 2(x + 3)$$
$$3x - 6 = 2x + 6$$
$$x - 6 = 6$$
$$x = 12$$
The solution set is $\{12\}$.

10.
$$\frac{x+2}{3}+\frac{x-1}{2}=2$$
$$6\left(\frac{x+2}{3}+\frac{x-1}{2}\right)=6(2)$$
$$2(x+2)+3(x-1)=12$$
$$2x+4+3x-3=12$$
$$5x+1=12$$
$$5x=11$$
$$x=\frac{11}{5}$$
The solution set is $\left\{\frac{11}{5}\right\}$.

11.
$$\frac{x-3}{6}-\frac{x-1}{8}=\frac{13}{24}$$
$$24\left(\frac{x-3}{6}-\frac{x-1}{8}\right)=24\left(\frac{13}{24}\right)$$
$$4(x-3)-3(x-1)=13$$
$$4x-12-3x+3=13$$
$$x-9=13$$
$$x=22$$
The solution set is $\{22\}$.

12.
$$-5(n-2)=-3(n+7)$$
$$-5n+10=-3n-21$$
$$-2n+10=-21$$
$$-2n=-31$$
$$n=\frac{31}{2}$$
The solution set is $\left\{\frac{31}{2}\right\}$.

13.
$$3x-2<13$$
$$3x<15$$
$$x<5$$
The solution set is $\{x|x<5\}$ or $(-\infty, 5)$.

14.
$$-2x+5\geq 3$$
$$-2x\geq -2$$
$$x\leq 1$$
The solution set is $\{x|x\leq 1\}$ or $(-\infty, 1]$.

15.
$$3(x-1)\leq 5(x+3)$$
$$3x-3\leq 5x+15$$
$$-2x-3\leq 15$$
$$-2x\leq 18$$
$$x\geq -9$$
The solution set is $\{x|x\geq -9\}$ or $[-9, \infty)$.

16.
$$-4>7(x-1)+3$$
$$-4>7x-7+3$$
$$-4>7x-4$$
$$0>7x$$
$$0>x$$
$$x<0$$
The solution set is $\{x|x<0\}$ or $(-\infty, 0)$.

17.
$$-2(x-1)+5(x-2)<5(x+3)$$
$$-2x+2+5x-10<5x+15$$
$$3x-8<5x+15$$
$$-2x-8<15$$
$$-2x<23$$
$$x>-\frac{23}{2}$$
The solution set is $\left\{x|x>-\frac{23}{2}\right\}$ or $\left(-\frac{23}{2}, \infty\right)$.

18.
$$\frac{1}{2}n+2\leq\frac{3}{4}n-1$$
$$8\left(\frac{1}{2}n+2\right)\leq 8\left(\frac{3}{4}n-1\right)$$
$$4n+16\leq 6n-8$$
$$-2n+16\leq -8$$
$$-2n\leq -24$$
$$n\geq 12$$
The solution set is $\{n|n\geq 12\}$ or $[12, \infty)$.

19. Since it is an "and" statement, we need to satisfy both inequalities at the same time. Thus, all numbers between and including -2 and 4 are solutions.

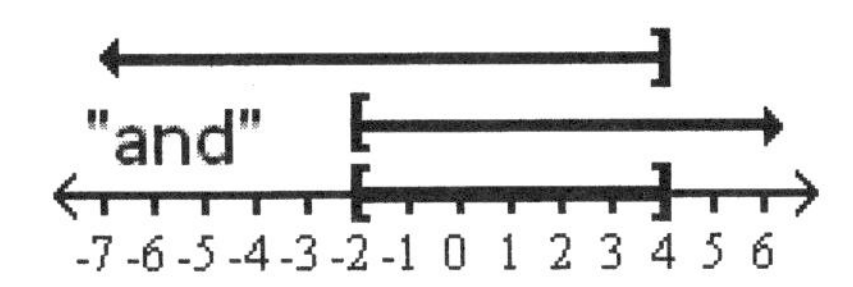

20. Since it is an "or" statement, the solution set consists of all numbers less than (but not including) 1 along with all numbers greater than (but not including) 3.

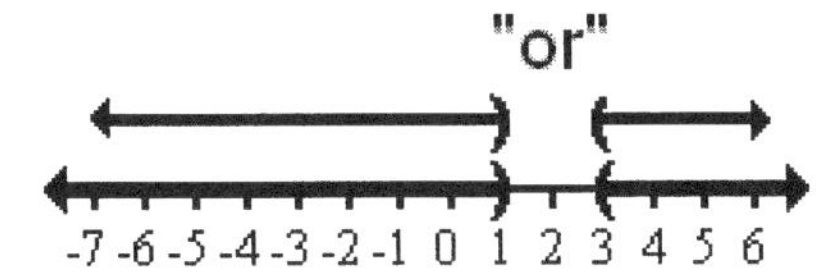

21. Let h represent the hourly rate for the labor.

$$153 + 4h = 441$$
$$4h = 288$$
$$h = 72$$

The hourly rate was \$72 per hour.

22. Let s represent the length of the shortest side, $2s$ the longest side, and $s + 10$ the third side.

$$s + 2s + (s + 10) = 70$$
$$4s + 10 = 70$$
$$4s = 60$$
$$s = 15$$

The shortest side is 15 meters, the longest side is 30 meters and the third side is 25 meters.

23. Let x represent the score on the fifth exam.

$$\frac{86 + 88 + 89 + 91 + x}{5} \geq 90$$
$$354 + x \geq 450$$
$$x \geq 96$$

She would need a score of 96 or better.

24. Let n represent the number of nickels, $2n - 1$ the number of dimes, and $3n + 2$ the number of quarters.

$$n + (2n - 1) + (3n + 2) = 103$$
$$6n + 1 = 103$$
$$6n = 102$$
$$n = 17$$

There are 17 nickels, $2(17) - 1 = 33$ dimes, and $3(17) + 2 = 53$ quarters.

25. Let a represent the measure of angle A; then $30 + a$ represents the measure of angle B and $\frac{1}{2}a$ represents the measure of angle C.

$$a + 30 + a + \frac{1}{2}a = 180$$
$$2\left[a + 30 + a + \frac{1}{2}a\right] = 2(180)$$
$$2a + 60 + 2a + a = 360$$
$$5a + 60 = 360$$
$$5a = 300$$
$$a = 60$$

Angle A measures 60°, angle B measures 90° and angle C measures 30°.

CHAPTERS 1-3 | Cumulative PRACTICE TEST

1.
$$3(-4) - 2 + (-3)(-6) - 1 =$$
$$-12 - 2 + 18 - 1 =$$
$$-14 + 18 - 1 = 3$$

2.
$$-(2)^7 = -(2 \cdot 2 \cdot 2 \cdot 2 \cdot 2 \cdot 2 \cdot 2)$$
$$= -128$$

3.
$$6.2 - 7.1 - 3.4 + 1.9 =$$
$$6.2 + 1.9 - 7.1 - 3.4 =$$
$$8.1 - 10.5 = -2.4$$

4.
$$-\frac{2}{3} + \frac{1}{2} - \frac{1}{4} =$$
$$-\frac{2}{3} \cdot \frac{4}{4} + \frac{1}{2} \cdot \frac{6}{6} - \frac{1}{4} \cdot \frac{3}{3} =$$
$$-\frac{8}{12} + \frac{6}{12} - \frac{3}{12} = -\frac{5}{12}$$

5. $-4x + 2y - xy$ for $x = -2$ and $y = 3$

$$-4(-2) + 2(3) - (-2)(3) =$$
$$8 + 6 - (-6) = 14 + 6 = 20$$

6. $\frac{1}{5}x - \frac{2}{3}y$ for $x = -\frac{1}{2}, y = \frac{1}{6}$

$\frac{1}{5}\left(-\frac{1}{2}\right) - \left(\frac{2}{3}\right)\left(\frac{1}{6}\right) =$

$-\frac{1}{10} - \frac{1}{9} = \left(-\frac{1}{10} \cdot \frac{9}{9}\right) - \left(\frac{1}{9} \cdot \frac{10}{10}\right) =$

$-\frac{9}{90} - \frac{10}{90} = -\frac{19}{90}$

7. $0.2(x - y) - 0.3(x + y) =$

$0.2x - 0.2y - 0.3x - 0.3y =$

$-0.1x - 0.5y$ for $x = 0.1, y = -0.2$

$-0.1(0.1) - 0.5(-0.2) =$

$-0.01 + 0.10 = 0.09$

8. $\left.\begin{array}{l} 48 = 2 \cdot 2 \cdot 2 \cdot 2 \cdot 3 \\ 60 = 2 \cdot 2 \cdot 3 \cdot 5 \\ 96 = 2 \cdot 2 \cdot 2 \cdot 2 \cdot 2 \cdot 3 \end{array}\right\}$

The greatest common factor is $2 \cdot 2 \cdot 3 = 12$.

9. $\left.\begin{array}{l} 9 = 3 \cdot 3 \\ 12 = 2 \cdot 2 \cdot 3 \end{array}\right\}$

The least common multiple is $2 \cdot 2 \cdot 3 \cdot 3 = 36$.

10. $\frac{3}{8}x + \frac{3}{7}y - \frac{5}{12}x - \frac{3}{4}y =$

$\frac{3}{8}x - \frac{5}{12}x + \frac{3}{7}y - \frac{3}{4}y =$

$\left(\frac{3}{8} - \frac{5}{12}\right)x + \left(\frac{3}{7} - \frac{3}{4}\right)y =$

$-\frac{1}{24}x + \left(-\frac{9}{28}\right)y = -\frac{1}{24}x - \frac{9}{28}y$

11. $-(x - 2) + 6(x + 4) - 2(x - 7) =$

$-x + 2 + 6x + 24 - 2x + 14 =$

$-x + 6x - 2x + 2 + 24 + 14 =$

$3x + 40$

12. $\left(\frac{5x^2y}{6xy^2}\right)\left(\frac{8x^2y^2}{30y}\right) = \frac{\overset{1}{\cancel{5}} \cdot \overset{4}{\cancel{8}} \cdot x^4 \cdot y^3}{\underset{6}{\cancel{30}} \cdot \underset{3}{\cancel{6}} \cdot x \cdot y^3} =$

$\frac{4x^3}{18} = \frac{2x^3}{9}$

13. $\frac{5}{x} - \frac{6}{y^2} = \left(\frac{5}{x} \cdot \frac{y^2}{y^2}\right) - \left(\frac{6}{y^2} \cdot \frac{x}{x}\right) =$

$\frac{5y^2}{xy^2} - \frac{6x}{xy^2} = \frac{5y^2 - 6x}{xy^2}$

14. $\left(\frac{ab^3}{3a^2}\right) \div \left(\frac{ab}{9b^2}\right) =$

$\frac{ab^3}{3a^2} \cdot \frac{9b^2}{ab} = \frac{9ab^5}{3a^3b} = \frac{3b^4}{a^2}$

15. $\frac{5}{3y} - \frac{2}{5y^2} =$

$\left(\frac{5}{3y} \cdot \frac{5y}{5y}\right) - \left(\frac{2}{5y^2} \cdot \frac{3}{3}\right) =$

$\frac{25y}{15y^2} - \frac{6}{15y^2} = \frac{25y - 6}{15y^2}$

16. $-3(x + 4) = -4(x - 1)$

$-3x - 12 = -4x + 4$

$x - 12 = 4$

$x = 16$

The solution set is $\{16\}$.

17. $\frac{x + 1}{4} - \frac{x - 2}{3} = -2$

$12\left(\frac{x + 1}{4} - \frac{x - 2}{3}\right) = 12(-2)$

$12\left(\frac{x + 1}{4}\right) - 12\left(\frac{x - 2}{3}\right) = -24$

$3(x + 1) - 4(x - 2) = -24$

$3x + 3 - 4x + 8 = -24$

$-x + 11 = -24$

$-x = -35$

$x = 35$

The solution set is $\{35\}$.

18. $2(2x - 1) + 3(x - 3) = -4(x + 7)$

$4x - 2 + 3x - 9 = -4x - 28$

$7x - 11 = -4x - 28$

$11x = -17$

$x = -\frac{17}{11}$

The solution set is $\left\{-\frac{17}{11}\right\}$.

19. $2(x-1) \geq 3(x-6)$
$2x-2 \geq 3x-18$
$-x \geq -16$
$x \leq 16$
The solution set is
$\{x|x \leq 16\}$ or $(-\infty, 16]$.

20. $-2 < -(x-1)-4$
$-2 < -x+1-4$
$-2 < -x-3$
$1 < -x$
$-1 > x$ so $x < -1$
The solution set is $\{x|x < -1\}$ or $(-\infty, -1)$.

21. $300 = 2 \cdot 2 \cdot 3 \cdot 5 \cdot 5$

22. $x \geq -1$ and $x < 3$
Because it is an "and" statement, the solution set consists of all numbers between -1 and 3, including -1, but not including 3.

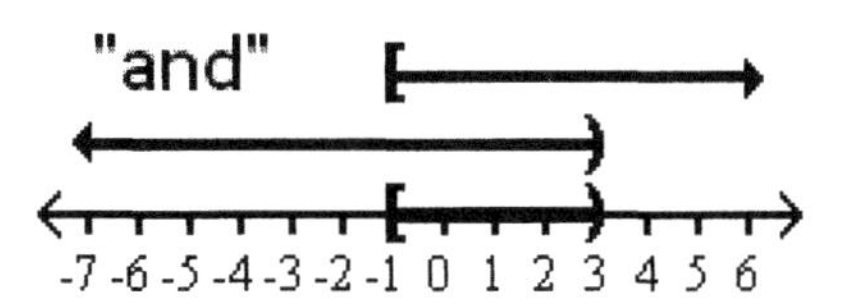

23. Let x represent the number of arrests on Friday night, then $6+3x$ represents the number of arrests on Saturday night.
$x+(6+3x) = 42$
$4x+6 = 42$
$4x = 36$
$x = 9$
On Friday night there were 9 arrests and on Saturday night there were $6+3(9) = 6+27 = 33$ arrests.

24. Let d represent the number of people who will attend the reception.
$125+35d < 2500$
$35d < 2375$
$d < 67.86$
The number of people attending the reception would be 67 or fewer people.

25. Let n and $n+2$ represent two consecutive odd numbers.
$n+5(n+2) = 76$
$n+5n+10 = 76$
$6n+10 = 76$
$6n = 66$
$n = 11$
The numbers are 11 and 13.

Chapter 4 Formulas and Problem Solving

PROBLEM SET 4.1 Ratio, Proportion, and Percent

1. $\frac{x}{6}=\frac{3}{2}$

$2x=18$ Cross products are equal.

$x=9$

The solution set is $\{9\}$.

3. $\frac{5}{12}=\frac{n}{24}$

$120=12n$ Cross products are equal.

$10=n$

The solution set is $\{10\}$.

5. $\frac{x}{3}=\frac{5}{2}$

$2x=15$ Cross products are equal.

$x=\frac{15}{2}$

The solution set is $\left\{\frac{15}{2}\right\}$.

7. $\frac{x-2}{4}=\frac{x+4}{3}$

Cross products are equal.

$3(x-2)=4(x+4)$

$3x-6=4x+16$

$-x=22$

$x=-22$

The solution set is $\{-22\}$.

9. $\frac{x+1}{6}=\frac{x+2}{4}$

Cross products are equal.

$4(x+1)=6(x+2)$

$4x+4=6x+12$

$-2x=8$

$x=-4$

The solution set is $\{-4\}$.

11. $\frac{h}{2}-\frac{h}{3}=1$

Be careful, this is not a proportion.

Multiply both sides by 6.

$6\left(\frac{h}{2}-\frac{h}{3}\right)=6(1)$

$3h-2h=6$

$h=6$

The solution set is $\{6\}$.

13. $\frac{x+1}{3}-\frac{x+2}{2}=4$

Be careful, this is not a proportion.

Multiply both sides by 6.

$6\left(\frac{x+1}{3}-\frac{x+2}{2}\right)=6(4)$

$2(x+1)-3(x+2)=24$

$2x+2-3x-6=24$

$-x-4=24$

$-x=28$

$x=-28$

The solution set is $\{-28\}$.

15. $\frac{-4}{x+2}=\frac{-3}{x-7}$

Cross products are equal.

$-4(x-7)=-3(x+2)$

$-4x+28=-3x-6$

$-x=-34$

$x=34$

The solution set is $\{34\}$.

17. $\frac{-1}{x-7}=\frac{5}{x-1}$

Cross products are equal.

$-1(x-1)=5(x-7)$

$-x+1=5x-35$

$-6x=-36$

$x=6$

The solution set is $\{6\}$.

19. $$\frac{3}{2x-1} = \frac{2}{3x+2}$$

Cross products are equal.

$$3(3x+2) = 2(2x-1)$$
$$9x + 6 = 4x - 2$$
$$5x = -8$$
$$x = -\frac{8}{5}$$

The solution set is $\left\{-\frac{8}{5}\right\}$.

21. $$\frac{n+1}{n} = \frac{8}{7}$$

Cross products are equal.

$$7(n+1) = 8n$$
$$7n + 7 = 8n$$
$$7 = n$$

The solution set is $\{7\}$.

23. $$\frac{x-1}{2} - 1 = \frac{3}{4}$$

Be careful, this is not a proportion.

Multiply both sides by 8.

$$8\left(\frac{x-1}{2} - 1\right) = 8\left(\frac{3}{4}\right)$$
$$4(x-1) - 8 = 2(3)$$
$$4x - 4 - 8 = 6$$
$$4x - 12 = 6$$
$$4x - 18$$
$$x = \frac{18}{4} = \frac{9}{2}$$

The solution set is $\left\{\frac{9}{2}\right\}$.

25. $$-3 - \frac{x+4}{5} = \frac{3}{2}$$

Be careful, this is not a proportion.

Multiply both sides by 10.

$$10\left(-3 - \frac{x+4}{5}\right) = 10\left(\frac{3}{2}\right)$$
$$-30 - 2(x+4) = 5(3)$$
$$-30 - 2x - 8 = 15$$
$$-2x - 38 = 15$$
$$-2x = 53$$
$$x = -\frac{53}{2}$$

The solution set is $\left\{\frac{-53}{2}\right\}$.

27. $$\frac{n}{150-n} = \frac{1}{2}$$

Cross products are equal.

$$2n = 150 - n$$
$$3n = 150$$
$$n = 50$$

The solution set is $\{50\}$.

29. $$\frac{300-n}{n} = \frac{3}{2}$$

Cross products are equal.

$$2(300-n) = 3n$$
$$600 - 2n = 3n$$
$$600 = 5n$$
$$120 = n$$

The solution set is $\{120\}$.

31. $$\frac{-1}{5x-1} = \frac{-2}{3x+7}$$

Cross products are equal.

$$-1(3x+7) = -2(5x-1)$$
$$-3x - 7 = -10x + 2$$
$$7x - 7 = 2$$
$$7x = 9$$
$$x = \frac{9}{7}$$

The solution set is $\left\{\frac{9}{7}\right\}$.

33. $\frac{11}{20} = \frac{n}{100}$
$20n = 1100$
$n = 55$
Therefore, $\frac{11}{20} = \frac{55}{100} = 55\%$

35. $\frac{3}{5} = \frac{n}{100}$
$5n = 300$
$n = 60$
Therefore, $\frac{3}{5} = \frac{60}{100} = 60\%$

37. $\frac{1}{6} = \frac{n}{100}$
$6n = 100$
$n = \frac{100}{6} = 16\frac{2}{3}$
Therefore, $\frac{1}{6} = \frac{16\frac{2}{3}}{100} = 16\frac{2}{3}\%$

39. $\frac{3}{8} = \frac{n}{100}$
$8n = 300$
$n = \frac{300}{8} = 37\frac{1}{2}$
Therefore, $\frac{3}{8} = \frac{37\frac{1}{2}}{100} = 37\frac{1}{2}\%$

41. $\frac{3}{2} = \frac{n}{100}$
$2n = 300$
$n = 150$
Therefore, $\frac{3}{2} = \frac{150}{100} = 150\%$

43. $\frac{12}{5} = \frac{n}{100}$
$5n = 1200$
$n = 240$
Therefore, $\frac{12}{5} = \frac{240}{100} = 240\%$

45. Let n represent the number.
$n = (7\%)(38)$
$n = 0.07(38)$
$n = 2.66$
Therefore, 2.66 is 7% of 38.

47. Let n represent the number.
$(15\%)(n) = 6.3$
$0.15n = 6.3$
Multiply both sides by 100.
$15n = 630$
$n = 42$
Therefore, 15% of 42 is 6.3.

49. Let r represent the percent to be found.
$76 = r(95)$
$\frac{76}{95} = r$
$0.80 = r$
Therefore, 76 is 80% of 95.

51. Let n represent the number.
$n = (120\%)(50)$
$n = 1.2(50)$
$n = 60$
Therefore, 60 is 120% of 50.

53. Let r represent the percent to be found.
$46 = r(40)$
$\frac{46}{40} = r$
$1.15 = r$
Therefore, 46 is 115% of 40.

55. Let n represent the number.
$(160\%)(n) = 144$
$1.6n = 144$
Multiply both sides by 10.
$16n = 1440$
$n = \frac{1440}{16} = 90$
Therefore, 160% of 90 is 144.

57. Let l and w represent the length and width of the room measured in feet.

$$\begin{array}{lcc} & \underline{\text{inches}} & \underline{\text{feet}} \\ \frac{\text{map}}{\text{room}} & \frac{1}{2\frac{1}{2}} & = \frac{6}{w} \end{array}$$

$$w = 6\left(2\frac{1}{2}\right)$$
$$w = 6\left(\frac{5}{2}\right)$$
$$w = 15$$

$$\begin{array}{lcc} & \underline{\text{inches}} & \underline{\text{feet}} \\ \frac{\text{map}}{\text{room}} & \frac{1}{3\frac{1}{4}} & = \frac{6}{l} \end{array}$$

$$l = 6\left(3\frac{1}{4}\right)$$
$$l = 6\left(\frac{13}{4}\right)$$
$$l = \frac{39}{2} = 19\frac{1}{2}$$

The room measures 15 feet by $19\frac{1}{2}$ feet.

59. Let m represent the number of miles traveled.

$$\frac{\text{miles}}{\text{gallons}} \quad \frac{264}{12} = \frac{m}{15}$$
$$12m = 3960$$
$$m = 330$$

The car will travel 330 miles.

61. Let l represent the length of the rectangle.

$$\frac{\text{length}}{\text{width}} \quad \frac{5}{2} = \frac{l}{24}$$
$$2l = 120$$
$$l = 60$$

The length of the rectangle is 60 centimeters.

63. Let s represent the number of pounds of salt needed.

$$\frac{\text{salt}}{\text{water}} \quad \frac{3}{10} = \frac{s}{25}$$
$$10s = 75$$
$$s = 7.5$$

It will take 7.5 pounds of salt.

65. Let p represent the number of pounds of fertilizer needed.

$$\frac{\text{fertilizer}}{\text{lawn}} \quad \frac{20}{1500} = \frac{p}{2500}$$
$$1500p = 50,000$$
$$p = 33\frac{1}{3}$$

It will take $33\frac{1}{3}$ pounds of fertilizer.

67. Let n represent the number of people who are expected to vote.

$$\frac{3}{7} = \frac{n}{210,000}$$
$$7n = 630,000$$
$$n = 90,000$$

It is expected that $90,000$ people will vote.

69. Let x represent the grams of carbohydrates.

$$\frac{\text{calories}}{\text{grams}} \quad \frac{16}{1} = \frac{2200}{x}$$
$$2200 = 16x$$
$$\frac{2200}{16} = x$$
$$137.5 = x$$

A plan with 2200 calories would have 137.5 grams of carbohydrates.

71. Let x represent the additional money to be invested.

$$\frac{\text{investment}}{\text{earnings}} \quad \frac{500}{45} = \frac{500 + x}{72}$$
$$45(500 + x) = 500(72)$$
$$22,500 + 45x = 36,000$$
$$45x = 13,500$$
$$x = 300$$

The additional investment would be \$300.

73. Let x represent the money the child will receive, then the cancer fund will receive $180,000 - x$.

$$\frac{\text{child}}{\text{fund}} \quad \frac{5}{1} = \frac{x}{180,000 - x}$$
$$x = 5(180,000 - x)$$
$$x = 900,000 - 5x$$
$$6x = 900,000$$
$$x = 150,000$$

The child would receive $\$150,000$.

77. $\dfrac{3}{x-2}=\dfrac{6}{2x-4}$

$3(2x-4)=6(x-2)$

$6x-12=6x-12$

$-12=-12$

(Always true except when $x=2$, because the denominator would become zero.)

The solution set is{all real numbers except 2}.

79. $\dfrac{5}{x-3}=\dfrac{10}{x-6}$

$5(x-6)=10(x-3)$

$5x-30=10x-30$

$-5x-30=-30$

$-5x=0$

$x=0$

The solution set is $\{0\}$.

81. $\dfrac{x-2}{2}=\dfrac{x}{2}-1$

$2\left(\dfrac{x-2}{2}\right)=2\left(\dfrac{x}{2}-1\right)$

$x-2=x-2$

$-2=-2$ (Always true.)

The solution set is {all real numbers}.

PROBLEM SET 4.2 More on Percents and Problem Solving

1. $x-0.36=0.75$

Add 0.36 to both sides.

$x-0.36+0.36=0.75+0.36$

$x=1.11$

The solution set is $\{1.11\}$.

3. $x+7.6=14.2$

Subtract 7.6 from both sides.

$x+7.6-7.6=14.2-7.6$

$x=6.6$

The solution set is $\{6.6\}$.

5. $0.62-y=0.14$

Subtract 0.62 from both sides.

$0.62-y-0.62=0.14-0.62$

$-y=-0.48$

$y=0.48$

The solution set is $\{0.48\}$.

7. $0.7t=56$

Multiply both sides by 10.

$10(0.7t)=10(56)$

$7t=560$

$t=80$

The solution set is $\{80\}$.

9. $x=3.36-0.12x$

Multiply both sides by 100.

$100(x)=100(3.36-0.12x)$

$100x=336-12x$

$112x=336$

$x=3$

The solution set is $\{3\}$.

11. $s=35+0.3s$

Multiply both sides by 10.

$10(s)=10(35+0.3s)$

$10s=350+3s$

$7s=350$

$s=50$

The solution set is $\{50\}$.

13. $s=42+0.4s$

Multiply both sides by 10.

$10(s)=10(42+0.4s)$

$10s=420+4s$

$6s=420$

$s=70$

The solution set is $\{70\}$.

15. $0.07x + 0.08(x + 600) = 78$
Multiply both sides by 100.
$100[0.07x + 0.08(x + 600)] = 100(78)$
$7x + 8(x + 600) = 7800$
$7x + 8x + 4800 = 7800$
$15x + 4800 = 7800$
$15x = 3000$
$x = 200$
The solution set is $\{200\}$.

17. $0.09x + 0.1(2x) = 130.5$
Multiply both sides by 100.
$100[0.09x + 0.1(2x)] = 100(130.5)$
$9x + 10(2x) = 13,050$
$9x + 20x = 13,050$
$29x = 13,050$
$x = 450$
The solution set is $\{450\}$.

19. $0.08x + 0.11(500 - x) = 50.5$
Multiply both sides by 100.
$100[0.08x + 0.11(500 - x)] = 100(50.5)$
$8x + 11(500 - x) = 5050$
$8x + 5500 - 11x = 5050$
$-3x + 5500 = 5050$
$-3x = -450$
$x = 150$
The solution set is $\{150\}$.

21. $0.09x = 550 - 0.11(5400 - x)$
Multiply both sides by 100.
$9x = 55,000 - 11(5400 - x)$
$9x = 55,000 - 59,400 + 11x$
$-2x = -4400$
$x = 2200$
The solution set is $\{2200\}$.

23. Let p represent the original price of the electric drill.
$(100\%)(p) - (30\%)(p) = 35$
$(70\%)(p) = 35$
$0.7p = 35$
$p = 50$
The original price of the electric drill was $50.

25. Let d represent the discount sale price of the wide-screen TV. Since the TV is on sale for 25% off, the discount price is 75% of the original price.
$d = (75\%)(4800)$
$d = 0.75(4800)$
$d = 3600$
The discount sale price is $3,600.

27. Let d represent the discount sale price of the putter. Since the putter is on sale for 35% off, the discount price is 65% of the original price.
$d = (65\%)(32)$
$d = 0.65(32)$
$d = 20.8$
The discount sale price is $20.80.

29. Let r represent the rate of discount.
$180 - r(180) = 126$
$180 - 180r = 126$
$-180r = -54$
$r = 0.3$
The rate of discount is 30%.

31. Let s represent the selling price.
Profit is a percent of cost.
Selling price = Cost + Profit
$s = 5 + (70\%)(5)$
$s = 5 + 0.7(5)$
$s = 5 + 3.5$
$s = 8.5$
The selling price would be $8.50.

33. Let s represent the selling price.
Profit is a percent of cost.
Selling price = Cost + Profit
$s = 8 + (55\%)(8)$
$s = 8 + 0.55(8)$
$s = 8 + 4.40$
$s = 12.40$
The selling price would be $12.40.

35. Let s represent the selling price.
Profit is a percent of selling price.
Selling price = Cost + Profit

$$s = 400 + (60\%)(s)$$
$$s = 400 + 0.6s$$
$$10s = 4000 + 6s$$
$$4s = 4000$$
$$s = 1000$$

The selling price would be \$1000.

37. Let r represent the rate of profit.
Profit is a percent of cost.
Selling price = Cost + Profit

$$44.8 = 32 + r(32)$$
$$448 = 320 + 320r$$
$$128 = 320r$$
$$0.4 = r$$

The rate of profit would be 40% of the cost.

39.

$$i = Prt$$
$$560 = 3500(r)(2)$$
$$560 = 7000r$$
$$\frac{560}{7000} = r$$
$$0.08 = r$$

The annual interest rate will be 8%.

41.

$$i = Prt$$
$$1000 = P(0.08)(3)$$
$$1000 = 0.24P$$
$$\frac{1000}{0.24} = P$$
$$4166.67 = P$$

The principal will be \$4166.67.

43.

$$i = Prt$$
$$i = (5000)(0.068)(10)$$
$$i = 3400$$

The interest earned will be \$3400.

45. $i = Prt$
Remember time must be in years.
So 1 month $= \frac{1}{12}$ year.

$$i = 95000(0.08)\left(\frac{1}{12}\right)$$
$$i = 633.33$$

The interest for a month will be \$633.33.

49. The profit in dollars on the item is \$50 − \$40 = \$10. Since 10 is 20% of 50, his claim would be correct that he makes a 20% profit if he calculates the profit as a percent of the selling price.

51. Use problem 50 and reverse the calculations to make a comparison.

$$d = (90\%)(60) = 0.9(60) = 54$$
$$a = (60\%)(54) = 0.6(54) = 32.40$$

Yes, both discount patterns would give the same result. This is the result of the commutative property of multiplication.

53.

$$2.4x + 5.7 = 9.6$$
$$2.4x = 3.9$$
$$x = 1.625$$

The solution set is $\{1.625\}$.

55.

$$0.08x + 0.09(800 - x) = 68.5$$
$$8x + 9(800 - x) = 6850$$
$$8x + 7200 - 9x = 6850$$
$$-x + 7200 = 6850$$
$$-x = -350$$
$$x = 350$$

The solution set is $\{350\}$.

57.

$$7x - 0.39 = 0.03$$
$$7x = 0.42$$
$$x = 0.06$$

The solution set is $\{0.06\}$.

59. $0.2(t + 1.6) = 3.4$
$2(t + 1.6) = 34$
$2t + 3.2 = 34$
$2t = 30.8$
$t = 15.4$
The solution set is $\{15.4\}$.

PROBLEM SET 4.3 Formulas: Geometric and Others

1. $d = 336,\ r = 48 : d = rt$
$336 = 48t$
$7 = t$

3. $i = 200,\ r = 0.08,\ t = 5 : i = Prt$
$200 = P(0.08)(5)$
$200 = 0.4P$
$500 = P$

5. $F = 68 : F = \frac{9}{5}C + 32$
$68 = \frac{9}{5}C + 32$
Multiply both sides by 5.
$340 = 9C + 160$
$180 = 9C$
$20 = C$

7. $V = 112,\ h = 7 : V = \frac{1}{3}Bh$
$112 = \frac{1}{3}B(7)$
Multiply both sides by 3.
$336 = 7B$
$48 = B$

9. $A = 5080, P = 4000, r = 0.03:$
$A = P + Prt$
$5080 = 4000 + 4000(0.03)t$
$5080 = 4000 + 120t$
$1080 = 120t$
$9 = t$

11. Substitute 14 for l and 9 for w in the formula for finding perimeter of a rectangle.

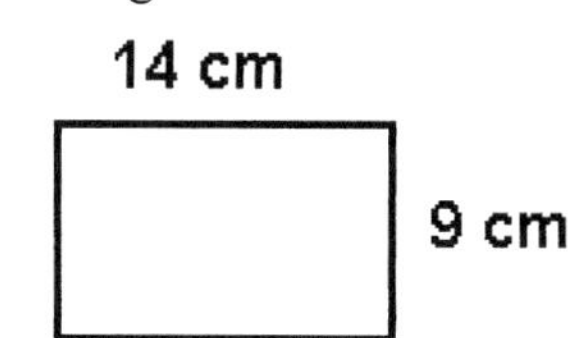

$P = 2l + 2w$
$P = 2(14) + 2(9)$
$P = 28 + 18$
$P = 46$
The perimeter of the rectangle is 46 centimeters.

13. Substitute $3\frac{1}{4}$ feet $= 39$ inches for l and 108 inches for P in the formula for finding perimeter of a rectangle.

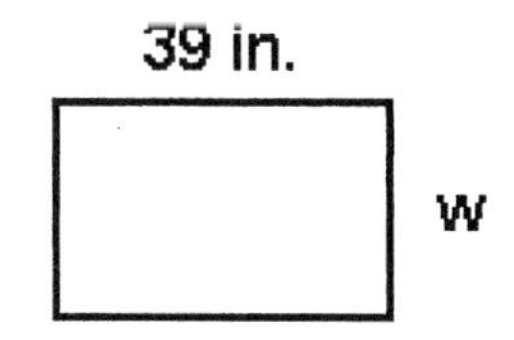

$P = 2l + 2w$
$108 = 2(39) + 2w$
$108 = 78 + 2w$
$30 = 2w$
$15 = w$
The width of the rectangle is 15 inches.

15. Subtract the area of the rectangular garden from the area of the garden plus the dirt path. The width of the larger rectangle is $17 + 4 + 4 = 25$ feet and the length of the larger rectangle is $38 + 4 + 4 = 46$ feet.

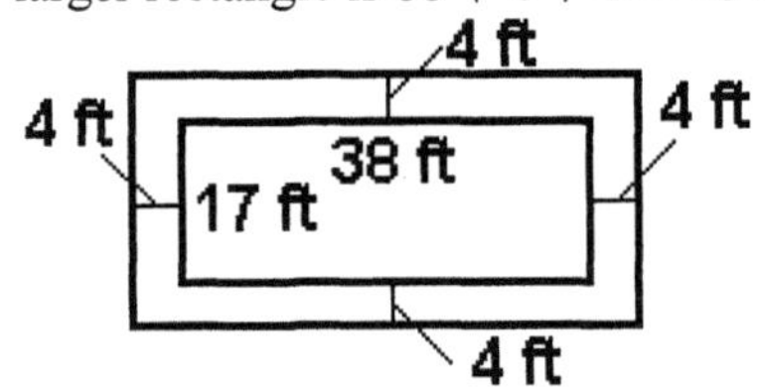

$A =$ large area $-$ garden area
$A = (46)(25) - (38)(17)$
$A = 1150 - 646$
$A = 504$
The area of the dirt path is 504 square feet.

17. Since the area is needed in square meters, it is easier to convert the measurements to meters first. Thus, the length of the wood piece is 60 centimeters $= 0.6$ meters and the width is 30 centimeters $= 0.3$ meters. The area of one piece is $(0.3)(0.6) = 0.18$ square meters. There are 50 pieces for a total area of $50(0.18) = 9$ square meters. One liter of paint would cover the total area for a cost of \$6.00.

19. Let h represent the length of the altitude of the trapezoid.

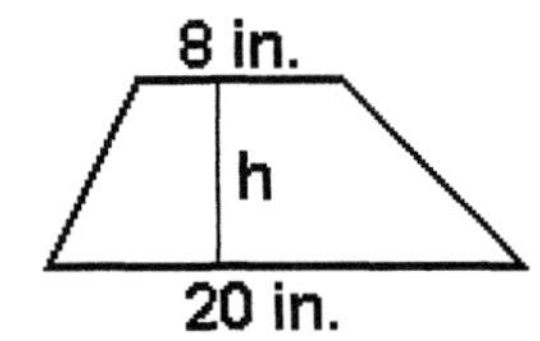

$$A = \frac{1}{2}h(b_1 + b_2)$$
$$98 = \frac{1}{2}h(8 + 20)$$
$$98 = \frac{1}{2}h(28)$$
$$98 = 14h$$
$$7 = h$$
The length of the altitude of the trapezoid is 7 inches.

21. The area of one washer can be computed by subtracting the area of the inner circle (diameter $= 2$ cm) from the area of the outer circle (diameter $= 4$ cm). The radius, which is one-half of the diameter, is used in the formula. See Figure 4.17 in the text for a drawing.
$A = \pi(2)^2 - \pi(1)^2 = 4\pi - 1\pi = 3\pi$
The area of one washer is 3π square centimeters, so 50 washers would have $50(3\pi) = 150\pi$ square centimeters of metal.

23. The radius of the circular region is $\frac{1}{2}$ yard.

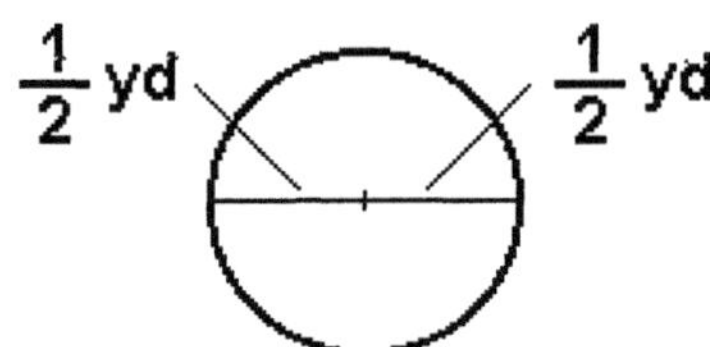

$$A = \pi r^2$$
$$A = \pi\left(\frac{1}{2}\right)^2 = \frac{1}{4}\pi$$
The area of the circular region is $\frac{1}{4}\pi$ square yards.

25. Substitute $r = 9$ into both formulas.
$$S = 4\pi r^2 = 4\pi(9)^2$$
$$= 324\pi \text{ square inches}$$
$$V = \frac{4}{3}\pi r^3 = \frac{4}{3}\pi(9)^3$$
$$= 972\pi \text{ cubic inches}$$

27. Substitute $r = 8$ and $h = 18$ into both formulas.
$$V = \pi r^2 h = \pi(8)^2(18)$$
$$= 1152\pi \text{ cubic inches}$$
$$S = 2\pi r^2 + 2\pi rh = 2\pi(8)^2 + 2\pi(8)(18)$$
$$= 128\pi + 288\pi$$
$$= 416\pi \text{ square inches}$$

29. Substitute $V = 324\pi$ and $r = 9$ in the formula.

$$V = \frac{1}{3}\pi r^2 h$$
$$324\pi = \frac{1}{3}\pi(9)^2 h$$
$$324\pi = \frac{81}{3}\pi h$$
$$324\pi = 27\pi h$$
$$12 = h$$

The height of the cone is 12 inches.

31. Substitute $S = 65\pi$ and $r = 5$ in the formula.

$$S = \pi r^2 + \pi rs$$
$$65\pi = \pi(5)^2 + \pi(5)s$$
$$65\pi = 25\pi + 5\pi s$$
$$40\pi = 5\pi s$$
$$8 = s$$

The slant height of the cone is 8 feet.

33. $V = Bh$

Divide both sides by B.

$$\frac{V}{B} = \frac{Bh}{B}$$
$$\frac{V}{B} = h$$

35. $V = \frac{1}{3}Bh$

Multiply both sides by 3.

$$3V = Bh$$

Divide both sides by h.

$$\frac{3V}{h} = B$$

37. $P = 2l + 2w$

Subtract $2l$ from both sides.

$$P - 2l = 2w$$

Divide both sides by 2.

$$\frac{P - 2l}{2} = w$$

39. $V = \frac{1}{3}\pi r^2 h$

Multiply both sides by 3.

$$3V = \pi r^2 h$$

Divide both sides by πr^2.

$$\frac{3V}{\pi r^2} = h$$

41. $F = \frac{9}{5}C + 32$

Subtract 32 from both sides.

$$F - 32 = \frac{9}{5}C$$

Multiply both sides by $\frac{5}{9}$.

$$\frac{5}{9}(F - 32) = C$$

43. $A = 2\pi r^2 + 2\pi rh$

Subtract $2\pi r^2$ from both sides.

$$A - 2\pi r^2 = 2\pi rh$$

Divide both sides by $2\pi r$.

$$\frac{A - 2\pi r^2}{2\pi r} = h$$

45. $3x + 7y = 9$

Subtract $7y$ from both sides.

$$3x = 9 - 7y$$

Divide both sides by 3.

$$x = \frac{9 - 7y}{3}$$

47. $9x - 6y = 13$

Subtract $9x$ from both sides.

$$-6y = 13 - 9x$$

Multiply both sides by -1.

$$6y = -13 + 9x$$

Apply commutative property.

$$6y = 9x - 13$$

Divide both sides by 6.

$$y = \frac{9x - 13}{6}$$

49. $-2x + 11y = 14$
Subtract $11y$ from both sides.
$-2x = 14 - 11y$
Multiply both sides by -1.
$2x = -14 + 11y$
Apply commutative property.
$2x = 11y - 14$
Divide both sides by 2.
$$x = \frac{11y - 14}{2}$$

51. $y = -3x - 4$
Add 4 to both sides.
$y + 4 = -3x$
Multiply both sides by -1.
$-y - 4 = 3x$
Divide both sides by 3.
$$\frac{-y - 4}{3} = x$$

53. $\frac{x - 2}{4} = \frac{y - 3}{6}$
Cross products are equal.
$4(y - 3) = 6(x - 2)$
Distributive property.
$4y - 12 = 6x - 12$
Add 12 to both sides.
$4y = 6x$
Divide both sides by 4.
$$y = \frac{6x}{4} = \frac{3x}{2}$$

55. $ax - by - c = 0$
Add by to both sides.
$ax - c = by$
Divide both sides by b.
$$\frac{ax - c}{b} = y$$

57. $\frac{x + 6}{2} = \frac{y + 4}{5}$
Cross products are equal.
$5(x + 6) = 2(y + 4)$
$5x + 30 = 2y + 8$
Subtract 30 from both sides.
$5x = 2y - 22$
Divide both sides by 5.
$$x = \frac{2y - 22}{5}$$

59. $m = \frac{y - b}{x}$
Multiply both sides by x.
$mx = y - b$
Add b to both sides.
$mx + b = y$

65. Find the larger area by substituting $r = 7$ into the area formula. Then substitute $r = 3$ into the formula to find the smaller area. Finally, subtract the smaller area from the larger area to find the area of the shaded ring.
Larger Area $= \pi r^2 = (3.14)(7)^2 = 153.86$
Smaller Area $= \pi r^2 = (3.14)(3)^2 = 28.26$
Shaded Area $= 153.86 - 28.26 =$
125.6 square centimeters

67. Right Circular Cylinder with $r = 3$ and $h = 10$:
$S = 2\pi r^2 + 2\pi rh$
$= 2(3.14)(3)^2 + 2(3.14)(3)(10)$
$= 56.52 + 188.4 = 244.92$
$S = 244.92$ rounds to 245 square centimeters

69. Sphere with diameter of 5 $(r = 2.5)$:
$V = \frac{4}{3}\pi r^3$
$= \frac{4}{3}(3.14)(2.5)^3$
$= 65.41666...$
$V = 65.41666...$ rounds to 65 cubic inches

PROBLEM SET 4.4 Problem Solving

1. $950(0.12)t = 950$
Divide both sides by 950.
$0.12t = 1$
Multiply both sides by 100.
$12t = 100$
Divide both sides by 12 and reduce.
$t = \frac{100}{12} = 8\frac{1}{3}$
The solution set is $\left\{8\frac{1}{3}\right\}$.

3. $l + \frac{1}{4}l - 1 = 19$
Add 1 to both sides.
$l + \frac{1}{4}l = 20$
Multiply both sides by 4.
$4l + l = 80$
$5l = 80$
Divide both sides by 5.
$l = 16$
The solution set is $\{16\}$.

5. $500(0.08)t = 1000$
Simplify the left side.
$40t = 1000$
Divide both sides by 40.
$t = 25$
The solution set is $\{25\}$.

7. $s + (2s - 1) + (3s - 4) = 37$
Combine like terms.
$6s - 5 = 37$
Add 5 to both sides.
$6s = 42$
Divide both sides by 6.
$s = 7$
The solution set is $\{7\}$.

9. $\frac{5}{2}r + \frac{5}{2}(r + 6) = 135$
Multiply both sides by 2.
$2\left[\frac{5}{2}r + \frac{5}{2}(r + 6)\right] = 2(135)$
$5r + 5(r + 6) = 270$
Apply distributive property.
$5r + 5r + 30 = 270$
Combine like terms.
$10r + 30 = 270$
Subtract 30 from both sides.
$10r = 240$
Divide both sides by 10.
$r = 24$
The solution set is $\{24\}$.

11. $24\left(t - \frac{2}{3}\right) = 18t + 8$
Apply distributive property.
$24t - 16 = 18t + 8$
Subtract $18t$ from both sides.
$6t - 16 = 8$
Add 16 to both sides.
$6t = 24$
Divide both sides by 6.
$t = 4$
The solution set is $\{4\}$.

13. To "double" the interest earned, i would equal 4000 when $P = 4000$ and $r = 8\%$.
$i = Prt$
$4000 = 4000(0.08)t$
Divide both sides by 4000.
$1 = 0.08t$
Multiply by 100.
$100 = 8t$
Divide both sides by 8.
$12\frac{1}{2} = t$
It would take $12\frac{1}{2}$ years for \$4000 to double itself.

15. To "triple itself" the interest earned, i would equal 16000 when $P = 8000$ and $r = 6\%$.

$16000 = 8000(6\%)t$

Divide both sides by 8000.

$2 = 0.06t$

Multiply by 100.

$200 = 6t$

Divide both sides by 6.

$33\frac{1}{3} = \frac{100}{3} = t$

It would take $33\frac{1}{3}$ years for $8000 to triple itself.

17. Let w represent the width of the rectangle. Then $3w$ represents the length and the perimeter is 112 inches.

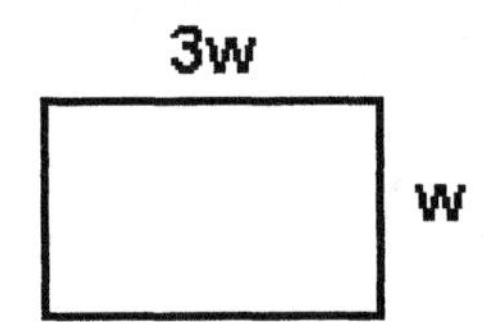

$P = 2l + 2w$

$112 = 2(3w) + 2w$

$112 = 6w + 2w$

$112 = 8w$

$14 = w$

The width is 14 inches and the length is $3(14) = 42$ inches.

19. Let w represent the width of the rectangle. Then $3w - 2$ represents the length and the perimeter is 92 cm.

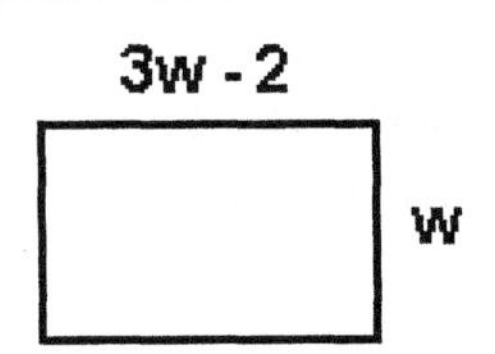

$P = 2l + 2w$

$92 = 2(3w - 2) + 2w$

$92 = 6w - 4 + 2w$

$96 = 8w$

$12 = w$

The width is 12 centimeters and the length is $3(12) - 2 = 34$ centimeters.

21. Let l represent the length of the rectangle; then $\frac{1}{2}l - 3$ represents the width when $P = 42$.

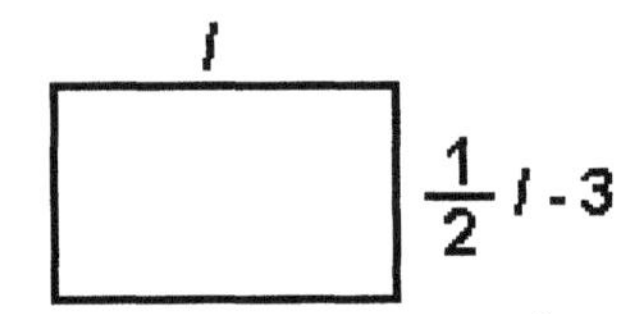

$P = 2l + 2w$

$42 = 2l + 2\left(\frac{1}{2}l - 3\right)$

$42 = 2l + l - 6$

$48 = 3l$

$16 = l$

The length is 16 inches and the width is $\frac{1}{2}(16) - 3 = 5$ inches. Therefore, the area $= lw = (16)(5) = 80$ square inches.

23. Let s represent the shortest side, $2s - 3$ the longest side, and $s + 7$ the third side of a triangle when the perimeter equals 100 feet.

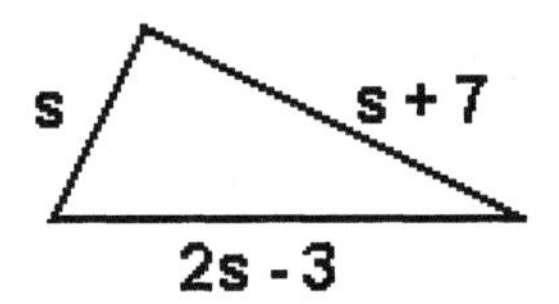

$s + (2s - 3) + (s + 7) = 100$

$4s + 4 = 100$

$4s = 96$

$s = 24$

The lengths of the sides are 24 feet, $2(24) - 3 = 45$ feet, and $24 + 7 = 31$ feet.

25. Let f represent the first side, $3f + 1$ the second side, and $3f + 3$ the third side of a triangle when the perimeter equals 46 centimeters.

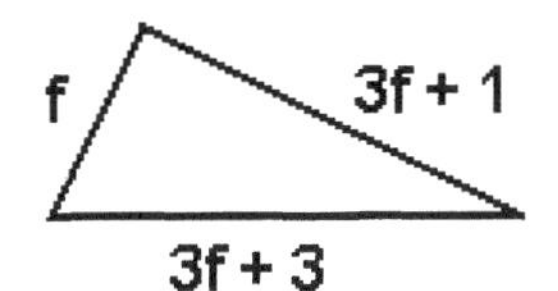

$$f + (3f + 1) + (3f + 3) = 46$$
$$7f + 4 = 46$$
$$7f = 42$$
$$f = 6$$

The lengths of the sides are 6 centimeters, 19 centimeters and 21 centimeters.

27. Let s represent each side of the equilateral triangle; then $s - 4$ represents each side of the square. The perimeter of the triangle is $3s$ and the perimeter of the rectangle is $4(s - 4)$. As stated in the problem, the perimeter of the triangle is 4 centimeters more than the perimeter of the rectangle. This gives the guideline for the equation to be solved.

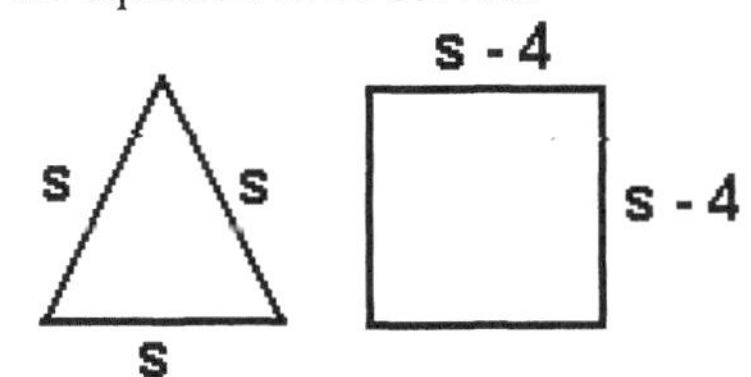

Triangle perimeter = Rectangle perimeter + 4

$$3s = 4(s - 4) + 4$$
$$3s = 4s - 16 + 4$$
$$-s = -12$$
$$s = 12$$

The length of a side of the triangle is 12 centimeters.

29. Let s represent the length of each side of the square; then s also represents the radius of the circle.

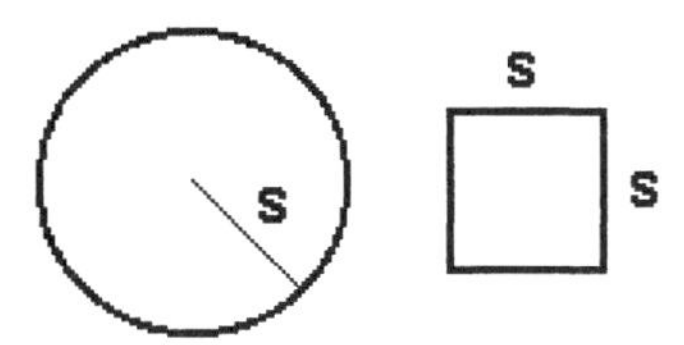

circle circumference = square perimeter + 15.96

$$2\pi r = 2s + 2s + 15.96$$
$$2(3.14)s = 4s + 15.96$$
$$6.28s = 4s + 15.96$$
$$2.28s = 15.96$$
$$s = 7$$

The length of the radius of the circle is 7 centimeters.

31. Let t represent Monica's time; then $t + 1$ represents Sandy's time as she travels one hour longer. A chart of the information would be as follows:

	Rate	Time	Distance ($d = rt$)
Sandy	45	$t+1$	$45(t+1)$
Monica	50	t	$50t$

A diagram of the problem would be as follows:

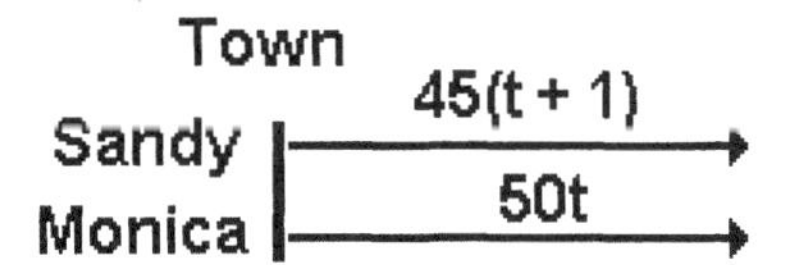

Set the distances equal to each other and solve.

$$45(t + 1) = 50t$$
$$45t + 45 = 50t$$
$$45 = 5t$$
$$9 = t$$

It would take 9 hours for Monica to overtake Sandy.

33. Let t represent the freight train's time; then t also represents the passenger train's time as both trains left at the same time. A chart of the information would be as follows:

	Rate	Time	Distance $(d = rt)$
Freight	40	t	$40t$
Passenger	90	t	$90t$

A diagram of the problem would be as follows:

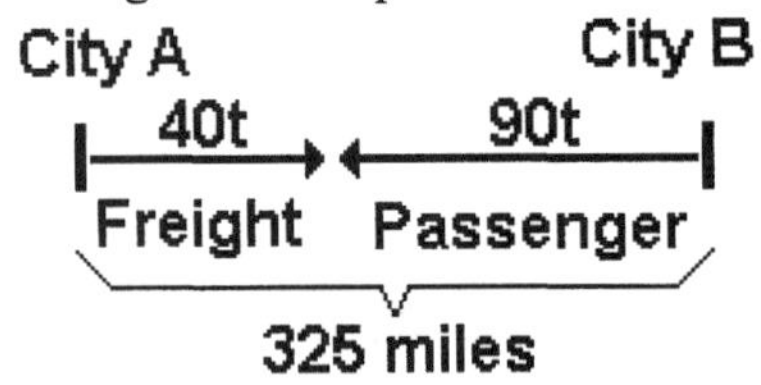

The sum of the two distances would equal the total distance of 325 miles.

$$40t + 90t = 325$$
$$130t = 325$$
$$t = 2\frac{1}{2}$$

The two trains will meet in $2\frac{1}{2}$ hours.

35. Let r represent the second car's rate. Add two hours to the second car's time to find the first car's time. The time also needs to be converted into hours only. A chart of the information would be as follows:

	Rate	Time	Distance $(d = rt)$
First car	40	$7\frac{1}{3} = \frac{22}{3}$	$40\left(\frac{22}{3}\right) = \frac{880}{3}$
Second car	r	$5\frac{1}{3} = \frac{16}{3}$	$\frac{16}{3}r$

A diagram of the problem would be as follows:

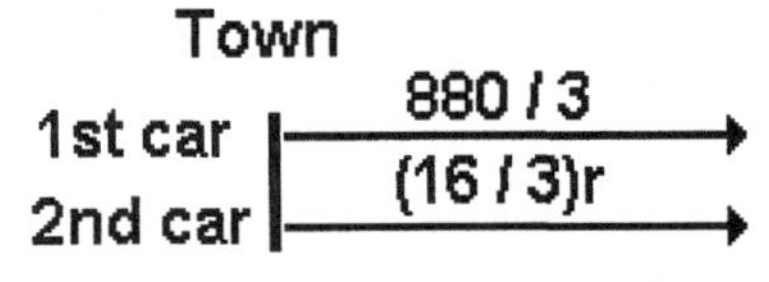

Distances are equal.

Set the two distances equal and solve.

$$\frac{16}{3}r = \frac{880}{3}$$
$$16r = 880$$
$$r = 55$$

The second car would be traveling 55 mph.

37. Let r represent the rate of the train traveling west; then $r + 8$ represents the rate of the train traveling east The time is the same for both trains. A chart of the information would be as follows:

	Rate	Time	Distance $(d = rt)$
West train	r	$9\frac{1}{2} = \frac{19}{2}$	$\frac{19}{2}r$
East train	$r + 8$	$9\frac{1}{2} = \frac{19}{2}$	$\frac{19}{2}(r + 8)$

A diagram of the problem would be as follows:

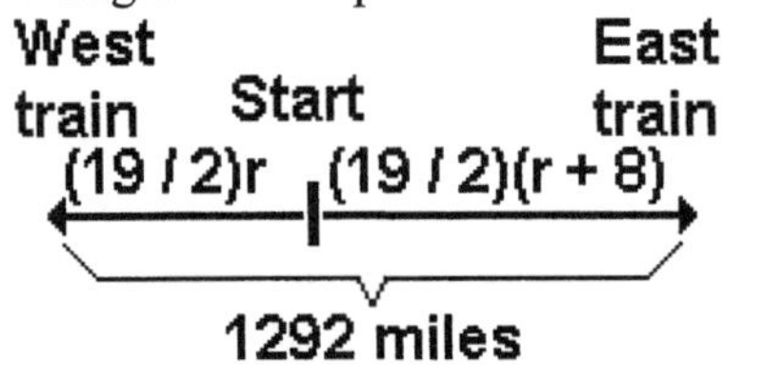

1292 miles

The sum of the two distances would be 1292 miles.

$$\frac{19}{2}r + \frac{19}{2}(r + 8) = 1292$$
$$19r + 19(r + 8) = 2(1292)$$
$$19r + 19r + 152 = 2584$$
$$38r = 2432$$
$$r = 64$$

The train traveling west has a rate of 64 mph and the train traveling east has a rate of 72 mph.

39. Let r represent Jeff's rate when he was leaving; then $r - 2$ represents his rate returning. A chart of the information would be as follows:

	Rate	Time	Distance $(d = rt)$
Leaving	r	3	$3r$
Returning	$r - 2$	$3\frac{3}{4} = \frac{15}{4}$	$\frac{15}{4}(r - 2)$

A diagram of the problem would be as follows:

Home
Leaving
3r
(15 / 4)(r - 2)
Returning
Distances are equal.

Set the distances equal to each other and solve for r.

$$3r = \frac{15}{4}(r-2)$$
$$12r = 15(r-2)$$
$$12r = 15r - 30$$
$$-3r = -30$$
$$r = 10$$

Jeff's rate leaving was 10 mph. Therefore, he travelled 30 miles leaving and 30 miles returning for a total of 60 miles.

PROBLEM SET 4.5 More about Problem Solving

1. $0.3x + 0.7(20 - x) = 0.4(20)$
$$0.3x + 14 - 0.7x = 8$$
$$14 - 0.4x = 8$$
$$-0.4x = -6$$
$$x = 15$$
The solution set is $\{15\}$.

3. $0.2(20) + x = 0.3(20 + x)$
$$4 + x = 6 + 0.3x$$
$$4 + 0.7x = 6$$
$$0.7x = 2$$
$$7x = 20$$
$$x = \frac{20}{7}$$
The solution set is $\left\{\frac{20}{7}\right\}$.

5. $0.7(15) - x = 0.6(15 - x)$
$$10[0.7(15)] - 10x = 10[0.6(15 - x)]$$
$$7(15) - 10x = 6(15 - x)$$
$$105 - 10x = 90 - 6x$$
$$105 - 4x = 90$$
$$-4x = -15$$
$$x = \frac{15}{4}$$
The solution set is $\left\{\frac{15}{4}\right\}$.

7. $0.4(10) - 0.4x + x = 0.5(10)$
$$4 - 0.4x + x = 5$$
$$4 + 0.6x = 5$$
$$0.6x = 1$$
$$6x = 10$$
$$x = \frac{10}{6} = \frac{5}{3}$$
The solution set is $\left\{\frac{5}{3}\right\}$.

9. $20x + 12\left(4\frac{1}{2} - x\right) = 70$
$$20x + 12\left(\frac{9}{2} - x\right) = 70$$
$$20x + 54 - 12x = 70$$
$$8x + 54 = 70$$
$$8x = 16$$
$$x = 2$$
The solution set is $\{2\}$.

11. $3t = \frac{11}{2}\left(t - \frac{3}{2}\right)$
$$3t = \frac{11}{2}t - \frac{33}{4}$$
$$4(3t) = 4\left(\frac{11}{2}t - \frac{33}{4}\right)$$
$$12t = 22t - 33$$
$$-10t = -33$$
$$t = \frac{33}{10}$$
The solution set is $\left\{\frac{33}{10}\right\}$.

13. Let x represent the amount of pure acid to be added; then $100 + x$ represents the amount of final solution.

$$\begin{pmatrix}\text{pure acid in}\\ \text{10\% solution}\end{pmatrix} + \begin{pmatrix}\text{pure acid}\\ \text{to be added}\end{pmatrix} = \begin{pmatrix}\text{pure acid in}\\ \text{final solution}\end{pmatrix}$$

$$\begin{aligned}(10\%)(100) + x &= (20\%)(100 + x)\\ 0.10(100) + x &= 0.20(100 + x)\\ 10[0.10(100) + x] &= 10[0.20(100 + x)]\\ 1(100) + 10x &= 2(100 + x)\\ 100 + 10x &= 200 + 2x\\ 100 + 8x &= 200\\ 8x &= 100\\ x &= 12.5\end{aligned}$$

We must add 12.5 milliliters of pure acid.

15. Let x represent the amount of distilled water to be added; then $10 + x$ represents the amount of final solution.

$$\begin{pmatrix}\text{pure acid in}\\ \text{50\% solution}\end{pmatrix} + \begin{pmatrix}\text{no acid}\\ \text{to be added}\end{pmatrix} = \begin{pmatrix}\text{pure acid in}\\ \text{final solution}\end{pmatrix}$$

$$\begin{aligned}(50\%)(10) + 0 &= (20\%)(10 + x)\\ 0.50(10) &= 0.20(10 + x)\\ 10[0.50(10)] &= 10[0.20(10 + x)]\\ 5(10) &= 2(10 + x)\\ 50 &= 20 + 2x\\ 30 &= 2x\\ 15 &= x\end{aligned}$$

We must add 15 centiliters of distilled water.

17. Let x represent the amount of 30% alcohol to be used; then $10 - x$ represents the amount of 50% solution.

$$\begin{pmatrix}\text{pure alcohol in}\\ \text{30\% solution}\end{pmatrix} + \begin{pmatrix}\text{pure alcohol in}\\ \text{50\% solution}\end{pmatrix} = \begin{pmatrix}\text{pure alcohol in}\\ \text{final 35\% solution}\end{pmatrix}$$

$$\begin{aligned}(30\%)(x) + (50\%)(10 - x) &= (35\%)(10)\\ 0.30(x) + 0.50(10 - x) &= 0.35(10)\\ 100[0.30(x) + 0.50(10 - x)] &= 100[0.35(10)]\\ 30(x) + 50(10 - x) &= 35(10)\\ 30x + 500 - 50x &= 350\\ 500 - 20x &= 350\\ -20x &= -150\\ x &= \frac{150}{20} = 7\frac{1}{2}\end{aligned}$$

We must add $7\frac{1}{2}$ quarts of 30% alcohol and $2\frac{1}{2}$ quarts of 50% alcohol.

19. Let x represent the amount of water to be removed; then $20 - x$ represents the amount of final salt solution.

$$\begin{pmatrix}\text{pure salt in}\\ \text{30\% solution}\end{pmatrix} - \begin{pmatrix}\text{no salt to}\\ \text{be removed}\end{pmatrix} = \begin{pmatrix}\text{pure salt in}\\ \text{final solution}\end{pmatrix}$$

$$\begin{aligned}(30\%)(20) - 0 &= (40\%)(20 - x)\\ 0.30(20) - 0 &= 0.40(20 - x)\\ 6 &= 8 - 0.4x\\ -2 &= -0.4x\\ -20 &= -4x\\ 5 &= x\end{aligned}$$

We must remove 5 gallons of water.

21. Let x represent the amount of pure antifreeze to be added; then x also represents the amount of solution to be drained.

$$\begin{pmatrix}\text{antifreeze in}\\ \text{20\% solution}\end{pmatrix} - \begin{pmatrix}\text{antifreeze to}\\ \text{be drained}\end{pmatrix} + \begin{pmatrix}\text{antifreeze to}\\ \text{be added}\end{pmatrix} = \begin{pmatrix}\text{antifreeze in}\\ \text{final solution}\end{pmatrix}$$

$$\begin{aligned}(20\%)(12) - (20\%)(x) + x &= (40\%)(12)\\ 0.20(12) - 0.20x + x &= 0.40(12)\\ 10[0.20(12) - 0.20x + x] &= 10[0.40(12)]\\ 2(12) - 2(x) + 10x &= 4(12)\\ 24 - 2x + 10x &= 48\\ 24 + 8x &= 48\\ 8x &= 24\\ x &= 3\end{aligned}$$

We must drain 3 quarts of the 20% solution and then replace with 3 quarts of pure antifreeze.

23. Let x represent the amount of 15% salt solution to be used; then $8 + x$ represents the amount of final solution.

$$\begin{pmatrix}\text{pure salt in}\\ \text{15\% solution}\end{pmatrix} + \begin{pmatrix}\text{pure salt in}\\ \text{20\% solution}\end{pmatrix} = \begin{pmatrix}\text{pure salt in}\\ \text{final solution}\end{pmatrix}$$

$$\begin{aligned}(15\%)(x) + (20\%)(8) &= (17\%)(8 + x)\\ 0.15(x) + 0.20(8) &= 0.17(8 + x)\\ 15x + 20(8) &= 17(8 + x)\\ 15x + 160 &= 136 + 17x\\ -2x &= -24\\ x &= 12\end{aligned}$$

We must add 12 gallons of the 15% salt solution.

Problem Set 4.5

25. Let p represent the percent of grapefruit juice in the resulting mixture.

$$\begin{pmatrix}\text{pure juice in}\\ \text{10\% solution}\end{pmatrix} + \begin{pmatrix}\text{pure juice in}\\ \text{20\% solution}\end{pmatrix} = \begin{pmatrix}\text{pure juice in}\\ \text{final solution}\end{pmatrix}$$

$$\begin{aligned}(10\%)(30) + (20\%)(50) &= (p)(30+50)\\ 0.10(30) + 0.20(50) &= p(80)\\ 3 + 10 &= 80p\\ 13 &= 80p\\ \frac{13}{80} &= p\\ p &= 0.1625 = 16.25\%\end{aligned}$$

The resulting mixture would be 16.25% grapefruit juice.

27. Let x represent the length of the side of the square; then $2x - 9$ represents the width of the rectangle and $2x - 3$ represents the length of the rectangle.

2x - 3
2x - 9
x
x
x
x

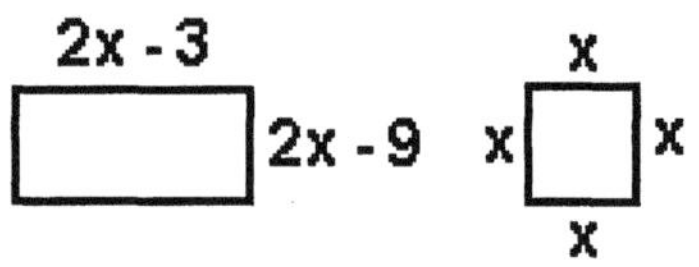

$$\text{Rectangular perimeter} = \text{Square perimeter}$$

$$\begin{aligned}2(2x-9) + 2(2x-3) &= 4x\\ 4x - 18 + 4x - 6 &= 4x\\ 8x - 24 &= 4x\\ -24 &= -4x\\ 6 &= x\end{aligned}$$

Each side of the square is 6 inches, the width of the rectangle is $2(6) - 9 = 3$ inches, and the length of the rectangle is $2(6) - 3 = 9$ inches.

29. Let t represent Aaron's time. Add one-half hour to Aaron's time to represent Andy's time. A chart of the information would be as follows:

	Rate	Time	Distance ($d = rt$)
Andy	2	$t + \frac{1}{2}$	$2\left(t + \frac{1}{2}\right) = 2t + 1$
Aaron	$3\frac{1}{2} = \frac{7}{2}$	t	$\frac{7}{2}t$

A diagram of the problem would be as follows:

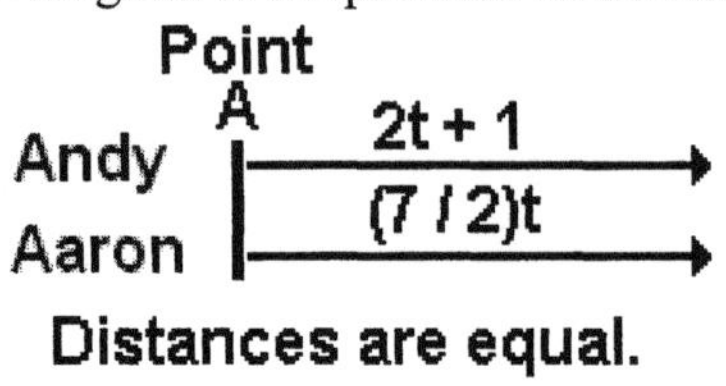

Point A
Andy
2t + 1
Aaron
(7 / 2)t
Distances are equal.

Set the two distances equal and solve.

$$2t + 1 = \frac{7}{2}t$$
$$4t + 2 = 7t$$
$$2 = 3t$$
$$t = \frac{2}{3}\text{hour} = 40 \text{ minutes}$$

Aaron would catch up with Andy in 40 minutes.

31. Let the ages be represented as follows:

$2x$: Bill's present age
x : Pam's present age
$2x - 6$: Bill's age 6 years ago
$x - 6$: Pam's age 6 years ago

$$(\text{Bill's age 6 years ago}) = (4)(\text{Pam's age 6 years ago})$$
$$2x - 6 = 4(x - 6)$$
$$2x - 6 = 4x - 24$$
$$-2x - 6 = -24$$
$$-2x = -18$$
$$x = 9$$

Pam is 9 years old and Bill is 18 years old.

33. Let x represent the amount of money invested at 3% interest; then $x + 750$ represents the amount invested at 5%.

$$\begin{pmatrix}\text{Interest earned}\\ \text{at 3\%}\end{pmatrix} + \begin{pmatrix}\text{Interest earned}\\ \text{at 5\%}\end{pmatrix} = \begin{pmatrix}\text{Total interest}\\ \text{earned}\end{pmatrix}$$

$$(3\%)(x) + (5\%)(x + 750) = 157.50$$
$$0.03(x) + 0.05(x + 750) = 157.50$$
$$100[0.03x + 0.05(x + 750)] = 100(157.50)$$
$$3x + 5x + 3750 = 15,750$$
$$8x + 3750 = 15,750$$
$$8x = 12,000$$
$$x = 1500$$

He invested \$1,500 at 3% and \$2,250 at 5%.

35. Let x represent the amount of money invested at 4% interest; then $1200 - x$ represents the amount of money invested at 6%.

$$\begin{pmatrix}\text{Interest earned}\\ \text{at 4\%}\end{pmatrix} + \begin{pmatrix}\text{Interest earned}\\ \text{at 6\%}\end{pmatrix} = \begin{pmatrix}\text{Total interest}\\ \text{earned}\end{pmatrix}$$

$$\begin{aligned}(4\%)(x) + (6\%)(1200 - x) &= 62\\ 0.04x + 0.06(1200 - x) &= 62\\ 100[0.04x + 0.06(1200 - x)] &= 100(62)\\ 4x + 6(1200 - x) &= 6200\\ 4x + 7200 - 6x &= 6200\\ -2x + 7200 &= 6200\\ -2x &= -1000\\ x &= 500\end{aligned}$$

He invested \$500 at 4% and \$700 at 6%.

37. Let x represent the amount of money invested at 7% interest.

$$\begin{pmatrix}\text{Interest earned}\\ \text{at 4\%}\end{pmatrix} + \begin{pmatrix}\text{Interest earned}\\ \text{at 7\%}\end{pmatrix} = \begin{pmatrix}\text{Total interest}\\ \text{earned}\end{pmatrix}$$

$$\begin{aligned}(4\%)(2000) + (7\%)(x) &= (6\%)(2000 + x)\\ 0.04(2000) + 0.07(x) &= 0.06(2000 + x)\\ 4(2000) + 7x &= 6(2000 + x)\\ 8,000 + 7x &= 12,000 + 6x\\ x &= 4000\end{aligned}$$

Invest \$4000 at 7%.

39. Let x represent the amount of money invested at 6% interest; then $2300 - x$ represents the amount of money invested at 8%.

$$\begin{pmatrix}\text{Interest earned}\\ \text{at 6\%}\end{pmatrix} + 100 = \begin{pmatrix}\text{Interest earned}\\ \text{at 8\%}\end{pmatrix}$$

$$\begin{aligned}(6\%)(x) + 100 &= (8\%)(2300 - x)\\ 0.06x + 100 &= 0.08(2300 - x)\\ 6x + 10,000 &= 8(2300 - x)\\ 6x + 10,000 &= 18,400 - 8x\\ 14x &= 8,400\\ x &= 600\end{aligned}$$

Invest \$600 at 6% and \$1700 at 8%.

41. Let x represent the amount of money invested at 8% interest; then $5400 - x$ represents the amount of money invested at 10%.

$$\begin{pmatrix}\text{Interest earned} \\ \text{at 8\%}\end{pmatrix} = \begin{pmatrix}\text{Interest earned} \\ \text{at 10\%}\end{pmatrix}$$

$$\begin{aligned} (8\%)x &= (10\%)(5400 - x) \\ 0.08x &= 0.10(5400 - x) \\ 8x &= 10(5400 - x) \\ 8x &= 54,000 - 10x \\ 18x &= 54,000 \\ x &= 3000 \end{aligned}$$

He invested \$3000 at 8% and \$2400 at 10%.

CHAPTER 4 | Review Problem Set

1. $0.5x + 0.7x = 1.7$

Combine like terms.

$1.2x = 1.7$

Multiplied both sides by 10.

$$\begin{aligned} 12x &= 17 \\ x &= \frac{17}{12} \end{aligned}$$

The solution set is $\left\{\frac{17}{12}\right\}$.

2. $0.07t + 0.12(t - 3) = 0.59$

Multiplied by 100.

$7t + 12(t - 3) = 59$

Distributive property

$7t + 12t - 36 = 59$

Combined like terms.

$19t - 36 = 59$

Added 36 to both sides.

$$\begin{aligned} 19t &= 95 \\ t &= 5 \end{aligned}$$

The solution set is $\{5\}$.

3. $0.1x + 0.12(1700 - x) = 188$

Multiplied both sides by 100.

$$\begin{aligned} 10x + 12(1700 - x) &= 18,800 \\ 10x + 20,400 - 12x &= 18,800 \\ -2x + 20,400 &= 18,800 \\ -2x &= -1600 \\ x &= 800 \end{aligned}$$

The solution set is $\{800\}$.

4. $x - 0.25x = 12$

Multiplied both sides by 100.

$$\begin{aligned} 100x - 25x &= 1200 \\ 75x &= 1200 \\ x &= 16 \end{aligned}$$

The solution set is $\{16\}$.

5. $0.2(x - 3) = 14$

Multiplied both sides by 10.

$$\begin{aligned} 2(x - 3) &= 140 \\ 2x - 6 &= 140 \\ 2x &= 146 \\ x &= 73 \end{aligned}$$

The solution set is $\{73\}$.

6. $P = 50,\ l = 19 : P = 2l + 2w$

$$\begin{aligned} 50 &= 2(19) + 2w \\ 50 &= 38 + 2w \\ 12 &= 2w \\ 6 &= w \end{aligned}$$

7. $F = 77 : F = \frac{9}{5}C + 32$

$77 = \frac{9}{5}C + 32$

Subtracted 32 from both sides.

$45 = \frac{9}{5}C$

Multiply both sides by $\frac{5}{9}$.

$\frac{5}{9}(45) = \frac{5}{9}\left(\frac{9}{5}\right)C$

$25 = C$

8. $A = P + Prt$

$A - P = Prt$

Divide both sides by Pr.

$\frac{A - P}{Pr} = t$

9. $2x - 3y = 13$

Add $3y$ to both sides.

$2x = 13 + 3y$

Divide both sides by 2.

$x = \frac{13 + 3y}{2}$

10. Substitute $b_1 = 8$, $b_2 = 14$, and $h = 7$ into the formula for the area of a trapezoid.

$A = \frac{1}{2}h(b_1 + b_2) = \frac{1}{2}(7)(8 + 14)$

$= \frac{1}{2}(7)(22) = 77$ square inches

11. Let h represent the altitude when $b = 9, A = 27$.

$A = \frac{1}{2}bh$

$27 = \frac{1}{2}(9)h$

$54 = 9h$

$6 = h$

The altitude of the triangle is 6 centimeters.

12. Let h represent the height when $r = 4$ and $S = 152\pi$.

$S = 2\pi r^2 + 2\pi rh$

$152\pi = 2\pi(4)^2 + 2\pi(4)h$

$152\pi = 32\pi + 8\pi h$

$120\pi = 8\pi h$

$15 = h$

The height of the trapezoid is 15 feet.

13. Let r represent the percent to be found.

$18 = r(30)$

$\frac{18}{30} = r$

$0.6 = r$

Therefore, 18 is 60% of 30.

14. Let n represent one of the numbers, then $96 - n$ represents the other number.

$\frac{5}{7} = \frac{n}{96 - n}$

$7n = 5(96 - n)$

$7n = 480 - 5n$

$12n = 480$

$n = 40$

Therefore, 40 and 56 are the numbers.

15. Let n represent the number.

$(15\%)(n) = 6$

$0.15n = 6$

$15n = 600$

$n = 40$

Therefore, 15% of 40 is 6.

16. Let w represent the width of the rectangle, then $2w + 5$ represents the length when the perimeter is 46 meters.

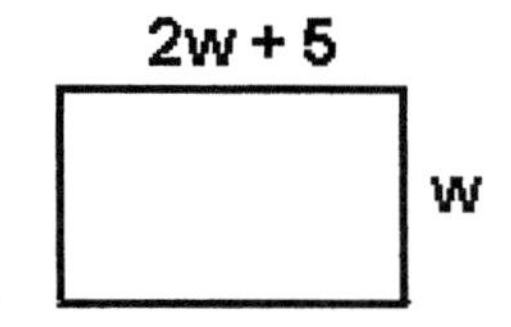

$2(2w + 5) + 2w = 46$

$4w + 10 + 2w = 46$

$6w = 36$

$w = 6$

The dimensions of the rectangle are 6 meters by 17 meters.

17. Let t represent the time for each airplane as both airplanes left at the same time. A chart of the information would be as follows:

	Rate	Time	Distance ($d = rt$)
Plane A	350	t	$350t$
Plane B	400	t	$400t$

A diagram of the problem would be as follows:

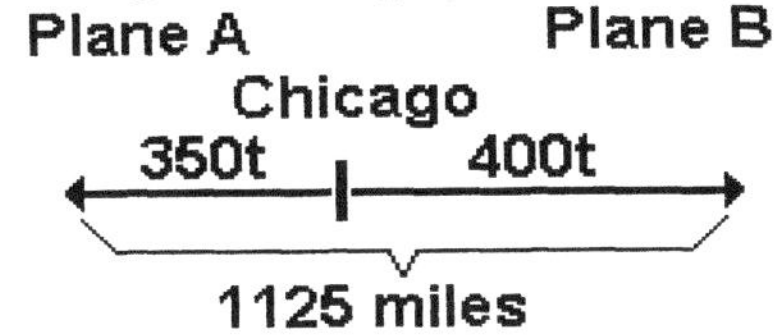

The sum of the two distances would equal the total distance of 1125 miles.

$$350t + 400t = 1125$$
$$750t = 1125$$
$$t = 1\frac{1}{2}$$

It would take $1\frac{1}{2}$ hours for the airplanes to be 1125 miles apart.

18. Let x represent the amount of pure alcohol to be added. Then $10 + x$ represents the amount of final solution.

$$\begin{pmatrix}\text{pure}\\\text{alcohol}\\\text{in 70\%}\\\text{solution}\end{pmatrix} + \begin{pmatrix}\text{pure}\\\text{alcohol}\\\text{to be}\\\text{added}\end{pmatrix} = \begin{pmatrix}\text{pure}\\\text{alcohol}\\\text{in final}\\\text{solution}\end{pmatrix}$$

$$(70\%)(10) + x = (90\%)(10 + x)$$
$$0.70(10) + x = 0.90(10 + x)$$
$$10[0.70(10) + x] = 10[0.90(10 + x)]$$
$$7(10) + 10x = 9(10 + x)$$
$$70 + 10x = 90 + 9x$$
$$70 + x = 90$$
$$x = 20$$

We must add 20 liters of pure alcohol.

19. Let w represent the width of the rectangle, then $2w + 10$ represents the length when the perimeter is 110.

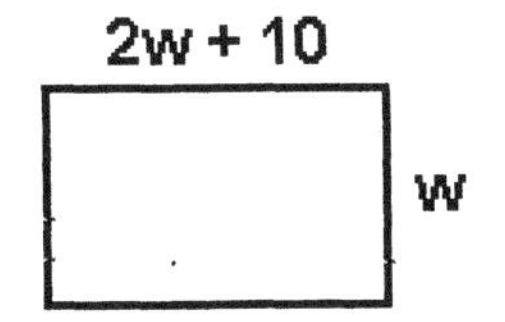

$$2(2w + 10) + 2w = 110$$
$$4w + 20 + 2w = 110$$
$$6w + 20 = 110$$
$$6w = 90$$
$$w = 15$$

The dimensions of the rectangle are 15 centimeters by 40 centimeters.

20. Let w represent the width of the rectangular garden, then $3w - 1$ represents the length when the perimeter is 78 yards.

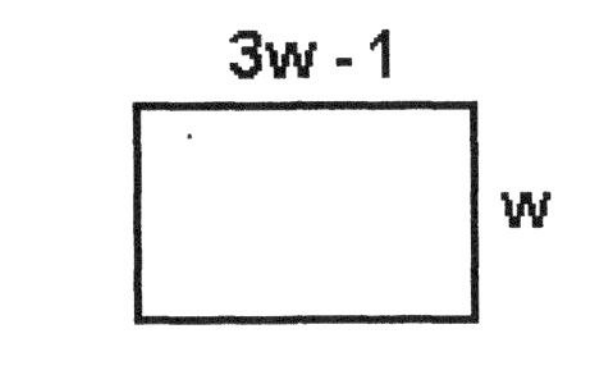

$$2(3w - 1) + 2w = 78$$
$$6w - 2 + 2w = 78$$
$$8w = 80$$
$$w = 10$$

The dimensions of the garden are 10 yards by 29 yards.

21. Let a represent the measure of the angle, then $90 - a$ represents the complement of the angle, and $180 - a$ represents the supplement.

$$\frac{\text{complement}}{\text{supplement}} \quad \frac{90 - a}{180 - a} = \frac{7}{16}$$
$$7(180 - a) = 16(90 - a)$$
$$1260 - 7a = 1440 - 16a$$
$$1260 + 9a = 1440$$
$$9a = 180$$
$$a = 20$$

The measure of the angle is 20°.

22. Let g represent the number of gallons required for the trip.

$$\frac{\text{gallons}}{\text{miles}} \quad \frac{18}{369} = \frac{g}{615}$$
$$369g = 18(615)$$
$$369g = 11{,}070$$
$$g = 30$$

Therefore, 30 gallons would be needed for the 615-mile trip.

23. Let x represent the amount of money invested at 3% interest; then $2100 - x$ represents the amount of money invested at 5%.

$$\begin{pmatrix}\text{Interest earned}\\ \text{at 3\%}\end{pmatrix} + 51 = \begin{pmatrix}\text{Interest earned}\\ \text{at 5\%}\end{pmatrix}$$

$$\begin{aligned}(3\%)x + 51 &= (5\%)(2100 - x)\\ 0.03x + 51 &= 0.05(2100 - x)\\ 3x + 5100 &= 5(2100 - x)\\ 3x + 5100 &= 10,500 - 5x\\ 8x &= 5,400\\ x &= 675\end{aligned}$$

He invested \$675 at 3% and \$1425 at 5%.

24. Let s represent the selling price. Profit is a percent of selling price.

Selling price = Cost + Profit

$$\begin{aligned}s &= 28 + (30\%)(s)\\ s &= 28 + 0.3s\\ 10s &= 280 + 3s\\ 7s &= 280\\ s &= 40\end{aligned}$$

The selling price would be \$40.

25. Let r represent the rate of discount.

$$\begin{aligned}60 - r(60) &= 39\\ 60 - 60r &= 39\\ -60r &= -21\\ r &= 0.35\end{aligned}$$

The percent of discount that she received was 35%.

26. Let a represent the measure of the second angle, then $3a - 3$ represents the third angle. The sum of the measures of the angles of a triangle are 180°.

$$\begin{aligned}47 + a + (3a - 3) &= 180\\ 44 + 4a &= 180\\ 4a &= 136\\ a &= 34\end{aligned}$$

The remaining angles are 34° and $3(34) - 3 = 99°$.

27. Let t represent Zak's time, then $t + 1$ represents Connie's time as she travels one hour longer. A chart of the information would be as follows:

	Rate	Time	Distance ($d = rt$)
Connie	10	$t + 1$	$10(t + 1)$
Zak	12	t	$12t$

A diagram of the problem would be as follows:

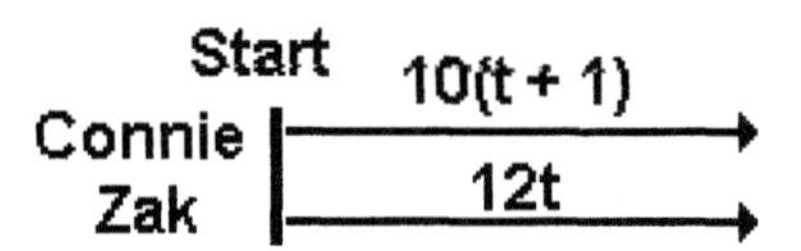

Distances are equal.

Set the distances equal to each other and solve.

$$\begin{aligned}10(t + 1) &= 12t\\ 10t + 10 &= 12t\\ 10 &= 2t\\ 5 &= t\end{aligned}$$

It would take 5 hours for Zak to catch up with Connie.

28. Let x represent the amount of 10% salt solution to be used, then $12 + x$ represents the amount of final solution.

$$\begin{pmatrix}\text{pure}\\ \text{salt}\\ \text{in 10\%}\\ \text{solution}\end{pmatrix} + \begin{pmatrix}\text{pure}\\ \text{salt}\\ \text{in 15\%}\\ \text{solution}\end{pmatrix} = \begin{pmatrix}\text{pure}\\ \text{salt}\\ \text{in final}\\ \text{solution}\end{pmatrix}$$

$$\begin{aligned}(10\%)(x) + (15\%)(12) &= (12\%)(12 + x)\\ 0.10(x) + 0.15(12) &= 0.12(12 + x)\\ 10x + 15(12) &= 12(12 + x)\\ 10x + 180 &= 144 + 12x\\ -2x &= -36\\ x &= 18\end{aligned}$$

We must add 18 gallons of the 10% salt solution.

29. Let p represent the percent of orange juice in the resulting mixture.

$$\begin{pmatrix}\text{pure}\\ \text{juice}\\ \text{in 20\%}\\ \text{solution}\end{pmatrix} + \begin{pmatrix}\text{pure}\\ \text{juice}\\ \text{in 30\%}\\ \text{solution}\end{pmatrix} = \begin{pmatrix}\text{pure}\\ \text{juice}\\ \text{in final}\\ \text{solution}\end{pmatrix}$$

$$\begin{aligned}(20\%)(20) + (30\%)(30) &= (p)(20 + 30)\\ 0.20(20) + 0.30(30) &= p(50)\\ 4 + 9 &= 50p\\ 13 &= 50p\\ \frac{13}{50} &= p\\ p &= 0.26 = 26\%\end{aligned}$$

The resulting mixture would be 26% orange juice.

30. $i = Prt$
$i = 3500(0.0525)(2)$
$i = 367.5$
The interest is \$367.50.

CHAPTER 4 Test

1. $\frac{x+2}{4} = \frac{x-3}{5}$
Cross products are equal.
$5(x+2) = 4(x-3)$
$5x + 10 = 4x - 12$
$x + 10 = -12$
$x = -22$
The solution set is $\{-22\}$.

2. $\frac{-4}{2x-1} = \frac{3}{3x+5}$
Cross products are equal.
$-4(3x+5) = 3(2x-1)$
$-12x - 20 = 6x - 3$
$-18x - 20 = -3$
$-18x = 17$
$x = -\frac{17}{18}$
The solution set is $\left\{-\frac{17}{18}\right\}$.

3. $\frac{x-1}{6} - \frac{x+2}{5} = 2$
This is NOT a proportion!
Multiply both sides by 30.
$30\left(\frac{x-1}{6} - \frac{x+2}{5}\right) = 30(2)$
$5(x-1) - 6(x+2) = 60$
$5x - 5 - 6x - 12 = 60$
$-x - 17 = 60$
$-x = 77$
$x = -77$
The solution set is $\{-77\}$.

4. $\frac{x+8}{7} - 2 = \frac{x-4}{4}$
This is NOT a proportion!
Multiply both sides by 28.
$28\left(\frac{x+8}{7} - 2\right) = 28\left(\frac{x-4}{4}\right)$
$4(x+8) - 28(2) = 7(x-4)$
$4x + 32 - 56 = 7x - 28$
$4x - 24 = 7x - 28$
$-3x - 24 = -28$
$-3x = -4$
$x = \frac{4}{3}$
The solution set is $\left\{\frac{4}{3}\right\}$.

5. $\frac{n}{20-n} = \frac{7}{3}$
Cross products are equal.
$7(20-n) = 3n$
$140 - 7n = 3n$
$140 - 10n$
$14 = n$
The solution set is $\{14\}$.

6. $\frac{h}{4} + \frac{h}{6} = 1$
This is NOT a proportion!
Multiply both sides by 12.
$12\left(\frac{h}{4} + \frac{h}{6}\right) = 12(1)$
$3h + 2h = 12$
$5h = 12$
$h = \frac{12}{5}$
The solution set is $\left\{\frac{12}{5}\right\}$.

7. $0.05n + 0.06(400 - n) = 23$

Multiplied both sides by 100.

$$5n + 6(400 - n) = 2300$$
$$5n + 2400 - 6n = 2300$$
$$-n + 2400 = 2300$$
$$-n = -100$$
$$n = 100$$

The solution set is $\{100\}$.

8. $s = 35 + 0.5s$

Multiplied both sides by 10.

$$10s = 350 + 5s$$
$$5s = 350$$
$$s = 70$$

The solution set is $\{70\}$.

9. $0.07n = 45.5 - 0.08(600 - n)$

Multiplied both sides by 100.

$$7n = 4550 - 8(600 - n)$$
$$7n = 4550 - 4800 + 8n$$
$$7n = -250 + 8n$$
$$-n = -250$$
$$n = 250$$

The solution set is $\{250\}$.

10. $12t + 8\left(\frac{7}{2} - t\right) = 50$

Distributive property.

$$12t + 28 - 8t = 50$$
$$4t + 28 = 50$$
$$4t = 22$$
$$t = \frac{22}{4} = \frac{11}{2}$$

The solution set is $\left\{\frac{11}{2}\right\}$.

11. $F = \frac{9C + 160}{5}$

Multiplied both sides by 5.

$$5F = 9C + 160$$

Subtracted 160 from both sides.

$$5F - 160 = 9C$$

Divided both sides by 9.

$$\frac{5F - 160}{9} = C$$

12. $y = 2(x - 4)$

Distributive Property.

$$y = 2x - 8$$

Added 8 to both sides.

$$y + 8 = 2x$$

Divided both sides by 2.

$$\frac{y + 8}{2} = x$$

13. $\frac{x + 3}{4} = \frac{y - 5}{9}$

Cross products are equal.

$$4(y - 5) = 9(x + 3)$$

Distributive property.

$$4y - 20 = 9x + 27$$

Added 20 to both sides.

$$4y = 9x + 47$$

Divided both sides by 4.

$$y = \frac{9x + 47}{4}$$

14. Find the radius by solving the formula for circumference for r when $C = 16\pi$.

$$C = 2\pi r$$
$$16\pi = 2\pi r$$
$$8 = r$$

When the radius is 8 centimeters,
the area will be $A = \pi(8)^2 = 64\pi$ sq cm.

15. Let w represent the width of the rectangle when $l = 32$ and $P = 100$.

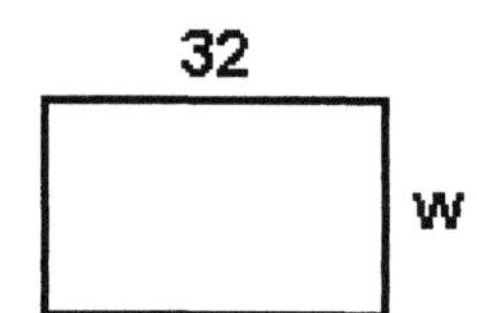

$$P = 2l + 2w$$
$$100 = 2(32) + 2w$$
$$100 = 64 + 2w$$
$$36 = 2w$$
$$18 = w$$

The width of the rectangle is 18 inches.
Therefore, the area of the rectangle is $(32)(18) = 576$ square inches.

16. Let h represent the altitude of the triangular plot when $A = 133$ and $b = 19$.

$$A = \frac{1}{2}bh$$
$$133 = \frac{1}{2}(19)h$$
$$266 = 19h$$
$$14 = h$$

The altitude of the triangular plot is 14 yards.

17. $\frac{5}{4} = \frac{n}{100}$

$$4n = 500$$
$$n = 125$$

Therefore, $\frac{5}{4} = \frac{125}{100} = 125\%$.

18. Let n represent the number.

$$(35\%)n = 24.5$$
$$0.35n = 24.5$$
$$35n = 2450$$
$$n = 70$$

Therefore, 35% of 70 is 24.5.

19. Let p represent the original price of the digital camera.

$$(100\%)(p) - (30\%)(p) = 132.30$$
$$(70\%)(p) = 132.30$$
$$0.7p = 132.30$$
$$7p = 1323$$
$$p = 189$$

The original price of the camera was $189.

20. Let s represent the selling price of the lamp. Profit is a percent of cost.

Selling price = Cost + Profit

$$s = 40 + (30\%)(40)$$
$$s = 40 + 0.3(40)$$
$$s = 40 + 12$$
$$s = 52$$

The selling price of the lamp should be $52.

21. Let r represent the rate of discount. he received.

$$80 - r(80) = 48$$
$$80 - 80r = 48$$
$$-80r = -32$$
$$r = 0.40$$

The rate of discount would be 40%.

22. Let f represent the number of females who voted, then $1500 - f$ represents the number of males who voted.

$$\frac{\text{female voters}}{\text{male voters}} \quad \frac{7}{5} = \frac{f}{1500 - f}$$
$$5f = 7(1500 - f)$$
$$5f = 10,500 - 7f$$
$$12f = 10,500$$
$$f = 875$$

There were 875 female voters.

23. Let t represent the time of the second car, then $t + 1$ represents the first car's time as it travels one hour longer. A chart of the information would be as follows:

	Rate	Time	Distance ($d = rt$)
First Car	50	$t+1$	$50(t+1)$
Second Car	55	t	$55t$

A diagram of the problem would be as follows:

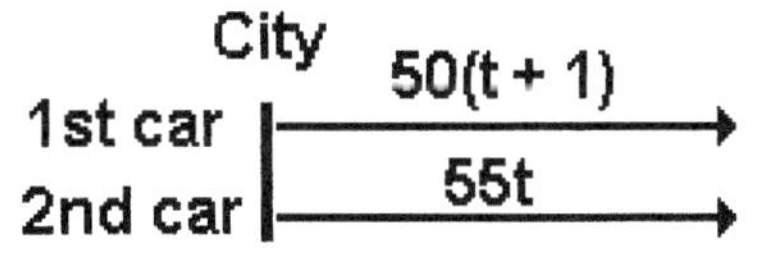

Set the distances equal to each other and solve.

$$50(t + 1) = 55t$$
$$50t + 50 = 55t$$
$$50 = 5t$$
$$10 = t$$

It would take 10 hours for the second car to overtake the first car.

24. Let x represent the amount of pure acid to be added. Then $6 + x$ represents the amount of final solution.

$$\begin{pmatrix}\text{pure}\\\text{acid}\\\text{in 50\%}\\\text{solution}\end{pmatrix} + \begin{pmatrix}\text{pure}\\\text{acid}\\\text{to be}\\\text{added}\end{pmatrix} = \begin{pmatrix}\text{pure}\\\text{acid}\\\text{in final}\\\text{solution}\end{pmatrix}$$

$$\begin{aligned}(50\%)(6) + x &= (70\%)(6 + x)\\ 0.50(6) + x &= 0.70(6 + x)\\ 10[0.50(6) + x] &= 10[0.70(6 + x)]\\ 5(6) + 10x &= 7(6 + x)\\ 30 + 10x &= 42 + 7x\\ 30 + 3x &= 42\\ 3x &= 12\\ x &= 4\end{aligned}$$

We must add 4 centiliters of pure acid.

25.

$$\begin{aligned}i &= Prt\\ 4000 &= 4000(0.09)t\\ 4000 &= 360t\\ \frac{4000}{360} &= t\\ 11.1 &= t\end{aligned}$$

It would take 11.1 years.

CHAPTERS 1-4 | Cumulative Review

1. $7x - 9x - 14x =$
$(7 - 9 - 14)x = -16x$

2. $-10a - 4 + 13a + a - 2 = 4a - 6$

3. $5(x - 3) + 7(x + 6) =$
$5x - 15 + 7x + 42 =$
$12x + 27$

4. $3(x - 1) - 4(2x - 1) =$
$3x - 3 - 8x + 4 = -5x + 1$

5. $-3n - 2(n - 1) + 5(3n - 2) - n =$
$-3n - 2n + 2 + 15n - 10 - n =$
$9n - 8$

6. $6n + 3(4n - 2) - 2(2n - 3) - 5 =$
$6n + 12n - 6 - 4n + 6 - 5 =$
$14n - 5$

7. $\frac{1}{2}x - \frac{3}{4}x + \frac{2}{3}x - \frac{1}{6}x =$
$\frac{6}{12}x - \frac{9}{12}x + \frac{8}{12}x - \frac{2}{12}x =$
$\frac{3}{12}x = \frac{1}{4}x$

8. $\frac{1}{3}n - \frac{4}{15}n + \frac{5}{6}n - n =$
$\frac{10}{30}n - \frac{8}{30}n + \frac{25}{30}n - \frac{30}{30}n =$
$-\frac{3}{30}n = -\frac{1}{10}n$

9. $0.4x - 0.7x - 0.8x + x = -0.1x$

10. $0.5(x - 2) + 0.4(x + 3) - 0.2x =$
$0.5x - 1.0 + 0.4x + 1.2 - 0.2x =$
$0.7x + 0.2$

11. $x = -2,\ y = 5:\ 5x - 7y + 2xy =$
$5(-2) - 7(5) + 2(-2)(5) =$
$-10 - 35 - 20 = -65$

12. $a = 3,\ b = -4:\ 2ab - a + 6b =$
$2(3)(-4) - (3) + 6(-4) =$
$-24 - 3 - 24 = -51$

13. $-3(x - 1) + 2(x + 6) =$
$-3x + 3 + 2x + 12 =$
$-x + 15$ for $x = -5$
$-(-5) + 15 = 5 + 15 = 20$

14. $5(n+3)-(n+4)-n=$
$5n+15-n-4-n=$
$3n+11$ for $n=7$
$3(7)+11=21+11=32$

15. $x=\frac{1}{2},\ y=\frac{1}{3}:\ \left(\frac{1}{x}+\frac{1}{y}\right)^2$
$\left(\frac{1}{\left(\frac{1}{2}\right)}+\frac{1}{\left(\frac{1}{3}\right)}\right)^2=\left(2+3\right)^2$
$=\left(5\right)^2=25$

16. $n=-\frac{2}{3}:\ \frac{3}{4}n-\frac{1}{3}n+\frac{5}{6}n=$
$\frac{9}{12}n-\frac{4}{12}n+\frac{10}{12}n=\frac{15}{12}n=\frac{5}{4}n=$
$\frac{5}{4}\left(-\frac{2}{3}\right)=-\frac{5}{6}$

17. $a=0.2,\ b=-0.3:\ 2a^2-4b^2=$
$2(0.2)^2-4(-0.3)^2=$
$2(0.04)-4(0.09)=$
$0.08-0.36=-0.28$

18. $x=\frac{1}{2},\ y=\frac{1}{4}:\ x^2-3xy-2y^2=$
$\left(\frac{1}{2}\right)^2-3\left(\frac{1}{2}\right)\left(\frac{1}{4}\right)-2\left(\frac{1}{4}\right)^2=$
$\frac{1}{4}-\frac{3}{8}-\frac{1}{8}=\frac{2}{8}-\frac{3}{8}-\frac{1}{8}=-\frac{2}{8}=-\frac{1}{4}$

19. $5x-7y-8x+3y=-3x-4y=$
$-3(9)-4(-8)$, when $x=9,\ y=-8$
$=-27+32=5$

20. $a=-1,\ b=3:\ \frac{3a-b-4a+3b}{a-6b-4b-3a}=$
$\frac{-a+2b}{-2a-10b}=\frac{-(-1)+2(3)}{-2(-1)-10(3)}=$
$\frac{1+6}{2-30}=-\frac{7}{28}=-\frac{1}{4}$

21. $3^4=3\cdot3\cdot3\cdot3=81$

22. $-2^6=-(2\cdot2\cdot2\cdot2\cdot2\cdot2)=-64$

23. $(0.4)^3=(0.4)(0.4)(0.4)=0.064$

24. $\left(-\frac{1}{2}\right)^5=$
$\left(-\frac{1}{2}\right)\left(-\frac{1}{2}\right)\left(-\frac{1}{2}\right)\left(-\frac{1}{2}\right)\left(-\frac{1}{2}\right)=$
$-\frac{1}{32}$

25. $\left(\frac{1}{2}+\frac{1}{3}\right)^2=\left(\frac{3}{6}+\frac{2}{6}\right)^2=$
$\left(\frac{5}{6}\right)^2=\frac{5}{6}\cdot\frac{5}{6}=\frac{25}{36}$

26. $\left(\frac{3}{4}-\frac{7}{8}\right)^3=\left(\frac{6}{8}-\frac{7}{8}\right)^3=\left(-\frac{1}{8}\right)^3=$
$\left(-\frac{1}{8}\right)\left(-\frac{1}{8}\right)\left(-\frac{1}{8}\right)=-\frac{1}{512}$

27. $-5x+2=22$
$-5x=20$
$x=-4$
The solution set is $\{-4\}$.

28. $3x-4=7x+4$
$-4x-4=4$
$-4x=8$
$x=-2$
The solution set is $\{-2\}$.

29. $7(n-3)=5(n+7)$
$7n-21=5n+35$
$2n-21=35$
$2n=56$
$n=28$
The solution set is $\{28\}$.

30.
$$\begin{aligned} 2(x-1)-3(x-2) &= 12 \\ 2x-2-3x+6 &= 12 \\ -x+4 &= 12 \\ -x &= 8 \\ x &= -8 \end{aligned}$$
The solution set is $\{-8\}$.

31.
$$\frac{2}{5}x-\frac{1}{3}=\frac{1}{3}x+\frac{1}{2}$$
Multiply both sides by 30.
$$\begin{aligned} 30\left(\frac{2}{5}x-\frac{1}{3}\right) &= 30\left(\frac{1}{3}x+\frac{1}{2}\right) \\ 12x-10 &= 10x+15 \\ 2x-10 &= 15 \\ 2x &= 25 \\ x &= \frac{25}{2} \end{aligned}$$
The solution set is $\left\{\frac{25}{2}\right\}$.

32.
$$\begin{aligned} \frac{t-2}{4}+\frac{t+3}{3} &= \frac{1}{6} \\ 12\left(\frac{t-2}{4}+\frac{t+3}{3}\right) &= 12\left(\frac{1}{6}\right) \\ 3(t-2)+4(t+3) &= 2 \\ 3t-6+4t+12 &= 2 \\ 7t+6 &= 2 \\ 7t &= -4 \\ t &= -\frac{4}{7} \end{aligned}$$
The solution set is $\left\{-\frac{4}{7}\right\}$.

33.
$$\frac{2n-1}{5}-\frac{n+2}{4}=1$$
Multiply both sides by 20.
$$\begin{aligned} 20\left(\frac{2n-1}{5}-\frac{n+2}{4}\right) &= 20(1) \\ 4(2n-1)-5(n+2) &= 20 \\ 8n-4-5n-10 &= 20 \\ 3n-14 &= 20 \\ 3n &= 34 \\ n &= \frac{34}{3} \end{aligned}$$
The solution set is $\left\{\frac{34}{3}\right\}$.

34.
$$0.09x+0.12(500-x)=54$$
Multiply both sides by 100.
$$\begin{aligned} 100[0.09x+0.12(500-x)] &= 100(54) \\ 9x+12(500-x) &= 5400 \\ 9x+6000-12x &= 5400 \\ 6000-3x &= 5400 \\ -3x &= -600 \\ x &= 200 \end{aligned}$$
The solution set is $\{200\}$.

35.
$$-5(n-1)-(n-2)=3(n-1)-2n$$
Distributive property.
$$-5n+5-n+2=3n-3-2n$$
Combined like terms.
$$\begin{aligned} -6n+7 &= n-3 \\ -7n &= -10 \\ n &= \frac{10}{7} \end{aligned}$$
The solution set is $\left\{\frac{10}{7}\right\}$.

36.
$$\frac{-2}{x-1}=\frac{-3}{x+4}$$
Cross products are equal.
$$\begin{aligned} -2(x+4) &= -3(x-1) \\ -2x-8 &= -3x+3 \\ x-8 &= 3 \\ x &= 11 \end{aligned}$$
The solution set is $\{11\}$.

37. $0.2x + 0.1(x - 4) = 0.7x - 1$

Multiplied both sides by 10.

$$2x + 1(x - 4) = 7x - 10$$
$$2x + x - 4 = 7x - 10$$
$$3x - 4 = 7x - 10$$
$$-4x - 4 = -10$$
$$-4x = -6$$
$$x = \frac{6}{4} = \frac{3}{2}$$

The solution set is $\left\{\frac{3}{2}\right\}$.

38. $-(t-2) + (t-4) = 2\left(t - \frac{1}{2}\right) - 3\left(t + \frac{1}{3}\right)$

Distributive Property.

$$-t + 2 + t - 4 = 2t - 1 - 3t - 1$$
$$-2 = -t - 2$$
$$0 = -t$$
$$0 = t$$

The solution set is $\{0\}$.

39. $4x - 6 > 3x + 1$

$$x - 6 > 1$$
$$x > 7$$

The solution set is $\{x|x > 7\}$ or $(7, \infty)$.

40. $-3x - 6 < 12$

$$-3x < 18$$

Reversed the inequality.

$$x > -6$$

The solution set is $\{x|x > -6\}$ or $(-6, \infty)$.

41. $-2(n-1) \leq 3(n-2) + 1$

Distributive property.

$$-2n + 2 \leq 3n - 6 + 1$$
$$-2n + 2 \leq 3n - 5$$
$$-5n + 2 \leq -5$$
$$-5n \leq -7$$

Reversed the inequality.

$$n \geq \frac{7}{5}$$

The solution set is $\left\{n|n \geq \frac{7}{5}\right\}$ or $\left[\frac{7}{5}, \infty\right)$.

42. $\frac{2}{7}x - \frac{1}{4} \geq \frac{1}{4}x + \frac{1}{2}$

Multiply both sides by 28.

$$28\left(\frac{2}{7}x - \frac{1}{4}\right) \geq 28\left(\frac{1}{4}x + \frac{1}{2}\right)$$
$$8x - 7 \geq 7x + 14$$
$$x - 7 \geq 14$$
$$x \geq 21$$

The solution set is $\{x|x \geq 21\}$ or $[21, \infty)$.

43. $0.08t + 0.1(300 - t) > 28$

Multiplied both sides by 100.

$$8t + 10(300 - t) > 2800$$
$$8t + 3000 - 10t > 2800$$
$$3000 - 2t > 2800$$
$$-2t > -200$$

Reversed the inequality.

$$t < 100$$

The solution set is $\{t|t < 100\}$ or $(-\infty, 100)$.

44. $-4 > 5x - 2 - 3x$

$$-4 > 2x - 2$$
$$-2 > 2x$$
$$-1 > x \text{ means } x < -1$$

The solution set is $\{x|x < -1\}$ or $(-\infty, -1)$.

45. $\frac{2}{3}n - 2 \geq \frac{1}{2}n + 1$

Multiply both sides by 6.

$6\left(\frac{2}{3}n - 2\right) \geq 6\left(\frac{1}{2}n + 1\right)$

$4n - 12 \geq 3n + 6$

$n - 12 \geq 6$

$n \geq 18$

The solution set is

$\{n|n \geq 18\}$ or $[18, \infty)$.

46. $-3 < -2(x - 1) - x$

$-3 < -2x + 2 - x$

$-3 < -3x + 2$

$-5 < -3x$

$\frac{5}{3} > x$ means $x < \frac{5}{3}$

The solution set is

$\left\{x|x < \frac{5}{3}\right\}$ or $\left(-\infty, \frac{5}{3}\right)$.

47. Let s represent her salary five years ago.

$2s + 2000 = 32,000$

$2s = 30,000$

$s = 15,000$

Five years ago, her salary was \$15,000.

48. Let a represent the measure of one angle; then $180 - a$ represents the supplementary angle.

$a = 4(180 - a) - 45$

$a = 720 - 4a - 45$

$a = 675 - 4a$

$5a = 675$

$a = 135$

One angle is 135° and the other is $180 - (135) = 45°$.

49. Let n represent the number of nickels; then $25 - n$ represents the number of dimes. The value of the nickels in cents is represented by $5n$ and the value of the dimes in cents is $10(25 - n)$.

$5n + 10(25 - n) = 210$

$5n + 250 - 10n = 210$

$-5n + 250 = 210$

$-5n = -40$

$n = 8$

Jasmal has 8 nickels and $25 - 8 = 17$ dimes.

50. Let x represent Hana's score in the third game.

$\frac{144 + 176 + x}{3} \geq 150$

$144 + 176 + x \geq 450$

$320 + x \geq 450$

$x \geq 130$

Hana must bowl 130 or higher in the third game.

51. Let x represent the shorter piece of the board; then $30 - x$ represents the longer piece.

$\frac{x}{30 - x} = \frac{2}{3}$

$3x = 2(30 - x)$

$3x = 60 - 2x$

$5x = 60$

$x = 12$

The pieces are 12 feet and $30 - 12 = 18$ feet.

52. Let x represent the selling price of the shoes. Profit is a percent of the selling price.

Selling price = Cost + Profit

$s = 32 + (20\%)(s)$

$s = 32 + 0.2s$

$10s = 320 + 2s$

$8s = 320$

$s = 40$

The selling price would be \$40.

53. Let r represent the rate of one car; then $r+5$ represents the other car's rate. Both cars travel for 6 hours. A chart of the information would be as follows:

	Rate	Time	Distance ($d=rt$)
Car A	r	6	$6r$
Car B	$r+5$	6	$6(r+5)$

A diagram of the problem would be as follows:

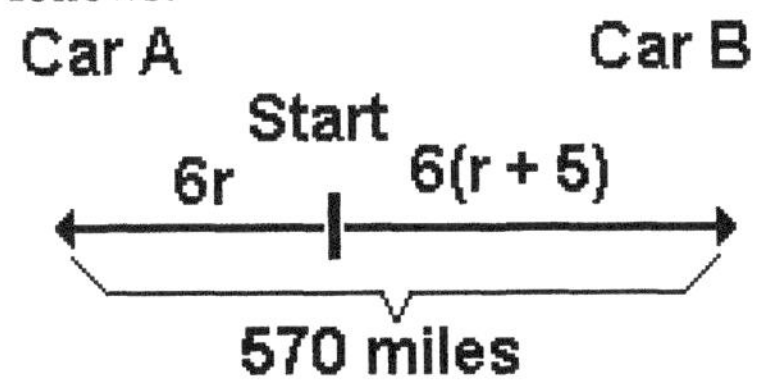

The sum of the two distances would equal the total distance of 570 miles.

$$6r+6(r+5)=570$$
$$6r+6r+30=570$$
$$12r+30=570$$
$$12r=540$$
$$r=45$$

The speeds for the cars were 45 mph and 50 mph.

54. Let x represent the amount of pure alcohol to be added. Then $15+x$ represents the amount of final solution.

$$\begin{pmatrix}\text{pure}\\\text{alcohol}\\\text{in 20\%}\\\text{solution}\end{pmatrix}+\begin{pmatrix}\text{pure}\\\text{alcohol}\\\text{to be}\\\text{added}\end{pmatrix}=\begin{pmatrix}\text{pure}\\\text{alcohol}\\\text{in final}\\\text{solution}\end{pmatrix}$$

$$(20\%)(15)+x=(40\%)(15+x)$$
$$0.20(15)+x=0.40(15+x)$$
$$10[0.20(15)+x]=10[0.40(15+x)]$$
$$2(15)+10x=4(15+x)$$
$$30+10x=60+4x$$
$$30+6x=60$$
$$6x=30$$
$$x=5$$

We must add 5 liters of pure alcohol.

Chapter 5 Coordinate Geometry and Linear Systems

PROBLEM SET 5.1 Cartesian Coordinate System

1. Six ordered pairs represent six days of change in light and water. Let x, the first coordinate, represent the change in light, and y, the second coordinate, represent the change in water. Then the ordered pairs would be as follows:

Monday	$(1, -3)$
Tuesday	$(-2, 4)$
Wednesday	$(-1, -1)$
Thursday	$(4, 0)$
Friday	$(-3, -5)$
Saturday	$(0, 1)$

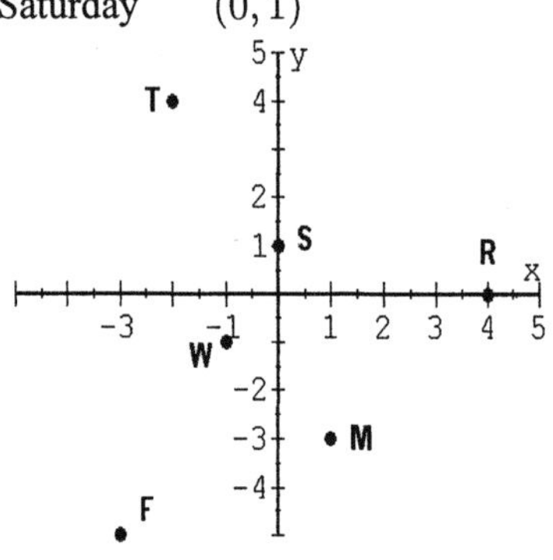

3.
$$3x + 7y = 13$$
$$7y = 13 - 3x$$
$$y = \frac{13 - 3x}{7}$$

5.
$$x - 3y = 9$$
$$x = 9 + 3y$$

7.
$$-x + 5y = 14$$
$$5y = x + 14$$
$$y = \frac{x + 14}{5}$$

9.
$$-3x + y = 7$$
$$-3x = -y + 7$$
$$x = \frac{-y + 7}{-3} \cdot \frac{-1}{-1} = \frac{y - 7}{3}$$

11.
$$-2x + 3y = -5$$
$$3y = 2x - 5$$
$$y = \frac{2x - 5}{3}$$

13. $y = x + 1$

$x = -2: \quad y = -2 + 1 = -1$

$x = 0: \quad y = 0 + 1 = 1$

$x = 2: \quad y = 2 + 1 = 3$

Point
$(-2, -1)$
$(0, 1)$
$(2, 3)$

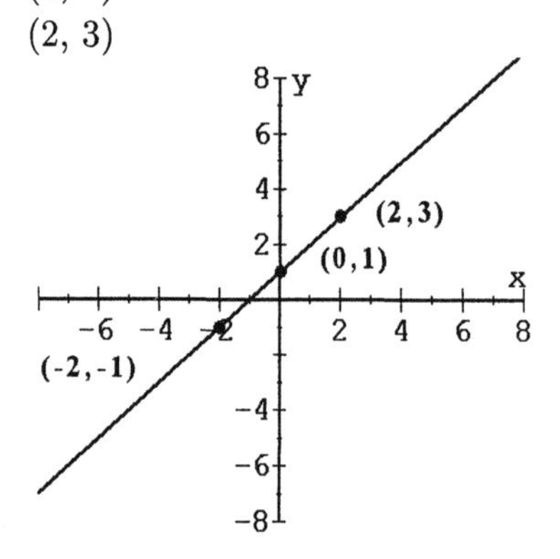

15. $y = x - 2$

$x = -2: \quad y = -2 - 2 = -4$

$x = 0: \quad y = 0 - 2 = -2$

$x = 2: \quad y = 2 - 2 = 0$

Point
$(-2, -4)$
$(0, -2)$
$(2, 0)$

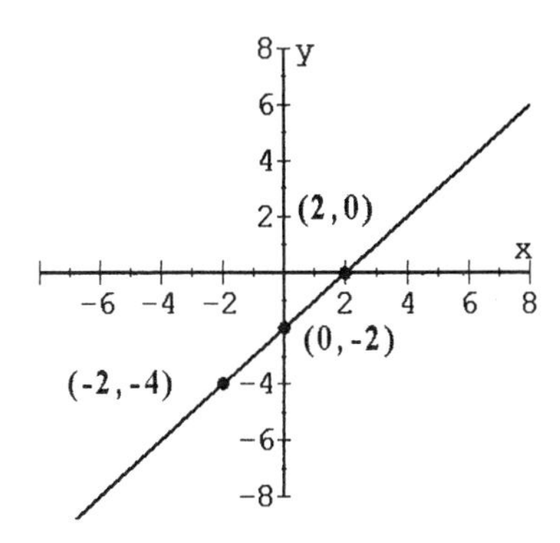

17. $y = (x-2)^2$

$x = -2:$	$y = (-2-2)^2 = (-4)^2 = 16$
$x = 0:$	$y = (0-2)^2 = (-2)^2 = 4$
$x = 2:$	$y = (2-2)^2 = (0)^2 = 0$
$x = 4:$	$y = (4-2)^2 = (2)^2 = 4$
$x = 6:$	$y = (6-2)^2 = (4)^2 = 16$

Point
$(-2, 16)$
$(0, 4)$
$(2, 0)$
$(4, 4)$
$(6, 16)$

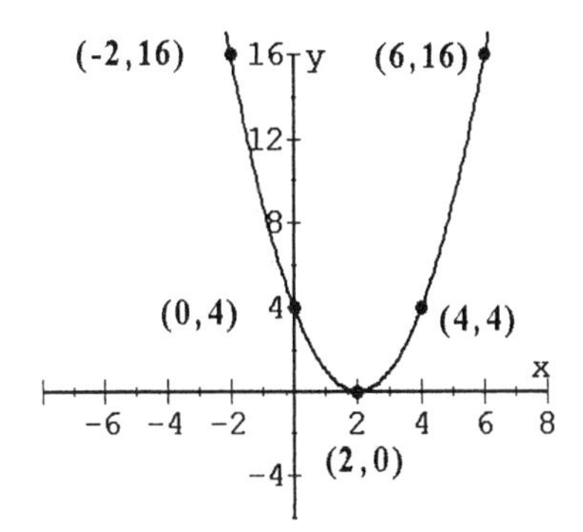

19. $y = x^2 - 2$

$x = -2:$	$y = (-2)^2 - 2 = 4 - 2 = 2$
$x = -1:$	$y = (-1)^2 - 2 = 1 - 2 = -1$
$x = 0:$	$y = (0)^2 - 2 = 0 - 2 = -2$
$x = 1:$	$y = (1)^2 - 2 = 1 - 2 = -1$
$x = 2:$	$y = (2)^2 - 2 = 4 - 2 = 2$

Point
$(-2, 2)$
$(-1, -1)$
$(0, -2)$
$(1, -1)$
$(2, 2)$

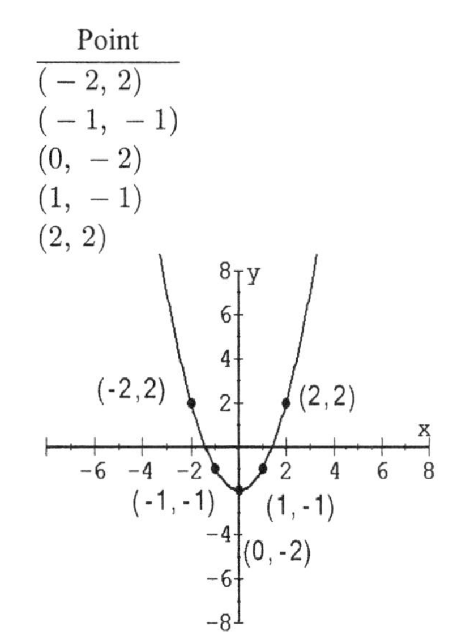

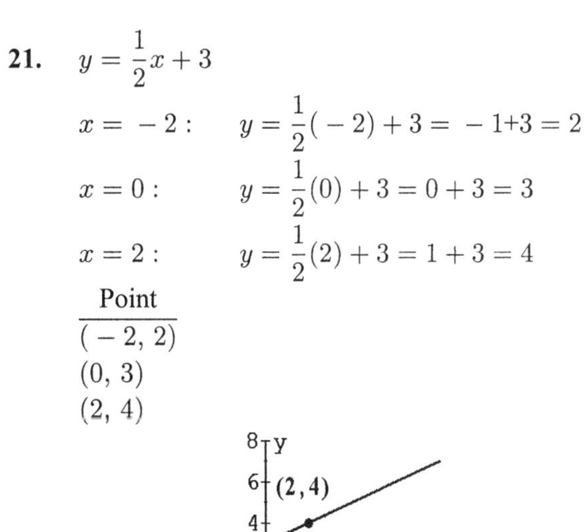

21. $y = \frac{1}{2}x + 3$

$x = -2:$	$y = \frac{1}{2}(-2) + 3 = -1 + 3 = 2$
$x = 0:$	$y = \frac{1}{2}(0) + 3 = 0 + 3 = 3$
$x = 2:$	$y = \frac{1}{2}(2) + 3 = 1 + 3 = 4$

Point
$(-2, 2)$
$(0, 3)$
$(2, 4)$

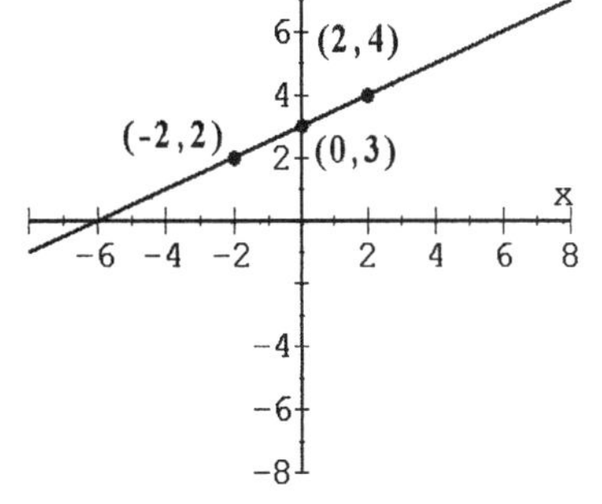

23. $x + 2y = 4$, so $y = \dfrac{4 - x}{2}$

(Use values of x that are divisible by 2.)

$x = -2: \quad y = \dfrac{4 - (-2)}{2} = \dfrac{6}{2} = 3$

$x = 0: \quad y = \dfrac{4 - (0)}{2} = \dfrac{4}{2} = 2$

$x = 2: \quad y = \dfrac{4 - (2)}{2} = \dfrac{2}{2} = 1$

Point
$(-2, 3)$
$(0, 2)$
$(2, 1)$

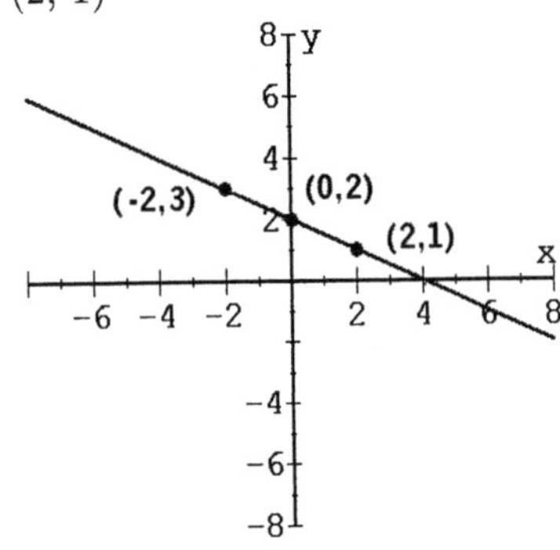

25. $2x - 5y = 10$, so $y = \dfrac{2x - 10}{5}$

(Use values of x that are divisible by 5.)

$x = -5: \quad y = \dfrac{2(-5) - 10}{5} = \dfrac{-10 - 10}{5} = \dfrac{-20}{5} = -4$

$x = 0: \quad y = \dfrac{2(0) - 10}{5} = \dfrac{-10}{5} = -2$

$x = 5: \quad y = \dfrac{2(5) - 10}{5} = \dfrac{10 - 10}{5} = \dfrac{0}{5} = 0$

Point
$(-5, -4)$
$(0, -2)$
$(5, 0)$

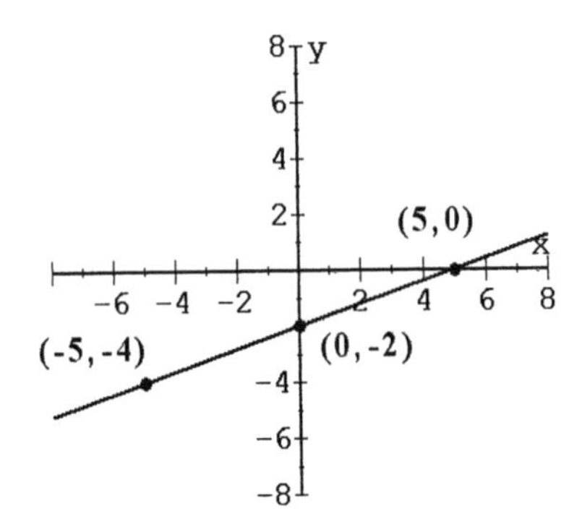

27. $y = x^3$

$x = -2: \quad y = (-2)^3 = -8$

$x = -1: \quad y = (-1)^3 = -1$

$x = 0: \quad y = (0)^3 = 0$

$x = 1: \quad y = (1)^3 = 1$

$x = 2: \quad y = (2)^3 = 8$

Point
$(-2, -8)$
$(-1, -1)$
$(0, 0)$
$(1, 1)$
$(2, 8)$

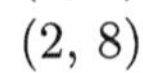

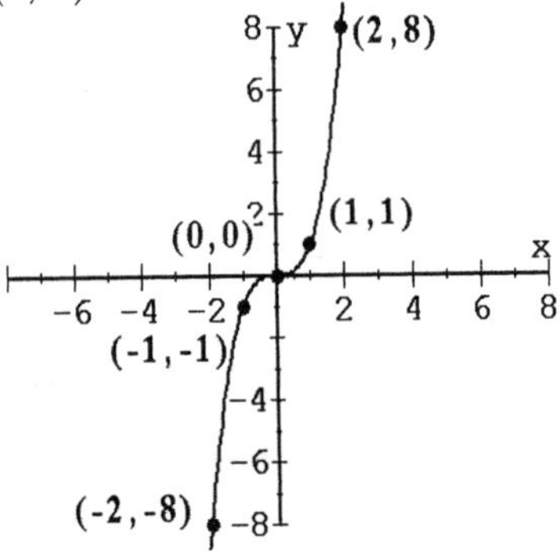

29. $y = -x^2$

$x = -2: \quad y = -(-2)^2 = -4$

$x = -1: \quad y = -(-1)^2 = -1$

$x = 0: \quad y = -(0)^2 = 0$

$x = 1: \quad y = -(1)^2 = -1$

$x = 2: \quad y = -(2)^2 = -4$

Point
$(-2,\ -4)$
$(-1,\ -1)$
$(0,\ 0)$
$(1,\ -1)$
$(2,\ -4)$

31. $y = x$

$x = -2: \quad y = -2$

$x = 0: \quad y = 0$

$x = 2: \quad y = 2$

Point
$(-2,\ -2)$
$(0,\ 0)$
$(2,\ 2)$

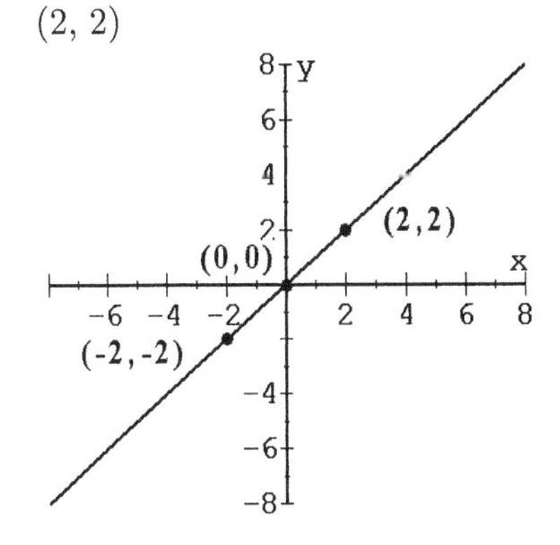

33. $y = -3x + 2$

$x = -2: \quad y = -3(-2) + 2 = 6+2 = 8$

$x = 0: \quad y = -3(0)+2 = 0+2 = 2$

$x = 2: \quad y = -3(2)+2 = -6+2 = -4$

Point
$(-2,\ 8)$
$(0,\ 2)$
$(2,\ -4)$

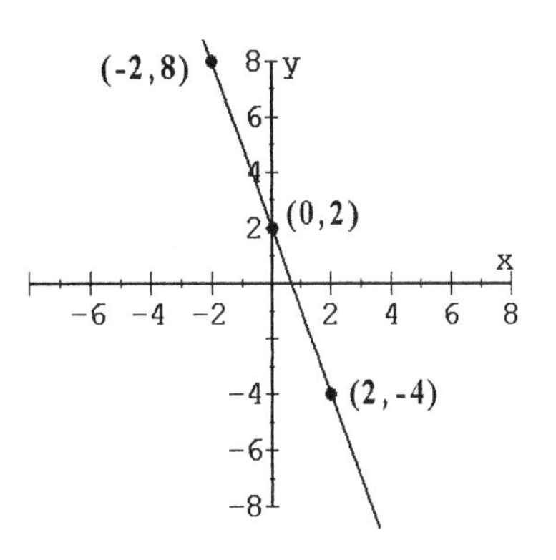

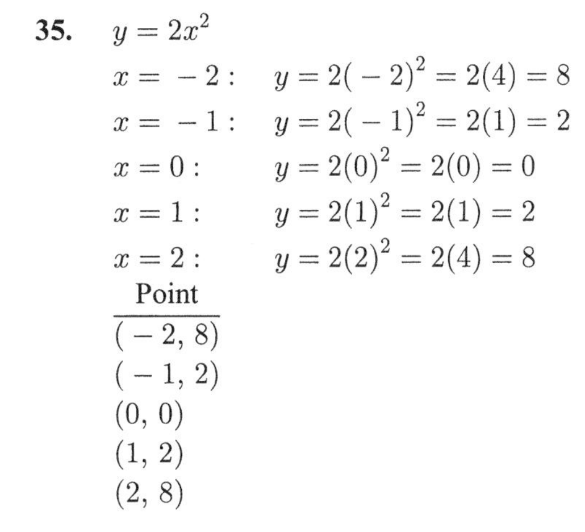

35. $y = 2x^2$

$x = -2: \quad y = 2(-2)^2 = 2(4) = 8$

$x = -1: \quad y = 2(-1)^2 = 2(1) = 2$

$x = 0: \quad y = 2(0)^2 = 2(0) = 0$

$x = 1: \quad y = 2(1)^2 = 2(1) = 2$

$x = 2: \quad y = 2(2)^2 = 2(4) = 8$

Point
$(-2,\ 8)$
$(-1,\ 2)$
$(0,\ 0)$
$(1,\ 2)$
$(2,\ 8)$

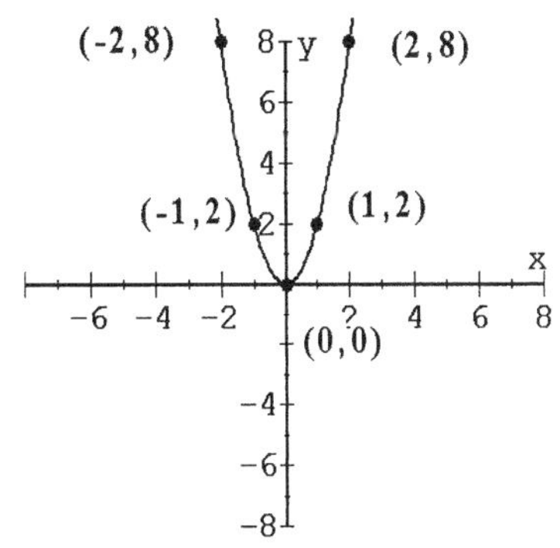

39 a. $y = x^2 + 2$

$x = -1: \quad y = (-1)^2 + 2 = 1 + 2 = 3$

$x = 0: \quad y = (0)^2 + 2 = 0 + 2 = 2$

$x = 1: \quad y = (1)^2 + 2 = 1 + 2 = 3$

Problem Set 5.1

$y = x^2 + 4$

$x = -1: \quad y = (-1)^2 + 4 = 1 + 4 = 5$

$x = 0: \quad y = (0)^2 + 4 = 0 + 4 = 4$

$x = 1: \quad y = (1)^2 + 4 = 1 + 4 = 5$

$y = x^2 - 3$

$x = -1: \quad y = (-1)^2 - 3 = 1 - 3 = -2$

$x = 0: \quad y = (0)^2 - 3 = 0 - 3 = -3$

$x = 1: \quad y = (1)^2 - 3 = 1 - 3 = -2$

$y = x^2 + 2$ Point	$y = x^2 + 4$ Point	$y = x^2 - 3$ Point
$(-1, 3)$	$(-1, 5)$	$(-1, -2)$
$(0, 2)$	$(0, 4)$	$(0, -3)$
$(1, 3)$	$(1, 5)$	$(1, -2)$

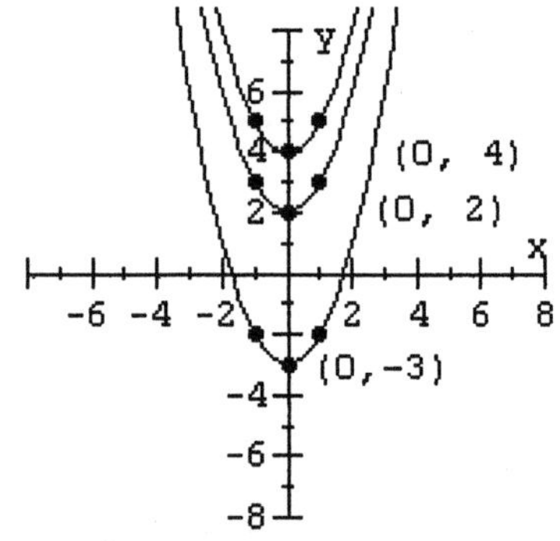

b. $y = x^2 - 1$

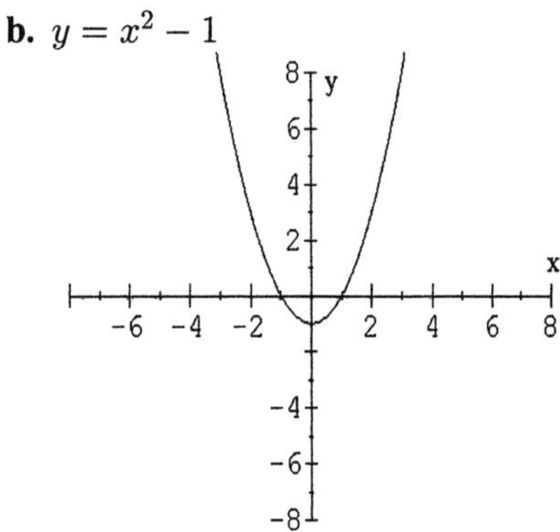

41a. $y = (x-1)^2 + 2$

$x = -1: \quad y = (-1-1)^2 + 2 = 4+2=6$

$x = 1: \quad y = (1-1)^2 + 2 = 0+2 = 2$

$x = 3: \quad y = (3-1)^2 + 2 = 4+2 = 6$

$y = (x-3)^2 - 2$

$x = 1: \quad y = (1-3)^2 - 2 = 4 - 2 = 2$

$x = 3: \quad y = (3-3)^2 - 2 = 0 - 2 = -2$

$x = 5: \quad y = (5-3)^2 - 2 = 4 - 2 = 2$

$y = (x+2)^2 + 3$

$x = -4: \quad y = (-4+2)^2 + 3 = 4+3 = 7$

$x = -2: \quad y = (-2+2)^2 + 3 = 0+3 = 3$

$x = 0: \quad y = (0+2)^2 + 3 = 4+3 = 7$

$y = (x-1)^2 + 2$ Point
$(-1, 6)$
$(1, 2)$
$(3, 6)$

$y = (x-3)^2 - 2$ Point
$(1, 2)$
$(3, -2)$
$(5, 2)$

$y = (x+2)^2 + 3$ Point
$(-4, 7)$
$(-2, 3)$
$(0, 7)$

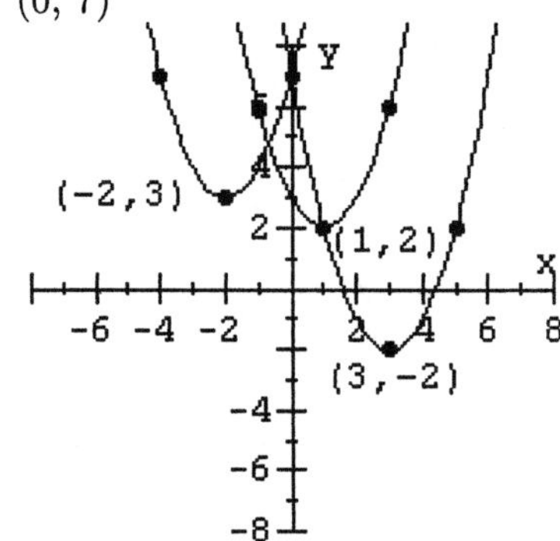

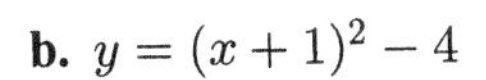
b. $y = (x+1)^2 - 4$

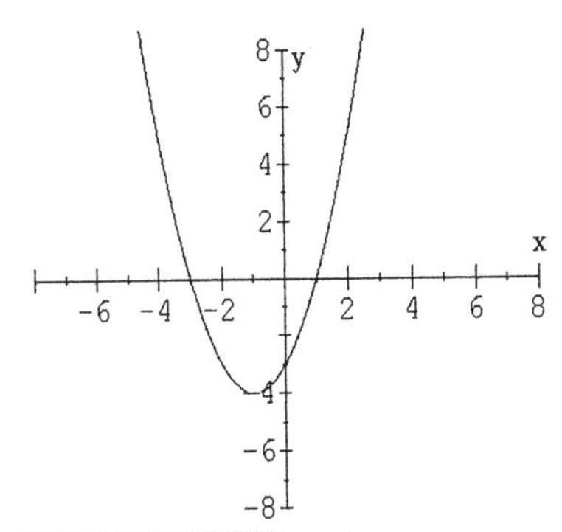

PROBLEM SET 5.2 Graphing Linear Equations

For Problems 1 – 36, the y-intercept, the x-intercept and one "check" point is given. If the line contains the origin or is parallel to an axis, two points in addition to the one intercept is given.

1.

		Point
$x + y = 2$		
$x = 0:$	$0 + y = 2$	(0, 2)
$y = 0:$	$x + 0 = 2$	(2, 0)
$x = 1:$	$1 + y = 2$	(1, 1)
	$y = 1$	

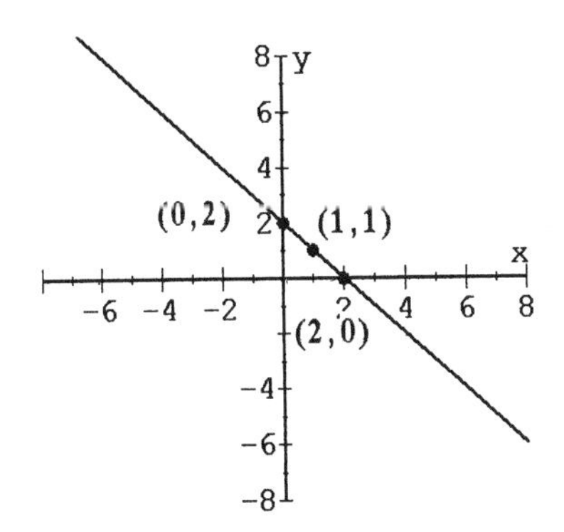

3.

		Point
$x - y = 3$		
$x = 0:$	$0 - y = 3$	(0, − 3)
	$-y = 3$	(3, 0)
	$y = -3$	(4, 1)
$y = 0:$	$x - 0 = 3$	
$y = 1:$	$x - 1 = 3$	
	$x = 4$	

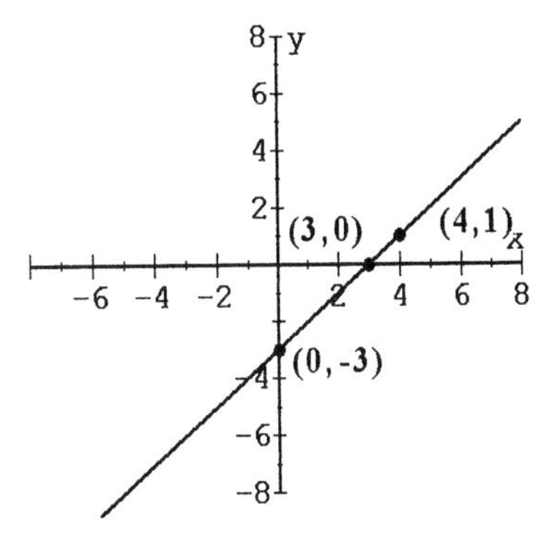

5.

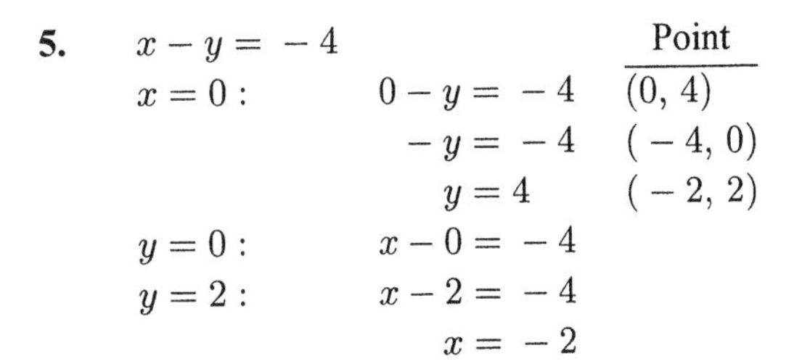

		Point
$x - y = -4$		
$x = 0:$	$0 - y = -4$	(0, 4)
	$-y = -4$	(− 4, 0)
	$y = 4$	(− 2, 2)
$y = 0:$	$x - 0 = -4$	
$y = 2:$	$x - 2 = -4$	
	$x = -2$	

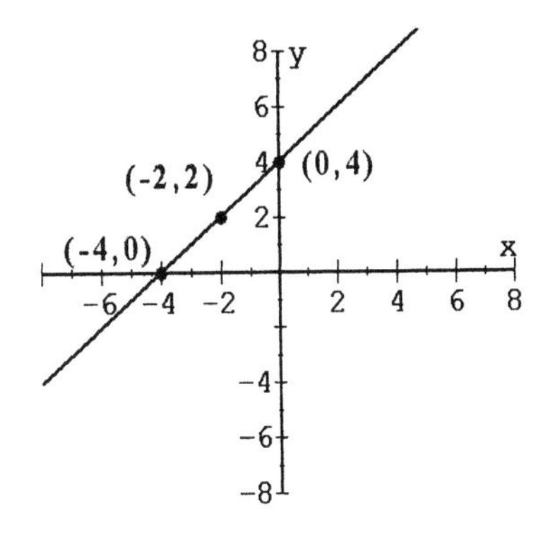

7. $x + 2y = 2$

		Point
$x = 0:$	$0 + 2y = 2$	(0, 1)
	$y = 1$	(2, 0)
$y = 0:$	$x + 0 = 2$	(− 2, 2)
$y = 2:$	$x + 4 = 2$	
	$x = -2$	

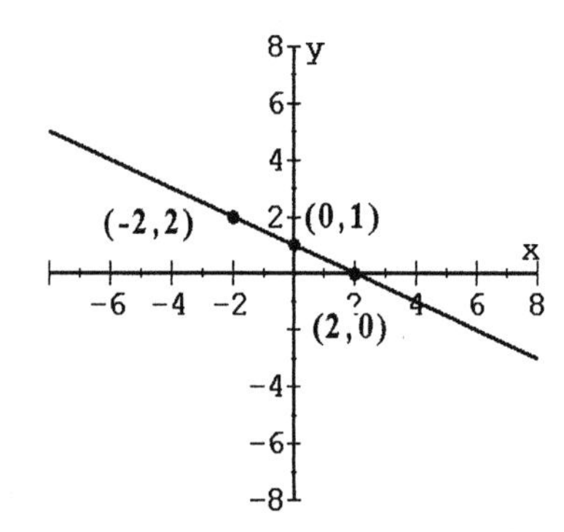

9. $3x - y = 6$

		Point
$x = 0:$	$0 - y = 6$	(0, − 6)
	$-y = 6$	(2, 0)
	$y = -6$	(1, − 3)
$y = 0:$	$3x - 0 = 6$	
	$x = 2$	
$x = 1:$	$3 - y = 6$	
	$-y = 3$	
	$y = -3$	

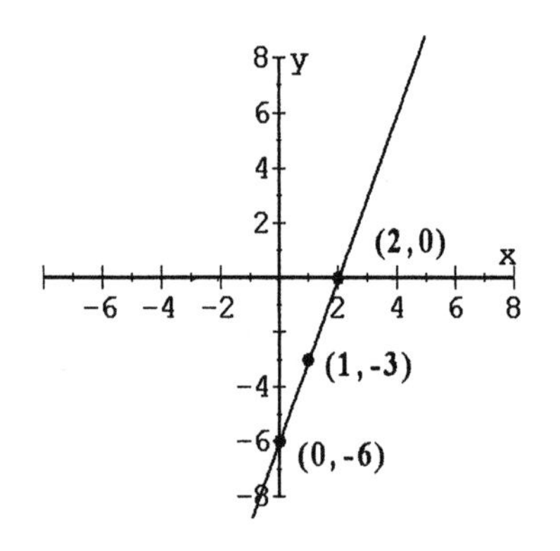

11. $3x - 2y = 6$

		Point
$x = 0:$	$0 - 2y = 6$	(0, − 3)
	$-2y = 6$	(2, 0)
	$y = -3$	(4, 3)
$y = 0:$	$3x - 0 = 6$	
	$x = 2$	
$x = 4:$	$12 - 2y = 6$	
	$-2y = -6$	
	$y = 3$	

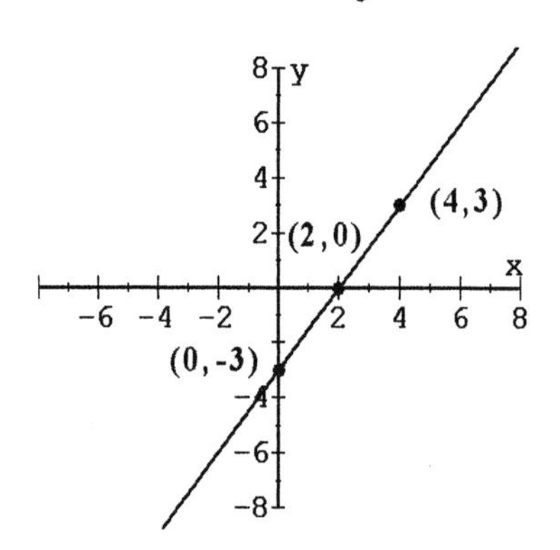

13. $x - y = 0$

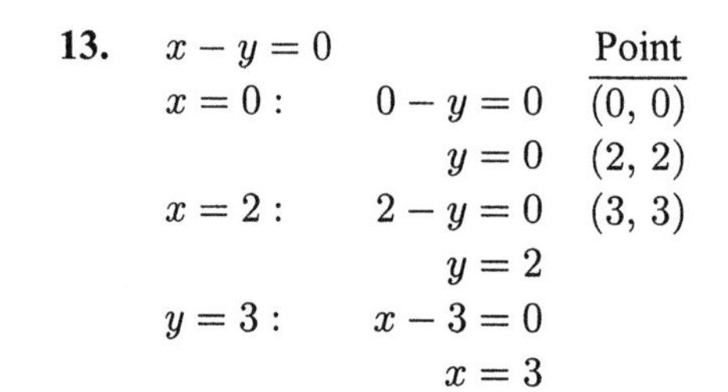

		Point
$x = 0:$	$0 - y = 0$	(0, 0)
	$y = 0$	(2, 2)
$x = 2:$	$2 - y = 0$	(3, 3)
	$y = 2$	
$y = 3:$	$x - 3 = 0$	
	$x = 3$	

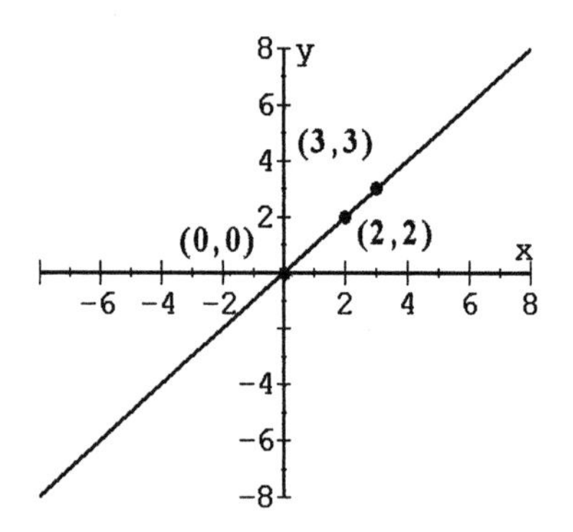

15. $y = 3x$

$x = 0:$	$y = 3(0) = 0$
$x = 1:$	$y = 3(1) = 3$
$x = -2:$	$y = 3(-2) = -6$

Point
(0, 0)
(1, 3)
(− 2, − 6)

8 y 6 4 2 (1,3) (0,0) x −6 −4 −2 2 4 6 8 −4 (-2,-6) −6 −8

17. $x = -2$

		Point
$y = 0:$	$x = -2$	(− 2, 0)
$y = 2:$	$x = -2$	(− 2, 2)
$y = -3:$	$x = -2$	(− 2, − 3)

8 y 6 4 (-2,2) 2 (-2,0) x −6 −4 −2 2 4 6 8 (-2,-3) −4 −6 −8

19. $y = 0$

		Point
$x = 0:$	$y = 0$	(0, 0)
$x = 2:$	$y = 0$	(2, 0)
$x = -3:$	$y = 0$	(− 3, 0)

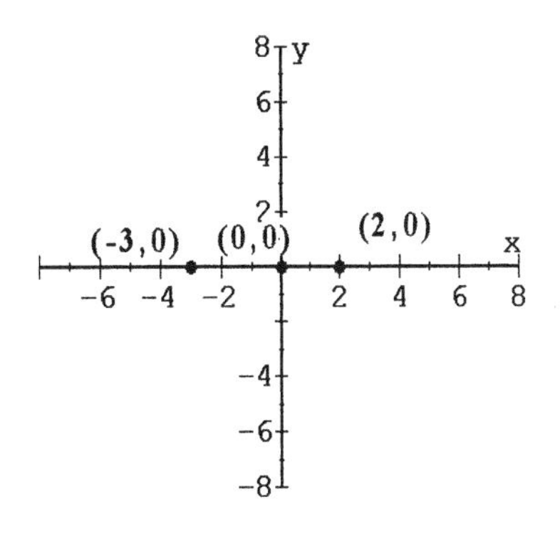

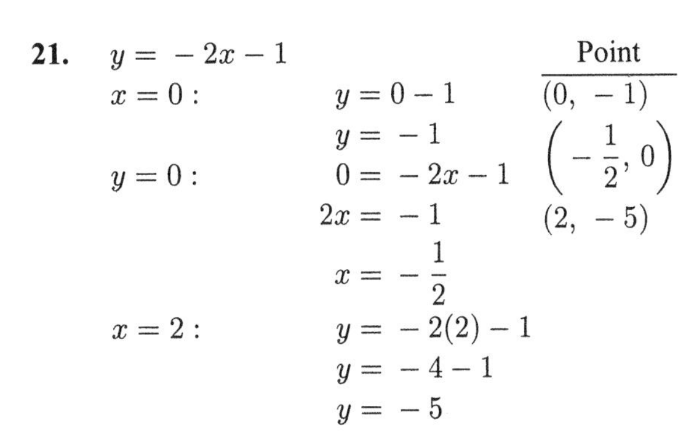

21. $y = -2x - 1$

$x = 0: \quad y = 0 - 1$

$y = -1$

$y = 0: \quad 0 = -2x - 1$

$2x = -1$

$x = -\frac{1}{2}$

$x = 2: \quad y = -2(2) - 1$

$y = -4 - 1$

$y = -5$

Point
(0, − 1)
$\left(-\frac{1}{2}, 0\right)$
(2, − 5)

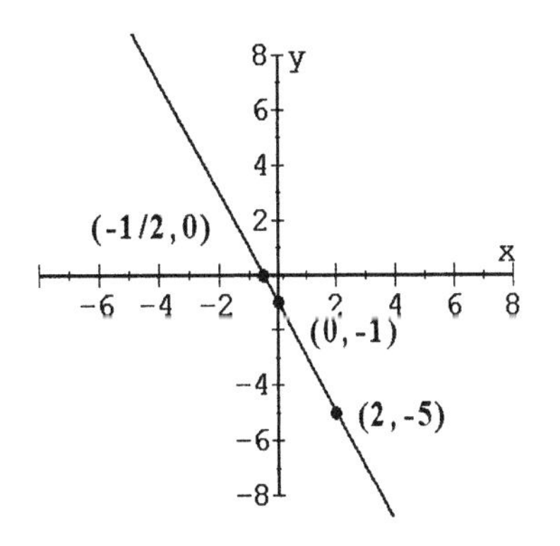

23. $y = \frac{1}{2}x + 1$

$x = 0: \quad y = 0 + 1$

$y = 1$

$y = 0: \quad 0 = \frac{1}{2}x + 1$

$0 = x + 2$

$x = -2$

$x = 2: \quad y = \frac{1}{2}(2) + 1$

$y = 1 + 1$

$y = 2$

Point
(0, 1)
(− 2, 0)
(2, 2)

Problem Set 5.2

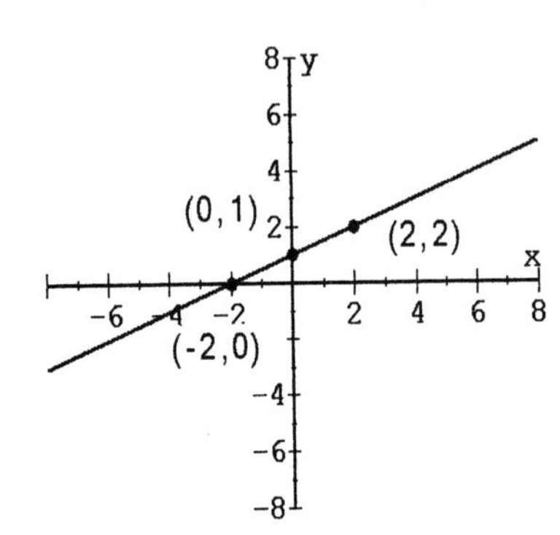

25. $y = -\frac{1}{3}x - 2$

$x = 0:$ $\quad y = 0 - 2$

$\quad y = -2$

$y = 0:$ $\quad 0 = -\frac{1}{3}x - 2$

$\quad 0 = -x - 6$

$\quad x = -6$

$x = 3:$ $\quad y = -\frac{1}{3}(3) - 2$

$\quad y = -1 - 2$

$\quad y = -3$

Point
$(0, -2)$
$(-6, 0)$
$(3, -3)$

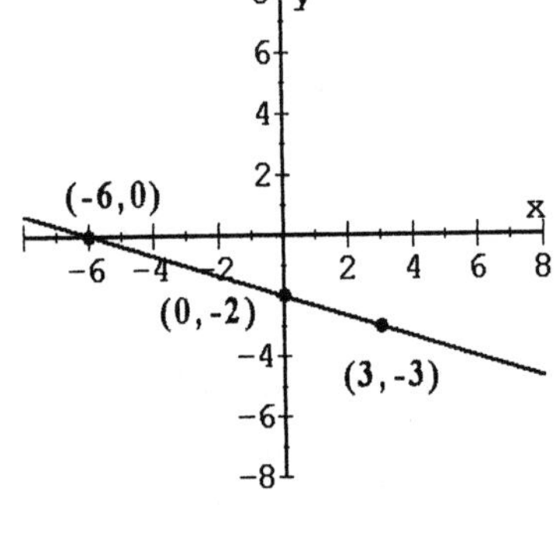

27. $4x + 5y = -10$

$x = 0:$ $\quad 0 + 5y = -10$

$\quad y = -2$

$y = 0:$ $\quad 4x + 0 = -10$

$\quad x = -\frac{10}{4}$

$\quad x = -\frac{5}{2}$

$x = 5:$ $\quad 20 + 5y = -10$

$\quad 5y = -30$

$\quad y = -6$

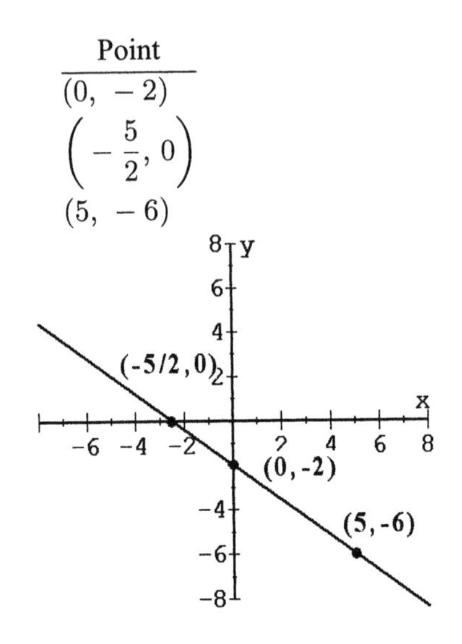

Point
$(0, -2)$
$\left(-\frac{5}{2}, 0\right)$
$(5, -6)$

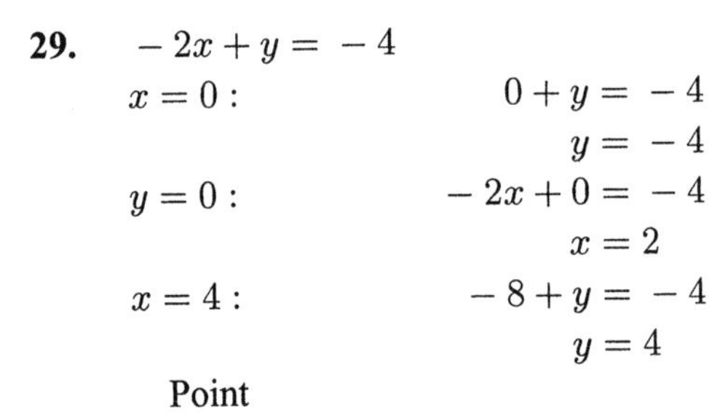

29. $-2x + y = -4$

$x = 0:$ $\quad 0 + y = -4$

$\quad y = -4$

$y = 0:$ $\quad -2x + 0 = -4$

$\quad x = 2$

$x = 4:$ $\quad -8 + y = -4$

$\quad y = 4$

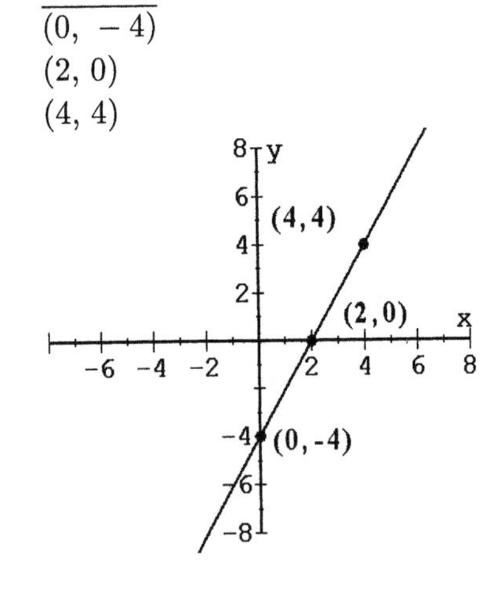

Point
$(0, -4)$
$(2, 0)$
$(4, 4)$

31. $3x - 4y = 7$

		Point
$x = 0:$	$0 - 4y = 7$	$\left(0,\ -\frac{7}{4}\right)$
	$y = -\frac{7}{4}$	$\left(\frac{7}{3},\ 0\right)$
$y = 0:$	$3x - 0 = 7$	$(-7,\ -7)$
	$x = \frac{7}{3}$	
$y = -7:$	$3x + 28 = 7$	
	$3x = -21$	
	$x = -7$	

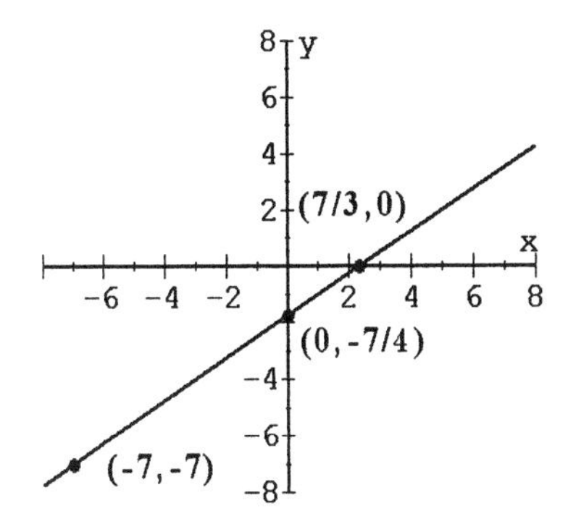

33. $y + 4x = 0$

		Point
$x = 0:$	$y + 0 = 0$	$(0,\ 0)$
	$y = 0$	$(1,\ -4)$
$x = 1:$	$y + 4 = 0$	$(-1,\ 4)$
	$y = -4$	
$x = -1:$	$y - 4 = 0$	
	$y = 4$	

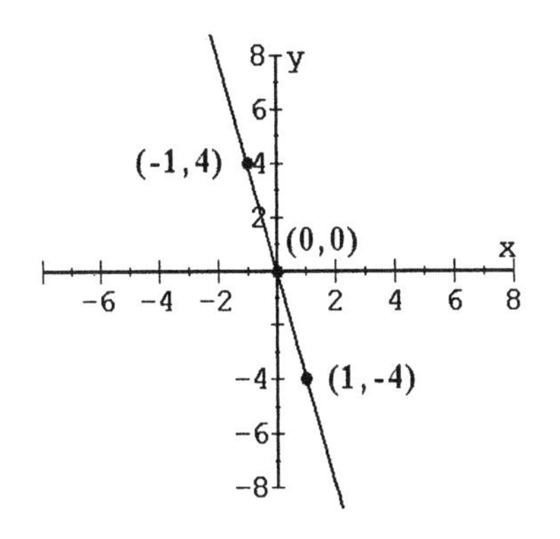

35. $x = 2y$

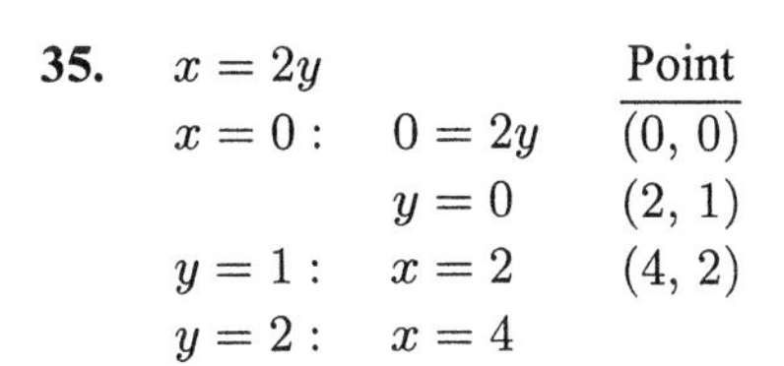

		Point
$x = 0:$	$0 = 2y$	$(0,\ 0)$
	$y = 0$	$(2,\ 1)$
$y = 1:$	$x = 2$	$(4,\ 2)$
$y = 2:$	$x = 4$	

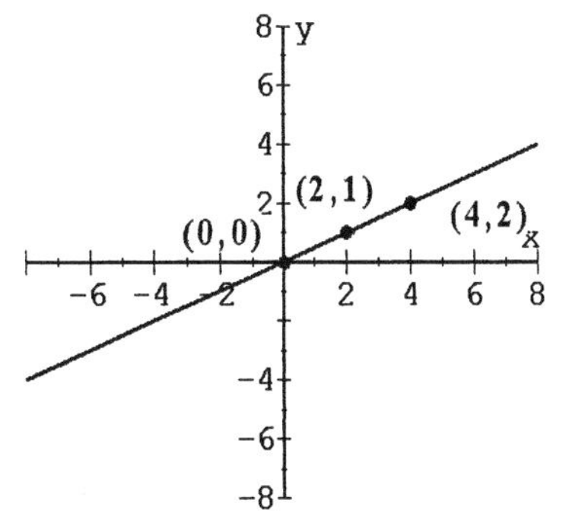

37. 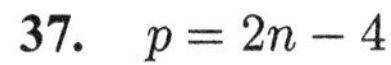
$p = 2n - 4$

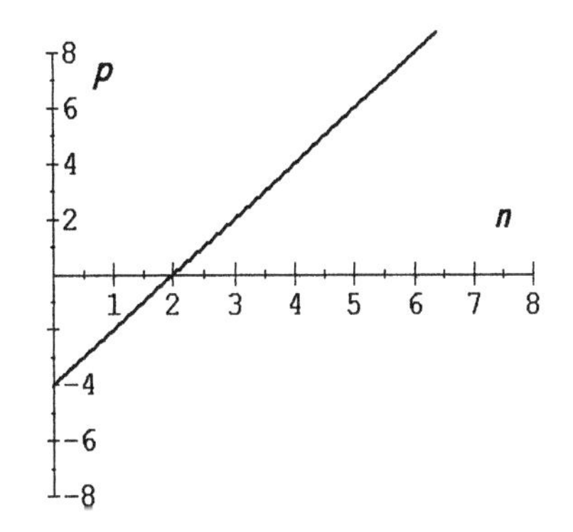

39. $A = 3l$

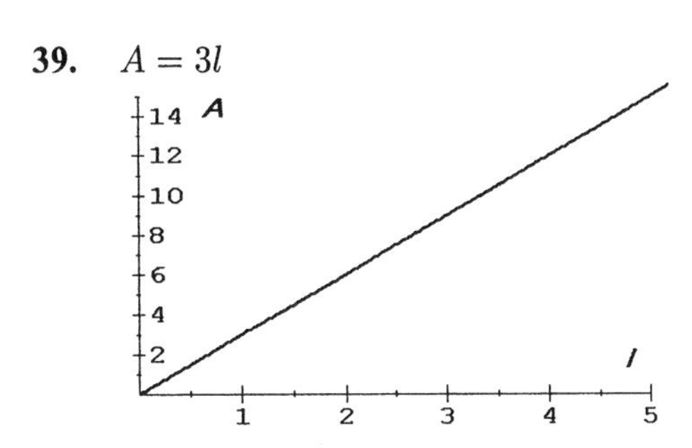

41. $A = 0.06t$

For $t = 696$, $A = 0.06(696) = \$41.76$

For $t = 720$, $A = 0.06(720) = \$43.20$

For $t = 740$, $A = 0.06(740) = \$44.40$

For $t = 775$, $A = 0.06(775) = \$46.50$

For $t = 782$, $A = 0.06(782) = \$46.92$

Problem Set 5.2

Hours, t	696	720	740	775	782
Dollars and cents, A	41.76	43.20	44.40	46.50	46.92

43a. $F = \frac{9}{5}C + 32$

For $C = 0$, $F = \frac{9}{5}(0) + 32 = 0 + 32 = 32$

For $C = 5$, $F = \frac{9}{5}(5) + 32 = 9 + 32 = 41$

For $C = 10$, $F = \frac{9}{5}(10) + 32 = 18 + 32 = 50$

For $C = 15$, $F = \frac{9}{5}(15) + 32 = 27 + 32 = 59$

For $C = 20$, $F = \frac{9}{5}(20) + 32 = 36 + 32 = 68$

For $C = -5$, $F = \frac{9}{5}(-5) + 32 = -9 + 32 = 23$

For $C = -10$, $F = \frac{9}{5}(-10) + 32 = -18 + 32 = 14$

For $C = -15$, $F = \frac{9}{5}(-15) + 32 = -27 + 32 = 5$

For $C = -20$, $F = \frac{9}{5}(-20) + 32 = -36 + 32 = -4$

For $C = -25$, $F = \frac{9}{5}(-25) + 32 = -45 + 32 = -13$

C	0	5	10	15	20	−5	−10	−15	−20	−25
F	32	41	50	59	68	23	14	5	−4	−13

b. 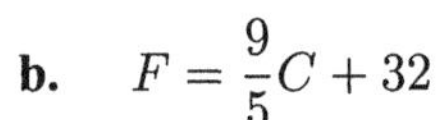
$F = \frac{9}{5}C + 32$

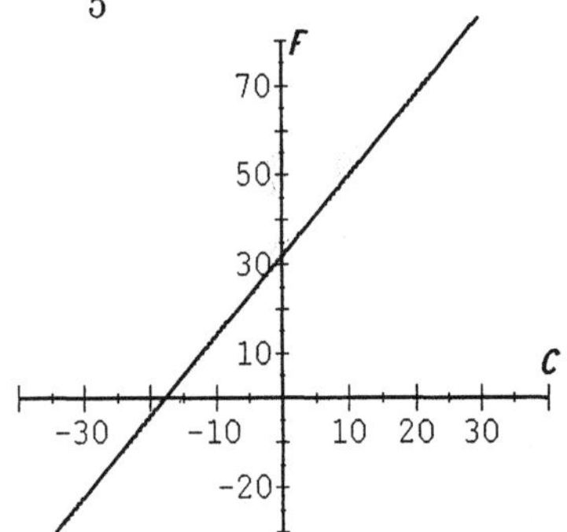

c. If $C = 25$, then $F \approx 70$.
If $C = 30$, then $F \approx 75$.
If $C = -30$, then $F \approx -20$.
If $C = -40$, then $F \approx -30$.

d. If $C = 25$, then $F = \frac{9}{5}(25) + 32 =$
$9(5) + 32 = 45 + 32 = 77°$.

If $C = 30$, then $F = \frac{9}{5}(30) + 32 =$
$9(6) + 32 = 54 + 32 = 86°$.

If $C = -30$, then $F = \frac{9}{5}(-30) + 32 =$
$9(-6) + 32 = -54 + 32 = -22°$.

If $C = -40$, then $F = \frac{9}{5}(-40) + 32 =$
$9(-8) + 32 = -72 + 32 = -40°$.

49. 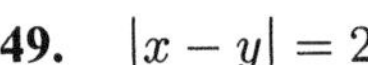
$|x - y| = 2$

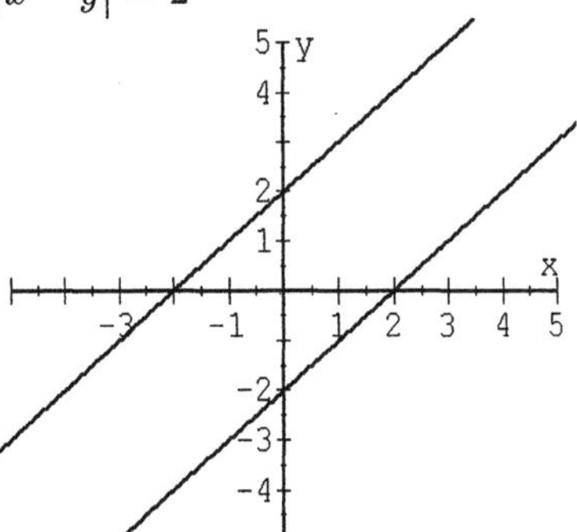

51. $|3x - y| = 6$

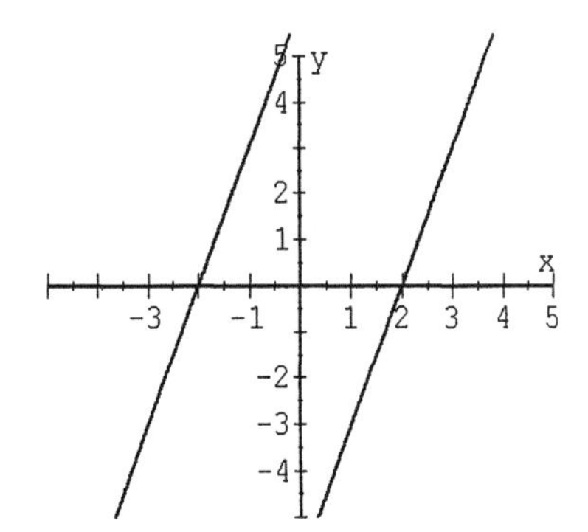

PROBLEM SET 5.3 Slope of a Line

1. Let (7, 5) be P_1 and (3, 2) be P_2.

$$m = \frac{y_2 - y_1}{x_2 - x_1} = \frac{2-5}{3-7} = \frac{-3}{-4} = \frac{3}{4}$$

3. Let (− 1, 3) be P_1 and (− 6, − 4) be P_2.

$$m = \frac{y_2 - y_1}{x_2 - x_1} = \frac{-4-3}{-6-(-1)}$$

$$m = \frac{-7}{-5} = \frac{7}{5}$$

5. Let (2, 8) be P_1 and (7, 2) be P_2.

$$m = \frac{y_2 - y_1}{x_2 - x_1} = \frac{2-8}{7-2} = \frac{-6}{5} = -\frac{6}{5}$$

7. Let (− 2, 5) be P_1 and (1, − 5) be P_2.

$$m = \frac{y_2 - y_1}{x_2 - x_1} = \frac{-5-5}{1-(-2)}$$

$$m = \frac{-10}{3} = -\frac{10}{3}$$

9. Let (4, − 1) be P_1 and (− 4, − 7) be P_2.

$$m = \frac{y_2 - y_1}{x_2 - x_1} = \frac{-7-(-1)}{-4-4}$$

$$m = \frac{-6}{-8} = \frac{3}{4}$$

11. Let (3, − 4) be P_1 and (2, − 4) be P_2.

$$m = \frac{y_2 - y_1}{x_2 - x_1} = \frac{-4-(-4)}{2-3} = \frac{0}{-1} = 0$$

13. Let (− 6, − 1) be P_1 and (− 2, − 7) be P_2.

$$m = \frac{y_2 - y_1}{x_2 - x_1} = \frac{-7-(-1)}{-2-(-6)}$$

$$m = \frac{-6}{4} = -\frac{3}{2}$$

15. Let (− 2, 4) be P_1 and (− 2, − 6) be P_2.
The slope is undefined because $x_1 = x_2$.

17. Let (− 1, 10) be P_1 and (− 9, 2) be P_2.

$$m = \frac{y_2 - y_1}{x_2 - x_1} = \frac{2-10}{-9-(-1)} = \frac{-8}{-8} = 1$$

19. Let (a, b) be P_1 and (c, d) be P_2.

$$m = \frac{y_2 - y_1}{x_2 - x_1} = \frac{d-b}{c-a}$$

21. Let (7, 8) be P_1 and (2, y) be P_2.

$$\frac{y-8}{2-7} = \frac{4}{5}$$

$$\frac{y-8}{-5} = \frac{4}{5}$$

$$5(y-8) = 4(-5)$$

$$5y - 40 = -20$$

$$5y = 20$$

$$y = 4$$

23. Let (− 2, − 4) be P_1 and (x, 2) be P_2.

$$\frac{2-(-4)}{x-(-2)} = -\frac{3}{2}$$

$$\frac{6}{x+2} = -\frac{3}{2}$$

$$-3(x+2) = 6(2)$$

$$-3x - 6 = 12$$

$$-3x = 18$$

$$x = -6$$

In Problems 25 − 32, the answers will vary but sample points are given.

25. From (3, 2) move 2 units up and 3 units right to (6, 4).
From (6, 4) move 2 units up and 3 units right to (9, 6).
From (9, 6) move 2 units up and 3 units right to (12, 8).

27. From (− 2, − 4) move 1 unit up and 2 units right to (0, − 3).
From (0, − 3) move 1 unit up and 2 units right to (2, − 2).
From (2, − 2) move 1 unit up and 2 units right to (4, − 1).

29. From $(-3, 4)$ move 3 units down and 4 units right to $(1, 1)$.
From $(1, 1)$ move 3 units down and 4 units right to $(5, -2)$.
From $(5, -2)$ move 3 units down and 4 units right to $(9, -5)$.

31. From $(4, -5)$ move 2 units down and 1 unit right to $(5, -7)$.
From $(5, -7)$ move 2 units down and 1 unit right to $(6, -9)$.
From $(6, -9)$ move 2 units down and 1 unit right to $(7, -11)$.

33. The line falls from left to right, so slope is negative.

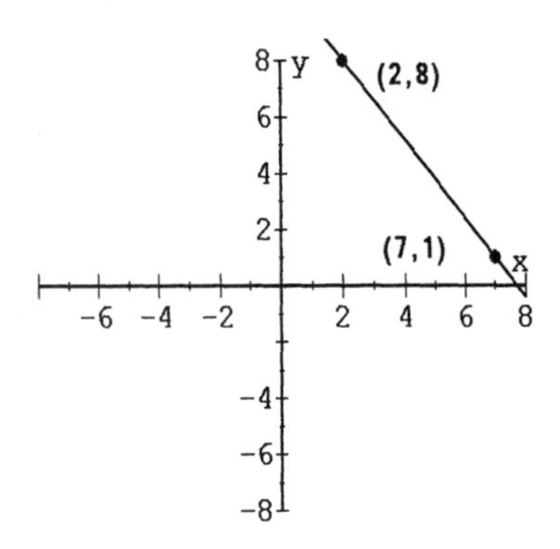

35. The line rises from left to right, so the slope is positive.

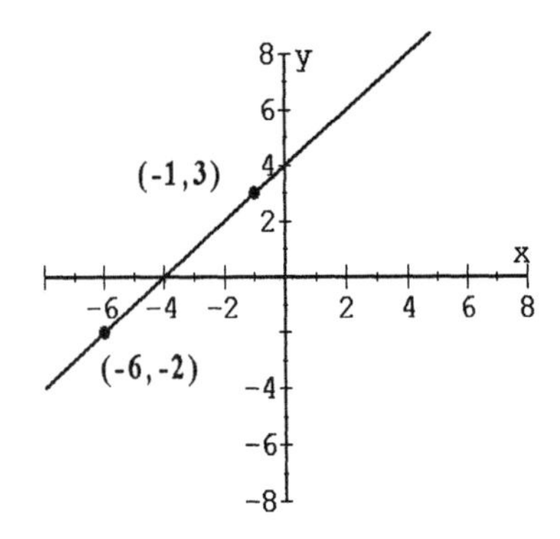

37. The line is horizontal, so the slope is zero.

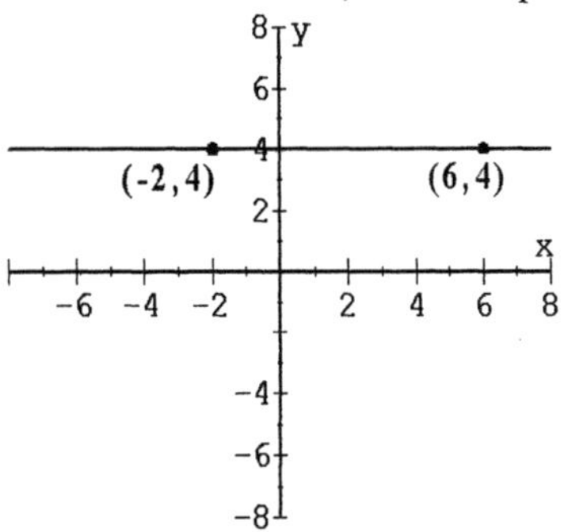

39. The line falls from left to right, so the slope is negative.

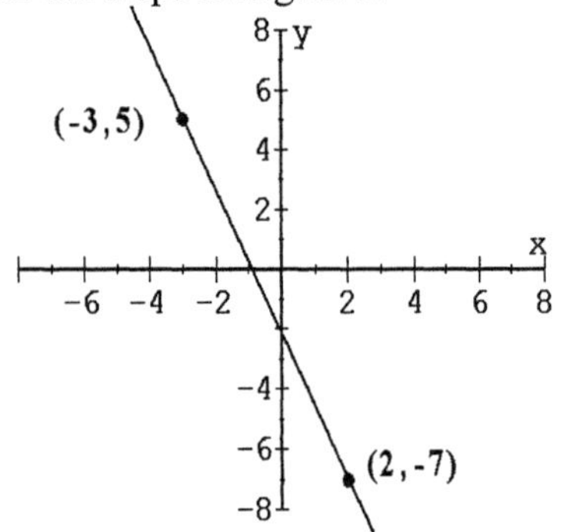

41. $(3,1)\ m = \dfrac{2}{3}$

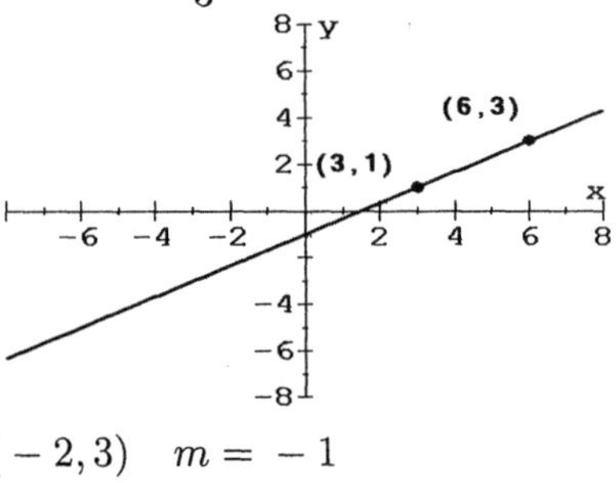

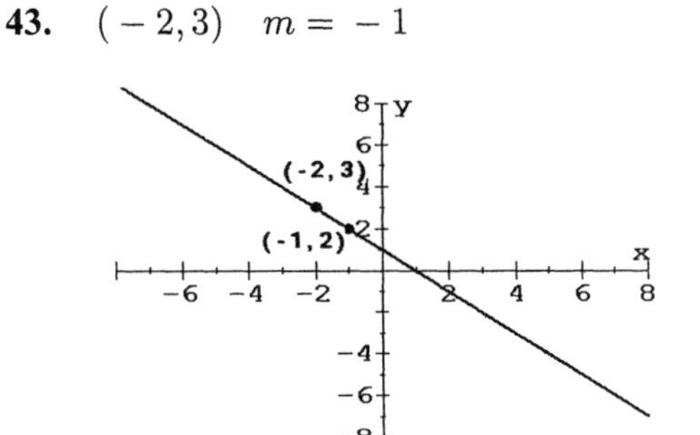

43. $(-2,3)\quad m = -1$

45. $(0,5)$ $\quad m = \dfrac{-1}{4}$

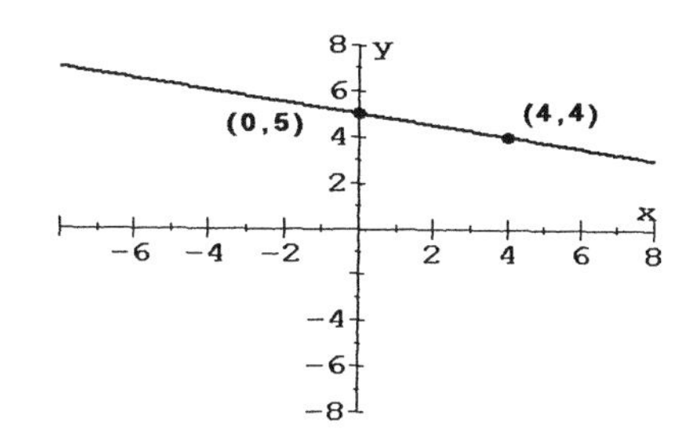

47. $(2, -2)$ $\quad m = \dfrac{3}{2}$

49. $3x + 2y = 6$

$x = 0: \quad 0 + 2y = 6$

$2y = 6$

$y = 3$

$y = 0: \quad 3x + 0 = 6$

$x = 2$

Point
(0, 3)
(2, 0)

$$m = \frac{0-3}{2-0} = \frac{-3}{2} = -\frac{3}{2}$$

51. $5x - 4y = 20$

$x = 0: \quad 0 - 4y = 20$

$-4y = 20$

$y = -5$

$y = 0: \quad 5x + 0 = 20$

$5x = 20$

$x = 4$

Point
(0, − 5)
(4, 0)

$$m = \frac{0-(-5)}{4-0} = \frac{5}{4}$$

53. $x + 5y = 6$

$y = 0: \quad x + 0 = 6$

$x = 6$

$y = 1: \quad x + 5 = 6$

$x = 1$

Point
(6, 0)
(1, 1)

$$m = \frac{1-0}{1-6} = \frac{1}{-5} = -\frac{1}{5}$$

55. $2x - y = -7$

$x = 0: \quad 0 - y = -7$

$-y = -7$

$y = 7$

$x = 1: \quad 2 - y = -7$

$-y = -9$

$y = 9$

Point
(0, 7)
(1, 9)

$$m = \frac{9-7}{1-0} = \frac{2}{1} = 2$$

57. $y = 3$

$x = 0: \quad y = 3$

$x = 2: \quad y = 3$

Point
(0, 3)
(2, 3)

$$m = \frac{3-3}{2-0} = \frac{0}{2} = 0$$

59. $-2x + 5y = 9$

$x = 0: \quad 0 + 5y = 9$

$5y = 9$

$y = \dfrac{9}{5}$

$y = 0: \quad -2x + 0 = 9$

$-2x = 9$

$x = -\dfrac{9}{2}$

Point
$\left(0, \frac{9}{5}\right)$
$\left(-\frac{9}{2}, 0\right)$

$$m = \frac{0 - \frac{9}{5}}{-\frac{9}{2} - 0} = \frac{-\frac{9}{5}}{-\frac{9}{2}}$$

$$m = \left(-\frac{9}{5}\right)\left(-\frac{2}{9}\right) = \frac{2}{5}$$

Problem Set 5.3

61. $6x - 5y = -30$

$x = 0: \quad 0 - 5y = -30$
$-5y = -30$
$y = 6$

$y = 0: \quad 6x + 0 = -30$
$6x = -30$
$x = -5$

Point
(0, 6)
(−5, 0)

$$m = \frac{0-6}{-5-0} = \frac{-6}{-5} = \frac{6}{5}$$

63. $y = -3x - 1$

$x = 0: \quad y = 0 - 1$
$y = -1$

$x = 1: \quad y = -3 - 1$
$y = -4$

Point
(0, −1)
(1, −4)

$$m = \frac{-4-(-1)}{1-0} = \frac{-3}{1} = -3$$

65. $y = 4x$

		Point
$x = 0:$	$y = 0$	(0, 0)
$x = 1:$	$y = 4$	(1, 4)

$$m = \frac{4-0}{1-0} = \frac{4}{1} = 4$$

67. $y = \frac{2}{3}x - \frac{1}{2}$

$x = 0: \quad y = 0 - \frac{1}{2}$
$y = -\frac{1}{2}$

$x = 3: \quad y = \frac{2}{3}(3) - \frac{1}{2}$
$y = 2 - \frac{1}{2}$
$y = \frac{4}{2} - \frac{1}{2}$
$y = \frac{3}{2}$

Point
$\left(0, -\frac{1}{2}\right)$
$\left(3, \frac{3}{2}\right)$

$$m = \frac{\frac{3}{2} - \left(-\frac{1}{2}\right)}{3-0} = \frac{\frac{4}{2}}{3} = \frac{2}{3}$$

69. The grade of the highway can be calculated as a ratio as follows:

$$\frac{\text{distance highway rises}}{\text{horizontal distance}} = \frac{135}{2640} = 0.051136...$$

The grade of the highway is 5.1%.

71. Let x represent the measure of the run of the stairs. Solve a proportion comparing the rise to the run of the stairs.

$$\frac{\text{rise}}{\text{run}} \quad \frac{3}{5} = \frac{19}{x}$$
$3x = 19(5)$
$3x = 95$
$x = 31.667$ or 32 to the nearest whole number.

The measure of the run of the stairs is 32 centimeters.

73. Let x represent the vertical drop for the sewage pipe.

$$\frac{\text{fall}}{100\text{ feet}} \quad \frac{2\frac{1}{4}}{100} = \frac{x}{45}$$
$100x = 45\left(\frac{9}{4}\right)$
$400x = 45(9)$
$400x = 405$
$x = 1.0125$ or 1.0 to the nearest tenth.

The vertical drop must be 1.0 feet for 45 feet.

PROBLEM SET 5.4 Systems of Two Linear Equations

1.

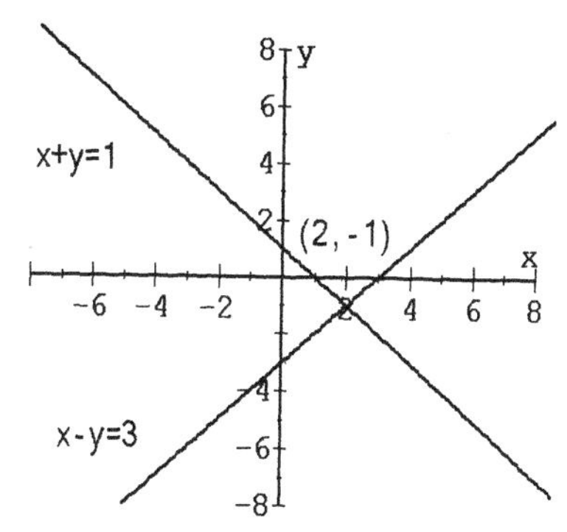

The solution set is $\{(2, -1)\}$.

3.

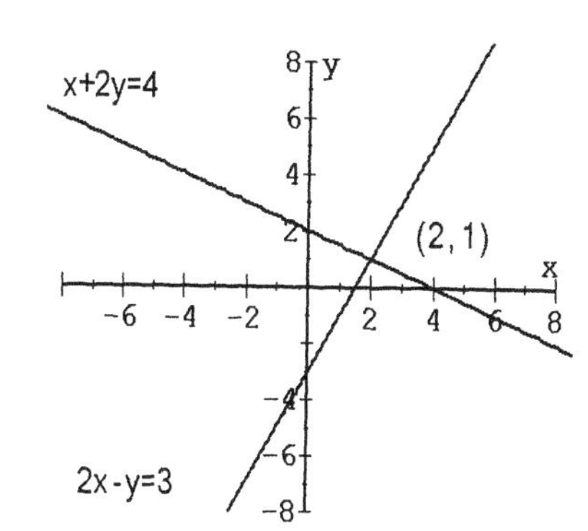

The solution set is $\{(2, 1)\}$.

5.

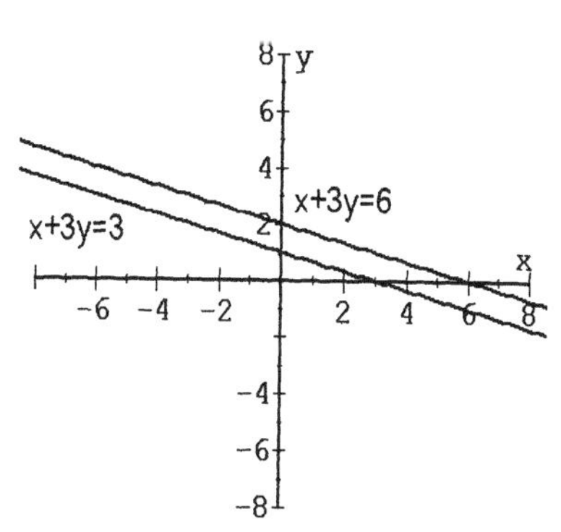

The solution set is $\emptyset$.

7.

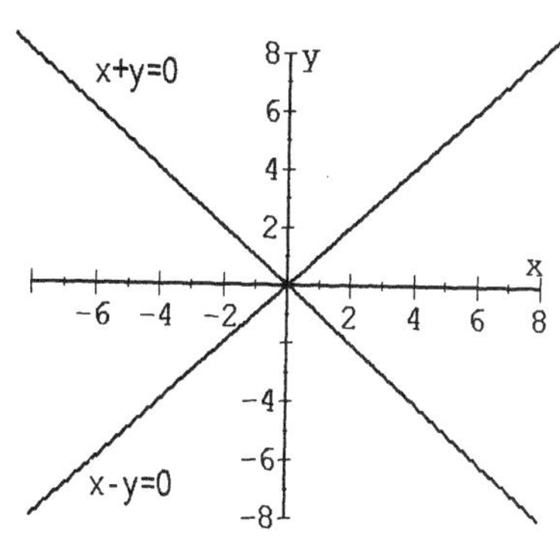

The solution set is $\{(0, 0)\}$.

9.

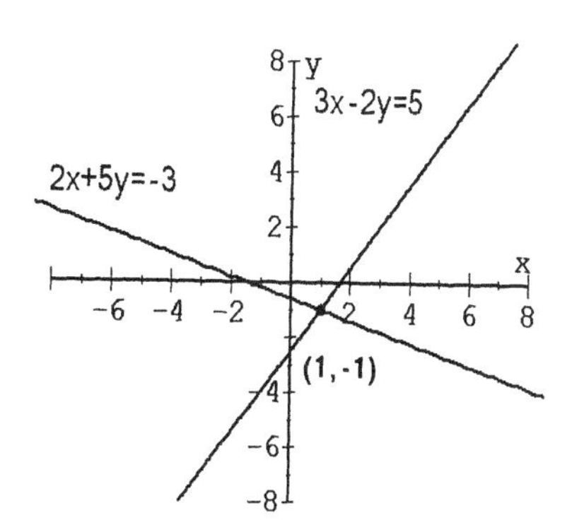

The solution set is $\{(1, -1)\}$.

11.

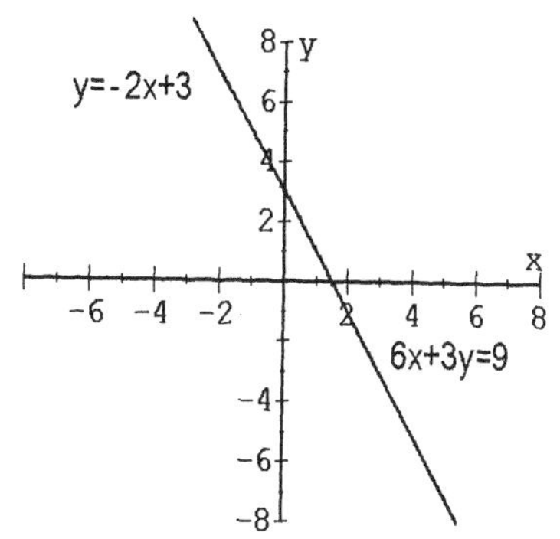

The solution set is $\{(x, y) \mid y = -2x + 3\}$.

Problem Set 5.4

13.

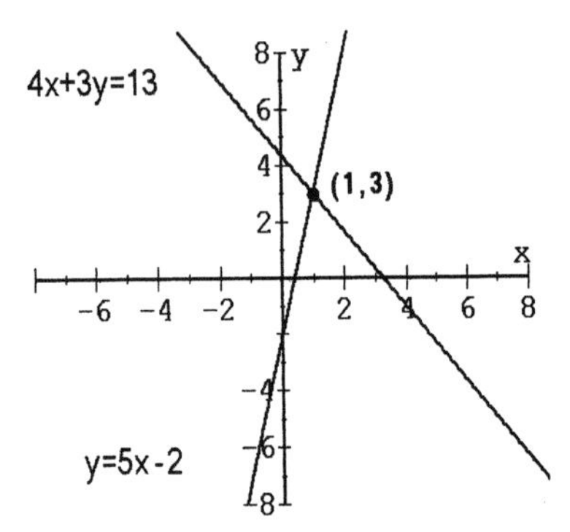

The solution set is $\{(1, 3)\}$.

15.

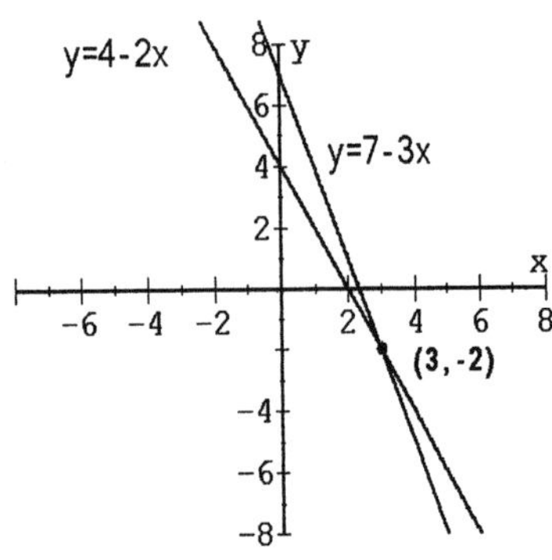

The solution set is $\{(3, -2)\}$.

17.

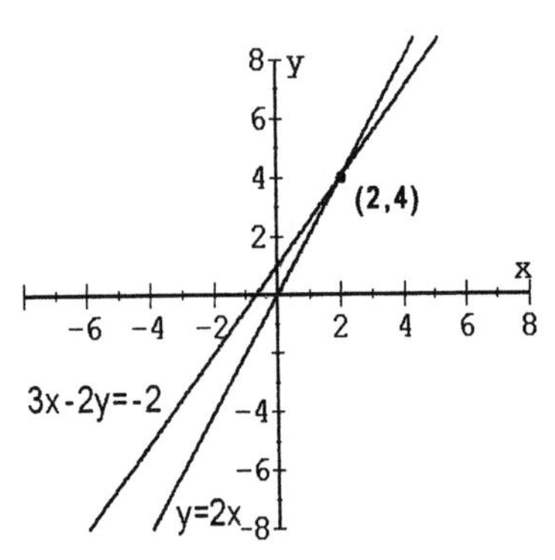

The solution set is $\{(2, 4)\}$.

19.

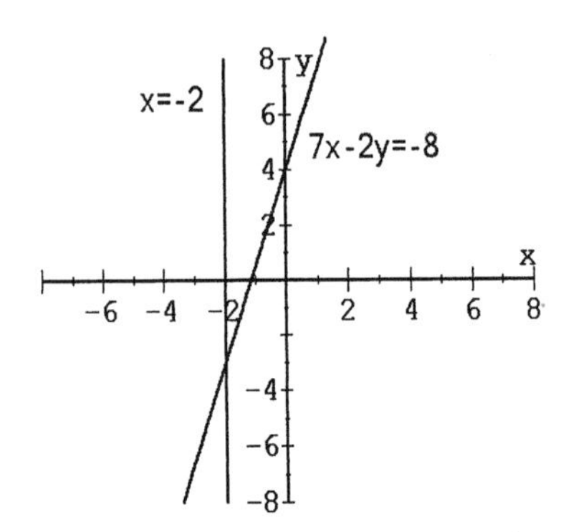

The solution set is $\{(-2, -3)\}$.

21. $\begin{pmatrix} x + y = 20 \\ x = y - 4 \end{pmatrix}$

$x + y = 20$
$(y - 4) + y = 20$
$2y - 4 = 20$
$2y = 24$
$y = 12$
$x = y - 4$
$x = 12 - 4$
$x = 8$
The solution set is $\{(8, 12)\}$.

23. $\begin{pmatrix} y = -3x - 18 \\ 5x - 2y = -8 \end{pmatrix}$

$5x - 2y = -8$
$5x - 2(-3x - 18) = -8$
$5x + 6x + 36 = -8$
$11x + 36 = -8$
$11x = -44$
$x = -4$
$y = -3x - 18$
$y = -3(-4) - 18$
$y = 12 - 18$
$y = -6$
The solution set is $\{(-4, -6)\}$.

25. $\begin{pmatrix} x = -3y \\ 7x - 2y = -69 \end{pmatrix}$

$7x - 2y = -69$
$7(-3y) - 2y = -69$
$-21y - 2y = -69$
$-23y = -69$
$y = 3$
$x = -3y$
$x = -3(3)$
$x = -9$
The solution set is $\{(-9, 3)\}$.

27. $\begin{pmatrix} x + 2y = 5 \\ 3x + 6y = -2 \end{pmatrix}$
Solve equation 1 for x.
$x + 2y = 5$
$x = 5 - 2y$
Substitute into equation 2.
$3x + 6y = -2$
$3(5 - 2y) + 6y = -2$
$15 - 6y + 6y = -2$
$15 = -2$

The false statement, $15 = -2$, implies that the system has no solution. Therefore, the solution set is Ø.

29. $\begin{pmatrix} 3x - 4y = 9 \\ x = 4y - 1 \end{pmatrix}$

$3x - 4y = 9$
$3(4y - 1) - 4y = 9$
$12y - 3 - 4y = 9$
$8y - 3 = 9$
$8y = 12$
$y = \frac{12}{8} = \frac{3}{2}$
$x = 4y - 1$
$x = 4\left(\frac{3}{2}\right) - 1$
$x = 6 - 1$
$x = 5$
The solution set is $\left\{\left(5, \frac{3}{2}\right)\right\}$.

31. $\begin{pmatrix} y = \frac{2}{5}x - 1 \\ 3x + 5y = 4 \end{pmatrix}$

$3x + 5y = 4$
$3x + 5\left(\frac{2}{5}x - 1\right) = 4$
$3x + 2x - 5 = 4$
$5x - 5 = 4$
$5x = 9$
$x = \frac{9}{5}$
$y = \frac{2}{5}x - 1$
$y = \frac{2}{5}\left(\frac{9}{5}\right) - 1$
$y = \frac{18}{25} - 1$
$y = -\frac{7}{25}$
The solution set is $\left\{\left(\frac{9}{5}, -\frac{7}{25}\right)\right\}$.

33. $\begin{pmatrix} 7x - 3y = -2 \\ x = \frac{3}{4}y + 1 \end{pmatrix}$
$7x - 3y = -2$
$7\left(\frac{3}{4}y + 1\right) - 3y = -2$
$\frac{21}{4}y + 7 - 3y = -2$
$\frac{9}{4}y + 7 = -2$
$\frac{9}{4}y = -9$
$\frac{4}{9}\left(\frac{9}{4}y\right) = \frac{4}{9}(-9)$
$y = -4$
$x = \frac{3}{4}y + 1$
$x = \frac{3}{4}(-4) + 1$
$x = -3 + 1 = -2$
The solution set is $\{(-2, -4)\}$.

35. $\begin{pmatrix} 2x+y=12 \\ 3x-y=13 \end{pmatrix}$

Solve equation 1 for y.

$2x+y=12$
$y=-2x+12$

Substitute into equation 2.

$3x-y=13$
$3x-(-2x+12)=13$
$3x+2x-12=13$
$5x-12=13$
$5x=25$
$x=5$
$y=-2x+12$
$y=-2(5)+12$
$y=-10+12$
$y=2$

The solution set is $\{(5, 2)\}$.

37. $\begin{pmatrix} 4x+3y=-40 \\ 5x-y=-12 \end{pmatrix}$

Solve equation 2 for y.

$5x-y=-12$
$-y=-5x-12$
$y=5x+12$

Substitute into equation 1.

$4x+3y=-40$
$4x+3(5x+12)=-40$
$4x+15x+36=-40$
$19x+36=-40$
$19x=-76$
$x=-4$
$y=5x+12$
$y=5(-4)+12$
$y=-20+12$
$y=-8$

The solution set is $\{(-4, -8)\}$.

39. $\begin{pmatrix} 3x+y=2 \\ 11x-3y=5 \end{pmatrix}$

Solve equation 1 for y.

$3x+y=2$
$y=-3x+2$

Substitute into equation 2.

$11x-3y=5$
$11x-3(-3x+2)=5$
$11x+9x-6=5$
$20x-6=5$
$20x=11$
$x=\frac{11}{20}$
$y=-3x+2$
$y=-3\left(\frac{11}{20}\right)+2$
$y=-\frac{33}{20}+2$
$y=\frac{7}{20}$

The solution set is $\left\{\left(\frac{11}{20}, \frac{7}{20}\right)\right\}$.

41. $\begin{pmatrix} 4x-8y=-12 \\ 3x-6y=-9 \end{pmatrix}$

Solve equation 2 for x.

$3x-6y=-9$
$3x=6y-9$
$x=2y-3$

Substitute into equation 1.

$4x-8y=-12$
$4(2y-3)-8y=-12$
$8y-12-8y=-12$
$-12=-12$

The true statement, $-12=-12$, implies that the system has infinitely many solutions. The solution set is expressed as $\{(x, y) | 4x-8y=-12\}$.

43. $\begin{pmatrix} 4x-5y=3 \\ 8x+15y=-24 \end{pmatrix}$

Solve equation 1 for x.

$4x-5y=3$
$4x=5y+3$
$x=\frac{5}{4}y+\frac{3}{4}$

Substitute into equation 2.

$8x+15y=-24$
$8\left(\frac{5}{4}y+\frac{3}{4}\right)+15y=-24$
$10y+6+15y=-24$
$25y+6=-24$
$25y=-30$
$y=-\frac{30}{25}=-\frac{6}{5}$
$x=\frac{5}{4}y+\frac{3}{4}$

$x = \frac{5}{4}\left(-\frac{6}{5}\right) + \frac{3}{4}$

$x = -\frac{6}{4} + \frac{3}{4}$

$x = -\frac{3}{4}$

The solution set is $\left\{\left(-\frac{3}{4}, -\frac{6}{5}\right)\right\}$.

45. $\begin{pmatrix} 6x - 3y = 4 \\ 5x + 2y = -1 \end{pmatrix}$

Solve equation 2 for y.

$5x + 2y = -1$

$2y = -5x - 1$

$y = -\frac{5}{2}x - \frac{1}{2}$

Substitute into equation 1.

$6x - 3y = 4$

$6x - 3\left(-\frac{5}{2}x - \frac{1}{2}\right) = 4$

$6x + \frac{15}{2}x + \frac{3}{2} = 4$

$\frac{27}{2}x + \frac{3}{2} = 4$

$\frac{27}{2}x = 4 - \frac{3}{2}$

$\frac{27}{2}x = \frac{5}{2}$

$\frac{2}{27}\left(\frac{27}{2}x\right) = \frac{2}{27}\left(\frac{5}{2}\right)$

$x = \frac{5}{27}$

$y = -\frac{5}{2}x - \frac{1}{2}$

$y = -\frac{5}{2}\left(\frac{5}{27}\right) - \frac{1}{2}$

$y = -\frac{25}{54} - \frac{1}{2}$

$y = -\frac{26}{27}$

The solution set is $\left\{\left(\frac{5}{27}, -\frac{26}{27}\right)\right\}$.

47. Let $x =$ money at 7% and $y =$ money at 8%.

$\begin{pmatrix} x + 6000 = y \\ 0.07x + 0.08y = 780 \end{pmatrix}$

$0.07x + 0.08y = 780$

$0.07x + 0.08(x + 6000) = 780$

$100[0.07x + 0.08(x + 6000)] = 100(780)$

$7x + 8(x + 6000) = 78000$

$7x + 8x + 48000 = 78000$

$15x = 30000$

$x = 2000$

$y = x + 6000$

$y = 2000 + 6000$

$y = 8000$

She invested \$2000 at 7% and \$8000 at 8%.

49. Let $x =$ one number and $y =$ other number.

$\begin{pmatrix} x + y = 131 \\ x = 3y - 5 \end{pmatrix}$

$x + y = 131$

$(3y - 5) + y = 131$

$4y - 5 = 131$

$4y = 136$

$y = 34$

$x = 3y - 5$

$x = 3(34) - 5$

$x = 102 - 5$

$x = 97$

The numbers are 34 and 97.

51. Let $x =$ number of women and $y =$ number of men.

$\begin{pmatrix} x + y = 50 \\ x = 5y + 2 \end{pmatrix}$

$x + y = 50$

$(5y + 2) + y = 50$

$6y + 2 = 50$

$6y = 48$

$y = 8$

$x = 5y + 2$

$x = 5(8) + 2$

$x = 42$

There are 42 women.

53. Let $x =$ width and $y =$ length.

$\begin{pmatrix} 2x + 2y = 94 \\ y = x + 7 \end{pmatrix}$

$2x + 2y = 94$
$2x + 2(x + 7) = 94$
$2x + 2x + 14 = 94$
$4x = 80$
$x = 20$
$y = x + 7$
$y = 20 + 7$
$y = 27$
The length is 27 inches and the width is 20 inches.

55. Let x = number of \$5 bills and y = number of \$10 bills.
$\begin{pmatrix} x + y = 100 \\ 5x + 10y = 700 \end{pmatrix}$
Solve equation 1 for x.
$x + y = 100$
$x = 100 - \text{y}$
Substitute into equation 2.
$5x + 10y = 700$
$5(100 - y) + 10y = 700$
$500 - 5y + 10y = 700$
$5y = 200$
$y = 40$
$x = 100 - y$
$x = 100 - 40$
$x = 60$
The are 60 five-dollar bills and 40 ten-dollar bills.

57. Let x = number of student tickets and y = number of non-student tickets.
$\begin{pmatrix} x + y = 3000 \\ 8x + 15y = 27500 \end{pmatrix}$
Solve equation 1 for x.
$x + y = 3000$
$x = 3000 - y$
Substitute into equation 2.
$8x + 15y = 27500$
$8(3000 - y) + 15y = 27500$
$24000 - 8y + 15y = 27500$
$7y = 3500$
$y = 500$
$x = 3000 - y = 3000 - 500 = 2500$
$x = 2500$
There were 2500 student tickets sold and 500 non-student tickets sold.

PROBLEM SET 5.5 Elimination-by-Addition Method

1. $\begin{pmatrix} 2x + 3y = -1 \\ 5x - 3y = 29 \end{pmatrix}$

$$\begin{array}{r} 2x + 3y = -1 \\ 5x - 3y = 29 \\ \hline 7x \quad = 28 \end{array}$$
$x = 4$
$2x + 3y = -1$
$2(4) + 3y = -1$
$8 + 3y = -1$
$3y = -9$
$y = -3$
The solution set is $\{(4, -3)\}$.

3. $\begin{pmatrix} 6x - 7y = 15 \\ 6x + 5y = -21 \end{pmatrix}$
Multiply equation 2 by -1, then add to equation 1.
$$\begin{array}{r} 6x - 7y = 15 \\ -6x - 5y = 21 \\ \hline -12y = 36 \end{array}$$
$y = -3$
$6x - 7y = 15$
$6x - 7(-3) = 15$
$6x + 21 = 15$
$6x = -6$
$x = -1$
The solution set is $\{(-1, -3)\}$.

5. $\begin{pmatrix} x - 2y = -12 \\ 2x + 9y = 2 \end{pmatrix}$
Multiply equation 1 by -2 and add to equation 2.
$$\begin{array}{r} -2x + 4y = 24 \\ 2x + 9y = 2 \\ \hline 13y = 26 \end{array}$$
$y = 2$

$x - 2y = -12$
$x - 2(2) = -12$
$x - 4 = -12$
$x = -8$
The solution set is $\{(-8, 2)\}$.

7. $\begin{pmatrix} 4x + 7y = -16 \\ 6x - y = -24 \end{pmatrix}$

Multiply equation 2 by 7 and add to equation 1.

$$\begin{aligned} 4x + 7y &= -16 \\ 42x - 7y &= -168 \\ \hline 46x \quad &= -184 \\ x &= -4 \end{aligned}$$

$4x + 7y = -16$
$4(-4) + 7y = -16$
$-16 + 7y = -16$
$7y = 0$
$y = 0$
The solution set is $\{(-4, 0)\}$.

9. $\begin{pmatrix} 3x - 2y = 5 \\ 2x + 5y = -3 \end{pmatrix}$

Multiply equation 1 by 5 and multiply equation 2 by 2. Then add the equations.

$$\begin{aligned} 15x - 10y &= 25 \\ 4x + 10y &= -6 \\ \hline 19x \quad &= 19 \\ x &= 1 \end{aligned}$$

$3x - 2y = 5$
$3(1) - 2y = 5$
$3 - 2y = 5$
$-2y = 2$
$y = -1$
The solution set is $\{(1, -1)\}$.

11. $\begin{pmatrix} 7x - 2y = 4 \\ 7x - 2y = 9 \end{pmatrix}$

Multiply equation 1 by -1, then add to equation 2.

$$\begin{aligned} -7x + 2y &= -4 \\ 7x - 2y &= 9 \\ \hline 0 &= 5 \end{aligned}$$

Since $0 \neq 5$, the system has no solution. The solution set is $\emptyset$.

13. $\begin{pmatrix} 5x + 4y = 1 \\ 3x - 2y = -1 \end{pmatrix}$

Multiply equation 2 by 2, then add to equation 1.

$$\begin{aligned} 5x + 4y &= 1 \\ 6x - 4y &= -2 \\ \hline 11x &= -1 \\ x &= -\frac{1}{11} \end{aligned}$$

$5x + 4y = 1$
$5\left(-\frac{1}{11}\right) + 4y = 1$
$-\frac{5}{11} + 4y = 1$
$4y = 1 + \frac{5}{11}$
$4y = \frac{16}{11}$
$\frac{1}{4}(4y) = \frac{1}{4}\left(\frac{16}{11}\right)$
$y = \frac{4}{11}$
The solution set is $\left\{\left(-\frac{1}{11}, \frac{4}{11}\right)\right\}$.

15. $\begin{pmatrix} 8x - 3y = 13 \\ 4x + 9y = 3 \end{pmatrix}$

Multiply equation 1 by 3 then add to equation 2.

$$\begin{aligned} 24x - 9y &= 39 \\ 4x + 9y &= 3 \\ \hline 28x \quad &= 42 \\ x &= \frac{42}{28} = \frac{3}{2} \end{aligned}$$

$8x - 3y = 13$
$8\left(\frac{3}{2}\right) - 3y = 13$
$12 - 3y = 13$
$-3y = 1$
$y = -\frac{1}{3}$
The solution set is $\left\{\left(\frac{3}{2}, -\frac{1}{3}\right)\right\}$.

Problem Set 5.5

17. $\begin{pmatrix} 5x+3y=-7 \\ 7x-3y=55 \end{pmatrix}$

$$\begin{array}{rl} 5x+3y= & -7 \\ 7x-3y= & 55 \\ \hline 12x \quad = & 48 \\ x = & 4 \end{array}$$

$5x+3y=-7$
$5(4)+3y=-7$
$20+3y=-7$
$3y=-27$
$y=-9$
The solution set is $\{(4,-9)\}$.

19. $\begin{pmatrix} x=5y+7 \\ 4x+9y=28 \end{pmatrix}$

$\begin{pmatrix} x-5y=7 \\ 4x+9y=28 \end{pmatrix}$

Multiply equation 1 by -4, then add to equation 2.

$$\begin{array}{rl} -4x+20y= & -28 \\ 4x+9y= & 28 \\ \hline 29y= & 0 \\ y= & 0 \end{array}$$

$x=5y+7$
$x=5(0)+7$
$x=7$
The solution set is $\{(7,0)\}$.

21. $\begin{pmatrix} x=-6y+79 \\ x=4y-41 \end{pmatrix}$

$\begin{pmatrix} x+6y=79 \\ x-4y=-41 \end{pmatrix}$

Multiiply equation 1 by -1, then add to equation 2.

$$\begin{array}{rl} -x-6y= & -79 \\ x-4y= & -41 \\ \hline -10y= & -120 \\ y= & 12 \end{array}$$

$x=-6y+79$
$x=-6(12)+79$
$x=-72+79$
$x=7$
The solution set is $\{(7,12)\}$.

23. $\begin{pmatrix} 4x-3y=2 \\ 5x-y=3 \end{pmatrix}$

Multiply equation 2 by -3, then add to equation 1.

$$\begin{array}{rl} 4x-3y= & 2 \\ -15x+3y= & -9 \\ \hline -11x \quad = & -7 \\ x= & \frac{7}{11} \end{array}$$

$4x-3y=2$
$4\left(\frac{7}{11}\right)-3y=2$
$\frac{28}{11}-3y=2$
$-3y=2-\frac{28}{11}$
$-\frac{1}{3}(-3y)=-\frac{1}{3}\left(-\frac{6}{11}\right)$
$y=\frac{2}{11}$
The solution set is $\left\{\left(\frac{7}{11},\frac{2}{11}\right)\right\}$.

25. $\begin{pmatrix} 5x-2y=1 \\ 10x-4y=7 \end{pmatrix}$

Multiply equation 1 by -2, then add to equation 2.

$$\begin{array}{rl} -10x+4y= & -2 \\ 10x-4y= & 7 \\ \hline 0= & 5 \end{array}$$

Since $0\neq 5$, the system has no solution. The solution set is $\emptyset$.

27. $\begin{pmatrix} 3x-2y=7 \\ 5x+7y=1 \end{pmatrix}$

Multiply equation 1 by 7 and multiply equation 2 by 2. Then add the equations.

$$\begin{array}{rl} 21x-14y= & 49 \\ 10x+14y= & 2 \\ \hline 31x \quad = & 51 \\ x= & \frac{51}{31} \end{array}$$

$3x - 2y = 7$

$3\left(\frac{51}{31}\right) - 2y = 7$

$\frac{153}{31} - 2y = 7$

$-2y = 7 - \frac{153}{31}$

$-2y = \frac{64}{31}$

$-\frac{1}{2}\left(-2y\right) = -\frac{1}{2}\left(\frac{64}{31}\right)$

$y = -\frac{32}{31}$

The solution set is $\left\{\left(\frac{51}{31}, -\frac{32}{31}\right)\right\}$.

29. $\begin{pmatrix} -2x + 5y = -16 \\ x = \frac{3}{4}y + 1 \end{pmatrix}$

Multiply equation 2 by 4.

$\begin{pmatrix} -2x + 5y = -16 \\ 4x = 3y + 4 \end{pmatrix}$

$\begin{pmatrix} -2x + 5y = -16 \\ 4x - 3y = 4 \end{pmatrix}$

Multiply equation 1 by 2, then add the equations.

$$\begin{aligned} -4x + 10y &= -32 \\ 4x - 3y &= 4 \\ \hline 7y &= -28 \\ y &= -4 \end{aligned}$$

$x = \frac{3}{4}y + 1$

$x = \frac{3}{4}\left(-4\right) + 1$

$x = -3 + 1$

$x = -2$

The solution set is $\{(-2, -4)\}$.

31. $\begin{pmatrix} y = \frac{2}{3}x - 4 \\ 5x - 3y = 9 \end{pmatrix}$

Multiply equation 1 by 12.

$\begin{pmatrix} 12y = 8x - 9 \\ 2x + 3y = 11 \end{pmatrix}$

$\begin{pmatrix} -8x + 12y = -9 \\ 2x + 3y = 11 \end{pmatrix}$

Multiply equation 2 by 4, then add the equations.

$$\begin{aligned} -8x + 12y &= -9 \\ 8x + 12y &= 44 \\ \hline 24y &= 35 \\ y &= \frac{35}{24} \end{aligned}$$

$2x + 3y = 11$

$2x + 3\left(\frac{35}{24}\right) = 11$

$8\left(2x + \frac{35}{8}\right) = 8(11)$

$16x + 35 = 88$

$16x = 53$

$x = \frac{53}{16}$

The solution set is $\left\{\left(\frac{53}{16}, \frac{35}{24}\right)\right\}$.

33. $\begin{pmatrix} \frac{x}{6} + \frac{y}{3} = 3 \\ \frac{5x}{2} - \frac{y}{6} = -17 \end{pmatrix}$

Multiply both equations by 6.

$\begin{pmatrix} x + 2y = 18 \\ 15x - y = -102 \end{pmatrix}$

Multiply equation 2 by 2, then add to equation 1.

$$\begin{aligned} x + 2y &= 18 \\ 30x - 2y &= -204 \\ \hline 31x &= -186 \\ x &= -6 \end{aligned}$$

$x + 2y = 18$

$-6 + 2y = 18$

$2y = 24$

$y = 12$

The solution set is $\{(-6, 12)\}$.

35. $\begin{pmatrix} -(x - 6) + 6(y + 1) = 58 \\ 3(x + 1) - 4(y - 2) = -15 \end{pmatrix}$

$\begin{pmatrix} -x + 6 + 6y + 6 = 58 \\ 3x + 3 - 4y + 8 = -15 \end{pmatrix}$

$\begin{pmatrix} -x + 6y = 46 \\ 3x - 4y = -26 \end{pmatrix}$

Multiply equation 1 by 3,
then add to equation 2.

$$\begin{array}{rl} -3x + 18y = & 138 \\ 3x - \ \ 4y = & -26 \\ \hline 14y = & 112 \end{array}$$

$y = 8$

$-x + 6y = 46$

$-x + 6(8) = 46$

$-x + 48 = 46$

$-x = -2$

$x = 2$

The solution set is $\{(2, 8)\}$.

37. $\begin{pmatrix} 5(x+1) - (y+3) = -6 \\ 2(x-2) + 3(y-1) = 0 \end{pmatrix}$

$\begin{pmatrix} 5x + 5 - y - 3 = -6 \\ 2x - 4 + 3y - 3 = 0 \end{pmatrix}$

$\begin{pmatrix} 5x - y = -8 \\ 2x + 3y = 7 \end{pmatrix}$

Multiply equation 1 by 3,
then add to equation 2.

$$\begin{array}{rl} 15x - 3y = & -24 \\ 2x + 3y = & 7 \\ \hline 17x \quad\quad = & -17 \end{array}$$

$x = -1$

$5x - y = -8$

$5(-1) - y = -8$

$-5 - y = -8$

$-y = -3$

$y = 3$

The solution set is $\{(-1, 3)\}$.

39. $\begin{pmatrix} \frac{1}{2}x - \frac{1}{3}y = 12 \\ \frac{3}{4}x + \frac{2}{3}y = 4 \end{pmatrix}$

Multiply equation 1 by 6 and
multiply equation 2 by 12.

$\begin{pmatrix} 3x - 2y = 72 \\ 9x + 8y = 48 \end{pmatrix}$

Multiply equation 1 by 4,
then add to equation 2.

$$\begin{array}{rl} 12x - 8y = & 288 \\ 9x + 8y = & 48 \\ \hline 21x \quad\quad = & 336 \end{array}$$

$x = 16$

$3x - 2y = 72$

$3(16) - 2y = 72$

$48 - 2y = 72$

$-2y = 24$

$y = -12$

The solution set is $\{(16, -12)\}$.

41. $\begin{pmatrix} \frac{2x}{3} - \frac{y}{2} = -\frac{5}{4} \\ \frac{x}{4} + \frac{5y}{6} = \frac{17}{16} \end{pmatrix}$

Multiply equation 1 by 12 and
multiply equation 2 by 48.

$\begin{pmatrix} 8x - 6y = -15 \\ 12x + 40y = 51 \end{pmatrix}$

Multiply equation 1 by -3 and
multiply equation 2 by 2.
Then add the equations.

$$\begin{array}{rl} -24x + 18y = & 45 \\ 24x + 80y = & 102 \\ \hline 98y = & 147 \end{array}$$

$y = \frac{147}{98} = \frac{3}{2}$

$8x - 6y = -15$

$8x - 6\left(\frac{3}{2}\right) = -15$

$8x - 9 = -15$

$8x = -6$

$x = -\frac{6}{8} = -\frac{3}{4}$

The solution set is $\left\{\left(-\frac{3}{4}, \frac{3}{2}\right)\right\}$.

43. $\begin{pmatrix} \frac{3x+y}{2} + \frac{x-2y}{5} = 8 \\ \frac{x-y}{3} - \frac{x+y}{6} = \frac{10}{3} \end{pmatrix}$

Multiply equation 1 by 10 and
multiply equation 2 by 6.

$\begin{pmatrix} 5(3x+y) + 2(x-2y) = 80 \\ 2(x-y) - (x+y) = 20 \end{pmatrix}$

$$\begin{pmatrix} 15x + 5y + 2x - 4y = 80 \\ 2x - 2y - x - y = 20 \end{pmatrix}$$

$$\begin{pmatrix} 17x + y = 80 \\ x - 3y = 20 \end{pmatrix}$$

Multiply equation 1 by 3, then add the equations

$$\begin{array}{r} 51x + 3y = 240 \\ x - 3y = 20 \\ \hline 52x \quad = 260 \end{array}$$

$x = 5$

Substitute into equation 1.

$17x + y = 80$

$17(5) + y = 80$

$85 + y = 80$

$y = -5$

The solution set is $\{(5, -5)\}$.

45. Let $x =$ gallons of 10% solution and $y =$ gallons of 20% solution.

$$\begin{pmatrix} x + y = 20 \\ 0.10x + 0.20y = 0.175(20) \end{pmatrix}$$

$$\begin{pmatrix} x + y = 20 \\ 0.10x + 0.20y = 3.5 \end{pmatrix}$$

Multiply equation 2 by -10, then add the equations.

$$\begin{array}{r} x + y = 20 \\ -x - 2y = -35 \\ \hline -y = -15 \end{array}$$

$y - 15$

$x = 20 - 15$

$x = 5$

There should be 5 gallons of 10% solution and 15 gallons of 20% solution.

47. Let $x =$ cost of tennis ball and $y =$ cost of golf ball.

$$\begin{pmatrix} 3x + 2y = 12 \\ 6x + 3y = 27 \end{pmatrix}$$

Multiply equation 1 by -2, then add to equation 2.

$$\begin{array}{r} -6x - 4y = -24 \\ 6x + 3y = 21 \\ \hline -y = -3 \end{array}$$

$y = 3$

$3x + 2y = 12$

$3x + 2(3) = 12$

$3x + 6 = 12$

$3x = 6$

$x = 2$

Tennis balls cost \$2 and golf balls cost \$3.

49. Let $x =$ the number of single rooms and $y =$ the number of double rooms.

$$\begin{pmatrix} x + y = 55 \\ 62x + 82y = 4210 \end{pmatrix}$$

Multiply equation 1 by -62, then add the equations.

$$\begin{array}{r} -62x - 62y = -3410 \\ 62x + 82y = 4210 \\ \hline 20y = 800 \end{array}$$

$y = 40$

$x + y = 55$

$x + 40 = 55$

$x = 15$

There were 15 single rooms and 40 double rooms.

51. Let $x =$ numerator and $y =$ denominator.

$$\begin{pmatrix} \dfrac{x+5}{y-1} = \dfrac{8}{3} \\ \dfrac{2x}{y+7} = \dfrac{6}{11} \end{pmatrix}$$

$$\begin{pmatrix} 3(x + 5) = 8(y - 1) \\ 11(2x) = 6(y + 7) \end{pmatrix}$$

$$\begin{pmatrix} 3x + 15 = 8y - 8 \\ 22x = 6y + 42 \end{pmatrix}$$

$$\begin{pmatrix} 3x - 8y = -23 \\ 22x - 6y = 42 \end{pmatrix}$$

Multiply equation 1 by 3 and equation 2 by -4. Then add the equations.

$$\begin{array}{r} 9x - 24y = -69 \\ -88x + 24y = -168 \\ \hline -79x \quad = -237 \end{array}$$

$x = 3$

$3x - 8y = -23$
$3(3) - 8y = -23$
$9 - 8y = -23$
$-8y = -32$
$y = 4$
The fraction is $\frac{3}{4}$.

53. Let x = width and
y = length.
$\begin{pmatrix} (x+2)(y-1) = xy + 28 \\ (x-1)(y+2) = xy + 10 \end{pmatrix}$

$\begin{pmatrix} xy - x + 2y - 2 = xy + 28 \\ xy + 2x - y - 2 = xy + 10 \end{pmatrix}$

$\begin{pmatrix} -x + 2y = 30 \\ 2x - y = 12 \end{pmatrix}$
Multiply equation 1 by 2,
then add to equation 2.

$$\begin{aligned} -2x + 4y &= 60 \\ 2x - y &= 12 \\ \hline 3y &= 72 \\ y &= 24 \end{aligned}$$

$-x + 2y = 30$
$-x + 2(24) = 30$
$-x + 48 = 30$
$-x = -18$
$x = 18$
The cover is 18 centimeters by 24 centimeters.

55. Let x = distance from 60 pound
weight to the fulcrum and
let y = distance from 100 pound
weight to the fulcrum.
$\begin{pmatrix} 60x = 100y \\ 60(x-1) = 80y \end{pmatrix}$
$\begin{pmatrix} 60x - 100y = 0 \\ 60x - 60 = 80y \end{pmatrix}$
$\begin{pmatrix} 60x - 100y = 0 \\ 60x - 80y = 60 \end{pmatrix}$
Multiply equation 2 by -1, then
add the equations.

$$\begin{aligned} 60x - 100y &= 0 \\ -60x + 80y &= -60 \\ \hline -20y &= -60 \\ y &= 3 \end{aligned}$$

$60x = 100y$
$60x = 100(3)$
$60x = 300$
$x = 5$
The distance between the
weights is 8 feet.

Further Investigations

59a) $\begin{pmatrix} 4x - 3y = 7 \\ 9x + 2y = 5 \end{pmatrix}$
$\frac{4}{9} \neq \frac{-3}{2}$
Consistent

b) $\begin{pmatrix} 5x - y = 6 \\ 10x - 2y = 19 \end{pmatrix}$
$\frac{5}{10} = \frac{-1}{-2} \neq \frac{6}{19}$
Inconsistent

c) $\begin{pmatrix} 5x - 4y = 11 \\ 4x + 5y = 12 \end{pmatrix}$
$\frac{5}{4} \neq -\frac{4}{5}$
Consistent

d) $\begin{pmatrix} x + 2y = 5 \\ x - 2y = 9 \end{pmatrix}$
$\frac{1}{1} \neq \frac{2}{-2}$
Consistent

e) $\begin{pmatrix} x - 3y = 5 \\ 3x - 9y = 15 \end{pmatrix}$
$\frac{1}{3} = \frac{-3}{-9} = \frac{5}{15}$
Dependent

f) $\begin{pmatrix} 4x + 3y = 7 \\ 2x - y = 10 \end{pmatrix}$
$\frac{4}{2} \neq \frac{3}{-1}$
Consistent

g) $\begin{pmatrix} 3x + 2y = 4 \\ y = -\frac{3}{2}x - 1 \end{pmatrix}$

$\begin{pmatrix} 3x + 2y = 4 \\ 3x + 2y = -2 \end{pmatrix}$

$\frac{3}{3} = \frac{2}{2} \neq \frac{4}{-2}$

Inconsistent

h) $\begin{pmatrix} y = \frac{4}{3}x - 2 \\ 4x - 3y = 6 \end{pmatrix}$

$\begin{pmatrix} -4x + 3y = -6 \\ 4x - 3y = 6 \end{pmatrix}$

$\frac{-4}{4} = \frac{3}{-3} = \frac{-6}{6}$

Dependent

61. $\begin{pmatrix} a_1x + b_1y = c_1 \\ a_2x + b_2y = c_2 \end{pmatrix}$

Multiply equation 1 by a_2 and multiply equation 2 by $-a_1$. Then add the equations.

$$\begin{aligned} a_1a_2x + a_2b_1y &= a_2c_1 \\ -a_1a_2x - a_1b_2y &= -a_1c_2 \\ \hline a_2b_1y - a_1b_2y &= a_2c_1 - a_1c_2 \\ y(a_2b_1 - a_1b_2) &= a_2c_1 - a_1c_2 \\ y &= \frac{a_2c_1 - a_1c_2}{a_2b_1 - a_1b_2} \end{aligned}$$

To solve for x:
Multiply equation 1 by b_2 and multiply equation 2 by $-b_1$. Then add the equations.

$$\begin{aligned} a_1b_2x + b_1b_2y &= b_2c_1 \\ -a_2b_1x - b_1b_2y &= -b_1c_2 \\ \hline a_1b_2x - a_2b_1x &= b_2c_1 - b_1c_2 \\ x(a_1b_2 - a_2b_1) &= b_2c_1 - b_1c_2 \\ x &= \frac{b_2c_1 - b_1c_2}{a_1b_2 - a_2b_1} \end{aligned}$$

PROBLEM SET 5.6 Graphing Linear Inequalities

1. Use the points $(0, 1)$ and $(1, 0)$ to graph a dashed line for $x + y = 1$. Use $(0, 0)$ as a test point. The given inequality $x + y > 1$ becomes $0 + 0 > 1$ which is a false statement. Therefore, the solution set is the half-plane that does not contain the origin.

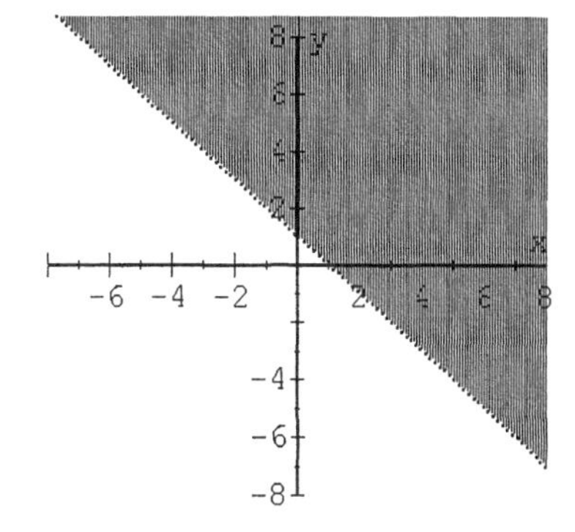

3. Use the points $(0, 3)$ and $(2, 0)$ to graph a dashed line for $3x + 2y = 6$. Use $(0, 0)$ as a test point. The given inequality $3x + 2y < 6$ becomes $0 + 0 < 6$ which is a true statement. Therefore, the solution set is the half-plane that does contain the origin.

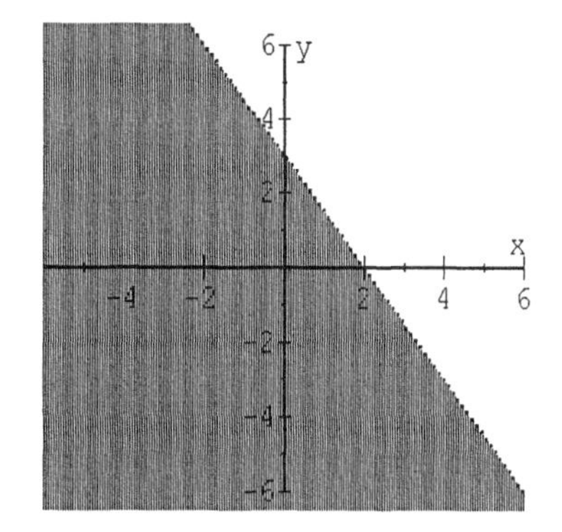

5. Use the points $(0,\ -4)$ and $(2,\ 0)$ to graph a solid line for $2x - y = 4$. Use $(0,\ 0)$ as a test point. The given inequality $2x - y \geq 4$ becomes $0 - 0 \geq 4$ which is a false statement. Therefore, the solution set is the half-plane that does not contain the origin.

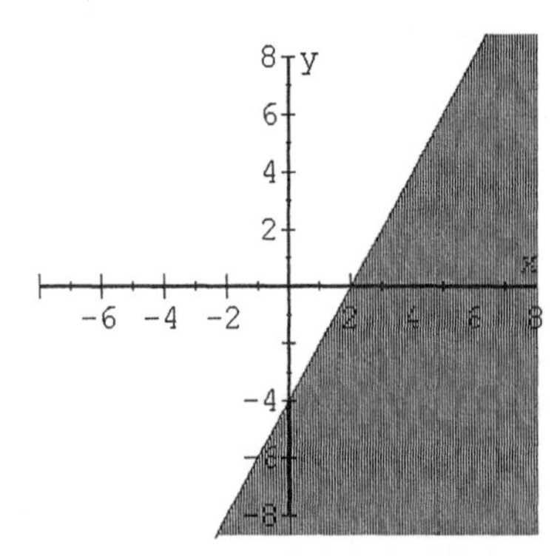

7. Use the points $(0,\ -4)$ and $(3,\ 0)$ to graph a solid line for $4x - 3y = 12$. Use $(0,\ 0)$ as a test point. The given inequality $4x - 3y \leq 12$ becomes $0 - 0 \leq 12$ which is a true statement. Therefore, the solution set is the half-plane that does contain the origin.

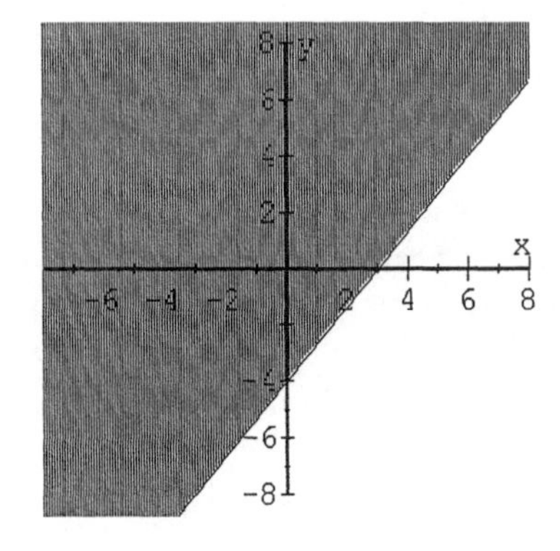

9. Use the points $(0,\ 0)$ and $(1,\ -1)$ to graph a dashed line for $y = -x$. Since the origin is on the line use $(2,\ 1)$ as a test point. The given inequality $y > -x$ becomes $1 > -2$ which is a true statement. Therefore, the solution set is the half-plane that contains the point $(2,\ 1)$.

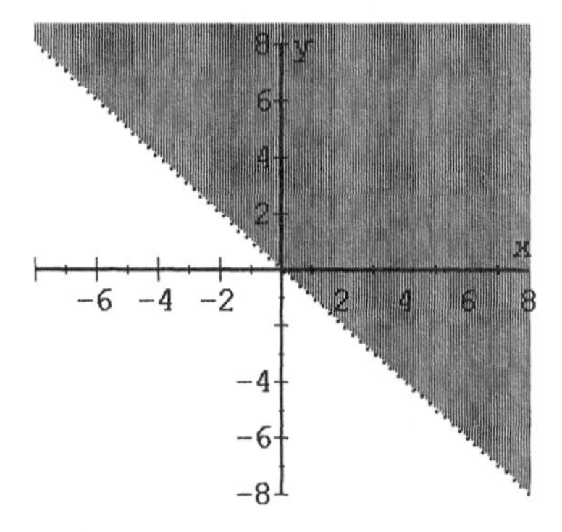

11. Use the points $(0,\ 0)$ and $(1,\ 2)$ to graph a solid line for $2x - y = 0$. Since the origin is on the line use $(2,\ -1)$ as a test point. The given inequality $2x - y \geq 0$ becomes $4 + 1 \geq 0$ which is a true statement. Therefore, the solution set is the half-plane that contains the point $(2,\ -1)$.

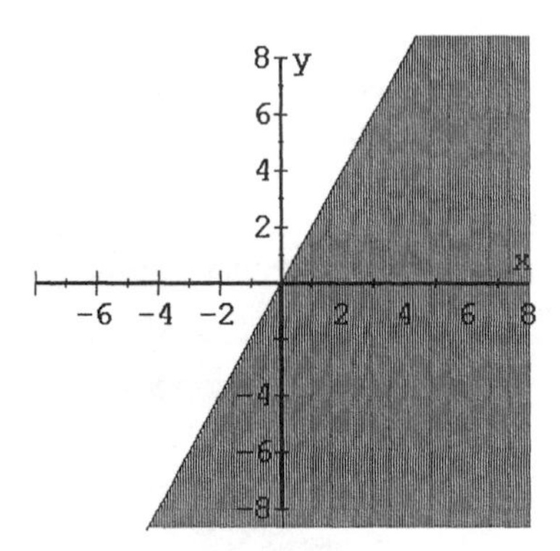

13. Use the points $(0,\ -1)$ and $(2,\ 0)$ to graph a dashed line for $-x+2y=-2$.
Use $(0,\ 0)$ as a test point. The given inequality $-x+2y<-2$ becomes $0+0<-2$ which is a false statement. Therefore, the solution set is the half-plane that does not contain the origin.

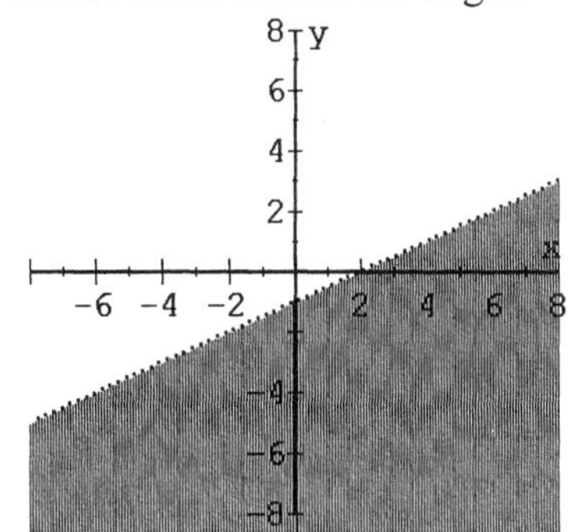

15. Use the points $(0,\ -2)$ and $(4,\ 0)$ to graph a solid line for $y=\frac{1}{2}x-2$. Use $(0,\ 0)$ as a test point. The given inequality $y\leq\frac{1}{2}x-2$ becomes $0\leq 0-2$ which is a false statement. Therefore, the solution set is the half-plane that does not contain the origin.

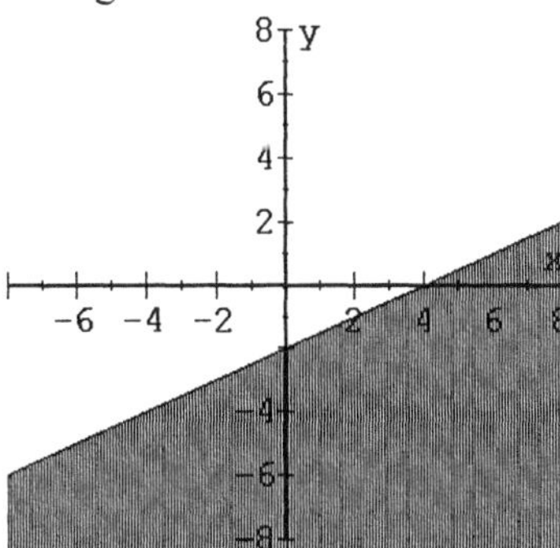

17. Use the points $(0,\ 4)$ and $(4,\ 0)$ to graph a solid line for $y=-x+4$. Use $(0,\ 0)$ as a test point. The given inequality $y\geq -x+4$ becomes $0\geq 0+4$ which is a false statement. Therefore, the solution set is the half-plane that does not contain the origin.

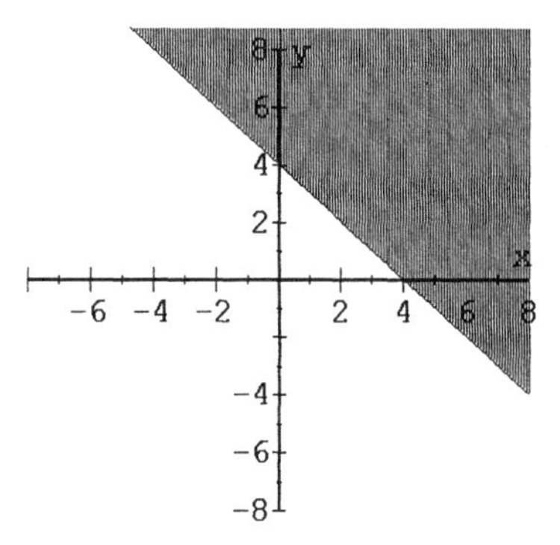

19. Use the points $(0,\ -3)$ and $(-4,\ 0)$ to graph a dashed line for $3x+4y=-12$. Use $(0,\ 0)$ as a test point. The given inequality $3x+4y>-12$ becomes $0+0>-12$ which is a true statement. Therefore, the solution set is the half-plane that does contain the origin.

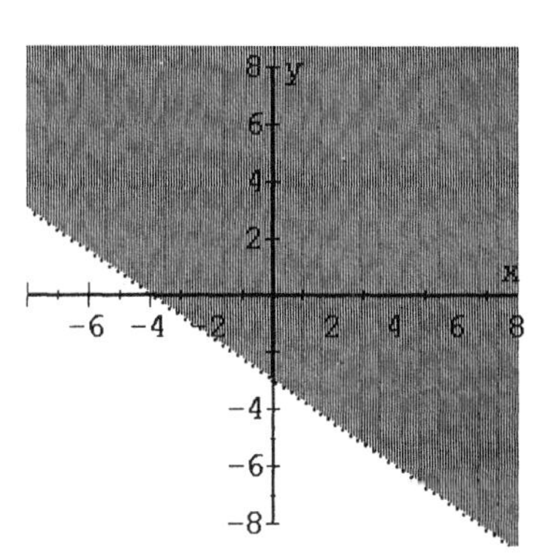

21. $\begin{pmatrix} 2x+3y>6 \\ x-y<2 \end{pmatrix}$

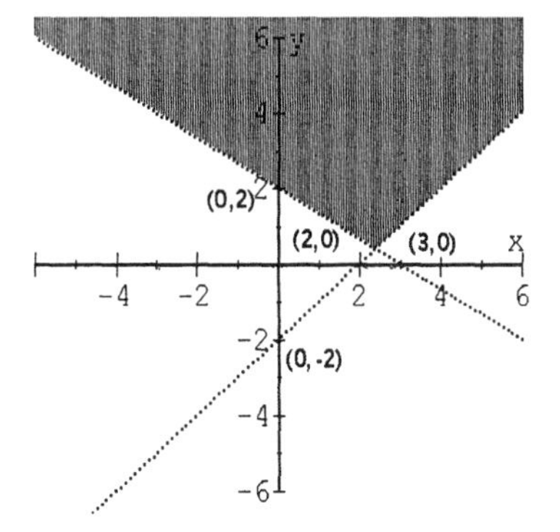

Problem Set 5.6

23. $\begin{pmatrix} x - 3y \geq 3 \\ 3x + y \leq 3 \end{pmatrix}$

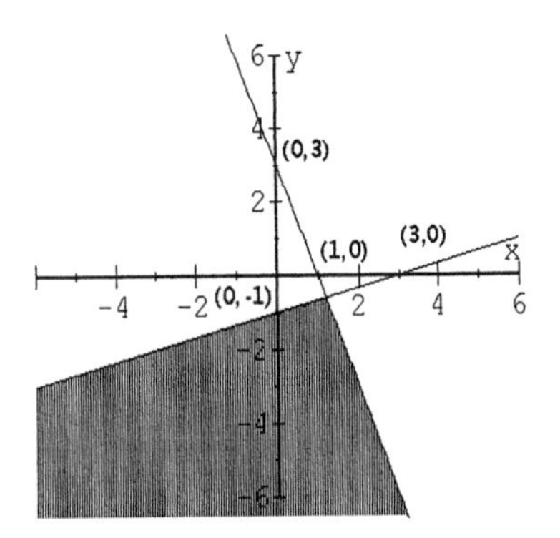

29. $\begin{pmatrix} y < \frac{1}{2}x + 2 \\ y < \frac{1}{2}x - 1 \end{pmatrix}$

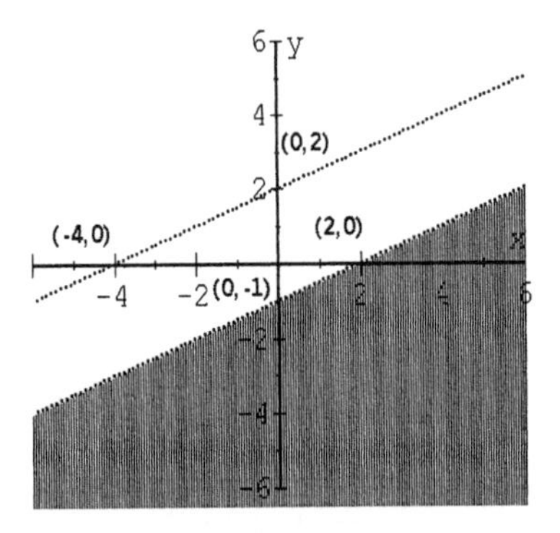

25. $\begin{pmatrix} y \geq 2x \\ y < x \end{pmatrix}$

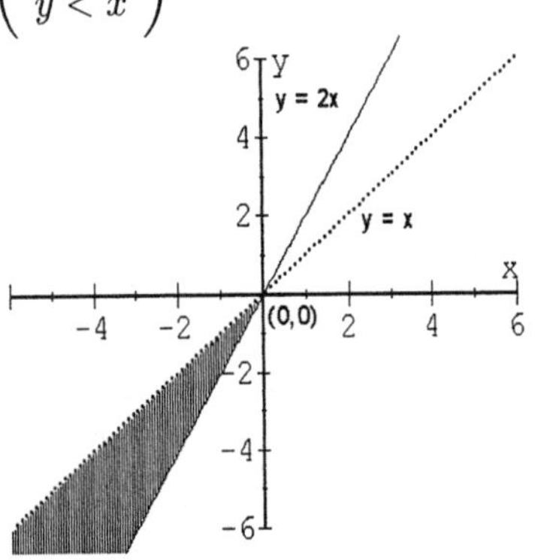

33. $\begin{pmatrix} y > x + 1 \\ y < x - 1 \end{pmatrix}$

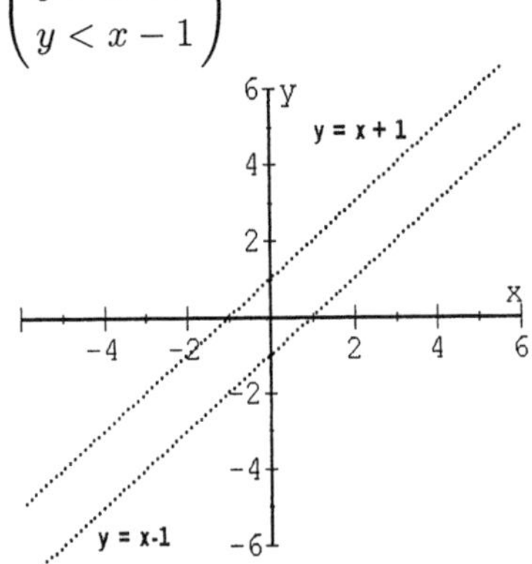

27. $\begin{pmatrix} y < -x + 1 \\ y > -x - 1 \end{pmatrix}$

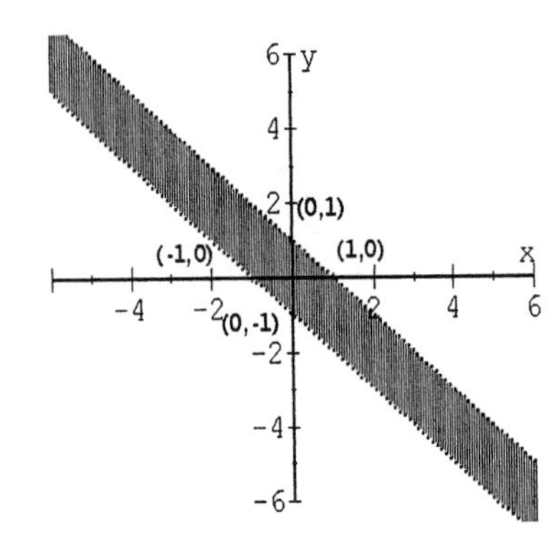

35. $\begin{pmatrix} x \geq 0 \\ y \geq 0 \\ 2x + y \leq 4 \\ 2x - 3y \leq 6 \end{pmatrix}$

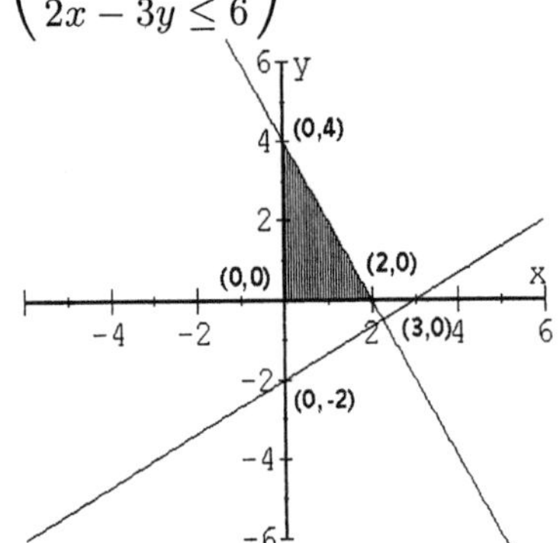

CHAPTER 5 | Review Problem Set

1. $2x - 5y = 10$

$x = 0$	$y = 0$
$2(0) - 5y = 10$	$2x - 5(0) = 10$
$-5y = 10$	$2x = 10$
$y = -2$	$x = 5$
$(0, -2)$	$(5, 0)$

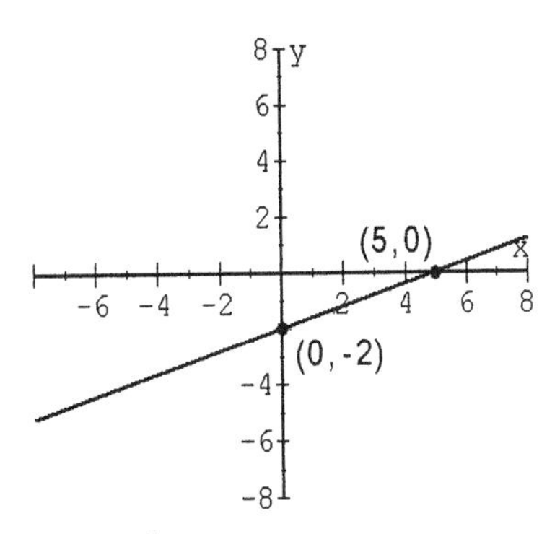

2. $y = -\frac{1}{3}x + 1$

$m = -\frac{1}{3} \quad b = 1$

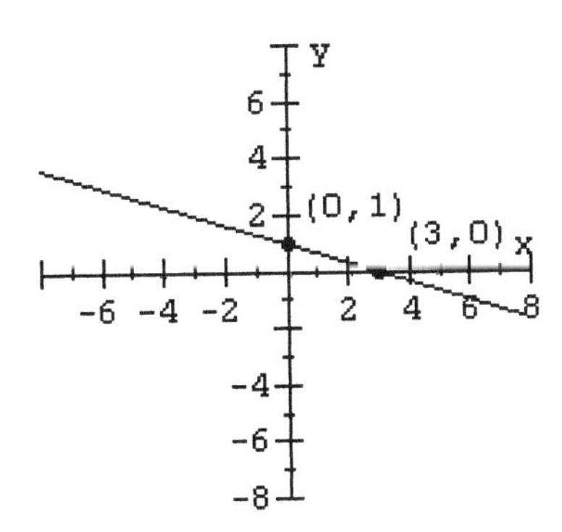

3. $y = 2x^2 + 1$

$x = 1$	$x = 2$	$x = 0$
$y = 2(1)^2 + 1$	$y = 2(2)^2 + 1$	$y = 2(0)^2+1$
$y = 2(1) + 1$	$y = 2(4) + 1$	$y = 2(0)+1$
$y = 2 + 1$	$y = 8 + 1$	$y = 0+1$
$y = 3$	$y = 9$	$y = 1$
$(1, 3)$	$(2, 9)$	$(0, 1)$

$x = -1$	$x = -2$
$y = 2(-1)^2 + 1$	$y = 2(-2)^2 + 1$
$y = 2(1) + 1$	$y = 2(4) + 1$
$y = 2 + 1$	$y = 8 + 1$
$y = 3$	$y = 9$
$(-1, 3)$	$(-2, 9)$

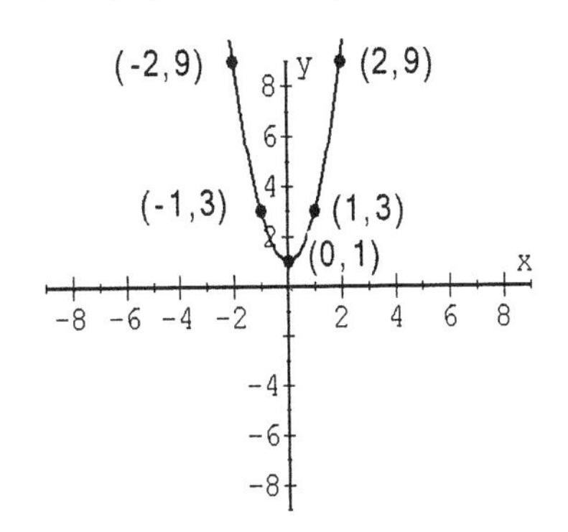

4. $y = -x^2 - 2$

$x = 0$	$x = 1$	$x = -1$
$y = -(0)^2 - 2$	$y = -(1)^2 - 2$	$y = -(-1)^2 - 2$
$y = 0 - 2$	$y = -1 - 2$	$y = -1 - 2$
$y = -2$	$y = -3$	$y = -3$
$(0, -2)$	$(1, -3)$	$(-1, -3)$

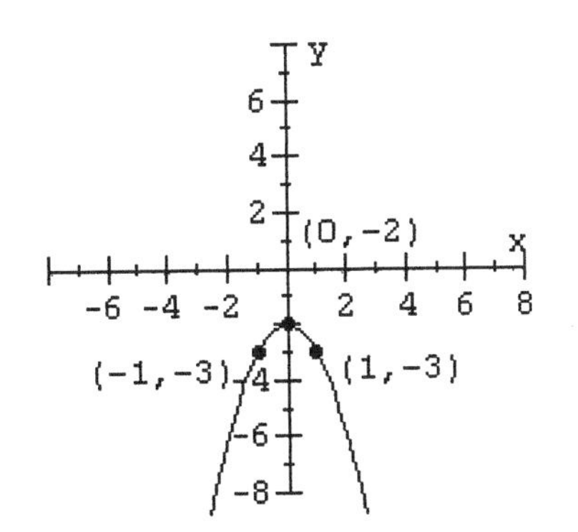

5. $2x - 3y = 0$

$x = 0$	$x = 3$
$2(0) - 3y = 0$	$2(3) - 3y = 0$
$-3y = \dfrac{0}{-3}$	$-3y = -6$
$y = 0$	$y = \dfrac{-6}{-3} = 2$
$(0, 0)$	$(3, 2)$

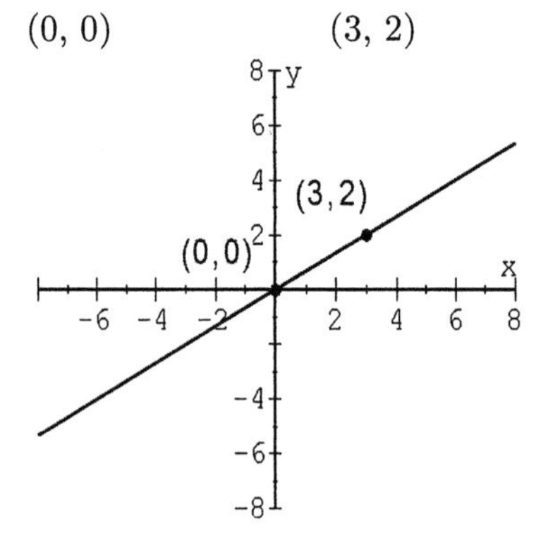

6. $y = -x^3$

$x = 0$	$x = 1$	$x = -1$
$y = -(0)^3$	$y = -(1)^3$	$y = -(-1)^3$
$y = 0$	$y = -1$	$y = 1$
$(0, 0)$	$(1, -1)$	$(-1, 1)$

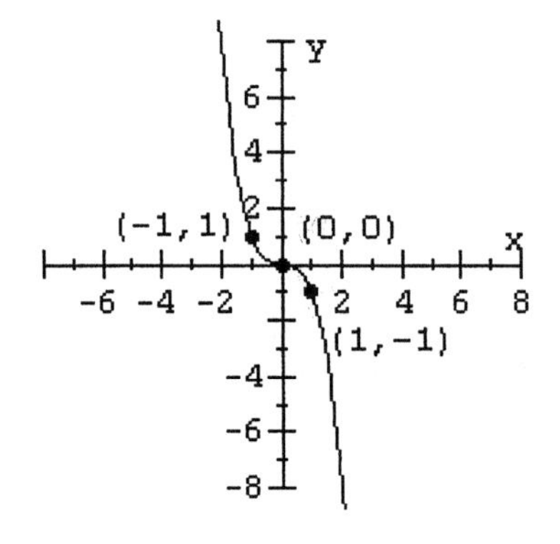

7. $4x + y = -4$

$x = 0$	$y = 0$
$4(0) + y = -4$	$4x + (0) = -4$
$y = -4$	$4x = -4$
	$x = -1$

The x-intercept is -1 and the y-intercept is -4.

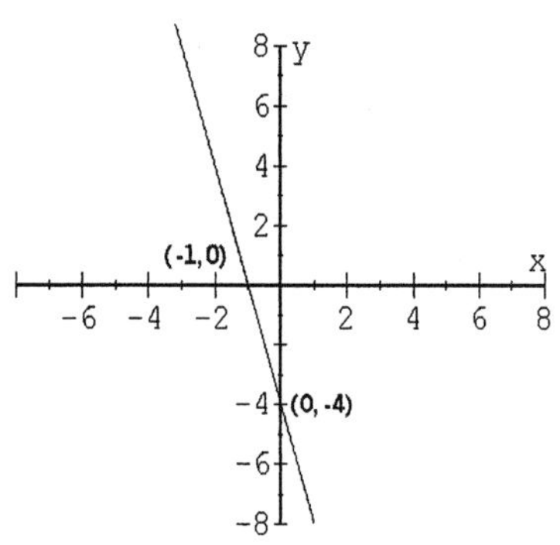

8. $x - 2y = 2$

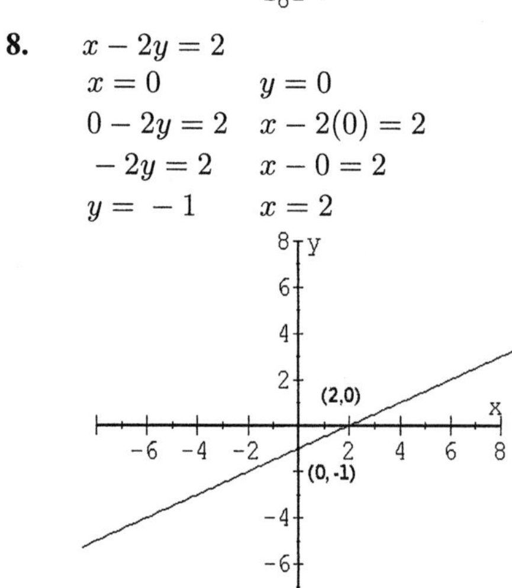

$x = 0$	$y = 0$
$0 - 2y = 2$	$x - 2(0) = 2$
$-2y = 2$	$x - 0 = 2$
$y = -1$	$x = 2$

9. $y = 3x + 6$

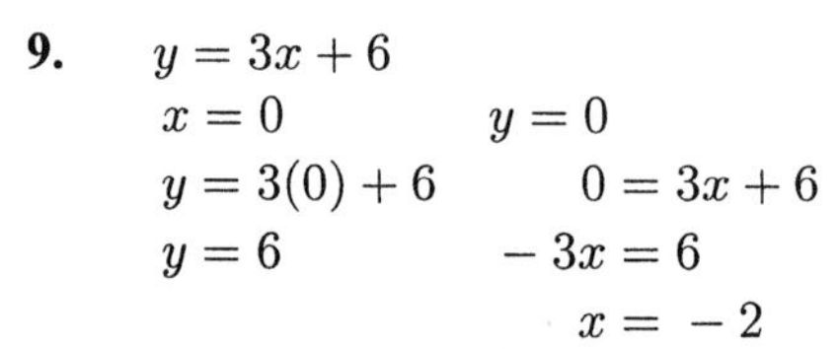

$x = 0$	$y = 0$
$y = 3(0) + 6$	$0 = 3x + 6$
$y = 6$	$-3x = 6$
	$x = -2$

The x-intercept is -2 and the y-intercept is 6.

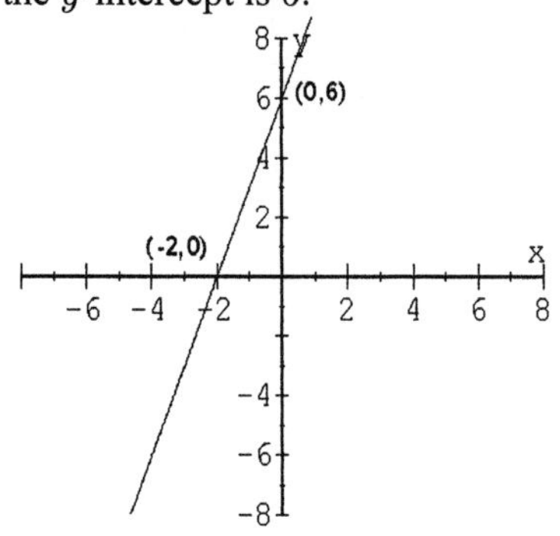

10. $y = -4x + 1$

$x = 0$ $\qquad$ $y = 0$

$y = -4(0) + 1$ $\qquad$ $0 = -4x + 1$

$y = 1$ $\qquad$ $4x = 1$

$x = \frac{1}{4}$

11. $(2, -5)$ and $(-1, 1)$

$$m = \frac{1 - (-5)}{-1 - 2} = \frac{1 + 5}{-3} = \frac{6}{-3} = -2$$

12. $(4, -3)$ and $(4, 2)$

$$m = \frac{2 - (-3)}{4 - 4} = \frac{2 + 3}{0} = \frac{5}{0}$$

Division by 0 is undefined, therefore the slope of the line determined by $(4, -3)$ and $(4, 2)$ is undefined.

13.

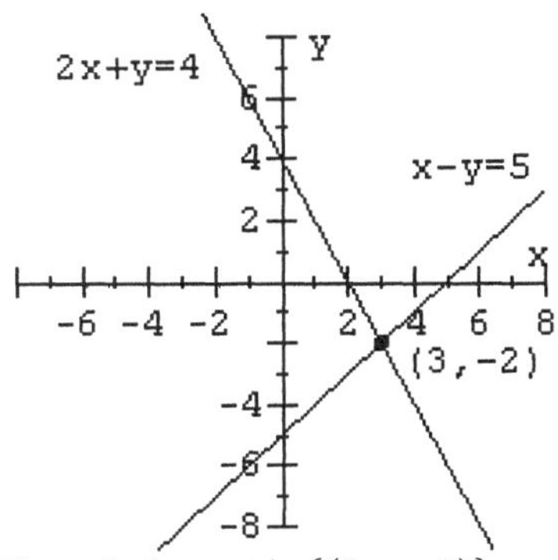

The solution set is $\{(3, -2)\}$.

14. $\begin{pmatrix} 2x - y = 1 \\ 3x - 2y = -5 \end{pmatrix}$

Multiply equation (1) by -2 and add the result to equation (2).

$$\begin{aligned} -4x + 2y &= -2 \\ 3x - 2y &= -5 \\ \hline -x \qquad &= -7 \\ x &= 7 \end{aligned}$$

Substitute $x = 7$ into equation (1).

$2(7) - y = 1$

$14 - y = 1$

$-y = -13$

$y = 13$

The solution set is $\{(7, 13)\}$.

15. Use the substitution method.

$\begin{pmatrix} 2x + 5y = 7 \\ x = -3y + 1 \end{pmatrix}$

Substitute $-3y + 1$ for x in $2x + 5y = 7$.

$2(-3y + 1) + 5y = 7$

$-6y + 2 + 5y = 7$

$-y + 2 = 7$

$-y = 5$

$y = -5$

Substitute -5 for y in $x = -3y + 1$.

$x = -3(-5) + 1$

$x = 15 + 1$

$x = 16$

The solution set is $\{(16, -5)\}$.

16. $\begin{pmatrix} 3x + 2y = 7 \\ 4x - 5y = 3 \end{pmatrix}$

Multiply equation (1) by -4.

$-12x - 8y = -28 \quad (3)$

Multiply equiation (2) by 3.

$12x - 15y = 9 \quad (4)$

Add equation (3) and equation (4).

$$\begin{aligned} -12x - 8y &= -28 \\ 12x - 15y &= 9 \\ \hline -23y &= -19 \\ y &= \frac{-19}{-23} = \frac{19}{23} \end{aligned}$$

Substitute $y = \frac{19}{23}$ into equation (1).

$$3x + 2\left(\frac{19}{23}\right) = 7$$
$$3x + \frac{38}{23} = 7$$
$$3x = 7 - \frac{38}{23}$$
$$3x = \frac{161}{23} - \frac{38}{23}$$
$$3x = \frac{123}{23}$$
$$\frac{1}{3}(3x) = \frac{1}{3}\left(\frac{123}{23}\right)$$
$$x = \frac{41}{23}$$

The solution set is $\left\{\left(\frac{41}{23}, \frac{19}{23}\right)\right\}$.

17. Use the elimination-by-addition method.

$$\begin{pmatrix} 9x + 2y = 140 \\ x + 5y = 135 \end{pmatrix}$$

$9x + 2y = 140$ Leave alone.

$x + 5y = 135$ Multiply by -9.

$$\begin{array}{rl} 9x + 2y = & 140 \\ -9x - 45y = & -1215 \\ \hline -43y = & -1075 \end{array}$$
$$y = 25$$

Substitute 25 for y in $x + 5y = 135$.

$$x + 5(25) = 135$$
$$x + 125 = 135$$
$$x = 10$$

The solution set is $\{(10, 25)\}$.

18. $$\begin{pmatrix} \frac{1}{2}x + \frac{1}{4}y = -5 \\ \frac{2}{3}x - \frac{1}{2}y = 0 \end{pmatrix}$$

Multiply equation (1) by 4.

$2x + y = -20$

Multiply equation (2) by 6.

$4x - 3y = 0$

The equivalent system is

$$\begin{pmatrix} 2x + y = -20 \\ 4x - 3y = 0 \end{pmatrix} \begin{matrix} (3) \\ (4) \end{matrix}$$

Multiply equation (3) by 3 and add the result to equation (4).

$$\begin{array}{rl} 6x + 3y = & -60 \\ 4x - 3y = & 0 \\ \hline 10x \quad\quad = & -60 \end{array}$$
$$x = -6$$

Substitute $x = -6$ into equation (3).

$$2(-6) + y = -20$$
$$-12 + y = -20$$
$$y = -8$$

The solution set is $\{(-6, -8)\}$.

19. Use the elimination-by-addition method.

$$\begin{pmatrix} x + y = 1000 \\ 0.07x + 0.09y = 82 \end{pmatrix}$$

$x + y = 1000$ Multiply by -7.

$0.07x + 0.09y = 82$ Multiply by 100.

$$\begin{array}{rl} -7x - 7y = & -7000 \\ 7x + 9y = & 8200 \\ \hline 2y = & 1200 \end{array}$$
$$y = 600$$

Substitute 600 for y in $x + y = 1000$.

$$x + 600 = 1000$$
$$x = 400$$

The solution set is $\{(400, 600)\}$.

20. $$\begin{pmatrix} y = 5x + 2 \\ 10x - 2y = 1 \end{pmatrix}$$

Substitute in equation (2) for y.

$$10x - 2(5x + 2) = 1$$
$$10x - 10x - 4 = 1$$
$$-4 = 1$$

Since $-4 \neq 1$ the system is inconsistent and the solution set is $\emptyset$.

21. Use the substitution method.

Substitute $3x - 2$ for y in $5x - 7y = 9$.

$$5x - 7(3x - 2) = 9$$
$$5x - 21x + 14 = 9$$
$$-16x + 14 = 9$$
$$-16x = -5$$
$$x = \frac{5}{16}$$

Substitute $\frac{5}{16}$ for x in $y = 3x - 2$.

$y = 3\left(\frac{5}{16}\right) - 2$

$y = \frac{15}{16} - \frac{32}{16}$

$y = -\frac{17}{16}$

The solution set is $\left\{\left(\frac{5}{16}, -\frac{17}{16}\right)\right\}$.

22. $\begin{pmatrix} 10t + u = 6u \\ t + u = 12 \end{pmatrix} \Rightarrow \begin{pmatrix} 10t - 5u = 0 \\ t + u = 12 \end{pmatrix} \begin{matrix} (1) \\ (2) \end{matrix}$

Multiply equation (2) by 5 and add the result to equation (1).

$$\begin{aligned} 5t + 5u &= 60 \\ 10t - 5u &= 0 \\ \hline 15t \quad &= 60 \\ t &= 4 \end{aligned}$$

Substitute $t = 4$ into equation (2).

$4 + u = 12$

$u = 8$

The solution set is $t = 4$ and $u = 8$.

23. Use the substitution method.

$\begin{pmatrix} t = 2u \\ 10t + u - 36 = 10u + t \end{pmatrix}$

Substitute $2u$ for t in $10t + u - 36 = 10u + t$.

$$\begin{aligned} 10(2u) + u - 36 &= 10u + (2u) \\ 20u + u - 36 &= 10u + 2u \\ 21u - 36 &= 12u \\ -36 &= -9u \\ 4 &= u \end{aligned}$$

Substitute 4 for u in $t = 2u$.

$t = 2(4) = 8$

The solution set is $t = 8$ and $u = 4$.

24. $\begin{pmatrix} u = 2t + 1 \\ 10t + u + 10u + t = 100 \end{pmatrix} \Rightarrow$

$\begin{pmatrix} u = 2t + 1 \\ 11t + 11u = 110 \end{pmatrix}$

Substitute for u into equation (2).

$$\begin{aligned} 11t + 11(2t + 1) &= 110 \\ 11t + 22t + 11 &= 110 \\ 33t + 11 &= 110 \\ 33t &= 99 \\ t &= 3 \end{aligned}$$

Substitute $t = 3$ into equation (1).

$u = 2(3) + 1$

$u = 6 + 1$

$u = 7$

The solution set is $t = 3$ and $u = 7$.

25. Use the substitution method.

$\begin{pmatrix} y = -\frac{2}{3}x \\ \frac{1}{3}x - y = -9 \end{pmatrix}$

Substitute $-\frac{2}{3}x$ for y.

in $\frac{1}{3}x - y = -9$.

$$\begin{aligned} \frac{1}{3}x - \left(-\frac{2}{3}x\right) &= -9 \\ \frac{1}{3}x + \frac{2}{3}x &= -9 \\ x &= -9 \end{aligned}$$

Substitute -9 for x in $y = -\frac{2}{3}x$.

$y = -\frac{2}{3}(-9) = 6$

The solution set is $\{(-9, 6)\}$.

26. Let x = one number and y = other number.

$\begin{pmatrix} x + y = 113 \\ x = 2y - 1 \end{pmatrix}$

Use substitution to solve.

$$\begin{aligned} x + y &= 113 \\ (2y - 1) + y &= 113 \\ 3y - 1 &= 113 \\ 3y &= 114 \\ y &= 38 \end{aligned}$$

Substitute $y = 38$ into equation (2).

$x = 2(38) - 1$

$x = 76 - 1$

$x = 75$

The numbers are 75 and 38.

27. Let x represent the money invested at 6%.
Let y represent the money invested at 8%.

$y = x + 50$
The money invested at 8% is $50 more than the money invested at 6%.

$0.06x + 0.08y = 39.00$
The yearly interest was $39.00.

Substitute $x + 50$ for y in
$0.06x + 0.08y = 39.00$

$0.06x + 0.08(x + 50) = 39.00$
Multiply by 100.

$$6x + 8(x + 50) = 3900$$
$$6x + 8x + 400 = 3900$$
$$14x + 400 = 3900$$
$$14x = 3500$$
$$x = 250$$

Substitute 250 for x in $y = x + 50$.
$y = 250 + 50 = 300$
The money invested was $250 at 6% and $300 at 8%.

28. Let n represent the number of nickels and d the number of dimes.
$n + d = 43$
The number of coins is 43.
$5n + 10d = 340$
The total value of the coins is 340 cents.

$n + d = 43$ Multiply by -5.
$5n + 10d = 340$ Leave alone.

$$\begin{aligned} -5n - 5d &= -215 \\ 5n + 10d &= 340 \\ \hline 5d &= 125 \\ d &= 25 \end{aligned}$$

Substitute 25 for d in $n + d = 43$.
$n + 25 = 43$
$n = 18$
She has 18 nickels and 25 dimes.

29. Let l represent the length of the rectangle.
Let w represent the width of the rectangle.
$l = 3w + 1$
The length is one more than three times the width.
$2l + 2w = 50$
The perimeter is 50 inches.

Substitute $3w + 1$ for l in $2l + 2w = 50$.

$$2(3w + 1) + 2w = 50$$
$$6w + 2 + 2w = 50$$
$$8w + 2 = 50$$
$$8w = 48$$
$$w = 6$$

Substitute 6 for w in $l = 3w + 1$.
$l = 3(6) + 1$
$l = 18 + 1$
$l = 19$
The rectangle has a length of 19 inches and a width of 6 inches.

30. Let a and b represent the angles.
$a + b = 90$
The angles are complementary.
$a = 2b - 6$
One angle is 6° less than twice the other angle.

Substitute $2b - 6$ for a in $a + b = 90$.

$$(2b - 6) + b = 90$$
$$3b - 6 = 90$$
$$3b = 96$$
$$b = 32$$

Substitute 32 for b in $a + b = 90$.
$a + 32 = 90$
$a = 58$
The two angles are 32° and 58°.

31. Let a represent the larger angle.
Let b represent the smaller angle.
$a + b = 180$
The angles are supplementary.
$a = 3b - 20$
The larger angle is 20° less than three times the smaller angle.
Substitute $3b - 20$ for a in $a + b = 180$.

$$(3b - 20) + b = 180$$
$$4b - 20 = 180$$
$$4b = 200$$
$$b = 50$$

Substitute 50 for b in $a + b = 180$.

$a + 50 = 180$
$a = 130$
The two angles are 50° and 130°.

32. Let c represent the cost of a cheeseburger and m the cost of a milkshake.
$4c + 5m = 25.50$
The cost of four cheeseburgers and five milkshakes is \$25.50.
$2m = c + 1.75$
Two milkshakes cost \$1.75 more than one cheeseburger.

$$\begin{aligned} 4c + 5m &= 25.50 && \text{Leave alone.} \\ -c + 2m &= 1.75 && \text{Multiply by 4.} \end{aligned}$$

$$\begin{aligned} 4c + 5m &= 25.50 \\ -4c + 8m &= 7.00 \\ \hline 13m &= 32.50 \\ m &= 2.50 \end{aligned}$$

Substitute 2.50 for m in $2m = c + 1.75$.
$2(2.50) = c + 1.75$
$5 = c + 1.75$
$3.25 = c$
The cost of a cheeseburger is \$3.25 and a milkshake is \$2.50.

33. Let j represent the cost of a bottle of orange juice and w the cost of a bottle of water.
$3j + 2w - 6.75$
The cost of three bottles of orange juice and two bottles of water cost \$6.75.
$2j + 3w = 6.15$
The cost of two bottles of orange juice and three bottles of water cost \$6.15.

$$\begin{aligned} 3j + 2w &= 6.75 && \text{Multiply by 2.} \\ 2j + 3w &= 6.15 && \text{Multiply by } -3. \end{aligned}$$

$6j + 4w = 13.50$
$-6j - 9w = -18.45$
Add the two above equations.
$-5w = -4.95$
$w = 0.99$
Substitute $w = 0.99$ into the first equation.
$3j + 2(0.99) = 6.75$
$3j + 1.98 = 6.75$
$3j = 4.77$
$j = 1.59$
The cost of a bottle of orange juice is \$1.59 and a bottle of water is \$0.99.

34. Graph $-2x + y < 4$.
Use the points $(0, 4)$ and $(-2, 0)$ to graph a dashed line for $-2x + y = 4$.
Use $(0, 0)$ as a test point. The given inequality $-2x + y < 4$ becomes $0 + 0 < 4$ which is a true statement. Therefore, the solution set is the half-plane that does contain the origin.

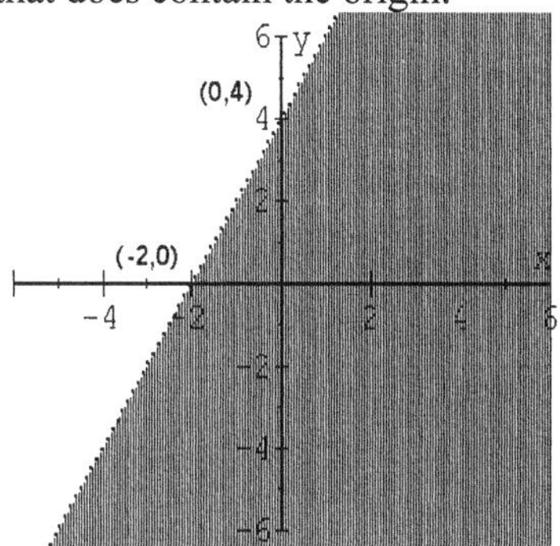

35. $3x + 2y \geq -6$
Use $(0, -3)$ and $(-2, 0)$ to graph a dashed line for $3x + 2y = -6$.
Use $(0, 0)$ for a test point.
$3x + 2y > -6$
$3(0) + 2(0) > -6$
$0 > -6$
Since $0 > -6$ is a true statement, the solution set is the half-plane that contains $(0, 0)$.

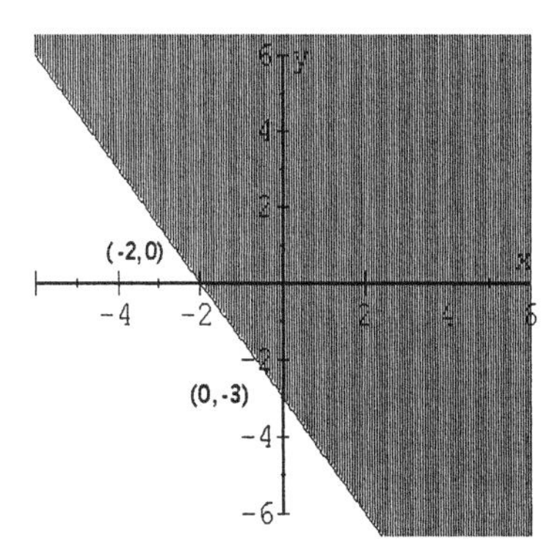

36. $\begin{pmatrix} -x + 2y > 2 \\ 3x - y > 3 \end{pmatrix}$

CHAPTER 5 Test

1. $7x - 2y = -8$
Substitute $(-2, -3)$ into the equation.
$7(-2) - 2(-3) = -8$
$-14 + 6 = -8$
$-8 = -8$
The result is a true statement so
$(-2, -3)$ is a solution of $7x - 2y = -8$.

2. $y = -x^2 - 4$
Substitute $(-1, -5)$ into the equation.
$-5 = -(-1)^2 - 4$
$-5 = -1 - 4$
$-5 = -5$
The result is a true statement so
$(-1, -5)$ is a solution of $y = -x^2 - 4$.

3. To find x-intercepts of the graph of
$y = -x^2 + 16$, set $y = 0$ and solve for x.

$0 = -x^2 + 16$
$x^2 = 16$
$x^2 - 16 = 0$
$(x + 4)(x - 4) = 0$
$x + 4 = 0$ or $x - 4 = 0$
$x = -4$ or $x = 4$
The x-intercepts are 4 and -4.

4. To find x-intercepts of the graph of
$-3x + 4y = -12$, set $y = 0$ and solve for x.
$-3x + 4y = -12$
$-3x + 4(0) = -12$
$-3x = -12$
$x = 4$
The x-intercept is 4.

5. $(3, -1)$ and $(-1, 1)$
$m = \dfrac{1 - (-1)}{-1 - 3} = \dfrac{2}{-4} = -\dfrac{1}{2}$
The slope of the line determined by
$(3, -1)$ and $(-1, 1)$ is $-\dfrac{1}{2}$.

6. $5x + 3y = 15$

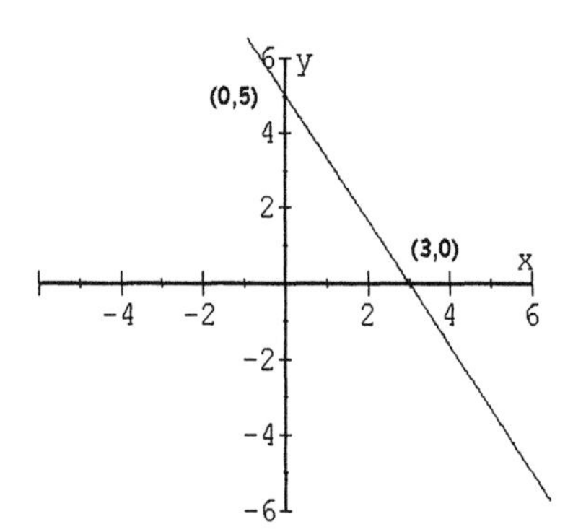

7. $-2x + y = -4$

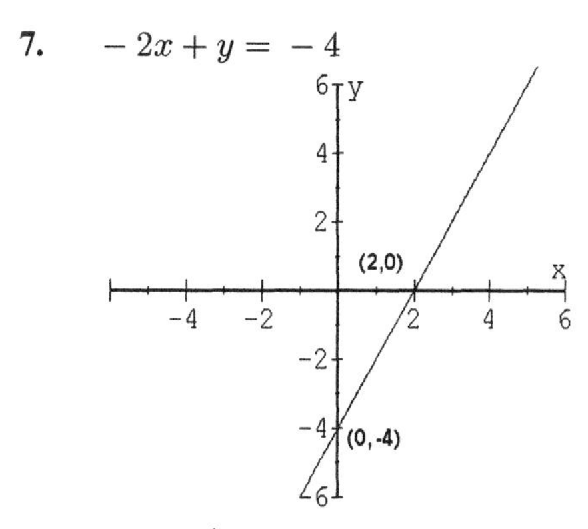

8. $y = -\frac{1}{2}x - 2$

9. $y = 2x^2 - 3$

10. $\begin{pmatrix} 3x - 2y = -14 \\ 5x + y = 14 \end{pmatrix}$

Substitute $(-2, 4)$ into the $1st$ equation.
$3(-2) - 2(4) = -14$
$-6 - 8 = -14$
$-14 = -14$ (true statement)
Substitute $(-2, 4)$ into the $2nd$ equation.
$5(-2) + (4) = -14$
$-10 + 4 = -14$
$-6 = -14$ (false statement)

$\{(-2, 4)\}$ is not the solution set of the above system.

11. $\begin{pmatrix} -x - 3y = 14 \\ 2x + 5y = -23 \end{pmatrix}$

Substitute $(1, -5)$ into the $1st$ equation.
$-(1) - 3(-5) = 14$
$-1 + 15 = 14$
$14 = 14$ (true statement)
Substitute $(1, -5)$ into the $2nd$ equation.
$2(1) + 5(-5) = -23$
$2 - 25 = -23$
$-23 = -23$ (true statement)

$\{(1, -5)\}$ is the solution set of the above system.

12.

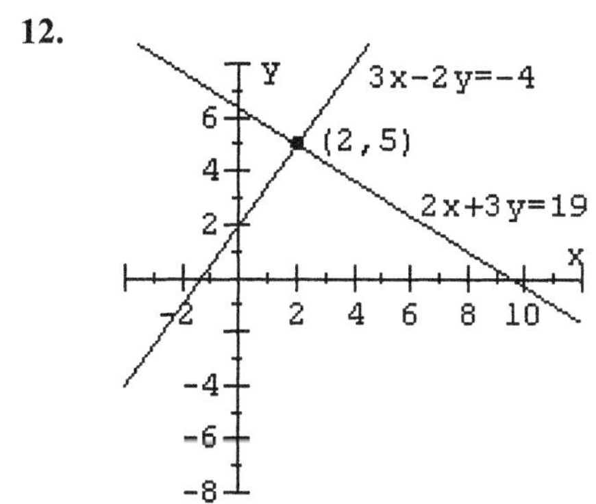

The solution set is $\{(2, 5)\}$.

13. $x - 3y = -9$ Multiply by -4.
$4x + 7y = 40$ Leave alone.

$$\begin{array}{r} -4x + 12y = 36 \\ 4x + \ \ 7y = 40 \\ \hline 19y = 76 \\ y = 4 \end{array}$$

Substitute 4 for y in $x - 3y = -9$.
$x - 3(4) = -9$
$x - 12 = -9$
$x = 3$
The solution set is $\{(3, 4)\}$.

14. Solve $5x + y = -14$ for y.

$5x + y = -14$

$y = -5x - 14$

Substitute $-5x - 14$ for y in $6x - 7y = -66$.

$6x - 7(-5x - 14) = -66$

$6x + 35x + 98 = -66$

$41x + 98 = -66$

$41x = -164$

$x = -4$

Substitute -4 for x in $y = -5x - 14$.

$y = -5(-4) - 14$

$y = 20 - 14$

$y = 6$

The solution set is $\{(-4, 6)\}$.

15. Use the elimination-by-addition method.

$2x - 7y = 26$ Multiply by 2.

$3x + 2y = -11$ Multiply by 7.

$$\begin{array}{r} 4x - 14y = \quad 52 \\ 21x + 14y = -77 \\ \hline 25x \qquad\quad = -25 \end{array}$$

$x = -1$

Substitute -1 for x in $2x - 7y = 26$.

$2(-1) - 7y = 26$

$-2 - 7y = 26$

$-7y = 28$

$y = -4$

The solution set is $\{(-1, -4)\}$.

16. Use the elimination-by-addition method.

$8x + 5y = -6$ Leave alone.

$4x - y = 18$ Multiply by 5.

$$\begin{array}{r} 8x + 5y = -6 \\ 20x - 5y = \ 90 \\ \hline 28x \qquad = 84 \end{array}$$

$x = 3$

Substitute 3 for x in $4x - y = 18$.

$4(3) - y = 18$

$12 - y = 18$

$-y = 6$

$y = -6$

The solution set is $\{(3, -6)\}$.

17. $3x - 2y > 6$

Substitute $(3, -2)$ into the inequality.

$3(3) - 2(-2) > 6$

$9 + 4 > 6$

$13 > 6$ (true statement)

Yes, $(3, -2)$ is a member of the solution set of the inequality.

18. $-2x - y \leq 4$

Substitute $(-3, -5)$ into the inequality.

$-2(-3) - (-5) \leq 4$

$6 + 5 \leq 4$

$11 \leq 4$ (false statement)

No, $(-3, -5)$ is not a member of the solution set of the inequality.

19. $\begin{pmatrix} 5x - y < 10 \\ 3x + 2y > 6 \end{pmatrix}$

Substitute $(2, 1)$ into the $1st$ inequality.

$5(2) - (1) < 10$

$5 - 1 < 10$

$4 < 10$ (true statement)

Substitute $(2, 1)$ into the $2nd$ inequality.

$3(2) + 2(1) > 6$

$6 + 2 > 6$

$8 > 6$ (true statement)

Yes $(2, 1)$ is a member of the solution set of the above system of inequalities.

20. Graph $3x - 2y > -6$.

Use $(0, 3)$ and $(-2, 0)$ to graph a dashed line for $3x - 2y = -6$.

Use $(0, 0)$ for a test point.

$3x - 2y > -6$

$3(0) + 2(0) > -6$

$0 > -6$

Since $0 > -6$ is a true statement, the solution set is the half-plane that contains $(0, 0)$.

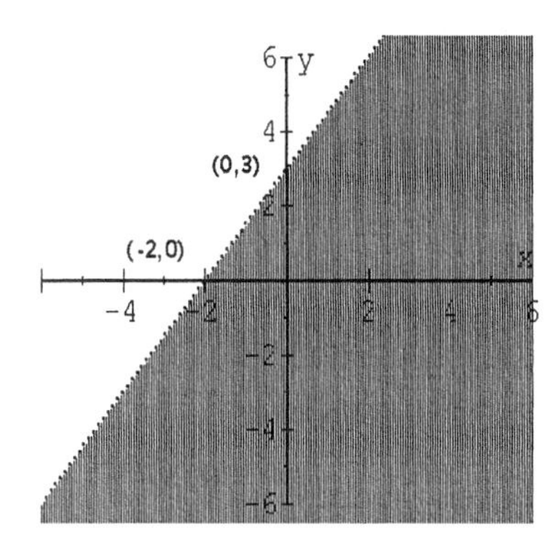

21. Graph $y \leq -3x$.
Use $(1, -3)$ and $(-1, 3)$ to graph a solid line for $y = -3x$. Use $(-2, 0)$ for a test point.
$y \leq -3x$
$(0) \leq -3(-2)$
$0 \leq 6$
Since $0 \leq 6$ is a true statement, the solution set is the half-plane that contains $(-2, 0)$.

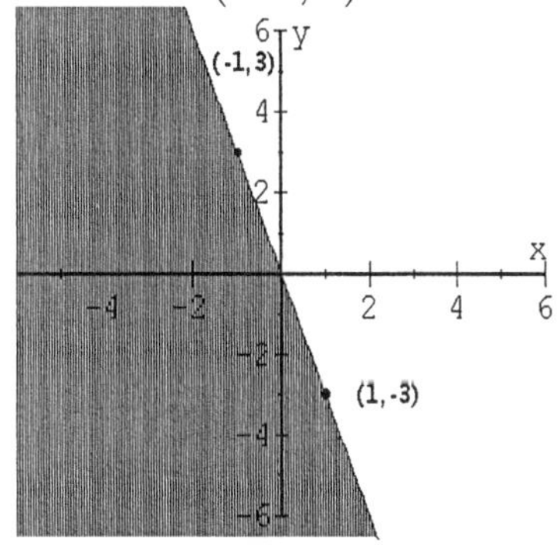

22. $\begin{pmatrix} x - 2y \leq 4 \\ 2x + y \leq 4 \end{pmatrix}$

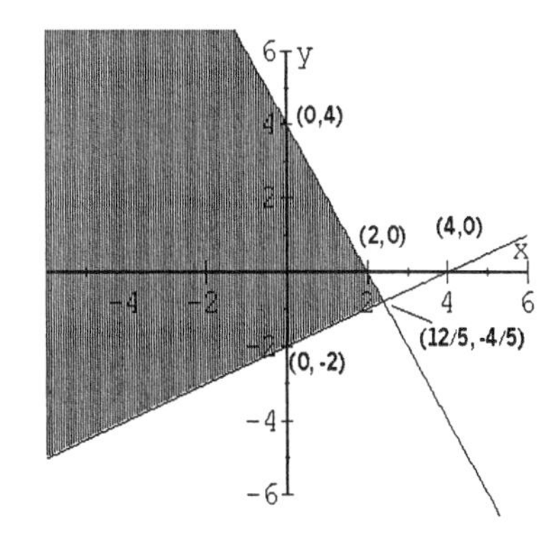

23. Let q represent the number of quarters and d the number of dimes.
$q + d = 40$
The number of coins is 40.
$25q + 10d = 760$
The total value of the coins is 760 cents.

$q + d = 40$ Multiply by -10.
$25q + 10d = 760$ Leave alone.

$$\begin{aligned} -10q - 10d &= -400 \\ 25q + 10d &= 760 \\ \hline 15q &= 360 \\ q &= 24 \end{aligned}$$

Substitute 24 for q in $q + d = 40$.
$24 + d = 40$
$d = 16$
She has 24 quarters and 16 dimes.

24. Let l represent the length of the rectangle.
Let w represent the width of the rectangle.
$l = 2w - 1$
The length is one less than twice the width.
$2l + 2w = 40$
The perimeter is 40 inches.
Substitute $2w - 1$ for l in $2l + 2w = 40$.
$2(2w - 1) + 2w = 40$
$4w - 2 + 2w = 40$
$6w - 2 = 40$
$6w = 42$
$w = 7$
Substitute 7 for w in $l = 2w - 1$.
$l = 2(7) - 1$
$l = 14 - 1$
$l = 13$
The rectangle has a length of 13 inches and a width of 7 inches.

25. Let $x =$ amount of 30% solution and $y =$ amount of 80% solution.

$$\begin{pmatrix} x + y = 5 \\ 0.30x + 0.80y = 0.60(5) \end{pmatrix}$$

Solve equation 1 for x.
$x + y = 5$
$x = 5 - y$

Substitute into equation 2:
$0.30x + 0.80y = 3$
$0.30(5 - y) + 0.80y = 3$
$10[0.30(5 - y) + 0.80y] = 10(3)$
$3(5 - y) + 8y = 30$
$15 - 3y + 8y = 30$
$5y = 15$
$y = 3$
There should be 3 liters of 80% solution.

CHAPTERS 1-5 | Cumulative Practice Test

1. $3.9 - 4.6 - 1.2 + 0.4$
$3.9 + 0.4 - 4.6 - 1.2$
$4.3 - 5.8 = -1.5$

2. $\left(\frac{1}{2}\right)^3 + \left(\frac{1}{4}\right)^2 - \frac{1}{8} =$
$\frac{1}{8} + \frac{1}{16} - \frac{1}{8} = \frac{1}{16}$

3. $(0.2)^3 - (0.4)^3 + (1.2)^2 =$
$0.008 - 0.064 + 1.44 = 1.384$

4. $\frac{1}{2}x - \frac{2}{5}y + xy$ for $x = -3$ and $y = \frac{1}{2}$
$\frac{1}{2}(-3) - \frac{2}{5}\left(\frac{1}{2}\right) + (-3)\left(\frac{1}{2}\right) =$
$-\frac{3}{2} - \frac{1}{5} - \frac{3}{2} = -\frac{6}{2} - \frac{1}{5} = -3 - \frac{1}{5} =$
$-3\frac{1}{5} = -\frac{16}{5}$

5. $1.1x - 2.3y + 2.5x + 1.6y =$
$1.1x + 2.5x - 2.3y + 1.6y =$
$3.6x - 0.7y$ for $x = 0.4$ and $y = 0.7$
$3.6(0.4) - 0.7(0.7) = 1.44 - 0.49 = 0.95$

6. $2(x + 6) - 3(x + 9) - 4(x - 5)$
$2x + 12 - 3x - 27 - 4x + 20 =$
$2x - 3x - 4x + 12 - 27 + 20 =$
$-5x + 5$ for $x = 17$
$-5(17) + 5 = -85 + 5 = -80$

7. $\frac{3}{x} + \frac{2}{y} - \frac{4}{xy} =$
$\frac{3}{x} \cdot \frac{y}{y} + \frac{2}{y} \cdot \frac{x}{x} - \frac{4}{xy} =$
$\frac{3y}{xy} + \frac{2x}{xy} - \frac{4}{xy} = \frac{3y + 2x + 4}{xy}$

8. $\left(\frac{7xy}{9xy^2}\right)\left(\frac{12x}{14y}\right) = \frac{\overset{1}{\cancel{7}} \cdot \overset{4}{\cancel{12}} \cdot x^2 \cdot y}{\underset{2}{\cancel{14}} \cdot \underset{3}{\cancel{9}} \cdot x \cdot y^3} =$
$\frac{4x}{6y^2} = \frac{2x}{3y^2}$

9. $\left(\frac{2ab^2}{7a}\right) \div \left(\frac{4b}{21a}\right) = \left(\frac{2ab^2}{7a}\right) \cdot \left(\frac{21a}{4b}\right) =$
$\frac{\overset{1}{\cancel{2}} \cdot \overset{3}{\cancel{21}} \cdot a^2 \cdot b^2}{\underset{2}{\cancel{4}} \cdot \underset{1}{\cancel{7}} \cdot a \cdot b} = \frac{3ab}{2}$

10. $2(x - 3) - (x + 4) = 3(x + 10)$
$2x - 6 - x - 4 = 3x + 30$
$x - 10 = 3x + 30$
$-2x - 10 = 30$
$-2x = 40$
$x = -20$
The solution set is $\{-20\}$.

11. $\frac{2}{3}x + \frac{1}{4} - \frac{1}{2}x = -1$
$12\left[\frac{2}{3}x + \frac{1}{4} - \frac{1}{2}x\right] = 12(-1)$
$12\left(\frac{2}{3}x\right) + 12\left(\frac{1}{4}\right) - 12\left(\frac{1}{2}x\right) = -12$
$8x + 3 - 6x = -12$
$2x + 3 = -12$
$2x = -15$
$x = -\frac{15}{2}$
The solution set is $\left\{-\frac{15}{2}\right\}$.

12. $\frac{x+1}{4} - \frac{x-2}{6} = \frac{3}{8}$

$24\left(\frac{x+1}{4} - \frac{x-2}{6}\right) = 24\left(\frac{3}{8}\right)$

$24\left(\frac{x+1}{4}\right) - 24\left(\frac{x-2}{6}\right) = 3(3)$

$6(x+1) - 4(x-2) = 9$

$6x + 6 - 4x + 8 = 9$

$2x + 14 = 9$

$2x = -5$

$x = -\frac{5}{2}$

The solution set is $\left\{-\frac{5}{2}\right\}$.

13. $\frac{x+6}{7} = \frac{x-2}{8}$

$8(x+6) = 7(x-2)$

$8x + 48 = 7x - 14$

$x + 48 = -14$

$x = -62$

The solution set is $\{-62\}$.

14. $0.06x + 0.07(1800 - x) = 120$

$100[0.06x + 0.07(1800 - x)] = 100(120)$

$6x + 7(1800 - x) = 12000$

$6x + 12600 - 7x = 12000$

$-x + 12600 = 12000$

$-x = -600$

$x = 600$

The solution set is $\{600\}$.

15. $2x - 3y = 13$

$-3y = -2x + 13$

$y = \frac{-2x+13}{-3} = \frac{2x-13}{3}$

16. $\begin{pmatrix} 5x - 2y = -20 \\ 2x + 3y = 11 \end{pmatrix}$

Use the elimination-by-addition method.

$5x - 2y = -20$ Multiply by 3.

$2x + 3y = 11$ Multiply by 2.

$15x - 6y = -60$

$4x + 6y = 22$

$19x = -38$

$x = -2$

Substitute -2 for x in $5x - 2y = -20$.

$5(-2) - 2y = -20$

$-10 - 2y = -20$

$-2y = -10$

$y = 5$

The solution set is $\{(-2, 5)\}$.

17. $\begin{pmatrix} 4x - 9y = 47 \\ 7x + y = 32 \end{pmatrix}$

Use the elimination-by-addition method.

$7x + y = 32$ Multiply by 9.

$4x - 9y = 47$

$63x + 9y = 288$

$67x = 335$

$x = 5$

Substitute 5 for x in $4x - 9y = 47$.

$4(5) - 9y = 47$

$20 - 9y = 47$

$-9y = 27$

$y = -3$

The solution set is $\{(5, -3)\}$.

18. $3(x-2) < 4(x+6)$

$3x - 6 < 4x + 24$

$-x - 6 < 24$

$-x < 30$

$x > -30$

The solution set is $\{x \mid x > -30\}$ or $(-30, \infty)$.

19.

$$\frac{1}{4}x - 2 \geq \frac{2}{3}x + 1$$
$$12\left(\frac{1}{4}x - 2\right) \geq 12\left(\frac{2}{3}x + 1\right)$$
$$12\left(\frac{1}{4}x\right) - 12(2) \geq 12\left(\frac{2}{3}x\right) + 12(1)$$
$$3x - 24 \geq 8x + 12$$
$$-5x \geq 36$$
$$\frac{-5x}{-5} \leq \frac{36}{-5}$$
$$x \leq -\frac{36}{5}$$

The solution set is $\left\{x \middle| x \leq -\frac{36}{5}\right\}$

or $\left(-\infty, -\frac{36}{5}\right]$.

20. Graph $y = -2x - 3$.

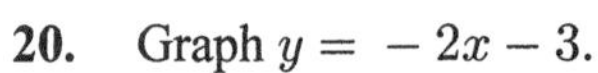

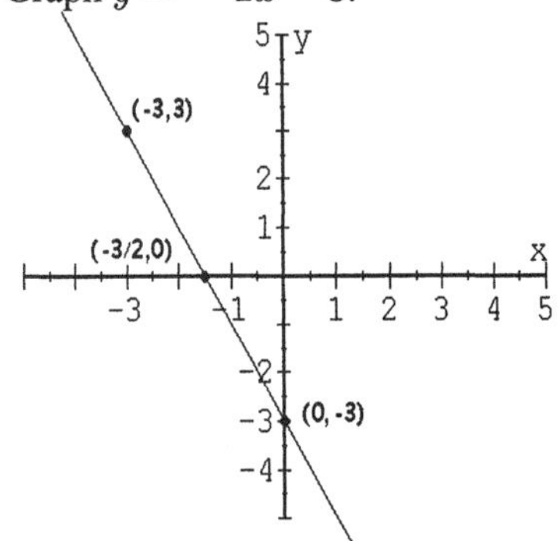

21. Graph $y = -2x^2 - 3$.

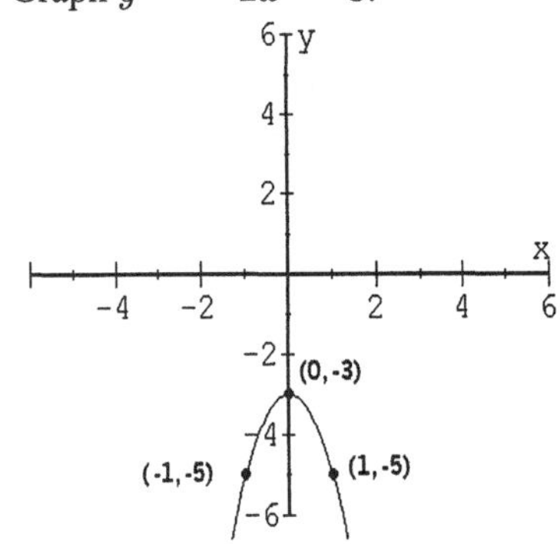

22. Graph $x - 2y > 4$.
Use $(0, -2)$ and $(4, 0)$ to graph a dashed line for $x - 2y = 4$.
Use $(0, 0)$ for a test point.

$$x - 2y > 4$$
$$(0) - 2(0) > 4$$
$$0 > 4$$

Since $0 > 4$ is a false statement, the solution set is the half-plane that does NOT contain $(0, 0)$.

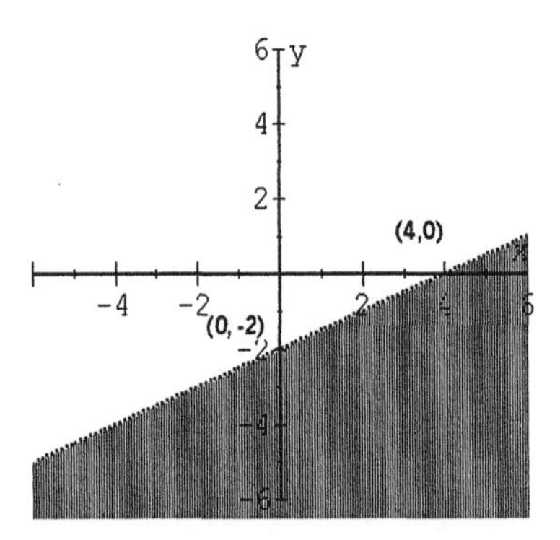

23. Let x represent the highest test grade and y represent the lowest test grade.

Highest grade = 3(lowest grade) − 9
$x = 3y - 9$
Sum of grades = 135.
$x + y = 135$

Solve the system by substitution.

$$\begin{pmatrix} x = 3y - 9 \\ x + y = 135 \end{pmatrix}$$

Substitute $3y - 9$ for x in equation 2.
$(3y - 9) + y = 135$
$3y - 9 + y = 135$
$4y - 9 = 135$
$4y = 144$
$y = 36$
$x = 3(36) - 9 = 108 - 9 = 99$
The grades are 36 and 99.

24. Let x represent golf score of the fifth day.
Average ≤ 80.

$$\frac{82+84+78+79+x}{5} \leq 80$$

$$\frac{323+x}{5} \leq 80$$

$323 + x \leq 400$
$x \leq 77$

On the fifth day of the tournament he must shoot 77 or less.

25. Let $x =$ gallons of 10% salt solution
$y =$ gallons of 15% salt solution

$$\begin{pmatrix} x+y=10 \\ 0.10x+0.15y=0.13(10) \end{pmatrix}$$

Solve equation 1 for x.
$x + y = 10$
$x = -y + 10$

Substitute into equation 2:
$0.10x + 0.15y = 1.3$
$0.10(-y+10) + 0.15y = 1.3$
$100[0.10(-y+10) + 0.15y] = 100(1.3)$
$10(-y+10) + 15y = 130$
$-10y + 100 + 15y = 130$
$5y + 100 = 130$
$5y = 30$
$y = 6$
$x = -y + 10$
$x = -6 + 10$
$x = 4$

There should be 4 gallons of 10% solution and 6 gallons of 15% solution.

Chapter 6 Exponents and Polynomials

PROBLEM SET 6.1 Addition and Subtraction of Polynomials

1. The degree of $7x^2y + 6xy$ is 3 because the degree of the term $7x^2y$ is 3.

3. The degree of $5x^2 - 9$ is 2 because the degree of the term $5x^2$ is 2.

5. The degree of $5x^3 - x^2 - x + 3$ is 3 because the degree of the term $5x^3$ is 3.

7. The degree of $5xy$ is 2 because the degree of the term $5xy$ is 2.

9. $(3x+4)+(5x+7)=$
$(3+5)x+(4+7)=$
$8x+11$

11. $(-5y-3)+(9y+13)=$
$(-5+9)y+(-3+13)=$
$4y+10$

13. $(-2x^2+7x-9)+(4x^2-9x-14)=$
$(-2+4)x^2+(7-9)x+(-9-14)=$
$2x^2-2x-23$

15. $(5x-2)+(3x-7)+(9x-10)=$
$(5+3+9)x+(-2-7-10)=$
$17x-19$

17. $(2x^2-x+4)+(-5x^2-7x-2)$
$+(9x^2+3x-6)=$
$(2-5+9)x^2+(-1-7+3)x$
$+(4-2-6)=$
$6x^2-5x-4$

19. $(-4n^2-n-1)+(4n^2+6n-5)=$
$(-4+4)n^2+(-1+6)n+(-1-5)=$
$5n-6$

21. $(2x^2-7x-10)+(-6x-2)+(-9x^2+5)=$
$(2-9)x^2+(-7-6)x+(-10-2+5)=$
$-7x^2-13x-7$

23. $(12x+6)-(7x+1)=$
$12x+6-7x-1=$
$(12-7)x+(6-1)=$
$5x+5$

25. $(3x-7)-(5x-2)=$
$3x-7-5x+2=$
$(3-5)x+(-7+2)=$
$-2x-5$

27. $(-4x+6)-(-x-1)=$
$-4x+6+x+1=$
$(-4+1)x+(6+1)=$
$-3x+7$

29. $(3x^2+8x-4)-(x^2-7x+2)=$
$3x^2+8x-4-x^2+7x-2=$
$(3-1)x^2+(8+7)x+(-4-2)=$
$2x^2+15x-6$

31. $(3n^2-n+7)-(-2n^2-3n+4)=$
$3n^2-n+7+2n^2+3n-4=$
$(3+2)n^2+(-1+3)n+(7-4)=$
$5n^2+2n+3$

33. $(-7x^3+x^2+6x-12)$
$-(-4x^3-x^2+6x-1)=$
$-7x^3+x^2+6x-12+4x^3$
$+x^2-6x+1=$
$(-7+4)x^3+(1+1)x^2+(6-6)x$
$+(-12+1)=$
$-3x^3+2x^2-11$

35.

$12x-4$	Add	$12x-4$
$\underline{3x-2}$	the	$\underline{-3x+2}$
	opposite.	$9x-2$

37.

$-3a+9$	Add	$-3a+9$
$\underline{-5a-6}$	the	$\underline{5a+6}$
	opposite.	$2a+15$

39. $\begin{array}{r} 6x^2 - x + 11 \\ \underline{8x^2 - x + \ 6} \end{array}$ Add the opposite. $\begin{array}{r} 6x^2 - x + 11 \\ \underline{-8x^2 + x - \ 6} \\ -2x^2 \qquad + \ 5 \end{array}$

41. $\begin{array}{r} 4x^3 + 6x^2 + 7x - 14 \\ \underline{-2x^3 - 6x^2 + 7x - \ 9} \end{array}$

Add the opposite.

$\begin{array}{r} 4x^3 + 6x^2 + 7x - 14 \\ \underline{2x^3 + 6x^2 - 7x + \ 9} \\ 6x^3 + 12x^2 \qquad - \ 5 \end{array}$

43. $\begin{array}{r} 4x^3 - 6x^2 + 7x - \ 2 \\ \underline{2x^2 - 6x - 14} \end{array}$

Add the opposite.

$\begin{array}{r} 4x^3 - 6x^2 + 7x - \ 2 \\ \underline{-2x^2 + 6x + 14} \\ 4x^3 - 8x^2 + 13x + 12 \end{array}$

45. $(5x + 3) - (7x - 2) + (3x + 6) =$
$5x + 3 - 7x + 2 + 3x + 6 =$
$(5 - 7 + 3)x + (3 + 2 + 6) =$
$x + 11$

47. $(-x - 1) - (-2x + 6) + (-4x - 7) =$
$-x - 1 + 2x - 6 - 4x - 7 =$
$(-1 + 2 - 4)x + (-1 - 6 - 7) =$
$-3x - 14$

49. $(x^2 - 7x - 4) + (2x^2 - 8x - 9)$
$\quad - (4x^2 - 2x - 1) =$
$x^2 - 7x - 4 + 2x^2 - 8x - 9$
$\quad - 4x^2 + 2x + 1 =$
$(1 + 2 - 4)x^2 + (-7 - 8 + 2)x$
$\quad + (-4 - 9 + 1) =$
$-x^2 - 13x - 12$

51. $(-x^2 - 3x + 4) + (-2x^2 - x - 2)$
$\quad - (-4x^2 + 7x + 10) =$
$-x^2 - 3x + 4 - 2x^2 - x - 2$
$\quad + 4x^2 - 7x - 10 =$
$(-1 - 2 + 4)x^2 + (-3 - 1 - 7)x$
$\quad + (4 - 2 - 10) =$
$x^2 - 11x - 8$

53. $(3a - 2b) - (7a + 4b) - (6a - 3b) =$
$3a - 2b - 7a - 4b - 6a + 3b =$
$(3 - 7 - 6)a + (-2 - 4 + 3)b =$
$-10a - 3b$

55. $(n - 6) - (2n^2 - n + 4) + (n^2 - 7) =$
$n - 6 - 2n^2 + n - 4 + n^2 - 7 =$
$(-2 + 1)n^2 + (1 + 1)n + (-6 - 4 - 7) =$
$-n^2 + 2n - 17$

57. $7x + [3x - (2x - 1)] =$
$7x + [3x - 2x + 1] =$
$7x + [x + 1] =$
$7x + x + 1 =$
$(7 + 1)x + 1 = 8x + 1$

59. $-7n - [4n - (6n - 1)] =$
$-7n - [4n - 6n + 1] =$
$-7n - [-2n + 1] =$
$-7n + 2n - 1 =$
$(-7 + 2)n - 1 =$
$-5n - 1$

61. $(5a - 1) - [3a + (4a - 7)] =$
$(5a - 1) - [3a + 4a - 7] =$
$(5a - 1) - [7a - 7] =$
$5a - 1 - 7a + 7 =$
$(5 - 7)a + (-1 + 7) =$
$-2a + 6$

63. $13x - [5x - [4x - (x - 6)]] =$
$13x - [5x - [4x - x + 6]] =$
$13x - [5x - [3x + 6]] =$
$13x - [5x - 3x - 6] =$
$13x - [2x - 6] =$
$13x - 2x + 6 =$
$11x + 6$

65. $[(4x - 2) + (7x + 6)] - (5x - 3) =$
$[4x - 2 + 7x + 6] - (5x - 3) =$
$[4x + 7x - 2 + 6] - (5x - 3) =$
$[11x + 4] - (5x - 3) =$
$11x + 4 - 5x + 3 =$
$11x - 5x + 4 + 3 =$
$6x + 7$

67. $(-8n+9)-[(-2n-5)+(-n+7)] =$
$(-8n+9)-[-2n-5-n+7] =$
$(-8n+9)-[-2n-n-5+7] =$
$(-8n+9)-[-3n+2] =$
$-8n+9+3n-2 =$
$-8n+3n+9-2 =$
$-5n+7$

69. Use the formula $P = 2l + 2w$ with $l = 3x+5$ and $w = x-3$.
$P = 2(3x+5)+2(x-2)$
$= 6x+10+2x-4$
$= 6x+2x+10-4$
$= 8x+6$

71. Use the formula $A = lw$ for all the figures. Remember, $x(x) = x^2$
$A = 3x(x)+4x(x)+2x(2x)+3x(3x)$
$= 3x^2+4x^2+4x^2+9x^2$
$= (3+4+4+9)x^2$
$= 20x^2$

PROBLEM SET 6.2 Multiplying Monomials

1. $(5x)(9x) = 5\cdot 9\cdot x\cdot x = 45x^{1+1} = 45x^2$

3. $(3x^2)(7x) = 3\cdot 7\cdot x^2\cdot x = 21x^{2+1} = 21x^3$

5. $(-3xy)(2xy) = -3\cdot 2\cdot x\cdot x\cdot y\cdot y =$
$-6x^{1+1}y^{1+1} = -6x^2y^2$

7. $(-2x^2y)(-7x) =$
$(-2)(-7)(x^2)(x)(y) =$
$14x^{2+1}y^1 = 14x^3y$

9. $(4a^2b^2)(-12ab) =$
$(4)(-12)(a^2)(a)(b^2)(b) =$
$-48a^{2+1}b^{2+1} = -48a^3b^3$

11. $(-xy)(-5x^3) =$
$(-1)(-5)(x)(x^3)(y) =$
$5x^{1+3}y^1 = 5x^4y$

13. $(8ab^2c)(13a^2c) =$
$8\cdot 13\cdot a\cdot a^2\cdot b^2\cdot c\cdot c =$
$104a^{1+2}b^2c^{1+1} = 104a^3b^2c^2$

15. $(5x^2)(2x)(3x^3) = 5\cdot 2\cdot 3\cdot x^2\cdot x\cdot x^3 =$
$30x^{2+1+3} = 30x^6$

17. $(4xy)(-2x)(7y^2) =$
$(4)(-2)(7)(x)(x)(y)(y^2) =$
$-56x^{1+1}y^{1+2} = -56x^2y^3$

19. $(-2ab)(-ab)(-3b) =$
$(-2)(-1)(-3)(a^{1+1})(b^{1+1+1}) =$
$-6a^2b^3$

21. $(6cd)(-3c^2d)(-4d) =$
$(6)(-3)(-4)(c^{1+2}d^{1+1+1}) =$
$72c^3d^3$

23. $\left(\frac{2}{3}xy\right)\left(\frac{3}{5}x^2y^4\right) =$
$\left(\frac{2}{3}\right)\left(\frac{3}{5}\right)x^{1+2}y^{1+4} =$
$\frac{2}{5}x^3y^5$

25. $\left(-\frac{7}{12}a^2b\right)\left(\frac{8}{21}b^4\right) =$
$\left(-\frac{\overset{1}{\cancel{7}}}{\underset{3}{\cancel{12}}}\right)\left(\frac{\overset{2}{\cancel{8}}}{\underset{3}{\cancel{21}}}\right)a^2b^{1+4} = -\frac{2}{9}a^2b^5$

27. $(0.4x^5)(0.7x^3) = (0.4)(0.7)x^{5+3} = 0.28x^8$

29. $(-4ab)(1.6a^3b) =$
$(-4)(1.6)a^{1+3}b^{1+1} = -6.4a^4b^2$

31. $(2x^4)^2 = (2)^2(x^4)^2 = 4x^{4\cdot 2} = 4x^8$

33. $(-3a^2b^3)^2 = (-3)^2(a^2)^2(b^3)^2 = 9a^4b^6$

35. $(3x^2)^3 = (3)^3(x^2)^3 = 27x^6$

37. $(-4x^4)^3 = (-4)^3(x^4)^3 = -64x^{12}$

39. $(9x^4y^5)^2 = (9)^2(x^4)^2(y^5)^2 = 81x^8y^{10}$

41. $(2x^2y)^4 = (2)^4(x^2)^4(y)^4 = 16x^8y^4$

43. $(-3a^3b^2)^4 = (-3)^4(a^3)^4(b^2)^4 =$
$81a^{12}b^8$

45. $(-x^2y)^6 = (-1)^6(x^2)^6(y)^6 =$
$1x^{12}y^6 = x^{12}y^6$

47. $5x(3x+2) = 5x(3x) + 5x(2) =$
$15x^2 + 10x$

49. $3x^2(6x-2) = 3x^2(6x) - 3x^2(2) =$
$18x^3 - 6x^2$

51. $-4x(7x^2-4) =$
$-4x(7x^2) - (-4x)(4) =$
$-28x^3 + 16x$

53. $2x(x^2-4x+6) =$
$2x(x^2) - 2x(4x) + 2x(6) =$
$2x^3 - 8x^2 + 12x$

55. $-6a(3a^2-5a-7) =$
$-6a(3a^2) - (-6a)(5a) - (-6a)(7) =$
$-18a^3 + 30a^2 + 42a$

57. $7xy(4x^2-x+5) =$
$7xy(4x^2) - 7xy(x) + 7xy(5) =$
$28x^3y - 7x^2y + 35xy$

59. $-xy(9x^2-2x-6) =$
$-xy(9x^2) - (-xy)(2x) - (-xy)(6) =$
$-9x^3y + 2x^2y + 6xy$

61. $5(x+2y) + 4(2x+3y) =$
$5x + 10y + 8x + 12y =$
$13x + 22y$

63. $4(x-3y) - 3(2x-y) =$
$4x - 12y - 6x + 3y =$
$-2x - 9y$

65. $2x(x^2-3x-4) + x(2x^2+3x-6) =$
$2x^3 - 6x^2 - 8x + 2x^3 + 3x^2 - 6x =$
$4x^3 - 3x^2 - 14x$

67. $3[2x-(x-2)] - 4(x-2) =$
$3[2x-x+2] - 4(x-2) =$
$3[x+2] - 4(x-2) =$
$3x + 6 - 4x + 8 =$
$-x + 14$

69. $-4(3x+2) - 5[2x-(3x+4)] =$
$-4(3x+2) - 5[2x-3x-4] =$
$-4(3x+2) - 5[-x-4] =$
$-12x - 8 + 5x + 20 =$
$-7x + 12$

71. $(3x)^2(2x^3) = (3)^2(x)^2(2)(x)^3 =$
$(9)(x^2)(2)(x^3) = 18x^5$

73. $(-3x)^3(-4x)^2 =$
$(-3)^3(x)^3(-4)^2(x)^2 =$
$(-27)(x^3)(16)(x^2) =$
$-432x^5$

75. $(5x^2y)^2(xy^2)^3 =$
$(5)^2(x^2)^2(y)^2(x)^3(y^2)^3 -$
$(25)(x^4)(y^2)(x^3)(y^6) =$
$25x^7y^8$

77. $(-a^2bc^3)^3(a^3b)^2 =$
$(-1)^3(a^2)^3(b)^3(c^3)^3(a^3)^2(b)^2 =$
$-1(a^6)(b^3)(c^9)(a^6)(b^2) =$
$-a^{12}b^5c^9$

79. $(-2x^2y^2)^4(-xy^3)^3 =$
$(-2)^4(x^2)^4(y^2)^4(-1)^3(x)^3(y^3)^3 =$
$(16)(x^8)(y^8)(-1)(x^3)(y^9) =$
$-16x^{11}y^{17}$

81. Use the formula $A = lw$ for both rectangles and then add.
Area = Left Area + Right Area
$$A = 3(x-1) + 4(x+2)$$
$$= 3x - 3 + 4x + 8$$
$$= 7x + 5$$

83. Use the formula $A = \pi r^2$ for the both circles, then subtract the area of the smaller circle from the area of the larger circle.
$$\text{Area} = \underset{\text{Larger Circle}}{\text{Area of}} - \underset{\text{Smaller Circle}}{\text{Area of}}$$
$$A = \pi(2x)^2 - \pi(x)^2$$
$$= \pi(4x^2) - \pi(x^2)$$
$$= 4\pi x^2 - \pi x^2 = 3\pi x^2$$
The area of the shaded region can be represented by $3\pi x^2$.

89. $(x^{2n})(x^{5n}) = x^{2n+5n} = x^{7n}$

91. $(x^{5n+2})(x^{n-1}) = x^{(5n+2)+(n-1)} = x^{6n+1}$

93. $(x^{6n-1})(x^4) = x^{(6n-1)+4} = x^{6n+3}$

95. $(4x^{3n})(-5x^{7n}) = -20x^{3n+7n} = -20x^{10n}$

97. $(-3x^{5n-2})(-4x^{2n+2}) =$
$12x^{(5n-2)+(2n+2)} =$
$12x^{7n}$

PROBLEM SET 6.3 Multiplying Polynomials

1. $(x+2)(y+3) =$
$x(y) + x(3) + 2(y) + 2(3) =$
$xy + 3x + 2y + 6$

3. $(x-4)(y+1) =$
$x(y) + x(1) - 4(y) - 4(1) =$
$xy + x - 4y - 4$

5. $(x-5)(y-6) =$
$x(y) + x(-6) - 5(y) - 5(-6) =$
$xy - 6x - 5y + 30$

7. $(x+2)(y+z+1) =$
$x(y) + x(z) + x(1) + 2(y) + 2(z) + 2(1) =$
$xy + xz + x + 2y + 2z + 2$

9. $(2x+3)(3y+1) =$
$2x(3y) + 2x(1) + 3(3y) + 3(1) =$
$6xy + 2x + 9y + 3$

11. $(x+3)(x+7) =$
$x(x) + x(7) + 3(x) + 3(7) =$
$x^2 + 7x + 3x + 21 =$
$x^2 + 10x + 21$

13. $(x+8)(x-3) =$
$x(x) + x(-3) + 8(x) + 8(-3) =$
$x^2 - 3x + 8x - 24 =$
$x^2 + 5x - 24$

15. $(x-7)(x+1) =$
$x(x) + x(1) - 7(x) - 7(1) =$
$x^2 + 1x - 7x - 7 = x^2 - 6x - 7$

17. $(n-4)(n-6) =$
$n(n) + n(-6) - 4(n) - 4(-6) =$
$n^2 - 6n - 4n + 24 =$
$n^2 - 10n + 24$

19. $(3n+1)(n+6) =$
$3n(n) + 3n(6) + 1(n) + 1(6) =$
$3n^2 + 18n + 1n + 6 = 3n^2 + 19n + 6$

21. $(5x-2)(3x+7) =$
$5x(3x) + 5x(7) - 2(3x) - 2(7) =$
$15x^2 + 35x - 6x - 14 =$
$15x^2 + 29x - 14$

23. $(x+3)(x^2+4x+9) =$
$x(x^2) + x(4x) + x(9) + 3(x^2)$
$+ 3(4x) + 3(9) =$
$x^3 + 4x^2 + 9x + 3x^2 + 12x + 27 =$
$x^3 + 7x^2 + 21x + 27$

25. $(x+4)(x^2-x-6) =$
$x(x^2)+x(-x)+x(-6)+4(x^2)$
$+4(-x)+4(-6) =$
$x^3-x^2-6x+4x^2-4x-24 =$
$x^3+3x^2-10x-24$

27. $(x-5)(2x^2+3x-7) =$
$x(2x^2)+x(3x)+x(-7)-5(2x^2)$
$-5(3x)-5(-7) =$
$2x^3+3x^2-7x-10x^2-15x+35 =$
$2x^3-7x^2-22x+35$

29. $(2a-1)(4a^2-5a+9) =$
$2a(4a^2)+2a(-5a)+2a(9)-1(4a^2)$
$-1(-5a)-1(9) =$
$8a^3-10a^2+18a-4a^2+5a-9 =$
$8a^3-14a^2+23a-9$

31. $(3a+5)(a^2-a-1) =$
$3a(a^2)+3a(-a)+3a(-1)+5(a^2)$
$+5(-a)+5(-1) =$
$3a^3-3a^2-3a+5a^2-5a-5 =$
$3a^3+2a^2-8a-5$

33. $(x^2+2x+3)(x^2+5x+4) =$
$x^2(x^2+5x+4)+2x(x^2+5x+4)$
$+3(x^2+5x+4) =$
$x^4+5x^3+4x^2+2x^3+10x^2+8x$
$+3x^2+15x+12 =$
$x^4+7x^3+17x^2+23x+12$

35. $(x^2-6x-7)(x^2+3x-9) =$
$x^2(x^2+3x-9)-6x(x^2+3x-9)$
$-7(x^2+3x-9) =$
$x^4+3x^3-9x^2-6x^3-18x^2+54x$
$-7x^2-21x+63 =$
$x^4-3x^3-34x^2+33x+63$

37. $(x+2)(x+9) =$
$x^2+(2+9)x+18 = x^2+11x+18$

39. $(x+6)(x-2) =$
$x^2+(6-2)x-12 =$
$x^2+4x-12$

41. $(x+3)(x-11) =$
$x^2+(3-11)x-33 =$
$x^2-8x-33$

43. $(n-4)(n-3) =$
$n^2+(-4-3)n+12 =$
$n^2-7n+12$

45. $(n+6)(n+12) =$
$n^2+(6+12)n+72 =$
$n^2+18n+72$

47. $(y+3)(y-7) =$
$y^2+(3-7)y-21 =$
$y^2-4y-21$

49. $(y-7)(y-12) =$
$y^2+(-7-12)y+84 =$
$y^2-19y+84$

51. $(x-5)(x+7) =$
$x^2+(-5+7)x-35 =$
$x^2+2x-35$

53. $(x-14)(x+8) =$
$x^2+(-14+8)x-112 =$
$x^2-6x-112$

55. $(a+10)(a-9) =$
$a^2+(10-9)a-90 =$
a^2+a-90

57. $(2a+1)(a+6) =$
$2a^2+(12+1)a+6 =$
$2a^2+13a+6$

59. $(5x-2)(x+7) =$
$5x^2+(35-2)x-14 =$
$5x^2+33x-14$

61. $(3x-7)(2x+1) =$
$6x^2+(3-14)x-7 =$
$6x^2-11x-7$

63. $(4a+3)(3a-4) =$
$12a^2+(-16+9)a-12 =$
$12a^2-7a-12$

65. $(6n-5)(2n-3) =$
$12n^2+(-18-10)n+15 =$
$12n^2-28n+15$

67. $(7x-4)(2x+3)=$
$14x^2+(21-8)x-12=$
$14x^2+13x-12$

69. $(5-x)(9-2x)=$
$45+(-10-9)x+2x^2=$
$45-19x+2x^2$

71. $(-2x+3)(4x-5)=$
$-8x^2+(10+12)x-15=$
$-8x^2+22x-15$

73. $(-3x-1)(3x-4)=$
$-9x^2+(12-3)x+4=$
$-9x^2+9x+4$

75. $(8n+3)(9n-4)=$
$72n^2+(-32+27)n-12=$
$72n^2-5n-12$

77. $(3-2x)(9-x)=$
$27+(-3-18)x+2x^2=$
$27-21x+2x^2$

79. $(-4x+3)(-5x-2)=$
$20x^2+(8-15)x-6=$
$20x^2-7x-6$

81. Use the pattern
$(a+b)^2=a^2+2ab+b^2$
$(x+7)^2=x^2+2(x)(7)+(7)^2=$
$x^2+14x+49$

83. Use the pattern
$(a+b)(a-b)=a^2-b^2$
$(5x-2)(5x+2)=(5x)^2-(2)^2=$
$25x^2-4$

85. Use the pattern
$(a-b)^2=a^2-2ab+b^2$
$(x-1)^2=x^2-2(x)(1)+(1)^2=$
x^2-2x+1

87. Use the pattern
$(a+b)^2=a^2+2ab+b^2$
$(3x+7)^2=(3x)^2+2(3x)(7)+(7)^2=$
$9x^2+42x+49$

89. Use the pattern
$(a-b)^2=a^2-2ab+b^2$
$(2x-3)^2=(2x)^2-2(2x)(3)+(3)^2=$
$4x^2-12x+9$

91. Use the pattern
$(a+b)(a-b)=a^2-b^2$
$(2x+3y)(2x-3y)=(2x)^2-(3y)^2=$
$4x^2-9y^2$

93. Use the pattern
$(a-b)^2=a^2-2ab+b^2$
$(1-5n)^2=1^2-2(1)(5n)+(5n)^2=$
$1-10n+25n^2$

95. Use the pattern
$(a+b)^2=a^2+2ab+b^2$
$(3x+4y)^2=(3x)^2+2(3x)(4y)+(4y)^2=$
$9x^2+24xy+16y^2$

97. Use the pattern
$(a+b)^2=a^2+2ab+b^2$
$(3+4y)^2=3^2+2(3)(4y)+(4y)^2=$
$9+24y+16y^2$

99. Use the pattern
$(a+b)(a-b)=a^2-b^2$
$(1+7n)(1-7n)=(1)^2-(7n)^2=$
$1-49n^2$

101. Use the pattern
$(a-b)^2=a^2-2ab+b^2$
$(4a-7b)^2=(4a)^2-2(4a)(7b)+(7b)^2=$
$16a^2-56ab+49b^2$

103. Use the pattern
$(a+b)^2=a^2+2ab+b^2$
$(x+8y)^2=x^2+2(x)(8y)+(8y)^2=$
$x^2+16xy+64y^2$

105. Use the pattern
$(a+b)(a-b)=a^2-b^2$
$(5x-11y)(5x+11y)=(5x)^2-(11y)^2=$
$25x^2-121y^2$

107. Use the pattern
$(a+b)(a-b)=a^2-b^2$
$x(8x+1)(8x-1)=x\left[(8x)^2-(1)^2\right]$
$x[64x^2-1]=64x^3-x$

109. Use the pattern
$(a+b)(a-b)=a^2-b^2$
$-2x(4x+y)(4x-y)=-2x\left[(4x)^2-(y)^2\right]$
$-2x[16x^2-y^2]=-32x^3+2xy^2$

111. $(x+2)^3=(x+2)(x+2)(x+2)=$
$(x+2)(x^2+4x+4)=$
$x(x^2+4x+4)+2(x^2+4x+4)=$
$x^3+4x^2+4x+2x^2+8x+8=$
$x^3+6x^2+12x+8$

113. $(x-3)^3=$
$(x-3)(x-3)(x-3)=$
$(x-3)(x^2-6x+9)=$
$x(x^2-6x+9)-3(x^2-6x+9)=$
$x^3-6x^2+9x-3x^2+18x-27=$
$x^3-9x^2+27x-27$

115. $(2n+1)^3=$
$(2n+1)(2n+1)(2n+1)=$
$(2n+1)(4n^2+4n+1)=$
$2n(4n^2+4n+1)+1(4n^2+4n+1)=$
$8n^3+8n^2+2n+4n^2+4n+1=$
$8n^3+12n^2+6n+1$

117. $(3n-2)^3=$
$(3n-2)(3n-2)(3n-2)=$
$(3n-2)(9n^2-12n+4)=$
$3n(9n^2-12n+4)-2(9n^2-12n+4)=$
$27n^3-36n^2+12n-18n^2+24n-8=$
$27n^3-54n^2+36n-8$

119. Let $x+3$ represent the width of the rectangle, then $x+5$ represents the length. Therefore, the area of the figure is represented by

	x	5
3	A	B
x	C	D

$\text{Area}=(\text{length})(\text{width})$
$=(x+3)(x+5)$
$=x^2+8x+15$

Geometrically, the sum of the area of each section would be

$\text{Area}=\text{Area of A}+\text{Area of B}+\text{Area of C}+\text{Area of D}$
$A=3x+15+x^2+5x$
$=x^2+8x+15$

121. Each side of the box will be $14-2x$ and the height will be x.

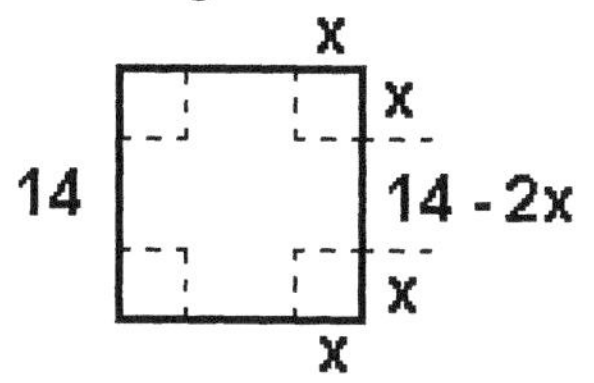

The volume is $V=lwh$
$=(14-2x)(14-2x)x$
$=\left(196-56x+4x^2\right)x$
$=196x-56x^2+4x^3$

The outside surface area is $S=\text{Original Area}-4\text{ Corners}$
$=(14)^2-4(x)^2$
$=196-4x^2$

125. **111)** $(x+2)^3=$
$(x)^3+3(x)^2 2+3(x)(2)^2+(2)^3=$
$x^3+6x^2+12x+8$

112) $(x+4)^3=$
$(x)^3+3(x)^2 4+3(x)(4)^2+(4)^3=$
$x^3+12x^2+48x+64$

113) $(x-3)^3=$
$(x)^3+3(x)^2(-3)+3(x)(-3)^2+(-3)^3=$
$x^3-9x^2+27x-27$

114) $(x-1)^3=$
$(x^3)-3(x)^2(1)+3(x)(1)^2-(1)^3=$
x^3-3x^2+3x-1

115) $(2n+1)^3=$
$(2n)^3+3(2n)^2(1)+3(2n)(1)^2+(1)^3=$
$8n^3+12n^2+6n+1$

116) $(3n+2)^3 =$
$(3n)^3+3(3n)^2(2)+3(3n)(2)^2+(2)^3 =$
$27n^3+54n^2+36n+8$

117) $(3n-2)^3 =$
$(3n)^3+3(3n)^2(-2)+3(3n)(-2)^2+(-2)^3 =$
$27n^3-54n^2+36n-8$

118) $(4n-3)^3 =$
$(4n)^3-3(4n)^2(3)+3(4n)(3)^2-(3)^3=$
$64n^3-144n^2+108n-27$

127 a. $21^2=(20+1)^2=20^2+2(20)(1)+1=$
$400+40+1=441$
b. $41^2=(40+1)^2=40^2+2(40)(1)+1=$
$1600+80+1=1681$
c. $71^2=(70+1)^2=70^2+2(70)(1)+1=$
$4900+140+1=5041$
d. $32^2=(30+2)^2=30^2+2(30)(2)+2^2=$
$900+120+4=1024$
e. $52^2=(50+2)^2=50^2+2(50)(2)+2^2=$
$2500+200+4=2704$
f. $82^2=(80+2)^2=80^2+2(80)(2)+2^2=$
$6400+320+4=6724$

129 a. $15^2=1(2)(100)+25=225$
b. $25^2=2(3)(100)+25=625$
c. $45^2=4(5)(100)+25=2025$
d. $55^2=5(6)(100)+25=3025$
e. $65^2=6(7)(100)+25=4225$
f. $75^2=7(8)(100)+25=5625$
g. $85^2=8(9)(100)+25=7225$
h. $95^2=9(10)(100)+25=9025$
i. $105^2=10(11)(100)+25=11025$

PROBLEM SET 6.4 Dividing by Monomials

1. $\dfrac{x^{10}}{x^2}=x^{10-2}=x^8$

3. $\dfrac{4x^3}{2x}=2x^{3-1}=2x^2$

5. $\dfrac{-16n^6}{2n^2}=-8n^{6-2}=-8n^4$

7. $\dfrac{72x^3}{-9x^3}=-8(1)=-8$

9. $\dfrac{65x^2y^3}{5xy}=13x^{2-1}y^{3-1}=13xy^2$

11. $\dfrac{-91a^4b^6}{-13a^3b^4}=7a^{4-3}b^{6-4}=7ab^2$

13. $\dfrac{18x^2y^6}{xy^2}=18x^{2-1}y^{6-2}=18xy^4$

15. $\dfrac{32x^6y^2}{-x}=-32x^{6-1}y^2=-32x^5y^2$

17. $\dfrac{-96x^5y^7}{12y^3}=-8x^5y^{7-3}=-8x^5y^4$

19. $\dfrac{-ab}{ab}=-1(1)(1)=-1$

21. $\dfrac{56a^2b^3c^5}{4abc}=14a^{2-1}b^{3-1}c^{5-1}=14ab^2c^4$

23. $\dfrac{-80xy^2z^6}{-5xyz^2}=16(1)y^{2-1}z^{6-2}=16yz^4$

25. $\dfrac{8x^4+12x^5}{2x^2}=\dfrac{8x^4}{2x^2}+\dfrac{12x^5}{2x^2}=4x^2+6x^3$

27. $\dfrac{9x^6-24x^4}{3x^3}=\dfrac{9x^6}{3x^3}-\dfrac{24x^4}{3x^3}=3x^3-8x$

29. $\dfrac{-28n^5+36n^2}{4n^2}=\dfrac{-28n^5}{4n^2}+\dfrac{36n^2}{4n^2}=$
$-7n^3+9$

31. $\dfrac{35x^6 - 56x^5 - 84x^3}{7x^2} =$

$\dfrac{35x^6}{7x^2} - \dfrac{56x^5}{7x^2} - \dfrac{84x^3}{7x^2} =$

$5x^4 - 8x^3 - 12x$

33. $\dfrac{-24n^8 + 48n^5 - 78n^3}{-6n^3} =$

$\dfrac{-24n^8}{-6n^3} + \dfrac{48n^5}{-6n^3} - \dfrac{78n^3}{-6n^3} =$

$4n^5 - 8n^2 + 13$

35. $\dfrac{-60a^7 - 96a^3}{-12a} = \dfrac{-60a^7}{-12a} - \dfrac{96a^3}{-12a} =$

$5a^6 + 8a^2$

37. $\dfrac{27x^2y^4 - 45xy^4}{-9xy^3} =$

$\dfrac{27x^2y^4}{-9xy^3} - \dfrac{45xy^4}{-9xy^3} =$

$-3xy + 5y$

39. $\dfrac{48a^2b^2 + 60a^3b^4}{-6ab} =$

$\dfrac{48a^2b^2}{-6ab} + \dfrac{60a^3b^4}{-6ab} =$

$-8ab - 10a^2b^3$

41. $\dfrac{12a^2b^2c^2 - 52a^2b^3c^5}{-4a^2bc} =$

$\dfrac{12a^2b^2c^2}{-4a^2bc} - \dfrac{52a^2b^3c^5}{-4a^2bc} =$

$-3bc + 13b^2c^4$

43. $\dfrac{9x^2y^3 - 12x^3y^4}{-xy} = \dfrac{9x^2y^3}{-xy} - \dfrac{12x^3y^4}{-xy} =$

$-9xy^2 + 12x^2y^3$

45. $\dfrac{-42x^6 - 70x^4 + 98x^2}{14x^2} =$

$\dfrac{-42x^6}{14x^2} - \dfrac{70x^4}{14x^2} + \dfrac{98x^2}{14x^2} =$

$-3x^4 - 5x^2 + 7$

47. $\dfrac{15a^3b - 35a^2b - 65ab^2}{-5ab} =$

$\dfrac{15a^3b}{-5ab} - \dfrac{35a^2b}{-5ab} - \dfrac{65ab^2}{-5ab} =$

$-3a^2 + 7a + 13b$

49. $\dfrac{-xy + 5x^2y^3 - 7x^2y^6}{xy} =$

$\dfrac{-xy}{xy} + \dfrac{5x^2y^3}{xy} - \dfrac{7x^2y^6}{xy} =$

$-1 + 5xy^2 - 7xy^5$

PROBLEM SET 6.5 Dividing by Binomials

1.

$$\begin{array}{r|l} & x + 12 \\ \hline x + 4 & x^2 + 16x + 48 \\ & \underline{x^2 + 4x} \\ & 12x + 48 \\ & \underline{12x + 48} \end{array}$$

3.

$$\begin{array}{r|l} & x + 2 \\ \hline x - 7 & x^2 - 5x - 14 \\ & \underline{x^2 - 7x} \\ & 2x - 14 \\ & \underline{2x - 14} \end{array}$$

Problem Set 6.5

5.

$$\begin{array}{r} x+8 \\ x+3\overline{)x^2+11x+28} \\ \underline{x^2+3x} \\ 8x+28 \\ \underline{8x+24} \\ 4 \end{array}$$

7.

$$\begin{array}{r} x+4 \\ x-8\overline{)x^2-4x-39} \\ \underline{x^2-8x} \\ 4x-39 \\ \underline{4x-32} \\ -7 \end{array}$$

9.

$$\begin{array}{r} 5n+4 \\ n-1\overline{)5n^2-n-4} \\ \underline{5n^2-5n} \\ 4n-4 \\ \underline{4n-4} \end{array}$$

11.

$$\begin{array}{r} 8y-3 \\ y+7\overline{)8y^2+53y-19} \\ \underline{8y^2+56y} \\ -3y-19 \\ \underline{-3y-21} \\ 2 \end{array}$$

13.

$$\begin{array}{r} 4x-7 \\ 5x+1\overline{)20x^2-31x-7} \\ \underline{20x^2+4x} \\ -35x-7 \\ \underline{-35x-7} \end{array}$$

15.

$$\begin{array}{r} 3x+2 \\ 2x+7\overline{)6x^2+25x+8} \\ \underline{6x^2+21x} \\ 4x+8 \\ \underline{4x+14} \\ -6 \end{array}$$

17.

$$\begin{array}{r} 2x^2+3x+4 \\ x-2\overline{)2x^3-x^2-2x-8} \\ \underline{2x^3-4x^2} \\ 3x^2-2x-8 \\ \underline{3x^2-6x} \\ 4x-8 \\ \underline{4x-8} \end{array}$$

19.

$$\begin{array}{r} 5n^2-4n-3 \\ n+3\overline{)5n^3+11n^2-15n-9} \\ \underline{5n^3+15n^2} \\ -4n^2-15n-9 \\ \underline{-4n^2-12n} \\ -3n-9 \\ \underline{-3n-9} \end{array}$$

21.

$$\begin{array}{r} n^2+6n-4 \\ n-6\overline{)n^3+0n^2-40n+24} \\ \underline{n^3-6n^2} \\ 6n^2-40n+24 \\ \underline{6n^2-36n} \\ -4n+24 \\ \underline{-4n+24} \end{array}$$

23.

$$\begin{array}{r} x^2+3x+9 \\ x-3\overline{)x^3+0x^2+0x-27} \\ \underline{x^3-3x^2} \\ 3x^2-0x-27 \\ \underline{3x^2-9x} \\ 9x-27 \\ \underline{9x-27} \end{array}$$

25.

$$\begin{array}{r} 9x^2+12x+16 \\ 3x-4\overline{)27x^3+0x^2+0x-64} \\ \underline{27x^3-36x^2} \\ 36x^2+0x-64 \\ \underline{36x^2-48x} \\ 48x-64 \\ \underline{48x-64} \end{array}$$

27.
$$\begin{array}{r} 3n - 8 \\ n+2\overline{\smash{)}\,3n^2 - 2n + 1} \\ \underline{3n^2 + 6n\phantom{{}+1}} \\ -8n + 1 \\ \underline{-8n - 16} \\ 17 \end{array}$$

29.
$$\begin{array}{r} 3t + 2 \\ 3t-1\overline{\smash{)}\,9t^2 + 3t + 4} \\ \underline{9t^2 - 3t\phantom{{}+4}} \\ 6t + 4 \\ \underline{6t - 2} \\ 6 \end{array}$$

31.
$$\begin{array}{r} 3n^2 - n - 4 \\ 2n-1\overline{\smash{)}\,6n^3 - 5n^2 - 7n + 4} \\ \underline{6n^3 - 3n^2\phantom{{}-7n+4}} \\ -2n^2 - 7n + 4 \\ \underline{-2n^2 + 1n\phantom{{}+4}} \\ -8n + 4 \\ \underline{-8n + 4} \end{array}$$

33.
$$\begin{array}{r} 4x^2 - 5x + 5 \\ x+7\overline{\smash{)}\,4x^3 + 23x^2 - 30x + 32} \\ \underline{4x^3 + 28x^2\phantom{{}-30x+32}} \\ -5x^2 - 30x + 32 \\ \underline{-5x^2 - 35x\phantom{{}+32}} \\ 5x + 32 \\ \underline{5x + 35} \\ -3 \end{array}$$

35.
$$\begin{array}{r} x + 4 \\ x^2-2x\overline{\smash{)}\,x^3 + 2x^2 - 3x - 1} \\ \underline{x^3 - 2x^2\phantom{{}-3x-1}} \\ 4x^2 - 3x - 1 \\ \underline{4x^2 - 8x\phantom{{}-1}} \\ 5x - 1 \end{array}$$

37.
$$\begin{array}{r} 2x - 12 \\ x^2+4x\overline{\smash{)}\,2x^3 - 4x^2 + x - 5} \\ \underline{2x^3 + 8x^2\phantom{{}+x-5}} \\ -12x^2 + x - 5 \\ \underline{-12x^2 - 48x\phantom{{}-5}} \\ 49x - 5 \end{array}$$

39.
$$\begin{array}{r} x^3 - 2x^2 + 4x - 8 \\ x+2\overline{\smash{)}\,x^4 + 0x^3 + 0x^2 + 0x - 16} \\ \underline{x^4 + 2x^3\phantom{{}+0x^2+0x-16}} \\ -2x^3 + 0x^2 + 0x - 16 \\ \underline{-2x^3 - 4x^2\phantom{{}+0x-16}} \\ 4x^2 + 0x - 16 \\ \underline{4x^2 + 8x\phantom{{}-16}} \\ -8x - 16 \\ \underline{-8x - 16} \end{array}$$

PROBLEM SET 6.6 Zero and Negative Integers as Exponents

1. $3^{-2} = \dfrac{1}{3^2} = \dfrac{1}{9}$

3. $4^{-3} = \dfrac{1}{4^3} = \dfrac{1}{64}$

5. $\left(\dfrac{3}{2}\right)^{-1} = \left(\dfrac{2}{3}\right)^{1} = \dfrac{2}{3}$

7. $\dfrac{1}{2^{-4}} = \dfrac{1}{\frac{1}{2^4}} = 2^4 = 16$

9. $\left(-\dfrac{4}{3}\right)^{0} = 1$

Problem Set 6.6

11. $\left(-\frac{2}{3}\right)^{-3} = \left(-\frac{3}{2}\right)^{3} = -\frac{27}{8}$

13. $(-2)^{-2} = \frac{1}{(-2)^2} = \frac{1}{4}$

15. $-\left(3^{-2}\right) = -\frac{1}{3^2} = -\frac{1}{9}$

17. $\frac{1}{\left(\frac{3}{4}\right)^{-3}} = \left(\frac{3}{4}\right)^{3} = \frac{27}{64}$

19. $2^6 \cdot 2^{-9} = 2^{6-9} = 2^{-3} = \frac{1}{2^3} = \frac{1}{8}$

21. $3^6 \cdot 3^{-3} = 3^{6-3} = 3^3 = 27$

23. $\frac{10^2}{10^{-1}} = 10^{2-(-1)} = 10^{2+1} = 10^3 = 1000$

25. $\frac{10^{-1}}{10^2} = 10^{-1-2} = 10^{-3} = \frac{1}{10^3} = \frac{1}{1000}$

27. $\left(2^{-1} \cdot 3^{-2}\right)^{-1} = \left(2^{-1}\right)^{-1} \cdot \left(3^{-2}\right)^{-1} =$
$2 \cdot 3^2 = 2 \cdot 9 = 18$

29. $\left(\frac{4^{-1}}{3}\right)^{-2} = \frac{\left(4^{-1}\right)^{-2}}{3^{-2}} = \frac{4^2}{3^{-2}} =$
$4^2 \cdot 3^2 = 16 \cdot 9 = 144$

31. $x^6 x^{-1} = x^{6-1} = x^5$

33. $n^{-4} n^2 = n^{-4+2} = n^{-2} = \frac{1}{n^2}$

35. $a^{-2} a^{-3} = a^{-2-3} = a^{-5} = \frac{1}{a^5}$

37. $(2x^3)(4x^{-2}) = 2 \cdot 4 \cdot x^{3-2} = 8x$

39. $\left(3x^{-6}\right)\left(9x^2\right) = 3 \cdot 9 \cdot x^{-6+2} = 27x^{-4} = \frac{27}{x^4}$

41. $\left(5y^{-1}\right)\left(-3y^{-2}\right) = (5)(-3)y^{-1-2} =$
$-15y^{-3} = -\frac{15}{y^3}$

43. $\left(8x^{-4}\right)\left(12x^4\right) = 8 \cdot 12 \cdot x^{-4+4} =$
$96x^0 = 96(1) = 96$

45. $\frac{x^7}{x^{-3}} = x^{7-(-3)} = x^{7+3} = x^{10}$

47. $\frac{n^{-1}}{n^3} = n^{-1-3} = n^{-4} = \frac{1}{n^4}$

49. $\frac{4n^{-1}}{2n^{-3}} = 2n^{-1-(-3)} = 2n^{-1+3} = 2n^2$

51. $\frac{-24x^{-6}}{8x^{-2}} = -3x^{-6-(-2)} =$
$-3x^{-6+2} = -3x^{-4} = -\frac{3}{x^4}$

53. $\frac{-52y^{-2}}{-13y^{-2}} = 4y^{-2-(-2)} = 4y^{-2+2} =$
$4y^0 = 4(1) = 4$

55. $\left(x^{-3}\right)^{-2} = x^{-3(-2)} = x^6$

57. $\left(x^2\right)^{-2} = x^{2(-2)} = x^{-4} = \frac{1}{x^4}$

59. $\left(x^3 y^4\right)^{-1} = \left(x^3\right)^{-1}\left(y^4\right)^{-1} =$
$x^{-3} y^{-4} = \frac{1}{x^3 y^4}$

61. $\left(x^{-2} y^{-1}\right)^3 = \left(x^{-2}\right)^3\left(y^{-1}\right)^3 =$
$x^{-6} y^{-3} = \frac{1}{x^6 y^3}$

63. $\left(2n^{-2}\right)^3 = (2)^3\left(n^{-2}\right)^3 = 8n^{-6} = \frac{8}{n^6}$

65. $\left(4n^3\right)^{-2} = (4)^{-2}\left(n^3\right)^{-2} = 4^{-2} n^{-6} =$
$\frac{1}{4^2 n^6} = \frac{1}{16n^6}$

67. $\left(3a^{-2}\right)^4 = 3^4\left(a^{-2}\right)^4 = 81a^{-8} = \frac{81}{a^8}$

69. $\left(5x^{-1}\right)^{-2} = 5^{-2}\left(x^{-1}\right)^{-2} = 5^{-2} x^2 =$
$\frac{x^2}{5^2} = \frac{x^2}{25}$

71. $\left(2x^{-2}y^{-1}\right)^{-1} = 2^{-1}\left(x^{-2}\right)^{-1}\left(y^{-1}\right)^{-1} = 2^{-1}x^2y^1 = \frac{x^2y}{2}$

73. $\left(\frac{x^2}{y}\right)^{-1} = \frac{\left(x^2\right)^{-1}}{y^{-1}} = \frac{x^{-2}}{y^{-1}} = \frac{y}{x^2}$

75. $\left(\frac{a^{-1}}{b^2}\right)^{-4} = \frac{a^4}{b^{-8}} = a^4b^8$

77. $\left(\frac{x^{-1}}{y^{-3}}\right)^{-2} = \frac{x^2}{y^6}$

79. $\left(\frac{x^2}{x^3}\right)^{-1} = \left(x^{2-3}\right)^{-1} = \left(x^{-1}\right)^{-1} = x$

81. $\left(\frac{2x^{-1}}{x^{-2}}\right)^{-3} = \left(2x^{-1-(-2)}\right)^{-3} = \left(2x^{-1+2}\right)^{-3} = (2x)^{-3} = \frac{1}{(2x)^3} = \frac{1}{8x^3}$

83. $\left(\frac{18x^{-1}}{9x}\right)^{-2} = \left(2x^{-1-1}\right)^{-2} = \left(2x^{-2}\right)^{-2} = 2^{-2}x^4 = \frac{x^4}{2^2} = \frac{x^4}{4}$

85. $321 = (3.21)(10^2)$
Number > 10, positive exponent.

87. $8000 = (8)(10^3)$
Number > 10, positive exponent.

89. $0.00246 = (2.46)(10^{-3})$
Number < 1, negative exponent.

91. $0.0000179 = (1.79)(10^{-5})$
Number < 1, negative exponent.

93. $87{,}000{,}000 = (8.7)(10^7)$
Number > 10, positive exponent.

95. $(8)(10^3) = 8000$
Positive exponent, move decimal right.

97. $(5.21)(10^4) = 52{,}100$
Positive exponent, move decimal right.

99. $(1.14)(10^7) = 11{,}400{,}000$
Positive exponent, move decimal right.

101. $(7)(10^{-2}) = 0.07$
Negative exponent, move decimal left.

103. $(9.87)(10^{-4}) = 0.000987$
Negative exponent, move decimal left.

105. $(8.64)(10^{-6}) = 0.00000864$
Negative exponent, move decimal left.

107. $(0.007)(120) = (7)(10^{-3})(1.2)(10^2) = (7)(1.2)(10^{-3})(10^2) = (8.4)(10^{-1}) = 0.84$

109. $(5{,}000{,}000)(0.00009) = (5)(10^6)(9)(10^{-5}) = (5)(9)(10^6)(10^{-5}) = (45)(10^1) = 450$

111. $\frac{6000}{0.0015} = \frac{(6)(10^3)}{(1.5)(10^{-3})} = (4)\left(10^{3-(-3)}\right) = (4)(10^6) = 4{,}000{,}000$

113. $\frac{0.00086}{4300} = \frac{(8.6)(10^{-4})}{(4.3)(10^3)} = (2)\left(10^{-4-3}\right) = (2)(10^{-7}) = 0.0000002$

115. $\frac{0.00039}{0.0013} = \frac{(3.9)(10^{-4})}{(1.3)(10^{-3})} = (3)\left(10^{-4-(-3)}\right) = (3)(10^{-1}) = 0.3$

117. $\frac{(0.0008)(0.07)}{(20{,}000)(0.0004)} = \frac{(8)(10^{-4})(7)(10^{-2})}{(2)(10^4)(4)(10^{-4})} = \frac{(56)(10^{-6})}{(8)(10^0)} =$

$(7)(10^{-6}) = 0.000007$

CHAPTER 6 Review Problem Set

1. $(5x^2 - 6x + 4) + (3x^2 - 7x - 2) =$
$(5 + 3)x^2 + (-6 - 7)x + (4 - 2) =$
$8x^2 - 13x + 2$

2. $(7y^2 + 9y - 3) - (4y^2 - 2y + 6) =$
$7y^2 + 9y - 3 - 4y^2 + 2y - 6 =$
$3y^2 + 11y - 9$

3. $(2x^2 + 3x - 4) + (4x^2 - 3x - 6)$
$- (3x^2 - 2x - 1) =$
$2x^2 + 3x - 4 + 4x^2 - 3x - 6 - 3x^2$
$+ 2x + 1 = 3x^2 + 2x - 9$

4. $(-3x^2 - 2x + 4) - (x^2 - 5x - 6)$
$- (4x^2 + 3x - 8) =$
$-3x^2 - 2x + 4 - x^2 + 5x + 6$
$- 4x^2 - 3x + 8 =$
$-8x^2 + 18$

5. $5(2x - 1) + 7(x + 3) - 2(3x + 4) =$
$10x - 5 + 7x + 21 - 6x - 8 =$
$11x + 8$

6. $3(2x^2 - 4x - 5) - 5(3x^2 - 4x + 1) =$
$6x^2 - 12x - 15 - 15x^2 + 20x - 5 =$
$-9x^2 + 8x - 20$

7. $6(y^2 - 7y - 3) - 4(y^2 + 3y - 9) =$
$6y^2 - 42y - 18 - 4y^2 - 12y + 36 =$
$2y^2 - 54y + 18$

8. $3(a - 1) - 2(3a - 4) - 5(2a + 7) =$
$3a - 3 - 6a + 8 - 10a - 35 =$
$-13a - 30$

9. $-(a + 4) + 5(-a - 2) - 7(3a - 1) =$
$-a - 4 - 5a - 10 - 21a + 7 =$
$-27a - 7$

10. $-2(3n - 1) - 4(2n + 6) + 5(3n + 4) =$
$-6n + 2 - 8n - 24 + 15n + 20 =$
$n - 2$

11. $3(n^2 - 2n - 4) - 4(2n^2 - n - 3) =$
$3n^2 - 6n - 12 - 8n^2 + 4n + 12 =$
$-5n^2 - 2n$

12. $-5(-n^2 + n - 1) + 3(4n^2 - 3n - 7) =$
$5n^2 - 5n + 5 + 12n^2 - 9n - 21 =$
$17n^2 - 14n - 16$

13. $(5x^2)(7x^4) = 5 \cdot 7 \cdot x^2 \cdot x^4$
$= 35x^{2+4} = 35x^6$

14. $(-6x^3)(9x^5) = -6 \cdot 9 \cdot x^3 \cdot x^5 =$
$-54x^{3+5} = -54x^8$

15. $(-4xy^2)(-6x^2y^3) =$
$(-4)(-6)(x^{1+2})(y^{2+3}) = 24x^3y^5$

16. $(2a^3b^4)(-3ab^5) =$
$(2)(-3)(a^{3+1})(b^{4+5}) =$
$-6a^4b^9$

17. $(2a^2b^3)^3 = (2)^3(a^2)^3(b^3)^3 = 8a^6b^9$

18. $(-3xy^2)^2 = (-3)^2(x)^2(y^2)^2 =$
$9x^2y^4$

19. $5x(7x + 3) = 35x^2 + 15x$

20. $(-3x^2)(8x - 1) = -24x^3 + 3x^2$

21. $(x + 9)(x + 8) =$
$x^2 + (9 + 8)x + 72 =$
$x^2 + 17x + 72$

22. $(3x + 7)(x + 1) =$
$3x^2 + (3 + 7)x + 7 =$
$3x^2 + 10x + 7$

23. $(x - 5)(x + 2) =$
$x^2 + (-5 + 2)x - 10 =$
$x^2 - 3x - 10$

24. $(y - 4)(y - 9) =$
$y^2 + (-9 - 4)y + 36 =$
$y^2 - 13y + 36$

25. $(2x - 1)(7x + 3) =$
$14x^2 + (6 - 7)x - 3 =$
$14x^2 - x - 3$

26. $(4a-7)(5a+8) =$
$20a^2 + (32-35)a - 56 =$
$20a^2 - 3a - 56$

27. $(3a-5)^2 =$
$(3a)^2 - 2(3a)(5) + (5)^2 =$
$9a^2 - 30a + 25$

28. $(x+6)(2x^2+5x-4) =$
$x(2x^2+5x-4) + 6(2x^2+5x-4) =$
$2x^3 + 5x^2 - 4x + 12x^2 + 30x - 24 =$
$2x^3 + 17x^2 + 26x - 24$

29. $(5n-1)(6n+5) =$
$30n^2 + (25-6)n - 5 =$
$30n^2 + 19n - 5$

30. $(3n+4)(4n-1) =$
$12n^2 + (-3+16)n - 4 =$
$12n^2 + 13n - 4$

31. $(2n+1)(2n-1) =$
$(2n)^2 - (1)^2 =$
$4n^2 - 1$

32. $(4n-5)(4n+5) =$
$(4n)^2 - (5)^2 =$
$16n^2 - 25$

33. $(2a+7)^2 =$
$(2a)^2 + 2(2a)(7) + (7)^2 =$
$4a^2 + 28a + 49$

34. $(3a+5)^2 =$
$(3a)^2 + 2(3a)(5) + (5)^2 =$
$9a^2 + 30a + 25$

35. $(x-2)(x^2-x+6) =$
$x(x^2-x+6) - 2(x^2-x+6) =$
$x^3 - x^2 + 6x - 2x^2 + 2x - 12 =$
$x^3 - 3x^2 + 8x - 12$

36. $(2x-1)(x^2+4x+7) =$
$2x(x^2+4x+7) - 1(x^2+4x+7) =$
$2x^3 + 8x^2 + 14x - x^2 - 4x - 7 =$
$2x^3 + 7x^2 + 10x - 7$

37. $(a+5)^3 = (a+5)(a+5)(a+5) =$
$(a+5)(a^2+10a+25) =$
$a(a^2+10a+25) + 5(a^2+10a+25) =$
$a^3 + 10a^2 + 25a + 5a^2 + 50a + 125 =$
$a^3 + 15a^2 + 75a + 125$

38. $(a-6)^3 = (a-6)(a-6)(a-6) =$
$(a-6)(a^2-12a+36) =$
$a(a^2-12a+36) - 6(a^2-12a+36) =$
$a^3 - 12a^2 + 36a - 6a^2 + 72a - 216 =$
$a^3 - 18a^2 + 108a - 216$

39. $(x^2-x-1)(x^2+2x+5) =$
$x^2(x^2+2x+5) - x(x^2+2x+5)$
$\quad - 1(x^2+2x+5) =$
$x^4 + 2x^3 + 5x^2 - x^3 - 2x^2 - 5x$
$\quad - x^2 - 2x - 5 =$
$x^4 + x^3 + 2x^2 - 7x - 5$

40. $(n^2+2n+4)(n^2-7n-1) =$
$n^2(n^2-7n-1) + 2n(n^2-7n-1)$
$\quad + 4(n^2-7n-1) =$
$n^4 - 7n^3 - n^2 + 2n^3 - 14n^2 - 2n$
$\quad + 4n^2 - 28n - 4 =$
$n^4 - 5n^3 - 11n^2 - 30n - 4$

41. $\dfrac{36x^4y^5}{-3xy^2} = -12x^{4-1}y^{5-2} = -12x^3y^3$

42. $\dfrac{-56a^5b^7}{-8a^2b^3} = 7a^{5-2}b^{7-3} = 7a^3b^4$

43. $\dfrac{-18x^4y^3 - 54x^6y^2}{6x^2y^2} =$
$\dfrac{-18x^4y^3}{6x^2y^2} - \dfrac{54x^6y^2}{6x^2y^2} =$
$-3x^2y - 9x^4$

44. $\dfrac{-30a^5b^{10} + 39a^4b^8}{-3ab} =$
$\dfrac{-30a^5b^{10}}{-3ab} + \dfrac{39a^4b^8}{-3ab} =$
$10a^4b^9 - 13a^3b^7$

Chapter 6 Review Problem Set

45. $\dfrac{56x^4 - 40x^3 - 32x^2}{4x^2} =$

$\dfrac{56x^4}{4x^2} - \dfrac{40x^3}{4x^2} - \dfrac{32x^2}{4x^2} =$

$14x^2 - 10x - 8$

46.

$$\begin{array}{r|l} & x + 4 \\ x+5 & x^2 + 9x - 1 \\ & \underline{x^2 + 5x} \\ & \quad 4x - 1 \\ & \quad \underline{4x + 20} \\ & \qquad -21 \end{array}$$

47.

$$\begin{array}{r|l} & 7x - 6 \\ 3x+2 & 21x^2 - 4x - 12 \\ & \underline{21x^2 + 14x} \\ & \quad -18x - 12 \\ & \quad \underline{-18x - 12} \end{array}$$

48.

$$\begin{array}{r|l} & 2x^2 + x + 4 \\ x-2 & 2x^3 - 3x^2 + 2x - 4 \\ & \underline{2x^3 - 4x^2} \\ & \quad x^2 + 2x - 4 \\ & \quad \underline{x^2 - 2x} \\ & \qquad 4x - 4 \\ & \qquad \underline{4x - 8} \\ & \qquad\quad 4 \end{array}$$

49. $3^2 + 2^2 = 9 + 4 = 13$

50. $(3 + 2)^2 = (5)^2 = 25$

51. $2^{-4} = \dfrac{1}{2^4} = \dfrac{1}{16}$

52. $(-5)^0 = 1$

53. $-5^0 = -1$

54. $\dfrac{1}{3^{-2}} = 3^2 = 9$

55. $\left(\dfrac{3}{4}\right)^{-2} = \left(\dfrac{4}{3}\right)^2 = \dfrac{16}{9}$

56. $\dfrac{1}{\left(\dfrac{1}{4}\right)^{-1}} = \left(\dfrac{1}{4}\right)^1 = \dfrac{1}{4}$

57. $\dfrac{1}{(-2)^{-3}} = (-2)^3 = -8$

58. $2^{-1} + 3^{-2} = \dfrac{1}{2} + \dfrac{1}{3^2} = \dfrac{1}{2} + \dfrac{1}{9} =$

$\dfrac{9}{18} + \dfrac{2}{18} = \dfrac{11}{18}$

59. $3^0 + 2^{-2} = 1 + \dfrac{1}{2^2} =$

$1 + \dfrac{1}{4} = \dfrac{4}{4} + \dfrac{1}{4} = \dfrac{5}{4}$

60. $(2 + 3)^{-2} = (5)^{-2} = \dfrac{1}{5^2} = \dfrac{1}{25}$

61. $x^5 x^{-8} = x^{5-8} = x^{-3} = \dfrac{1}{x^3}$

62. $(3x^5)(4x^{-2}) = 12x^{5-2} = 12x^3$

63. $\dfrac{x^{-4}}{x^{-6}} = x^{-4-(-6)} = x^{-4+6} = x^2$

64. $\dfrac{x^{-6}}{x^{-4}} = x^{-6-(-4)} = x^{-6+4} = x^{-2} = \dfrac{1}{x^2}$

65. $\dfrac{24a^5}{3a^{-1}} = 8a^{5-(-1)} = 8a^6$

66. $\dfrac{48n^{-2}}{12n^{-1}} = 4n^{-2-(-1)} =$

$4n^{-2+1} = 4n^{-1} = \dfrac{4}{n}$

67. $(x^{-2}y)^{-1} = (x^{-2})^{-1}(y)^{-1} = x^2y^{-1} = \dfrac{x^2}{y}$

68. $(a^2b^{-3})^{-2} = (a^2)^{-2}(b^{-3})^{-2} =$

$a^{-4}b^6 = \dfrac{b^6}{a^4}$

69. $(2x)^{-1} = \frac{1}{2x}$

70. $(3n^2)^{-2} = 3^{-2}(n^2)^{-2} = 3^{-2}n^{-4} = \frac{1}{3^2n^4} = \frac{1}{9n^4}$

71. $(2n^{-1})^{-3} = (2^{-3})(n^{-1})^{-3} = 2^{-3}n^3 = \frac{n^3}{2^3} = \frac{n^3}{8}$

72. $(4ab^{-1})(-3a^{-1}b^2) = -12a^{1-1}b^{-1+2} = -12a^0b^1 = -12b$

73. $(6.1)(10^2) = 610$

74. $(5.6)(10^4) = 56,000$

75. $(8)(10^{-2}) = 0.08$

76. $(9.2)(10^{-4}) = 0.00092$

77. $9000 = (9)(10^3)$

78. $47 = (4.7)(10^1)$

79. $0.047 = (4.7)(10^{-2})$

80. $0.00021 = (2.1)(10^{-4})$

81. $(0.00004)(12,000) = (4)(10^{-5})(1.2)(10^4) = (4)(1.2)(10^{-5})(10^4) = (4.8)(10^{-1}) = 0.48$

82. $(0.0021)(2000) = (2.1)(10^{-3})(2)(10^3) = (2.1)(2)(10^{-3})(10^3) = (4.2)(10^0) = 4.2$

83. $\frac{0.0056}{0.0000028} = \frac{(5.6)(10^{-3})}{(2.8)(10^{-6})} = (2)(10^{-3-(-6)}) = (2)(10^3) = 2000$

84. $\frac{0.00078}{39,000} = \frac{(7.8)(10^{-4})}{(3.9)(10^4)} = (2)(10^{-4-4}) = (2)(10^{-8}) = 0.00000002$

CHAPTER 6 | Test

1. $(-7x^2 + 6x - 2) + (5x^2 - 8x + 7) = -2x^2 - 2x + 5$

2. $(-4x^2 + 3x + 6) - (-x^2 + 9x - 14) = -4x^2 + 3x + 6 + x^2 - 9x + 14 = -3x^2 - 6x + 20$

3. $3(2x - 1) - 6(3x - 2) - (x + 7) = 6x - 3 - 18x + 12 - x - 7 = -13x + 2$

4. $(-4xy^2)(7x^2y^3) = (-4)(7)(x^{1+2})(y^{2+3}) = -28x^3y^5$

5. $(2x^2y)^2(3xy^3) = (4x^4y^2)(3xy^3) = 12x^5y^5$

6. $(x - 9)(x + 2) = x^2 + (2 - 9)x - 18 = x^2 - 7x - 18$

7. $(n + 14)(n - 7) = n^2 + (14 - 7)n - 98 = n^2 + 7n - 98$

8. $(5a + 3)(8a + 7) = 40a^2 + (35 + 24)a + 21 = 40a^2 + 59a + 21$

9. $(3x - 7y)^2 = 9x^2 - 2(3x)(7y) + 49y^2 = 9x^2 - 42xy + 49y^2$

10. $(x + 3)(2x^2 - 4x - 7) = x(2x^2 - 4x - 7) + 3(2x^2 - 4x - 7) = 2x^3 - 4x^2 - 7x + 6x^2 - 12x - 21 = 2x^3 + 2x^2 - 19x - 21$

11. $(9x - 5y)(9x + 5y) = 81x^2 - 25y^2$

Chapter 6 Test

12. $(3x-7)(5x-11)=$
$15x^2+(-33-35)x+77=$
$15x^2-68x+77$

13. $\dfrac{-96x^4y^5}{-12x^2y}=8x^{4-2}y^{5-1}=8x^2y^4$

14. $\dfrac{56x^2y-72xy^2}{-8xy}=\dfrac{56x^2y}{-8xy}-\dfrac{72xy^2}{-8xy}=$
$-7x+9y$

15.

$$
\begin{array}{r|l}
 & x^2+\ 4x-5 \\
2x-3 & 2x^3+5x^2-22x+15 \\
 & \underline{2x^3-3x^2} \\
 & 8x^2-22x+15 \\
 & \underline{8x^2-12x} \\
 & -10x+15 \\
 & \underline{-10x+15}
\end{array}
$$

16.

$$
\begin{array}{r|l}
 & 4x^2-\ x+6 \\
x+6 & 4x^3+23x^2+0x+36 \\
 & \underline{4x^3+24x^2} \\
 & -1x^2-0x+36 \\
 & \underline{-1x^2-6x} \\
 & 6x+36 \\
 & \underline{6x+36}
\end{array}
$$

17. $\left(\dfrac{2}{3}\right)^{-3}=\left(\dfrac{3}{2}\right)^{3}=\dfrac{27}{8}$

18. $4^{-2}+4^{-1}+4^0=\dfrac{1}{4^2}+\dfrac{1}{4}+1=$
$\dfrac{1}{16}+\dfrac{1}{4}+1=\dfrac{1}{16}+\dfrac{4}{16}+\dfrac{16}{16}=\dfrac{21}{16}$

19. $\dfrac{1}{2^{-4}}=2^4=16$

20. $(-6x^{-4})(4x^2)=-24x^{-2}=-\dfrac{24}{x^2}$

21. $\left(\dfrac{8x^{-1}}{2x^2}\right)^{-1}=\left(4x^{-1-2}\right)^{-1}=$
$\left(4x^{-3}\right)^{-1}=4^{-1}x^3=\dfrac{x^3}{4}$

22. $(x^{-3}y^5)^{-2}=(x^{-3})^{-2}(y^5)^{-2}=$
$x^6y^{-10}=\dfrac{x^6}{y^{10}}$

23. $0.00027=(2.7)(10^{-4})$

24. $(9.2)(10^6)=9{,}200{,}000$

25. $(0.000002)(3000)=$
$(2)(10^{-6})(3)(10^3)=$
$(6)(10^{-3})=0.006$

CHAPTERS 1-6 | Cumulative Review

1. $5+3(2-7)^2\div 3\cdot 5$
$5+3(-5)^2\div 3\cdot 5$
$5+3(25)\div 3\cdot 5$
$5+75\div 3\cdot 5$
$5+25\cdot 5$
$5+125$
130

2. $8\div 2\cdot(-1)+3$
$4\cdot(-1)+3$
$-4+3$
-1

3. $7-2^2\cdot 5\div(-1)$
$7-4\cdot 5\div(-1)$
$7-20\div(-1)$
$7+20$
27

4. $4+(-2)-3(6)$
$4+(-2)-18$
$2-18$
-16

5. $(-3)^4=(-3)(-3)(-3)(-3)=81$

6. $-2^5 = -(2 \cdot 2 \cdot 2 \cdot 2 \cdot 2) = -32$

7. $\left(\frac{2}{3}\right)^{-1} = \frac{2^{-1}}{3^{-1}} = \frac{3^1}{2^1} = \frac{3}{2}$

8. $\frac{1}{4^{-2}} = 4^2 = 16$

9. $\left(\frac{1}{2} - \frac{1}{3}\right)^{-2} = \left(\frac{3}{6} - \frac{2}{6}\right)^{-2} = \left(\frac{1}{6}\right)^{-2}$
$\frac{1^{-2}}{6^{-2}} = \frac{6^2}{1^2} = \frac{36}{1} = 36$

10. $2^0 + 2^{-1} + 2^{-2} = 1 + \frac{1}{2} + \frac{1}{2^2} =$
$1 + \frac{1}{2} + \frac{1}{4} = \frac{4}{4} + \frac{2}{4} + \frac{1}{4} = \frac{7}{4}$

11. $x = \frac{1}{2},\ y = -\frac{1}{3}:\ \frac{2x + 3y}{x - y} =$
$\frac{2\left(\frac{1}{2}\right) + 3\left(-\frac{1}{3}\right)}{\frac{1}{2} - \left(-\frac{1}{3}\right)} = \frac{1 - 1}{\frac{3}{6} + \frac{2}{6}} = \frac{0}{\frac{5}{6}} = 0$

12. $n = -\frac{3}{4}:\ \frac{2}{5}n - \frac{1}{3}n - n + \frac{1}{2}n =$
$\frac{12}{30}n - \frac{10}{30}n - \frac{30}{30}n + \frac{15}{30}n =$
$\frac{-13}{30}n = -\frac{13}{30}\left(-\frac{3}{4}\right) = \frac{13}{40}$

13. $a = -1,\ b = -\frac{1}{3}:\ \frac{3a - 2b - 4a + 7b}{-a - 3a + b - 2b} =$
$\frac{-a + 5b}{-4a - b} = \frac{-(-1) + 5\left(-\frac{1}{3}\right)}{-4(-1) - \left(-\frac{1}{3}\right)} =$
$\frac{1 - \frac{5}{3}}{4 + \frac{1}{3}} = \frac{\frac{3}{3} - \frac{5}{3}}{\frac{12}{3} + \frac{1}{3}} = \frac{-\frac{2}{3}}{\frac{13}{3}} = -\frac{2}{13}$

14. $x = -2:\ -2(x - 4) + 3(2x - 1) - (3x - 2) =$
$-2x + 8 + 6x - 3 - 3x + 2 =$
$x + 7 = (-2) + 7 = 5$

15. $x = -1:\ (x^2 + 2x - 4) - (x^2 - x - 2)$
$+ (2x^2 - 3x - 1) =$
$x^2 + 2x - 4 - x^2 + x + 2 + 2x^2 - 3x - 1 =$
$2x^2 - 3 =$
$2(-1)^2 - 3 = 2(1) - 3 = 2 - 3 = -1$

16. $n = 3:\ 2(n^2 - 3n - 1) - (n^2 + n + 4)$
$- 3(2n - 1) =$
$2n^2 - 6n - 2 - n^2 - n - 4 - 6n + 3 =$
$n^2 - 13n - 3 =$
$(3)^2 - 13(3) - 3 = 9 - 39 - 3 = -33$

17. $(3x^2y^3)(-5xy^4) =$
$(3)(-5)(x^{2+1})(y^{3+4}) =$
$-15x^3y^7$

18. $(-6ab^4)(-2b^3) =$
$(-6)(-2)(a^1)(b^{4+3}) =$
$12ab^7$

19. $(-2x^2y^5)^3 =$
$(-2)^3(x^2)^3(y^5)^3 =$
$-8x^6y^{15}$

20. $-3xy(2x - 5y) = -6x^2y + 15xy^2$

21. $(5x - 2)(3x - 1) =$
$15x^2 + (-5 - 6)x + 2 =$
$15x^2 - 11x + 2$

22. $(7x - 1)(3x + 4) =$
$21x^2 + (28 - 3)x - 4 =$
$21x^2 + 25x - 4$

23. $(-x - 2)(2x + 3) =$
$-2x^2 + (-3 - 4)x - 6 =$
$-2x^2 - 7x - 6$

24. $(7 - 2y)(7 + 2y) =$
$49 + (14 - 14)y - 4y^2 =$
$49 - 4y^2$

25. $(x - 2)(3x^2 - x - 4) =$
$x(3x^2 - x - 4) - 2(3x^2 - x - 4) =$
$3x^3 - x^2 - 4x - 6x^2 + 2x + 8 =$
$3x^3 - 7x^2 - 2x + 8$

26. $(2x-5)(x^2+x-4) =$
$2x(x^2+x-4) - 5(x^2+x-4) =$
$2x^3+2x^2-8x-5x^2-5x+20 =$
$2x^3-3x^2-13x+20$

27. $(2n+3)^3 = (2n+3)(2n+3)(2n+3) =$
$(2n+3)(4n^2+12n+9) =$
$2n(4n^2+12n+9)+3(4n^2+12n+9) =$
$8n^3+24n^2+18n+12n^2+36n+27 =$
$8n^3+36n^2+54n+27$

28. $(1-2n)^3 = (1-2n)(1-2n)(1-2n) =$
$(1-2n)(1-4n+4n^2) =$
$1(1-4n+4n^2)-2n(1-4n+4n^2) =$
$1-4n+4n^2-2n+8n^2-8n^3 =$
$1-6n+12n^2-8n^3$

29. $(x^2-2x+6)(2x^2+5x-6) =$
$x^2(2x^2+5x-6)-2x(2x^2+5x-6)$
$\quad +6(2x^2+5x-6) =$
$2x^4+5x^3-6x^2-4x^3-10x^2+12x$
$\quad +12x^2+30x-36 =$
$2x^4+x^3-4x^2+42x-36$

30. $\dfrac{-52x^3y^4}{13xy^2} = -4x^{3-1}y^{4-2} = -4x^2y^2$

31. $\dfrac{-126a^3b^5}{-9a^2b^3} = 14a^{3-2}b^{5-3}$
$= 14a^1b^2 = 14ab^2$

32. $\dfrac{56xy^2-64x^3y-72x^4y^4}{8xy} =$
$\dfrac{56xy^2}{8xy} - \dfrac{64x^3y}{8xy} - \dfrac{72x^4y^4}{8xy} =$
$7y-8x^2-9x^3y^3$

33.

$$\begin{array}{r}
2x^2 - 4x - 7 \\
x+3\overline{\big)\,2x^3+2x^2-19x-21} \\
\underline{2x^3+6x^2} \qquad\qquad \\
-4x^2-19x-21 \\
\underline{-4x^2-12x} \qquad \\
-7x-21 \\
\underline{-7x-21}
\end{array}$$

34.

$$\begin{array}{r}
x^2 + 6x + 4 \\
3x-1\overline{\big)\,3x^3+17x^2+6x-4} \\
\underline{3x^3-x^2} \qquad\qquad \\
18x^2+6x-4 \\
\underline{18x^2-6x} \qquad \\
12x-4 \\
\underline{12x-4}
\end{array}$$

35. $(-2x^3)(3x^{-4}) = -6x^{3-4} =$
$-6x^{-1} = -\dfrac{6}{x}$

36. $\dfrac{4x^{-2}}{2x^{-1}} = 2x^{-2-(-1)} = 2x^{-2+1} =$
$2x^{-1} = \dfrac{2}{x}$

37. $(3x^{-1}y^{-2})^{-1} = 3^{-1}(x^{-1})^{-1}(y^{-2})^{-1} =$
$3^{-1}x^1y^2 = \dfrac{xy^2}{3}$

38. $(xy^2z^{-1})^{-2} = x^{-2}(y^2)^{-2}(z^{-1})^{-2} =$
$x^{-2}y^{-4}z^2 = \dfrac{z^2}{x^2y^4}$

39. $(0.00003)(4000) =$
$(3)(10^{-5})(4)(10^3) =$
$(12)(10^{-5+3}) =$
$(12)(10^{-2}) = 0.12$

40. $(0.0002)(0.003)^2 =$
$(2)(10^{-4})(3)^2(10^{-3})^2 =$
$(2)(9)(10^{-4})(10^{-6}) =$
$(18)(10^{-4-6}) =$
$(18)(10^{-10}) = 0.0000000018$

41. $\dfrac{0.00034}{0.0000017} = \dfrac{(3.4)(10^{-4})}{(1.7)(10^{-6})} =$
$(2)(10^{-4-(-6)}) = (2)(10^2) = 200$

42.
$$\begin{aligned} 5x + 8 &= 6x - 3 \\ -x + 8 &= -3 \\ -x &= -11 \\ x &= 11 \end{aligned}$$
The solution set is $\{11\}$.

43.
$$\begin{aligned} -2(4x - 1) &= -5x + 3 - 2x \\ -8x + 2 &= -5x + 3 - 2x \\ -8x + 2 &= -7x + 3 \\ -x + 2 &= 3 \\ -x &= 1 \\ x &= -1 \end{aligned}$$
The solution set is $\{-1\}$.

44.
$$\begin{aligned} \frac{y}{2} - \frac{y}{3} &= 8 \\ 6\left(\frac{y}{2} - \frac{y}{3}\right) &= 6(8) \\ 6\left(\frac{y}{2}\right) - 6\left(\frac{y}{3}\right) &= 48 \\ 3y - 2y &= 48 \\ y &= 48 \end{aligned}$$
The solution set is $\{48\}$.

45.
$$\begin{aligned} 6x + 8 - 4x &= 10(3x + 2) \\ 6x + 8 - 4x &= 30x + 20 \\ 2x + 8 &= 30x + 20 \\ -28x + 8 &= 20 \\ -28x &= 12 \\ x &= \frac{12}{-28} = -\frac{3}{7} \end{aligned}$$
The solution set is $\left\{-\frac{3}{7}\right\}$.

46.
$$\begin{aligned} 1.6 - 2.4x &= 5x - 65 \\ 10(1.6 - 2.4x) &= 10(5x - 65) \\ 16 - 24x &= 50x - 650 \\ 16 - 74x &= -650 \\ -74x &= -666 \\ x &= 9 \end{aligned}$$
The solution set is $\{9\}$.

47.
$$\begin{aligned} -3(x - 1) + 2(x + 3) &= -4 \\ -3x + 3 + 2x + 6 &= -4 \\ -x + 9 &= -4 \\ -x &= -13 \\ x &= 13 \end{aligned}$$
The solution set is $\{13\}$.

48.
$$\begin{aligned} \frac{3n + 1}{5} + \frac{n - 2}{3} &= \frac{2}{15} \\ 15\left(\frac{3n + 1}{5} + \frac{n - 2}{3}\right) &= 15\left(\frac{2}{15}\right) \\ 15\left(\frac{3n + 1}{5}\right) + 15\left(\frac{n - 2}{3}\right) &= 2 \\ 3(3n + 1) + 5(n - 2) &= 2 \\ 9n + 3 + 5n - 10 &= 2 \\ 14n - 7 &= 2 \\ 14n &= 9 \\ n &= \frac{9}{14} \end{aligned}$$
The solution set is $\left\{\frac{9}{14}\right\}$.

49.
$$\begin{aligned} 0.06x + 0.08(1500 - x) &= 110 \\ 100[0.06x + 0.08(1500 - x)] &= 100(110) \\ 6x + 8(1500 - x) &= 11000 \\ 6x + 12000 - 8x &= 11000 \\ -2x + 12000 &= 11000 \\ -2x &= -1000 \\ x &= 500 \end{aligned}$$
The solution set is $\{500\}$.

50.
$$\begin{aligned} 2x - 7 &\le -3(x + 4) \\ 2x - 7 &\le -3x - 12 \\ 5x - 7 &\le -12 \\ 5x &\le -5 \\ x &\le -1 \end{aligned}$$
The solution set is
$\{x | x \le -1\}$ or $(-\infty, -1]$.

51. $$\begin{aligned} 6x + 5 - 3x &> 5 \\ 3x + 5 &> 5 \\ 3x &> 0 \\ x &> \frac{0}{3} \\ x &> 0 \end{aligned}$$
The solution set is
$\{x|x > 0\}$ or $(0, \infty)$.

52. $$\begin{aligned} 4(x-5) + 2(3x+6) &< 0 \\ 4x - 20 + 6x + 12 &< 0 \\ 10x - 8 &< 0 \\ 10x &< 8 \\ x &< \frac{8}{10} \\ x &< \frac{4}{5} \end{aligned}$$
The solution set is
$\left\{x|x < \frac{4}{5}\right\}$ or $\left(-\infty, \frac{4}{5}\right)$.

53. $$\begin{aligned} -5x + 3 &> -4x + 5 \\ -x + 3 &> 5 \\ -x &> 2 \\ \frac{-x}{-1} &< \frac{2}{-1} \\ x &< -2 \end{aligned}$$
The solution set is
$\{x|x < -2\}$ or $(-\infty, -2)$.

54. $$\begin{aligned} \frac{3x}{4} - \frac{x}{2} &\le \frac{5x}{6} - 1 \\ 12\left(\frac{3x}{4} - \frac{x}{2}\right) &\le 12\left(\frac{5x}{6} - 1\right) \\ 12\left(\frac{3x}{4}\right) - 12\left(\frac{x}{2}\right) &\le 12\left(\frac{5x}{6}\right) - 12(1) \\ 3(3x) - 6(x) &\le 2(5x) - 12 \\ 9x - 6x &\le 10x - 12 \\ 3x &\le 10x - 12 \\ -7x &\le -12 \\ \frac{-7x}{-7} &\ge \frac{-12}{-7} \\ x &\ge \frac{12}{7} \end{aligned}$$
The solution set is
$\left\{x|x \ge \frac{12}{7}\right\}$ or $\left[\frac{12}{7}, \infty\right)$.

55. $$\begin{aligned} 0.08(700 - x) + 0.11x &\ge 65 \\ 100[0.08(700 - x) + 0.11x] &\ge 100(65) \\ 8(700 - x) + 11x &\ge 6500 \\ 5600 - 8x + 11x &\ge 6500 \\ 5600 + 3x &\ge 6500 \\ 3x &\ge 900 \\ x &\ge 300 \end{aligned}$$
The solution set is
$\{x|x \ge 300\}$ or $[300, \infty)$.

56. Graph $y = x^2 - 1$.

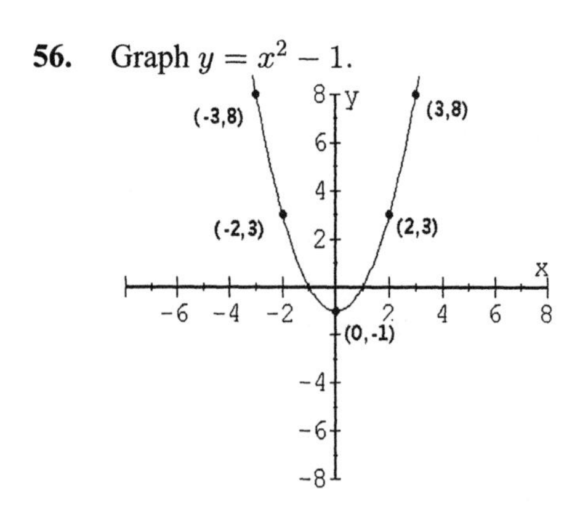

57. Graph $y = 2x + 3$.

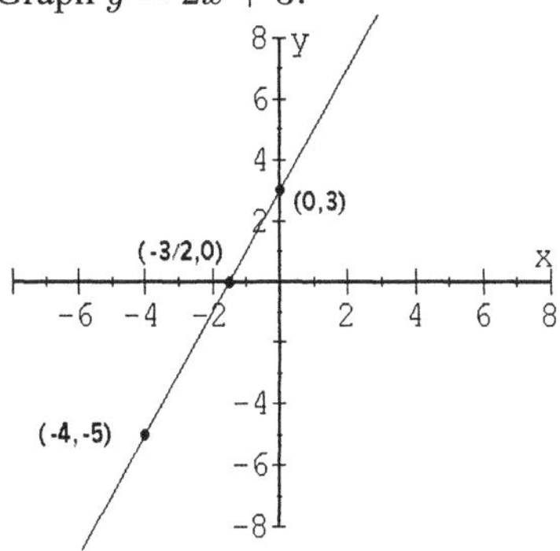

58. Graph $y = -5x$.

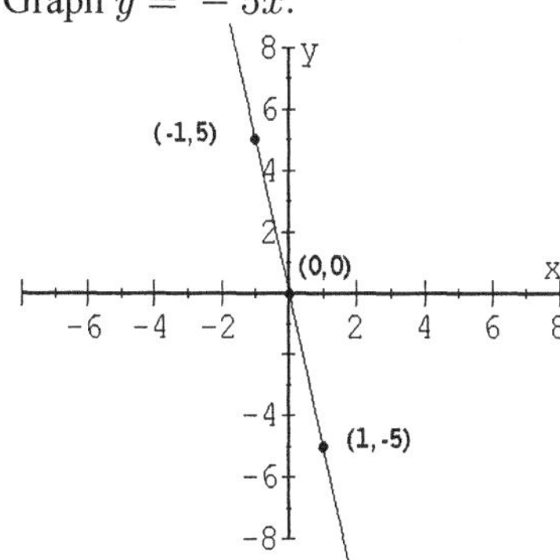

59. Graph $x - 2y = 6$.

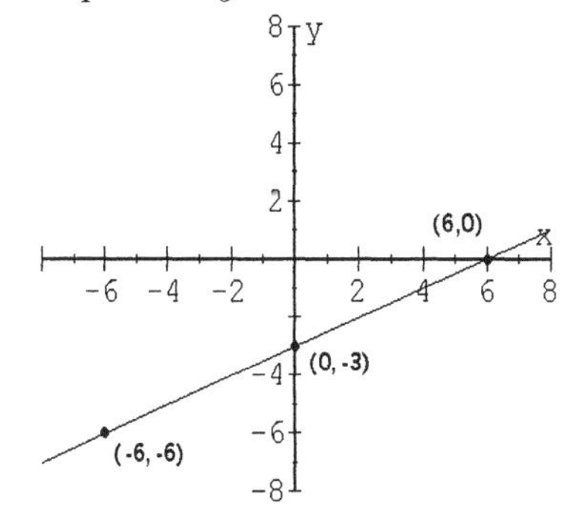

60. Graph $y = -\frac{1}{2}x + 2$.

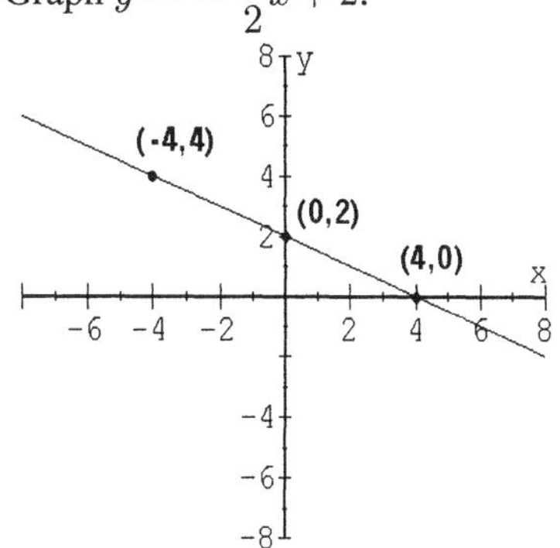

61. Let n represent the number.

$$\begin{aligned} 4 + 3n &= n + 10 \\ 4 + 2n &= 10 \\ 2n &= 6 \\ n &= 3 \end{aligned}$$

The number is 3.

62. Let n represent the number.

$$\begin{aligned} (15\%)(n) &= 6 \\ 0.15n &= 6 \\ 15n &= 600 \\ n &= 40 \end{aligned}$$

Therefore, 15% of 40 = 6.

63. Let d represent the number of dimes, then $18 - d$ represents the number of quarters. The value of the dimes in cents is 10d and the value of the quarters in cents is $25(18 - d)$.

$$\begin{aligned} 10d + 25(18 - d) &= 330 \\ 10d + 450 - 25d &= 330 \\ -15d + 450 &= 330 \\ -15d &= -120 \\ d &= 8 \end{aligned}$$

He would have 8 dimes and 10 quarters.

64. Let x represent the amount of money invested at 8% interest; then $1500 - x$ represents the amount of money invested at 9%.

$$\begin{pmatrix}\text{Interest}\\ \text{earned}\\ \text{at}\\ 8\%\end{pmatrix} + \begin{pmatrix}\text{Interest}\\ \text{earned}\\ \text{at}\\ 9\%\end{pmatrix} = \begin{pmatrix}\text{Total}\\ \text{interest}\\ \text{earned}\end{pmatrix}$$

$$\begin{aligned}(8\%)(x) + (9\%)(1500 - x) &= 128\\ 0.08x + 0.09(1500 - x) &= 128\\ 8x + 9(1500 - x) &= 12{,}800\\ 8x + 13{,}500 - 9x &= 12{,}800\\ 13{,}500 - x &= 12{,}800\\ -x &= -700\\ x &= 700\end{aligned}$$

The amounts invested are \$700 at 8% and \$800 at 9%.

65. Let x represent the amount of water to be added, then $15 + x$ represents the amount of final salt solution.

$$\begin{pmatrix}\text{Pure}\\ \text{salt in}\\ \text{water}\\ \text{added}\end{pmatrix} + \begin{pmatrix}\text{Pure}\\ \text{salt}\\ \text{in 12\%}\\ \text{solution}\end{pmatrix} = \begin{pmatrix}\text{Pure}\\ \text{salt}\\ \text{in final}\\ \text{solution}\end{pmatrix}$$

$$\begin{aligned}0 + (12\%)(15) &= (10\%)(15 + x)\\ 0 + 0.12(15) &= 0.10(15 + x)\\ 12(15) &= 10(15 + x)\\ 180 &= 150 + 10x\\ 30 &= 10x\\ 3 &= x\end{aligned}$$

The amount of water to be added would be 3 gallons.

66. Let t represent the time for both airplanes as they left Atlanta at the same time. A chart of the information would be as follows:

	Rate	Time	Distance ($d = rt$)
Plane A	400	t	$400t$
Plane B	450	t	$450t$

A diagram of the problem would be as follows:

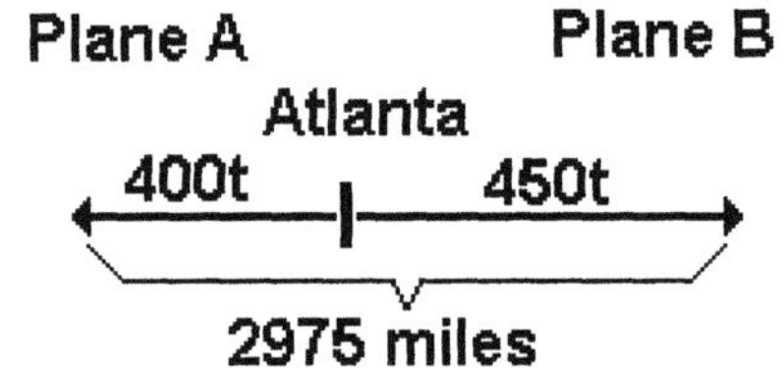

The sum of the two distances would be 2975 miles.

$$\begin{aligned}400t + 450t &= 2975\\ 850t &= 2975\\ t &= \frac{2975}{850} = 3\frac{1}{2} \text{ hours}\end{aligned}$$

It would take $3\frac{1}{2}$ hours for the airplanes to be 2975 miles apart.

67. Let w represent the width of the rectangle, then $2w + 1$ represents the length of the rectangle.

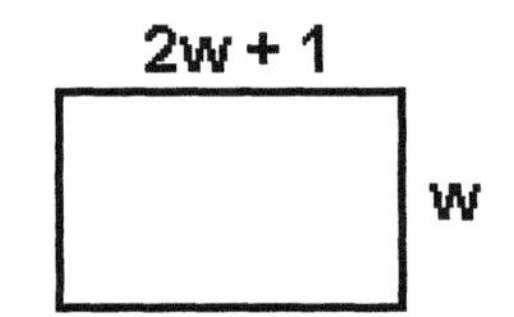

$$\begin{aligned}2(w) + 2(2w + 1) &= 44\\ 2w + 4w + 2 &= 44\\ 6w + 2 &= 44\\ 6w &= 42\\ w &= 7\end{aligned}$$

The width of the rectangle is 7 meters and the length is 15 meters.

Chapter 7 Factoring, Solving Equations, and Problem Solving

PROBLEM SET 7.1 Factoring by Using the Distributive Property

1. $24y = 2 \cdot 2 \cdot 2 \cdot 3 \cdot y$
$30xy = 2 \cdot 3 \cdot 5 \cdot x \cdot y$
The greatest common factor is
$2 \cdot 3 \cdot y = 6y.$

3. $60x^2y = 2 \cdot 2 \cdot 3 \cdot 5 \cdot x \cdot x \cdot y$
$84xy^2 = 2 \cdot 2 \cdot 3 \cdot 7 \cdot x \cdot y \cdot y$
The greatest common factor is
$2 \cdot 2 \cdot 3 \cdot x \cdot y = 12xy.$

5. $42ab^3 = 2 \cdot 3 \cdot 7 \cdot a \cdot b \cdot b \cdot b$
$70a^2b^2 = 2 \cdot 5 \cdot 7 \cdot a \cdot a \cdot b \cdot b$
The greatest common factor is
$2 \cdot 7 \cdot a \cdot b \cdot b = 14ab^2.$

7. $6x^3 = 2 \cdot 3 \cdot x \cdot x \cdot x$
$8x = 2 \cdot 2 \cdot 2 \cdot x$
$24x^2 = 2 \cdot 2 \cdot 2 \cdot 3 \cdot x \cdot x$
The greatest common factor is
$2 \cdot x = 2x.$

9. $16a^2b^2 = 2 \cdot 2 \cdot 2 \cdot 2 \cdot a \cdot a \cdot b \cdot b$
$40a^2b^3 = 2 \cdot 2 \cdot 2 \cdot 5 \cdot a \cdot a \cdot b \cdot b \cdot b$
$56a^3b^4 = 2 \cdot 2 \cdot 2 \cdot 7 \cdot a \cdot a \cdot a \cdot b \cdot b \cdot b \cdot b$
The greatest common factor is
$2 \cdot 2 \cdot 2 \cdot a \cdot a \cdot b \cdot b = 8a^2b^2.$

11. $8x + 12y =$
$4(2x) + 4(3y) =$
$4(2x + 3y)$

13. $14xy - 21y =$
$7y(2x) - 7y(3) =$
$7y(2x - 3)$

15. $18x^2 + 45x =$
$9x(2x) + 9x(5) =$
$9x(2x + 5)$

17. $12xy^2 - 30x^2y =$
$6xy(2y) - 6xy(5x) =$
$6xy(2y - 5x)$

19. $36a^2b - 60a^3b^4 =$
$12a^2b(3) - 12a^2b(5ab^3) =$
$12a^2b(3 - 5ab^3)$

21. $16xy^3 + 25x^2y^2 =$
$xy^2(16y) + xy^2(25x) =$
$xy^2(16y + 25x)$

23. $64ab - 72cd =$
$8(8ab) - 8(9cd) =$
$8(8ab - 9cd)$

25. $9a^2b^4 - 27a^2b =$
$9a^2b(b^3) - 9a^2b(3) =$
$9a^2b(b^3 - 3)$

27. $52x^4y^2 + 60x^6y =$
$4x^4y(13y) + 4x^4y(15x^2) =$
$4x^4y(13y + 15x^2)$

29. $40x^2y^2 + 8x^2y =$
$8x^2y(5y) + 8x^2y(1) =$
$8x^2y(5y + 1)$

31. $12x + 15xy + 21x^2 =$
$3x(4) + 3x(5y) + 3x(7x) =$
$3x(4 + 5y + 7x)$

33. $2x^3 - 3x^2 + 4x =$
$x(2x^2) - x(3x) + x(4) =$
$x(2x^2 - 3x + 4)$

35. $44y^5 - 24y^3 - 20y^2 =$
$4y^2(11y^3) - 4y^2(6y) - 4y^2(5) =$
$4y^2(11y^3 - 6y - 5)$

37. $14a^2b^3 + 35ab^2 - 49a^3b =$
$7ab(2ab^2) + 7ab(5b) - 7ab(7a^2) =$
$7ab(2ab^2 + 5b - 7a^2)$

39. $x(y + 1) + z(y + 1) =$
$(y + 1)(x + z)$

41. $a(b-4)-c(b-4)=$
$(b-4)(a-c)$

43. $x(x+3)+6(x+3)=$
$(x+3)(x+6)$

45. $2x(x+1)-3(x+1)=$
$(x+1)(2x-3)$

47. $5x+5y+bx+by=$
$5(x+y)+b(x+y)=$
$(x+y)(5+b)$

49. $bx-by-cx+cy=$
$b(x-y)-c(x-y)=$
$(x-y)(b-c)$

51. $ac+bc+a+b=$
$c(a+b)+1(a+b)=$
$(a+b)(c+1)$

53. $x^2+5x+12x+60=$
$x(x+5)+12(x+5)=$
$(x+5)(x+12)$

55. $x^2-2x-8x+16=$
$x(x-2)-8(x-2)=$
$(x-2)(x-8)$

57. $2x^2+x-10x-5=$
$x(2x+1)-5(2x+1)=$
$(2x+1)(x-5)$

59. $6n^2-3n-8n+4=$
$3n(2n-1)-4(2n-1)=$
$(2n-1)(3n-4)$

61. $x^2-8x=0$
$x(x-8)=0$
$x=0$ or $x-8=0$
$x=0$ or $x=8$
The solution set is $\{0, 8\}$.

63. $x^2+x=0$
$x(x+1)=0$
$x=0$ or $x+1=0$
$x=0$ or $x=-1$
The solution set is $\{-1, 0\}$.

65. $n^2=5n$
$n^2-5n=0$
$n(n-5)=0$
$n=0$ or $n-5=0$
$n=0$ or $n=5$
The solution set is $\{0, 5\}$.

67. $2y^2-3y=0$
$y(2y-3)=0$
$y=0$ or $2y-3=0$
$y=0$ or $2y=3$
$y=0$ or $y=\frac{3}{2}$
The solution set is $\left\{0, \frac{3}{2}\right\}$.

69. $7x^2=-3x$
$7x^2+3x=0$
$x(7x+3)=0$
$x=0$ or $7x+3=0$
$x=0$ or $7x=-3$
$x=0$ or $x=-\frac{3}{7}$
The solution set is $\left\{-\frac{3}{7}, 0\right\}$.

71. $3n^2+15n=0$
$3n(n+5)=0$
$3n=0$ or $n+5=0$
$n=0$ or $n=-5$
The solution set is $\{-5, 0\}$.

73. $4x^2=6x$
$2x^2=3x$
$2x^2-3x=0$
$x(2x-3)=0$
$x=0$ or $2x-3=0$
$x=0$ or $2x=3$
$x=0$ or $x=\frac{3}{2}$
The solution set is $\left\{0, \frac{3}{2}\right\}$.

75. $7x - x^2 = 0$
$x(7 - x) = 0$
$x = 0 \quad \text{or} \quad 7 - x = 0$
$x = 0 \quad \text{or} \quad 7 = x$
The solution set is $\{0, 7\}$.

77. $13x = x^2$
$13x - x^2 = 0$
$x(13 - x) = 0$
$x = 0 \quad \text{or} \quad 13 - x = 0$
$x = 0 \quad \text{or} \quad 13 = x$
The solution set is $\{0, 13\}$.

79. $5x = -2x^2$
$2x^2 + 5x = 0$
$x(2x + 5) = 0$
$x = 0 \quad \text{or} \quad 2x + 5 = 0$
$x = 0 \quad \text{or} \quad 2x = -5$
$x = 0 \quad \text{or} \quad x = -\frac{5}{2}$
The solution set is $\left\{-\frac{5}{2}, 0\right\}$.

81. $x(x + 5) - 4(x + 5) = 0$
$(x + 5)(x - 4) = 0$
$x + 5 = 0 \quad \text{or} \quad x - 4 = 0$
$x = -5 \quad \text{or} \quad x = 4$
The solution set is $\{-5, 4\}$.

83. $4(x - 6) - x(x - 6) = 0$
$(x - 6)(4 - x) = 0$
$x - 6 = 0 \quad \text{or} \quad 4 - x = 0$
$x = 6 \quad \text{or} \quad 4 = x$
The solution set is $\{4, 6\}$.

85. Let n represent the number, then n^2 represents the square of the number.
$n^2 = 9n$
$n^2 - 9n = 0$
$n(n - 9) = 0$
$n = 0 \quad \text{or} \quad n - 9 = 0$
$n = 0 \quad \text{or} \quad n = 9$
The number is 0 or 9.

87. Let s represent the length of a side of the square.

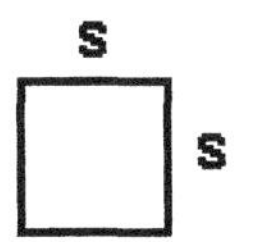

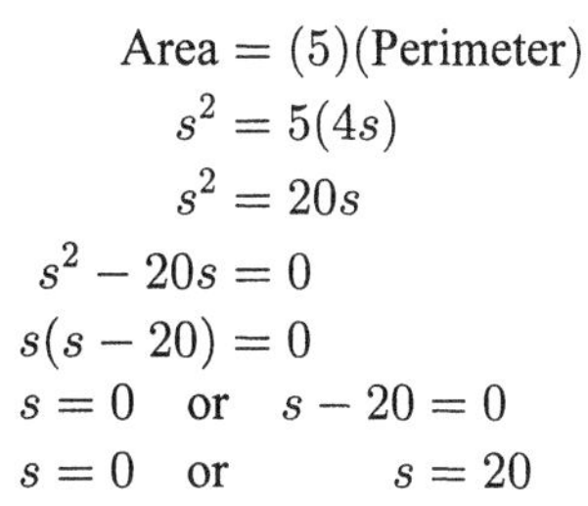

Area = (5)(Perimeter)
$s^2 = 5(4s)$
$s^2 = 20s$
$s^2 - 20s = 0$
$s(s - 20) = 0$
$s = 0 \quad \text{or} \quad s - 20 = 0$
$s = 0 \quad \text{or} \quad s = 20$
Since 0 is not a reasonable answer to the problem, the length of the side of the square must be 20 units.

89. Let s represent the length of a side of the square, then s also represents the length of a radius of the circle.

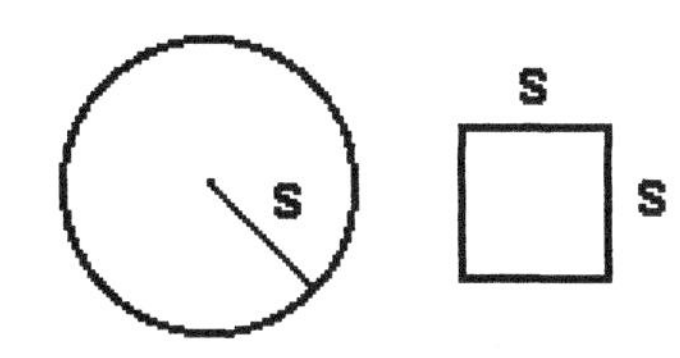

Circle Area = Square Perimeter
$\pi s^2 = 4s$
$\pi s^2 - 4s = 0$
$s(\pi s - 4) = 0$
$s = 0 \quad \text{or} \quad \pi s - 4 = 0$
$s = 0 \quad \text{or} \quad \pi s = 4$
$s = 0 \quad \text{or} \quad s = \frac{4}{\pi}$
The answer of 0 must be discarded, so the length of a side of the square is $\frac{4}{\pi}$ units.

91. Let w represent the width of the rectangle, then w also represents the length of a side of the square.

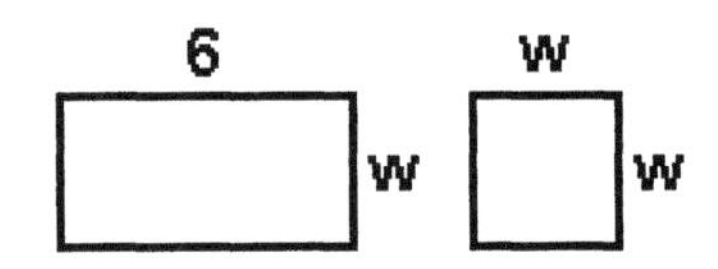

Rectangle Area = (2)(Square Area)

$$6w = 2(w^2)$$
$$6w - 2w^2 = 0$$
$$2w(3 - w) = 0$$
$$2w = 0 \quad \text{or} \quad 3 - w = 0$$
$$w = 0 \quad \text{or} \quad 3 = w$$

The answer of 0 must be discarded, so the dimensions of the rectangle are 3 inches by 6 inches and the dimensions of the square are 3 inches by 3 inches.

95. $A = P(1 + rt)$

a. $P = \$100,\ r = 8\%,\ t = 2:$
$A = 100[1 + 0.08(2)] =$
$100(1 + 0.16) = 100(1.16) = \$116$

b. $P = \$200,\ r = 9\%,\ t = 3:$
$A = 200[1 + 0.09(3)] =$
$200(1 + 0.27) = 200(1.27) = \$254$

c. $P = \$500,\ r = 10\%,\ t = 5:$
$A = 500[1 + 0.10(5)] =$
$500(1 + 0.50) = 500(1.5) = \$750$

d. $P = \$1000,\ r = 10\%,\ t = 10:$
$A = 1000[1 + 0.10(10)] =$
$1000(1 + 1.00) = 1000(2) = \$2000$

97.
$$b^2x^2 - cx = 0$$
$$x(b^2x - c) = 0$$
$$x = 0 \quad \text{or} \quad b^2x - c = 0$$
$$x = 0 \quad \text{or} \quad b^2x = c$$
$$x = 0 \quad \text{or} \quad x = \frac{c}{b^2}$$

99.
$$y + ay - by - c = 0$$
$$y + ay - by = c$$
$$y(1 + a - b) = c$$
$$y = \frac{c}{1 + a - b}$$

PROBLEM SET 7.2 Factoring the Difference of Two Squares

1. $x^2 - 1 =$
$(x)^2 - (1)^2 =$
$(x - 1)(x + 1)$

3. $x^2 - 100 =$
$(x)^2 - (10)^2 =$
$(x - 10)(x + 10)$

5. $x^2 - 4y^2 =$
$(x)^2 - (2y)^2 =$
$(x - 2y)(x + 2y)$

7. $9x^2 - y^2 =$
$(3x)^2 - (y)^2 =$
$(3x - y)(3x + y)$

9. $36a^2 - 25b^2 =$
$(6a)^2 - (5b)^2 =$
$(6a - 5b)(6a + 5b)$

11. $1 - 4n^2 =$
$(1)^2 - (2n)^2 =$
$(1 - 2n)(1 + 2n)$

13. $5x^2 - 20 =$
$5(x^2 - 4) =$
$5(x - 2)(x + 2)$

15. $8x^2 + 32 = 8(x^2 + 4)$

17. $2x^2 - 18y^2 =$
$2(x^2 - 9y^2) =$
$2(x - 3y)(x + 3y)$

19. $x^3 - 25x =$
$x(x^2 - 25) =$
$x(x - 5)(x + 5)$

21. $x^2 + 9y^2$ is not factorable.

23. $45x^2 - 36xy = 9x(5x - 4y)$

25. $36 - 4x^2 =$
$4(9 - x^2) =$
$4(3 - x)(3 + x)$

27. $4a^4 + 16a^2 = 4a^2(a^2 + 4)$

29. $x^4 - 81 =$
$(x^2 + 9)(x^2 - 9) =$
$(x^2 + 9)(x + 3)(x - 3)$

31. $x^4 + x^2 = x^2(x^2 + 1)$

33. $3x^3 + 48x = 3x(x^2 + 16)$

35. $5x - 20x^3 =$
$5x(1 - 4x^2) =$
$5x(1 - 2x)(1 + 2x)$

37. $4x^2 - 64 =$
$4(x^2 - 16) =$
$4(x - 4)(x + 4)$

39. $75x^3y - 12xy^3 =$
$3xy(25x^2 - 4y^2) =$
$3xy(5x + 2y)(5x - 2y)$

41. $x^2 = 9$
$x^2 - 9 = 0$
$(x - 3)(x + 3) = 0$
$x - 3 = 0$ or $x + 3 = 0$
$x = 3$ or $x = -3$
The solution set is $\{-3, 3\}$.

43. $4 = n^2$
$4 - n^2 = 0$
$(2 - n)(2 + n) = 0$
$2 - n = 0$ or $2 + n = 0$
$2 = n$ or $n = -2$
The solution set is $\{-2, 2\}$.

45. $9x^2 = 16$
$9x^2 - 16 = 0$
$(3x - 4)(3x + 4) = 0$
$3x - 4 = 0$ or $3x + 4 = 0$
$3x = 4$ or $3x = -4$
$x = \frac{4}{3}$ or $x = -\frac{4}{3}$
The solution set is $\left\{-\frac{4}{3}, \frac{4}{3}\right\}$.

47. $n^2 - 121 = 0$
$(n - 11)(n + 11) = 0$
$n - 11 = 0$ or $n + 11 = 0$
$n = 11$ or $n = -11$
The solution set is $\{-11, 11\}$.

49. $25x^2 = 4$
$25x^2 - 4 = 0$
$(5x - 2)(5x + 2) = 0$
$5x - 2 = 0$ or $5x + 2 = 0$
$5x = 2$ or $5x = -2$
$x = \frac{2}{5}$ or $x = -\frac{2}{5}$
The solution set is $\left\{-\frac{2}{5}, \frac{2}{5}\right\}$.

51. $3x^2 = 75$
Divide both sides by 3.
$x^2 = 25$
$x^2 - 25 = 0$
$(x - 5)(x + 5) = 0$
$x - 5 = 0$ or $x + 5 = 0$
$x = 5$ or $x = -5$
The solution set is $\{-5, 5\}$.

53. $3x^3 - 48x = 0$
Divide both sides by 3.
$x^3 - 16x = 0$
$x(x^2 - 16) = 0$
$x(x - 4)(x + 4) = 0$
$x = 0$ or $x - 4 = 0$ or $x + 4 = 0$
$x = 0$ or $x = 4$ or $x = -4$
The solution set is $\{-4, 0, 4\}$.

55. $n^3 = 16n$
$n^3 - 16n = 0$
$n(n^2 - 16) = 0$
$n(n - 4)(n + 4) = 0$
$n = 0$ or $n - 4 = 0$ or $n + 4 = 0$
$n = 0$ or $n = 4$ or $n = -4$
The solution set is $\{-4, 0, 4\}$.

Problem Set 7.2

57. $5 - 45x^2 = 0$

Divide by 5.

$1 - 9x^2 = 0$

$(1 - 3x)(1 + 3x) = 0$

$1 - 3x = 0$ or $1 + 3x = 0$

$1 = 3x$ or $3x = -1$

$x = \frac{1}{3}$ or $x = -\frac{1}{3}$

The solution set is $\left\{-\frac{1}{3}, \frac{1}{3}\right\}$.

59. $4x^3 - 400x = 0$

Divide by 4.

$x^3 - 100x = 0$

$x(x^2 - 100) = 0$

$x(x - 10)(x + 10) = 0$

$x = 0$ or $x - 10 = 0$ or $x + 10 = 0$

$x = 0$ or $x = 10$ or $x = -10$

The solution set is $\{-10, 0, 10\}$.

61. $64x^2 = 81$

$64x^2 - 81 = 0$

$(8x - 9)(8x + 9) = 0$

$8x - 9 = 0$ or $8x + 9 = 0$

$8x = 9$ or $8x = -9$

$x = \frac{9}{8}$ or $x = -\frac{9}{8}$

The solution set is $\left\{-\frac{9}{8}, \frac{9}{8}\right\}$.

63. $36x^3 = 9x$

Divide both sides by 9.

$4x^3 = x$

$4x^3 - x = 0$

$x(4x^2 - 1) = 0$

$x(2x - 1)(2x + 1) = 0$

$x = 0$ or $2x - 1 = 0$ or $2x + 1 = 0$

$x = 0$ or $2x = 1$ or $2x = -1$

$x = 0$ or $x = \frac{1}{2}$ or $x = -\frac{1}{2}$

The solution set is $\left\{-\frac{1}{2}, 0, \frac{1}{2}\right\}$.

65. Let n represent the number.

$n^2 - 49 = 0$

$(n - 7)(n + 7) = 0$

$n - 7 = 0$ or $n + 7 = 0$

$n = 7$ or $n = -7$

The number could be either -7 or 7.

67. Let n represent the number.

$5n^3 = 80n$

Divide both sides by 5.

$n^3 = 16n$

$n^3 - 16n = 0$

$n(n^2 - 16) = 0$

$n(n - 4)(n + 4) = 0$

$n = 0$ or $n - 4 = 0$ or $n + 4 = 0$

$n = 0$ or $n = 4$ or $n = -4$

The number could be -4, 0 or 4.

69. Let x represent the length of a side of the smaller square, then $5x$ represents the length of a side of the larger square.

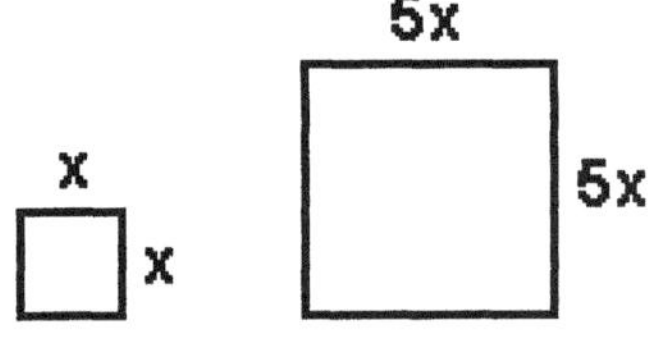

Larger Area + Smaller Area = 234

$(5x)^2 + x^2 = 234$

$25x^2 + x^2 = 234$

$26x^2 = 234$

$x^2 = 9$

$x^2 - 9 = 0$

$(x - 3)(x + 3) = 0$

$x - 3 = 0$ or $x + 3 = 0$

$x = 3$ or $x = -3$

The solution -3 must be discarded. Thus, the length of a side of the smaller square is 3 inches and the length of a side of the larger square is 15 inches.

71. Let w represent the width of the rectangle, then $2\frac{1}{2}w = \frac{5}{2}w$ represents the length of the rectangle.

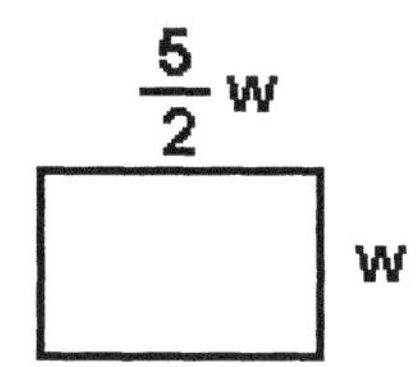

$$\text{Rectangular Area} = 160$$
$$w\left(\frac{5}{2}w\right) = 160$$
$$\frac{5}{2}w^2 = 160$$

Multiply both sides by $\frac{2}{5}$.

$$w^2 = 64$$
$$w^2 - 64 = 0$$
$$(w-8)(w+8) = 0$$
$$w - 8 = 0 \quad \text{or} \quad w + 8 = 0$$
$$w = 8 \quad \text{or} \quad w = -8$$

The solution of -8 must be discarded. Thus, the width of the rectangle is 8 centimeters and the length is $\frac{5}{2}(8) = 20$ centimeters.

73. Let r represent the length of a radius of the smaller circle, then $2r$ represents the length of a radius of the larger circle.

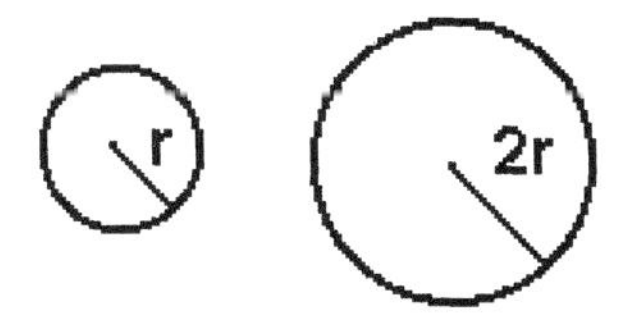

$$\text{Smaller Area} + \text{Larger Area} = 80\pi$$
$$\pi r^2 + \pi(2r)^2 = 80\pi$$
$$\pi r^2 + 4\pi r^2 = 80\pi$$
$$5\pi r^2 = 80\pi$$
$$r^2 = 16$$
$$r^2 - 16 = 0$$
$$(r-4)(r+4) = 0$$
$$r - 4 = 0 \quad \text{or} \quad r + 4 = 0$$
$$r = 4 \quad \text{or} \quad r = -4 \left(\text{Discard this solution.}\right)$$

The length of a radius of the smaller circle is 4 meters and the length of a radius of the larger circle is 8 meters.

75. Let x represent a radius of the base, then x also represents the altitude of the cylinder.

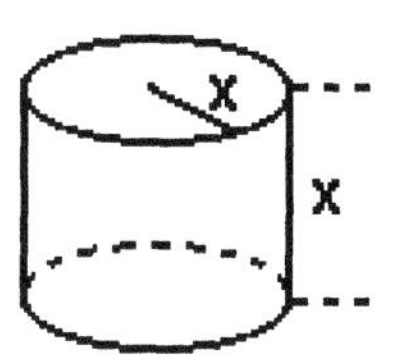

$$\text{Surface Area} = 100\pi$$
$$2\pi x^2 + 2\pi x(x) = 100\pi$$
$$2\pi x^2 + 2\pi x^2 = 100\pi$$
$$4\pi x^2 = 100\pi$$
$$x^2 = 25$$
$$x^2 - 25 = 0$$
$$(x-5)(x+5) = 0$$
$$x - 5 = 0 \quad \text{or} \quad x + 5 = 0$$
$$x = 5 \quad \text{or} \quad x = -5 \left(\text{Discard this solution.}\right)$$

The length of a radius is 5 centimeters.

81. $x^3 - 8 =$
$(x)^3 - (2)^3 =$
$(x-2)(x^2 + 2x + 4)$

83. $n^3 + 64 =$
$(n)^3 + (4)^3 =$
$(n+4)(n^2 - 4n + 16)$

85. $27a^3 - 64b^3 =$
$(3a)^3 - (4b)^3 =$
$(3a - 4b)(9a^2 + 12ab + 16b^2)$

87. $1 + 27a^3 =$
$(1)^3 + (3a)^3 =$
$(1 + 3a)(1 - 3a + 9a^2)$

89. $8x^3 - y^3 =$
$(2x)^3 - (y)^3 =$
$(2x - y)(4x^2 + 2xy + y^2)$

91. $27x^3 - 8y^3 =$
$(3x)^3 - (2y)^3 =$
$(3x - 2y)(9x^2 + 6xy + 4y^2)$

93. $125x^3 + 8y^3 =$
$(5x)^3 + (2y)^3 =$
$(5x + 2y)(25x^2 - 10xy + 4y^2)$

95. $64 + x^3 =$
$(4)^3 + (x)^3 =$
$(4 + x)(16 - 4x + x^2)$

PROBLEM SET 7.3 Factoring Trinomials of the Form $x^2 + bx + c$

1. We need two integers whose product is 24 and whose sum is 10. They are 4 and 6.
$x^2 + 10x + 24 = (x + 4)(x + 6)$

3. We need two integers whose product is 40 and whose sum is 13. They are 5 and 8.
$x^2 + 13x + 40 = (x + 5)(x + 8)$

5. We need two integers whose product is 18 and whose sum is -11. They are -2 and -9.
$x^2 - 11x + 18 = (x - 2)(x - 9)$

7. We need two integers whose product is 28 and whose sum is -11. They are -4 and -7.
$n^2 - 11n + 28 = (n - 4)(n - 7)$

9. We need two integers whose product is -27 and whose sum is 6. They are 9 and -3.
$n^2 + 6n - 27 = (n - 3)(n + 9)$

11. We need two integers whose product is -40 and whose sum is -6. They are 4 and -10.
$n^2 - 6n - 40 = (n + 4)(n - 10)$

13. We need two integers whose product is 24 and whose sum is 12. No such integers exist, therefore $t^2 + 12t + 24$ is not factorable.

15. We need two integers whose product is 72 and whose sum is -18. They are -6 and -12.
$x^2 - 18x + 72 = (x - 6)(x - 12)$

17. We need two integers whose product is -66 and whose sum is 5. They are -6 and 11.
$x^2 + 5x - 66 = (x - 6)(x + 11)$

19. We need two integers whose product is -72 and whose sum is -1. They are 8 and -9.
$y^2 - y - 72 = (y + 8)(y - 9)$

21. We need two integers whose product is 80 and whose sum is 21. They are 5 and 16.
$x^2 + 21x + 80 = (x + 5)(x + 16)$

23. We need two integers whose product is -72 and whose sum is 6. They are -6 and 12.
$x^2 + 6x - 72 = (x - 6)(x + 12)$

25. We need two integers whose product is -48 and whose sum is -10. No such integers exist, therefore $x^2 - 10x - 48$ is not factorable.

27. We need two integers whose product is -10 and whose sum is 3. They are -2 and 5.
$x^2 + 3xy - 10y^2 = (x - 2y)(x + 5y)$

29. We need two integers whose product is -32 and whose sum is -4. They are 4 and -8.
$a^2 - 4ab - 32b^2 = (a + 4b)(a - 8b)$

31. $x^2 + 10x + 21 = 0$
$(x + 3)(x + 7) = 0$
$x + 3 = 0$ or $x + 7 = 0$
$x = -3$ or $x = -7$
The solution set is $\{-7, -3\}$.

33. $x^2 - 9x + 18 = 0$
$(x - 3)(x - 6) = 0$
$x - 3 = 0$ or $x - 6 = 0$
$x = 3$ or $x = 6$
The solution set is $\{3, 6\}$.

35. $x^2 - 3x - 10 = 0$
$(x - 5)(x + 2) = 0$
$x - 5 = 0$ or $x + 2 = 0$
$x = 5$ or $x = -2$
The solution set is $\{-2, 5\}$.

37. $n^2 + 5n - 36 = 0$
$(n + 9)(n - 4) = 0$
$n + 9 = 0$ or $n - 4 = 0$
$n = -9$ or $n = 4$
The solution set is $\{-9, 4\}$

39. $n^2 - 6n - 40 = 0$
$(n - 10)(n + 4) = 0$
$n - 10 = 0$ or $n + 4 = 0$
$n = 10$ or $n = -4$
The solution set is $\{-4, 10\}$.

41. $t^2 + t - 56 = 0$
$(t + 8)(t - 7) = 0$
$t + 8 = 0$ or $t - 7 = 0$
$t = -8$ or $t = 7$
The solution set is $\{-8, 7\}$.

43. $x^2 - 16x + 28 = 0$
$(x - 14)(x - 2) = 0$
$x - 14 = 0$ or $x - 2 = 0$
$x = 14$ or $x = 2$
The solution set is $\{2, 14\}$.

45. $x^2 + 11x = 12$
$x^2 + 11x - 12 = 0$
$(x + 12)(x - 1) = 0$
$x + 12 = 0$ or $x - 1 = 0$
$x = -12$ or $x = 1$
The solution set is $\{-12, 1\}$.

47. $x(x - 10) = -16$
$x^2 - 10x = -16$
$x^2 - 10x + 16 = 0$
$(x - 8)(x - 2) = 0$
$x - 8 = 0$ or $x - 2 = 0$
$x = 8$ or $x = 2$
The solution set is $\{2, 8\}$.

49. $-x^2 - 2x + 24 = 0$
Multiply both sides by -1.
$x^2 + 2x - 24 = 0$
$(x + 6)(x - 4) = 0$
$x + 6 = 0$ or $x - 4 = 0$
$x = -6$ or $x = 4$
The solution set is $\{-6, 4\}$.

51. Let n represent one integer, then $n + 1$ represents the next integer.
$n(n + 1) = 56$
$n^2 + n = 56$
$n^2 + n - 56 = 0$
$(n + 8)(n - 7) = 0$
$n + 8 = 0$ or $n - 7 = 0$
$n = -8$ or $n = 7$
If $n = -8$, then $n + 1 = -7$. If $n = 7$, then $n + 1 = 8$. Thus, the consecutive integers are either -8 and -7, or 7 and 8.

53. Let n represent an even whole number, then $n + 2$ represents the next consecutive even whole number.
$n(n + 2) = 168$
$n^2 + 2n = 168$
$n^2 + 2n - 168 = 0$
$(n + 14)(n - 12) = 0$
$n + 14 = 0$ or $n - 12 = 0$
$n = -14$ or $n = 12$
$n + 2 = -12$ or $n + 2 = 14$
Since -14 and -12 are not whole numbers, the consecutive even whole numbers would be 12 and 14.

55. Let n represent the first integer, then $n+1, n+2$, and $n+3$ represent the other three consecutive integers.

$$(n+2)(n+3)=2[n(n+1)]-22$$
$$n^2+5n+6=2(n^2+n)-22$$
$$n^2+5n+6=2n^2+2n-22$$
$$-n^2+3n+28=0$$
$$n^2-3n-28=0$$
$$(n-7)(n+4)=0$$
$$n-7=0 \quad \text{or} \quad n+4=0$$
$$n=7 \quad \text{or} \quad n=-4$$

If the first integer is -4, then the four consecutive integers are $-4, -3, -2$, and -1. If the first integer is 7, then the four consecutive integers are $7, 8, 9$, and 10.

57. Let n represent the larger number, then $n-3$ is the smaller number.

$$n^2=10(n-3)+9$$
$$n^2=10n-30+9$$
$$n^2=10n-21$$
$$n^2-10n+21=0$$
$$(n-7)(n-3)=0$$
$$n-7=0 \quad \text{or} \quad n-3=0$$
$$n=7 \quad \text{or} \quad n=3$$

If $n=7$, then $n-3=4$. If $n=3$, then $n-3=0$. Thus the numbers are 4 and 7, or 0 and 3.

59. Let l represent the length of the rectangle, then $l-3$ represents the width of the rectangle.

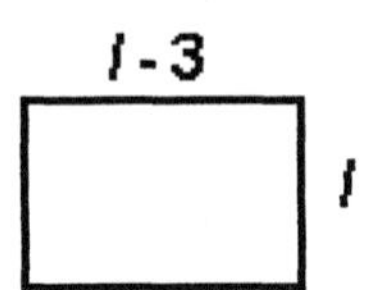

$$\text{Area}=(2)\text{Perimeter}-6$$
$$l(l-3)=2[2l+2(l-3)]-6$$
$$l^2-3l=2[2l+2l-6]-6$$
$$l^2-3l=2[4l-6]-6$$
$$l^2-3l=8l-12-6$$
$$l^2-3l=8l-18$$
$$l^2-11l+18=0$$
$$(l-2)(l-9)=0$$
$$l-2=0 \quad \text{or} \quad l-9=0$$
$$l=2 \quad \text{or} \quad l=9$$

If $l=2$, then $l-3=-1$ which is not possible. If $l=9$, then $l-3=6$. Therefore, the rectangle is 9 inches by 6 inches.

61. Let w represent the width of the rectangle, then one-half the perimeter less the width, $15-w$, represents the length of the rectangle.

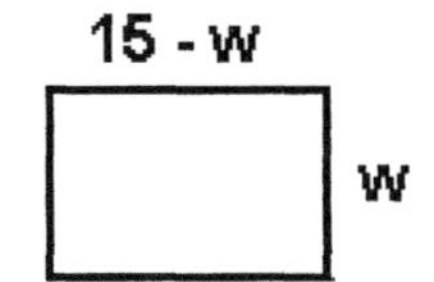

$$\text{Area}=54$$
$$w(15-w)=54$$
$$15w-w^2=54$$
$$-w^2+15w-54=0$$
$$w^2-15w+54=0$$
$$(w-6)(w-9)=0$$
$$w-6=0 \quad \text{or} \quad w-9=0$$
$$w=6 \quad \text{or} \quad w=9$$

If the width is 6, then $15-6=9$. If the width is 9, then $15-9=6$. Thus, the length and width of the rectangle must be 6 centimeters by 9 centimeters.

63. Let r represent the number of rows in the orchard, then $r+5$ represents the number of trees per row.

$$(\text{trees per row})(\text{rows})=\text{number of trees}$$
$$(r+5)(r)=84$$
$$r^2+5r=84$$
$$r^2+5r-84=0$$
$$(r+12)(r-7)=0$$

$r + 12 = 0$ or $r - 7 = 0$
$r = -12$ or $r = 7$
The number of rows cannot be negative, so $r = -12$ is discarded. Thus there would be 7 rows of trees.

65. Let x represent the length of one leg of the triangle, then $x - 7$ represents the other leg and $x + 2$ represents the length of the hypotenuse.

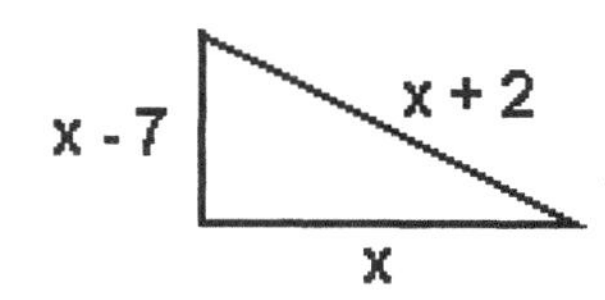

$$x^2 + (x - 7)^2 = (x + 2)^2$$
Pythagorean Theorem
$$x^2 + x^2 - 14x + 49 = x^2 + 4x + 4$$
$$2x^2 - 14x + 49 = x^2 + 4x + 4$$
$$x^2 - 18x + 45 = 0$$
$$(x - 3)(x - 15) = 0$$
$x - 3 = 0$ or $x - 15 = 0$
$x = 3$ or $x = 15$
If $x = 3$, then $x - 7 = -4$ which is not possible. If $x = 15$ then $x - 7 = 8$ and $x + 2 = 17$. Thus, the sides of the right triangle are 8 feet, 15 feet and 17 feet.

67. Let x represent the length of one leg of the triangle, then $x - 2$ represents the other leg.

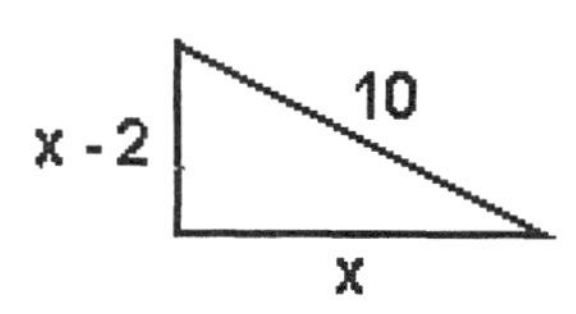

$$x^2 + (x - 2)^2 = 10^2$$
Pythagorean Theorem
$$x^2 + x^2 - 4x + 4 = 100$$
$$2x^2 - 4x + 4 = 100$$
$$2x^2 - 4x - 96 = 0$$
$$x^2 - 2x - 48 = 0$$
$$(x - 8)(x + 6) = 0$$
$x - 8 = 0$ or $x + 6 = 0$
$x = 8$ or $x = -6$
The solution of $x = -6$ is not possible. If $x = 8$, then $x - 2 = 6$. Thus, the legs of the triangle are 6 inches and 8 inches.

73. We need two integers whose product is 40 and whose sum is 13. They are 5 and 8.
$x^{2a} + 13x^a + 40 = (x^a + 5)(x^a + 8)$

75. We need two integers whose product is -27 and whose sum is 6. They are -3 and 9.
$x^{2a} + 6x^a - 27 = (x^a - 3)(x^a + 9)$

PROBLEM SET 7.4 Factoring Trinomials of the Form $ax^2 + bx + c$

1. We need two integers whose product is 6 and whose sum is 7. They are 6 and 1.
$3x^2 + 7x + 2 =$
$3x^2 + 1x + 6x + 2 =$
$x(3x + 1) + 2(3x + 1) =$
$(3x + 1)(x + 2)$

3. We need two integers whose product is 60 and whose sum is 19. They are 4 and 15.
$6x^2 + 19x + 10 =$
$6x^2 + 4x + 15x + 10 =$
$2x(3x + 2) + 5(3x + 2) =$
$(3x + 2)(2x + 5)$

5. We need two integers whose product is 24 and whose sum is -25. They are -1 and -24.
$4x^2 - 25x + 6 =$
$4x^2 - 1x - 24x + 6 =$
$x(4x - 1) - 6(4x - 1) =$
$(4x - 1)(x - 6)$

Problem Set 7.4

7. We need two integers whose product is 240 and whose sum is -31. They are -15 and -16.
$12x^2 - 31x + 20 =$
$12x^2 - 15x - 16x + 20 =$
$3x(4x - 5) - 4(4x - 5) =$
$(4x - 5)(3x - 4)$

9. We need two integers whose product is -70 and whose sum is -33. They are 2 and -35.
$5y^2 - 33y - 14 =$
$5y^2 + 2y - 35y - 14 =$
$y(5y + 2) - 7(5y + 2) =$
$(5y + 2)(y - 7)$

11. $4n^2 + 26n - 48 = 2(2n^2 + 13n - 24)$
We need two integers whose product is -48 and whose sum is 13. They are -3 and 16.
$2n^2 + 13n - 24 =$
$2n^2 - 3n + 16n - 24 =$
$n(2n - 3) + 8(2n - 3) = (2n - 3)(n + 8)$
$4n^2 + 26n - 48 = 2(2n - 3)(n + 8)$

13. We need two integers whose product is 14 and whose sum is 1. No such integers exist, therefore $2x^2 + x + 7$ is not factorable.

15. We need two integers whose product is 126 and whose sum is 45. They are 3 and 42.
$18x^2 + 45x + 7 =$
$18x^2 + 3x + 42x + 7 =$
$3x(6x + 1) + 7(6x + 1) = (6x + 1)(3x + 7)$

17. $21x^2 - 90x + 24 = 3(7x^2 - 30x + 8)$
We need two integers whose product is 56 and whose sum is -30. They are -2 and -28.
$7x^2 - 30x + 8 =$
$7x^2 - 2x - 28x + 8 =$
$x(7x - 2) - 4(7x - 2) = (7x - 2)(x - 4)$
$21x^2 - 90x + 24 = 3(7x - 2)(x - 4)$

19. We need two integers whose product is -168 and whose sum is 2. They are -12 and 14.
$8x^2 + 2x - 21 =$
$8x^2 - 12x + 14x - 21 =$
$4x(2x - 3) + 7(2x - 3) =$
$(2x - 3)(4x + 7)$

21. We need two integers whose product is -126 and whose sum is -15. They are 6 and -21.
$9t^2 - 15t - 14 =$
$9t^2 + 6t - 21t - 14 =$
$3t(3t + 2) - 7(3t + 2) =$
$(3t + 2)(3t - 7)$

23. We need two integers whose product is -420 and whose sum is 79. They are -5 and 84.
$12y^2 + 79y - 35 =$
$12y^2 - 5y + 84y - 35 =$
$y(12y - 5) + 7(12y - 5) =$
$(12y - 5)(y + 7)$

25. We need two integers whose product is -30 and whose sum is 2. No such integers exist, therefore $6n^2 + 2n - 5$ is not factorable.

27. We need two integers whose product is 294 and whose sum is 55. They are 6 and 49.
$14x^2 + 55x + 21 =$
$14x^2 + 6x + 49x + 21 =$
$2x(7x + 3) + 7(7x + 3) =$
$(7x + 3)(2x + 7)$

29. We need two integers whose product is 240 and whose sum is -31. They are -15 and -16.
$20x^2 - 31x + 12 =$
$20x^2 - 15x - 16x + 12 =$
$5x(4x - 3) - 4(4x - 3) =$
$(4x - 3)(5x - 4)$

31. $2x^2 + 13x + 6 = 0$
$(2x + 1)(x + 6) = 0$
$2x + 1 = 0$ or $x + 6 = 0$
$2x = -1$ or $x = -6$
$x = -\frac{1}{2}$ or $x = -6$
The solution set is $\left\{-6, -\frac{1}{2}\right\}$.

33. $12x^2 + 11x + 2 = 0$
$(4x + 1)(3x + 2) = 0$
$4x + 1 = 0$ or $3x + 2 = 0$
$4x = -1$ or $3x = -2$
$x = -\frac{1}{4}$ or $x = -\frac{2}{3}$
The solution set is $\left\{-\frac{2}{3}, -\frac{1}{4}\right\}$.

35. $3x^2 - 25x + 8 = 0$
$(3x - 1)(x - 8) = 0$
$3x - 1 = 0$ or $x - 8 = 0$
$3x = 1$ or $x = 8$
$x = \frac{1}{3}$ or $x = 8$
The solution set is $\left\{\frac{1}{3}, 8\right\}$

37. $15n^2 - 41n + 14 = 0$
$(3n - 7)(5n - 2) = 0$
$3n - 7 = 0$ or $5n - 2 = 0$
$3n = 7$ or $5n = 2$
$n = \frac{7}{3}$ or $n = \frac{2}{5}$
The solution set is $\left\{\frac{2}{5}, \frac{7}{3}\right\}$.

39. $6t^2 + 37t - 35 = 0$
$(6t - 5)(t + 7) = 0$
$6t - 5 = 0$ or $t + 7 = 0$
$6t = 5$ or $t = -7$
$t = \frac{5}{6}$ or $t = -7$
The solution set is $\left\{-7, \frac{5}{6}\right\}$.

41. $16y^2 - 18y - 9 = 0$
$(8y + 3)(2y - 3) = 0$
$8y + 3 = 0$ or $2y - 3 = 0$
$8y = -3$ or $2y = 3$
$y = -\frac{3}{8}$ or $y = \frac{3}{2}$
The solution set is $\left\{-\frac{3}{8}, \frac{3}{2}\right\}$.

43. $9x^2 - 6x - 8 = 0$
$(3x - 4)(3x + 2) = 0$
$3x - 4 = 0$ or $3x + 2 = 0$
$3x = 4$ or $3x = -2$
$x = \frac{4}{3}$ or $x = -\frac{2}{3}$
The solution set is $\left\{-\frac{2}{3}, \frac{4}{3}\right\}$.

45. $10x^2 - 29x + 10 = 0$
$(2x - 5)(5x - 2) = 0$
$2x - 5 = 0$ or $5x - 2 = 0$
$2x = 5$ or $5x = 2$
$x = \frac{5}{2}$ or $x = \frac{2}{5}$
The solution set is $\left\{\frac{2}{5}, \frac{5}{2}\right\}$.

47. $6x^2 + 19x = -10$
$6x^2 + 19x + 10 = 0$
$(2x + 5)(3x + 2) = 0$
$2x + 5 = 0$ or $3x + 2 = 0$
$2x = -5$ or $3x = -2$
$x = -\frac{5}{2}$ or $x = -\frac{2}{3}$
The solution set is $\left\{-\frac{5}{2}, -\frac{2}{3}\right\}$.

49. $16x(x + 1) = 5$
$16x^2 + 16x = 5$
$16x^2 + 16x - 5 = 0$
$(4x - 1)(4x + 5) = 0$
$4x - 1 = 0$ or $4x + 5 = 0$
$4x = 1$ or $4x = -5$
$x = \frac{1}{4}$ or $x = -\frac{5}{4}$
The solution set is $\left\{-\frac{5}{4}, \frac{1}{4}\right\}$.

PROBLEM SET 7.5 Factoring, Solving Equations, and Problem Solving

1. $x^2 + 4x + 4 =$
$(x)^2 + 2(x)(2) + (2)^2 = (x+2)^2$

3. $x^2 - 10x + 25 =$
$(x)^2 - 2(x)(5) + (5)^2 = (x-5)^2$

5. $9n^2 + 12n + 4 =$
$(3n)^2 + 2(3n)(2) + (2)^2 = (3n+2)^2$

7. $16a^2 - 8a + 1 =$
$(4a)^2 - 2(4a)(1) + (1)^2 = (4a-1)^2$

9. $4 + 36x + 81x^2 =$
$(2)^2 + 2(2)(9x) + (9x)^2 = (2+9x)^2$

11. $16x^2 - 24xy + 9y^2 =$
$(4x)^2 - 2(4x)(3y) + (3y)^2 = (4x-3y)^2$

13. We need two integers whose product is 16 and whose sum is 17. They are 1 and 16.
$2x^2 + 17x + 8 =$
$2x^2 + 1x + 16x + 8 =$
$x(2x+1) + 8(2x+1) = (2x+1)(x+8)$

15. $2x^3 - 72x = 2x(x^2 - 36) =$
$2x(x-6)(x+6)$

17. We need two integers whose product is -60 and whose sum is -7. They are 5 and -12.
$n^2 - 7n - 60 = (n+5)(n-12)$

19. We need two integers whose product is -12 and whose sum is -7. No such integers exist, therefore $3a^2 - 7a - 4$ is not factorable.

21. $8x^2 + 72 = 8(x^2 + 9)$

23. $9x^2 + 30x + 25 =$
$(3x)^2 + 2(3x)(5) + (5)^2 = (3x+5)^2$

25. $15x^2 + 65x + 70 = 5(3x^2 + 13x + 14)$
We need two integers whose product is 42 and whose sum is 13. They are 6 and 7.
$5(3x^2 + 6x + 7x + 14) =$
$5[3x(x+2) + 7(x+2)] =$
$5(x+2)(3x+7)$

27. We need two integers whose product is -360 and whose sum is 2. They are -18 and 20.
$24x^2 + 2x - 15 =$
$24x^2 - 18x + 20x - 15 =$
$6x(4x-3) + 5(4x-3) =$
$(4x-3)(6x+5)$

29. $xy + 5y - 8x - 40 =$
$y(x+5) - 8(x+5) = (x+5)(y-8)$

31. We need two integers whose product is -140 and whose sum is 31. They are -4 and 35.
$20x^2 + 31xy - 7y^2 =$
$20x^2 - 4xy + 35xy - 7y^2 =$
$4x(5x-y) + 7y(5x-y) =$
$(5x-y)(4x+7y)$

33. $24x^2 + 18x - 81 = 3(8x^2 + 6x - 27)$
We need two integers whose product is -216 and whose sum is 6. They are -12 and 18.
$3(8x^2 + 6x - 27) =$
$3(8x^2 - 12x + 18x - 27) =$
$3[4x(2x-3) + 9(2x-3)] =$
$3(2x-3)(4x+9)$

35. $2x^2 + 6x + 30 = 6(2x^2 + x + 5)$
$[2x^2 + x + 5$ is not factorable.$]$

37. $5x^4 - 80 = 5(x^4 - 16) =$
$5[(x^2-4)(x^2+4)] =$
$5(x-2)(x+2)(x^2+4)$

39. $x^2 + 12xy + 36y^2 =$
$(x)^2 + 2(x)(6y) + (6y)^2 = (x+6y)^2$

41. $4x^2 - 20x = 0$
$4x(x - 5) = 0$
$4x = 0$ or $x - 5 = 0$
$x = 0$ or $x = 5$
The solution set is $\{0, 5\}$.

43. $x^2 - 9x - 36 = 0$
$(x - 12)(x + 3) = 0$
$x - 12 = 0$ or $x + 3 = 0$
$x = 12$ or $x = -3$
The solution set is $\{-3, 12\}$.

45. $-2x^3 + 8x = 0$
Divide by -2.
$x^3 - 4x = 0$
$x(x^2 - 4) = 0$
$x(x - 2)(x + 2) = 0$
$x = 0$ or $x - 2 = 0$ or $x + 2 = 0$
$x = 0$ or $x = 2$ or $x = -2$
The solution set is $\{-2, 0, 2\}$.

47. $6n^2 - 29n - 22 = 0$
$(2n - 11)(3n + 2) = 0$
$2n - 11 = 0$ or $3n + 2 = 0$
$2n = 11$ or $3n = -2$
$n = \frac{11}{2}$ or $n = -\frac{2}{3}$
The solution set is $\left\{-\frac{2}{3}, \frac{11}{2}\right\}$.

49. $(3n - 1)(4n - 3) = 0$
$3n - 1 = 0$ or $4n - 3 = 0$
$3n = 1$ or $4n = 3$
$n = \frac{1}{3}$ or $n = \frac{3}{4}$
The solution set is $\left\{\frac{1}{3}, \frac{3}{4}\right\}$.

51. $(n - 2)(n + 6) = -15$
$n^2 + 4n - 12 = -15$
$n^2 + 4n + 3 = 0$
$(n + 3)(n + 1) = 0$
$n + 3 = 0$ or $n + 1 = 0$
$n = -3$ or $n = -1$
The solution set is $\{-3, -1\}$.

53. $2x^2 = 12x$
Divide by 2.
$x^2 = 6x$
$x^2 - 6x = 0$
$x(x - 6) = 0$
$x = 0$ or $x - 6 = 0$
$x = 0$ or $x = 6$
The solution set is $\{0, 6\}$.

55. $t^3 - 2t^2 - 24t = 0$
$t(t^2 - 2t - 24) = 0$
$t(t - 6)(t + 4) = 0$
$t = 0$ or $t - 6 = 0$ or $t + 4 = 0$
$t = 0$ or $t = 6$ or $t = -4$
The solution set is $\{-4, 0, 6\}$.

57. $12 - 40x + 25x^2 = 0$
$(2 - 5x)(6 - 5x) = 0$
$2 - 5x = 0$ or $6 - 5x = 0$
$-5x = -2$ or $-5x = -6$
$x = \frac{2}{5}$ or $x = \frac{6}{5}$
The solution set is $\left\{\frac{2}{5}, \frac{6}{5}\right\}$.

59. $n^2 - 28n + 192 = 0$
$(n - 12)(n - 16) = 0$
$n - 12 = 0$ or $n - 16 = 0$
$n = 12$ or $n = 16$
The solution set is $\{12, 16\}$.

61. $(3n + 1)(n + 2) = 12$
$3n^2 + 7n + 2 = 12$
$3n^2 + 7n - 10 = 0$
$(3n + 10)(n - 1) = 0$
$3n + 10 = 0$ or $n - 1 = 0$
$3n = -10$ or $n = 1$
$n = -\frac{10}{3}$ or $n = 1$
The solution set is $\left\{-\frac{10}{3}, 1\right\}$.

63.
$$x^3 = 6x^2$$
$$x^3 - 6x^2 = 0$$
$$x^2(x-6) = 0$$
$$x(x)(x-6) = 0$$
$$x = 0 \quad \text{or} \quad x = 0 \quad \text{or} \quad x - 6 = 0$$
$$x = 0 \quad \text{or} \quad x = 0 \quad \text{or} \quad x = 6$$
The solution set is $\{0, 6\}$.

65.
$$9x^2 - 24x + 16 = 0$$
$$(3x-4)(3x-4) = 0$$
$$3x - 4 = 0 \quad \text{or} \quad 3x - 4 = 0$$
$$3x = 4 \quad \text{or} \quad 3x = 4$$
$$x = \frac{4}{3} \quad \text{or} \quad x = \frac{4}{3}$$
The solution set is $\left\{\frac{4}{3}\right\}$.

67.
$$x^3 + 10x^2 + 25x = 0$$
$$x(x^2 + 10x + 25) = 0$$
$$x(x+5)(x+5) = 0$$
$$x = 0 \quad \text{or} \quad x + 5 = 0 \quad \text{or} \quad x + 5 = 0$$
$$x = 0 \quad \text{or} \quad x = -5 \quad \text{or} \quad x = -5$$
The solution set is $\{-5, 0\}$.

69.
$$24x^2 + 17x - 20 = 0$$
$$(3x+4)(8x-5) = 0$$
$$3x + 4 = 0 \quad \text{or} \quad 8x - 5 = 0$$
$$3x = -4 \quad \text{or} \quad 8x = 5$$
$$x = -\frac{4}{3} \quad \text{or} \quad x = \frac{5}{8}$$
The solution set is $\left\{-\frac{4}{3}, \frac{5}{8}\right\}$.

71. Let n represent one of the numbers, then $4n + 7$ represents the other number.
$$n(4n+7) = 15$$
$$4n^2 + 7n = 15$$
$$4n^2 + 7n - 15 = 0$$
$$(4n-5)(n+3) = 0$$
$$4n - 5 = 0 \quad \text{or} \quad n + 3 = 0$$
$$4n = 5 \quad \text{or} \quad n = -3$$
$$n = \frac{5}{4} \quad \text{or} \quad n = -3$$
If $n = \frac{5}{4}$, then $4n + 7 = 4\left(\frac{5}{4}\right) + 7 = 12$.
If $n = -3$, then $4n + 7 = 4(-3) + 7 = -5$.
Thus, the numbers are $\frac{5}{4}$ and 12, or -3 and -5.

73. Let n represent one number, then $2n + 3$ represents the other number.
$$n(2n+3) = -1$$
$$2n^2 + 3n = -1$$
$$2n^2 + 3n + 1 = 0$$
$$(2n+1)(n+1) = 0$$
$$2n + 1 = 0 \quad \text{or} \quad n + 1 = 0$$
$$2n = -1 \quad \text{or} \quad n = -1$$
$$n = -\frac{1}{2} \quad \text{or} \quad n = -1$$
If $n = -\frac{1}{2}$, then $2n + 3 = 2\left(-\frac{1}{2}\right) + 3 = 2$.
If $n = -1$, then $2n + 3 = 2(-1) + 3 = 1$.
Thus, the numbers are $-\frac{1}{2}$ and 2, or -1 and 1.

75. Let n represent one number, then $2n + 1$ represents the other number.
$$n^2 + (2n+1)^2 = 97$$
$$n^2 + 4n^2 + 4n + 1 = 97$$
$$5n^2 + 4n + 1 = 97$$
$$5n^2 + 4n - 96 = 0$$
$$(5n+24)(n-4) = 0$$
$$5n + 24 = 0 \quad \text{or} \quad n - 4 = 0$$
$$5n = -24 \quad \text{or} \quad n = 4$$
$$n = -\frac{24}{5} \quad \text{or} \quad n = 4$$
If $n = -\frac{24}{5}$, then $2n + 1 =$
$2\left(-\frac{24}{5}\right) + 1 = -\frac{48}{5} + \frac{5}{5} = -\frac{43}{5}$.
If $n = 4$, then $2n + 1 = 2(4) + 1 = 9$
Thus, the numbers are $-\frac{24}{5}$ and $-\frac{43}{5}$, or 4 and 9.

77. Let r represent the number of rows, then $2r-3$ represents the number of chairs per row.

$$\begin{aligned}(\text{rows})(\text{chairs per row}) &= 54\\ r(2r-3) &= 54\\ 2r^2-3r &= 54\\ 2r^2-3r-54 &= 0\\ (2r+9)(r-6) &= 0\end{aligned}$$

$2r+9=0$ or $r-6=0$

$2r=-9$ or $r=6$

$r=-\frac{9}{2}$ or $r=6$

The solution of $r=-\frac{9}{2}$ is not reasonable.

If $r=6$, then $2r-3=2(6)-3=9$.Thus, there would be 6 rows with 9 chairs per row.

79. Let s represent the length of a side of the smaller square, then $3s$ represents the length of a side of the larger square.

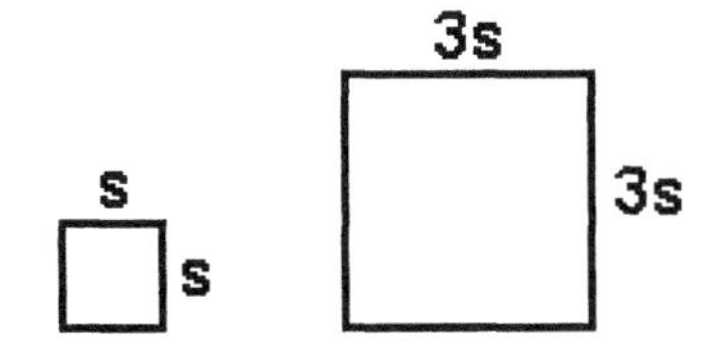

$$\begin{aligned}\text{Smaller Area} + \text{Larger Area} &= 360\\ s^2+(3s)^2 &= 360\\ s^2+9s^2 &= 360\\ 10s^2 &= 360\\ s^2 &= 36\\ s^2-36 &= 0\\ (s-6)(s+6) &= 0\end{aligned}$$

$s-6=0$ or $s+6=0$

$s=6$ or $s=-6$

The negative solution must be discarded. Therefore, the smaller square is 6 feet by 6 feet and the larger square is 18 feet by 18 feet.

81. Let w represent the width of the rectangle, then $2w+1$ represents the length.

$$\begin{aligned}w(2w+1) &= 55\\ 2w^2+w &= 55\\ 2w^2+w-55 &= 0\\ (2w+11)(w-5) &= 0\end{aligned}$$

$2w+11=0$ or $w-5=0$

$2w=-11$ or $w=5$

$w=-\frac{11}{2}$ or $w=5$

The negative solution must be discarded. Therefore, the rectangle is 5 centimeters by 11 centimeters.

83. Let h represent the length of an altitude to a side of the triangle, then $3h-1$ represents the side.

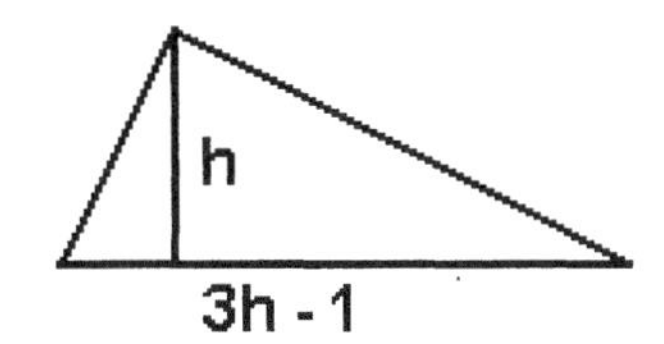

$$\begin{aligned}\frac{1}{2}h(3h-1) &= 51\\ h(3h-1) &= 102\\ 3h^2-h &= 102\\ 3h^2-h-102 &= 0\\ (3h+17)(h-6) &= 0\end{aligned}$$

$3h+17=0$ or $h-6=0$

$3h=-17$ or $h=6$

$h=-\frac{17}{3}$ or $h=6$

The negative solution must be discarded. Therefore, the length of the side is 17 inches and the altitude to that side is 6 inches long.

85. Let x represent the width of the strip, then $8-2x$ represents the reduced width of the paper, and $11-2x$ represents the reduced length of the paper.

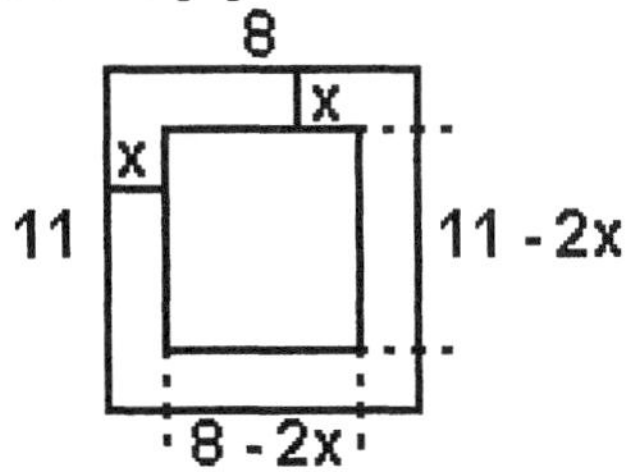

$(8-2x)(11-2x)=40$
$88-38x+4x^2=40$
$48-38x+4x^2=0$
$24-19x+2x^2=0$
$(3-2x)(8-x)=0$
$3-2x=0$ or $8-x=0$
$-2x=-3$ or $8=x$
$x=\frac{3}{2}$ or $x=8$

The solution $x=8$ must be discarded since the width of the paper is 8 inches.
Therefore, the width of the strip is $1\frac{1}{2}$ inches.

87. Let r represent a radius of the larger circle, then $r-6$ represents a radius of the smaller circle.

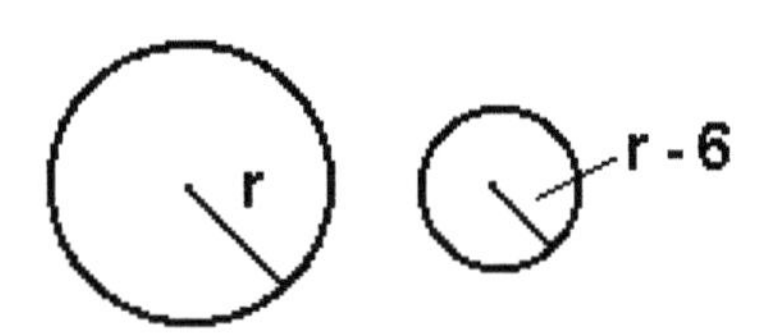

Larger Area + Smaller Area $=180\pi$
$\pi r^2+\pi(r-6)^2=180\pi$
$\pi r^2+\pi(r^2-12r+36)=180\pi$
$\pi r^2+\pi r^2-12\pi r+36\pi=180\pi$
$2\pi r^2-12\pi r+36\pi=180\pi$
$2\pi r^2-12\pi r-144\pi=0$
Divide both sides by 2π.
$r^2-6r-72=0$
$(r-12)(r+6)=0$
$r-12=0$ or $r+6=0$
$r=12$ or $r=-6$
Discard the negative solution. The length of a radius of the larger circle is 12 inches and a radius of the smaller circle is 6 inches.

CHAPTER 7 | Review Problem Set

1. We need two integers whose product is 14 and whose sum is -9. They are -2 and -7.
$x^2-9x+14=(x-2)(x-7)$

2. $3x^2+21x=3x(x+7)$

3. $9x^2-4=(3x)^2-(2)^2=$
$(3x-2)(3x+2)$

4. We need two integers whose product is -20 and whose sum is 8. They are -2 and 10.
$4x^2+8x-5=$
$4x^2-2x+10x-5=$
$2x(2x-1)+5(2x-1)=$
$(2x-1)(2x+5)$

5. $25x^2-60x+36=$
$(5x)^2-2(5x)(6)+(6)^2=(5x-6)^2$

6. $n^3+13n^2+40n=n(n^2+13n+40)$
We need two integers whose product is 40 and whose sum is 13. They are 5 and 8.
$n(n^2+13n+40)=n(n+5)(n+8)$

7. We need two integers whose product is -12 and whose sum is 11. They are -1 and 12.
$y^2+11y-12=(y-1)(y+12)$

8. $3xy^2+6x^2y=3xy(y+2x)$

9. $x^4-1=(x^2)^2-(1)^2=$
$(x^2-1)(x^2+1)=$
$(x-1)(x+1)(x^2+1)$

10. We need two integers whose product is -90 and whose sum is 9. They are -6 and 15.
$18n^2+9n-5=$
$18n^2-6n+15n-5=$
$6n(3n-1)+5(3n-1)=$
$(3n-1)(6n+5)$

11. We need two integers whose product is 24 and whose sum is 7. No such integers exist, therefore $x^2+7x+24$ is not factorable.

12. We need two integers whose product is -28 and whose sum is -3. They are 4 and -7.
$4x^2 - 3x - 7 = 4x^2 + 4x - 7x - 7 =$
$4x(x+1) - 7(x+1) =$
$(x+1)(4x-7)$

13. $3n^2 + 3n - 90 = 3(n^2 + n - 30)$
We need two integers whose product is -30 and whose sum is 1. They are -5 and 6.
$3(n^2 + n - 30) = 3(n-5)(n+6)$

14. $x^3 - xy^2 = x(x^2 - y^2) = x(x-y)(x+y)$

15. We need two integers whose product is -4 and whose sum is 3. They are -1 and 4.
$2x^2 + 3xy - 2y^2 =$
$2x^2 - xy + 4xy - 2y^2 =$
$x(2x-y) + 2y(2x-y) = (2x-y)(x+2y)$

16. $4n^2 - 6n - 40 = 2(2n^2 - 3n - 20)$
We need two integers whose product is -40 and whose sum is -3. They are 5 and -8.
$2(2n^2 - 3n - 20) =$
$2(2n^2 + 5n - 8n - 20) =$
$2[n(2n+5) - 4(2n+5)] =$
$2(2n+5)(n-4)$

17. $5x + 5y + ax + ay =$
$5(x+y) + a(x+y) = (x+y)(5+a)$

18. We need two integers whose product is -84 and whose sum is -5. They are 7 and -12.
$21t^2 - 5t - 4 =$
$21t^2 + 7t - 12t - 4 =$
$7t(3t+1) - 4(3t+1) =$
$(3t+1)(7t-4)$

19. $2x^3 - 2x = 2x(x^2 - 1) =$
$2x(x-1)(x+1)$

20. $3x^3 - 108x =$
$3x(x^2 - 36) =$
$3x(x-6)(x+6)$

21. $16x^2 + 40x + 25 =$
$(4x)^2 + 2(4x)(5) + (5)^2 = (4x+5)^2$

22. $xy - 3x - 2y + 6 =$
$x(y-3) - 2(y-3) = (y-3)(x-2)$

23. We need two integers whose product is -30 and whose sum is -7. They are 3 and -10.
$15x^2 - 7xy - 2y^2 =$
$15x^2 + 3xy - 10xy - 2y^2 =$
$3x(5x+y) - 2y(5x+y) =$
$(5x+y)(3x-2y)$

24. $6n^4 - 5n^3 + n^2 = n^2(6n^2 - 5n + 1)$
We need two integers whose product is 6 and whose sum is -5. They are -2 and -3.
$n^2(6n^2 - 5n + 1) =$
$n^2(6n^2 - 2n - 3n + 1) =$
$n^2[2n(3n-1) - 1(3n-1)] =$
$n^2(3n-1)(2n-1)$

25. $x^2 + 4x - 12 = 0$
$(x+6)(x-2) = 0$
$x + 6 = 0 \quad \text{or} \quad x - 2 = 0$
$x = -6 \quad \text{or} \quad x = 2$
The solution set is $\{-6, 2\}$.

26. $x^2 - 11x$
$x^2 - 11x = 0$
$x(x-11) = 0$
$x = 0 \quad \text{or} \quad x - 11 = 0$
$x = 0 \quad \text{or} \quad x = 11$
The solution set is $\{0, 11\}$.

27. $2x^2 + 3x - 20 = 0$
$(2x-5)(x+4) = 0$
$2x - 5 = 0 \quad \text{or} \quad x + 4 = 0$
$2x = 5 \quad \text{or} \quad x = -4$
$x = \frac{5}{2} \quad \text{or} \quad x = -4$
The solution set is $\left\{-4, \frac{5}{2}\right\}$.

28. $9n^2 + 21n - 8 = 0$
$(3n - 1)(3n + 8) = 0$
$3n - 1 = 0$ or $3n + 8 = 0$
$3n = 1$ or $3n = -8$
$n = \frac{1}{3}$ or $n = -\frac{8}{3}$
The solution set is $\left\{-\frac{8}{3}, \frac{1}{3}\right\}$.

29. $6n^2 = 24$
$6n^2 - 24 = 0$
$n^2 - 4 = 0$
$(n - 2)(n + 2) = 0$
$n - 2 = 0$ or $n + 2 = 0$
$n = 2$ or $n = -2$
The solution set is $\{-2, 2\}$.

30. $16y^2 + 40y + 25 = 0$
$(4y + 5)(4y + 5) = 0$
$4y + 5 = 0$ or $4y + 5 = 0$
$4y = -5$ or $4y = -5$
$y = -\frac{5}{4}$ or $y = -\frac{5}{4}$
The solution set is $\left\{-\frac{5}{4}\right\}$.

31. $t^3 - t = 0$
$t(t^2 - 1) = 0$
$t(t - 1)(t + 1) = 0$
$t = 0$ or $t - 1 = 0$ or $t + 1 = 0$
$t = 0$ or $t = 1$ or $t = -1$
The solution set is $\{-1, 0, 1\}$.

32. $28x^2 + 71x + 18 = 0$
$(4x + 9)(7x + 2) = 0$
$4x + 9 = 0$ or $7x + 2 = 0$
$4x = -9$ or $7x = -2$
$x = -\frac{9}{4}$ or $x = -\frac{2}{7}$
The solution set is $\left\{-\frac{9}{4}, -\frac{2}{7}\right\}$.

33. $x^2 + 3x - 28 = 0$
$(x + 7)(x - 4) = 0$
$x + 7 = 0$ or $x - 4 = 0$
$x = -7$ or $x = 4$
The solution set is $\{-7, 4\}$.

34. $(x - 2)(x + 2) = 21$
$x^2 - 4 = 21$
$x^2 - 25 = 0$
$(x + 5)(x - 5) = 0$
$x + 5 = 0$ or $x - 5 = 0$
$x = -5$ or $x = 5$
The solution set is $\{-5, 5\}$.

35. $5n^2 + 27n = 18$
$5n^2 + 27n - 18 = 0$
$(5n - 3)(n + 6) = 0$
$5n - 3 = 0$ or $n + 6 = 0$
$5n = 3$ or $n = -6$
$n = \frac{3}{5}$ or $n = -6$
The solution set is $\left\{-6, \frac{3}{5}\right\}$.

36. $4n^2 + 10n = 14$
$4n^2 + 10n - 14 = 0$
$2n^2 + 5n - 7 = 0$
$(2n + 7)(n - 1) = 0$
$2n + 7 = 0$ or $n - 1 = 0$
$2n = -7$ or $n = 1$
$n = -\frac{7}{2}$ or $n = 1$
The solution set is $\left\{-\frac{7}{2}, 1\right\}$.

37. $2x^3 - 8x = 0$
Divide both sides by 2.
$x^3 - 4x = 0$
$x(x^2 - 4) = 0$
$x(x - 2)(x + 2) = 0$
$x = 0$ or $x - 2 = 0$ or $x + 2 = 0$
$x = 0$ or $x = 2$ or $x = -2$
The solution set is $\{-2, 0, 2\}$.

38. $x^2 - 20x + 96 = 0$
$(x - 8)(x - 12) = 0$
$x - 8 = 0$ or $x - 12 = 0$
$x = 8$ or $x = 12$
The solution set is $\{8, 12\}$.

39. $4t^2 + 17t - 15 = 0$
$(4t - 3)(t + 5) = 0$
$4t - 3 = 0$ or $t + 5 = 0$
$4t = 3$ or $t = -5$
$t = \frac{3}{4}$ or $t = -5$
The solution set is $\left\{ -5, \frac{3}{4} \right\}$.

40. $3(x + 2) - x(x + 2) = 0$
$(x + 2)(3 - x) = 0$
$x + 2 = 0$ or $3 - x = 0$
$x = -2$ or $3 = x$
The solution set is $\{ -2, 3\}$.

41. $(2x - 5)(3x + 7) = 0$
$2x - 5 = 0$ or $3x + 7 = 0$
$2x = 5$ or $3x = -7$
$x = \frac{5}{2}$ or $t = -\frac{7}{3}$
The solution set is $\left\{ -\frac{7}{3}, \frac{5}{2} \right\}$.

42. $(x + 4)(x - 1) = 50$
$x^2 + 3x - 4 = 50$
$x^2 + 3x - 54 = 0$
$(x + 9)(x - 6) = 0$
$x + 9 = 0$ or $x - 6 = 0$
$x - -9$ or $x - 6$
The solution set is $\{ -9, 6\}$.

43. $-7n - 2n^2 = -15$
$-2n^2 - 7n = -15$
$-2n^2 - 7n + 15 = 0$
$2n^2 + 7n - 15 = 0$
$(2n - 3)(n + 5) = 0$
$2n - 3 = 0$ or $n + 5 = 0$
$2n = 3$ or $n = -5$
$n = \frac{3}{2}$ or $n = -5$
The solution set is $\left\{ -5, \frac{3}{2} \right\}$.

44. $-23x + 6x^2 = -20$
$6x^2 - 23x + 20 = 0$
$(2x - 5)(3x - 4) = 0$
$2x - 5 = 0$ or $3x - 4 = 0$
$2x = 5$ or $3x = 4$
$x = \frac{5}{2}$ or $x = \frac{4}{3}$
The solution set is $\left\{ \frac{4}{3}, \frac{5}{2} \right\}$.

45. Let n represent the smaller number, then $2n - 1$ is the larger number.
$(2n - 1)^2 - n^2 = 33$
$4n^2 - 4n + 1 - n^2 = 33$
$3n^2 - 4n + 1 = 33$
$3n^2 - 4n - 32 = 0$
$(3n + 8)(n - 4) = 0$
$3n + 8 = 0$ or $n - 4 = 0$
$3n = -8$ or $n = 4$
$n = -\frac{8}{3}$ or $n = 4$
If $n = -\frac{8}{3}$, then $2n - 1 = 2\left(-\frac{8}{3} \right) - 1 = -\frac{16}{3} - \frac{3}{3} = -\frac{19}{3}$.
If $n = 4$, then $2n - 1 = 2(4) - 1 = 7$
The numbers are $-\frac{8}{3}$ and $-\frac{19}{3}$, or 4 and 7.

46. Let w represent the width of the rectangle, then $5w - 2$ represents the length of the rectangle.

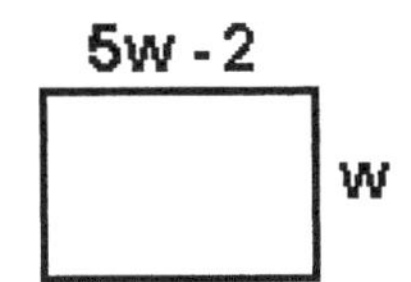

$w(5w - 2) = 16$
$5w^2 - 2w = 16$
$5w^2 - 2w - 16 = 0$
$(5w + 8)(w - 2) = 0$
$5w + 8 = 0$ or $w - 2 = 0$
$5w = -8$ or $w = 2$
$w = -\frac{8}{5}$ or $w = 2$

Discard the negative solution. The width of the rectangle is 2 centimeters and the length is $5(2) - 2 = 8$ centimeters.

47. Let s represent the length of a side of the smaller square, then $5s$ represents the length of a side of the larger square.

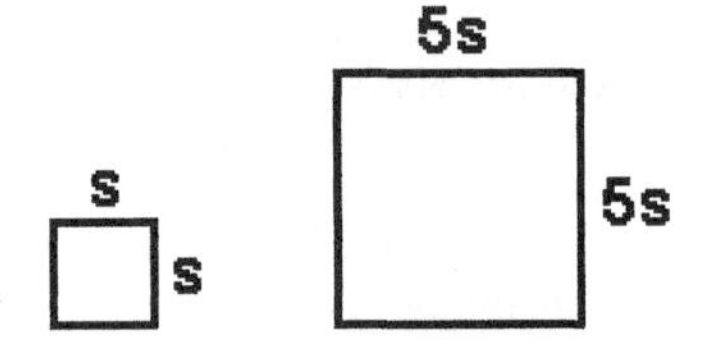

Smaller Area + Larger Area = 104

$$s^2 + (5s)^2 = 104$$
$$s^2 + 25s^2 = 104$$
$$26s^2 = 104$$
$$s^2 = 4$$
$$s^2 - 4 = 0$$
$$(s-2)(s+2) = 0$$
$$s - 2 = 0 \quad \text{or} \quad s + 2 = 0$$
$$s = 2 \quad \text{or} \quad s = -2$$

Discard the negative solution. The smaller square is 2 inches by 2 inches and the larger square is 10 inches by 10 inches.

48. Let $x =$ the length of the shorter leg, then $2x - 1 =$ the length of the longer leg, and $2x + 1 =$ the the length of the hypotenuse.

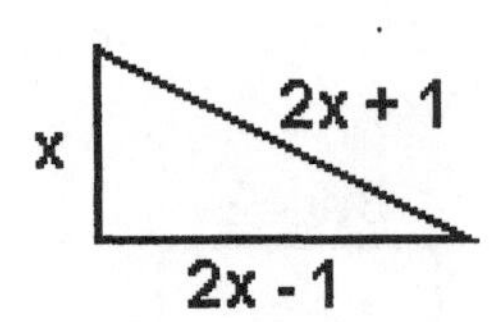

$$x^2 + (2x-1)^2 = (2x+1)^2$$

Pythagorean Theorem

$$x^2 + 4x^2 - 4x + 1 = 4x^2 + 4x + 1$$
$$5x^2 - 4x + 1 = 4x^2 + 4x + 1$$
$$x^2 - 8x = 0$$
$$x(x-8) = 0$$
$$x = 0 \quad \text{or} \quad x - 8 = 0$$
$$x = 0 \quad \text{or} \quad x = 8$$

Discard the zero solution. The sides of the triangle are 8 units, 15 units, and 17 units.

49. Let n represent one number, then $6n + 1$ represents the other number.

$$n(6n+1) = 26$$
$$6n^2 + n = 26$$
$$6n^2 + n - 26 = 0$$
$$(6n+13)(n-2) = 0$$
$$6n + 13 = 0 \quad \text{or} \quad n - 2 = 0$$
$$6n = -13 \quad \text{or} \quad n = 2$$
$$n = -\frac{13}{6} \quad \text{or} \quad n = 2$$

If $n = -\frac{13}{6}$, then $6n + 1 =$

$$6\left(-\frac{13}{6}\right) + 1 = -13 + 1 = -12.$$

If $n = 2$, then $6n + 1 = 6(2) + 1 = 13$.

The numbers are $-\frac{13}{6}$ and -12, or 2 and 13.

50. Let n represent the first whole number, then $n + 2$ and $n + 4$ represent the next two consecutive odd whole numbers.

$$n^2 + (n+2)^2 = (n+4)^2 + 9$$
$$n^2 + n^2 + 4n + 4 = n^2 + 8n + 16 + 9$$
$$2n^2 + 4n + 4 = n^2 + 8n + 25$$
$$n^2 - 4n - 21 = 0$$
$$(n+3)(n-7) = 0$$
$$n + 3 = 0 \quad \text{or} \quad n - 7 = 0$$
$$n = -3 \quad \text{or} \quad n = 7$$

Discard the negative solution because only positive whole numbers are wanted. Thus, the positive odd whole numbers are 7, 9, and 11.

51. Let s represent the number of shelves, then $9s - 1$ represents the number of books per shelf.

$$\binom{\text{Number of}}{\text{shelves}}\binom{\text{books per}}{\text{shelf}} = 140$$
$$s(9s-1) = 140$$
$$9s^2 - s = 140$$
$$9s^2 - s - 140 = 0$$
$$(9s+35)(s-4) = 0$$
$$9s + 35 = 0 \quad \text{or} \quad s - 4 = 0$$
$$9s = -35 \quad \text{or} \quad s = 4$$
$$s = -\frac{35}{9} \quad \text{or} \quad s = 4$$

Discard the negative solution. There would be 4 shelves in the bookcase.

52. Let w represent the width of the rectangle, then $8w$ represents the length of the rectangle and w represents a side of the square.

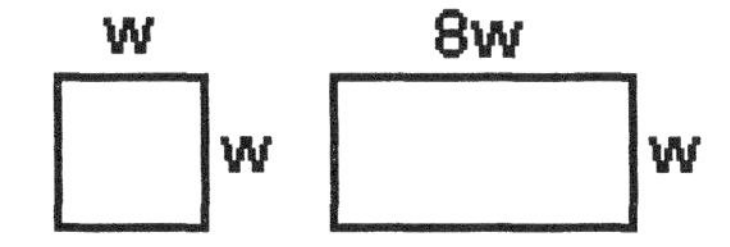

Square Area + Rectangle Area = 225

$$w^2 + w(8w) = 225$$
$$w^2 + 8w^2 = 225$$
$$9w^2 = 225$$
$$w^2 = 25$$
$$w^2 - 25 = 0$$
$$(w+5)(w-5) = 0$$
$$w+5=0 \quad \text{or} \quad w-5=0$$
$$w=-5 \quad \text{or} \quad w=5$$

Discard the negative solution. The square would be 5 yards by 5 yards and the rectangle would be 5 yards by 40 yards.

53. Let n represent the first integer, then $n+1$ represents the next integer.

$$n^2 + (n+1)^2 = 613$$
$$n^2 + n^2 + 2n + 1 = 613$$
$$2n^2 + 2n - 612 = 0$$
$$n^2 + n - 306 = 0$$
$$(n+18)(n-17) = 0$$
$$n+18=0 \quad \text{or} \quad n-17=0$$
$$n=-18 \quad \text{or} \quad n=17$$

The numbers would be -18 and -17, or 17 and 18.

54. Let x represent the length of a side of the cube.

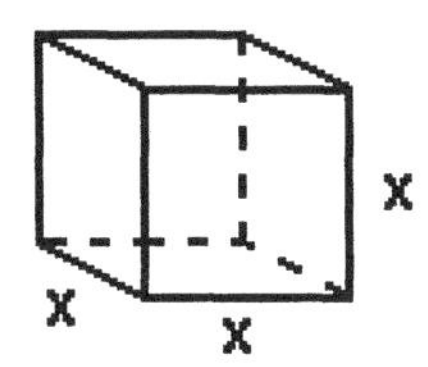

Volume of Cube = Surface Area

$$x^3 = 2x(x) + 2x(x) + 2x(x)$$
$$x^3 = 2x^2 + 2x^2 + 2x^2$$
$$x^3 = 6x^2$$
$$x^3 - 6x^2 = 0$$
$$x^2(x-6) = 0$$
$$x=0 \quad \text{or} \quad x=0 \quad \text{or} \quad x-6=0$$
$$x=0 \quad \text{or} \quad x=0 \quad \text{or} \quad x=6$$

Discard the zero solution. The length of a side of the cube is 6 units.

55. Let r represent a radius of the smaller circle, then $3r+1$ represents a radius of the larger circle.

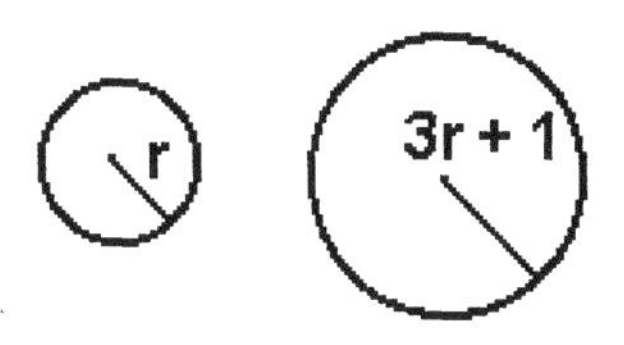

Smaller Area + Larger Area = 53π

$$\pi r^2 + \pi(3r+1)^2 = 53\pi$$
$$\pi r^2 + \pi(9r^2 + 6r + 1) = 53\pi$$
$$\pi r^2 + 9\pi r^2 + 6\pi r + \pi = 53\pi$$
$$10\pi r^2 + 6\pi r - 52\pi = 0$$

Divide both sides by 2π.

$$5r^2 + 3r - 26 = 0$$
$$(5r+13)(r-2) = 0$$
$$5r+13=0 \quad \text{or} \quad r-2=0$$
$$5r=-13 \quad \text{or} \quad r=2$$
$$r=-\frac{13}{5} \quad \text{or} \quad r=2$$

Discard the negative solution. The length of a radius of the smaller circle is 2 meters and a radius of the larger circle is 7 meters.

56. Let n represent the first odd whole number, then $n+2$ represents the next odd whole number. The sum of the two numbers is $n + (n+2) = 2n+2$.

$$n(n+2) = 5(2n+2) - 1$$
$$n^2 + 2n = 10n + 10 - 1$$
$$n^2 + 2n = 10n + 9$$
$$n^2 - 8n - 9 = 0$$
$$(n+1)(n-9) = 0$$
$$n+1=0 \quad \text{or} \quad n-9=0$$
$$n = -1 \quad \text{or} \quad n = 9$$

Discard the negative solution because whole numbers are positive. The numbers would be 9 and 11.

57. Let x represent the amount that the width and length are both reduced. The original area of the photograph is $8(14) = 112$ square centimeters. The new area is $112 - 40 = 72$ square centimeters.

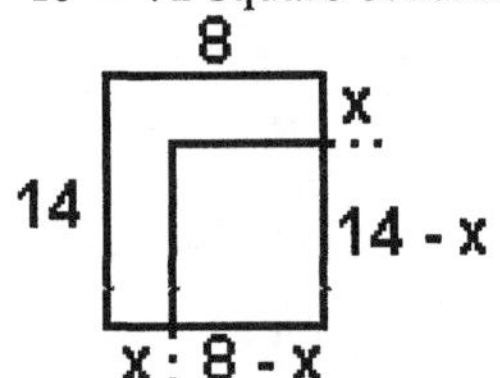

$$(8-x)(14-x) = 72$$
$$112 - 22x + x^2 = 72$$
$$40 - 22x + x^2 = 0$$
$$(20-x)(2-x) = 0$$
$$20 - x = 0 \quad \text{or} \quad 2 - x = 0$$
$$20 = x \quad \text{or} \quad 2 = x$$

The solution $x = 20$ is not reasonable. Thus, the length and the width must be reduced by 2 centimeters.

58. Let x represent the width of the plowed strip. Then $120 - 2x$ represents the length of the unplowed garden, and $90 - 2x$ represents the width. The original area of the garden is $90(120) = 10,800$ square feet. Thus, the unplowed garden is 5400 square feet.

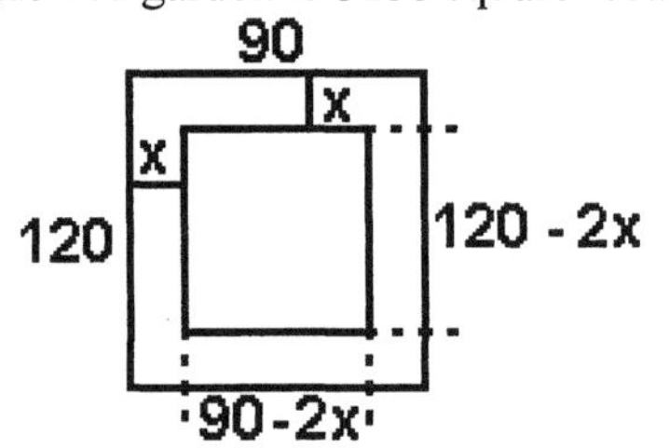

$$(90-2x)(120-2x) = 5400$$
$$10,800 - 420x + 4x^2 = 5400$$
$$5400 - 420x + 4x^2 = 0$$
$$1350 - 105x + x^2 = 0$$
$$(90-x)(15-x) = 0$$
$$90 - x = 0 \quad \text{or} \quad 15 - x = 0$$
$$90 = x \quad \text{or} \quad 15 = x$$

The solution $x = 90$ is not reasonable. Thus, the width of the strip must be 15 feet.

CHAPTER 7 Test

1. We need two integers whose product is -10 and whose sum is 3. They are -2 and 5.
$x^2 + 3x - 10 = (x-2)(x+5)$

2. We need two integers whose product is -24 and whose sum is -5. They are 3 and -8.
$x^2 - 5x - 24 = (x+3)(x-8)$

3. $2x^3 - 2x = 2x(x^2 - 1) = 2x(x-1)(x+1)$

4. We need two integers whose product is 108 and whose sum is 21. They are 9 and 12.
$x^2 + 21x + 108 = (x+9)(x+12)$

5. $18n^2 + 21n + 6 = 3(6n^2 + 7n + 2)$
We need two integers whose product is 12 and whose sum is 7. They are 3 and 4.
$3(6n^2 + 7n + 2) =$
$3(6n^2 + 3n + 4n + 2) =$
$3[3n(2n+1) + 2(2n+1)] =$
$3(2n+1)(3n+2)$

6. $ax + ay + 2bx + 2by =$
$a(x+y) + 2b(x+y) = (x+y)(a+2b)$

7. We need two integers whose product is -60 and whose sum is 17. They are -3 and 20.
$4x^2 + 17x - 15 = 4x^2 - 3x + 20x - 15 =$
$x(4x-3) + 5(4x-3) = (4x-3)(x+5)$

8. $6x^2 + 24 = 6(x^2 + 4)$

9. $30x^3 - 76x^2 + 48x = 2x(15x^2 - 38x + 24)$
We need two integers whose product is 360 and whose sum is -38. They are -18 and -20.
$2x(15x^2 - 38x + 24) =$
$2x(15x^2 - 18x - 20x + 24) =$
$2x[3x(5x - 6) - 4(5x - 6)] =$
$2x(5x - 6)(3x - 4)$

10. We need two integers whose product is -168 and whose sum is 13. They are -8 and 21.
$28 + 13x - 6x^2 = 28 - 8x + 21x - 6x^2 =$
$4(7 - 2x) + 3x(7 - 2x) = (7 - 2x)(4 + 3x)$

11. $7x^2 = 63$
Divide both sides by 7.
$x^2 = 9$
$x^2 - 9 = 0$
$(x - 3)(x + 3) = 0$
$x - 3 = 0$ or $x + 3 = 0$
$x = 3$ or $x = -3$
The solution set is $\{-3, 3\}$.

12. $x^2 + 5x - 6 = 0$
$(x - 1)(x + 6) = 0$
$x - 1 = 0$ or $x + 6 = 0$
$x = 1$ or $x = -6$
The solution set is $\{-6, 1\}$.

13. $4n^2 = 32n$
$4n^2 - 32n = 0$
$n^2 - 8n = 0$
$n(n - 8) = 0$
$n = 0$ or $n - 8 = 0$
$n = 0$ or $n = 8$
The solution set is $\{0, 8\}$.

14. $(3x - 2)(2x + 5) = 0$
$3x - 2 = 0$ or $2x + 5 = 0$
$3x = 2$ or $2x = -5$
$x = \frac{2}{3}$ or $x = -\frac{5}{2}$
The solution set is $\left\{-\frac{5}{2}, \frac{2}{3}\right\}$.

15. $(x - 3)(x + 7) = -9$
$x^2 + 4x - 21 = -9$
$x^2 + 4x - 12 = 0$
$(x + 6)(x - 2) = 0$
$x + 6 = 0$ or $x - 2 = 0$
$n = -6$ or $x = 2$
The solution set is $\{-6, 2\}$.

16. $x^3 + 16x^2 + 48x = 0$
$x(x + 4)(x + 12) = 0$
$x = 0$ or $x + 4 = 0$ or $x + 12 = 0$
$x = 0$ or $x = -4$ or $x = -12$
The solution set is $\{-12, -4, 0\}$.

17. $9(x - 5) - x(x - 5) = 0$
$(x - 5)(9 - x) = 0$
$x - 5 = 0$ or $9 - x = 0$
$x = 5$ or $9 = x$
The solution set is $\{5, 9\}$.

18. $3t^2 + 35t = 12$
$3t^2 + 35t - 12 = 0$
$(3t - 1)(t + 12) = 0$
$3t - 1 = 0$ or $t + 12 = 0$
$3t = 1$ or $t = -12$
$t = \frac{1}{3}$ or $t = -12$
The solution set is $\left\{-12, \frac{1}{3}\right\}$.

19. $8 - 10x - 3x^2 = 0$
$(4 + x)(2 - 3x) = 0$
$4 + x = 0$ or $2 - 3x = 0$
$x = -4$ or $-3x = -2$
$x = -4$ or $x = \frac{2}{3}$
The solution set is $\left\{-4, \frac{2}{3}\right\}$.

20. $3x^3 = 75x$
$x^3 = 25x$
$x^3 - 25x = 0$
$x(x^2 - 25) = 0$
$x(x + 5)(x - 5) = 0$

$x = 0$ or $x + 5 = 0$ or $x - 5 = 0$
$x = 0$ or $x = -5$ or $x = 5$
The solution set is $\{-5, 0, 5\}$.

21. $25n^2 - 70n + 49 = 0$
$(5n - 7)(5n - 7) = 0$
$5n - 7 = 0$ or $5n - 7 = 0$
$5n = 7$ or $5n = 7$
$x = \frac{7}{5}$ or $x = \frac{7}{5}$
The solution set is $\left\{\frac{7}{5}\right\}$.

22. Let w represent the width of the rectangle, then $2w - 2$ represents the length of the rectangle.

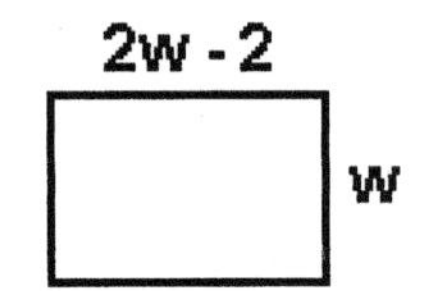

$w(2w - 2) = 112$
$2w^2 - 2w = 112$
$2w^2 - 2w - 112 = 0$
$w^2 - w - 56 = 0$
$(w - 8)(w + 7) = 0$
$w - 8 = 0$ or $w + 7 = 0$
$w = 8$ or $w = -7$
Discard the negative solution. If $w = 8$, then the length of the rectangle is 2(8)-2=14 inches.

23. Let x represent the length of the shorter leg, then $x + 4$ represents the length of the longer leg and $x + 8$ represents the length of the hypotenuse.

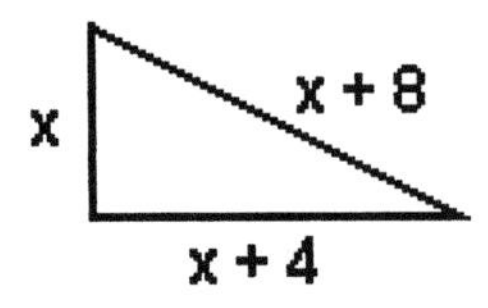

$x^2 + (x + 4)^2 = (x + 8)^2$
Pythagorean Theorem
$x^2 + x^2 + 8x + 16 = x^2 + 16x + 64$
$2x^2 + 8x + 16 = x^2 + 16x + 64$
$x^2 - 8x - 48 = 0$
$(x - 12)(x + 4) = 0$
$x - 12 = 0$ or $x + 4 = 0$
$x = 12$ or $x = -4$
Discard the negative solution. The length of the shorter leg is 12 centimeters.

24. Let r represent the number of rows, then $3r - 5$ represents the number of chairs per row.
$(\text{rows})(\text{chairs per rows}) = 112$
$r(3r - 5) = 112$
$3r^2 - 5r = 112$
$3r^2 - 5r - 112 = 0$
$(3r + 16)(r - 7) = 0$
$3r + 16 = 0$ or $r - 7 = 0$
$3r = -16$ or $r = 7$
$r = -\frac{16}{3}$ or $r = 7$
Discard the negative solution. There are 7 rows with $3(7) - 5 = 16$ chairs per row.

25. Let x represent the length of a side of the cube.

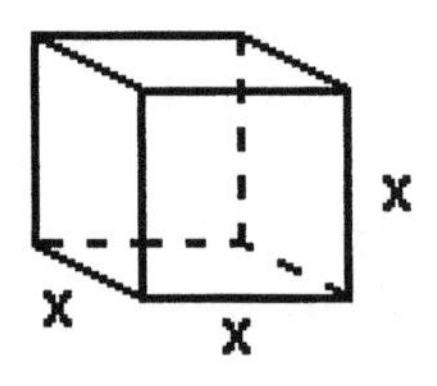

Volume of Cube = 2[Surface Area]
$x^3 = 2[2x(x) + 2x(x) + 2x(x)]$
$x^3 = 2\left[2x^2 + 2x^2 + 2x^2\right]$
$x^3 = 2\left[6x^2\right]$
$x^3 = 12x^2$
$x^3 - 12x^2 = 0$
$x^2(x - 12) = 0$
$x = 0$ or $x = 0$ or $x - 12 = 0$
$x = 0$ or $x = 0$ or $x = 12$
Discard the zero solution. The length of a side of the cube is 12 units.

CHAPTERS 1-7 | Cumulative Review

1. $3x - 2xy - 7x + 5xy = -4x + 3xy$
when $x = \frac{1}{2},\ y = 3:\ -4\left(\frac{1}{2}\right) + 3\left(\frac{1}{2}\right)(3)$
$= \frac{-4}{2} + \frac{9}{2} = \frac{5}{2}$

2. $a = -3,\ b = -5:$
$7(a-b) - 3(a-b) - (a-b) =$
$3(a-b) = 3[(-3) - (-5)] =$
$3[2] = 6$

3. $a = 0.4,\ b = 0.6:\ ab + b^2 =$
$(0.4)(0.6) + (0.6)^2 = 0.24 + 0.36 = 0.6$

4. $x = -6,\ y = 4:\ x^2 - y^2 =$
$(-6)^2 - (4)^2 = 36 - 16 = 20$

5. $x = -10:\ x^2 + x - 72 =$
$(x+9)(x-8) = (-10+9)(-10-8) =$
$(-1)(-18) = 18$

6. $x = -6: 3(x-2) - (x+3) - 5(x+6) =$
$3x - 6 - x - 3 - 5x - 30 =$
$-3x - 39 = -3(-6) - 39 =$
$18 - 39 = -21$

7. $3^{-3} = \frac{1}{3^3} = \frac{1}{27}$

8. $\left(\frac{2}{3}\right)^{-1} = \left(\frac{3}{2}\right)^{1} = \frac{3}{2}$

9. $\left(\frac{1}{2} + \frac{1}{3}\right)^{0} = 1$

10. $\left(\frac{1}{3} + \frac{1}{4}\right)^{-1} = \left(\frac{1\cdot 4}{3\cdot 4} + \frac{1\cdot 3}{4\cdot 3}\right)^{-1} =$
$\left(\frac{4}{12} + \frac{3}{12}\right)^{-1} = \left(\frac{7}{12}\right)^{-1} = \frac{12}{7}$

11. $-4^{-2} = -\frac{1}{4^2} = -\frac{1}{16}$

12. $-4^2 = -(4\cdot 4) = -16$

13. $\frac{1}{\left(\frac{2}{5}\right)^2} = \left(\frac{2}{5}\right)^{-2} = \left(\frac{5}{2}\right)^{2} = \frac{25}{4}$

14. $(-3)^{-3} = \frac{1}{(-3)^3} = \frac{1}{-27} = -\frac{1}{27}$

15. $\frac{7}{5x} + \frac{2}{x} - \frac{3}{2x} =$
$\frac{7}{5x}\cdot\frac{2}{2} + \frac{2}{x}\cdot\frac{10}{10} - \frac{3}{2x}\cdot\frac{5}{5} =$
$\frac{14 + 20 - 15}{10x} = \frac{19}{10x}$

16. $\frac{4x}{5y} \div \frac{12x^2}{10y^2} = \frac{4x}{5y}\cdot\frac{10y^2}{12x^2} =$
$\frac{\cancel{4}\cdot\overset{2}{\cancel{10}}\cdot\cancel{x}\cdot\overset{y}{\cancel{y^2}}}{\cancel{5}\cdot\underset{3}{\cancel{12}}\cdot\underset{x}{\cancel{x^2}}\cdot\cancel{y}} = \frac{2y}{3x}$

17. $(-5x^2y)(7x^3y^4) =$
$(-5)(7)(x^{2+3})(y^{1+4}) = -35x^5y^5$

18. $(9ab^3)^2 = 9^2a^2(b^3)^2 = 81a^2b^6$

19. $(-3n^2)(5n^2 + 6n - 2) =$
$(-3n^2)(5n^2) + (-3n^2)(6n) - (-3n^2)(2) =$
$-15n^4 - 18n^3 + 6n^2$

20. $(5x-1)(3x+4) =$
$(5x)(3x) + (20-3)x - 4 =$
$15x^2 + 17x - 4$

21. $(2x+5)^2 = (2x)^2 + 2(2x)(5) + (5)^2 =$
$4x^2 + 20x + 25$

22. $(x+2)(2x^2 - 3x - 1) =$
$x(2x^2 - 3x - 1) + 2(2x^2 - 3x - 1) =$
$2x^3 - 3x^2 - x + 4x^2 - 6x - 2 =$
$2x^3 + x^2 - 7x - 2$

23. $(x^2 - x - 1)(x^2 + 2x - 3) =$
$x^2(x^2 + 2x - 3) - x(x^2 + 2x - 3)$
$\quad - 1(x^2 + 2x - 3) =$
$x^4 + 2x^3 - 3x^2 - x^3 - 2x^2 + 3x$
$\quad - x^2 - 2x + 3 =$
$x^4 + x^3 - 6x^2 + x + 3$

24. $(-2x - 1)(3x - 7) =$
$(-2x)(3x) + (14 - 3)x + 7 =$
$-6x^2 + 11x + 7$

25. $\dfrac{24x^2y^3 - 48x^4y^5}{8xy^2} =$
$\dfrac{24x^2y^3}{8xy^2} - \dfrac{48x^4y^5}{8xy^2} = 3xy - 6x^3y^3$

26.
$$\begin{array}{r|l} & 7x + 4 \\ 4x - 5 & 28x^2 - 19x - 20 \\ & \underline{28x^2 - 35x\quad\quad} \\ & \quad\quad 16x - 20 \\ & \quad\quad \underline{16x - 20} \end{array}$$

27. $3x^3 + 15x + 27x = 3x(x^2 + 5x + 9)$
($x^2 + 5x + 9$ is not factorable.)

28. $x^2 - 100 = (x - 10)(x + 10)$

29. $5x^2 - 22x + 8 =$
$5x^2 - 20x - 2x + 8 =$
$5x(x - 4) - 2(x - 4) =$
$(x - 4)(5x - 2)$

30. $8x^2 - 22x - 63 =$
$8x^2 + 14x - 36x - 63 =$
$2x(4x + 7) - 9(4x + 7) =$
$(4x + 7)(2x - 9)$

31. $n^2 + 25n + 144 =$
$n^2 + 16n + 9n + 144 =$
$n(n + 16) + 9(n + 16) =$
$(n + 16)(n + 9)$

32. $nx + ny - 2x - 2y =$
$n(x + y) - 2(x + y) = (x + y)(n - 2)$

33. $3x^3 - 3x = 3x(x^2 - 1) =$
$3x(x - 1)(x + 1)$

34. $2x^3 - 6x^2 - 108x =$
$2x(x^2 - 3x - 54) = 2x(x - 9)(x + 6)$

35. $36x^2 - 60x + 25 =$
$(6x)^2 - 2(6x)(5) + (5)^2 = (6x - 5)^2$

36. $3x^2 - 5xy - 2y^2 =$
$3x^2 - 6xy + xy - 2y^2 =$
$3x(x - 2y) + y(x - 2y) =$
$(x - 2y)(3x + y)$

37.
$$\begin{aligned} 3(x - 2) - 2(x + 6) &= -2(x + 1) \\ 3x - 6 - 2x - 12 &= -2x - 2 \\ x - 18 &= -2x - 2 \\ 3x - 18 &= -2 \\ 3x &= 16 \\ x &= \frac{16}{3} \end{aligned}$$
The solution set is $\left\{\frac{16}{3}\right\}$.

38.
$$\begin{aligned} x^2 &= -11x \\ x^2 + 11x &= 0 \\ x(x + 11) &= 0 \end{aligned}$$
$x = 0$ or $x + 11 = 0$
$x = 0$ or $x = -11$
The solution set is $\{-11, 0\}$.

39. $0.2x - 3(x - 0.4) = 1$
Multiply by 10.
$$\begin{aligned} 2x - 30(x - 0.4) &= 10 \\ 2x - 30x + 12 &= 10 \\ -28x + 12 &= 10 \\ -28x &= -2 \\ x &= \frac{2}{28} = \frac{1}{14} \end{aligned}$$
The solution set is $\left\{\frac{1}{14}\right\}$.

40. $5n^2 - 5 = 0$

Divide by 5.

$n^2 - 1 = 0$

$(n-1)(n+1) = 0$

$n - 1 = 0$ or $n + 1 = 0$

$n = 1$ or $n = -1$

The solution set is $\{-1, 1\}$.

41. $x^2 + 5x - 6 = 0$

$(x-1)(x+6) = 0$

$x - 1 = 0$ or $x + 6 = 0$

$x = 1$ or $x = -6$

The solution set is $\{-6, 1\}$.

42. $\frac{2x+1}{2} + \frac{3x-4}{3} = 1$

$6\left(\frac{2x+1}{2} + \frac{3x-4}{3}\right) = 6(1)$

$3(2x+1) + 2(3x-4) = 6$

$6x + 3 + 6x - 8 = 6$

$12x - 5 = 6$

$12x = 11$

$x = \frac{11}{12}$

The solution set is $\left\{\frac{11}{12}\right\}$.

43. $2(x-1) - x(x-1) = 0$

$(x-1)(2-x) = 0$

$x - 1 = 0$ or $2 - x = 0$

$x = 1$ or $2 = x$

The solution set is $\{1, 2\}$.

44. $6x^2 + 19x - 7 = 0$

$(3x-1)(2x+7) = 0$

$3x - 1 = 0$ or $2x + 7 = 0$

$3x = 1$ or $2x = -7$

$x = \frac{1}{3}$ or $x = -\frac{7}{2}$

The solution set is $\left\{-\frac{7}{2}, \frac{1}{3}\right\}$.

45. $(2x-1)(x-8) = 0$

$2x - 1 = 0$ or $x - 8 = 0$

$2x = 1$ or $x = 8$

$x = \frac{1}{2}$ or $x = 8$

The solution set is $\left\{\frac{1}{2}, 8\right\}$.

46. $(x+1)(x+6) = 24$

$x^2 + 7x + 6 = 24$

$x^2 + 7x - 18 = 0$

$(x+9)(x-2) = 0$

$x + 9 = 0$ or $x - 2 = 0$

$x = -9$ or $x = 2$

The solution set is $\{-9, 2\}$.

47. $-3x - 2 \geq 1$

$-3x \geq 3$

$\frac{-3x}{-3} \leq \frac{3}{-3}$

$x \leq -1$

The solution set is $\{x | x \leq -1\}$.
or $(-\infty, -1]$.

48. $18 < 2(x-4)$

$18 < 2x - 8$

$26 < 2x$

$\frac{26}{2} < \frac{2x}{2}$

$13 < x$

$x > 13$

The solution set is $\{x | x > 13\}$ or $(13, \infty)$.

49. $3(x-2) - 2(x+1) \leq -(x+5)$

$3x - 6 - 2x - 2 \leq -x - 5$

$x - 8 \leq -x - 5$

$2x \leq 3$

$\frac{2x}{2} \leq \frac{3}{2}$

$x \leq \frac{3}{2}$

The solution set is $\left\{x \middle| x \leq \frac{3}{2}\right\}$

or $\left(-\infty, \frac{3}{2}\right]$.

50.

$$\frac{2}{3}x - \frac{1}{4}x - 1 > 3$$
$$\frac{2}{3}x - \frac{1}{4}x > 4$$
$$12\left(\frac{2}{3}x - \frac{1}{4}x\right) > 12(4)$$
$$12\left(\frac{2}{3}x\right) - 12\left(\frac{1}{4}x\right) > 48$$
$$4(2x) - 3(x) > 48$$
$$8x - 3x > 48$$
$$5x > 48$$
$$x > \frac{48}{5}$$

The solution set is $\left\{x \middle| x > \frac{48}{5}\right\}$ or $\left(\frac{48}{5}, \infty\right)$.

51.

$$0.08x + 0.09(2x) \leq 130$$
$$100[0.08x + 0.09(2x)] \leq 100(130)$$
$$8x + 9(2x) \leq 13000$$
$$8x + 18x \leq 13000$$
$$26x \leq 13000$$
$$x \leq 500$$

The solution set is $\{x| x \leq 500\}$ or $(-\infty, 500]$.

52. Graph $y = -3x + 5$.

53. Graph $y = \frac{1}{4}x + 2$.

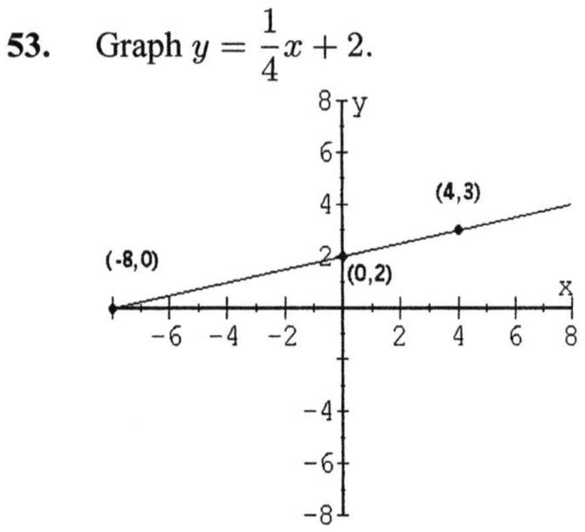

54. Graph $3x - y = 3$.

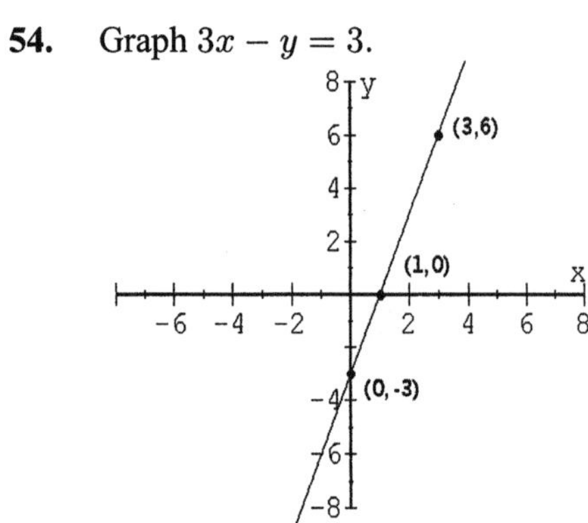

55. Graph $y = 2x^2 - 4$.

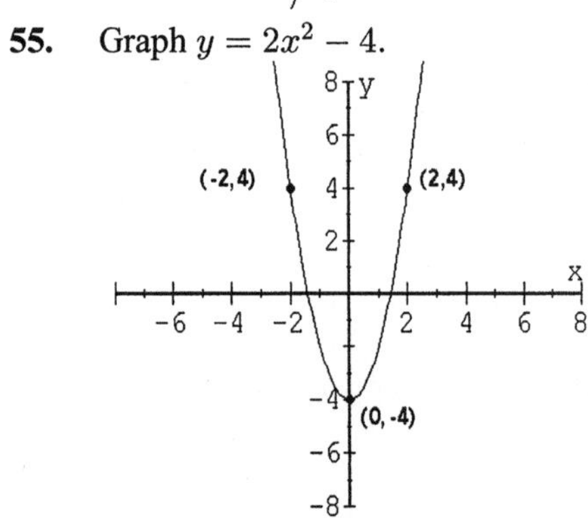

56.
$$\begin{pmatrix} 7x - 2y = -34 \\ x + 2y = -14 \end{pmatrix}$$

Add the equations.

$8x = -48$

$x = -6$

Substitute into equation 2.

$x + 2y = -14$

$-6 + 2y = -14$

$2y = -8$

$y = -4$

The solution set is $\{(-6, -4)\}$.

57. $\begin{pmatrix} 5x + 3y = -9 \\ 3x - 5y = 15 \end{pmatrix}$

Multiply equation 1by 5
and multiply equation 2 by 3.

$\begin{pmatrix} 25x + 15y = -45 \\ 9x - 15y = 45 \end{pmatrix}$

Add the equations.

$34x = 0$
$x = 0$
$5x + 3y = -9$
$5(0) + 3y = -9$
$3y = -9$
$y = -3$

The solution set is $\{(0, -3)\}$.

58. $\begin{pmatrix} 6x - 11y = -58 \\ 8x + y = 1 \end{pmatrix}$

Solve equation 2 for y.

$8x + y = 1$
$y = -8x + 1$

Substitute into equation 1.

$6x - 11(-8x + 1) = -58$
$6x + 88x - 11 = -58$
$94x - 11 = -58$
$94x = -47$
$x = \frac{-47}{94} = -\frac{1}{2}$
$y = -8(-\frac{1}{2}) + 1$
$y = 4 + 1 = 5$

The solution set is $\left\{\left(-\frac{1}{2}, 5\right)\right\}$.

59. $\begin{pmatrix} \frac{1}{2}x + \frac{2}{3}y = -1 \\ \frac{2}{5}x - \frac{1}{3}y = -6 \end{pmatrix}$

Multiply equation 1 by 6 and
multiply equation 2 by 15.

$\begin{pmatrix} 3x + 4y = -6 \\ 6x - 5y = -90 \end{pmatrix}$

Multiply equation 1 by -2,
then add to equation 2.

$$\begin{array}{r} -6x - 8y = 12 \\ 6x - 5y = -90 \\ \hline -13y = -78 \end{array}$$

$y = 6$
$3x + 4y = -6$
$3x + 4(6) = -6$
$3x + 24 = -6$
$3x = -30$
$x = -10$

The solution set is $\{(-10, 6)\}$.

60. Since $91 = 7 \cdot 13$,
91 is composite because it has factors other than 1 and itself.

61. $\left.\begin{array}{l} 18 = 2 \cdot 3 \cdot 3 \\ 48 = 2 \cdot 2 \cdot 2 \cdot 2 \cdot 3 \end{array}\right\}$

The greatest common factor is $2 \cdot 3 = 6$.

62. $\left.\begin{array}{l} 6 = 2 \cdot 3 \\ 8 = 2 \cdot 2 \cdot 2 \\ 9 = 3 \cdot 3 \end{array}\right\}$

The least common multiple is $2 \cdot 2 \cdot 2 \cdot 3 \cdot 3 = 72$.

63. $\frac{7}{4} = \frac{n}{100}$
$4n = 700$
$n = 175$

Therefore, $\frac{7}{4} = \frac{175}{100} = 175\%$.

64. $0.0024 = (2.4)(10^{-3})$

65. $(3.14)(10^3) = 3140$

66. Because it is an "or" statement, the solution set consists of all numbers less than 0 along with all numbers greater than 3.

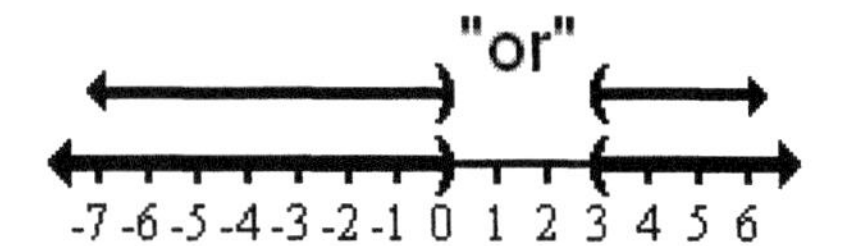

67. Let r represent the radius of the circular region.
$C = 8\pi; C = 2\pi r$
$8\pi = 2\pi r$
$4 = \frac{8\pi}{2\pi} = r$
Area of a circle $= \pi r^2$
Area of a circle $= \pi(4)^2 = 16\pi$
Area of a circle is 16π square centimeters.

68. 30% of what number is 5.4.
$0.30(x) = 5.4$
$x = \frac{5.4}{0.30} = 18$
30% of 18 is 5.4.

69. Graph $3x - 2y < -6$.
Use the points $(0, 3)$ and $(-2, 0)$ to graph a dashed line for $3x - 2y = -6$. Use $(0, 0)$ as a test point. The given inequality $3x + 2y < -6$ becomes $0 + 0 < -6$ which is a false statement. Therefore, the solution set is the half-plane that does NOT contain the origin.

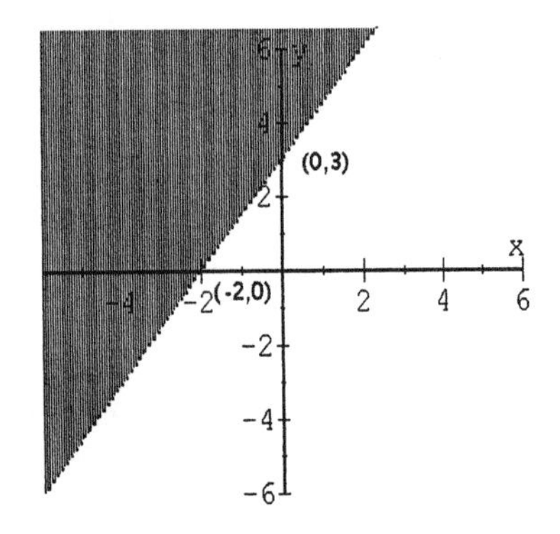

70.

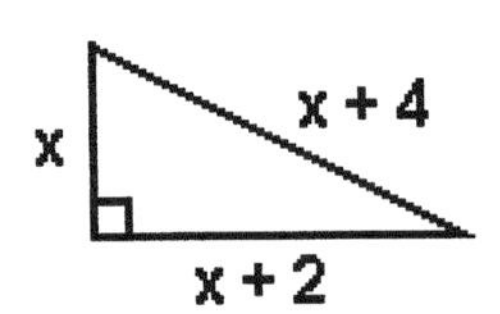

Use the Pythagorean Theorem.

$x^2 + (x+2)^2 = (x+4)^2$
$x^2 + x^2 + 4x + 4 = x^2 + 8x + 16$
$2x^2 + 4x + 4 = x^2 + 8x + 16$
$x^2 - 4x - 12 = 0$
$(x-6)(x+2) = 0$
$x - 6 = 0$ or $x + 2 = 0$
$x = 6$ or $x = -2$
Since length can not be negative, discard the root $x = -2$. The length of the legs are 6 inches and 8 inches. The hypotenuse is 10 inches.

71. Let $x =$ milliliters of 65% HCl acid.
$0.65x + 0.30(40) = 0.55(x + 40)$
$100[0.65x + 0.30(40)] = 100[0.55(x + 40)]$
$65x + 30(40) = 55(x + 40)$
$65x + 1200 = 55x + 2200$
$10x = 1000$
$x = 100$
100 ml of 65% hydrochloric acid must be added.

72. Let $x =$ width of rectangle and then $x + 2 =$ length of rectangle.
$P = 2L + 2W$
$28 = 2(x+2) + 2x$
$28 = 2x + 4 + 2x$
$28 = 4x + 4$
$24 = 4x$
$6 = x$
The dimensions will be 6 feet by 8 feet.

73. Let $t =$ time traveling.
$55t + 65t = 300$
$120t = 300$
$t = \frac{300}{120} = 2.5$
It will take 2.5 hours to be 300 miles apart.

74.
$$A = \frac{1}{2}h(b_1 + b_2)$$
$$120 = \frac{1}{2}h(10 + 22)$$
$$120 = \frac{1}{2}h(32)$$
$$120 = 16h$$
$$\frac{120}{16} = h$$
$$7.5 = h$$
The altitude is 7.5 cm.

75. Let x represent the gallons used on 594 mile trip.
$$\frac{16}{352} = \frac{x}{594}$$
$$352x = 16(594)$$
$$352x = 9504$$
$$x = \frac{9504}{352} = 27$$
The consumption would be 27 gallons.

76. Let a represent the smaller angle. Then $3a - 20$ represents the larger angle. Since they are supplementary angles, the sum of their measures is 180°.
$$a + (3a - 20) = 180$$
$$4a - 20 = 180$$
$$4a = 200$$
$$a = 50$$
The measures of the angles are 50°, and $3(50) - 20 = 130°$.

77. Let a represent the smallest of the three angles; then $2a + 10$ represents the largest angle, and the other angle is $a + 10$. The sum of the measures of the three angles of a triangle is 180°.
$$a + (2a + 10) + (a + 10) = 180$$
$$4a + 20 = 180$$
$$4a = 160$$
$$a = 40$$
The measures of the angles are 40°, $2(40) + 10 = 90°$, and $(40) + 10 = 50°$.

78. Let p represent the number of pennies, $2p + 5$ the number of dimes, and $p + 10$ the number of nickels. Total number of coins is 175.
$$p + (2p + 5) + (p + 10) = 175$$
$$p + 2p + 5 + p + 10 = 175$$
$$4p + 15 = 175$$
$$4p = 160$$
$$p = 40$$
There would be 40 pennies, $2(40) + 5 = 85$ dimes, and $(40) + 10 = 50$ nickels.

79. Let x represent the number of dimes, then $2x - 3$ represents the number of quarters.
$$0.10x + 0.25(2x - 3) = 7.65$$
$$100[0.10x + 0.25(2x - 3)] = 100(7.65)$$
$$10x + 25(2x - 3) = 765$$
$$10x + 50x - 75 = 765$$
$$60x = 840$$
$$x = 14$$
There are 14 dimes and $2(14) - 3 = 25$ quarters.

80. Let x represent the score on the fifth exam.
$$\frac{85 + 87 + 90 + 91 + x}{5} \geq 90$$
$$\frac{353 + x}{5} \geq 90$$
$$353 + x \geq 450$$
$$x \geq 97$$
She must have a score of 97 or better.

81. Let x represent the number of boys and $1650 - x$ represent the number of girls in the school.
$$\frac{\text{boys}}{\text{girls}} \quad \frac{6}{5} = \frac{x}{1650 - x}$$
$$5x = 6(1650 - x)$$
$$5x = 9900 - 6x$$
$$11x = 9900$$
$$x = 900$$
There would be 900 boys and $1650 - 900 = 750$ girls in the school.

82. Let s represent the selling price.
Profit is a percent of selling price.
Selling price = Cost + Profit

$$\begin{aligned} s &= 750 + (70\%)(s) \\ s &= 750 + 0.7s \\ 10s &= 7500 + 7s \\ 3s &= 7500 \\ s &= 2500 \end{aligned}$$

The selling price would be $\$2,500$.

83. Let s represent the selling price.
Profit is a percent of cost.
Selling price = Cost + Profit

$$\begin{aligned} s &= 750 + (70\%)(750) \\ s &= 750 + 0.70(750) \\ s &= 750 + 525 \\ s &= 1275 \end{aligned}$$

The selling price would be $\$1,275$.

84. Let x represent the amount of pure alcohol to be added. Then $20 + x$ represents the amount of final solution.

$$\begin{pmatrix}\text{pure alcohol in} \\ \text{30\% solution}\end{pmatrix} + \begin{pmatrix}\text{pure alcohol} \\ \text{to be added}\end{pmatrix} = \begin{pmatrix}\text{pure alcohol} \\ \text{in final solution}\end{pmatrix}$$

$$\begin{aligned} (30\%)(6) + x &= (40\%)(6 + x) \\ 0.30(6) + x &= 0.40(6 + x) \\ 10[0.30(6) + x] &= 10[0.40(6 + x)] \\ 3(6) + 10x &= 4(6 + x) \\ 18 + 10x &= 24 + 4x \\ 18 + 6x &= 24 \\ 6x &= 6 \\ x &= 1 \end{aligned}$$

We must add 1 quart of pure alcohol.

85. Let x = cost of tennis ball and y = cost of golf ball.

$$\begin{pmatrix} 5x + 4y = 17.00 \\ 3x + 7y = 20.55 \end{pmatrix}$$

Multiply equation 1 by -3.
Multiply equation 2 by 5,
then add the equations.

$$\begin{aligned} -15x - 12y &= -51.00 \\ 15x + 35y &= 102.75 \\ \hline 23y &= 51.75 \\ y &= 2.25 \end{aligned}$$

$5x + 4y = 17$
$5x + 4(2.25) = 17$
$5x + 9 = 17$
$5x = 8$
$x = 1.6$

Tennis balls cost $\$1.60$ and golf balls cost $\$2.25$.

Chapter 8 : A Transition from Elementary Algebra to Intermediate Algebra

PROBLEM SET 8.1 Equations: A Brief Review

1.
$$5x - 4 = 16$$
$$5x = 20$$
$$x = \frac{20}{5} = 4$$
The solution set is $\{4\}$.

3.
$$-6 = 7x + 1$$
$$-7 = 7x$$
$$-1 = x$$
The solution set is $\{-1\}$.

5.
$$-2x + 8 = -3x + 14$$
$$-2x = -3x + 6$$
$$x = 6$$
The solution set is $\{6\}$.

7.
$$4x - 6 - 5x = 3x + 1 - x$$
$$-x - 6 = 2x + 1$$
$$-x = 2x + 7$$
$$-3x = 7$$
$$x = \frac{7}{-3} = -\frac{7}{3}$$
The solution set is $\left\{-\frac{7}{3}\right\}$.

9.
$$6(2x + 5) = 5(3x - 4)$$
$$12x + 30 = 15x - 20$$
$$12x = 15x - 50$$
$$-3x = -50$$
$$x = \frac{-50}{-3} = \frac{50}{3}$$
The solution set is $\left\{\frac{50}{3}\right\}$.

11.
$$-2(3x - 1) - 3(2x + 5) = -5(2x - 8)$$
$$-6x + 2 - 6x - 15 = -10x + 40$$
$$-12x - 13 = -10x + 40$$
$$-12x = -10x + 53$$
$$-2x = 53$$
$$x = \frac{53}{-2} = -\frac{53}{2}$$
The solution set is $\left\{-\frac{53}{2}\right\}$.

13.
$$\frac{2}{3}x - \frac{1}{2} + \frac{1}{4}x = \frac{5}{8}$$
$$24\left(\frac{2}{3}x - \frac{1}{2} + \frac{1}{4}x\right) = 24\left(\frac{5}{8}\right)$$
$$24\left(\frac{2}{3}x\right) - 24\left(\frac{1}{2}\right) + 24\left(\frac{1}{4}x\right) = 3(5)$$
$$8(2x) - 12(1) + 6(1x) = 15$$
$$16x - 12 + 6x = 15$$
$$22x - 12 = 15$$
$$22x = 27$$
$$x = \frac{27}{22}$$
The solution set is $\left\{\frac{27}{22}\right\}$.

15.
$$\frac{2x - 1}{3} - \frac{3x + 2}{5} = -2$$
$$15\left(\frac{2x - 1}{3} - \frac{3x + 2}{5}\right) = 15(-2)$$
$$15\left(\frac{2x - 1}{3}\right) - 15\left(\frac{3x + 2}{5}\right) = -30$$
$$5(2x - 1) - 3(3x + 2) = -30$$
$$10x - 5 - 9x - 6 = -30$$
$$x - 11 = -30$$
$$x = -19$$
The solution set is $\{-19\}$.

17.
$$\frac{-3}{2x-1}=\frac{2}{4x+7}$$
$$-3(4x+7)=2(2x-1)$$
$$-12x-21=4x-2$$
$$-16x=19$$
$$x=\frac{19}{-16}=-\frac{19}{16}$$
The solution set is $\left\{-\frac{19}{16}\right\}$.

19.
$$\frac{x+2}{3}+1=\frac{x-2}{3}$$
$$3\left(\frac{x+2}{3}+1\right)=3\left(\frac{x-2}{3}\right)$$
$$3\left(\frac{x+2}{3}\right)+3(1)=1(x-2)$$
$$1(x+2)+3=x-2$$
$$x+5=x-2$$
$$0=-7$$
Because $0=-7$ is a false statement, there is no solution. The solution set is Ø.

21.
$$0.09x+0.11(x+125)=68.75$$
$$100[0.09x+0.11(x+125)]=100(68.75)$$
$$9x+11(x+125)=6875$$
$$9x+11x+1375=6875$$
$$20x+1375=6875$$
$$20x=5500$$
$$x=275$$
The solution set is $\{275\}$.

23.
$$\frac{x-1}{5}-2=\frac{x-11}{5}$$
$$5\left(\frac{x-1}{5}-2\right)=5\left(\frac{x-11}{5}\right)$$
$$5\left(\frac{x-1}{5}\right)-5(2)=x-11$$
$$x-1-10=x-11$$
$$x-11=x-11$$
$$0=0$$
Because $0=0$ is a true statement, the solution set is $\{x|\ x \text{ is a real number}\}$.

25.
$$0.05x-4(x+0.5)=1.2$$
$$100[0.05x-4(x+0.5)]=100(1.2)$$
$$5x-400(x+0.5)=120$$
$$5x-400x-200=120$$
$$-395x=320$$
$$x=\frac{320}{-395}=-\frac{64}{79}$$
The solution set is $\left\{-\frac{64}{79}\right\}$.

27.
$$\begin{pmatrix} x-2y=10 \\ 3x+7y=-22 \end{pmatrix}$$
Solve equation 1 for x.
$x-2y=10$
$x=2y+10$
Substitute into equation 2.
$3x+7y=-22$
$3(2y+10)+7y=-22$
$6y+30+7y=-22$
$13y=-52$
$y=-4$
$x=2(-4)+10$
$x=-8+10=2$
The solution set is $\{(2,-4)\}$.

29.
$$\begin{pmatrix} 5x-2y=-3 \\ 3x-4y=15 \end{pmatrix}$$
Multiply equation 1 by -2, then add to equation 2.
$$\begin{array}{r} -10x+4y=6 \\ 3x-4y=15 \\ \hline -7x\qquad=21 \end{array}$$
$$x=-3$$
$5x-2y=-3$
$5(-3)-2y=-3$
$-15-2y=-3$
$-2y=12$
$y=-6$
The solution set is $\{(-3,-6)\}$.

31. $\begin{pmatrix} 2x - y = 6 \\ 4x - 2y = -1 \end{pmatrix}$

Multiply equation 1 by -2, then add to equation 2.

$$\begin{aligned} -4x + 2y &= -12 \\ 4x - 2y &= -1 \\ \hline 0 &= -13 \end{aligned}$$

Since $0 \neq -13$, the system has no solution. The solution set is $\emptyset$.

33. $\begin{pmatrix} 3x + 4y = -18 \\ 5x - 7y = -30 \end{pmatrix}$

Multiply equation 1 by 7, multiply equation 2 by 4, then add equations.

$$\begin{aligned} 21x + 28y &= -126 \\ 20x - 28y &= -120 \\ \hline 41x &= -246 \\ x &= -6 \end{aligned}$$

$$\begin{aligned} 3x + 4y &= -18 \\ 3(-6) + 4y &= -18 \\ -18 + 4y &= -18 \\ 4y &= 0 \\ y &= 0 \end{aligned}$$

The solution set is $\{(-6, 0)\}$.

35. $\begin{pmatrix} \frac{1}{2}x - \frac{2}{3}y = -8 \\ \frac{3}{2}x + \frac{1}{3}y = -10 \end{pmatrix}$

Multiply both equations by 6.

$\begin{pmatrix} 3x - 4y = -48 \\ 9x + 2y = -60 \end{pmatrix}$

Multiply equation 2 by 2, then add to equation 1.

$$\begin{aligned} 3x - 4y &= -48 \\ 18x + 4y &= -120 \\ \hline 21x &= -168 \\ x &= -8 \end{aligned}$$

$$\begin{aligned} 3x - 4y &= -48 \\ 3(-8) - 4y &= -48 \\ -24 - 4y &= -48 \\ -4y &= -24 \\ y &= 6 \end{aligned}$$

The solution set is $\{(-8, 6)\}$.

37. Let n, $n + 1$, $n + 2$ represent the consecutive integers.

$$\begin{aligned} n + (n + 1) + (n + 2) &= -45 \\ 3n + 3 &= -45 \\ 3n &= -48 \\ n &= -16 \end{aligned}$$

The integers are -16, -15, and -14.

39. Let x represent the number of males in class, then $3x - 5$ represents the number of females.

$$\begin{aligned} x + 3x - 5 &= 51 \\ 4x - 5 &= 51 \\ 4x &= 56 \\ x &= 14 \end{aligned}$$

There are 14 males and $51 - 14 = 37$ females.

41. Let x represent the cost of the skirt, so $\frac{8}{7}x$ represents the cost of the sweater, and $x - 8$ represents the cost of the shoes.

$$x + \left(\frac{8}{7}x\right) + (x - 8) = 124$$

$$7\left[x + \left(\frac{8}{7}x\right) + (x - 8)\right] = 7(124)$$

$$\begin{aligned} 7x + 8x + 7(x - 8) &= 868 \\ 7x + 8x + 7x - 56 &= 868 \\ 22x &= 924 \\ x &= 42 \end{aligned}$$

The cost of the skirt is \$42; the cost of the sweater is $\frac{8}{7}(42) = \$48$; the cost of the shoes is $(42) - 8 = \$34$.

43. Let x represent the length the side of the small square and $x - 2$ and $x + 3$ represent lengths of opposite sides of the larger figure.

x
x
x + 3
x - 2

$$\left(\text{Area of Small Square}\right) + 8 = \left(\text{Area of new larger rectangle}\right)$$

$$(x)^2 + 8 = (x - 2)(x + 3)$$

$$x^2 + 8 = x^2 + 3x - 2x - 6$$

$8 = x - 6$
$14 = x$
The length of the side of the square is 14 centimeters.

45. Let s represent the selling price. Profit is a percent of selling price.
Selling price = Cost + Profit

$$s = 50 + (50\%)(s)$$
$$s = 50 + 0.5s$$
$$10s = 500 + 5s$$
$$5s = 500$$
$$s = 100$$

The selling price would be \$100 if profit is based on selling price.
Profit = Selling price − Cost
Profit = 100 − 50 = 50

Let s represent the selling price. Profit is a percent of cost.
Selling price = Cost + Profit

$$s = 50 + (50\%)(50)$$
$$s = 50 + 0.50(50)$$
$$s = 50 + 25$$
$$s = 75$$

The selling price would be \$75 if profit is based on cost.
Profit = Selling price − Cost
Profit = 75 − 50 = 25

$$\begin{pmatrix}\text{Profit based}\\ \text{on selling price}\end{pmatrix} - \begin{pmatrix}\text{Profit based}\\ \text{on cost}\end{pmatrix} = \begin{pmatrix}\text{Difference}\\ \text{in profit}\end{pmatrix}$$
$$50 - 25 = 25$$

The difference in profit is \$25.

47. Let x represent the weight of sodium, then $200 - x$ represents the weight of chlorine.

$$\frac{\text{sodium}}{\text{chlorine}} \quad \frac{5}{3} = \frac{x}{200 - x}$$
$$3x = 5(200 - x)$$
$$3x = 1000 - 5x$$
$$8x = 1000$$
$$x = 125$$

There is 125 pounds of sodium and $200 - 125 = 75$ pounds of chlorine.

49. Let x represent the length of one leg of a right triangle, then $x + 7$ represents the length of the other leg and the hypotenuse has length of 7.

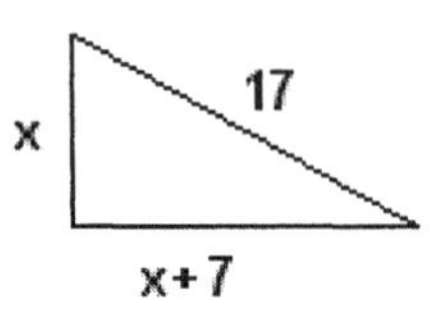

Use the Pythagorean Theorem to find x.
$x^2 + (x + 7)^2 = 17^2$
$x^2 + x^2 + 14x + 49 = 289$
$2x^2 + 14x - 240 = 0$
$x^2 + 7x - 120 = 0$
$(x + 15)(x - 8) = 0$
$x + 15 = 0$ or $x - 8 = 0$
$x = -15$ or $x = 8$
Discard $x = -15$ because length cannot be negative. Therefore, $x = 8$ and the lengths of the two legs are 8 meters and $8 + 7 = 15$ meters.

51. Let r represent the rate for Javier and $r + 3$ represent the rate for Domenica. Chart of the information would be as follows:

	Rate	Time	Distance($d = rt$)
Javier	r	4	$4r$
Domenica	$r + 3$	4	$4(r + 3)$

A diagram of the problem would be as follows:

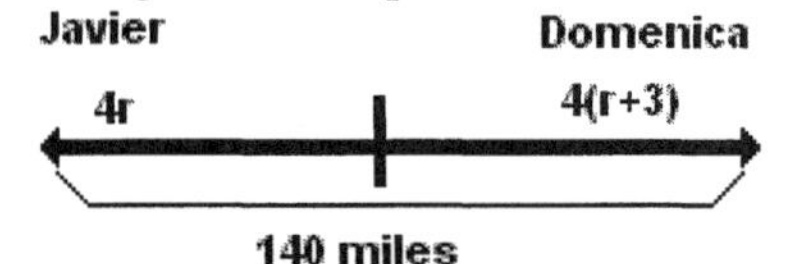

The sum of the two distances would be 140 miles.

$$4r + 4(r + 3) = 140$$
$$4r + 4r + 12 = 140$$
$$8r + 12 = 140$$
$$8r = 128$$
$$r = 16$$

Javier rides at 16 miles per hour and Domenica rides at $16 + 3 = 19$ miles per hour.

PROBLEM SET 8.2 Inequalities: A Brief Review

1. $6x - 2 > 4x - 14$
$6x > 4x - 12$
$2x > -12$
$\frac{1}{2}(2x) > \frac{1}{2}(-12)$
$x > -6$
The solution set is $(-6, \infty)$.

3. $2x - 7 < 6x + 13$
$2x < 6x + 20$
$-4x < 20$
$-\frac{1}{4}(-4x) > -\frac{1}{4}(20)$
$x > -5$
The solution set is $(-5, \infty)$.

5. $4(x - 3) \leq -2(x + 1)$
$4x - 12 \leq -2x - 2$
$4x \leq -2x + 10$
$6x \leq 10$
$\frac{1}{6}(6x) \leq \frac{1}{6}(10)$
$x \leq \frac{10}{6}$
$x \leq \frac{5}{3}$
The solution set is $\left(-\infty, \frac{5}{3}\right]$.

7. $5(x - 4) - 6(x + 2) < 4$
$5x - 20 - 6x - 12 < 4$
$-x - 32 < 4$
$-x < 36$
$-1(-x) > -1(36)$
$x > -36$
The solution set is $(-36, \infty)$.

9. $-3(3x + 2) - 2(4x + 1) \geq 0$
$-9x - 6 - 8x - 2 \geq 0$
$-17x - 8 \geq 0$
$-17x \geq 8$
$-\frac{1}{17}(-17x) \leq -\frac{1}{17}(8)$
$x \leq -\frac{8}{17}$
The solution set is $\left(-\infty, -\frac{8}{17}\right]$.

11. $-(x - 3) + 2(x - 1) < 3(x + 4)$
$-x + 3 + 2x - 2 < 3x + 12$
$x + 1 < 3x + 12$
$x < 3x + 11$
$-2x < 11$
$-\frac{1}{2}(-2x) > -\frac{1}{2}(11)$
$x > -\frac{11}{2}$
The solution set is $\left(-\frac{11}{2}, \infty\right)$.

13. $7(x + 1) - 8(x - 2) < 0$
$7x + 7 - 8x + 16 < 0$
$-x + 23 < 0$
$-x < -23$
$-1(-x) > -1(-23)$
$x > 23$
The solution set is $(23, \infty)$.

15. $\frac{x+3}{8} - \frac{x+5}{5} \geq \frac{3}{10}$
$40\left(\frac{x+3}{8} - \frac{x+5}{5}\right) \geq 40\left(\frac{3}{10}\right)$
$40\left(\frac{x+3}{8}\right) - 40\left(\frac{x+5}{5}\right) \geq 4(3)$
$5(x + 3) - 8(x + 5) \geq 12$
$5x + 15 - 8x - 40 \geq 12$
$-3x - 25 \geq 12$
$-3x \geq 37$
$-\frac{1}{3}(-3x) \leq -\frac{1}{3}(37)$
$x \leq -\frac{37}{3}$
The solution set is $\left(-\infty, -\frac{37}{3}\right]$.

17. $\frac{4x-3}{6} - \frac{2x-1}{12} < -2$
$12\left(\frac{4x-3}{6} - \frac{2x-1}{12}\right) < 12(-2)$
$12\left(\frac{4x-3}{6}\right) - 12\left(\frac{2x-1}{12}\right) < -24$
$2(4x - 3) - (2x - 1) < -24$
$8x - 6 - 2x + 1 < -24$
$6x - 5 < -24$

$6x < -19$

$x < -\frac{19}{6}$

The solution set is $\left(-\infty, -\frac{19}{6}\right)$.

19. $0.06x + 0.08(250 - x) \geq 19$

$100[0.06x + 0.08(250 - x)] \geq 100(19)$

$100(0.06x) + 100[0.08(250 - x)] \geq 1900$

$6x + 8(250 - x) \geq 1900$

$6x + 2000 - 8x \geq 1900$

$-2x + 2000 \geq 1900$

$-2x \geq -100$

$-\frac{1}{2}(-2x) \leq -\frac{1}{2}(-100)$

$x \leq 50$

The solution set is $(-\infty, 50]$.

21. $0.09x + 0.1(x + 200) > 77$

$100[0.09x + 0.1(x + 200)] > 100(77)$

$100(0.09x) + 100[0.1(x + 200)] > 7700$

$9x + 10(x + 200) > 7700$

$9x + 10x + 2000 > 7700$

$19x + 2000 > 7700$

$19x > 5700$

$x > 300$

The solution set is $(300, \infty)$.

23. $x \geq 3.4 + 0.15x$

$100(x) \geq 100(3.4 + 0.15x)$

$100x \geq 100(3.4) + 100(0.15x)$

$100x \geq 340 + 15x$

$85x \geq 340$

$x \geq 4$

The solution set is $[4, \infty)$.

25. $2x - 1 \geq 5$ and $x > 0$

$2x \geq 6$ and $x > 0$

$x \geq 3$ and $x > 0$

The solution set is $[3, \infty)$.

27. $5x - 2 < 0$ and $3x - 1 > 0$

$5x < 2$ and $3x > 1$

$x < \frac{2}{5}$ and $x > \frac{1}{3}$

The solution set is $\left(\frac{1}{3}, \frac{2}{5}\right)$.

29. $3x + 2 < -1$ or $3x + 2 > 1$

$3x < -3$ or $3x > -1$

$x < -1$ or $x > -\frac{1}{3}$

The solution set is $(-\infty, -1) \cup \left(-\frac{1}{3}, \infty\right)$.

31. $-6 < 4x - 5 < 6$

$-1 < 4x < 11$

$-\frac{1}{4} < x < \frac{11}{4}$

The solution set is $\left(-\frac{1}{4}, \frac{11}{4}\right)$.

33. $-4 \leq \frac{x-1}{3} \leq 4$

$3(-4) \leq 3\left(\frac{x-1}{3}\right) \leq 3(4)$

$-12 \leq x - 1 \leq 12$

$-11 \leq x \leq 13$

The solution set is $[-11, 13]$.

35. $-3 < 2 - x < 3$

$-5 < -x < 1$

$-1(-5) > -1(-x) > -1(1)$

$5 > x > -1$

$-1 < x < 5$

The solution set is $(-1, 5)$.

37. Let x = amount invested at 9%.

$(8\%)(1000) + (9\%)(x) > 98$

$0.08(1000) + 0.09x > 98$

$100[0.08(1000) + 0.09x] > 100(98)$

$8(1000) + 9x > 9800$
$8000 + 9x > 9800$
$9x > 1800$
$x > 200$
She must invest more than \$200.

39. Let x = score on 5th exam.
$\frac{95 + 82 + 93 + 84 + x}{5} \geq 90$
$\frac{354 + x}{5} \geq 90$
$5\left(\frac{354 + x}{5}\right) \geq 5(90)$
$354 + x \geq 450$
$x \geq 96$
A score of 96 or better.

41. $-4 \leq \frac{9}{5}C + 32 \leq 23$
$-36 \leq \frac{9}{5}C \leq -9$
$\frac{5}{9}(-36) \leq \frac{5}{9}\left(\frac{9}{5}C\right) \leq \frac{5}{9}(-9)$
$-20 \leq C \leq -5$
The temperature would be between -20°C and -5°C.

43. $80 \leq \frac{100M}{C} \leq 140$
$80 \leq \frac{100M}{11} \leq 140$
$\frac{11}{100}(80) \leq \frac{11}{100}\left(\frac{100M}{11}\right) \leq \frac{11}{100}(140)$
$8.8 \leq M \leq 15.4$
The mental age ranges from 8.8 to 15.4 years inclusive.

PROBLEM SET 8.3 Equations and Inequalities Involving Absolute Value

1. $|x| < 5$
$-5 < x < 5$
The solution set is $(-5, 5)$.

3. $|x| \leq 2$
$-2 \leq x \leq 2$
The solution set is $[-2, 2]$.

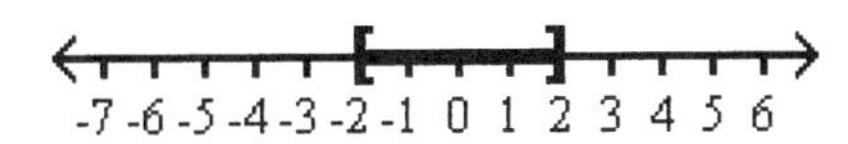

5. $|x| > 2$
$x < -2$ or $x > 2$
The solution set is $(-\infty, -2) \cup (2, \infty)$.

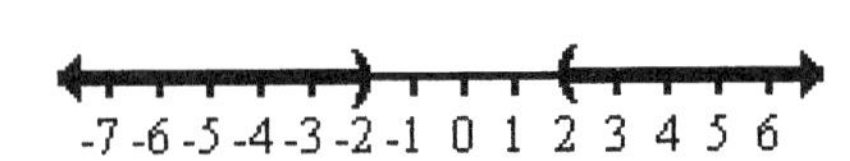

7. $|x - 1| < 2$
$-2 < x - 1 < 2$
$-1 < x < 3$
The solution set is $(-1, 3)$.

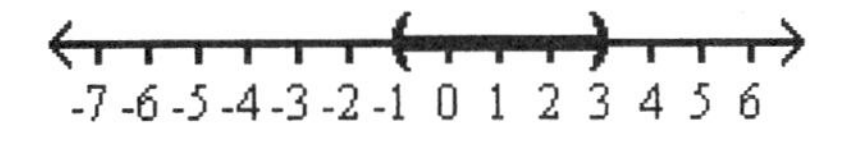

9. $|x + 2| \leq 4$
$-4 \leq x + 2 \leq 4$
$-6 \leq x \leq 2$
The solution set is $[-6, 2]$.

11. $|x + 2| > 1$
$x + 2 < -1$ or $x + 2 > 1$
$x < -3$ or $x > -1$
The solution set is $(-\infty, -3) \cup (-1, \infty)$.

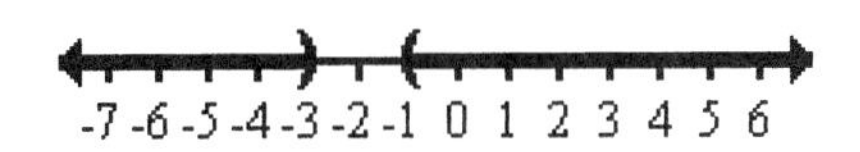

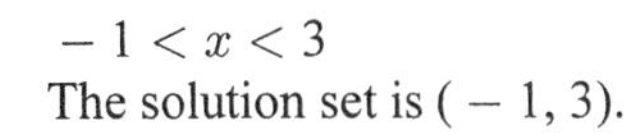

Problem Set 8.3

13. $|x - 3| \geq 2$

$x - 3 \leq -2$ or $x - 3 \geq 2$

$x \leq 1$ or $x \geq 5$

The solution set is $(-\infty, 1] \cup [5, \infty)$.

-7 -6 -5 -4 -3 -2 -1 0 1 2 3 4 5 6

15. $|x - 1| = 8$

$x - 1 = 8$ or $x - 1 = -8$

$x = 9$ or $x = -7$

The solution set is $\{-7, 9\}$.

17. $|x - 2| > 6$

$x - 2 < -6$ or $x - 2 > 6$

$x < -4$ or $x > 8$

The solution set is $(-\infty, -4) \cup (8, \infty)$.

19. $|x + 3| < 5$

$-5 < x + 3 < 5$

$-8 < x < 2$

The solution set is $(-8, 2)$.

21. $|2x - 4| = 6$

$2x - 4 = 6$ or $2x - 4 = -6$

$2x = 10$ or $2x = -2$

$x = 5$ or $x = -1$

The solution set is $\{-1, 5\}$.

23. $|2x - 1| \leq 9$

$-9 \leq 2x - 1 \leq 9$

$-8 \leq 2x \leq 10$

$-4 \leq x \leq 5$

The solution set is $[-4, 5]$.

25. $|4x + 2| \geq 12$

$4x + 2 \leq -12$ or $4x + 2 \geq 12$

$4x \leq -14$ or $4x \geq 10$

$x \leq -\dfrac{14}{4}$ or $x \geq \dfrac{10}{4}$

$x \leq -\dfrac{7}{2}$ or $x \geq \dfrac{5}{2}$

The solution set is $\left(-\infty, -\dfrac{7}{2}\right] \cup \left[\dfrac{5}{2}, \infty\right)$.

27. $|3x + 4| = 11$

$3x + 4 = 11$ or $3x + 4 = -11$

$3x = 7$ or $3x = -15$

$x = \dfrac{7}{3}$ or $x = -5$

The solution set is $\left\{-5, \dfrac{7}{3}\right\}$.

29. $|4 - 2x| = 6$

$4 - 2x = 6$ or $4 - 2x = -6$

$-2x = 2$ or $-2x = -10$

$x = -1$ or $x = 5$

The solution set is $\{-1, 5\}$.

31. $|2 - x| > 4$

$2 - x < -4$ or $2 - x > 4$

$-x < -6$ or $-x > 2$

$-1(-x) > -1(-6)$ or $-1(-x) < -1(2)$

$x > 6$ or $x < -2$

The solution set is $(-\infty, -2) \cup (6, \infty)$.

33. $|1 - 2x| < 2$

$-2 < 1 - 2x < 2$

$-3 < -2x < 1$

$-\dfrac{1}{2}(-3) > -\dfrac{1}{2}(-2x) > -\dfrac{1}{2}(1)$

$\dfrac{3}{2} > x > -\dfrac{1}{2}$

$-\dfrac{1}{2} < x < \dfrac{3}{2}$

The solution set is $\left(-\dfrac{1}{2}, \dfrac{3}{2}\right)$.

35. $|5x + 9| \leq 16$

$-16 \leq 5x + 9 \leq 16$

$-25 \leq 5x \leq 7$

$-5 \leq x \leq \dfrac{7}{5}$

The solution set is $\left[-5, \dfrac{7}{5}\right]$.

37. $\left|x - \dfrac{3}{4}\right| = \dfrac{2}{3}$

$x - \dfrac{3}{4} = \dfrac{2}{3}$ or $x - \dfrac{3}{4} = -\dfrac{2}{3}$

$x = \dfrac{2}{3} + \dfrac{3}{4}$ or $x = -\dfrac{2}{3} + \dfrac{3}{4}$

$x = \dfrac{8}{12} + \dfrac{9}{12}$ or $x = -\dfrac{8}{12} + \dfrac{9}{12}$

$x = \frac{17}{12}$ or $x = \frac{1}{12}$
The solution set is $\left\{\frac{1}{12}, \frac{17}{12}\right\}$.

39. $|-2x + 7| \leq 13$
$-13 \leq -2x + 7 \leq 13$
$-20 \leq -2x \leq 6$
$-\frac{1}{2}(-20) \geq -\frac{1}{2}(-2x) \geq -\frac{1}{2}(6)$
$10 \geq x \geq -3$
$-3 \leq x \leq 10$
The solution set is $[-3, 10]$.

41. $\left|\frac{x-3}{4}\right| < 2$
$-2 < \frac{x-4}{4} < 2$
$4(-2) < 4\left(\frac{x-3}{4}\right) < 4(2)$
$-8 < x - 3 < 8$
$-5 < x < 11$
The solution set is $(-5, 11)$.

43. $\left|\frac{2x+1}{2}\right| > 1$
$\frac{2x+1}{2} < -1$ or $\frac{2x+1}{2} > 1$
$2\left(\frac{2x+1}{2}\right) < -1(2)$ or $2\left(\frac{2x+1}{2}\right) > 1(2)$
$2x + 1 < -2$ or $2x + 1 > 2$
$2x < -3$ or $2x > 1$
$x < -\frac{3}{2}$ or $x > \frac{1}{2}$
The solution set is $\left(-\infty, -\frac{3}{2}\right) \cup \left(\frac{1}{2}, \infty\right)$.

45. $|2x - 3| + 2 = 5$
$|2x - 3| = 3$
$2x - 3 = 3$ or $2x - 3 = -3$
$2x = 6$ or $2x = 0$
$x = 3$ or $x = 0$
The solution set is $\{0, 3\}$.

47. $|x + 7| - 3 \geq 4$
$|x + 7| \geq 7$
$x + 7 \leq -7$ or $x + 7 \geq 7$
$x \leq -14$ or $x \geq 0$
The solution set is $(-\infty, -14] \cup [0, \infty)$.

49. $|2x - 1| + 1 \leq 6$
$|2x - 1| \leq 5$
$-5 \leq 2x - 1 \leq 5$
$-4 \leq 2x \leq 6$
$-2 \leq x \leq 3$
The solution set is $[-2, 3]$.

51. $|2x + 1| = -4$
The solution set is $\emptyset$.
Absolute value is never negative.

53. $|3x - 1| > -2$
The solution set is $(-\infty, \infty)$.
Absolute value is always greater than a negative value.

55. $|5x - 2| = 0$
$5x - 2 = 0$
$5x = 2$
$x = \frac{2}{5}$
The solution set is $\left\{\frac{2}{5}\right\}$.

57. $|4x - 6| < -1$
The solution set is $\emptyset$.
Absolute value is never less than a negative value.

59. $|x + 4| < 0$
The solution set is $\emptyset$.
Absolute value is never less than zero.

Further Investigations

65. $|-2x - 3| = |x + 1|$
$-2x - 3 = x + 1$ or $-2x - 3 = -(x + 1)$
$-3x = 4$ or $-2x - 3 = -x - 1$
$x = -\frac{4}{3}$ or $-x = 2$
$x = -2$
The solution set is $\left\{-2, -\frac{4}{3}\right\}$.

67. $|x-2|=|x+6|$
$x-2=x+6$ or $x-2=-(x+6)$
$x=x+8$ or $x-2=-x-6$
$0=8$ or $x=-x-4$
contradiction $\quad 2x=-4$
$x=-2$
The solution set is $\{-2\}$.

69. $|x+1|=|x-1|$
$x+1=x-1$ or $x+1=-(x-1)$
$x=x-2$ or $x+1=-x+1$
$0=-2$ or $x=-x+0$
contradiction $\quad 2x=0$
$x=0$
The solution set is $\{0\}$.

71. $|ax+b|<k$
If $ax+b\geq 0$, then $|ax+b|=ax+b$
so, $ax+b<k$
If $ax+b<0$, then $|ax+b|=-(ax+b)$
so, $-(ax+b)<k$
$-1[-(ax+b)]>-1(k)$
$ax+b>-k$
Therefore $ax+b<k$ and $ax+b>-k$
is $-k<ax+b<-k$.

PROBLEM SET 8.4 Polynomials: A Brief Review and Binomial Expansions

1. $\left(-\frac{1}{2}xy\right)\left(\frac{1}{3}x^2y^3\right)$
$-\frac{1}{6}x^{1+2}y^{1+3}$
$-\frac{1}{6}x^3y^4$

3. $(3x)(-2x^2)(-5x^3)$
$30x^{1+2+3}$
$30x^6$

5. $(-6x^2)(3x^3)(x^4)$
$-18x^{2+3+4}$
$-18x^9$

7. $(x^2y)(-3xy^2)(x^3y^3)$
$-3x^{2+1+3}y^{1+2+3}$
$-3x^6y^6$

9. $(-3ab^3)^4$
$(-3)^4a^{1(4)}b^{3(4)}$
$81a^4b^{12}$

11. $-(2ab)^4$
$-[2^4a^{1(4)}b^{1(4)}]$
$-(16a^4b^4)$
$-16a^4b^4$

13. $(4x+5)(x+7)$
$4x(x+7)+5(x+7)$
$4x^2+28x+5x+35$
$4x^2+33x+35$

15. $(3y-1)(3y+1)$
$3y(3y+1)-1(3y+1)$
$9y^2+3y-3y-1$
$9y^2-1$

17. $(7x-2)(2x+1)$
$7x(2x+1)-2(2x+1)$
$14x^2+7x-4x-2$
$14x^2+3x-2$

19. $(1+t)(5-2t)$
$1(5-2t)+t(5-2t)$
$5-2t+5t-2t^2$
$5+3t-2t^2$

21. $(3t+7)^2$
$(3t)^2+2(3t)(7)+(7)^2$
$9t^2+42t+49$

23. $(2-5x)(2+5x)$
$2(2+5x)-5x(2+5x)$
$4+10x-10x-25x^2$
$4-25x^2$

25. $(x+1)(x-2)(x-3)$
$(x+1)[x(x-3)-2(x-3)]$
$(x+1)[x^2-3x-2x+6]$
$(x+1)(x^2-5x+6)$
$x(x^2-5x+6)+1(x^2-5x+6)$
$x^3-5x^2+6x+x^2-5x+6$
x^3-4x^2+x+6

27. $(x-3)(x+3)(x-1)$
$(x-3)[x(x-1)+3(x-1)]$
$(x-3)[x^2-x+3x-3]$
$(x-3)(x^2+2x-3)$
$x(x^2+2x-3)-3(x^2+2x-3)$
$x^3+2x^2-3x-3x^2-6x+9$
x^3-x^2-9x+9

29. $(x-4)(x^2+5x-4)$
$x(x^2+5x-4)-4(x^2+5x-4)$
$x^3+5x^2-4x-4x^2-20x+16$
$x^3+x^2-24x+16$

31. $(2x-3)(x^2+6x+10)$
$2x(x^2+6x+10)-3(x^2+6x+10)$
$2x^3+12x^2+20x-3x^2-18x-30$
$2x^3+9x^2+2x-30$

33. $(4x-1)(3x^2-x+6)$
$4x(3x^2-x+6)-1(3x^2-x+6)$
$12x^3-4x^2+24x-3x^2+x-6$
$12x^3-7x^2+25x-6$

35. $(x^2+2x+1)(x^2+3x+4)$

$x^2(x^2+3x+4)+2x(x^2+3x+4)$
$+1(x^2+3x+4)$

$x^4+3x^3+4x^2+2x^3+6x^2+8x$
$+x^2+3x+4$

$x^4+5x^3+11x^2+11x+4$

37. $(4x-1)^3$
$(4x)^3+3(4x)^2(-1)+3(4x)(-1)^2+(-1)^3$
$64x^3+3(16x^2)(-1)+3(4x)(1)-1$
$64x^3-48x^2+12x-1$

39. $(5x+2)^3$
$(5x)^3+3(5x)^2(2)+3(5x)(2)^2+(2)^3$
$125x^3+3(25x^2)(2)+3(5x)(4)+8$
$125x^3+150x^2+60x+8$

41. $(x^n-4)(x^n+4)$
$x^n(x^n+4)-4(x^n+4)$
$x^{2n}+4x^n-4x^n-16$
$x^{2n}-16$

43. $(x^a+6)(x^a-2)$
$x^a(x^a-2)+6(x^a-2)$
$x^{2a}-2x^a+6x^a-12$
$x^{2a}+4x^a-12$

45. $(2x^n+5)(3x^n-7)$
$2x^n(3x^n-7)+5(3x^n-7)$
$6x^{2n}-14x^n+15x^n-35$
$6x^{2n}+x^n-35$

47. $(x^{2a}-7)(x^{2a}-3)$
$x^{2a}(x^{2a}-3)-7(x^{2a}-3)$
$x^{4a}-3x^{2a}-7x^{2a}+21$
$x^{4a}-10x^{2a}+21$

49. $(2x^n+5)^2$
$(2x^n)^2+2(2x^n)(5)+(5)^2$
$4x^{2n}+20x^n+25$

51. $\dfrac{9x^4y^5}{3xy^2}$
$3x^{4-1}y^{5-2}$
$3x^3y^3$

53. $\dfrac{25x^5y^6}{-5x^2y^4}$
$-5x^{5-2}y^{6-4}$
$-5x^3y^2$

55. $\dfrac{-54ab^2c^3}{-6abc}$
$9a^{1-1}b^{2-1}c^{3-1}$
$9a^0bc^2$
$9(1)bc^2$
$9bc^2$

Problem Set 8.4

57. $\dfrac{-18x^2y^2z^6}{xyz^2}$

$-18x^{2-1}y^{2-1}z^{6-2}$

$-18xyz^4$

59. $\dfrac{a^3b^4c^7}{-abc^5}$

$-a^{3-1}b^{4-1}c^{7-5}$

$-a^2b^3c^2$

61. $(a+b)^7 = a^7 + \binom{7}{1}a^6(b) + \binom{7}{2}a^5(b)^2 + \binom{7}{3}a^4(b)^3 + \binom{7}{4}a^3(b)^4 + \binom{7}{5}a^2(b)^5 + \binom{7}{6}a(b)^6 + b^7$

$= a^7 + 7a^6b + 21a^5b^2 + 35a^4b^3 + 35a^3a^4 + 21a^2b^5 + 7ab^6 + b^7$

63. $(x-y)^5 = x^5 + \binom{5}{1}x^4(-y) + \binom{5}{2}x^3(-y)^2 + \binom{5}{3}x^2(-y)^3 + \binom{5}{4}x(-y)^4 + (-y)^5$

$= x^5 - 5x^4y + 10x^3y^2 - 10x^2y^3 + 5xy^4 - y^5$

65. $(x+2y)^4 = x^4 + \binom{4}{1}x^3(2y) + \binom{4}{2}x^2(2y)^2 + \binom{4}{3}x(2y)^3 + (2y)^4$

$= x^4 + 8x^3y + 24x^2y^2 + 32xy^3 + 16y^4$

67. $(2a-b)^6 = (2a)^6 + \binom{6}{1}(2a)^5(-b) + \binom{6}{2}(2a)^4(-b)^2 + \binom{6}{3}(2a)^3(-b)^3 + \binom{6}{4}(2a)^2(-b)^4 + \binom{6}{5}(2a)(-b)^5 + (-b)^6$

$= 64a^6 - 192a^5b + 240a^4b^2 - 160a^3b^3 + 60a^2b^4 - 12ab^5 + b^6$

69. $(x^2+y)^7 = (x^2)^7 + \binom{7}{1}(x^2)^6(y) + \binom{7}{2}(x^2)^5(y)^2 + \binom{7}{3}(x^2)^4(y)^3 + \binom{7}{4}(x^2)^3(y)^4 + \binom{7}{5}(x^2)^2(y)^5 + \binom{7}{6}(x^2)(y)^6 + y^7$

$= x^{14} + 7x^{12}y + 21x^{10}y^2 + 35x^8y^3 + 35x^6y^4 + 21x^4y^5 + 7x^2y^6 + y^7$

71. $(2a-3b)^5 = (2a)^5 + \binom{5}{1}(2a)^4(-3b) + \binom{5}{2}(2a)^3(-3b)^2 + \binom{5}{3}(2a)^2(-3b)^3 + \binom{5}{4}(2a)(-3b)^4 + (-3b)^5$

$= 32a^5 - 240a^4b + 720a^3b^2 - 1080a^2b^3 + 810ab^4 - 243b^5$

PROBLEM SET 8.5 Dividing Polynomials: Synthetic Division

1. $\dfrac{9x^4 + 18x^3}{3x}$

$\dfrac{9x^4}{3x} + \dfrac{18x^3}{3x}$

$3x^3 + 6x^2$

3. $\dfrac{-24x^6 + 36x^8}{4x^2}$

$\dfrac{-24x^6}{4x^2} + \dfrac{36x^8}{4x^2}$

$-6x^4 + 9x^6$

5. $\dfrac{15a^3 - 25a^2 - 40a}{5a}$

$\dfrac{15a^3}{5a} + \dfrac{-25a^2}{5a} + \dfrac{-40a}{5a}$

$3a^2 - 5a - 8$

7. $\dfrac{13x^3 - 17x^2 + 28x}{-x}$

$\dfrac{13x^3}{-x} + \dfrac{-17x^2}{-x} + \dfrac{28x}{-x}$

$-13x^2 + 17x - 28$

9. $\dfrac{-18x^2y^2 + 24x^3y^2 - 48x^2y^3}{6xy}$

$\dfrac{-18x^2y^2}{6xy} + \dfrac{24x^3y^2}{6xy} + \dfrac{-48x^2y^3}{6xy}$

$-3xy + 4x^2y - 8xy^2$

11.

$$\begin{array}{r|l} & 4x + 5 \\ \hline 3x - 2 & 12x^2 + 7x - 10 \\ & 12x^2 - 8x \\ \hline & 15x - 10 \\ & 15x - 10 \\ \hline & 0 \end{array}$$

Q: $4x + 5$ R: 0

13.

$$\begin{array}{r|l} & t^2 + 2t - 4 \\ \hline 3t + 1 & 3t^3 + 7t^2 - 10t - 4 \\ & 3t^3 + t^2 \\ \hline & 6t^2 - 10t - 4 \\ & 6t^2 + 2t \\ \hline & -12t - 4 \\ & -12t - 4 \\ \hline & 0 \end{array}$$

Q: $t^2 + 2t - 4$ R: 0

15.

$$\begin{array}{r|l} & 2x + 5 \\ \hline 3x + 2 & 6x^2 + 19x + 11 \\ & 6x^2 + 4x \\ \hline & 15x + 11 \\ & 15x + 10 \\ \hline & 1 \end{array}$$

Q: $2x + 5$ R: 1

17.

$$\begin{array}{r|l} & 3x - 4 \\ \hline x^2 + 2x & 3x^3 + 2x^2 - 5x - 1 \\ & 3x^3 + 6x^2 \\ \hline & -4x^2 - 5x - 1 \\ & -4x^2 - 8x \\ \hline & 3x - 1 \end{array}$$

Q: $3x - 4$ R: $3x - 1$

19.

$$\begin{array}{r|l} & 5y - 1 \\ \hline y^2 - y & 5y^3 - 6y^2 - 7y - 2 \\ & 5y^3 - 5y^2 \\ \hline & -y^2 - 7y - 2 \\ & -y^2 + y \\ \hline & -8y - 2 \end{array}$$

Q: $5y - 1$ R: $-8y - 2$

21.

$$\begin{array}{r|l} & 4a + 6 \\ \hline a^2 - 2a + 3 & 4a^3 - 2a^2 + 7a - 1 \\ & 4a^3 - 8a^2 + 12a \\ \hline & 6a^2 - 5a - 1 \\ & 6a^2 - 12a + 18 \\ \hline & 7a - 19 \end{array}$$

Q: $4a + 6$ R: $7a - 19$

23. $(3x^2 + x - 4) \div (x - 1)$

$$\begin{array}{r|rrr} 1 & 3 & 1 & -4 \\ & & 3 & 4 \\ \hline & 3 & 4 & 0 \end{array}$$

Q: $3x + 4$ R: 0

25. $(x^2 + 2x - 10) \div (x - 4)$

$$\begin{array}{r|rrr} 4 & 1 & 2 & -10 \\ & & 4 & 24 \\ \hline & 1 & 6 & 14 \end{array}$$

Q: $x + 6$ R: 14

27. $(4x^2 + 5x - 4) \div (x + 2)$

$$\begin{array}{r|rrr} -2 & 4 & 5 & -4 \\ & & -8 & 6 \\ \hline & 4 & -3 & 2 \end{array}$$

Q: $4x - 3$ R: 2

29. $(x^3 - 2x^2 - x + 2) \div (x - 2)$

2	1	-2	-1	2
		2	0	-2
	1	0	-1	0

Q: $x^2 - 1$ R: 0

31. $(3x^4 - x^3 + 2x^2 - 7x - 1) \div (x + 1)$

-1	3	-1	2	-7	-1
		-3	4	-6	13
	3	-4	6	-13	12

Q: $3x^3 - 4x^2 + 6x - 13$ R: 12

33. $(x^3 - 7x - 6) \div (x + 2)$

-2	1	0	-7	-6
		-2	4	6
	1	-2	-3	0

Q: $x^2 - 2x - 3$ R: 0

35. $(x^4 + 4x^3 - 7x - 1) \div (x - 3)$

3	1	4	0	-7	-1
		3	21	63	168
	1	7	21	56	167

Q: $x^3 + 7x^2 + 21x + 56$ R: 167

37. $(x^3 + 6x^2 + 11x + 6) \div (x + 3)$

-3	1	6	11	6
		-3	-9	-6
	1	3	2	0

Q: $x^2 + 3x + 2$ R: 0

39. $(x^5 - 1) \div (x - 1)$

1	1	0	0	0	0	-1
		1	1	1	1	1
	1	1	1	1	1	0

Q: $x^4 + x^3 + x^2 + x + 1$ R: 0

41. $(x^5 + 1) \div (x - 1)$

1	1	0	0	0	0	1
		1	1	1	1	1
	1	1	1	1	1	2

Q: $x^4 + x^3 + x^2 + x + 1$ R: 2

43. $(2x^3 + 3x^2 - 2x + 3) \div \left(x + \frac{1}{2}\right)$

$-\frac{1}{2}$	2	3	-2	3
		-1	-1	$\frac{3}{2}$
	2	2	-3	$\frac{9}{2}$

Q: $2x^2 + 2x - 3$ R: $\frac{9}{2}$

45. $(4x^4 - 5x^2 + 1) \div \left(x - \frac{1}{2}\right)$

$\frac{1}{2}$	4	0	-5	0	1
		2	1	-2	-1
	4	2	-4	-2	0

Q: $4x^3 + 2x^2 - 4x - 2$ R: 0

PROBLEM SET 8.6 Factoring: A Brief Review and a Step Further

1. $6xy - 8xy^2 =$
$2xy(3) - 2xy(4y) = 2xy(3 - 4y)$

3. $x(z + 3) + y(z + 3) =$
$(x + 3)(z + y)$

5. $3x + 3y + ax + ay =$
$3(x + y) + a(x + y) = (x + y)(3 + a)$

7. $ax - ay - bx + by =$
$a(x - y) - b(x - y) = (x - y)(a - b)$

9. $9x^2 - 25 =$
$(3x - 5)(3x + 5)$

11. $1 - 81n^2 =$
$(1 - 9n)(1 + 9n)$

13. $(x + 4)^2 - y^2$
$[(x + 4) - y][(x + 4) + y]$
$(x + 4 - y)(x + 4 + y)$

15. $9s^2 - (2t - 1)^2$
$[3s - (2t - 1)][3s + (2t - 1)]$
$(3s - 2t + 1)(3s + 2t - 1)$

17. We need two integers whose product is -14 and whose sum is -5. They are 2 and -7.
$x^2 - 5x - 14 = (x - 7)(x + 2)$

19. We need two integers whose product is -15 and whose sum is -2. They are -5 and 3.
$15-2x-x^2=15-5x+3x-x^2=$
$5(3-x)+x(3-x)=(3-x)(5+x)$

21. We need two integers whose product is -36 and whose sum is 7. No such numbers exist. $x^2+7x-36$ is not factorable.

23. We need two integers whose product is 30 and whose sum is -11. They are -6 and -5.
$3x^2-11x+10=$
$3x^2-6x-5x+10=$
$3x(x-2)-5(x-2)$
$(x-2)(3x-5)$

25. We need two integers whose product is -70 and whose sum is -33. They are -35 and 2.
$10x^2-33x-7=$
$10x^2-35x+2x-7=$
$5x(2x-7)+1(2x-7)=$
$(2x-7)(5x+1)$

27. $x^3-8=$
$(x)^3-(2)^3=$
$(x-2)(x^2+2x+4)$

29. $64x^3+27y^3=$
$(4x)^3+(3y)^3=$
$(4x+3y)(16x^2-12xy+9y^2)$

31. $4x^2+16=4(x^2+4)$

33. $x^3-9x=$
$x(x^2-9)=$
$x(x-3)(x+3)$

35. $9a^2-42a+49=$
$(3a)^2-2(3a)(7)+(7)^2=$
$(3a-7)^2$

37. $2n^3+6n^2+10n=2n(n^2+3n+5)$
[n^2+3n+5 is not factorable.]

39. We need two integers whose product is -270 and whose sum is 39. They are -6 and 45.
$10x^2+39x-27=$
$10x^2+45x-6x-27=$
$5x(2x+9)-3(2x+9)=$
$(2x+9)(5x-3)$

41. $36a^2-12a+1=$
$(6a)^2-2(6a)(1)+(1)^2=$
$(6a-1)^2$

43. We need two integers whose product is -8 and whose sum is 2. They are 4 and -2.
$8x^2+2xy-y^2=8x^2+4xy-2xy-y^2=$
$4x(2x+y)-y(2x+y)=$
$(2x+y)(4x-y)$

45. We need two integers whose product is -10 and whose sum is -1. There are no such integers. $2n^2-n-5$ is not factorable.

47. $2n^3+14n^2-20n=2n(n^2+7n-10)$
We need two integers whose product is -10 and whose sum is 7. There are no such integers. Therefore the final factored form is $2n(n^2+7n-10)$.

49. $4x^3+32=4(x^3+8)$
$(x^3+8)=(x)^3+(2)^3=$
$(x+2)(x^2-2x+4)$
$4x^3+32=4(x+2)(x^2-2x+4)$

51. $x^2+4x+3=0$
$(x+3)(x+1)=0$
$x+3=0$ or $x+1=0$
$x=-3$ or $x=-1$
The solution set is $\{-3,-1\}$.

53. $x^2+18x+72=0$
$(x+12)(x+6)=0$
$x+12=0$ or $x+6=0$
$x=-12$ or $x=-6$
The solution set is $\{-12,-6\}$.

55. $n^2-13n+36=0$
$(n-9)(n-4)=0$
$n-9=0$ or $n-4=0$
$n=9$ or $n=4$
The solution set is $\{4, 9\}$.

Problem Set 8.6

57. $x^2 + 4x - 12 = 0$
$(x + 6)(x - 2) = 0$
$x + 6 = 0$ or $x - 2 = 0$
$x = -6$ or $x = 2$
The solution set is $\{-6, 2\}$.

59. $w^2 - 4w = 5$
$w^2 - 4w - 5 = 0$
$(w - 5)(w + 1) = 0$
$w - 5 = 0$ or $w + 1 = 0$
$w = 5$ or $w = -1$
The solution set is $\{-1, 5\}$.

61. $n^2 + 25n + 156 = 0$
$(n + 12)(n + 13) = 0$
$n + 12 = 0$ or $n + 13 = 0$
$n = -12$ or $n = -13$
The solution set is $\{-13, -12\}$.

63. $3t^2 + 14t - 5 = 0$
$(3t - 1)(t + 5) = 0$
$3t - 1 = 0$ or $t + 5 = 0$
$3t = 1$ or $t = -5$
$t = \frac{1}{3}$ or $t = -5$
The solution set is $\left\{-5, \frac{1}{3}\right\}$.

65. $6x^2 + 25x + 14 = 0$
$(2x + 7)(3x + 2) = 0$
$2x + 7 = 0$ or $3x + 2 = 0$
$2x = -7$ or $3x = -2$
$x = -\frac{7}{2}$ or $x = -\frac{2}{3}$
The solution set is $\left\{-\frac{7}{2}, -\frac{2}{3}\right\}$.

67. $3t(t - 4) = 0$
$3t = 0$ or $t - 4 = 0$
$t = 0$ or $t = 4$
The solution set is $\{0, 4\}$.

69. $-6n^2 + 13n - 2 = 0$
$-1(-6n^2 + 13n - 2) = -1(0)$
$6n^2 - 13n + 2 = 0$
$(6n - 1)(n - 2) = 0$
$6n - 1 = 0$ or $n - 2 = 0$
$6n = 1$ or $n = 2$
$n = \frac{1}{6}$ or $n = 2$
The solution set is $\left\{\frac{1}{6}, 2\right\}$.

71. $2n^3 = 72n$
$2n^3 - 72n = 0$
$2n(n^2 - 36) = 0$
$2n(n + 6)(n - 6) = 0$
$2n = 0$ or $n + 6 = 0$ or $n - 6 = 0$
$n = 0$ or $n = -6$ or $n = 6$
The solution set is $\{-6, 0, 6\}$.

73. $(x - 5)(x + 3) = 9$
$x^2 - 2x - 15 = 9$
$x^2 - 2x - 24 = 0$
$(x - 6)(x + 4) = 0$
$x - 6 = 0$ or $x + 4 = 0$
$x = 6$ or $x = -4$
The solution set is $\{-4, 6\}$.

75. $9x^2 - 6x + 1 = 0$
$(3x - 1)(3x - 1) = 0$
$3x - 1 = 0$ or $3x - 1 = 0$
$3x = 1$ or $3x = 1$
$x = \frac{1}{3}$ or $x = \frac{1}{3}$
The solution set is $\left\{\frac{1}{3}\right\}$.

77. $n^2 + 7n - 44 = 0$
$(n + 11)(n - 4) = 0$
$n + 11 = 0$ or $n - 4 = 0$
$n = -11$ or $n = 4$
The solution set is $\{-11, 4\}$.

79. $3x^2 = 75$
$3x^2 - 75 = 0$
$3(x^2 - 25) = 0$
$3(x + 5)(x - 5) = 0$
$x + 5 = 0$ or $x - 5 = 0$
$x = -5$ or $x = 5$
The solution set is $\{-5, 5\}$.

81. $\begin{pmatrix}\text{Volume}\\ \text{of Sphere}\end{pmatrix} = 2 \cdot \begin{pmatrix}\text{Surface Area}\\ \text{of Sphere}\end{pmatrix}$
$\frac{4}{3}\pi r^3 = 2(4\pi r^2)$

$\frac{4}{3}\pi r^3 = 8\pi r^2$
$3\left(\frac{4}{3}\pi r^3\right) = 3(8\pi r^2)$
$4\pi r^3 = 24\pi r^2$
$\frac{4\pi r^3}{4\pi} = \frac{24\pi r^2}{4\pi}$
$r^3 = 6r^2$
$r^3 - 6r^2 = 0$
$r^2(r-6) = 0$
$r^2 = 0$ or $r - 6 = 0$
$r = 0$ or $r = 6$
Discard $r = 0$ because length cannot be 0. Therefore the radius is 6 units.

83. Let $x =$ 1st integer and
$2x - 3 =$ 2nd integer.
$x(2x-3) = 104$
$2x^2 - 3x = 104$
$2x^2 - 3x - 104 = 0$
$(2x+13)(x-8) = 0$
$2x + 13 = 0$ or $x - 8 = 0$
$2x = -13$ or $x = 8$
$x = -\frac{13}{2}$
Discard the root $x = -\frac{13}{2}$ because it is not an integer.
1st integer $= 8$
2nd integer $= 2(8) - 3 = 13$
The integers are 8 and 13.

85. Let $x =$ 1st leg,
$x + 2 =$ 2nd leg, and
$x + 4 =$ hypotenuse.
Use the Pythagorean Theorem.
$x^2 + (x+2)^2 = (x+4)^2$
$x^2 + x^2 + 4x + 4 = x^2 + 8x + 16$
$2x^2 + 4x + 4 = x^2 + 8x + 16$
$x^2 - 4x - 12 = 0$
$(x-6)(x+2) = 0$
$x - 6 = 0$ or $x + 2 = 0$
$x = 6$ or $x = -2$
Discard the root $x = -2$.
The sides are 6, 8, and 10 units.

87. Let r = the radius of the base of the cylinder, then the altitude $= 2r$.

$\text{Surface Area of Right Circular Cylinder} = 2\pi r^2 + 2\pi rh$
$54\pi = 2\pi r^2 + 2\pi rh$
$54\pi = 2\pi r^2 + 2\pi r(2r)$
$54\pi = 2\pi r^2 + 4\pi r^2$
$54\pi = 6\pi r^2$
$0 = 6\pi r^2 - 54\pi$
$0 = 6\pi(r^2 - 9)$
$0 = r^2 - 9$
$0 = (r+3)(r-3)$
$r + 3 = 0$ or $r - 3 = 0$
$r = -3$ or $r = 3$
Discard $r = -3$ because length cannot be negative. Therefore the radius is 6 inches.

89. Let x = side of square,
$x + 2$ = width of rectangle, and
$x + 4$ = length of rectangle
Area of square $= x^2$
Area of rectangle$=(x+2)(x+4)=x^2+6x+8$
$x^2 + x^2 + 6x + 8 = 64$
$2x^2 + 6x + 8 = 64$
$2x^2 + 6x - 56 = 0$
$2(x^2 + 3x - 28) = 0$
$2(x+7)(x-4) = 0$
$x + 7 = 0$ or $x - 4 = 0$
$x = -7$ or $x = 4$
Discard the root $x = -7$.
The length of the side of the square is 4 centimeters. The dimensions of the rectangle are 6 centimeters by 8 centimeters.

CHAPTER 8 Review Problem Set

1.
$$\begin{aligned} 5(x-6) &= 3(x+2) \\ 5x-30 &= 3x+6 \\ 2x-30 &= 6 \\ 2x &= 36 \\ x &= 18 \end{aligned}$$
The solution set is $\{18\}$.

2.
$$\begin{aligned} 2(2x+1)-(x-4) &= 4(x+5) \\ 4x+2-x+4 &= 4x+20 \\ 3x+6 &= 4x+20 \\ -x+6 &= 20 \\ -x &= 14 \\ x &= -14 \end{aligned}$$
The solution set is $\{-14\}$.

3.
$$\begin{aligned} -(2n-1)+3(n+2) &= 7 \\ -2n+1+3n+6 &= 7 \\ n+7 &= 7 \\ n &= 0 \end{aligned}$$
The solution set is$\{0\}$.

4.
$$\begin{aligned} 2(3n-4)+3(2n-3) &= -2(n+5) \\ 6n-8+6n-9 &= -2n-10 \\ 12n-17 &= -2n-10 \\ 14n &= 7 \\ n &= \frac{7}{14} = \frac{1}{2} \end{aligned}$$
The solution set is $\left\{\frac{1}{2}\right\}$.

5.
$$\begin{aligned} \frac{3t-2}{4} &= \frac{2t+1}{3} \\ 3(3t-2) &= 4(2t+1) \\ 9t-6 &= 8t+4 \\ t-6 &= 4 \\ t &= 10 \end{aligned}$$
The solution set is $\{10\}$.

6.
$$\begin{aligned} \frac{x+6}{5}+\frac{x-1}{4} &= 2 \\ 20\left(\frac{x+6}{5}+\frac{x-1}{4}\right) &= 20(2) \\ 4(x+6)+5(x-1) &= 40 \\ 4x+24+5x-5 &= 40 \\ 9x+19 &= 40 \\ 9x &= 21 \\ x &= \frac{21}{9} = \frac{7}{3} \end{aligned}$$
The solution set is $\left\{\frac{7}{3}\right\}$.

7.
$$\begin{aligned} 1-\frac{2x-1}{6} &= \frac{3x}{8} \\ 24\left(1-\frac{2x-1}{6}\right) &= 24\left(\frac{3x}{8}\right) \\ 24-4(2x-1) &= 3(3x) \\ 24-8x+4 &= 9x \\ 28-8x &= 9x \\ 28 &= 17x \\ \frac{28}{17} &= x \end{aligned}$$
The solution set is $\left\{\frac{28}{17}\right\}$.

8.
$$\begin{aligned} \frac{2x+1}{3}+\frac{3x-1}{5} &= \frac{1}{10} \\ 30\left(\frac{2x+1}{3}+\frac{3x-1}{5}\right) &= 30\left(\frac{1}{10}\right) \\ 10(2x+1)+6(3x-1) &= 3(1) \\ 20x+10+18x-6 &= 3 \\ 38x+4 &= 3 \\ 38x &= -1 \\ x &= \left\{-\frac{1}{38}\right\} \end{aligned}$$
The solution set is $\left\{-\frac{1}{38}\right\}$.

9.

$$\frac{3n-1}{2}-\frac{2n+3}{7}=1$$
$$14\left(\frac{3n-1}{2}-\frac{2n+3}{7}\right)=14(1)$$
$$7(3n-1)-2(2n+3)=14$$
$$21n-7-4n-6=14$$
$$17n-13=14$$
$$17n=27$$
$$n=\frac{27}{17}$$

The solution set is $\left\{\frac{27}{17}\right\}$.

10. $|3x-1|=11$ is equivalent to $3x-1=11$ or $3x-1=-11$.

$3x-1=-11$ or $3x-1=11$

$3x=-10$ or $3x=12$

$x=-\frac{10}{3}$ or $x=4$

The solution set is $\left\{-\frac{10}{3},4\right\}$.

11.

$$0.06x+0.08(x+100)=15$$

Multiply both sides by 100.

$$100[0.06x+0.08(x+100)]=100(15)$$
$$6x+8(x+100)=1500$$
$$6x+8x+800=1500$$
$$14x+800=1500$$
$$14x=700$$
$$x=50$$

The solution set is $\{50\}$.

12.

$$0.4(t-6)=0.3(2t+5)$$

Multiply both sides by 100.

$$10[0.4(t-6)]=10[0.3(2t+5)]$$
$$4(t-6)=3(2t+5)$$
$$4t-24=6t+15$$
$$-2t=39$$
$$t=-\frac{39}{2}$$

The solution set is $\left\{-\frac{39}{2}\right\}$.

13.

$$0.1(n+300)=0.09n+32$$

Multiply both sides by 100.

$$100[0.1(n+300)]=100(0.09n+32)$$
$$10(n+300)=9n+3200$$
$$10n+3000=9n+3200$$
$$n+3000=3200$$
$$n=200$$

The solution set is $\{200\}$.

14.

$$0.2(x-0.5)-0.3(x+1)=0.4$$

Multiply both sides by 10.

$$10[0.2(x-0.5)-0.3(x+1)]=10(0.4)$$
$$2(x-0.5)-3(x+1)=4$$
$$2x-1-3x-3=4$$
$$-x-4=4$$
$$-x=8$$
$$x=-8$$

The solution set is $\{-8\}$.

15.

$$4x^2-36=0$$
$$4(x^2-9)=0$$
$$x^2-9=0$$
$$(x-3)(x+3)=0$$

$x-3=0$ or $x+3=0$

$x=3$ or $x=-3$

The solution set is $\{-3, 3\}$.

16.

$$x^2+5x-6=0$$
$$(x-1)(x+6)=0$$

$x-1=0$ or $x+6=0$

$x=1$ or $x=-6$

The solution set is $\{-6,1\}$.

17.

$$49n^2-28n+4=0$$
$$(7n-2)(7n-2)=0$$

$7n-2=0$ or $7n-2=0$

$7n=2$ or $7n=2$

$n=\frac{2}{7}$ or $n=\frac{2}{7}$

The solution set is $\left\{\frac{2}{7}\right\}$.

18. $(3x-1)(5x+2)=0$
$3x-1=0$ or $5x+2=0$
$3x=1$ or $5x=-2$
$x=\frac{1}{3}$ or $x=-\frac{2}{5}$
The solution set is $\left\{-\frac{2}{5},\frac{1}{3}\right\}$.

19. $(3x-4)^2-25=0$
$[(3x-4)-5][(3x-4)+5]=0$
$(3x-9)(3x+1)=0$
$3x-9=0$ or $3x+1=0$
$3x=9$ or $3x=-1$
$x=3$ or $x=-\frac{1}{3}$
The solution set is $\left\{-\frac{1}{3},3\right\}$.

20. $6a^3=54a$
$6a^3-54a=0$
$6a\left(a^2-9\right)=0$
$6a(a-3)(a+3)=0$
$6a=0$ or $a-3=0$ or $a+3=0$
$a=0$ or $a=3$ or $a=-3$
The solution set is $\{-3,0,3\}$.

21. $7n(7n+2)=8$
$49n^2+14n=8$
$49n^2+14n-8=0$
$(7n+4)(7n-2)=0$
$7n+4=0$ or $7n-2=0$
$7n=-4$ or $7n=2$
$n=-\frac{4}{7}$ or $n=\frac{2}{7}$
The solution set is $\left\{-\frac{4}{7},\frac{2}{7}\right\}$.

22. $30w^2-w-20=0$
$(6w-5)(5w+4)=0$
$6w-5=0$ or $5w+4=0$
$6w=5$ or $5w=-4$
$w=\frac{5}{6}$ or $x=-\frac{4}{5}$
The solution set is $\left\{-\frac{4}{5},\frac{5}{6}\right\}$.

23. $3t^3-27t^2+24t=0$
$t^3-9t^2+8t=0$
$t(t-8)(t-1)=0$
$t=0$ or $t-8=0$ or $t-1=0$
$t=0$ or $t=8$ or $t=1$
The solution set is $\{0,1,8\}$.

24.. $-4n^2-39n+10=0$
$4n^2+39n-10=0$
$(4n-1)(n+10)=0$
$4n-1=0$ or $n+10=0$
$4n=1$ or $n=-10$
$n=\frac{1}{4}$ or
The solution set is $\left\{-10,\frac{1}{4}\right\}$.

25. $ax-b=b+2$
Add b to both sides.
$ax=2b+2$
Divide both sides by a.
$x=\frac{2b+2}{a}$

26. $ax=bx+c$
Subtract bx from both sides.
$ax-bx=c$
Factor x on the left side.
$x(a-b)=c$
Divide both sides by $(a-b)$.
$x=\frac{c}{a-b}$

27. $m(x+a)=p(x+b)$
Distribute on both sides.
$mx+ma=px+pb$
Subtract px from both sides.
$mx+ma-px=pb$
Subtract ma from both sides.
$mx-px=pb-ma$
Factor x on the left side.
$x(m-p)=pb-ma$
Divide both sides by $(m-p)$.
$x=\frac{pb-ma}{m-p}$

28. $5x - 7y = 11$

Add $7y$ to both sides.

$$5x = 11 + 7y$$

Divide both sides by 5.

$$x = \frac{11 + 7y}{5}$$

29. $\frac{x - a}{b} = \frac{y + 1}{c}$

Cross products are equal.

$c(x - a) = b(y + 1)$

Distributive property.

$cx - ac = by + b$

Add ac to both sides.

$$cx = by + b + ac$$

Divide both sides by c.

$$x = \frac{by + b + ac}{c}$$

30. $5x - 2 \geq 4x - 7$

$x - 2 \geq -7$

$x \geq -5$

The solution set is

$\{x|x \geq -5\}$ or $[-5, \infty)$.

31. $3 - 2x < -5$

$-2x < -8$

$x > 4$

The solution set is

$\{x|x > 4\}$ or $(4, \infty)$.

32. $2(3x - 1) - 3(x - 3) > 0$

$6x - 2 - 3x + 9 > 0$

$3x + 7 > 0$

$3x > -7$

$x > -\frac{7}{3}$

The solution set is

$\left\{x|x > -\frac{7}{3}\right\}$ or $\left(-\frac{7}{3}, \infty\right)$.

33. $3(x + 4) \leq 5(x - 1)$

$3x + 12 \leq 5x - 5$

$-2x + 12 \leq -5$

$-2x \leq -17$

$x \geq \frac{-17}{-2}$

$x \geq \frac{17}{2}$

The solution set is

$\left\{x|x \geq \frac{17}{2}\right\}$ or $\left[\frac{17}{2}, \infty\right)$.

34. $\frac{5}{6}n - \frac{1}{3}n < \frac{1}{6}$

$6\left(\frac{5}{6}n - \frac{1}{3}n\right) < 6\left(\frac{1}{6}\right)$

$5n - 2n < 1$

$3n < 1$

$\frac{3n}{3} < \frac{1}{3}$

$n < \frac{1}{3}$

The solution set is

$\left\{n|n < \frac{1}{3}\right\}$ or $\left(-\infty, \frac{1}{3}\right)$.

35. $\frac{n - 4}{5} + \frac{n - 3}{6} > \frac{7}{15}$

$30\left(\frac{n - 4}{5} + \frac{n - 3}{6}\right) > 30\left(\frac{7}{15}\right)$

$6(n - 4) + 5(n - 3) > 2(7)$

$6n - 24 + 5n - 15 > 14$

$11n - 39 > 14$

$11n > 53$

$n > \frac{53}{11}$

The solution set is

$\left\{n|n > \frac{53}{11}\right\}$ or $\left(\frac{53}{11}, \infty\right)$.

36. $s \geq 4.5 + 0.25s$
$100(s) \geq 100(4.5 + 0.25s)$
$100s \geq 450 + 25s$
$75s \geq 450$
$s \geq 6$
The solution set is
$\{s|s > 6\}$ or $[6, \infty)$.

37. $0.07x + 0.09(500 - x) \geq 43$
$100[0.07x + 0.09(500 - x)] \geq 100(43)$
$7x + 9(500 - x) \geq 4300$
$7x + 4500 - 9x \geq 4300$
$-2x + 4500 \geq 4300$
$-2x \geq -200$
$x \leq 100$
The solution set is
$\{x|x \leq 100\}$ or $(-\infty, 100]$.

38. $|2x - 1| < 11$
$-11 < 2x - 1 < 11$
$-10 < 2x < 12$
$-5 < x < 6$
The solution set is $(-5, 6)$.

39. $|3x + 1| > 10$
$3x + 1 < -10$ or $3x + 1 > 10$
$3x < -11$ or $3x > 9$
$x < -\frac{11}{3}$ or $x > 3$
The solution set is $\left(-\infty, -\frac{11}{3}\right) \cup (3, \infty)$.

40. $x > -1$ and $x < 1$

-7 -6 -5 -4 -3 -2 -1 0 1 2 3 4 5 6

41. $x > 2$ or $x \leq -3$

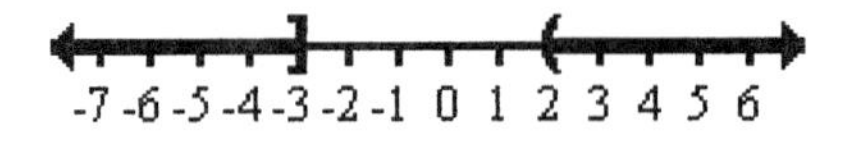

42. $x > 2$ and $x > 3$

-7 -6 -5 -4 -3 -2 -1 0 1 2 3 4 5 6

43. $x < 2$ or $x > -1$

-7 -6 -5 -4 -3 -2 -1 0 1 2 3 4 5 6

44. $(3x - 2) + (4x - 6) + (-2x + 5)$
$3x - 2 + 4x - 6 - 2x + 5$
$3x + 4x - 2x - 2 - 6 + 5$
$5x - 3$

45. $(8x^2 + 9x - 3) - (5x^2 - 3x - 1)$
$8x^2 + 9x - 3 - 5x^2 + 3x + 1$
$3x^2 + 12x - 2$

46. $(6x^2 - 2x - 1) + (4x^2 + 2x + 5)$
$-(-2x^2 + x - 1)$

$6x^2 - 2x - 1 + 4x^2 + 2x + 5$
$+ 2x^2 - x + 1$

$6x^2 + 4x^2 + 2x^2 - 2x + 2x - x$
$-1 + 5 + 1$

$12x^2 - x + 5$

47. $(-5x^2y^3)(4x^3y^4)$
$-20x^{2+3}y^{3+4}$
$-20x^5y^7$

48. $(-2a^2)(3ab^2)(a^2b^3)$
$-6a^{2+1+2}b^{2+3}$
$-6a^5b^5$

49. $5a^2(3a^2 - 2a - 1)$
$15a^4 - 10a^3 - 5a^2$

50. $(4x - 3y)(6x + 5y)$
$24x^2 + 20xy - 18xy - 15y^2$
$24x^2 + 2xy - 15y^2$

51. $(x+4)(3x^2-5x-1)$
$x(3x^2-5x-1)+4(3x^2-5x-1)$
$3x^3-5x^2-x+12x^2-20x-4$
$3x^3+7x^2-21x-4$

52. $(4x^2y^3)^4$
$4^4x^{2(4)}y^{3(4)}$
$256x^8y^{12}$

53. $(3x-2y)^2$
$(3x)^2+2(3x)(-2y)+(-2y)^2$
$9x^2-12xy+4y^2$

54. $(-2x^2y^3z)^3$
$(-2)^3x^{2(3)}y^{3(3)}z^3$
$-8x^6y^9z^3$

55. $\dfrac{-39x^3y^4}{3xy^3}$
$-13x^{3-1}y^{4-3}$
$-13x^2y$

56. $[3x-(2x-3y+1)]-[2y-(x-1)]$
$[3x-2x+3y-1]-[2y-x+1]$
$[x+3y-1]-2y+x-1$
$x+3y-1-2y+x-1$
$2x+y-2$

57. $(x^2-2x-5)(x^2+3x-7)$
$x^2(x^2+3x-7)-2x(x^2+3x-7)-5(x^2+3x-7)$
$x^4+3x^3-7x^2-2x^3-6x^2+14x-5x^2-15x+35$
$x^4+x^3-18x^2-x+35$

58. $(7-3x)(3+5x)$
$21+35x-9x-15x^2$
$21+26x-15x^2$

59. $-(3ab)(2a^2b^3)^2$
$-(3ab)(2^2a^4b^6)$
$-(3ab)(4a^4b^6)$
$-12a^5b^7$

60. $\left(\dfrac{1}{2}ab\right)(8a^3b^2)(-2a^3)$
$-8a^{1+3+3}b^{1+2}$
$-8a^7b^3$

61. $(7x-9)(x+4)$
$7x^2+28x-9x-36$
$7x^2+19x-36$

62. $(3x+2)(2x^2-5x+1)$
$3x(2x^2-5x+1)+2(2x^2-5x+1)$
$6x^3-15x^2+3x+4x^2-10x+2$
$6x^3-11x^2-7x+2$

63. $(3x^{n+1})(2x^{3n-1})$
$6x^{n+1+3n-1}$
$6x^{4n}$

64. $(2x+5y)^2$
$(2x)^2+2(2x)(5y)+(5y)^2$
$4x^2+20xy+25y^2$

65. $(x-2)^3$
$(x-2)[(x-2)(x-2)]$
$(x-2)(x^2-4x+4)$
$x(x^2-4x+4)-2(x^2-4x+4)$
$x^3-4x^2+4x-2x^2+8x-8$
$x^3-6x^2+12x-8$

66. $\left(3x^3-10x^2+2x+41\right)\div\left(x-2\right)$

$$\begin{array}{r|rrrr} 2 & 3 & -10 & 2 & 41 \\ & & 6 & -8 & -12 \\ \hline & 3 & -4 & -6 & 29 \end{array}$$

Q: $3x^2-4x-6$ R: 29

67. $\left(5x^3+8x^2+x+6\right)\div\left(x+1\right)$

$$\begin{array}{r|rrrr} -1 & 5 & 8 & 1 & 6 \\ & & -5 & -3 & 2 \\ \hline & 5 & 3 & -2 & 8 \end{array}$$

Q: $5x^2+3x-2$ R: 8

68. $\left(x^4-x^3-19x^2-22x-3\right)\div\left(x+3\right)$

$$\begin{array}{r|rrrrr} -3 & 1 & -1 & -19 & -22 & -3 \\ & & -3 & 12 & 21 & 3 \\ \hline & 1 & -4 & -7 & -1 & 0 \end{array}$$

Q: x^3-4x^2-7x-1 R: 0

69. $\left(2x^4-5x^3-16x^2+14x+8\right)\div\left(x-4\right)$

$$\begin{array}{r|rrrrr} 4 & 2 & -5 & -16 & 14 & 8 \\ & & 8 & 12 & -16 & -8 \\ \hline & 2 & 3 & -4 & -2 & 0 \end{array}$$

Q: $2x^3+3x^2-4x-2$ R: 0

70. $x^2 + 3x - 28$
$(x + 7)(x - 4)$

71. $2t^2 - 18$
$2(t^2 - 9)$
$2(t + 3)(t - 3)$

72. $4n^2 + 9$
Not factorable

73. $12n^2 - 7n + 1$
$12(1) = 12$
We need two integers whose product is 12 and whose sum is -7. They are -3 and -4.
$12n^2 - 3n - 4n + 1$
$3n(4n - 1) - 1(4n - 1)$
$(4n - 1)(3n - 1)$

74. $x^6 - x^2$
$x^2(x^4 - 1)$
$x^2(x^2 + 1)(x^2 - 1)$
$x^2(x^2 + 1)(x + 1)(x - 1)$

75. $x^3 - 6x^2 - 72x$
$x(x^2 - 6x - 72)$
$x(x - 12)(x + 6)$

76. $6a^3b + 4a^2b^2 - 2a^2bc$
$2a^2b(3a + 2b - c)$

77. $x^2 - (y - 1)^2$
$[x + (y - 1)][x - (y - 1)]$
$(x + y - 1)(x - y + 1)$

78. $8x^2 + 12$
$4(2x^2 + 3)$
[$2x^2 + 3$ is not factorable.]

79. $12x^2 + x - 35$
$12(-35) = -420$
We need two integers whose product is -420 and whose sum is 1. They are 21 and -20.
$12x^2 + 21x - 20x - 35$
$3x(4x + 7) - 5(4x + 7)$
$(4x + 7)(3x - 5)$

80. $16n^2 - 40n + 25$
$(4n)^2 - 2(4n)(5) + (5)^2$
$(4n - 5)^2$

81. $4n^2 - 8n$
$4n(n - 2)$

82. $3w^3 + 18w^2 - 24w$
$3w(w^2 + 6w - 8)$
[$w^2 + 6w - 8$ is not factorable.]

83. $20x^2 + 3xy - 2y^2$
$20(-2) = -40$
We need two integers whose product is -40 and whose sum is 3. They are 8 and -5.
$20x^2 + 8xy - 5xy - 2y^2$
$4x(5x + 2y) - y(5x + 2y)$
$(5x + 2y)(4x - y)$

84. $16a^2 - 64a$
$16a(a - 4)$

85. $3x^3 - 15x^2 - 18x$
$3x(x^2 - 5x - 6)$
$3x(x - 6)(x + 1)$

86. $n^2 - 8n - 128$
$(n + 8)(n - 16)$

87. $t^4 - 22t^2 - 75$
$(t^2 - 25)(t^2 + 3)$
$(t + 5)(t - 5)(t^2 + 3)$

88. $35x^2 - 11x - 6$
We need two integers whose product is -210 and whose sum is -11.
They are -21 and 10.
$35x^2 - 21x + 10x - 6$
$7x(5x - 3) + 2(5x - 3)$
$(5x - 3)(7x + 2)$

89. $15 - 14x + 3x^2$
$15(3) = 45$
We need two integers whose product is 45 and whose sum is -14. They are -9 and -5.
$15 - 9x - 5x + 3x^2$
$3(5 - 3x) - x(5 - 3x)$
$(5 - 3x)(3 - x)$

90. $64n^3 - 27$
$(4n)^3 - (3)^3$
$(4n - 3)(16n^2 + 12n + 9)$

91. $16x^3 + 250$
$2(8x^3 + 125)$
$2(2x + 5)(4x^2 - 10x + 25)$

92. Let l represent the length of the rectangle.
Let w represent the width of the rectangle.

$w = 2 + \frac{1}{3}l$ The width is 2 more than one-third of the length.

$2l + 2w = 44$ The perimeter is 44 meters.

Solve the system.

$$\begin{pmatrix} 2l + 2w = 44 \\ w = 2 + \frac{1}{3}l \end{pmatrix}$$

Substitute $2 + \frac{1}{3}l$ for w in $2l + 2w = 44$.

$$2l + 2\left(2 + \frac{1}{3}l\right) = 44$$
$$2l + 4 + \frac{2}{3}l = 44$$
$$2l + \frac{2}{3}l = 40$$
$$3\left(2l + \frac{2}{3}l\right) = 3(40)$$
$$6l + 2l = 120$$
$$8l - 120$$
$$l = 15$$

Substitute 15 for l in $w = 2 + \frac{1}{3}l$.

$$w = 2 + \frac{1}{3}\left(15\right) = 2 + 5 = 7$$

The length is 15 meters and the width is 7 meters.

93. Let x represent the money invested at 7%.
Let y represent the money invested at 8%.

$x + y = 5000$ Total \$5000 was invested.

$0.07x + 0.08y = 380$ The total yearly interest was \$380.

Solve first equation for y and substitute into second equation.

$y = 5000 - x$
$0.07x + 0.08y = 380$

$$0.07x + 0.08(5000 - x) = 380$$
$$100\left[0.07x + 0.08(5000 - x)\right] = 100(380)$$
$$7x + 8(5000 - x) = 38000$$
$$7x + 40000 - 8x = 38000$$
$$-x + 40000 = 38000$$
$$-x = -2000$$
$$x = 2000$$

\$2,000 was invested at 7% and
$5000 - 2000 = \$3,000$ was invested at 8%.

94. Let x represent the score on the fourth exam.

$$\frac{3(84) + x}{4} \geq 85$$
$$4\left(\frac{252 + x}{4}\right) \geq 4(85)$$
$$252 + x \geq 340$$
$$x \geq 88$$

She would need a score of 88 or better.

95. Let $x =$ 1st integer,
$x + 1 =$ 2nd integer, and
$x + 2 =$ 3rd integer.

$$\frac{1}{2}x + \frac{1}{3}(x + 2) = x + 1 - 1$$
$$\frac{1}{2}x + \frac{1}{3}(x + 2) = x$$
$$6\left[\frac{1}{2}x + \frac{1}{3}(x + 2)\right] = 6(x)$$
$$6\left(\frac{1}{2}x\right) + 6\left[\frac{1}{3}(x + 2)\right] - 6x$$
$$3x + 2(x + 2) = 6x$$
$$3x + 2x + 4 = 6x$$
$$5x + 4 = 6x$$
$$4 = x$$

The first integer is 4, the second integer is 5, and the third integer is 6.

96. Let $x =$ normal hourly rate and
$\frac{3}{2}x =$ over 36 hours rate.

Pat worked 36 hours at the normal hourly rate and 6 hours at the overtime rate.

$36x + 6\left(\frac{3}{2}x\right) = 472.50$
$36x + 9x = 472.50$
$45x = 472.50$
$x = 10.50$
The normal hourly rate is \$10.50.

97. Let $x =$ the number of nickels,
$2x + 10 =$ the number of dimes, and
$2x + 10 + 25 =$ the number of quarters.

$0.05(x)$+$0.10(2x$+$10)$+$0.25(2x$+10+$25) = 24.75$

$0.05(x)$+$0.10(2x$+$10)$+$0.25(2x$+$35) = 24.75$

$100[0.05$+$0.10(2x$+$10)$+$.025(2x$+$35)]$=$100(24.75)$

$100(0.05x)$+$100[0.10(2x$+$10)]$+$100[0.25(2x$+$35)]$=2475

$5x + 10(2x + 10) + 25(2x + 35) = 2475$
$5x + 20x + 100 + 50x + 875 = 2475$
$75x + 975 = 2475$
$75x = 1500$
$x = 20$
The number of nickels $= 20$, the number of dimes $= 2(20) + 10 = 50$, and the number of quarters $= 2(20) + 10 + 25 = 75$.
There are 20 nickels, 50 dimes, and 75 quarters.

98. Let $x =$ angle
$90 - x =$ complement of the angle.
$180 - x =$ supplement of the angle.
$90 - x = \frac{1}{10}(180 - x)$
$10(90 - x) = 10\left[\frac{1}{10}(180 - x)\right]$
$900 - 10x = 1(180 - x)$
$900 - 10x = 180 - x$
$-10x = -720 - x$
$-9x = -720$
$x = 80$
The angle is 80°.

99. Price = cost + 20%(cost)
Price $= 38 + (20\%)(38)$
Price $= 38 + 0.20(38)$
Price $= 38 + 7.6$
Price $= 45.60$
The selling price should be \$45.60.

100. Let x represent the score on 5th game.
$\frac{16 + 22 + 18 + 14 + x}{5} \geq 20$
$\frac{70 + x}{5} \geq 20$
$5\left(\frac{70 + x}{5}\right) \geq 5(20)$
$70 + x \geq 100$
$x \geq 30$
She should score 30 points or more.

101. Reena's time $= 5$ hours 20 minutes $= 5\frac{1}{3}$ hr.
Gladys's time $= 5\frac{1}{3} + 2 = 7\frac{1}{3}$ hr.
Let $x =$ Reena's rate.

Distance = rate · time.
Reena's distance $= 5\left(\frac{1}{3}\right)(x) = \frac{16}{3}x$.
Gladys's distance $= \left(7\frac{1}{3}\right)(40) = \frac{22}{3}(40)$.
Reena's distance = Gladys's distance.
$\frac{16}{3}x = \frac{22}{3}(40)$
$3\left(\frac{16}{3}x\right) = 3\left[\frac{22}{3}(40)\right]$
$16x = 22(40)$
$16x = 880$
$x = 55$
Reena's rate is 55 miles per hour.

102. Let $t =$ time Sonya rode then
$t + \frac{5}{4} =$ time Rita rode.
Distance = rate · time.
Sonya's distance = 16t.
Rita's distance $= 12\left(t + \frac{5}{4}\right)$.
Rita's distance = Sonya's distance + 2

$12\left(t+\frac{5}{4}\right)=16t+2$
$12t+12\left(\frac{5}{4}\right)=16t+2$
$12t+15=16t+2$
$12t=16t-13$
$-4t=-13$
$t=\frac{-13}{-4}=3\frac{1}{4}$
Sonya's time $=3\frac{1}{4}$ hours.
Rita's time $=\frac{13}{4}+\frac{5}{4}=\frac{18}{4}=4\frac{1}{2}$ hours.
Sonya rides for $3\frac{1}{4}$ hours and
Rita rides for $4\frac{1}{2}$ hours.

103. Let $x =$ number of cups of orange juice.
$(100\%)(x)+(10\%)(50)=(20\%)(x+50)$
$1.00x+0.10(50)=0.20(x+50)$
$100[1x+0.10(50)]=100[0.20(x+50)]$
$100(1x)+100[0.10(50)]=20(x+50)$
$100x+10(50)=20x+1000$
$100x+500=20x+1000$
$100x=20x+500$
$80x=500$
$x=6.25$
$x=6\frac{1}{4}$
There should be $6\frac{1}{4}$ cups of orange juice.

104. Let $x =$ distance of the northbound car and $x+4 =$ distance of the eastbound car.

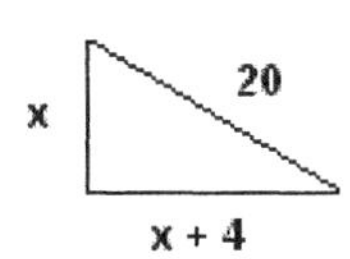

Use the Pythagorean theorem.
$x^2+(x+4)^2=20^2$
$x^2+x^2+8x+16=400$
$2x^2+8x-384=0$
$2(x^2+4x-192)=0$
$2(x+16)(x-12)=0$
$x+16=0$ or $x-12=0$
$x=-16$ or $x=12$
Discard the root $x=-16$.
The northbound car traveled 12 miles and the eastbound car traveled 16 miles.

105. Use $P=2l+2w$.
Let $x =$ width.
$32=2l+2x$
$32-2x=2l$
$16-x=l=$ length
Use $A=lw$.
$(16-x)x=48$
$16x-x^2=48$
$x^2-16x+48=0$
$(x-4)(x-12)=0$
$x-4=0$ or $x-12=0$
$x=4$ or $x=12$
The width is 4 meters and the length is 12 meters.

106. Let $x =$ number of rows and $2x-2 =$ number of chairs per row.
$x(2x-2)=144$
$2x^2-2x=144$
$2x^2-2x-144=0$
$2(x^2-x-72)=0$
$2(x-9)(x+8)=0$
$x-9=0$ or $x+8=0$
$x=9$ or $x=-8$
Discard the root $x=-8$.
Number of rows $=9$.
Number of chairs per row $=2(9)-2=16$.
There are 9 rows of 16 chairs per row.

107. Let $x =$ altitude and $2x+1 =$ side (base).
Use $A=\frac{1}{2}bh$.
$\frac{1}{2}x\left(2x+1\right)=39$
$2\left[\frac{1}{2}x(2x+1)\right]=2(39)$
$x(2x+1)=78$
$2x^2+x=78$
$2x^2+x-78=0$
$(2x+13)(x-6)=0$
$2x+13=0$ or $x-6=0$
$2x=-13$ or $x=6$
$x=-\frac{13}{2}$

Discard the root $x = -\frac{13}{2}$.
The length of the altitude is 6 feet and the side is $2(6) + 1 = 13$ feet.

CHAPTER 8 | Test

1. $(-3x - 1) + (9x - 2) - (4x + 8)$
$-3x - 1 + 9x - 2 - 4x - 8$
$-3x + 9x - 4x - 1 - 2 - 8$
$2x - 11$

2. $(5x - 7)(4x + 9)$
$20x^2 + 45x - 28x - 63$
$20x^2 + 17x - 63$

3. $(x + 6)(2x^2 - x - 5)$
$x(2x^2 - x - 5) + 6(2x^2 - x - 5)$
$2x^3 - x^2 - 5x + 12x^2 - 6x - 30$
$2x^3 + 11x^2 - 11x - 30$

4. $(x - 4y)^3$
$(x - 4y)^2(x - 4y)$
$[(x)^2 + 2(x)(-4y) + (-4y)^2](x - 4y)$
$(x^2 - 8xy + 16y^2)(x - 4y)$
$(x - 4y)(x^2 - 8xy + 16y^2)$
$x(x^2 - 8xy + 16y^2) - 4y(x^2 - 8xy + 16y^2)$
$x^3 - 8x^2y + 16xy^2 - 4x^2y + 32xy^2 - 64y^3$
$x^3 - 12x^2y + 48xy^2 - 64y^3$

5.

$$\begin{array}{r|l} & 2x^2 - 3x - 4 \\ \hline 3x - 5 & 6x^3 - 19x^2 + 3x + 20 \\ & 6x^3 - 10x^2 \\ \hline & -9x^2 + 3x \\ & -9x^2 + 15x \\ \hline & -12x + 20 \\ & -12x + 20 \\ \hline & 0 \end{array}$$

Q: $2x^2 - 3x - 4$ R: 0

6.

$$\begin{array}{r|l} & 3x^3 - x^2 - 2x - 6 \\ \hline x + 3 & 3x^4 + 8x^3 - 5x^2 - 12x - 15 \\ & 3x^4 + 9x^3 \\ \hline & -x^3 - 5x^2 \\ & -x^3 - 3x^2 \\ \hline & -2x^2 - 12x \\ & -2x^2 - 6x \\ \hline & -6x - 15 \\ & -6x - 18 \\ \hline & 3 \end{array}$$

Q: $3x^3 - x^2 - 2x - 6$ R: 3

7. $x^2 - xy + 4x - 4y$
$x(x - y) + 4(x - y)$
$(x - y)(x + 4)$

8. $12x^2 - 3$
$3(4x^2 - 1)$
$3(2x - 1)(2x + 1)$

9. $3(2x - 1) - 2(x + 5) = -(x - 3)$
$6x - 3 - 2x - 10 = -x + 3$
$4x - 13 = -x + 3$
$5x - 13 = 3$
$5x = 16$
$x = \frac{16}{5}$
The solution set is $\left\{\frac{16}{5}\right\}$.

10. $\frac{3t - 2}{4} = \frac{5t + 1}{5}$
$5(3t - 2) = 4(5t + 1)$
$15t - 10 = 20t + 4$
$-5t = 14$
$x = -\frac{14}{5}$
The solution set is $\left\{-\frac{14}{5}\right\}$.

11. $|4x-3|=9$ is equivalent to

$$4x-3=-9 \quad \text{or} \quad 4x-3=9.$$
$$4x=-6 \quad \text{or} \quad 4x-3=9$$
$$x=-\frac{6}{4}=-\frac{3}{2} \quad \text{or} \quad 4x=12$$
$$\text{or} \quad x=3$$

The solution set is $\left\{-\frac{3}{2}, 3\right\}$.

12.
$$\frac{1-3x}{4}+\frac{2x+3}{3}=1$$
$$12\left(\frac{1-3x}{4}+\frac{2x+3}{3}\right)=12(1)$$
$$3(1-3x)+4(2x+3)=12$$
$$3-9x+8x+12=12$$
$$-x+15=12$$
$$-x=-3$$
$$x=3$$

The solution set is $\{3\}$.

13.
$$0.05x+0.06(1500-x)=83.5$$
$$100[0.05x+0.06(1500-x)]=100(83.5)$$
$$5x+6(1500-x)=8350$$
$$5x+9000-6x=8350$$
$$-x+9000=8350$$
$$-x=-650$$
$$x=650$$

The solution set is $\{650\}$.

14.
$$4n^2=n$$
$$4n^2-n=0$$
$$n(4n-1)=0$$
$$n=0 \quad \text{or} \quad 4n-1=0$$
$$n=0 \quad \text{or} \quad 4n=1$$
$$n=0 \quad \text{or} \quad n=\frac{1}{4}$$

The solution set is $\left\{0, \frac{1}{4}\right\}$.

15.
$$4x^2-12x+9=0$$
$$(2x-3)(2x-3)=0$$
$$2x-3=0 \quad \text{or} \quad 2x-3=0$$
$$2x=3 \quad \text{or} \quad 2x=3$$
$$x=\frac{3}{2} \quad \text{or} \quad x=\frac{3}{2}$$

The solution set is $\left\{\frac{3}{2}\right\}$.

16.
$$3x^3+21x^2-54x=0$$
$$3x(x^2+7x-18)=0$$
$$3x(x+9)(x-2)=0$$
$$3x=0 \quad \text{or} \quad x+9=0 \quad \text{or} \quad x-2=0$$
$$x=0 \quad \text{or} \quad x=-9 \quad \text{or} \quad x=2$$

The solution set is $\{-9, 0, 2\}$.

17.
$$12+13x-35x^2=0$$
$$(3+7x)(4-5x)=0$$
$$3+7x=0 \quad \text{or} \quad 4-5x=0$$
$$7x=-3 \quad \text{or} \quad 4=5x$$
$$x=-\frac{3}{7} \quad \text{or} \quad \frac{4}{5}=x$$

The solution set is $\left\{-\frac{3}{7}, \frac{4}{5}\right\}$.

18.
$$n(3n-5)=2$$
$$3n^2-5n=2$$
$$3n^2-5n-2=0$$
$$(3n+1)(n-2)=0$$
$$3n+1=0 \quad \text{or} \quad n-2=0$$
$$3n=-1 \quad \text{or} \quad n=2$$
$$n=-\frac{1}{3} \quad \text{or} \quad n=2$$

The solution set is $\left\{-\frac{1}{3}, 2\right\}$.

19.
$$|6x-4|<10$$
$$-10<6x-4<10$$
$$-6<6x<14$$
$$-1<x<\frac{14}{6}$$
$$-1<x<\frac{7}{3}$$

The solution set is $\left(-1, \frac{7}{3}\right)$.

20. $\frac{x-2}{6} - \frac{x+3}{9} > -\frac{1}{2}$

$18\left(\frac{x-2}{6} - \frac{x+3}{9}\right) > 18\left(-\frac{1}{2}\right)$

$18\left(\frac{x-2}{6}\right) - 18\left(\frac{x+3}{9}\right) > -9$

$3(x-2) - 2(x+3) > -9$

$3x - 6 - 2x - 6 > -9$

$x - 12 > -9$

$x > 3$

The solution set is $(3, \infty)$.

21. $2(x-1) - 3(3x+1) \geq -6(x-5)$

$2x - 2 - 9x - 3 \geq -6x + 30$

$-7x - 5 \geq -6x + 30$

$-x - 5 \geq 30$

$-x \geq 35$

$x \leq -35$

The solution set is $(-\infty, -35]$.

22. Let x = number of cups of grapefruit juice.

$100\%(x) + (8\%)(30) = (10\%)(x+30)$

$1.00x + 0.08(30) = 0.10(x+30)$

$100[1.00x + 0.08(30)] = 100[0.10(x+30)]$

$100(1.00x) + 100[0.08(30)] = 10(x+30)$

$100x + 8(30) = 10x + 300$

$100x + 240 = 10x + 300$

$100x = 10x + 60$

$90x = 60$

$x = \frac{60}{90} = \frac{2}{3}$

There should be $\frac{2}{3}$ cup of grapefruit juice added.

23. Let x = score on 6th exam.

$\frac{85 + 92 + 87 + 88 + 91 + x}{6} \geq 90$

$\frac{443 + x}{6} \geq 90$

$6\left(\frac{443 + x}{6}\right) \geq 6(90)$

$443 + x \geq 540$

$x \geq 97$

The score should be 97 or better.

24. Let x = angle,
$90 - x$ = complement of the angle, and
$180 - x$ = supplement of the angle.

$90 - x = \frac{2}{11}(180 - x)$

$11(90 - x) = 11\left[\frac{2}{11}(180 - x)\right]$

$11(90) - 11(x) = 2(180 - x)$

$990 - 11x = 360 - 2x$

$-11x = -630 - 2x$

$-9x = -630$

$x = 70$

The angle is 70°.

25. Let x = side of square,
$x + 3$ = width of rectangle, and
$x + 5$ = length of rectangle.

$x^2 + (x+3)(x+5) = 57$

$x^2 + x^2 + 8x + 15 = 57$

$2x^2 + 8x - 42 = 0$

$2(x^2 + 4x - 21) = 0$

$2(x+7)(x-3) = 0$

$x + 7 = 0$ or $x - 3 = 0$

$x = -7$ or $x = 3$

Discard the root $x = -7$.

The length of the rectangle is $3 + 5 = 8$ feet.

Chapter 9 Rational Expressions

PROBLEM SET 9.1 Simplifying Rational Expressions

1. $\frac{27}{36} = \frac{9 \bullet 3}{9 \bullet 4} = \frac{3}{4}$

3. $\frac{45}{54} = \frac{9 \bullet 5}{9 \bullet 6} = \frac{5}{6}$

5. $\frac{24}{-60} = \frac{12 \bullet 2}{12 \bullet -5} = -\frac{2}{5}$

7. $\frac{-16}{-56} = \frac{-8 \bullet 2}{-8 \bullet 7} = \frac{2}{7}$

9. $\frac{12xy}{42y} = \frac{6y}{6y} \bullet \frac{2x}{7} = \frac{2x}{7}$

11. $\frac{18a^2}{45ab} = \frac{9a}{9a} \bullet \frac{2a}{5b} = \frac{2a}{5b}$

13. $\frac{-14y^3}{56xy^2} = \frac{14y^2}{14y^2} \bullet \frac{-y}{4x} = -\frac{y}{4x}$

15. $\frac{54c^2d}{-78cd^2} = \frac{6cd}{6cd} \bullet \frac{9c}{-13d} = -\frac{9c}{13d}$

17. $\frac{-40x^3y}{-24xy^4} = \frac{-8xy}{-8xy} \bullet \frac{5x^2}{3y^3} = \frac{5x^2}{3y^3}$

19. $\frac{x^2-4}{x^2+2x} = \frac{(x+2)(x-2)}{x(x+2)} = \frac{x-2}{x}$

21. $\frac{18x+12}{12x-6}$

$\frac{6(3x+2)}{6(2x-1)}$

$\frac{3x+2}{2x-1}$

23. $\frac{a^2+7a+10}{a^2-7a-18}$

$\frac{(a+5)(a+2)}{(a-9)(a+2)}$

$\frac{a+5}{a-9}$

25. $\frac{2n^2+n-21}{10n^2+33n-7}$

$\frac{(2n+7)(n-3)}{(2n+7)(5n-1)}$

$\frac{n-3}{5n-1}$

27. $\frac{5x^2+7}{10x}$

29. $\frac{6x^2+x-15}{8x^2-10x-3}$

$\frac{(3x+5)(2x-3)}{(4x+1)(2x-3)}$

$\frac{3x+5}{4x+1}$

31. $\frac{3x^2-12x}{x^3-64}$

$\frac{3x(x-4)}{(x-4)(x^2+4x+16)}$

$\frac{3x}{x^2+4x+16}$

33. $\frac{3x^2+17x-6}{9x^2-6x+1}$

$\frac{(3x-1)(x+6)}{(3x-1)(3x-1)}$

$\frac{x+6}{3x-1}$

Problem Set 9.1

35. $\dfrac{2x^3 + 3x^2 - 14x}{x^2y + 7xy - 18y}$

$\dfrac{x(2x^2 + 3x - 14)}{y(x^2 + 7x - 18)}$

$\dfrac{x(2x + 7)(x - 2)}{y(x + 9)(x - 2)}$

$\dfrac{x(2x + 7)}{y(x + 9)}$

37. $\dfrac{5y^2 + 22y + 8}{25y^2 - 4}$

$\dfrac{(5y + 2)(y + 4)}{(5y + 2)(5y - 2)}$

$\dfrac{y + 4}{5y - 2}$

39. $\dfrac{15x^3 - 15x^2}{5x^3 + 5x}$

$\dfrac{15x^2(x - 1)}{5x(x^2 + 1)}$

$\dfrac{3x(x - 1)}{x^2 + 1}$

41. $\dfrac{4x^2y + 8xy^2 - 12y^3}{18x^3y - 12x^2y^2 - 6xy^3}$

$\dfrac{4y(x^2 + 2xy - 3y^2)}{6xy(3x^2 - 2xy - y^2)}$

$\dfrac{4y(x + 3y)(x - y)}{6xy(3x + y)(x - y)}$

$\dfrac{2(x + 3y)}{3x(3x + y)}$

43. $\dfrac{3n^2 + 16n - 12}{7n^2 + 44n + 12}$

$\dfrac{(3n - 2)(n + 6)}{(7n + 2)(n + 6)}$

$\dfrac{3n - 2}{7n + 2}$

45. $\dfrac{8 + 18x - 5x^2}{10 + 31x + 15x^2}$

$\dfrac{(4 - x)(2 + 5x)}{(5 + 3x)(2 + 5x)}$

$\dfrac{4 - x}{5 + 3x}$

47. $\dfrac{27x^4 - x}{6x^3 + 10x^2 - 4x}$

$\dfrac{x(27x^3 - 1)}{2x(3x^2 + 5x - 2)}$

$\dfrac{x(3x - 1)(9x^2 + 3x + 1)}{2x(3x - 1)(x + 2)}$

$\dfrac{9x^2 + 3x + 1}{2(x + 2)}$

49. $\dfrac{-40x^3 + 24x^2 + 16x}{20x^3 + 28x^2 + 8x}$

$\dfrac{-8x(5x^2 - 3x - 2)}{4x(5x^2 + 7x + 2)}$

$\dfrac{-8x(5x + 2)(x - 1)}{4x(5x + 2)(x + 1)}$

$\dfrac{-2(x - 1)}{x + 1}$

51. $\dfrac{xy + ay + bx + ab}{xy + ay + cx + ac}$

$\dfrac{y(x + a) + b(x + a)}{y(x + a) + c(x + a)}$

$\dfrac{(x + a)(y + b)}{(x + a)(y + c)}$

$\dfrac{y + b}{y + c}$

53. $\dfrac{ax - 3x + 2ay - 6y}{2ax - 6x + ay - 3y}$

$\dfrac{x(a-3) + 2y(a-3)}{2x(a-3) + y(a-3)}$

$\dfrac{(a-3)(x+2y)}{(a-3)(2x+y)}$

$\dfrac{x+2y}{2x+y}$

55. $\dfrac{5x^2 + 5x + 3x + 3}{5x^2 + 3x - 30x - 18}$

$\dfrac{5x(x+1) + 3(x+1)}{x(5x+3) - 6(5x+3)}$

$\dfrac{(x+1)(5x+3)}{(5x+3)(x-6)}$

$\dfrac{x+1}{x-6}$

57. $\dfrac{2st - 30 - 12s + 5t}{3st - 6 - 18s + t}$

$\dfrac{2st - 12s + 5t - 30}{3st - 18s + t - 6}$

$\dfrac{2s(t-6) + 5(t\ \ \ 6)}{3s(t-6) + 1(t-6)}$

$\dfrac{(t-6)(2s+5)}{(t-6)(3s+1)}$

$\dfrac{2s+5}{3s+1}$

59. $\dfrac{5x-7}{7-5x}$

$\dfrac{-1(-5x+7)}{(7-5x)}$

$\dfrac{-1(7-5x)}{(7-5x)}$

-1

61. $\dfrac{n^2-49}{7-n}$

$\dfrac{(n+7)(n-7)}{-1(-7+n)}$

$\dfrac{(n+7)(n-7)}{-1(n-7)}$

$\dfrac{n+7}{-1}$

$-n-7$

63. $\dfrac{2y-2xy}{x^2y-y}$

$\dfrac{2y(1-x)}{y(x^2-1)}$

$\dfrac{2y(1-x)}{y(x+1)(x-1)}$

$\dfrac{2y(-1)(-1+x)}{y(x+1)(x-1)}$

$\dfrac{-2y(x-1)}{y(x+1)(x-1)}$

$\dfrac{-2y}{y(x+1)}$

$\dfrac{-2}{x+1}$

65. $\dfrac{2x^3-8x}{4x-x^3}$

$\dfrac{2x(x^2-4)}{-x(-4+x^2)}$

$\dfrac{2x(x^2-4)}{-x(x^2-4)}$

-2

67. $\dfrac{n^2 - 5n - 24}{40 + 3n - n^2}$

$\dfrac{(n-8)(n+3)}{(8-n)(5+n)}$

$\dfrac{(n-8)(n+3)}{-1(-8+n)(5+n)}$

$\dfrac{(n-8)(n+3)}{-1(n-8)(n+5)}$

$\dfrac{-(n+3)}{n+5}$

PROBLEM SET 9.2 Multiplying and Dividing Rational Expressions

1. $\dfrac{7}{12} \bullet \dfrac{6}{35}$

$\dfrac{7}{2 \bullet 6} \bullet \dfrac{6}{5 \bullet 7}$

$\dfrac{1}{10}$

3. $\dfrac{-4}{9} \bullet \dfrac{18}{30}$

$\dfrac{-4}{9} \bullet \dfrac{9 \bullet 2}{2 \bullet 15}$

$-\dfrac{4}{15}$

5. $\dfrac{3}{-8} \bullet \dfrac{-6}{12}$

$\dfrac{3}{-8} \bullet \dfrac{-6}{2 \bullet 6}$

$\dfrac{3}{16}$

7. $\left(-\dfrac{5}{7}\right) \div \dfrac{6}{7}$

$-\dfrac{5}{7} \bullet \dfrac{7}{6}$

$-\dfrac{5}{6}$

9. $-\dfrac{9}{5} \div \dfrac{27}{10}$

$-\dfrac{9}{5} \bullet \dfrac{10}{27}$

$-\dfrac{9}{5} \bullet \dfrac{5 \bullet 2}{3 \bullet 9}$

$-\dfrac{2}{3}$

11. $\dfrac{4}{9} \bullet \dfrac{6}{11} \div \dfrac{4}{15}$

$\dfrac{4}{9} \bullet \dfrac{6}{11} \bullet \dfrac{15}{4}$

$\dfrac{4}{3 \bullet 3} \bullet \dfrac{2 \bullet 3}{11} \bullet \dfrac{3 \bullet 5}{4}$

$\dfrac{10}{11}$

13. $\dfrac{6xy}{9y^4} \bullet \dfrac{30x^3y}{-48x}$

$\dfrac{6(30)x^4y^2}{9(-48)xy^4}$

$\dfrac{-5x^3}{12y^2}$

15. $\dfrac{5a^2b^2}{11ab} \bullet \dfrac{22a^3}{15ab^2}$

$\dfrac{5(22)a^5b^2}{11(15)a^2b^3}$

$\dfrac{2a^3}{3b}$

17. $\dfrac{5xy}{8y^2} \bullet \dfrac{18x^2y}{15}$

$\dfrac{5(18)x^3y^2}{8(15)y^2}$

$\dfrac{3x^3}{4}$

19. $\dfrac{5x^4}{12x^2y^3} \div \dfrac{9}{5xy}$

$\dfrac{5x^4}{12x^2y^3} \bullet \dfrac{5xy}{9}$

$$\frac{5(5)x^5y}{12(9)x^2y^3}$$

$$\frac{25x^3}{108y^2}$$

21. $$\frac{9a^2c}{12bc^2} \div \frac{21ab}{14c^3}$$

$$\frac{9a^2c}{12bc^2} \bullet \frac{14c^3}{21ab}$$

$$\frac{9(14)a^2c^4}{12(21)ab^2c^2}$$

$$\frac{3(2)ac^2}{4(3)b^2}$$

$$\frac{ac^2}{2b^2}$$

23. $$\frac{9x^2y^3}{14x} \bullet \frac{21y}{15xy^2} \bullet \frac{10x}{12y^3}$$

$$\frac{9(21)(10)x^3y^4}{14(15)(12)x^2y^5}$$

$$\frac{21}{14} \bullet \frac{9}{12} \bullet \frac{10}{15} \bullet \frac{x}{y}$$

$$\frac{3}{2} \bullet \frac{3}{4} \bullet \frac{2}{3} \bullet \frac{x}{y}$$

$$\frac{3x}{4y}$$

25. $$\frac{3x+6}{5y} \bullet \frac{x^2+4}{x^2+10x+16}$$

$$\frac{3(x+2)}{5y} \bullet \frac{x^2+4}{(x+8)(x+2)}$$

$$\frac{3(x^2+4)}{5y(x+8)}$$

27. $$\frac{5a^2+20a}{a^3-2a^2} \bullet \frac{a^2-a-12}{a^2-16}$$

$$\frac{5a(a+4)}{a^2(a-2)} \bullet \frac{(a-4)(a+3)}{(a-4)(a+4)}$$

$$\frac{5(a+3)}{a(a-2)}$$

29. $$\frac{3n^2+15n-18}{3n^2+10n-48} \bullet \frac{12n^2-17n-40}{4n^2+6n-10}$$

$$\frac{3(n^2+5n-6)}{3n^2+10n-48} \bullet \frac{12n^2-17n-40}{2(2n^2+3n-5)}$$

$$\frac{3(n+6)(n-1)}{(3n-8)(n+6)} \bullet \frac{(4n+5)(3n-8)}{2(4n+5)(n-1)}$$

$$\frac{3}{2}$$

31. $$\frac{9y^2}{x^2+12x+36} \div \frac{12y}{x^2+6x}$$

$$\frac{9y^2}{x^2+12x+36} \bullet \frac{x^2+6x}{12y}$$

$$\frac{9y^2}{(x+6)(x+6)} \bullet \frac{x(x+6)}{12y}$$

$$\frac{3xy}{4(x+6)}$$

33. $$\frac{x^2-4xy+4y^2}{7xy^2} \div \frac{4x^2-3xy-10y^2}{20x^2y+25xy^2}$$

$$\frac{x^2-4xy+4y^2}{7xy^2} \bullet \frac{20x^2y+25xy^2}{4x^2-3xy-10y^2}$$

$$\frac{(x-2y)(x-2y)}{7xy^2} \bullet \frac{5xy(4x+5y)}{(4x+5y)(x-2y)}$$

$$\frac{5(x-2y)}{7y}$$

35. $$\frac{5-14n-3n^2}{1-2n-3n^2} \bullet \frac{9+7n-2n^2}{27-15n+2n^2}$$

$$\frac{(5+n)(1-3n)}{(1+n)(1-3n)} \bullet \frac{(9-2n)(1+n)}{(9-2n)(3-n)}$$

$$\frac{5+n}{3-n}$$

37. $$\frac{3x^4+2x^2-1}{3x^4+14x^2-5} \bullet \frac{x^4-2x^2-35}{x^4-17x^2+70}$$

$$\frac{(3x^2-1)(x^2+1)}{(3x^2-1)(x^2+5)} \bullet \frac{(x^2-7)(x^2+5)}{(x^2-7)(x^2-10)}$$

$$\frac{x^2+1}{x^2-10}$$

39. $$\frac{3x^2-20x+25}{2x^2-7x-15} \div \frac{9x^2-3x-20}{12x^2+28x+15}$$

$$\frac{3x^2-20x+25}{2x^2-7x-15} \bullet \frac{12x^2+28x+15}{9x^2-3x-20}$$

$\frac{(3x-5)(x-5)}{(2x+3)(x-5)} \bullet \frac{(2x+3)(6x+5)}{(3x+4)(3x-5)}$

$\frac{6x+5}{3x+4}$

41. $\frac{10t^3+25t}{20t+10} \bullet \frac{2t^2-t-1}{t^5-t}$

$\frac{5t(2t^2+5)}{10(2t+1)} \bullet \frac{(2t+1)(t-1)}{t(t^4-1)}$

$\frac{5t(2t^2+5)}{10(2t+1)} \bullet \frac{(2t+1)(t-1)}{t(t^2+1)(t^2-1)}$

$\frac{5t(2t^2+5)}{10(2t+1)} \bullet \frac{(2t+1)(t-1)}{t(t^2+1)(t+1)(t-1)}$

$\frac{2t^2+5}{2(t^2+1)(t+1)}$

43. $\frac{4t^2+t-5}{t^3-t^2} \bullet \frac{t^4+6t^3}{16t^2+40t+25}$

$\frac{(4t+5)(t-1)}{t^2(t-1)} \bullet \frac{t^3(t+6)}{(4t+5)(4t+5)}$

$\frac{t(t+6)}{4t+5}$

45. $\frac{nr+3n+2r+6}{nr+3n-3r-9} \bullet \frac{n^2-9}{n^3-4n}$

$\frac{n(r+3)+2(r+3)}{n(r+3)-3(r+3)} \bullet \frac{(n+3)(n-3)}{n(n^2-4)}$

$\frac{(r+3)(n+2)}{(r+3)(n-3)} \bullet \frac{(n+3)(n-3)}{n(n+2)(n-2)}$

$\frac{n+3}{n(n-2)}$

47. $\frac{x^2-x}{4y} \bullet \frac{10xy^2}{2x-2} \div \frac{3x^2+3x}{15x^2y^2}$

$\frac{x^2-x}{4y} \bullet \frac{10xy^2}{2x-2} \bullet \frac{15x^2y^2}{3x^2+3x}$

$\frac{x(x-1)}{4y} \bullet \frac{10xy^2}{2(x-1)} \bullet \frac{15x^2y^2}{3x(x+1)}$

$\frac{10 \cdot 15 \cdot x^4y^4(x-1)}{8 \cdot 3 \cdot (x+1)(x-1)}$

$\frac{25x^3y^3}{4(x+1)}$

49. $\frac{a^2-4ab+4b^2}{6a^2-4ab} \bullet \frac{3a^2+5ab-2b^2}{6a^2+ab-b^2} \div \frac{a^2-4b^2}{8a+4b}$

$\frac{a^2-4ab+4b^2}{6a^2-4ab} \bullet \frac{3a^2+5ab-2b^2}{6a^2+ab-b^2} \bullet \frac{8a+4b}{a^2-4b^2}$

$\frac{(a-2b)(a-2b)}{2a(3a-2b)} \bullet \frac{(3a-b)(a+2b)}{(3a-b)(2a+b)} \bullet \frac{4(2a+b)}{(a+2b)(a-2b)}$

$\frac{2(a-2b)}{a(3a-2b)}$

PROBLEM SET 9.3 Adding and Subtracting Rational Expressions

1. $\frac{1}{4}+\frac{5}{6}$; LCD = 12

$\frac{1}{4}\left(\frac{3}{3}\right)+\frac{5}{6}\left(\frac{2}{2}\right)$

$\frac{3}{12}+\frac{10}{12}$

$\frac{13}{12}$

3. $\frac{7}{8}-\frac{3}{5}$; LCD = 40

$\frac{7}{8}\left(\frac{5}{5}\right)-\frac{3}{5}\left(\frac{8}{8}\right)$

$\frac{35}{40}-\frac{24}{40}$

$\frac{11}{40}$

5. $\frac{6}{5}+\frac{1}{-4}$; LCD = 20

$\frac{6}{5}\left(\frac{4}{4}\right)+\left(-\frac{1}{4}\right)\left(\frac{5}{5}\right)$

$\frac{24}{20}+\left(-\frac{5}{20}\right)$

$\frac{19}{20}$

7. $\frac{8}{15}+\frac{3}{25}$; LCD = 75

$\frac{8}{15}\left(\frac{5}{5}\right)+\frac{3}{25}\left(\frac{3}{3}\right)$

$\frac{40}{75}+\frac{9}{75}$

$\frac{49}{75}$

9. $\frac{1}{5}+\frac{5}{6}-\frac{7}{15}$; LCD = 30

$\frac{1}{5}\left(\frac{6}{6}\right)+\frac{5}{6}\left(\frac{5}{5}\right)-\frac{7}{15}\left(\frac{2}{2}\right)$

$\frac{6}{30}+\frac{25}{30}-\frac{14}{30}$

$\frac{31}{30}-\frac{14}{30}$

$\frac{17}{30}$

11. $\frac{1}{3}-\frac{1}{4}-\frac{3}{14}$; LCD = 84

$\frac{1}{3}\left(\frac{28}{28}\right)-\frac{1}{4}\left(\frac{21}{21}\right)-\frac{3}{14}\left(\frac{6}{6}\right)$

$\frac{28}{84}-\frac{21}{84}-\frac{18}{84}$

$\frac{7}{84}-\frac{18}{84}$

$-\frac{11}{84}$

13. $\frac{2x}{x-1}+\frac{4}{x-1}$

$\frac{2x+4}{x-1}$

15. $\frac{4a}{a+2}+\frac{8}{a+2}$

$\frac{4a+8}{a+2}$

$\frac{4(a+2)}{a+2}$

4

17. $\frac{3(y-2)}{7y}+\frac{4(y-1)}{7y}$

$\frac{3(y-2)+4(y-1)}{7y}$

$\frac{3y-6+4y-4}{7y}$

$\frac{7y-10}{7y}$

19. $\frac{x-1}{2}+\frac{x+3}{3}$; LCD = 6

$\frac{(x-1)}{2}\left(\frac{3}{3}\right)+\frac{(x+3)}{3}\left(\frac{2}{2}\right)$

$\frac{3(x-1)}{6}+\frac{2(x+3)}{6}$

$\frac{3(x-1)+2(x+3)}{6}$

$\frac{3x-3+2x+6}{6}$

$\frac{5x+3}{6}$

21. $\frac{2a-1}{4}+\frac{3a+2}{6}$; LCD = 12

$\frac{(2a-1)}{4}\left(\frac{3}{3}\right)+\frac{(3a+2)}{6}\left(\frac{2}{2}\right)$

$\frac{3(2a-1)}{12}+\frac{2(3a+2)}{12}$

$\frac{3(2a-1)+2(3a+2)}{12}$

$\frac{6a-3+6a+4}{12}$

$\frac{12a+1}{12}$

23. $\frac{n+2}{6}-\frac{n-4}{9}$; LCD = 18

$\frac{(n+2)}{6}\left(\frac{3}{3}\right)-\frac{(n-4)}{9}\left(\frac{2}{2}\right)$

$\frac{3(n+2)}{18}-\frac{2(n-4)}{18}$

$\frac{3(n+2)-2(n-4)}{18}$

$\frac{3n+6-2n+8}{18}$

$\frac{n+14}{18}$

Problem Set 9.3

25. $\frac{3x-1}{3} - \frac{5x+2}{5}$; LCD = 15

$\frac{(3x-1)}{3}\left(\frac{5}{5}\right) - \frac{(5x+2)}{5}\left(\frac{3}{3}\right)$

$\frac{5(3x-1)}{15} - \frac{3(5x+2)}{15}$

$\frac{5(3x-1)-3(5x+2)}{15}$

$\frac{15x-5-15x-6}{15}$

$-\frac{11}{15}$

27. $\frac{x-2}{5} - \frac{x+3}{6} + \frac{x+1}{15}$; LCD = 30

$\frac{(x-2)}{5}\left(\frac{6}{6}\right) - \frac{(x+3)}{6}\left(\frac{5}{5}\right) + \frac{(x+1)}{15}\left(\frac{2}{2}\right)$

$\frac{6(x-2)}{30} - \frac{5(x+3)}{30} + \frac{2(x+1)}{30}$

$\frac{6(x-2)-5(x+3)+2(x+1)}{30}$

$\frac{6x-12-5x-15+2x+2}{30}$

$\frac{3x-25}{30}$

29. $\frac{3}{8x} + \frac{7}{10x}$; LCD = $40x$

$\frac{3}{8x}\left(\frac{5}{5}\right) + \frac{7}{10x}\left(\frac{4}{4}\right)$

$\frac{15}{40x} + \frac{28}{40x}$

$\frac{43}{40x}$

31. $\frac{5}{7x} - \frac{11}{4y}$; LCD = $28xy$

$\frac{5}{7x}\left(\frac{4y}{4y}\right) - \frac{11}{4y}\left(\frac{7x}{7x}\right)$

$\frac{20y}{28xy} - \frac{77x}{28xy}$

$\frac{20y-77x}{28xy}$

33. $\frac{4}{3x} + \frac{5}{4y} - 1$; LCD = $12xy$

$\frac{4}{3x}\left(\frac{4y}{4y}\right) + \frac{5}{4y}\left(\frac{3x}{3x}\right) - \left(\frac{12xy}{12xy}\right)$

$\frac{16y}{12xy} + \frac{15x}{12xy} - \frac{12xy}{12xy}$

$\frac{16y+15x-12xy}{12xy}$

35. $\frac{7}{10x^2} + \frac{11}{15x}$; LCD = $30x^2$

$\frac{7}{10x^2}\left(\frac{3}{3}\right) + \frac{11}{15x}\left(\frac{2x}{2x}\right)$

$\frac{21}{30x^2} + \frac{22x}{30x^2}$

$\frac{21+22x}{30x^2}$

37. $\frac{10}{7n} - \frac{12}{4n^2}$; LCD = $28n^2$

$\frac{10}{7n}\left(\frac{4n}{4n}\right) - \frac{12}{4n^2}\left(\frac{7}{7}\right)$

$\frac{40n}{28n^2} - \frac{84}{28n^2}$

$\frac{40n-84}{28n^2}$

$\frac{4(10n-21)}{28n^2}$

$\frac{10n-21}{7n^2}$

39. $\frac{3}{n^2} - \frac{2}{5n} + \frac{4}{3}$; LCD = $15n^2$

$\frac{3}{n^2}\left(\frac{15}{15}\right) - \frac{2}{5n}\left(\frac{3n}{3n}\right) + \frac{4}{3}\left(\frac{5n^2}{5n^2}\right)$

$\frac{45}{15n^2} - \frac{6n}{15n^2} + \frac{20n^2}{15n^2}$

$\frac{45-6n+20n^2}{15n^2}$

41. $\frac{3}{x} - \frac{5}{3x^2} - \frac{7}{6x}$; LCD = $6x^2$

$\frac{3}{x}\left(\frac{6x}{6x}\right) - \frac{5}{3x^2}\left(\frac{2}{2}\right) - \frac{7}{6x}\left(\frac{x}{x}\right)$

$$\frac{18x}{6x^2} - \frac{10}{6x^2} - \frac{7x}{6x^2}$$
$$\frac{18x - 10 - 7x}{6x^2}$$
$$\frac{11x - 10}{6x^2}$$

43. $\frac{6}{5t^2} - \frac{4}{7t^3} + \frac{9}{5t^3}$; LCD $= 35t^3$
$$\frac{6}{5t^2}\left(\frac{7t}{7t}\right) - \frac{4}{7t^3}\left(\frac{5}{5}\right) + \frac{9}{5t^3}\left(\frac{7}{7}\right)$$
$$\frac{42t}{35t^3} - \frac{20}{35t^3} + \frac{63}{35t^3}$$
$$\frac{42t - 20 + 63}{35t^3}$$
$$\frac{42t + 43}{35t^3}$$

45. $\frac{5b}{24a^2} - \frac{11a}{32b}$; LCD $= 96a^2b$
$$\frac{5b}{24a^2}\left(\frac{4b}{4b}\right) - \frac{11a}{32b}\left(\frac{3a^2}{3a^2}\right)$$
$$\frac{20b^2}{96a^2b} - \frac{33a^3}{96a^2b}$$
$$\frac{20b^2 - 33a^3}{96a^2b}$$

47. $\frac{7}{9xy^3} - \frac{4}{3x} + \frac{5}{2y^2}$; LCD $= 18xy^3$
$$\frac{7}{9xy^3}\left(\frac{2}{2}\right) - \frac{4}{3x}\left(\frac{6y^3}{6y^3}\right) + \frac{5}{2y^2}\left(\frac{9xy}{9xy}\right)$$
$$\frac{14}{18xy^3} - \frac{24y^3}{18xy^3} + \frac{45xy}{18xy^3}$$
$$\frac{14 - 24y^3 + 45xy}{18xy^3}$$

49. $\frac{2x}{x-1} + \frac{3}{x}$; LCD $= x(x-1)$
$$\frac{2x}{(x-1)}\left(\frac{x}{x}\right) + \frac{3}{x}\left(\frac{x-1}{x-1}\right)$$
$$\frac{2x(x)}{x(x-1)} + \frac{3(x-1)}{x(x-1)}$$
$$\frac{2x(x) + 3(x-1)}{x(x-1)}$$
$$\frac{2x^2 + 3x - 3}{x(x-1)}$$

51. $\frac{a-2}{a} - \frac{3}{a+4}$; LCD $= a(a+4)$
$$\frac{(a-2)}{a}\left(\frac{a+4}{a+4}\right) - \frac{3}{(a+4)}\left(\frac{a}{a}\right)$$
$$\frac{(a-2)(a+4)}{a(a+4)} - \frac{3a}{a(a+4)}$$
$$\frac{(a-2)(a+4) - 3a}{a(a+4)}$$
$$\frac{a^2 + 2a - 8 - 3a}{a(a+4)}$$
$$\frac{a^2 - a - 8}{a(a+4)}$$

53. $\frac{-3}{4n+5} - \frac{8}{3n+5}$;LCD $= (4n+5)(3n+5)$
$$\frac{-3}{(4n+5)}\left(\frac{3n+5}{3n+5}\right) - \frac{8}{(3n+5)}\left(\frac{4n+5}{4n+5}\right)$$
$$\frac{-3(3n+5)}{(4n+5)(3n+5)} - \frac{8(4n+5)}{(4n+5)(3n+5)}$$
$$\frac{-3(3n+5) - 8(4n+5)}{(4n+5)(3n+5)}$$
$$\frac{-9n - 15 - 32n - 40}{(4n+5)(3n+5)}$$
$$\frac{-41n - 55}{(4n+5)(3n+5)}$$

55. $\frac{-1}{x+4} + \frac{4}{7x-1}$; LCD $= (x+4)(7x-1)$
$$\frac{-1}{(x+4)}\left(\frac{7x-1}{7x-1}\right) + \frac{4}{(7x-1)}\left(\frac{x+4}{x+4}\right)$$
$$\frac{-1(7x-1)}{(x+4)(7x-1)} + \frac{4(x+4)}{(7x-1)(x+4)}$$
$$\frac{-1(7x-1) + 4(x+4)}{(7x-1)(x+4)}$$
$$\frac{-7x + 1 + 4x + 16}{(7x-1)(x+4)}$$
$$\frac{-3x + 17}{(7x-1)(x+4)}$$

57. $\frac{7}{3x-5}-\frac{5}{2x+7}$;

LCD $=(3x-5)(2x+7)$

$$\frac{7}{(3x-5)}\left(\frac{2x+7}{2x+7}\right)-\frac{5}{(2x+7)}\left(\frac{3x-5}{3x-5}\right)$$

$$\frac{7(2x+7)}{(3x-5)(2x+7)}-\frac{5(3x-5)}{(2x+7)(3x-5)}$$

$$\frac{7(2x+7)-5(3x-5)}{(3x-5)(2x+7)}$$

$$\frac{14x+49-15x+25}{(3x-5)(2x+7)}$$

$$\frac{-x+74}{(3x-5)(2x+7)}$$

59. $\frac{5}{3x-2}+\frac{6}{4x+5}$;

LCD $=(3x-2)(4x+5)$

$$\frac{5}{(3x-2)}\left(\frac{4x+5}{4x+5}\right)+\frac{6}{(4x+5)}\left(\frac{3x-2}{3x-2}\right)$$

$$\frac{5(4x+5)}{(3x-2)(4x+5)}+\frac{6(3x-2)}{(4x+5)(3x-2)}$$

$$\frac{5(4x+5)+6(3x-2)}{(3x-2)(4x+5)}$$

$$\frac{20x+25+18x-12}{(3x-2)(4x+5)}$$

$$\frac{38x+13}{(3x-2)(4x+5)}$$

61. $\frac{3x}{2x+5}+1$; LCD $=2x+5$

$$\frac{3x}{2x+5}+1\left(\frac{2x+5}{2x+5}\right)$$

$$\frac{3x}{2x+5}+\frac{2x+5}{2x+5}$$

$$\frac{3x+2x+5}{2x+5}$$

$$\frac{5x+5}{2x+5}$$

63. $\frac{4x}{x-5}-3$; LCD $=x-5$

$$\frac{4x}{x-5}-3\left(\frac{x-5}{x-5}\right)$$

$$\frac{4x}{x-5}-\frac{3(x-5)}{x-5}$$

$$\frac{4x-3(x-5)}{x-5}$$

$$\frac{4x-3x+15}{x-5}$$

$$\frac{x+15}{x-5}$$

65. $-1-\frac{3}{2x+1}$; LCD $=2x+1$

$$-\left(\frac{2x+1}{2x+1}\right)-\frac{3}{2x+1}$$

$$\frac{-1(2x+1)}{2x+1}-\frac{3}{2x+1}$$

$$\frac{-1(2x+1)-3}{2x+1}$$

$$\frac{-2x-1-3}{2x+1}$$

$$\frac{-2x-4}{2x+1}$$

67a. $\frac{1}{x-1}-\frac{x}{x-1}$

$$\frac{1-x}{x-1}$$

$$-1$$

b. $\frac{3}{2x-3}-\frac{2x}{2x-3}$

$$\frac{3-2x}{2x-3}$$

$$-1$$

c. $\frac{4}{x-4}-\frac{x}{x-4}+1$

$$\frac{4-x}{x-4}+1$$

$$-1+1$$

$$0$$

d. $-1+\frac{2}{x-2}-\frac{x}{x-2}$

$-1+\frac{2-x}{x-2}$

$-1+(-1)$

-2

PROBLEM SET 9.4 More on Rational Expressions and Complex Fractions

1. $\frac{2x}{x^2+4x}+\frac{5}{x}$

$\frac{2x}{x(x+4)}+\frac{5}{x}$; LCD $= x(x+4)$

$\frac{2x}{x(x+4)}+\frac{5}{x}\left(\frac{x+4}{x+4}\right)$

$\frac{2x+5(x+4)}{x(x+4)}$

$\frac{2x+5x+20}{x(x+4)}$

$\frac{7x+20}{x(x+4)}$

3. $\frac{4}{x^2+7x}-\frac{1}{x}$

$\frac{4}{x(x+7)}-\frac{1}{x}$; LCD $= x(x+7)$

$\frac{4}{x(x+7)}-\frac{1}{x}\left(\frac{x+7}{x+7}\right)$

$\frac{4-1(x+7)}{x(x+7)}$

$\frac{4-x-7}{x(x+7)}$

$\frac{-x-3}{x(x+7)}$

5. $\frac{x}{x^2-1}+\frac{5}{x+1}$

$\frac{x}{(x+1)(x-1)}+\frac{5}{x+1}$

LCD $= (x+1)(x-1)$

$\frac{x}{(x+1)(x-1)}+\frac{5}{(x+1)}\left(\frac{x-1}{x-1}\right)$

$\frac{x+5(x-1)}{(x+1)(x-1)}$

$\frac{x+5x-5}{(x+1)(x-1)}$

$\frac{6x-5}{(x+1)(x-1)}$

7. $\frac{6a+4}{a^2-1}-\frac{5}{a-1}$

$\frac{6a+4}{(a+1)(a-1)}-\frac{5}{a-1}$

LCD $= (a+1)(a-1)$

$\frac{6a+4}{(a+1)(a-1)}-\frac{5}{(a-1)}\left(\frac{a+1}{a+1}\right)$

$\frac{6a+4-5(a+1)}{(a+1)(a-1)}$

$\frac{6a+4-5a-5}{(a+1)(a-1)}$

$\frac{a-1}{(a+1)(a-1)}$

$\frac{1}{a+1}$

9. $\frac{2n}{n^2-25}-\frac{3}{4n+20}$

$\frac{2n}{(n+5)(n-5)}-\frac{3}{4(n+5)}$

LCD $= 4(n+5)(n-5)$

$\frac{2n}{(n+5)(n-5)}\left(\frac{4}{4}\right)-\frac{3}{4(n+5)}\left(\frac{n-5}{n-5}\right)$

$\frac{2n(4)-3(n-5)}{4(n+5)(n-5)}$

Problem Set 9.4

$$\frac{8n-3n+15}{4(n+5)(n-5)}$$
$$\frac{5n+15}{4(n+5)(n-5)}$$

11. $$\frac{5}{x}-\frac{5x-30}{x^2+6x}+\frac{x}{x+6}$$
$$\frac{5}{x}-\frac{5x-30}{x(x+6)}+\frac{x}{x+6}$$
$$\text{LCD}=x(x+6)$$
$$\frac{5}{x}\left(\frac{x+6}{x+6}\right)-\frac{5x-30}{x(x+6)}+\frac{x}{x+6}\left(\frac{x}{x}\right)$$
$$\frac{5(x+6)-(5x-30)+x^2}{x(x+6)}$$
$$\frac{5x+30-5x+30+x^2}{x(x+6)}$$
$$\frac{x^2+60}{x(x+6)}$$

13. $$\frac{3}{x^2+9x+14}+\frac{5}{2x^2+15x+7}$$
$$\frac{3}{(x+2)(x+7)}+\frac{5}{(2x+1)(x+7)}$$
$$\text{LCD}=(x+2)(x+7)(2x+1)$$
$$\frac{3}{(x+2)(x+7)}\left(\frac{2x+1}{2x+1}\right)+\frac{5}{(2x+1)(x+7)}\left(\frac{x+2}{x+2}\right)$$
$$\frac{3(2x+1)+5(x+2)}{(x+2)(2x+1)(x+7)}$$
$$\frac{6x+3+5x+10}{(x+2)(2x+1)(x+7)}$$
$$\frac{11x+13}{(x+2)(2x+1)(x+7)}$$

15. $$\frac{1}{a^2-3a-10}-\frac{4}{a^2+4a-45}$$
$$\frac{1}{(a-5)(a+2)}-\frac{4}{(a+9)(a-5)}$$
$$\text{LCD}=(a-5)(a+2)(a+9)$$
$$\frac{1}{(a-5)(a+2)}\left(\frac{a+9}{a+9}\right)-\frac{4}{(a+9)(a-5)}\left(\frac{a+2}{a+2}\right)$$
$$\frac{1(a+9)-4(a+2)}{(a-5)(a+2)(a+9)}$$
$$\frac{a+9-4a-8}{(a-5)(a+2)(a+9)}$$
$$\frac{-3a+1}{(a-5)(a+2)(a+9)}$$

17. $$\frac{3a}{20a^2-11a-3}+\frac{1}{12a^2+7a-12}$$
$$\frac{3a}{(5a+1)(4a-3)}+\frac{1}{(4a-3)(3a+4)}$$
$$\text{LCD}=(5a+1)(4a-3)(3a+4)$$
$$\frac{3a}{(5a+1)(4a-3)}\left(\frac{3a+4}{3a+4}\right)+\frac{1}{(4a-3)(3a+4)}\left(\frac{5a+1}{5a+1}\right)$$
$$\frac{3a(3a+4)+1(5a+1)}{(5a+1)(4a-3)(3a+4)}$$
$$\frac{9a^2+12a+5a+1}{(5a+1)(4a-3)(3a+4)}$$
$$\frac{9a^2+17a+1}{(2a+1)(4a-3)(3a+4)}$$

19. $$\frac{5}{x^2+3}-\frac{2}{x^2+4x-21}$$
$$\frac{5}{x^2+3}-\frac{2}{(x+7)(x-3)}$$
$$\text{LCD}=(x^2+3)(x+7)(x-3)$$
$$\frac{5}{(x^2+3)}\left(\frac{x+7}{x+7}\right)\left(\frac{x-3}{x-3}\right)-\frac{2}{(x+7)(x-3)}\left(\frac{x^2+3}{x^2+3}\right)$$
$$\frac{5(x+7)(x-3)-2(x^2+3)}{(x^2+3)(x+7)(x-3)}$$
$$\frac{5(x^2+4x-21)-2x^2-6}{(x^2+3)(x+7)(x-3)}$$
$$\frac{5x^2+20x-105-2x^2-6}{(x^2+3)(x+7)(x-3)}$$
$$\frac{3x^2+20x-111}{(x^2+3)(x+7)(x-3)}$$

21. $\frac{2}{y^2+6y-16}-\frac{4}{y+8}-\frac{3}{y-2}$

$\frac{2}{(y+8)(y-2)}-\frac{4}{y+8}-\frac{3}{y-2}$

LCD $=(y+8)(y-2)$

$\frac{2}{(y+8)(y-2)}-\frac{4}{(y+8)}\left(\frac{y-2}{y-2}\right)-\frac{3}{(y-2)}\left(\frac{y+8}{y+8}\right)$

$\frac{2-4(y-2)-3(y+8)}{(y+8)(y-2)}$

$\frac{2-4y+8-3y-24}{(y+8)(y-2)}$

$\frac{-7y-14}{(y+8)(y-2)}$

23. $x-\frac{x^2}{x-2}+\frac{3}{x^2-4}$

$x-\frac{x^2}{x-2}+\frac{3}{(x+2)(x-2)}$

LCD $=(x+2)(x-2)$

$x\left(\frac{x+2}{x+2}\right)\left(\frac{x-2}{x-2}\right)-\frac{x^2}{(x-2)}\left(\frac{x+2}{x+2}\right)+\frac{3}{(x+2)(x-2)}$

$\frac{x(x+2)(x-2)-x^2(x+2)+3}{(x+2)(x-2)}$

$\frac{x(x^2-4)-x^3-2x^2+3}{(x+2)(x-2)}$

$\frac{x^3-4x-x^3-2x^2+3}{(x+2)(x-2)}$

$\frac{-2x^2-4x+3}{(x+2)(x-2)}$

25. $\frac{x+3}{x+10}+\frac{4x-3}{x^2+8x-20}+\frac{x-1}{x-2}$

$\frac{x+3}{x+10}+\frac{4x-3}{(x+10)(x-2)}+\frac{x-1}{x-2}$

LCD $=(x+10)(x-2)$

$\frac{(x+3)}{(x+10)}\left(\frac{x-2}{x-2}\right)+\frac{4x-3}{(x+10)(x-2)}+\frac{(x-1)}{(x-2)}\left(\frac{x+10}{x+10}\right)$

$\frac{(x+3)(x-2)+4x-3+(x-1)(x+10)}{(x+10)(x-2)}$

$\frac{x^2+x-6+4x-3+x^2+9x-10}{(x+10)(x-2)}$

$\frac{2x^2+14x-19}{(x+10)(x-2)}$

27. $\frac{n}{n-6}+\frac{n+3}{n+8}+\frac{12n+26}{n^2+2n-48}$

$\frac{n}{n-6}+\frac{n+3}{n+8}+\frac{12n+26}{(n+8)(n-6)}$

LCD $=(n-6)(n+8)$

$\frac{n}{(n-6)}\left(\frac{n+8}{n+8}\right)+\frac{(n+3)}{(n+8)}\left(\frac{n-6}{n-6}\right)+\frac{12n+26}{(n+8)(n-6)}$

$\frac{n(n+8)+(n+3)(n-6)+12n+26}{(n-6)(n+8)}$

$\frac{n^2+8n+n^2-3n-18+12n+26}{(n-6)(n+8)}$

$\frac{2n^2+17n+8}{(n-6)(n+8)}$

$\frac{(2n+1)(n+8)}{(n-6)(n+8)}$

$\frac{2n+1}{n-6}$

Problem Set 9.4

29. $\dfrac{4x-3}{2x^2+x-1}-\dfrac{2x+7}{3x^2+x-2}-\dfrac{3}{3x-2}$

$$\frac{4x-3}{(2x-1)(x+1)}-\frac{2x+7}{(3x-2)(x+1)}-\frac{3}{3x-2}$$

$$\text{LCD}=(2x-1)(x+1)(3x-2)$$

$$\frac{(4x-3)}{(2x-1)(x+1)}\left(\frac{3x-2}{3x-2}\right)-\frac{(2x+7)}{(3x-2)(x+1)}\left(\frac{2x-1}{2x-1}\right)-\frac{3}{(3x-2)}\left(\frac{2x-1}{2x-1}\right)\left(\frac{x+1}{x+1}\right)$$

$$\frac{(4x-3)(3x-2)-(2x+7)(2x-1)-3(2x-1)(x+1)}{(2x-1)(3x-2)(x+1)}$$

$$\frac{12x^2-17x+6-(4x^2+12x-7)-3(2x^2+x-1)}{(2x-1)(3x-2)(x+1)}$$

$$\frac{12x^2-17x+6-4x^2-12x+7-6x^2-3x+3}{(2x-1)(3x-2)(x+1)}$$

$$\frac{2x^2-32x+16}{(2x-1)(3x-2)(x+1)}$$

31. $\dfrac{n}{n^2+1}+\dfrac{n^2+3n}{n^4-1}-\dfrac{1}{n-1}$

$$\frac{n}{n^2+1}+\frac{n^2+3n}{(n^2+1)(n+1)(n-1)}-\frac{1}{n-1}$$

$$\text{LCD}=(n^2+1)(n+1)(n-1)$$

$$\frac{n}{(n^2+1)}\left(\frac{n+1}{n+1}\right)\left(\frac{n-1}{n-1}\right)+\frac{n^2+3n}{(n^2+1)(n+1)(n-1)}-\frac{1}{(n-1)}\left(\frac{n^2+1}{n^2+1}\right)\left(\frac{n+1}{n+1}\right)$$

$$\frac{n(n+1)(n-1)+n^2+3n-(n^2+1)(n+1)}{(n^2+1)(n+1)(n-1)}$$

$$\frac{n(n^2-1)+n^2+3n-(n^3+n^2+n+1)}{(n^2+1)(n+1)(n-1)}$$

$$\frac{n^3-n+n^2+3n-n^3-n^2-n-1}{(n^2+1)(n+1)(n-1)}$$

$$\frac{n-1}{(n^2+1)(n+1)(n-1)}$$

$$\frac{1}{(n^2+1)(n+1)}$$

33. $$\frac{15x^2-10}{5x^2-7x+2}-\frac{3x+4}{x-1}-\frac{2}{5x-2}$$

$$\frac{15x^2-10}{(5x-2)(x-1)}-\frac{3x+4}{x-1}-\frac{2}{5x-2}$$

LCD $= (5x-2)(x-1)$

$$\frac{15x^2-10}{(5x-2)(x-1)}-\frac{(3x+4)}{(x-1)}\left(\frac{5x-2}{5x-2}\right)-\frac{2}{(5x-2)}\left(\frac{x-1}{x-1}\right)$$

$$\frac{15x^2-10-(3x+4)(5x-2)-2(x-1)}{(5x-2)(x-1)}$$

$$\frac{15x^2-10-(15x^2+14x-8)-2x+2}{(5x-2)(x-1)}$$

$$\frac{15x^2-10-15x^2-14x+8-2x+2}{(5x-2)(x-1)}$$

$$\frac{-16x}{(5x-2)(x-1)}$$

35. $$\frac{t+3}{3t-1}+\frac{8t^2+8t+2}{3t^2-7t+2}-\frac{2t+3}{t-2}$$

$$\frac{t+3}{3t-1}+\frac{8t^2+8t+2}{(3t-1)(t-2)}-\frac{2t+3}{t-2}$$

LCD $= (3t-1)(t-2)$

$$\frac{(t+3)}{(3t-1)}\left(\frac{t-2}{t-2}\right)+\frac{8t^2+8t+2}{(3t-1)(t-2)}-\frac{(2t+3)}{(t-2)}\left(\frac{3t-1}{3t-1}\right)$$

$$\frac{(t+3)(t-2)+8t^2+8t+2-(2t+3)(3t-1)}{(3t-1)(t-2)}$$

$$\frac{t^2+t-6+8t^2+8t+2-(6t^2+7t-3)}{(3t-1)(t-2)}$$

$$\frac{t^2+t-6+8t^2+8t+2-6t^2-7t+3}{(3t-1)(t-2)}$$

$$\frac{3t^2+2t-1}{(3t-1)(t-2)}$$

$$\frac{(3t-1)(t+1)}{(3t-1)(t-2)}$$

$$\frac{t+1}{t-2}$$

Problem Set 9.4

37. $\dfrac{\frac{1}{2}-\frac{1}{4}}{\frac{5}{8}+\frac{3}{4}}=\dfrac{8}{8}\bullet\dfrac{\left(\frac{1}{2}-\frac{1}{4}\right)}{\left(\frac{5}{8}+\frac{3}{4}\right)}=$

$$\frac{8\left(\frac{1}{2}\right)-8\left(\frac{1}{4}\right)}{8\left(\frac{5}{8}\right)+8\left(\frac{3}{4}\right)}=\frac{4-2}{5+6}=\frac{2}{11}$$

39. $\dfrac{\frac{3}{28}-\frac{5}{14}}{\frac{5}{7}+\frac{1}{4}}$

$$\frac{28}{28}\bullet\frac{\left(\frac{3}{28}-\frac{5}{14}\right)}{\left(\frac{5}{7}+\frac{1}{4}\right)}$$

$$\frac{28\left(\frac{3}{28}\right)-28\left(\frac{5}{14}\right)}{28\left(\frac{5}{7}\right)+28\left(\frac{1}{4}\right)}$$

$$\frac{3-10}{20+7}=\frac{-7}{27}=-\frac{7}{27}$$

41. $\dfrac{\frac{5}{6y}}{\frac{10}{3xy}}$

$$\frac{5}{6y}\bullet\frac{3xy}{10}$$

$$\frac{15xy}{60y}$$

$$\frac{x}{4}$$

43. $\dfrac{\frac{3}{x}-\frac{2}{y}}{\frac{4}{y}-\frac{7}{xy}}$

$$\frac{xy}{xy}\bullet\frac{\left(\frac{3}{x}-\frac{2}{y}\right)}{\left(\frac{4}{y}-\frac{7}{xy}\right)}$$

$$\frac{xy\left(\frac{3}{x}\right)-xy\left(\frac{2}{y}\right)}{xy\left(\frac{4}{y}\right)-xy\left(\frac{7}{xy}\right)}$$

$$\frac{3y-2x}{4x-7}$$

45. $\dfrac{\frac{6}{a}-\frac{5}{b^2}}{\frac{12}{a^2}+\frac{2}{b}}$

$$\frac{a^2b^2}{a^2b^2}\bullet\frac{\left(\frac{6}{a}-\frac{5}{b^2}\right)}{\left(\frac{12}{a^2}+\frac{2}{b}\right)}$$

$$\frac{a^2b^2\left(\frac{6}{a}\right)-a^2b^2\left(\frac{5}{b^2}\right)}{a^2b^2\left(\frac{12}{a^2}\right)+a^2b^2\left(\frac{2}{b}\right)}$$

$$\frac{6ab^2-5a^2}{12b^2+2a^2b}$$

47. $\dfrac{\frac{2}{x}-3}{\frac{3}{y}+4}$

$$\frac{xy}{xy}\bullet\frac{\left(\frac{2}{x}-3\right)}{\left(\frac{3}{y}+4\right)}$$

$$\frac{xy\left(\frac{2}{x}\right)-xy(3)}{xy\left(\frac{3}{y}\right)+xy(4)}$$

$$\frac{2y-3xy}{3x+4xy}$$

49. $$\frac{3+\frac{2}{n+4}}{5-\frac{1}{n+4}}$$

$$\frac{(n+4)}{(n+4)}\bullet\frac{\left(3+\frac{2}{n+4}\right)}{\left(5-\frac{1}{n+4}\right)}$$

$$\frac{(n+4)(3)+(n+4)\left(\frac{2}{n+4}\right)}{(n+4)(5)-(n+4)\left(\frac{1}{n+4}\right)}$$

$$\frac{3n+12+2}{5n+20-1}$$

$$\frac{3n+14}{5n+19}$$

51. $$\frac{5-\frac{2}{n-3}}{4-\frac{1}{n-3}}$$

$$\frac{(n-3)}{(n-3)}\bullet\frac{\left(5-\frac{2}{n-3}\right)}{\left(4-\frac{1}{n-3}\right)}$$

$$\frac{(n-3)(5)-(n-3)\left(\frac{2}{n-3}\right)}{(n-3)(4)-(n-3)\left(\frac{1}{n-3}\right)}$$

$$\frac{5n-15-2}{4n-12-1}$$

$$\frac{5n-17}{4n-13}$$

53. $$\frac{\frac{-1}{y-2}+\frac{5}{x}}{\frac{3}{x}-\frac{4}{xy-2x}}$$

$$\frac{-\frac{1}{y-2}+\frac{5}{x}}{\frac{3}{x}-\frac{4}{x(y-2)}}$$

$$\frac{x(y-2)}{x(y-2)}\bullet\frac{\left(-\frac{1}{y-2}+\frac{5}{x}\right)}{\left[\frac{3}{x}-\frac{4}{x(y-2)}\right]}$$

$$\frac{x(y-2)\left(\frac{-1}{y-2}\right)+x(y-2)\left(\frac{5}{x}\right)}{x(y-2)\left(\frac{3}{x}\right)-x(y-2)\left[\frac{4}{x(y-2)}\right]}$$

$$\frac{-x+5(y-2)}{3(y-2)-4}$$

$$\frac{-x+5y-10}{3y-6-4}$$

$$\frac{-x+5y-10}{3y-10}$$

55. $$\frac{\frac{2}{x-3}-\frac{3}{x+3}}{\frac{5}{x^2-9}-\frac{2}{x-3}}$$

$$\frac{\frac{2}{x-3}-\frac{3}{x+3}}{\frac{5}{(x+3)(x-3)}-\frac{2}{x-3}}$$

Problem Set 9.4

$$\frac{(x+3)(x-3)}{(x+3)(x-3)} \bullet \frac{\left(\frac{2}{x-3} - \frac{3}{x+3}\right)}{\left[\frac{5}{(x+3)(x-3)} - \frac{2}{x-3}\right]}$$

$$\frac{(x+3)(x-3)\left(\frac{2}{x-3}\right) - (x+3)(x-3)\left(\frac{3}{x+3}\right)}{(x+3)(x-3)\left[\frac{5}{(x+3)(x-3)}\right] - (x+3)(x-3)\left(\frac{2}{x-3}\right)}$$

$$\frac{2(x+3) - 3(x-3)}{5 - 2(x+3)}$$

$$\frac{2x + 6 - 3x + 9}{5 - 2x - 6}$$

$$\frac{-x + 15}{-2x - 1}$$

57. $$\frac{3a}{2 - \frac{1}{a}} - 1$$

$$\frac{a}{a} \bullet \frac{(3a)}{\left(2 - \frac{1}{a}\right)} - 1$$

$$\frac{a(3a)}{a(2) - a\left(\frac{1}{a}\right)} - 1$$

$$\frac{3a^2}{2a - 1} - 1$$

$$\frac{3a^2}{2a - 1} - 1\left(\frac{2a - 1}{2a - 1}\right)$$

$$\frac{3a^2}{2a - 1} - \frac{2a - 1}{2a - 1}$$

$$\frac{3a^2 - (2a - 1)}{2a - 1}$$

$$\frac{3a^2 - 2a + 1}{2a - 1}$$

59. $$2 - \frac{x}{3 - \frac{2}{x}}$$

$$2 - \frac{x}{x} \bullet \frac{(x)}{\left(3 - \frac{2}{x}\right)}$$

$$2 - \frac{x(x)}{x(3) - x\left(\frac{2}{x}\right)}$$

$$2 - \frac{x^2}{3x - 2}$$

$$2\left(\frac{3x - 2}{3x - 2}\right) - \frac{x^2}{3x - 2}$$

$$\frac{6x - 4}{3x - 2} - \frac{x^2}{3x - 2}$$

$$\frac{6x - 4 - x^2}{3x - 2}$$

$$\frac{-x^2 + 6x - 4}{3x - 2}$$

PROBLEM SET 9.5 Equations Containing Rational Expressions

1. $\frac{x+1}{4}+\frac{x-2}{6}=\frac{3}{4}$

$12\left(\frac{x+1}{4}+\frac{x-2}{6}\right)=12\left(\frac{3}{4}\right)$

$12\left(\frac{x+1}{4}\right)+12\left(\frac{x-2}{6}\right)=3(3)$

$3(x+1)+2(x-2)=9$

$3x+3+2x-4=9$

$5x-1=9$

$5x=10$

$x=2$

The solution set is $\{2\}$.

3. $\frac{x+3}{2}-\frac{x-4}{7}=1$

$14\left(\frac{x+3}{2}-\frac{x-4}{7}\right)=14(1)$

$14\left(\frac{x+3}{2}\right)-14\left(\frac{x-4}{7}\right)=14$

$7(x+3)-2(x-4)=14$

$7x+21-2x+8=14$

$5x+29=14$

$5x=-15$

$x=-3$

The solution set is $\{-3\}$.

5. $\frac{5}{n}+\frac{1}{3}=\frac{7}{n};\ n\neq 0$

$3n\left(\frac{5}{n}+\frac{1}{3}\right)=3n\left(\frac{7}{n}\right)$

$3n\left(\frac{5}{n}\right)+3n\left(\frac{1}{3}\right)=3(7)$

$15+n=21$

$n=6$

The solution set is $\{6\}$.

7. $\frac{7}{2x}+\frac{3}{5}=\frac{2}{3x};\ x\neq 0$

$30x\left(\frac{7}{2x}+\frac{3}{5}\right)=30x\left(\frac{2}{3x}\right)$

$30x\left(\frac{7}{2x}\right)+30x\left(\frac{3}{5}\right)=10(2)$

$15(7)+6x(3)=20$

$105+18x=20$

$18x=-85$

$x=-\frac{85}{18}$

The solution set is $\left\{-\frac{85}{18}\right\}$.

9. $\frac{3}{4x}+\frac{5}{6}=\frac{4}{3x};\ x\neq 0$

$12x\left(\frac{3}{4x}+\frac{5}{6}\right)=12x\left(\frac{4}{3x}\right)$

$12x\left(\frac{3}{4x}\right)+12x\left(\frac{5}{6}\right)=4(4)$

$3(3)+2x(5)=16$

$9+10x=16$

$10x=7$

$x=\frac{7}{10}$

The solution set is $\left\{\frac{7}{10}\right\}$.

11. $\frac{47-n}{n}=8+\frac{2}{n};\ n\neq 0$

$n\left(\frac{47-n}{n}\right)=n\left(8+\frac{2}{n}\right)$

$47-n=n(8)+n\left(\frac{2}{n}\right)$

$47-n=8n+2$

$47=9n+2$

$45=9n$

$5=n$

The solution set is $\{5\}$.

13. $\frac{n}{65-n}=8+\frac{2}{65-n};\ n\neq 65$

$(65-n)\left(\frac{n}{65-n}\right)=(65-n)\left(8+\frac{2}{65-n}\right)$

$n=(65-n)(8)+(65-n)\left(\frac{2}{65-n}\right)$

$n=520-8n+2$

$9n=522$

$n=58$

The solution set is $\{58\}$.

Problem Set 9.5

15. $n+\frac{1}{n}=\frac{17}{4}; n\neq 0$

$4n\left(n+\frac{1}{n}\right)=4n\left(\frac{17}{4}\right)$

$4n(n)+4n\left(\frac{1}{n}\right)=17n$

$4n^2+4=17n$

$4n^2-17n+4=0$

$(4n-1)(n-4)=0$

$4n-1=0$ or $n-4=0$

$4n=1$ or $n=4$

$n=\frac{1}{4}$ or $n=4$

The solution set is $\left\{\frac{1}{4}, 4\right\}$.

17. $n-\frac{2}{n}=\frac{23}{5}; n\neq 0$

$5n\left(n-\frac{2}{n}\right)=5n\left(\frac{23}{5}\right)$

$5n(n)-5n\left(\frac{2}{n}\right)=23n$

$5n^2-10=23n$

$5n^2-23n-10=0$

$(5n+2)(n-5)=0$

$5n+2=0$ or $n-5=0$

$5n=-2$ or $n=5$

$n=-\frac{2}{5}$ or $n=5$

The solution set is $\left\{-\frac{2}{5}, 5\right\}$.

19. $\frac{5}{7x-3}=\frac{3}{4x-5}; x\neq\frac{3}{7}, x\neq\frac{5}{4}$

$5(4x-5)=3(7x-3)$

$20x-25=21x-9$

$20x=21x+16$

$-x=16$

$x=-16$

The solution set is $\{-16\}$.

21. $\frac{-2}{x-5}=\frac{1}{x+9}; x\neq 5, x\neq -9$

$-2(x+9)=1(x-5)$

$-2x-18=x-5$

$-2x=x+13$

$-3x=13$

$x=-\frac{13}{3}$

The solution set is $\left\{-\frac{13}{3}\right\}$.

23. $\frac{x}{x+1}-2=\frac{3}{x-3}; x\neq -1, x\neq 3$

$(x+1)(x-3)\left(\frac{x}{x+1}-2\right)=(x+1)(x-3)\left(\frac{3}{x-3}\right)$

$(x+1)(x-3)\left(\frac{x}{x+1}\right)-2(x+1)(x-3)=3(x+1)$

$x(x-3)-2(x^2-2x-3)=3x+3$

$x^2-3x-2x^2+4x+6=3x+3$

$-x^2+x+6=3x+3$

$-x^2-2x+3=0$

$x^2+2x-3=0$

$(x+3)(x-1)=0$

$x+3=0$ or $x-1=0$

$x=-3$ or $x=1$

The solution set is $\{-3, 1\}$.

25. $\frac{a}{a+5}-2=\frac{3a}{a+5}; a\neq -5$

$(a+5)\left(\frac{a}{a+5}-2\right)=(a+5)\left(\frac{3a}{a+5}\right)$

$(a+5)\left(\frac{a}{a+5}\right)-2(a+5)=3a$

$a-2a-10=3a$

$-a-10=3a$

$-10=4a$

$-\frac{10}{4}=a$

$-\frac{5}{2}=a$

The solution set is $\left\{-\frac{5}{2}\right\}$.

27. $\frac{5}{x+6}=\frac{6}{x-3}; x\neq -6, x\neq 3$

$5(x-3)=6(x+6)$

$5x-15=6x+36$

$5x=6x+51$

$-x=51$

$x=-51$

The solution set is $\{-51\}$

29. $\frac{3x-7}{10} = \frac{2}{x};\ x \neq 0$

$x(3x-7) = 10(2)$

$3x^2 - 7x = 20$

$3x^2 - 7x - 20 = 0$

$(3x+5)(x-4) = 0$

$3x + 5 = 0$ or $x - 4 = 0$

$3x = -5$ or $x = 4$

$x = -\frac{5}{3}$ or $x = 4$

The solution set is $\left\{-\frac{5}{3}, 4\right\}$.

31. $\frac{x}{x-6} - 3 = \frac{6}{x-6};\ x \neq 6$

$(x-6)\left(\frac{x}{x-6} - 3\right) = (x-6)\left(\frac{6}{x-6}\right)$

$(x-6)\left(\frac{x}{x-6}\right) - 3(x-6) = 6$

$x - 3(x-6) = 6$

$x - 3x + 18 = 6$

$-2x + 18 = 6$

$-2x = -12$

$x = 6$

Discard the solution $x = 6$.

The solution set is $\emptyset$.

33. $\frac{3s}{s+2} + 1 = \frac{35}{2(3s+1)};\ s \neq -\frac{1}{3}, s \neq -2$

$2(s+2)(3s+1)\left[\frac{3s}{s+2} + 1\right] = 2(s+2)(3s+1)\left[\frac{35}{2(3s+1)}\right]$

$2(s+2)(3s+1)\left(\frac{3s}{s+2}\right) + 2(s+2)(3s+1)(1) = 35(s+2)$

$6s(3s+1) + 2(3s^2 + 7s + 2) = 35s + 70$

$18s^2 + 6s + 6s^2 + 14s + 4 = 35s + 70$

$24s^2 + 20s + 4 = 35s + 70$

$24s^2 - 15s - 66 = 0$

$3(8s^2 - 5s - 22) = 0$

$3(8s+11)(s-2) = 0$

$8s + 11 = 0$ or $s - 2 = 0$

$8s = -11$ or $s = 2$

$s = -\frac{11}{8}$ or $s = 2$

The solution set is $\left\{-\frac{11}{8}, 2\right\}$.

35. $2 - \frac{3x}{x-4} = \frac{14}{x+7};\ x \neq 4, x \neq -7$

$(x-4)(x+7)\left(2 - \frac{3x}{x-4}\right) = (x-4)(x+7)\left(\frac{14}{x+7}\right)$

$(x-4)(x+7)(2) - (x-4)(x+7)\left(\frac{3x}{x-4}\right) = 14(x-4)$

$2(x^2 + 3x - 28) - 3x(x+7) = 14x - 56$

$2x^2 + 6x - 56 - 3x^2 - 21x = 14x - 56$

$-x^2 - 15x - 56 = 14x - 56$

$0 = x^2 + 29x$

$0 = x(x+29)$

$x = 0$ or $x + 29 = 0$

$x = 0$ or $x = -29$

The solution set is $\{-29, 0\}$.

Problem Set 9.5

37. $\frac{n+6}{27} = \frac{1}{n}; n \neq 0$

$n(n+6) = 1(27)$

$n^2 + n = 27$

$n^2 + 6n - 27 = 0$

$(n+9)(n-3) = 0$

$n + 9 = 0 \quad \text{or} \quad n - 3 = 0$

$n = -9 \quad \text{or} \quad n = 3$

The solution set is $\{-9, 3\}$.

39. $\frac{3n}{n-1} - \frac{1}{3} = \frac{-40}{3n-18}$

$\frac{3n}{n-1} - \frac{1}{3} = \frac{-40}{3(n-6)}; n \neq 1, n \neq 6$

$3(n-1)(n-6)\left(\frac{3n}{n-1} - \frac{1}{3}\right) = 3(n-1)(n-6)\left[\frac{-40}{3(n-6)}\right]$

$3(n-1)(n-6)\left(\frac{3n}{n-1}\right) - 3(n-1)(n-6)\left(\frac{1}{3}\right) = -40(n-1)$

$9n(n-6) - (n-1)(n-6) = -40n + 40$

$9n^2 - 54n - (n^2 - 7n + 6) = -40n + 40$

$9n^2 - 54n - n^2 + 7n - 6 = -40n + 40$

$8n^2 - 47n - 6 = -40n + 40$

$8n^2 - 7n - 46 = 0$

$(8n - 23)(n + 2) = 0$

$8n - 23 = 0 \quad \text{or} \quad n + 2 = 0$

$8n = 23 \quad \text{or} \quad n = -2$

$n = \frac{23}{8} \quad \text{or} \quad n = -2$

The solution set is $\left\{-2, \frac{23}{8}\right\}$.

41. $\frac{-3}{4x+5} = \frac{2}{5x-7}; x \neq -\frac{5}{4}, x \neq \frac{7}{5}$

$-3(5x-7) = 2(4x+5)$

$-15x + 21 = 8x + 10$

$-15x = 8x - 11$

$-23x = -11$

$x = \frac{-11}{-23} = \frac{11}{23}$

The solution set is $\left\{\frac{11}{23}\right\}$.

43. Let d = the first amount and $1750 - d$ = the second amount.

$\frac{d}{1750-d} = \frac{3}{4}; d \neq 1750$

$4d = 3(1750 - d)$

$4d = 5250 - 3d$

$7d = 5250$

$d = 750$

The amounts are \$750 and \$1000.

45. Let one angle $= x$ and the other angle $= 180 - x - 60 = 120 - x$.

$$\frac{x}{120 - x} = \frac{2}{3}; x \neq 120$$
$$3x = 2(120 - x)$$
$$3x = 240 - 2x$$
$$5x = 240$$
$$x = 48$$

The angles are 48° and 72°.

47. Let the angle $= x$, the complement of angle $= 90 - x$, and the supplement of angle $= 180 - x$.

$$\frac{90 - x}{180 - x} = \frac{1}{4}; x \neq 90, x \neq 180$$
$$4(90 - x) = 1(180 - x)$$
$$360 - 4x = 180 - x$$
$$-4x = -180 - x$$
$$-3x = -180$$
$$x = 60$$

The measure of the angle is 60°.

49. Let $x =$ the taxes on a home valued at \$180,000.

$$\frac{150,000}{1,900} = \frac{180,000}{x}; x \neq 0$$
$$150000x = 180000(1900)$$
$$150000x = 342000000$$
$$x = 2280$$

The taxes are \$2,280.

51. Let $x =$ Laura's sales and $210 - x =$ Tammy's sales.

$$\frac{210 - x}{x} = \frac{4}{3}; x \neq 0$$
$$4x = 3(210 - x)$$
$$4x = 630 - 3x$$
$$7x = 630$$
$$x = 90$$

Laura's sales are \$90 and Tammy's sales are \$120.

53. Let $x =$ the smaller number and $90 - x =$ the larger number.

$$\frac{90 - x}{x} = 10 + \frac{2}{x}; x \neq 0$$
$$x\left(\frac{90 - x}{x}\right) = x\left(10 + \frac{2}{x}\right)$$
$$90 - x = x(10) + x\left(\frac{2}{x}\right)$$
$$90 - x = 10x + 2$$
$$-x = 10x - 88$$
$$-11x = -88$$
$$x = 8$$

The numbers are 8 and 82.

55. Let $x =$ the first piece and $20 - x =$ the second piece.

$$\frac{x}{20 - x} = \frac{7}{3}; x \neq 20$$
$$3x = 7(20 - x)$$
$$3x = 140 - 7x$$
$$10x = 140$$
$$x = 14$$

The lengths are 14 feet and 6 feet.

57. Let $x =$ the number of female voters and $21150 - x =$ the number of male voters.

$$\frac{x}{21150 - x} = \frac{3}{2}; x \neq 21150$$
$$2x = 3(21150 - x)$$
$$2x = 63450 - 3x$$
$$5x = 63450$$
$$x = 12690$$

They were 12,690 female voters and 8,460 male voters.

PROBLEM SET 9.6 More on Rational Equations and Applications

1. $$\frac{x}{4x - 4} + \frac{5}{x^2 - 1} = \frac{1}{4}$$
$$\frac{x}{4(x - 1)} + \frac{5}{(x + 1)(x - 1)} = \frac{1}{4}; x \neq 1, x \neq -1$$

$$4(x-1)(x+1)\left[\frac{x}{4(x-1)}+\frac{5}{(x+1)(x-1)}\right]=4(x+1)(x-1)\left(\frac{1}{4}\right)$$

$$4(x-1)(x+1)\left[\frac{x}{4(x-1)}\right]+4(x-1)(x+1)\left[\frac{5}{(x+1)(x-1)}\right]=(x+1)(x-1)$$

$x(x+1)+4(5)=x^2-1$

$x^2+x+20=x^2-1$

$x+20=-1$

$x=-21$

The solution set is $\{-21\}$.

3. $3+\dfrac{6}{t-3}=\dfrac{6}{t^2-3t}$

$3+\dfrac{6}{t-3}=\dfrac{6}{t(t-3)};\ t\neq 0, t\neq 3$

$t(t-3)\left(3+\dfrac{6}{t-3}\right)=t(t-3)\left[\dfrac{6}{t(t-3)}\right]$

$3t(t-3)+6t=6$

$3t^2-9t+6t=6$

$3t^2-3t=6$

$3t^2-3t-6=0$

$3(t^2-t-2)=0$

$3(t-2)(t+1)=0$

$t-2=0$ or $t+1=0$

$t=2$ or $t=-1$

The solution set is $\{-1, 2\}$.

5. $\dfrac{3}{n-5}+\dfrac{4}{n+7}=\dfrac{2n+11}{n^2+2n-35}$

$\dfrac{3}{n-5}+\dfrac{4}{n+7}=\dfrac{2n+11}{(n+7)(n-5)};\ n\neq -7, n\neq 5$

$(n+7)(n-5)\left(\dfrac{3}{n-5}+\dfrac{4}{n+7}\right)=(n+7)(n-5)\left[\dfrac{2n+11}{(n+7)(n-5)}\right]$

$(n+7)(n-5)\left(\dfrac{3}{n-5}\right)+(n+7)(n-5)\left(\dfrac{4}{n+7}\right)=2n+11$

$3(n+7)+4(n-5)=2n+11$

$3n+21+4n-20=2n+11$

$7n+1=2n+11$

$7n=2n+10$

$5n=10$

$n=2$

The solution set is $\{2\}$.

7. $\dfrac{5x}{2x+6}-\dfrac{4}{x^2-9}=\dfrac{5}{2}$

$\dfrac{5x}{2(x+3)}-\dfrac{4}{(x+3)(x-3)}=\dfrac{5}{2};\ x\neq 3, x\neq -3$

$$2(x+3)(x-3)\left[\frac{5x}{2(x+3)}-\frac{4}{(x+3)(x-3)}\right]=2(x+3)(x-3)\left(\frac{5}{2}\right)$$
$$2(x+3)(x-3)\left[\frac{5x}{2(x+3)}\right]-2(x+3)(x-3)\left[\frac{4}{(x+3)(x-3)}\right]=5(x+3)(x-3)$$
$$5x(x-3)-8=5(x^2-9)$$
$$5x^2-15x-8=5x^2-45$$
$$-15x-8=-45$$
$$-15x=-37$$
$$x=\frac{37}{15}$$
The solution set is $\left\{\frac{37}{15}\right\}$.

9. $$1+\frac{1}{n-1}=\frac{1}{n^2-n}$$
$$1+\frac{1}{n-1}=\frac{1}{n(n-1)};\ n\neq 0,\ n\neq 1$$
$$n(n-1)\left(1+\frac{1}{n-1}\right)=n(n-1)\left[\frac{1}{n(n-1)}\right]$$
$$n(n-1)+n(n-1)\left(\frac{1}{n-1}\right)=1$$
$$n^2-n+n=1$$
$$n^2=1$$
$$n^2-1=0$$
$$(n+1)(n-1)=0$$
$n+1=0$ or $n-1=0$
$n=-1$ or $n=1$
Discard the solution $n=1$.
The solution set is $\{-1\}$.

11. $$\frac{2}{n-2}-\frac{n}{n+5}=\frac{10n+15}{n^2+3n-10}$$
$$\frac{2}{n-2}-\frac{n}{n+5}=\frac{10n+15}{(n+5)(n-2)};\ n\neq -5,\ n\neq 2$$
$$(n+5)(n-2)\left(\frac{2}{n-2}-\frac{n}{n+5}\right)=(n+5)(n-2)\left[\frac{10n+15}{(n+5)(n-2)}\right]$$
$$(n+5)(n-2)\left(\frac{2}{n-2}\right)-(n+5)(n-2)\left(\frac{n}{n+5}\right)=10n+15$$
$$2(n+5)-n(n-2)=10n+15$$
$$2n+10-n^2+2n=10n+15$$
$$-n^2+4n+10=10n+15$$
$$0=n^2+6n+5$$
$$0=(n+5)(n+1)$$
$n+5=0$ or $n+1=0$
$n=-5$ or $n=-1$
Discard the solution $x=-5$. The solution set is $\{-1\}$.

Problem Set 9.6

13. $\frac{2}{2x-3} - \frac{2}{10x^2-13x-3} = \frac{x}{5x+1}$

$\frac{2}{2x-3} - \frac{2}{(5x+1)(2x-3)} = \frac{x}{5x+1}; x \neq \frac{3}{2}, x \neq -\frac{1}{5}$

$(2x-3)(5x+1)\left[\frac{2}{2x-3} - \frac{2}{(5x+1)(2x-3)}\right] = (2x-3)(5x+1)\left(\frac{x}{5x+1}\right)$

$(2x-3)(5x+1)\left(\frac{2}{2x-3}\right) - (2x+3)(5x+1)\left[\frac{2}{(5x+1)(2x-3)}\right] = x(2x-3)$

$2(5x+1) - 2 = 2x^2 - 3x$

$10x + 2 - 2 = 2x^2 - 3x$

$0 = 2x^2 - 13x$

$0 = x(2x-13)$

$x = 0 \quad \text{or} \quad 2x - 13 = 0$

$x = 0 \quad \text{or} \quad 2x = 13$

$x = 0 \quad \text{or} \quad x = \frac{13}{2}$

The solution set is $\left\{0, \frac{13}{2}\right\}$.

15. $\frac{2x}{x+3} - \frac{3}{x-6} = \frac{29}{x^2-3x-18}$

$\frac{2x}{x+3} - \frac{3}{x-6} = \frac{29}{(x-6)(x+3)}; x \neq 6, x \neq -3$

$(x-6)(x+3)\left(\frac{2x}{x+3} - \frac{3}{x-6}\right) = (x-6)(x+3)\left[\frac{29}{(x-6)(x+3)}\right]$

$(x-6)(x+3)\left(\frac{2x}{x+3}\right) - (x-6)(x+3)\left(\frac{3}{x-6}\right) = 29$

$2x(x-6) - 3(x+3) = 29$

$2x^2 - 12x - 3x - 9 = 29$

$2x^2 - 15x - 38 = 0$

$(2x-19)(x+2) = 0$

$2x - 19 = 0 \quad \text{or} \quad x + 2 = 0$

$2x = 19 \quad \text{or} \quad x = -2$

$x = \frac{19}{2} \quad \text{or} \quad x = -2$

The solution set is $\left\{-2, \frac{19}{2}\right\}$.

17. $\frac{a}{a-5} + \frac{2}{a-6} = \frac{2}{a^2-11a+30}$

$\frac{a}{a-5} + \frac{2}{a-6} = \frac{2}{(a-6)(a-5)}; a \neq 6, a \neq 5$

$(a-6)(a-5)\left(\frac{a}{a-5} + \frac{2}{a-6}\right) = (a-6)(a-5)\left[\frac{2}{(a-6)(a-5)}\right]$

$(a-6)(a-5)\left(\frac{a}{a-5}\right)+(a-6)(a-5)\left(\frac{2}{a-6}\right)=2$

$a(a-6)+2(a-5)=2$

$a^2-6a+2a-10=2$

$a^2-4a-12=0$

$(a-6)(a+2)=0$

$a-6=0$ or $a+2=0$

$a=6$ or $a=-2$

Discard the solution $a=6$. The solution set is $\{-2\}$.

19. $\frac{-1}{2x-5}+\frac{2x-4}{4x^2-25}=\frac{5}{6x+15}$

$\frac{-1}{2x-5}+\frac{2x-4}{(2x+5)(2x-5)}=\frac{5}{3(2x+5)}; x\neq\frac{5}{2}, x\neq-\frac{5}{2}$

$3(2x+5)(2x-5)\left[\frac{-1}{2x-5}+\frac{2x-4}{(2x+5)(2x-5)}\right]=3(2x+5)(2x-5)\left[\frac{5}{3(2x+5)}\right]$

$3(2x+5)(2x-5)\left(\frac{-1}{2x-5}\right)+3(2x+5)(2x-5)\left[\frac{2x-4}{(2x+5)(2x-5)}\right]=5(2x-5)$

$-3(2x+5)+3(2x-4)=10x-25$

$-6x-15+6x-12=10x-25$

$-27=10x-25$

$-2=10x$

$-\frac{2}{10}=x$

$-\frac{1}{5}=x$

The solution set is $\left\{-\frac{1}{5}\right\}$.

21. $\frac{7y+2}{12y^2+11y-15}-\frac{1}{3y+5}=\frac{2}{4y-3}$

$\frac{7y+2}{(3y+5)(4y-3)}-\frac{1}{3y+5}=\frac{2}{4y-3}; y\neq-\frac{5}{3}, y\neq\frac{3}{4}$

$(3y+5)(4y-3)\left[\frac{7y+2}{(3y+5)(4y-3)}-\frac{1}{3y+5}\right]=(3y+5)(4y-3)\left(\frac{2}{4y-3}\right)$

$(3y+5)(4y-3)\left[\frac{7y+2}{(3y+5)(4y-3)}\right]-(3y+5)(4y-3)\left(\frac{1}{3y+5}\right)=2(3y+5)$

$7y+2-(4y-3)=2(3y+5)$

$7y+2-4y+3=6y+10$

$3y+5=6y+10$

$3y=6y+5$

$-3y=5$

$y=-\frac{5}{3}$

Discard the solution $y=-\frac{5}{3}$. The solution set is $\emptyset$.

Problem Set 9.6

23. $\frac{2n}{6n^2+7n-3} - \frac{n-3}{3n^2+11n-4} = \frac{5}{2n^2+11n+12}$

$\frac{2n}{(3n-1)(2n+3)} - \frac{n-3}{(3n-1)(n+4)} = \frac{5}{(2n+3)(n+4)}; n \neq \frac{1}{3}, n \neq -\frac{3}{2}, n \neq -4$

$(3n-1)(2n+3)(n+4)\left[\frac{2n}{(3n-1)(2n+3)} - \frac{n-3}{(3n-1)(n+4)}\right] = (3n-1)(2n+3)(n+4)\left[\frac{5}{(2n+3)(n+4)}\right]$

$(3n\text{-}1)(2n+3)(n+4)\left[\frac{2n}{(3n\text{-}1)(2n+3)}\right] - (3n\text{-}1)(2n+3)(n+4)\left[\frac{n-3}{(3n-1)(n+4)}\right] = (3n\text{-}1)(2n+3)(n+4)\left[\frac{5}{(2n+3)(n+4)}\right]$

$2n(n+4) - (2n+3)(n-3) = 5(3n-1)$

$2n^2 + 8n - (2n^2 - 3n - 9) = 15n - 5$

$2n^2 + 8n - 2n^2 + 3n + 9 = 15n - 5$

$11n + 9 = 15n - 5$

$11n = 15n - 14$

$-4n = -14$

$n = \frac{-14}{-4} = \frac{7}{2}$

The solution set is $\left\{\frac{7}{2}\right\}$.

25. $\frac{1}{2x^2-x-1} + \frac{3}{2x^2+x} = \frac{2}{x^2-1}$

$\frac{1}{(2x+1)(x-1)} + \frac{3}{x(2x+1)} = \frac{2}{(x+1)(x-1)}; x \neq -\frac{1}{2}, x \neq 1, x \neq -1, x \neq 0$

$x(2x+1)(x+1)(x-1)\left[\frac{1}{(2x+1)(x-1)} + \frac{3}{x(2x+1)}\right] = x(2x+1)(x-1)(x+1)\left[\frac{2}{(x+1)(x-1)}\right]$

$x(2x+1)(x+1)(x-1)\left[\frac{1}{(2x+1)(x-1)}\right] + x(2x+1)(x+1)(x-1)\left[\frac{3}{x(2x+1)}\right] = 2x(2x+1)$

$x(x+1) + 3(x+1)(x-1) = 2x(2x+1)$

$x^2 + x + 3(x^2-1) = 4x^2 + 2x$

$x^2 + x + 3x^2 - 3 = 4x^2 + 2x$

$4x^2 + x - 3 = 4x^2 + 2x$

$x - 3 = 2x$

$-3 = x$

The solution set is $\{-3\}$.

27. $\frac{x+1}{x^3-9x} - \frac{1}{2x^2+x-21} = \frac{1}{2x^2+13x+21}$

$\frac{x+1}{x(x+3)(x-3)} - \frac{1}{(2x+7)(x-3)} = \frac{1}{(2x+7)(x+3)}; x \neq 0, x \neq 3, x \neq -3, x \neq -\frac{7}{2}$

$x(2x+7)(x+3)(x-3)\left[\frac{x+1}{x(x+3)(x-3)} - \frac{1}{(2x+7)(x-3)}\right] = x(2x+7)(x+3)(x-3)\left[\frac{1}{(2x+7)(x+3)}\right]$

$x(2x+7)(x+3)(x-3)\left[\frac{x+1}{x(x+3)(x-3)}\right] - x(2x+7)(x+3)(x-3)\left[\frac{1}{(2x+7)(x-3)}\right] = x(x-3)$

$(2x+7)(x+1) - x(x+3) = x(x-3)$

$2x^2 + 9x + 7 - x^2 - 3x = x^2 - 3x$

$x^2 + 6x + 7 = x^2 - 3x$

$6x + 7 = -3x$

$7 = -9x$

$-\frac{7}{9} = x$

The solution set is $\left\{-\frac{7}{9}\right\}$.

29. $\frac{4t}{4t^2 - t - 3} + \frac{2 - 3t}{3t^2 - t - 2} = \frac{1}{12t^2 + 17t + 6}$

$\frac{4t}{(4t + 3)(t - 1)} + \frac{2 - 3t}{(3t + 2)(t - 1)} = \frac{1}{(4t + 3)(3t + 2)}; t \neq -\frac{3}{4}, t \neq 1, t \neq -\frac{2}{3}$

$(4t+3)(t - 1)(3t+2)\left[\frac{4t}{(4t+3)(t - 1)} + \frac{2 - 3t}{(3t+2)(t - 1)}\right] = (4t+3)(t - 1)(3t+2)\left[\frac{1}{(4t+3)(3t+2)}\right]$

$(4t+3)(t - 1)(3t+2)\left[\frac{4t}{(4t+3)(t - 1)}\right] + (4t+3)(t - 1)(3t+2)\left[\frac{2 - 3t}{(3t+2)(t - 1)}\right] = (4t+3)(t - 1)(3t+2)\left[\frac{1}{(4t+3)(3t+2)}\right]$

$4t(3t + 2) + (4t + 3)(2 - 3t) = t - 1$

$12t^2 + 8t - 12t^2 - t + 6 = t - 1$

$7t + 6 = t - 1$

$7t = t - 7$

$6t = -7$

$t = -\frac{7}{6}$

The solution set is $\left\{-\frac{7}{6}\right\}$.

31. $y = \frac{5}{6}x + \frac{2}{9}$ for x

$18(y) = 18\left(\frac{5}{6}x + \frac{2}{9}\right)$

$18y = 18\left(\frac{5}{6}x\right) + 18\left(\frac{2}{9}\right)$

$18y = 15x + 4$

$18y - 4 = 15x$

$\frac{1}{15}(18y - 4) = \frac{1}{15}(15x)$

$\frac{18y - 4}{15} = x$

33. $\frac{-2}{x - 4} = \frac{5}{y - 1}$ for y

$-2(y - 1) = 5(x - 4)$

$-2y + 2 = 5x - 20$

$-2y = 5x - 22$

$-\frac{1}{2}(-2y) = -\frac{1}{2}(5x - 22)$

$y = \frac{-5x + 22}{2}$

35. $I = \frac{100M}{C}$ for M

$\frac{CI}{100} = M$

$M = \frac{CI}{100}$

Problem Set 9.6

37. $\dfrac{R}{S} = \dfrac{T}{S+T}$ for R

$S\left(\dfrac{R}{S}\right) = S\left(\dfrac{T}{S+T}\right)$

$R = \dfrac{ST}{S+T}$

39. $\dfrac{y-1}{x-3} = \dfrac{b-1}{a-3}$ for y

$(y-1)(a-3) = (x-3)(b-1)$

$y(a-3) - 1(a-3) = x(b-1) - 3(b-1)$

$y(a-3) - a + 3 = bx - x - 3b + 3$

$y(a-3) = bx - x - 3b + 3 + a - 3$

$y(a-3) = bx - x - 3b + a$

$y = \dfrac{bx - x - 3b + a}{a-3}$

41. $\dfrac{x}{a} + \dfrac{y}{b} = 1$ for y

$ab\left(\dfrac{x}{a} + \dfrac{y}{b}\right) = ab(1)$

$ab\left(\dfrac{x}{a}\right) + ab\left(\dfrac{y}{b}\right) = ab$

$bx + ay = ab$

$ay = ab - bx$

$y = \dfrac{ab - bx}{a}$

43. $\dfrac{y-1}{x+6} = \dfrac{-2}{3}$ for y

$3(y-1) = -2(x+6)$

$3y - 3 = -2x - 12$

$3y = -2x - 9$

$y = \dfrac{-2x-9}{3}$

45.

	Rate	Time	Distance
Kent	$x+4$	$\dfrac{270}{x+4}$	270
Dave	x	$\dfrac{250}{x}$	250

$\dfrac{270}{x+4} = \dfrac{250}{x}$

$270x = 250(x+4)$

$270x = 250x + 1000$

$20x = 1000$

$x = 50$

Dave's rate is 50 mph and Kent's rate is 54 mph.

47.

	Time in Minutes	Rate
Inlet	10	$\dfrac{1}{10}$
Outlet	12	$\dfrac{1}{12}$
Together	x	$\dfrac{1}{x}$

$\dfrac{1}{10} - \dfrac{1}{12} = \dfrac{1}{x}$

$60x\left(\dfrac{1}{10} - \dfrac{1}{12}\right) = 60x\left(\dfrac{1}{x}\right)$

$60x\left(\dfrac{1}{10}\right) - 60x\left(\dfrac{1}{12}\right) = 60$

$6x - 5x = 60$

$x = 60$

The tank will overflow in 60 minutes.

49.

	Rate	Time	Words
Connie	$x+20$	$\dfrac{600}{x+20}$	600
Katie	x	$\dfrac{600}{x}$	600

$\dfrac{600}{x+20} + 5 = \dfrac{600}{x}$

$x(x+20)\left(\dfrac{600}{x+20} + 5\right) = x(x+20)\left(\dfrac{600}{x}\right)$

$x(x+20)\left(\dfrac{600}{x+20}\right) + 5x(x+20) = 600(x+20)$

$600x + 5x^2 + 100x = 600x + 12000$

$5x^2 + 700x = 600x + 12000$

$5x^2 + 100x - 12000 = 0$

$5(x^2 + 20x - 2400) = 0$

$5(x+60)(x-40) = 0$

$x + 60 = 0$ or $x - 40$

$x = -60$ or $x = 40$

Discard the root $x = -60$. Katie's rate is 40 words per minute and Connie's rate is 60 words per minute.

51.

	Rate	Time	Distance
Plane A	x	$\frac{1400}{x}$	1400
Plane B	$x+50$	$\frac{2000}{x+50}$	2000

$$\frac{1400}{x}+1=\frac{2000}{x+50}$$
$$x(x+50)\left(\frac{1400}{x}+1\right)=x(x+50)\left(\frac{2000}{x+50}\right)$$
$$x(x+50)\left(\frac{1400}{x}\right)+x(x+50)=2000x$$
$$1400(x+50)+x^2+50x=2000x$$
$$1400x+70000+x^2+50x=2000x$$
$$x^2+1450x+70000=2000x$$
$$x^2-550x+70000=0$$
$$(x-350)(x-200)=0$$
$$x-350=0 \quad \text{or} \quad x-200=0$$
$$x=350 \quad \text{or} \quad x=200$$

Case 1: Plane B travels at 400 mph for 5 hours and Plane A travels at 350 mph for 4 hours.

Case 2: Plane B travels at 250mph for 8 hours and Plane A travels at 200 mph for 7 hours.

53.

	Time in Minutes	Rate
Amy	$2x$	$\frac{1}{2x}$
Nancy	x	$\frac{1}{x}$
Together	40	$\frac{1}{40}$

$$\frac{1}{2x}+\frac{1}{x}=\frac{1}{40}$$
$$40x\left(\frac{1}{2x}+\frac{1}{x}\right)=40x\left(\frac{1}{40}\right)$$
$$40x\left(\frac{1}{2x}\right)+40x\left(\frac{1}{x}\right)=x$$
$$20+40=x$$
$$60=x$$

It would take Nancy 60 minutes and Amy would take 120 minutes.

55.

	Rate	Time	Money
Anticipated	$\frac{12}{x}$	x	12
Actual	$\frac{12}{x+1}$	$x+1$	12

$$\frac{12}{x}-1=\frac{12}{x+1}$$
$$x(x+1)\left(\frac{12}{x}-1\right)=x(x+1)\left(\frac{12}{x+1}\right)$$
$$x(x+1)\left(\frac{12}{x}\right)-x(x+1)=12x$$
$$12(x+1)-x^2-x=12x$$
$$12x+12-x^2-x=12x$$
$$-x^2+11x+12=12x$$
$$0=x^2+x-12$$
$$(x+4)(x-3)=0$$
$$x+4=0 \quad \text{or} \quad x-3=0$$
$$x=-4 \quad \text{or} \quad x=3$$

Discard the root $x=-4$.
The anticipated time was 3 hours.

57.

	Rate	Time	Distance
Out	$x+4$	$\frac{24}{x+4}$	24
Back	x	$\frac{12}{x}$	12

$$\frac{24}{x+4}-\frac{1}{2}=\frac{12}{x}$$
$$2x(x+4)\left(\frac{24}{x+4}-\frac{1}{2}\right)=2x(x+4)\left(\frac{12}{x}\right)$$
$$2x(x+4)\left(\frac{24}{x+4}\right)-2x(x+4)\left(\frac{1}{2}\right)=24(x+4)$$
$$2x(24)-x(x+4)=24x+96$$
$$48x-x^2-4x=24x+96$$
$$-x^2+44x=24x+96$$
$$0=x^2-20x+96$$
$$0=(x-12)(x-8)$$
$$x-12=0 \quad \text{or} \quad x-8=0$$
$$x=12 \quad \text{or} \quad x=8$$

Case 1: 16 mph on the way out and 12 mph back

Case 2: 12 mph on the way out and 8 mph back

CHAPTER 9 Review Problem Set

1. $\dfrac{26x^2y^3}{39x^4y^2} = \dfrac{2 \bullet 13x^2y^3}{3 \bullet 13x^4y^2} = \dfrac{2y}{3x^2}$

2. $\dfrac{a^2 - 9}{a^2 + 3a}$

$\dfrac{(a+3)(a-3)}{a(a+3)}$

$\dfrac{a-3}{a}$

3. $\dfrac{n^2 - 3n - 10}{n^2 + n - 2}$

$\dfrac{(n-5)(n+2)}{(n-1)(n+2)}$

$\dfrac{n-5}{n-1}$

4. $\dfrac{x^4 - 1}{x^3 - x}$

$\dfrac{(x^2+1)(x+1)(x-1)}{x(x+1)(x-1)}$

$\dfrac{x^2+1}{x}$

5. $\dfrac{8x^3 - 2x^2 - 3x}{12x^2 - 9x}$

$\dfrac{x(8x^2 - 2x - 3)}{3x(4x-3)}$

$\dfrac{x(4x-3)(2x+1)}{3x(4x-3)}$

$\dfrac{2x+1}{3}$

6. $\dfrac{x^4 - 7x^2 - 30}{2x^4 + 7x^2 + 3}$

$\dfrac{(x^2-10)(x^2+3)}{(2x^2+1)(x^2+3)}$

$\dfrac{x^2-10}{2x^2+1}$

7. $\dfrac{\frac{5}{8} - \frac{1}{2}}{\frac{1}{6} + \frac{3}{4}} = \dfrac{24}{24} \bullet \dfrac{\left(\frac{5}{8} - \frac{1}{2}\right)}{\left(\frac{1}{6} + \frac{3}{4}\right)}$

$\dfrac{24\left(\frac{5}{8}\right) - 24\left(\frac{1}{2}\right)}{24\left(\frac{1}{6}\right) + 24\left(\frac{3}{4}\right)} = \dfrac{15 - 12}{4 + 18} = \dfrac{3}{22}$

8. $\dfrac{\frac{3}{2x} + \frac{5}{3y}}{\frac{4}{x} - \frac{3}{4y}}$

$\dfrac{12xy}{12xy} \bullet \dfrac{\left(\frac{3}{2x} + \frac{5}{3y}\right)}{\left(\frac{4}{x} - \frac{3}{4y}\right)}$

$\dfrac{12xy\left(\frac{3}{2x}\right) + 12xy\left(\frac{5}{3y}\right)}{12xy\left(\frac{4}{x}\right) - 12xy\left(\frac{3}{4y}\right)}$

$\dfrac{18y + 20x}{48y - 9x}$

9. $\dfrac{\frac{3}{x-2} - \frac{4}{x^2-4}}{\frac{2}{x+2} + \frac{1}{x-2}}$

$\dfrac{\frac{3}{x-2} - \frac{4}{(x+2)(x-2)}}{\frac{2}{x+2} + \frac{1}{x-2}}$

$$\frac{(x+2)(x-2)}{(x+2)(x-2)} \bullet \frac{\left[\frac{3}{x-2} - \frac{4}{(x+2)(x-2)}\right]}{\left(\frac{2}{x+2} + \frac{1}{x-2}\right)}$$

$$\frac{(x+2)(x-2)\left(\frac{3}{x-2}\right) - (x+2)(x-2)\left[\frac{4}{(x+2)(x-2)}\right]}{(x+2)(x-2)\left(\frac{2}{x+2}\right) + (x+2)(x-2)\left(\frac{1}{x-2}\right)}$$

$$\frac{3(x+2)-4}{2(x-2)+1(x+2)}$$

$$\frac{3x+6-4}{2x-4+x+2}$$

$$\frac{3x+2}{3x-2}$$

10. $1 - \dfrac{1}{2 - \frac{1}{x}}$

$$1 - \frac{x}{x} \bullet \frac{(1)}{\left(2 - \frac{1}{x}\right)}$$

$$1 - \frac{x}{2(x) - x\left(\frac{1}{x}\right)}$$

$$1 - \frac{x}{2x-1}$$

$$\frac{2x-1}{2x-1} - \frac{x}{2x-1}$$

$$\frac{2x-1-x}{2x-1}$$

$$\frac{x-1}{2x-1}$$

11. $\dfrac{6xy^2}{7y^3} \div \dfrac{15x^2y}{5x^2}$

$$\frac{6xy^2}{7y^3} \bullet \frac{5x^2}{15x^2y}$$

$$\frac{6(5)x^3y^2}{7(15)x^2y^4}$$

$$\frac{2x}{7y^2}$$

12. $\dfrac{9ab}{3a+6} \bullet \dfrac{a^2-4a-12}{a^2-6a}$

$$\frac{9ab}{3(a+2)} \bullet \frac{(a-6)(a+2)}{a(a-6)}$$

$$3b$$

13. $\dfrac{n^2+10n+25}{n^2-n} \bullet \dfrac{5n^3-3n^2}{5n^2+22n-15}$

$$\frac{(n+5)(n+5)}{n(n-1)} \bullet \frac{n^2(5n-3)}{(5n-3)(n+5)}$$

$$\frac{n(n+5)}{n-1}$$

14. $\dfrac{x^2-2xy-3y^2}{x^2+9y^2} \div \dfrac{2x^2+xy-y^2}{2x^2-xy}$

$$\frac{x^2-2xy-3y^2}{x^2+9y^2} \bullet \frac{2x^2-xy}{2x^2+xy-y^2}$$

$$\frac{(x-3y)(x+y)}{x^2+9y^2} \bullet \frac{x(2x-y)}{(2x-y)(x+y)}$$

$$\frac{x(x-3y)}{x^2+9y^2}$$

15. $\dfrac{2x+1}{5} + \dfrac{3x-2}{4}$; LCD $= 20$

$$\frac{(2x+1)}{5} \bullet \frac{4}{4} + \frac{(3x-2)}{4} \bullet \frac{5}{5}$$

$$\frac{4(2x+1)}{20} + \frac{5(3x-2)}{20}$$

$$\frac{4(2x+1)+5(3x-2)}{20}$$

$$\frac{8x+4+15x-10}{20}$$

$$\frac{23x-6}{20}$$

16. $\frac{3}{2n} + \frac{5}{3n} - \frac{1}{9}$; LCD = $18n$

$\frac{3}{2n} \bullet \frac{9}{9} + \frac{5}{3n} \bullet \frac{6}{6} - \frac{1}{9} \bullet \frac{2n}{2n}$

$\frac{27}{18n} + \frac{30}{18n} - \frac{2\text{n}}{18n}$

$\frac{57 - 2n}{18n}$

17. $\frac{3x}{x+7} - \frac{2}{x}$; LCD = $x(x+7)$

$\frac{3x}{(x+7)} \bullet \frac{x}{x} - \frac{2}{x} \bullet \frac{(x+7)}{(x+7)}$

$\frac{3x^2}{x(x+7)} - \frac{2(x+7)}{x(x+7)}$

$\frac{3x^2 - 2(x+7)}{x(x+7)}$

$\frac{3x^2 - 2x - 14}{x(x+7)}$

18. $\frac{10}{x^2 - 5x} + \frac{2}{x}$

$\frac{10}{x(x-5)} + \frac{2}{x}$; LCD = $x(x-5)$

$\frac{10}{x(x-5)} + \frac{2}{x} \bullet \frac{(x-5)}{(x-5)}$

$\frac{10 + 2x - 10}{x(x-5)}$

$\frac{2x}{x(x-5)}$

$\frac{2}{x-5}$

19. $\frac{3}{n^2 - 5n - 36} + \frac{2}{n^2 + 3n - 4}$

$\frac{3}{(n-9)(n+4)} + \frac{2}{(n+4)(n-1)}$

LCD = $(n-9)(n+4)(n-1)$

$\frac{3}{(n-9)(n+4)} \bullet \frac{(n-1)}{(n-1)} + \frac{2}{(n+4)(n-1)} \bullet \frac{(n-9)}{(n-9)}$

$\frac{3(n-1)}{(n-9)(n+4)(n-1)} + \frac{2(n-9)}{(n-9)(n+4)(n-1)}$

$\frac{3(n-1) + 2(n-9)}{(n-9)(n+4)(n-1)}$

$\frac{3n - 3 + 2n - 18}{(n-9)(n+4)(n-1)}$

$\frac{5n - 21}{(n-9)(n+4)(n-1)}$

20. $\frac{3}{2y+3} + \frac{5y-2}{2y^2 - 9y - 18} - \frac{1}{y-6}$

$\frac{3}{2y+3} + \frac{5y-2}{(2y+3)(y-6)} - \frac{1}{y-6}$

LCD = $(2y+3)(y-6)$

$\frac{3}{(2y+3)} \bullet \frac{(y-6)}{(y-6)} + \frac{5y-2}{(2y+3)(y-6)} - \frac{1}{(y-6)} \bullet \frac{(2y+3)}{(2y+3)}$

$\frac{3(y-6) + 5y - 2 - 1(2y+3)}{(2y+3)(y-6)}$

$\frac{3y - 18 + 5y - 2 - 2y - 3}{(2y+3)(y-6)}$

$\frac{6y - 23}{(2y+3)(y-6)}$

21. $\frac{3}{x+1} + \frac{x+5}{x^2-1} - \frac{3}{x-1}$

$\frac{3}{x+1} + \frac{x+5}{(x+1)(x-1)} - \frac{3}{x-1}$

LCD = $(x+1)(x-1)$

$\frac{3}{x+1} \bullet \frac{x-1}{x-1} + \frac{x+5}{(x+1)(x-1)} - \frac{3}{x-1} \bullet \frac{x+1}{x+1}$

$$\frac{3(x-1)+x+5-3(x+1)}{(x+1)(x-1)}$$

$$\frac{3x-3+x+5-3x-3}{(x+1)(x-1)}$$

$$\frac{x-1}{(x+1)(x-1)} = \frac{1}{(x+1)}$$

22. $\dfrac{3n}{n^2+6n+5} + \dfrac{4}{n^2-7n-8}$

$\dfrac{3n}{(n+5)(n+1)} + \dfrac{4}{(n-8)(n+1)}$

LCD $= (n-8)(n+5)(n+1)$

$$\frac{3n}{(n+5)(n+1)} \bullet \frac{(n-8)}{(n-8)} + \frac{4}{(n-8)(n+1)} \bullet \frac{(n+5)}{(n+5)}$$

$$\frac{3n(n-8)}{(n+5)(n+1)(n-8)} + \frac{4(n+5)}{(n-8)(n+1)(n+5)}$$

$$\frac{3n(n-8)+4(n+5)}{(n-8)(n+5)(n+1)}$$

$$\frac{3n^2-24n+4n+20}{(n-8)(n+5)(n+1)}$$

$$\frac{3n^2-20n+20}{(n-8)(n+5)(n+1)}$$

23. $\dfrac{4x+5}{3} + \dfrac{2x-1}{5} = 2$

$15\left(\dfrac{4x+5}{3} + \dfrac{2x-1}{5}\right) = 15(2)$

$15\left(\dfrac{4x+5}{3}\right) + 15\left(\dfrac{2x-1}{5}\right) = 30$

$5(4x+5)+3(2x-1) = 30$

$20x+25+6x-3 = 30$

$26x = 8$

$x = \dfrac{8}{26} = \dfrac{4}{13}$

The solution set is $\left\{\dfrac{4}{13}\right\}$.

24. $\dfrac{3}{4x} + \dfrac{4}{5} = \dfrac{9}{10x}$; $x \neq 0$

$20x\left(\dfrac{3}{4x} + \dfrac{4}{5}\right) = 20x\left(\dfrac{9}{10x}\right)$

$20x\left(\dfrac{3}{4x}\right) + 20x\left(\dfrac{4}{5}\right) = 18$

$15 + 16x = 18$

$16x = 3$

$x = \dfrac{3}{16}$

The solution set is $\left\{\dfrac{3}{16}\right\}$.

25. $\dfrac{a}{a-2} - \dfrac{3}{2} = \dfrac{2}{a-2}$; $a \neq 2$

$2(a-2)\left(\dfrac{a}{a-2} - \dfrac{3}{2}\right) = 2(a-2)\left(\dfrac{2}{a-2}\right)$

$2(a-2)\left(\dfrac{a}{a-2}\right) - 2(a-2)\left(\dfrac{3}{2}\right) = 4$

$2a - 3(a-2) = 4$

$2a - 3a + 6 = 4$

$-a = -2$

$a = 2$

Discard the solution $a = 2$.

The solution set is $\emptyset$.

26. $\dfrac{4}{5y-3} = \dfrac{2}{3y+7}$;$y \neq \dfrac{3}{5}$, $y \neq -\dfrac{7}{3}$

$4(3y+7) = 2(5y-3)$

$12y + 28 = 10y - 6$

$2y = -34$

$y = -17$

The solution set is $\{-17\}$.

27. $n + \dfrac{1}{n} = \dfrac{53}{14}$; $n \neq 0$

$14n\left(n + \dfrac{1}{n}\right) = 14n\left(\dfrac{53}{14}\right)$

$14n(n) + 14n\left(\dfrac{1}{n}\right) = 53n$

$14n^2 + 14 = 53n$

$14n^2 - 53n + 14 = 0$

$(2n-7)(7n-2) = 0$

$2n - 7 = 0$ or $7n - 2 = 0$

$2n = 7$ or $7n = 2$

$n = \dfrac{7}{2}$ or $n = \dfrac{2}{7}$

The solution set is $\left\{\dfrac{2}{7}, \dfrac{7}{2}\right\}$.

28. $\dfrac{1}{2x-7}+\dfrac{x-5}{4x^2-49}=\dfrac{4}{6x-21}$

$\dfrac{1}{2x-7}+\dfrac{x-5}{(2x+7)(2x-7)}=\dfrac{4}{3(2x-7)};\ x\neq\dfrac{7}{2},\ x\neq-\dfrac{7}{2}$

$3(2x+7)(2x-7)\left[\dfrac{1}{2x-7}+\dfrac{x-5}{(2x+7)(2x-7)}\right]=3(2x+7)(2x-7)\left[\dfrac{4}{3(2x-7)}\right]$

$3(2x+7)+3(x-5)=4(2x+7)$

$6x+21+3x-15=8x+28$

$9x+6=8x+28$

$x=22$ The solution set is $\{22\}$.

29. $\dfrac{x}{2x+1}-1=\dfrac{-4}{7(x-2)};\ x\neq-\dfrac{1}{2},\ x\neq 2$

$7(2x+1)(x-2)\left(\dfrac{x}{2x+1}-1\right)=7(2x+1)(x-2)\left[\dfrac{-4}{7(x-2)}\right]$

$7(2x+1)(x-2)\left(\dfrac{x}{2x+1}\right)-1(7)(2x+1)(x-2)=-4(2x+1)$

$7x(x-2)-7(2x^2-3x-2)=-8x-4$

$7x^2-14x-14x^2+21x+14=-8x-4$

$-7x^2+7x+14=-8x-4$

$0=7x^2-15x-18$

$0=(7x+6)(x-3)$

$7x+6=0$ or $x-3=0$

$7x=-6$ or $x=3$

$x=-\dfrac{6}{7}$ or $x=3$

The solution set is $\left\{-\dfrac{6}{7},3\right\}$.

30. $\dfrac{2x}{-5}=\dfrac{3}{4x-13};\ x\neq\dfrac{13}{4}$

$2x(4x-13)=-5(3)$

$8x^2-26x=-15$

$8x^2-26x+15=0$

$(4x-3)(2x-5)=0$

$4x-3=0$ or $2x-5=0$

$4x=3$ or $2x=5$

$x=\dfrac{3}{4}$ or $x=\dfrac{5}{2}$

The solution set is $\left\{\dfrac{3}{4},\dfrac{5}{2}\right\}$.

31. $\dfrac{2n}{2n^2+11n-21}-\dfrac{n}{n^2+5n-14}=\dfrac{3}{n^2+5n-14}$

$\dfrac{2n}{(2n-3)(n+7)}-\dfrac{n}{(n+7)(n-2)}=\dfrac{3}{(n+7)(n-2)};\ n\neq\dfrac{3}{2},\ n\neq-7,\ n\neq 2$

$$(2n-3)(n+7)(n-2)\left[\frac{2n}{(2n-3)(n+7)}-\frac{n}{(n+7)(n-2)}\right]=(2n-3)(n+7)(n-2)\left[\frac{3}{(n+7)(n-2)}\right]$$

$$(2n-3)(n+7)(n-2)\left[\frac{2n}{(2n-3)(n+7)}\right]-(2n-3)(n+7)(n-2)\left[\frac{n}{(n+7)(n-2)}\right]=3(2n-3)$$

$2n(n-2)-n(2n-3)=3(2n-3)$

$2n^2-4n-2n^2+3n=6n-9$

$-n=6n-9$

$-7n=-9$

$n=\frac{-9}{-7}=\frac{9}{7}$

The solution set is $\left\{\frac{9}{7}\right\}$.

32. $\frac{2}{t^2-t-6}+\frac{t+1}{t^2+t-12}=\frac{t}{t^2+6t+8}$

$\frac{2}{(t-3)(t+2)}+\frac{t+1}{(t+4)(t-3)}=\frac{t}{(t+4)(t+2)}$ $;t\neq 3, t\neq -2, t\neq -4$

$$(t-3)(t+2)(t+4)\left[\frac{2}{(t-3)(t+2)}+\frac{t+1}{(t+4)(t-3)}\right]=(t-3)(t+2)(t+4)\left[\frac{t}{(t+4)(t+2)}\right]$$

$2(t+4)+(t+2)(t+1)=t(t-3)$

$2t+8+t^2+3t+2=t^2-3t$

$t^2+5t+10=t^2-3t$

$5t+10=-3t$

$10=-8t$

$-\frac{10}{8}=t$

$-\frac{5}{4}=t$ The solution set is $\left\{-\frac{5}{4}\right\}$.

33. $\frac{y-6}{x+1}=\frac{3}{4}$ for y

$4(y-6)=3(x+1)$

$4y-24=3x+3$

$4y=3x+27$

$y=\frac{3x+27}{4}$

34. $\frac{x}{a}-\frac{y}{b}=1$ for y

$ab\left(\frac{x}{a}-\frac{y}{b}\right)=ab(1)$

$bx-ay=ab$

$-ay=-bx+ab$

$y=\frac{-bx+ab}{-a}$

$y=\frac{bx-ab}{a}$

35. Let $x=$ one part and $1400-x=$ the other part.

$\frac{x}{1400-x}=\frac{3}{5}$

$5x=3(1400-x)$

$5x=4200-3x$

$8x=4200$

$x=525$

One part is $525 and the other part is $875.

36.

	Time in Minutes	Rate
Dan	x	$\frac{1}{x}$
Julio	$x - 10$	$\frac{1}{x-10}$
Together	12	$\frac{1}{12}$

$$\frac{1}{x} + \frac{1}{x-10} = \frac{1}{12}$$

$$12x(x-10)\left(\frac{1}{x} + \frac{1}{x-10}\right) = 12x(x-10)\left(\frac{1}{12}\right)$$

$$12x(x-10)\left(\frac{1}{x}\right) + 12x(x-10)\left(\frac{1}{x-10}\right) = x(x-10)$$

$$12(x-10) + 12x = x^2 - 10x$$
$$12x - 120 + 12x = x^2 - 10x$$
$$24x - 120 = x^2 - 10x$$
$$0 = x^2 - 34x + 120$$
$$0 = (x-30)(x-4)$$
$$x - 30 = 0 \quad \text{or} \quad x - 4 = 0$$
$$x = 30 \quad \text{or} \quad x = 4$$

Discard the root $x = 4$ since $x - 10 = 4 - 10 = -6$ and time cannot be negative.
Dan's time is 30 minutes and Julio's time is 20 minutes.

37.

	Rate	Time	Distance
Car A	x	$\frac{250}{x}$	250
Car B	$x + 5$	$\frac{440}{x+5}$	440

$$\frac{250}{x} + 3 = \frac{440}{x+5}$$

$$x(x+5)\left(\frac{250}{x} + 3\right) = x(x+5)\left(\frac{440}{x+5}\right)$$

$$x(x+5)\left(\frac{250}{x}\right) + 3x(x+5) = 440x$$

$$250(x+5) + 3x^2 + 15x = 440x$$
$$250x + 1250 + 3x^2 + 15x = 440x$$
$$3x^2 - 175x + 1250 = 0$$
$$(3x - 25)(x - 50) = 0$$
$$3x - 25 = 0 \quad \text{or} \quad x - 50 = 0$$
$$3x = 25 \quad \text{or} \quad x = 50$$
$$x = \frac{25}{3} \quad \text{or} \quad x = 50$$

Case 1: Car A would travel at $\frac{25}{3} = 8\frac{1}{3}$ mph and Car B would travel at $13\frac{1}{3}$ mph.

Case 2: Car A would travel at 50 mph and Car B would travel at 55 mph.

38.

	Time in Hours	Rate
Mark	20	$\frac{1}{20}$
Phil	30	$\frac{1}{30}$

In 5 hours Mark would finish

$5\left(\frac{1}{20}\right) = \frac{5}{20} = \frac{1}{4}$ of the job.

Then $\frac{3}{4}$ of the job is remaining.

Let t = the hours worked together.

$$\frac{1}{20}t + \frac{1}{30}t = \frac{3}{4}$$

$$60\left(\frac{1}{20}t\right) + 60\left(\frac{1}{30}t\right) = 60\left(\frac{3}{4}\right)$$

$$3t + 2t = 45$$
$$5t = 45$$
$$t = 9$$

It would take 9 hours.

39.

	Rate	Time	Pay
Anticipated	$\frac{640}{x}$	x	640
Actual	$\frac{640}{x+20}$	$x + 20$	640

$$\frac{640}{x} - 1.60 = \frac{640}{x+20}$$

$$10x(x+20)\left(\frac{640}{x} - 1.6\right) = 10x(x+20)\left(\frac{640}{x+20}\right)$$

$$10x(x+20)\left(\frac{640}{x}\right) - 10x(x+20)(1.6) = 10x(640)$$

$$6400(x+20) - 16x(x+20) = 6400x$$
$$6400x + 128000 - 16x^2 - 320x = 6400x$$
$$-16x^2 + 6080x + 128000 = 6400x$$
$$-16x^2 - 320x + 128000 = 0$$
$$16x^2 + 320x - 128000 = 0$$
$$16(x^2 + 20x - 8000) = 0$$
$$16(x - 80)(x + 100) = 0$$

$x - 80 = 0$ or $x + 100 = 0$
$x = 80$ or $x = -100$
Discard the root $x = -100$.
The anticipated time would take 80 hours.

40.

	Rate	Time	Distance
1st part	x	$\frac{40}{x}$	40
2nd part	$x-3$	$\frac{26}{x-3}$	26

$\frac{40}{x} + \frac{26}{x-3} = 4\frac{1}{2}$

$2x(x-3)\left(\frac{40}{x} + \frac{26}{x-3}\right) = 2x(x-3)\left(\frac{9}{2}\right)$
$80(x-3) + 52x = 9x(x-3)$
$80x - 240 + 52x = 9x^2 - 27x$
$132x - 240 = 9x^2 - 27x$
$0 = 9x^2 - 159x + 240$
$0 = 3(3x^2 - 53x + 80)$
$0 = 3(3x-5)(x-16)$
$3x - 5 = 0$ or $x - 16 = 0$
$3x = 5$ or $x = 16$
$x = \frac{5}{3}$
Discard the root $x = \frac{5}{3}$ since
$x - 3 = \frac{5}{3} - 3 = -\frac{4}{3}$ and rate cannot be negative. He travels 13 mph for the last part.

CHAPTER 9 | Test

1. $\frac{39x^2y^3}{72x^3y} = \frac{3(13)x^2y^3}{3(24)x^3y} = \frac{13y^2}{24x}$

2. $\frac{3x^2 + 17x - 6}{x^3 - 36x}$

$\frac{(3x-1)(x+6)}{x(x+6)(x-6)}$

$\frac{3x-1}{x(x-6)}$

3. $\frac{6n^2 - 5n - 6}{3n^2 + 14n + 8}$

$\frac{(2n-3)(3n+2)}{(3n+2)(n+4)}$

$\frac{2n-3}{n+4}$

4. $\frac{2x - 2x^2}{x^2 - 1}$

$\frac{2x(1-x)}{(x+1)(x-1)}$

$\frac{-2x(x-1)}{(x+1)(x-1)}$

$\frac{-2x}{x+1}$

5. $\frac{5x^2y}{8x} \bullet \frac{12y^2}{20xy} = \frac{5(12)x^2y^3}{8(20)x^2y} = \frac{3y^2}{8}$

6. $\frac{5a + 5b}{20a + 10b} \bullet \frac{a^2 - ab}{2a^2 + 2ab}$

$\frac{5(a+b)}{10(2a+b)} \bullet \frac{a(a-b)}{2a(a+b)}$

$\frac{a-b}{4(2a+b)}$

7. $\frac{3x^2 + 10x - 8}{5x^2 + 19x - 4} \div \frac{3x^2 - 23x + 14}{x^2 - 3x - 28}$

$\frac{3x^2 + 10x - 8}{5x^2 + 19x - 4} \bullet \frac{x^2 - 3x - 28}{3x^2 - 23x + 14}$

$\frac{(3x-2)(x+4)}{(5x-1)(x+4)} \bullet \frac{(x-7)(x+4)}{(3x-2)(x-7)}$

$\frac{x+4}{5x-1}$

8. $\dfrac{3x-1}{4} + \dfrac{2x+5}{6}$; LCD $= 12$

$\dfrac{(3x-1)}{4} \bullet \dfrac{3}{3} + \dfrac{(2x+5)}{6} \bullet \dfrac{2}{2}$

$\dfrac{3(3x-1)+2(2x+5)}{12}$

$\dfrac{9x-3+4x+10}{12}$

$\dfrac{13x+7}{12}$

9. $\dfrac{5x-6}{3} - \dfrac{x-12}{6}$; LCD $= 6$

$\dfrac{(5x-6)}{3} \bullet \dfrac{2}{2} - \dfrac{x-12}{6}$

$\dfrac{2(5x-6)}{6} - \dfrac{x-12}{6}$

$\dfrac{2(5x-6)-(x-12)}{6}$

$\dfrac{10x-12-x+12}{6}$

$\dfrac{9x}{6} = \dfrac{3x}{2}$

10. $\dfrac{3}{5n} + \dfrac{2}{3} - \dfrac{7}{3n}$; LCD $= 15n$

$\dfrac{3}{5n} \bullet \dfrac{3}{3} + \dfrac{2}{3} \bullet \dfrac{5n}{5n} - \dfrac{7}{3n} \bullet \dfrac{5}{5}$

$\dfrac{9}{15n} + \dfrac{10n}{15n} - \dfrac{35}{15n}$

$\dfrac{10n-26}{15n}$

11. $\dfrac{3x}{x-6} + \dfrac{2}{x}$; LCD $= x(x-6)$

$\dfrac{3x}{(x-6)} \bullet \dfrac{x}{x} + \dfrac{2}{x} \bullet \dfrac{(x-6)}{(x-6)}$

$\dfrac{3x^2}{x(x-6)} + \dfrac{2(x-6)}{x(x-6)}$

$\dfrac{3x^2+2(x-6)}{x(x-6)}$

$\dfrac{3x^2+2x-12}{x(x-6)}$

12. $\dfrac{9}{x^2-x} - \dfrac{2}{x}$

$\dfrac{9}{x(x-1)} - \dfrac{2}{x}$; LCD $= x(x-1)$

$\dfrac{9}{x(x-1)} - \dfrac{2}{x} \bullet \dfrac{(x-1)}{(x-1)}$

$\dfrac{9-2(x-1)}{x(x-1)}$

$\dfrac{9-2x+2}{x(x-1)}$

$\dfrac{-2x+11}{x(x-1)}$

13. $\dfrac{3}{2n^2+n+10} + \dfrac{5}{n^2+5n-14}$

$\dfrac{3}{(2n+5)(n-2)} + \dfrac{5}{(n+7)(n-2)}$

LCD $= (2n+5)(n-2)(n+7)$

$\dfrac{3}{(2n+5)(n-2)} \bullet \dfrac{(n+7)}{(n+7)} + \dfrac{5}{(n+7)(n-2)} \bullet \dfrac{(2n+5)}{(2n+5)}$

$\dfrac{3(n+7)}{(2n+5)(n-2)(n+7)} + \dfrac{5(2n+5)}{(n+7)(n+2)(2n+5)}$

$\dfrac{3(n+7)+5(2n+5)}{(2n+5)(n-2)(n+7)}$

$\dfrac{3n+21+10n+25}{(2n+5)(n-2)(n+7)}$

$\dfrac{13n+46}{(2n+5)(n-2)(n+7)}$

14. $\dfrac{5}{2x^2-6x} + \dfrac{4}{3x^2+6x}$

$\dfrac{5}{2x(x-3)} + \dfrac{4}{3x(x+2)}$

LCD $= 6x(x-3)(x+2)$

$$\frac{5}{2x(x-3)} \bullet \frac{3(x+2)}{3(x+2)} + \frac{4}{3x(x+2)} \bullet \frac{2(x-3)}{2(x-3)}$$

$$\frac{15(x+2)}{6x(x-3)(x+2)} + \frac{8(x-3)}{6x(x-3)(x+2)}$$

$$\frac{15x+30+8x-24}{6x(x-3)(x+2)}$$

$$\frac{23x+6}{6x(x-3)(x+2)}$$

15. $$\frac{\frac{3}{2x} - \frac{1}{6}}{\frac{2}{3x} + \frac{3}{4}}$$

$$\frac{12x}{12x} \bullet \frac{\left(\frac{3}{2x} - \frac{1}{6}\right)}{\left(\frac{2}{3x} + \frac{3}{4}\right)}$$

$$\frac{12x\left(\frac{3}{2x}\right) - 12x\left(\frac{1}{6}\right)}{12x\left(\frac{2}{3x}\right) + 12x\left(\frac{3}{4}\right)}$$

$$\frac{18-2x}{8+9x}$$

16. $\frac{x+2}{y-4} = \frac{3}{4}$ for y

$3(y-4) = 4(x+2)$
$3y - 12 = 4x + 8$
$3y = 4x + 20$

$$y = \frac{4x+20}{3}$$

17. $\frac{x-1}{2} - \frac{x+2}{5} = -\frac{3}{5}$

$$10\left(\frac{x-1}{2} - \frac{x+2}{5}\right) = 10\left(-\frac{3}{5}\right)$$

$$10\left(\frac{x-1}{2}\right) - 10\left(\frac{x+2}{5}\right) = 2(-3)$$

$5(x-1) - 2(x+2) = -6$
$5x - 5 - 2x - 4 = -6$
$3x - 9 = -6$
$3x = 3$
$x = 1$
The solution set is $\{1\}$.

18. $\frac{5}{4x} + \frac{3}{2} = \frac{7}{5x}$; $x \neq 0$

$$20x\left(\frac{5}{4x} + \frac{3}{2}\right) = 20x\left(\frac{7}{5x}\right)$$

$$20x\left(\frac{5}{4x}\right) + 20x\left(\frac{3}{2}\right) = 28$$

$25 + 30x = 28$
$30x = 3$
$x = \frac{3}{30} = \frac{1}{10}$
The solution set is $\left\{\frac{1}{10}\right\}$.

19. $\frac{-3}{4n-1} = \frac{-2}{3n+11}$; $n \neq \frac{1}{4}, n \neq -\frac{11}{3}$

$-3(3n+11) = -2(4n-1)$
$-9n - 33 = -8n + 2$
$-9n = -8n + 35$
$-n = 35$
$n = -35$
The solution set is $\{-35\}$.

20. $n - \frac{5}{n} = 4$; $n \neq 0$

$$n\left(n - \frac{5}{n}\right) = n(4)$$

$$n(n) - n\left(\frac{5}{n}\right) = 4n$$

$n^2 - 5 = 4n$
$n^2 - 4n - 5 = 0$
$(n-5)(n+1) = 0$
$n - 5 = 0$ or $n + 1 = 0$
$n = 5$ or $n = -1$
The solution set is $\{-1, 5\}$.

Chapter 9 Test

21. $\frac{6}{x-4} - \frac{4}{x+3} = \frac{8}{x-4}; x \neq 4, x \neq -3$

$(x\text{-}4)(x\text{+}3)\left(\frac{6}{x\text{-}4} - \frac{4}{x\text{+}3}\right) = (x\text{-}4)(x\text{+}3)\left(\frac{8}{x\text{-}4}\right)$

$(x\text{-}4)(x\text{+}3)\left(\frac{6}{x\text{-}4}\right)\text{-}(x\text{-}4)(x\text{+}3)\left(\frac{4}{x\text{+}3}\right) = 8(x\text{+}3)$

$6(x+3) - 4(x-4) = 8x + 24$

$6x + 18 - 4x + 16 = 8x + 24$

$2x + 34 = 8x + 24$

$2x = 8x - 10$

$-6x = -10$

$x = \frac{-10}{-6} = \frac{5}{3}$

The solution set is $\left\{\frac{5}{3}\right\}$.

22. $\frac{1}{3x-1} + \frac{x-2}{9x^2-1} = \frac{7}{6x-2}$

$\frac{1}{3x-1} + \frac{x-2}{(3x+1)(3x-1)} = \frac{7}{2(3x-1)}; x \neq \frac{1}{3}, x \neq -\frac{1}{3}$

$2(3x-1)(3x+1)\left[\frac{1}{3x-1} + \frac{x-2}{(3x+1)(3x-1)}\right] = 2(3x+1)(3x-1)\left[\frac{7}{2(3x-1)}\right]$

$2(3x+1) + 2(x-2) = 7(3x+1)$

$6x + 2 + 2x - 4 = 21x + 7$

$8x - 2 = 21x + 7$

$8x = 21x + 9$

$-13x = 9$

$x = -\frac{9}{13}$

The solution set is $\left\{-\frac{9}{13}\right\}$.

23. Let $x =$ the numerator and $3x - 9 =$ the denominator.

$\frac{x}{3x-9} = \frac{3}{8}; x \neq 3$

$8x = 3(3x - 9)$

$8x = 9x - 27$

$-x = -27$

$x = 27$

The number is $\frac{27}{3(27)-9} = \frac{27}{81-9} = \frac{27}{72}$.

24.

	Time in Minutes	Rate
Jodi	$3x$	$\frac{1}{3x}$
Jannie	x	$\frac{1}{x}$
Together	15	$\frac{1}{15}$

$\frac{1}{3x} + \frac{1}{x} = \frac{1}{15}$

$15x\left(\frac{1}{3x} + \frac{1}{x}\right) = 15x\left(\frac{1}{15}\right)$

$5 + 15 = x$

$20 = x$

It would take Jodi $3(20) = 60$ minutes.

25.

	Rate	Time	Distance
Rene	$x+3$	$\frac{60}{x+3}$	60
Sue	x	$\frac{60}{x}$	60

$$\frac{60}{x+3}+1=\frac{60}{x}$$

$$x(x+3)\left(\frac{60}{x+3}+1\right)=x(x+3)\left(\frac{60}{x}\right)$$

$$x(x+3)\left(\frac{60}{x+3}\right)+1x(x+3)=60(x+3)$$

$60x+x^2+3x=60x+180$

$x^2+63x=60x+180$

$x^2+3x-180=0$

$(x+15)(x-12)=0$

$x+15=0$ or $x-12=0$

$x=-15$ or $x=12$

Discard the root $x=-15$.

Rene's rate is $12+3=15$ mph.

CHAPTERS 1-9 | Cumulative Practice Test

1. $3(x-4)-4(2x+1)-6(4-3x)=$

$3x-12-8x-4-24+18x=$

$13x-40$ for $x=-19$

$13(-19)-40=-247-40=-287$

2. $$\frac{3x^2+2x-1}{3x^2-4x+1}=\frac{(3x-1)(x+1)}{(3x-1)(x-1)}=$$

$\frac{(x+1)}{(x-1)}$ for $x=15$

$$\frac{(15+1)}{(15-1)}=\frac{16}{14}=\frac{8}{7}$$

3. $$\frac{9x^2y^3}{3x^{-2}y^{-1}}=3x^{2-(-2)}y^{3-(-1)}=$$

$3x^4y^4$ for $x=-1$ and $y=-2$

$3(-1)^4(-2)^4=3(1)(16)=48$

4.

$$\begin{array}{r|l} & 2x^3+5x^2-7x-6 \\ \hline x-3 & 2x^4-x^3-22x^2+15x+21 \\ & 2x^4-6x^3 \\ \hline & 5x^3-22x^2 \\ & 5x^3-15x^2 \\ \hline & -7x^2+15x \\ & -7x^2+21x \\ \hline & -6x+21 \\ & -6x+18 \\ \hline & 3 \end{array}$$

Q: $2x^3+5x^2-7x-6$ R: 3

5. $$\begin{pmatrix} 3x-4y=-25 \\ 2x+5y=14 \end{pmatrix}$$

Multiply equation 1 by 5 and multiply equation 2 by 4. Then add the equations.

$$\begin{array}{l} 15x-20y=-125 \\ 8x+20y=56 \\ \hline 23x=-69 \end{array}$$

$x=-3$

$3x-4y=-25$

$3(-3)-4y=-25$

$-9-4y=-25$

$-4y=-16$

$y=4$

The solution set is $\{(-3,4)\}$.

6. $$\begin{pmatrix} 7x-4y=59 \\ 3x+y=9 \end{pmatrix}$$

Multiply equation 2 by 4, and add to equation 1.

$$\begin{array}{l} 7x-4y=59 \\ 12x+4y=36 \\ \hline 19x=95 \end{array}$$

$x=5$

$7x-4y=59$

$7(5)-4y=59$

$35-4y=59$

$-4y=24$

$y=-6$

The solution set is $\{(5,-6)\}$.

7. $y = -x^2 + 1$

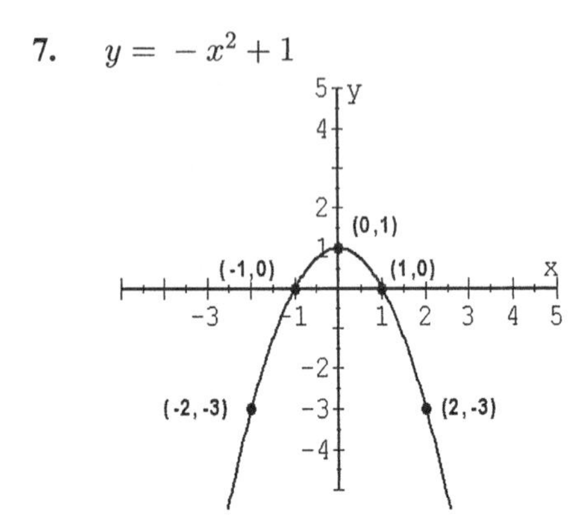

8. $y = -x^3 + 1$

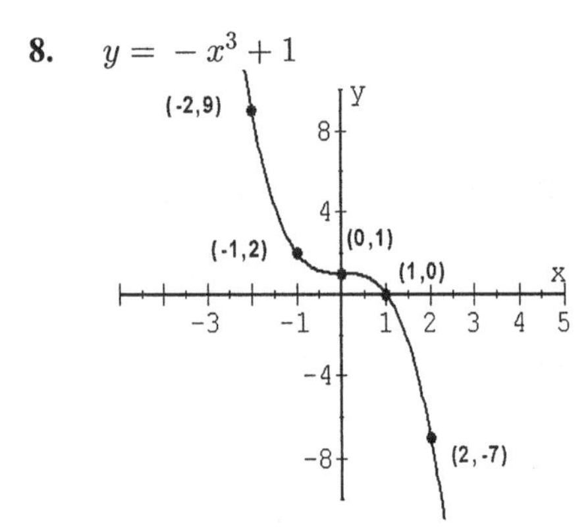

9. Graph $-2x - 4y \geq -4$

10. $\left(\frac{1}{2} - \frac{1}{3}\right)^{-2} = \left(\frac{3}{6} - \frac{2}{6}\right)^{-2} =$

$\left(\frac{1}{6}\right)^{-2} = \left(\frac{6}{1}\right)^{2} = 36$

11. $\frac{n}{100} = \frac{9}{4}$

$4n = 900$

$n = 225$

$\frac{9}{4} = \frac{225}{100} = 225\%$

12. $0.00013 = (1.3)(10)^{-4}$

13.
$$
\begin{aligned}
-3(2x-1)-(x+4) &= -5(x-3)\\
-6x+3-x-4 &= -5x+15\\
-7x-1 &= -5x+15\\
-2x-1 &= 15\\
-2x &= 16\\
x &= -8
\end{aligned}
$$

The solution set is $\{-8\}$.

14. $\frac{5-x}{2-x} - \frac{3-2x}{2x} = 1;\ x \neq 2 \text{ or } x \neq 0$

$2x(2-x)\left[\frac{5-x}{2-x} - \frac{3-2x}{2x}\right] = 2x(2-x)(1)$

$2x(2\text{-}x)\left(\frac{5\text{-}x}{2\text{-}x}\right) - (2x)(2\text{-}x)\left(\frac{3\text{-}2x}{2x}\right) = 4x\text{-}2x^2$

$2x(5-x) - (2-x)(3-2x)) = 4x - 2x^2$

$10x - 2x^2 - (6 - 7x + 2x^2) = 4x - 2x^2$

$10x - 2x^2 - 6 + 7x - 2x^2 = 4x - 2x^2$

$-2x^2 + 13x - 6 = 0$

$2x^2 - 13x + 6 = 0$

$(2x-1)(x-6) = 0$

$2x - 1 = 0$ or $x - 6 = 0$

$2x = 1$ or $x = 6$

$x = \frac{1}{2}$ or $x = 6$

The solution set is $\left\{\frac{1}{2}, 6\right\}$.

15. $15x^3 + x^2 - 2x = 0$

$x(15x^2 + x - 2) = 0$

$x(5x+2)(3x-1) = 0$

$x = 0$ or $5x + 2 = 0$ or $3x - 1 = 0$

$x = 0$ or $5x = -2$ or $3x = 1$

$x = 0$ or $x = -\frac{2}{5}$ or $x = \frac{1}{3}$

The solution set is $\left\{-\frac{2}{5}, 0, \frac{1}{3}\right\}$.

16. $|5x-1|=7$ is equivalent to

$5x-1=-7$ or $5x-1=7$.

$$\begin{aligned} 5x-1&=-7 &\text{or}\quad 5x-1&=7\\ 5x&=-6 &\text{or}\quad 5x&=8\\ x&=-\frac{6}{5} &\text{or}\quad x&=\frac{8}{5}\end{aligned}$$

The solution set is $\left\{-\frac{6}{5},\frac{8}{5}\right\}$.

17. $(3x-1)^2=16$

$$\begin{aligned} 3x-1&=\pm\sqrt{16}\\ 3x-1&=\pm 4\end{aligned}$$

$$\begin{aligned} 3x-1&=4 &\text{or}\quad 3x-1&=-4\\ 3x&=5 &\text{or}\quad 3x&=-3\\ x&=\frac{5}{3} &\text{or}\quad x&=-1\end{aligned}$$

The solution set is $\left\{-1,\frac{5}{3}\right\}$.

18. $-16\le 7x-2\le 5$

$-14\le 7x\le 7$

$-2\le x\le 1$

The solution set is $[-2,1]$.

19. $\frac{x-1}{3}-\frac{2x+1}{4}>\frac{1}{6}$

$12\left(\frac{x-1}{3}-\frac{2x+1}{4}\right)>12\left(\frac{1}{6}\right)$

$12\left(\frac{x-1}{3}\right)-12\left(\frac{2x+1}{4}\right)>2$

$4(x-1)-3(2x+1)>2$

$4x-4-6x-3>2$

$-2x-7>2$

$-2x>9$

$-\frac{1}{2}\left(-2x\right)<-\frac{1}{2}\left(9\right)$

$x<-\frac{9}{2}$

The solution set is $\left(-\infty,-\frac{9}{2}\right)$.

20. $|2x-1|>1$

$$\begin{aligned} 2x-1&<-1 &\text{or}\quad 2x-1&>1\\ 2x&<0 &\text{or}\quad 2x&>2\\ x&<0 &\text{or}\quad x&>1\end{aligned}$$

The solution set is $(-\infty,0)\cup(1,\infty)$.

21. Let s represent the selling price.
Profit is a percent of selling price.
Selling price = Cost + Profit

$$\begin{aligned} s&=14+(30\%)(s)\\ s&=14+0.3s\\ 10s&=140+3s\\ 7s&=140\\ s&=20\end{aligned}$$

The selling price would be \$20.

22. Let x represent the gallons of 55% glycerine solution to be used.

$$\begin{aligned} (55\%)(x)+(20\%)(15)&=(40\%)(x+15)\\ 0.55(x)+0.20(15)&=0.40(x+15)\\ 100[0.55(x)+0.20(15)]&=100[0.40(x+15)]\\ 55x+20(15)&=40(x+15)\\ 55x+300&=40x+600\\ 15x&=300\\ x&=20\end{aligned}$$

We must add 20 gallons of the 55% glycerine solution.

23.

	Time in Minutes	Rate
Russ	40	$\frac{1}{40}$
Jay	x	$\frac{1}{x}$
Together	10	$\frac{1}{10}$

Let x represent the time it takes Jay to mow the lawn alone. Then $\frac{1}{x}$ represents Jay's rate.

Russ did $15\left(\frac{1}{40}\right)=\frac{15}{40}=\frac{3}{8}$ of the job.

Then $\frac{5}{8}$ of the job is remaining.

Russ and Jay together did $\frac{5}{8}$ of the job together in 10 minutes.

$10\left(\frac{1}{40}\right)+10\left(\frac{1}{x}\right)=\frac{5}{8}$

$\frac{1}{4}+\frac{10}{x}=\frac{5}{8}$

$$8x\left(\frac{1}{4}+\frac{10}{x}\right)=8x\left(\frac{5}{8}\right)$$
$$2x+80=5x$$
$$80=3x$$
$$x=\frac{80}{3}=26\frac{2}{3}$$

It would have taken Jay $26\frac{2}{3}$ minutes to mow the lawn by himself.

24. Let x represent the length of one leg of a right triangle. Then $x+5$ represents the length of the other leg.

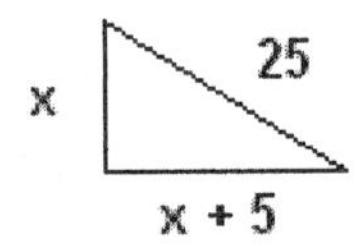

Use Pythagorean Theorem

$$c^2=a^2+b^2$$
$$(25)^2=x^2+(x+5)^2$$
$$625=x^2+x^2+10x+25$$
$$0=2x^2+10x-600$$
$$0=x^2+5x-300$$
$$0=(x+20)(x-15)$$
$$x+20=0 \text{ or } x-15=0$$
$$x=-20 \text{ or } x=15$$

Discard $x=-20$ because length cannot be negative. If $x=15$, then $x+5=20$, so the lengths of the legs are 15 centimeters and 20 centimeters.

25. Let x represent the score on the fifth exam.

$$\frac{93+88+89+95+x}{5}\geq 92$$
$$\frac{365+x}{5}\geq 92$$
$$5\left(\frac{365+x}{5}\right)\geq 5(92)$$
$$365+x\geq 460$$
$$x\geq 95$$

She would need a score of 95 or better.

Chapter 10 Exponents and Radicals

PROBLEM SET 10.1 Integral Exponents and Scientific Notation Revisited

1. $3^{-3} = \frac{1}{3^3} = \frac{1}{27}$

3. $-10^{-2} = \frac{-1}{10^2} = -\frac{1}{100}$

5. $\frac{1}{3^{-4}} = 3^4 = 81$

7. $-\left(\frac{1}{3}\right)^{-3} = -\left(\frac{3}{1}\right)^3 = -\frac{3^3}{1^3} =$
$-\frac{27}{1} = -27$

9. $\left(-\frac{1}{2}\right)^{-3} = \left(-\frac{2}{1}\right)^3 =$
$\frac{(2)^3}{(-1)^3} = \frac{8}{-1} = -8$

11. $\left(-\frac{3}{4}\right)^0 = 1$

13. $\frac{1}{\left(\frac{3}{7}\right)^{-2}} = \left(\frac{3}{7}\right)^2 = \frac{3^2}{7^2} = \frac{9}{49}$

15. $2^7 \bullet 2^{-3}$
2^{7-3}
$2^4 = 16$

17. $10^{-5} \bullet 10^2$
$10^{-5+2} = 10^{-3}$
$\frac{1}{10^3} = \frac{1}{1000}$

19. $10^{-1} \bullet 10^{-2}$
$10^{-1-2} = 10^{-3}$
$\frac{1}{10^3} = \frac{1}{1000}$

21. $(3^{-1})^{-3} = 3^{-1(-3)}$
$3^3 = 27$

23. $(5^3)^{-1} = 5^{3(-1)}$
$5^{-3} = \frac{1}{5^3} = \frac{1}{125}$

25. $(2^3 \bullet 3^{-2})^{-1}$
$2^{3(-1)} \bullet 3^{-2(-1)}$
$2^{-3} \bullet 3^2$
$\frac{3^2}{2^3} = \frac{9}{8}$

27. $(4^2 \bullet 5^{-1})^2$
$4^{2(2)} \bullet 5^{-1(2)}$
$4^4 \bullet 5^{-2}$
$\frac{4^4}{5^2} = \frac{256}{25}$

29. $\left(\frac{2^{-1}}{5^{-2}}\right)^{-1}$

$\frac{2^{-1(-1)}}{5^{-2(-1)}}$

$\frac{2^1}{5^2} = \frac{2}{25}$

31. $\left(\frac{2^{-1}}{3^{-2}}\right)^2$

$\frac{2^{-1(2)}}{3^{-2(2)}}$

$\frac{2^{-2}}{3^{-4}} = \frac{3^4}{2^2} = \frac{81}{4}$

33. $\frac{3^3}{3^{-1}}$

$3^{3-(-1)}$

$3^4 = 81$

35. $\dfrac{10^{-2}}{10^{2}}$

$10^{-2-2} = 10^{-4}$

$\dfrac{1}{10^{4}} = \dfrac{1}{10{,}000}$

37. $2^{-2} + 3^{-2}$

$\dfrac{1}{2^{2}} + \dfrac{1}{3^{2}}$

$\dfrac{1}{4} + \dfrac{1}{9}$; LCD = 36

$\dfrac{1}{4} \bullet \dfrac{9}{9} + \dfrac{1}{9} \bullet \dfrac{4}{4}$

$\dfrac{9}{36} + \dfrac{4}{36}$

$\dfrac{13}{36}$

39. $\left(\dfrac{1}{3}\right)^{-1} - \left(\dfrac{2}{5}\right)^{-1}$

$\left(\dfrac{3}{1}\right)^{1} - \left(\dfrac{5}{2}\right)^{1}$

$\dfrac{3}{1} - \dfrac{5}{2}$; LCD = 2

$\dfrac{3}{1} \bullet \dfrac{2}{2} - \dfrac{5}{2}$

$\dfrac{6}{2} - \dfrac{5}{2}$

$\dfrac{1}{2}$

41. $(2^{-3} + 3^{-2})^{-1}$

$\left(\dfrac{1}{2^{3}} + \dfrac{1}{3^{2}}\right)^{-1}$

$\left(\dfrac{1}{8} + \dfrac{1}{9}\right)^{-1}$

$\left(\dfrac{9}{72} + \dfrac{8}{72}\right)^{-1}$

$\left(\dfrac{17}{72}\right)^{-1} = \left(\dfrac{72}{17}\right)^{1} = \dfrac{72}{17}$

43. $x^{2} \bullet x^{-8}$

x^{2-8}

$x^{-6} = \dfrac{1}{x^{6}}$

45. $a^{3} \bullet a^{-5} \bullet a^{-1}$

a^{3-5-1}

$a^{-3} = \dfrac{1}{a^{3}}$

47. $(a^{-4})^{2}$

$a^{-4(2)}$

$a^{-8} = \dfrac{1}{a^{8}}$

49. $(x^{2}y^{-6})^{-1}$

$x^{2(-1)}y^{-6(-1)}$

$x^{-2}y^{6} = \dfrac{y^{6}}{x^{2}}$

51. $(ab^{3}c^{-2})^{-4}$

$a^{1(-4)}b^{3(-4)}c^{-2(-4)}$

$a^{-4}b^{-12}c^{8}$

$\dfrac{c^{8}}{a^{4}b^{12}}$

53. $(2x^{3}y^{-4})^{-3}$

$2^{-3}x^{3(-3)}y^{-4(-3)}$

$2^{-3}x^{-9}y^{12}$

$\dfrac{y^{12}}{2^{3}x^{9}} = \dfrac{y^{12}}{8x^{9}}$

55. $\left(\dfrac{x^{-1}}{y^{-4}}\right)^{-3}$

$\dfrac{x^{-1(-3)}}{y^{-4(-3)}} = \dfrac{x^{3}}{y^{12}}$

57. $\left(\frac{3a^{-2}}{2b^{-1}}\right)^{-2}$

$\frac{3^{-2}a^{-2(-2)}}{2^{-2}b^{-1(-2)}}$

$\frac{3^{-2}a^{4}}{2^{-2}b^{2}}$

$\frac{2^{2}a^{4}}{3^{2}b^{2}} = \frac{4a^{4}}{9b^{2}}$

59. $\frac{x^{-6}}{x^{-4}}$

$x^{-6-(-4)}$

$x^{-2} = \frac{1}{x^{2}}$

61. $\frac{a^{3}b^{-2}}{a^{-2}b^{-4}}$

$a^{3-(-2)}b^{-2-(-4)}$

$a^{5}b^{2}$

63. $(2xy^{-1})(3x^{-2}y^{4})$

$6x^{1-2}y^{-1+4}$

$6x^{-1}y^{3}$

$\frac{6y^{3}}{x}$

65. $(-7a^{2}b^{-5})(-a^{-2}b^{7})$

$7a^{2-2}b^{-5+7}$

$7a^{0}b^{2}$

$7(1)b^{2}$

$7b^{2}$

67. $\frac{28x^{-2}y^{-3}}{4x^{-3}y^{-1}}$

$7x^{-2-(-3)}y^{-3-(-1)}$

$7x^{1}y^{-2}$

$\frac{7x}{y^{2}}$

69. $\frac{-72a^{2}b^{-4}}{6a^{3}b^{-7}}$

$-12a^{2-3}b^{-4-(-7)}$

$-12a^{-1}b^{3}$

$\frac{-12b^{3}}{a}$

71. $\left(\frac{35x^{-1}y^{-2}}{7x^{4}y^{3}}\right)^{-1}$

$(5x^{-1-4}y^{-2-3})^{-1}$

$(5x^{-5}y^{-5})^{-1}$

$5^{-1}x^{-5(-1)}y^{-5(-1)}$

$5^{-1}x^{5}y^{5}$

$\frac{x^{5}y^{5}}{5}$

73. $\left(\frac{-36a^{-1}b^{-6}}{4a^{-1}b^{4}}\right)^{-2}$

$[-9a^{-1-(-1)}b^{-6-4}]^{-2}$

$(-9a^{0}b^{-10})^{-2}$

$(-9b^{-10})^{-2}$

$(-9)^{-2}b^{-10(-2)}$

$(-9)^{-2}b^{20}$

$\frac{b^{20}}{(-9)^{2}} = \frac{b^{20}}{81}$

75. $x^{-2} + x^{-3}$

$\frac{1}{x^{2}} + \frac{1}{x^{3}}$; LCD $= x^{3}$

$\frac{1}{x^{2}} \bullet \frac{x}{x} + \frac{1}{x^{3}}$

$\frac{x}{x^{3}} + \frac{1}{x^{3}}$

$\frac{x+1}{x^{3}}$

77. $x^{-3} - y^{-1}$

$\frac{1}{x^3} - \frac{1}{y}$; LCD $= x^3y$

$\frac{1}{x^3} \bullet \frac{y}{y} - \frac{1}{y} \bullet \frac{x^3}{x^3}$

$\frac{y}{x^3y} - \frac{x^3}{x^3y}$

$\frac{y - x^3}{x^3y}$

79. $3a^{-2} + 4b^{-1}$

$\frac{3}{a^2} + \frac{4}{b}$; LCD $= a^2b$

$\frac{3}{a^2} \bullet \frac{b}{b} + \frac{4}{b} \bullet \frac{a^2}{a^2}$

$\frac{3b}{a^2b} + \frac{4a^2}{a^2b}$

$\frac{3b + 4a^2}{a^2b}$

81. $x^{-1}y^{-2} - xy^{-1}$

$\frac{1}{xy^2} - \frac{x}{y}$; LCD $= xy^2$

$\frac{1}{xy^2} - \frac{x}{y} \bullet \frac{xy}{xy}$

$\frac{1}{xy^2} - \frac{x^2y}{xy^2}$

$\frac{1 - x^2y}{xy^2}$

83. $2x^{-1} - 3x^{-2}$

$\frac{2}{x} - \frac{3}{x^2}$; LCD $= x^2$

$\frac{2}{x} \bullet \frac{x}{x} - \frac{3}{x^2}$

$\frac{2x}{x^2} - \frac{3}{x^2}$

$\frac{2x - 3}{x^2}$

85. $40,000,000 = (4)(10)^7$

87. $376.4 = (3.764)(10)^2$

89. $0.347 = (3.47)(10)^{-1}$

91. $0.0214 = (2.14)(10)^{-2}$

93. $(3.14)(10)^{10} = 31,400,000,000$

95. $(4.3)(10)^{-1} = 0.43$

97. $(9.14)(10)^{-4} = 0.000914$

99. $(5.123)(10)^{-8} = 0.00000005123$

101. $\frac{(60,000)(0.006)}{(0.0009)(400)} = \frac{(6)(10)^4(6)(10)^{-3}}{(9)(10)^{-4}(4)(10)^2} =$

$\frac{(36)(10)^1}{(36)(10)^{-2}} = (1)(10)^{1-(-2)} =$

$(1)(10)^3 = 1000$

103. $\frac{(0.0045)(60,000)}{(1800)(0.00015)} = \frac{(4.5)(10)^{-3}(6)(10)^4}{(1.8)(10)^3(1.5)(10)^{-4}} =$

$\frac{(27)(10)^1}{(2.7)(10)^{-1}} = \frac{(27)}{(2.7)} \cdot \frac{(10)^1}{(10)^{-1}} =$

$(10)(10)^2 = 1000$

PROBLEM SET 10.2 Roots and Radicals

1. $\sqrt{64} = 8$

3. $-\sqrt{100} = -(10) = -10$

5. $\sqrt[3]{27} = 3$

7. $\sqrt[3]{-64} = -4$

9. $\sqrt[4]{81} = 3$

11. $\sqrt{\frac{16}{25}} = \frac{4}{5}$

13. $-\sqrt{\frac{36}{49}} = -\left(\frac{6}{7}\right) = -\frac{6}{7}$

15. $\sqrt{\frac{9}{36}} = \frac{3}{6} = \frac{1}{2}$

17. $\sqrt[3]{\frac{27}{64}} = \frac{3}{4}$

19. $\sqrt[3]{8^3} = 8$

21. $\sqrt{27} = \sqrt{9}\sqrt{3} = 3\sqrt{3}$

23. $\sqrt{32} = \sqrt{16}\sqrt{2} = 4\sqrt{2}$

25. $\sqrt{80} = \sqrt{16}\sqrt{5} = 4\sqrt{5}$

27. $\sqrt{160} = \sqrt{16}\sqrt{10} = 4\sqrt{10}$

29. $4\sqrt{18} = 4\sqrt{9}\sqrt{2} = 4(3)\sqrt{2} = 12\sqrt{2}$

31. $-6\sqrt{20} = -6\sqrt{4}\sqrt{5} = -6(2)\sqrt{5} = -12\sqrt{5}$

33. $\frac{2}{5}\sqrt{75} = \frac{2}{5}\sqrt{25}\sqrt{3} = \frac{2}{5}(5)\sqrt{3} = 2\sqrt{3}$

35. $\frac{3}{2}\sqrt{24} = \frac{3}{2}\sqrt{4}\sqrt{6} = \frac{3}{2}(2)\sqrt{6} = 3\sqrt{6}$

37. $-\frac{5}{6}\sqrt{28} = -\frac{5}{6}\sqrt{4}\sqrt{7} = -\frac{5}{6}(2)\sqrt{7} = -\frac{5}{3}\sqrt{7}$

39. $\sqrt{\frac{19}{4}} = \frac{\sqrt{19}}{\sqrt{4}} = \frac{\sqrt{19}}{2}$

41. $\sqrt{\frac{27}{16}} = \frac{\sqrt{27}}{\sqrt{16}} = \frac{\sqrt{9}\sqrt{3}}{4} = \frac{3\sqrt{3}}{4}$

43. $\sqrt{\frac{75}{81}} = \frac{\sqrt{75}}{\sqrt{81}} = \frac{\sqrt{25}\sqrt{3}}{9} = \frac{5\sqrt{3}}{9}$

45. $\sqrt{\frac{2}{7}} = \frac{\sqrt{2}}{\sqrt{7}} \bullet \frac{\sqrt{7}}{\sqrt{7}} = \frac{\sqrt{14}}{\sqrt{49}} = \frac{\sqrt{14}}{7}$

47. $\sqrt{\frac{2}{3}} = \frac{\sqrt{2}}{\sqrt{3}} \bullet \frac{\sqrt{3}}{\sqrt{3}} = \frac{\sqrt{6}}{\sqrt{9}} = \frac{\sqrt{6}}{3}$

49. $\frac{\sqrt{5}}{\sqrt{12}} = \frac{\sqrt{5}}{\sqrt{4}\sqrt{3}} = \frac{\sqrt{5}}{2\sqrt{3}} =$

$\frac{\sqrt{5}}{2\sqrt{3}} \bullet \frac{\sqrt{3}}{\sqrt{3}} = \frac{\sqrt{15}}{2\sqrt{9}} = \frac{\sqrt{15}}{2(3)} = \frac{\sqrt{15}}{6}$

51. $\frac{\sqrt{11}}{\sqrt{24}} = \frac{\sqrt{11}}{\sqrt{4}\sqrt{6}} = \frac{\sqrt{11}}{2\sqrt{6}} = \frac{\sqrt{11}}{2\sqrt{6}} \bullet \frac{\sqrt{6}}{\sqrt{6}} =$

$\frac{\sqrt{66}}{2\sqrt{36}} = \frac{\sqrt{66}}{2(6)} = \frac{\sqrt{66}}{12}$

53. $\frac{\sqrt{18}}{\sqrt{27}} = \sqrt{\frac{18}{27}} = \sqrt{\frac{2}{3}} = \frac{\sqrt{2}}{\sqrt{3}} =$

$\frac{\sqrt{2}}{\sqrt{3}} \bullet \frac{\sqrt{3}}{\sqrt{3}} = \frac{\sqrt{6}}{\sqrt{9}} = \frac{\sqrt{6}}{3}$

55. $\frac{\sqrt{35}}{\sqrt{7}} = \sqrt{\frac{35}{7}} = \sqrt{5}$

57. $\dfrac{2\sqrt{3}}{\sqrt{7}} = \dfrac{2\sqrt{3}}{\sqrt{7}} \bullet \dfrac{\sqrt{7}}{\sqrt{7}} = \dfrac{2\sqrt{21}}{\sqrt{49}} = \dfrac{2\sqrt{21}}{7}$

59. $\dfrac{-4\sqrt{12}}{\sqrt{5}} = \dfrac{-4\sqrt{4}\sqrt{3}}{\sqrt{5}} = \dfrac{-4(2)\sqrt{3}}{\sqrt{5}} =$

$\dfrac{-8\sqrt{3}}{\sqrt{5}} = \dfrac{-8\sqrt{3}}{\sqrt{5}} \bullet \dfrac{\sqrt{5}}{\sqrt{5}} =$

$\dfrac{-8\sqrt{15}}{\sqrt{25}} = \dfrac{-8\sqrt{15}}{5}$

61. $\dfrac{3\sqrt{2}}{4\sqrt{3}} = \dfrac{3\sqrt{2}}{4\sqrt{3}} \bullet \dfrac{\sqrt{3}}{\sqrt{3}} = \dfrac{3\sqrt{6}}{4(3)} = \dfrac{\sqrt{6}}{4}$

63. $\dfrac{-8\sqrt{18}}{10\sqrt{50}} = \dfrac{-4}{5}\sqrt{\dfrac{18}{50}} = \dfrac{-4}{5}\sqrt{\dfrac{9}{25}} =$

$-\dfrac{4}{5}\left(\dfrac{3}{5}\right) = -\dfrac{12}{25}$

65. $\sqrt[3]{16} = \sqrt[3]{8}\sqrt[3]{2} = 2\sqrt[3]{2}$

67. $2\sqrt[3]{81} = 2\sqrt[3]{27}\sqrt[3]{3} = 2(3)\sqrt[3]{3} = 6\sqrt[3]{3}$

69. $\dfrac{2}{\sqrt[3]{9}} = \dfrac{2}{\sqrt[3]{9}} \bullet \dfrac{\sqrt[3]{3}}{\sqrt[3]{3}} = \dfrac{2\sqrt[3]{3}}{\sqrt[3]{27}} = \dfrac{2\sqrt[3]{3}}{3}$

71. $\dfrac{\sqrt[3]{27}}{\sqrt[3]{4}} = \dfrac{3}{\sqrt[3]{4}} = \dfrac{3}{\sqrt[3]{4}} \bullet \dfrac{\sqrt[3]{2}}{\sqrt[3]{2}} =$

$\dfrac{3\sqrt[3]{2}}{\sqrt[3]{8}} = \dfrac{3\sqrt[3]{2}}{2}$

73. $\dfrac{\sqrt[3]{6}}{\sqrt[3]{4}} = \sqrt[3]{\dfrac{6}{4}} = \sqrt[3]{\dfrac{3}{2}} = \dfrac{\sqrt[3]{3}}{\sqrt[3]{2}} \bullet \dfrac{\sqrt[3]{4}}{\sqrt[3]{4}} =$

$\dfrac{\sqrt[3]{12}}{\sqrt[3]{8}} = \dfrac{\sqrt[3]{12}}{2}$

75. $S = \sqrt{30Df}$

$S = \sqrt{30(150)(0.4)} = \sqrt{1800} = 42$ mph

$S = \sqrt{30(200)(0.4)} = \sqrt{2400} = 49$ mph

$S = \sqrt{30(350)(0.4)} = \sqrt{4200} = 65$ mph

77. $K = \sqrt{s(s-a)(s-b)(s-c)}$

$s = \dfrac{a+b+c}{2} = \dfrac{14+16+18}{2} = \dfrac{48}{2} = 24$

$K = \sqrt{24(24-14)(24-16)(24-18)}$

$K = \sqrt{24(10)(8)(6)}$

$K = \sqrt{11520} = 107$ square centimeters

79. $K = \sqrt{s(s-a)(s-b)(s-c)}$

$s = \dfrac{a+b+c}{2} = \dfrac{18+18+18}{2} = \dfrac{54}{2} = 27$

$K = \sqrt{27(27-18)(27-18)(27-18)}$

$K = \sqrt{27(9)(9)(9)}$

$K = \sqrt{19683} = 140$ square inches.

Calculator Activities

85a) $\sqrt{2} = 1.414$
b) $\sqrt{75} = 8.660$
c) $\sqrt{156} = 12.490$
d) $\sqrt{691} = 26.287$
e) $\sqrt{3249} = 57.000$
f) $\sqrt{45123} = 212.422$
g) $\sqrt{0.14} = 0.374$
h) $\sqrt{0.023} = 0.152$
i) $\sqrt{0.8649} = 0.930$

PROBLEM SET 10.3 Simplifying and Combining Radicals

1. $5\sqrt{18} - 2\sqrt{2}$

$5\sqrt{9}\sqrt{2} - 2\sqrt{2}$

$5(3)\sqrt{2} - 2\sqrt{2}$

$15\sqrt{2} - 2\sqrt{2}$

$13\sqrt{2}$

3. $7\sqrt{12} + 10\sqrt{48}$

$7\sqrt{4}\sqrt{3} + 10\sqrt{16}\sqrt{3}$

$7(2)\sqrt{3} + 10(4)\sqrt{3}$

$14\sqrt{3} + 40\sqrt{3}$

$54\sqrt{3}$

5. $-2\sqrt{50}-5\sqrt{32}$
$-2\sqrt{25}\sqrt{2}-5\sqrt{16}\sqrt{2}$
$-2(5)\sqrt{2}-5(4)\sqrt{2}$
$-10\sqrt{2}-20\sqrt{2}$
$-30\sqrt{2}$

7. $3\sqrt{20}-\sqrt{5}-2\sqrt{45}$
$3\sqrt{4}\sqrt{5}-\sqrt{5}-2\sqrt{9}\sqrt{5}$
$3(2)\sqrt{5}-\sqrt{5}-2(3)\sqrt{5}$
$6\sqrt{5}-\sqrt{5}-6\sqrt{5}$
$-\sqrt{5}$

9. $-9\sqrt{24}+3\sqrt{54}-12\sqrt{6}$
$-9\sqrt{4}\sqrt{6}+3\sqrt{9}\sqrt{6}-12\sqrt{6}$
$-9(2)\sqrt{6}+3(3)\sqrt{6}-12\sqrt{6}$
$-18\sqrt{6}+9\sqrt{6}-12\sqrt{6}$
$-21\sqrt{6}$

11. $\frac{3}{4}\sqrt{7}-\frac{2}{3}\sqrt{28}$
$\frac{3}{4}\sqrt{7}-\frac{2}{3}\sqrt{4}\sqrt{7}$
$\frac{3}{4}\sqrt{7}-\frac{2}{3}(2)\sqrt{7}$
$\frac{3}{4}\sqrt{7}-\frac{4}{3}\sqrt{7}$
$\frac{9}{12}\sqrt{7}-\frac{16}{12}\sqrt{7}$
$-\frac{7\sqrt{7}}{12}$

13. $\frac{3}{5}\sqrt{40}+\frac{5}{6}\sqrt{90}$
$\frac{3}{5}\sqrt{4}\sqrt{10}+\frac{5}{6}\sqrt{9}\sqrt{10}$
$\frac{3}{5}(2)\sqrt{10}+\frac{5}{6}(3)\sqrt{10}$
$\frac{6}{5}\sqrt{10}+\frac{5}{2}\sqrt{10}$
$\frac{12}{10}\sqrt{10}+\frac{25}{10}\sqrt{10}$
$\frac{37\sqrt{10}}{10}$

15. $\frac{3\sqrt{18}}{5}-\frac{5\sqrt{72}}{6}+\frac{3\sqrt{98}}{4}$
$\frac{3\sqrt{9}\sqrt{2}}{5}-\frac{5\sqrt{36}\sqrt{2}}{6}+\frac{3\sqrt{49}\sqrt{2}}{4}$
$\frac{3(3)\sqrt{2}}{5}-\frac{5(6)\sqrt{2}}{6}+\frac{3(7)\sqrt{2}}{4}$
$\frac{9\sqrt{2}}{5}-5\sqrt{2}+\frac{21\sqrt{2}}{4}$
$\frac{(9\sqrt{2})}{5}\bullet\frac{4}{4}-(5\sqrt{2})\bullet\frac{20}{20}+\frac{(21\sqrt{2})}{4}\bullet\frac{5}{5}$
$\frac{36\sqrt{2}}{20}-\frac{100\sqrt{2}}{20}+\frac{105\sqrt{2}}{20}$
$\frac{41\sqrt{2}}{20}$

17. $5\sqrt[3]{3}+2\sqrt[3]{24}-6\sqrt[3]{81}$
$5\sqrt[3]{3}+2\sqrt[3]{8}\sqrt[3]{3}-6\sqrt[3]{27}\sqrt[3]{3}$
$5\sqrt[3]{3}+2(2)\sqrt[3]{3}-6(3)\sqrt[3]{3}$
$5\sqrt[3]{3}+4\sqrt[3]{3}-18\sqrt[3]{3}$
$-9\sqrt[3]{3}$

19. $-\sqrt[3]{16}+7\sqrt[3]{54}-9\sqrt[3]{2}$
$-\sqrt[3]{8}\sqrt[3]{2}+7\sqrt[3]{27}\sqrt[3]{2}-9\sqrt[3]{2}$
$-2\sqrt[3]{2}+7(3)\sqrt[3]{2}-9\sqrt[3]{2}$
$-2\sqrt[3]{2}+21\sqrt[3]{2}-9\sqrt[3]{2}$
$10\sqrt[3]{2}$

21. $\sqrt{32x}$
$\sqrt{16}\sqrt{2x}$
$4\sqrt{2x}$

23. $\sqrt{75x^2}$
$\sqrt{25x^2}\sqrt{3}$
$5x\sqrt{3}$

25. $\sqrt{20x^2y}$
$\sqrt{4x^2}\sqrt{5y}$
$2x\sqrt{5y}$

27. $\sqrt{64x^3y^7}$
$\sqrt{64x^2y^6}\sqrt{xy}$
$8xy^3\sqrt{xy}$

29. $\sqrt{54a^4b^3}$
$\sqrt{9a^4b^2}\sqrt{6b}$
$3a^2b\sqrt{6b}$

31. $\sqrt{63x^6y^8}$
$\sqrt{9x^6y^8}\sqrt{7}$
$3x^3y^4\sqrt{7}$

33. $2\sqrt{40a^3}$
$2\sqrt{4a^2}\sqrt{10a}$
$2(2a)\sqrt{10a}$
$4a\sqrt{10a}$

35. $\frac{2}{3}\sqrt{96xy^3}$
$\frac{2}{3}\sqrt{16y^2}\sqrt{6xy}$
$\frac{2}{3}(4y)\sqrt{6xy}$
$\frac{8y}{3}\sqrt{6xy}$

37. $\sqrt{\frac{2x}{5y}} = \frac{\sqrt{2x}}{\sqrt{5y}} \bullet \frac{\sqrt{5y}}{\sqrt{5y}} = \frac{\sqrt{10xy}}{\sqrt{25y^2}} =$
$\frac{\sqrt{10xy}}{5y}$

39. $\sqrt{\frac{5}{12x^4}} = \frac{\sqrt{5}}{\sqrt{4x^4}\sqrt{3}} = \frac{\sqrt{5}}{2x^2\sqrt{3}} \bullet \frac{\sqrt{3}}{\sqrt{3}} =$
$\frac{\sqrt{15}}{2x^2\sqrt{9}} = \frac{\sqrt{15}}{2x^2(3)} = \frac{\sqrt{15}}{6x^2}$

41. $\frac{5}{\sqrt{18y}} = \frac{5}{\sqrt{9}\sqrt{2y}} = \frac{5}{3\sqrt{2y}} \bullet \frac{\sqrt{2y}}{\sqrt{2y}} =$
$\frac{5\sqrt{2y}}{3\sqrt{4y^2}} = \frac{5\sqrt{2y}}{3(2y)} = \frac{5\sqrt{2y}}{6y}$

43. $\frac{\sqrt{7x}}{\sqrt{8y^5}} = \frac{\sqrt{7x}}{\sqrt{4y^4}\sqrt{2y}} = \frac{\sqrt{7x}}{2y^2\sqrt{2y}} \bullet \frac{\sqrt{2y}}{\sqrt{2y}} =$
$\frac{\sqrt{14xy}}{2y^2\sqrt{4y^2}} = \frac{\sqrt{14xy}}{(2y^2)(2y)} = \frac{\sqrt{14xy}}{4y^3}$

45. $\frac{\sqrt{18y^3}}{\sqrt{16x}} = \frac{\sqrt{9y^2}\sqrt{2y}}{\sqrt{16}\sqrt{x}} = \frac{3y\sqrt{2y}}{4\sqrt{x}} \bullet \frac{\sqrt{x}}{\sqrt{x}} =$
$\frac{3y\sqrt{2xy}}{4\sqrt{x^2}} = \frac{3y\sqrt{2xy}}{4x}$

47. $\frac{\sqrt{24a^2b^3}}{\sqrt{7ab^6}} = \frac{\sqrt{4a^2b^2}\sqrt{6b}}{\sqrt{b^6}\sqrt{7a}} =$
$\frac{2ab\sqrt{6b}}{b^3\sqrt{7a}} \bullet \frac{\sqrt{7a}}{\sqrt{7a}} = \frac{2a\sqrt{42ab}}{b^2\sqrt{49a^2}} =$
$\frac{2a\sqrt{42ab}}{b^2(7a)} = \frac{2a\sqrt{42ab}}{7ab^2} = \frac{2\sqrt{42ab}}{7b^2}$

49. $\sqrt[3]{24y} = \sqrt[3]{8}\sqrt[3]{3y} = 2\sqrt[3]{3y}$

51. $\sqrt[3]{16x^4} = \sqrt[3]{8x^3}\sqrt[3]{2x} = 2x\sqrt[3]{2x}$

53. $\sqrt[3]{56x^6y^8} = \sqrt[3]{8x^6y^6}\sqrt[3]{7y^2} = 2x^2y^2\sqrt[3]{7y^2}$

55. $\sqrt[3]{\frac{7}{9x^2}} = \frac{\sqrt[3]{7}}{\sqrt[3]{9x^2}} \bullet \frac{\sqrt[3]{3x}}{\sqrt[3]{3x}} = \frac{\sqrt[3]{21x}}{\sqrt[3]{27x^3}} =$
$\frac{\sqrt[3]{21x}}{3x}$

57. $\frac{\sqrt[3]{3y}}{\sqrt[3]{16x^4}} = \frac{\sqrt[3]{3y}}{\sqrt[3]{8x^3}\sqrt[3]{2x}} =$
$\frac{\sqrt[3]{3y}}{2x\sqrt[3]{2x}} \bullet \frac{\sqrt[3]{4x^2}}{\sqrt[3]{4x^2}} = \frac{\sqrt[3]{12x^2y}}{2x\sqrt[3]{8x^3}} =$
$\frac{\sqrt[3]{12x^2y}}{2x(2x)} = \frac{\sqrt[3]{12x^2y}}{4x^2}$

59. $\frac{\sqrt[3]{12xy}}{\sqrt[3]{3x^2y^5}} = \sqrt[3]{\frac{12xy}{3x^2y^5}} = \sqrt[3]{\frac{4}{xy^4}} = \frac{\sqrt[3]{4}}{\sqrt[3]{xy^4}} =$
$\frac{\sqrt[3]{4}}{\sqrt[3]{y^3}\sqrt[3]{xy}} = \frac{\sqrt[3]{4}}{y\sqrt[3]{xy}} \bullet \frac{\sqrt[3]{x^2y^2}}{\sqrt[3]{x^2y^2}} =$
$\frac{\sqrt[3]{4x^2y^2}}{y\sqrt[3]{x^3y^3}} = \frac{\sqrt[3]{4x^2y^2}}{y(xy)} = \frac{\sqrt[3]{4x^2y^2}}{xy^2}$

61. $\sqrt{8x+12y}$
$\sqrt{4(2x+3y)}$
$\sqrt{4}\sqrt{2x+3y}$
$2\sqrt{2x+3y}$

63. $\sqrt{16x+48y}$
$\sqrt{16(x+3y)}$
$\sqrt{16}\sqrt{x+3y}$
$4\sqrt{x+3y}$

65. $-3\sqrt{4x}+5\sqrt{9x}+6\sqrt{16x}$
$-3\sqrt{4}\sqrt{x}+5\sqrt{9}\sqrt{x}+6\sqrt{16}\sqrt{x}$
$-3(2)\sqrt{x}+5(3)\sqrt{x}+6(4)\sqrt{x}$
$-6\sqrt{x}+15\sqrt{x}+24\sqrt{x}$
$33\sqrt{x}$

67. $2\sqrt{18x}-3\sqrt{8x}-6\sqrt{50x}$
$2\sqrt{9}\sqrt{2x}-3\sqrt{4}\sqrt{2x}-6\sqrt{25}\sqrt{2x}$
$2(3)\sqrt{2x}-3(2)\sqrt{2x}-6(5)\sqrt{2x}$
$6\sqrt{2x}-6\sqrt{2x}-30\sqrt{2x}$
$-30\sqrt{2x}$

69. $5\sqrt{27n}-\sqrt{12n}-6\sqrt{3n}$
$5\sqrt{9}\sqrt{3n}-\sqrt{4}\sqrt{3n}-6\sqrt{3n}$
$5(3)\sqrt{3n}-2\sqrt{3n}-6\sqrt{3n}$
$15\sqrt{3n}-2\sqrt{3n}-6\sqrt{3n}$
$7\sqrt{3n}$

71. $7\sqrt{4ab}-\sqrt{16ab}-10\sqrt{25ab}$
$7\sqrt{4}\sqrt{ab}-\sqrt{16}\sqrt{ab}-10\sqrt{25}\sqrt{ab}$
$7(2)\sqrt{ab}-4\sqrt{ab}-10(5)\sqrt{ab}$
$14\sqrt{ab}-4\sqrt{ab}-50\sqrt{ab}$
$-40\sqrt{ab}$

73. $-3\sqrt{2x^3}+4\sqrt{8x^3}-3\sqrt{32x^3}$
$-3\sqrt{x^2}\sqrt{2x}+4\sqrt{4x^2}\sqrt{2x}-3\sqrt{16x^2}\sqrt{2x}$
$-3(x)\sqrt{2x}+4(2x)\sqrt{2x}-3(4x)\sqrt{2x}$
$-3x\sqrt{2x}+8x\sqrt{2x}-12x\sqrt{2x}$
$-7x\sqrt{2x}$

PROBLEM SET 10.4 Products and Quotients of Radicals

1. $\sqrt{6}\sqrt{12}$
$\sqrt{72}=\sqrt{36}\sqrt{2}=6\sqrt{2}$

3. $(3\sqrt{3})(2\sqrt{6})$
$6\sqrt{18}$
$6\sqrt{9}\sqrt{2}=6(3)\sqrt{2}=18\sqrt{2}$

5. $(4\sqrt{2})(-6\sqrt{5})$
$-24\sqrt{10}$

7. $(-3\sqrt{3})(-4\sqrt{8})$
$12\sqrt{24}$
$12\sqrt{4}\sqrt{6}=12(2)\sqrt{6}=24\sqrt{6}$

9. $(5\sqrt{6})(4\sqrt{6})$
$20\sqrt{36}$
$20(6)=120$

11. $\left(2\sqrt[3]{4}\right)\left(6\sqrt[3]{2}\right)$
$12\sqrt[3]{8}$
$12(2)=24$

13. $\left(4\sqrt[3]{6}\right)\left(7\sqrt[3]{4}\right)$
$28\sqrt[3]{24}$
$28\sqrt[3]{8}\sqrt[3]{3}=28(2)\sqrt[3]{3}=56\sqrt[3]{3}$

15. $\sqrt{2}(\sqrt{3}+\sqrt{5})$
$\sqrt{6}+\sqrt{10}$

17. $3\sqrt{5}(2\sqrt{2}-\sqrt{7})$
$6\sqrt{10}-3\sqrt{35}$

19. $2\sqrt{6}(3\sqrt{8}-5\sqrt{12})$
$6\sqrt{48}-10\sqrt{72}$
$6\sqrt{16}\sqrt{3}-10\sqrt{36}\sqrt{2}$
$6(4)\sqrt{3}-10(6)\sqrt{2}$
$24\sqrt{3}-60\sqrt{2}$

21. $-4\sqrt{5}(2\sqrt{5}+4\sqrt{12})$
$-8\sqrt{25}-16\sqrt{60}$
$-8(5)-16\sqrt{4}\sqrt{15}$
$-8(5)-16(2)\sqrt{15}$
$-40-32\sqrt{15}$

23. $3\sqrt{x}\left(5\sqrt{2}+\sqrt{y}\right)$
$15\sqrt{2x}+3\sqrt{xy}$

Problem Set 10.4

25. $\sqrt{xy}(5\sqrt{xy} - 6\sqrt{x})$
$5\sqrt{x^2y^2} - 6\sqrt{x^2y}$
$5xy - 6\sqrt{x^2}\sqrt{y}$
$5xy - 6x\sqrt{y}$

27. $\sqrt{5y}\left(\sqrt{8x} + \sqrt{12y^2}\right)$
$\sqrt{40xy} + \sqrt{60y^3}$
$\sqrt{4}\sqrt{10xy} + \sqrt{4y^2}\sqrt{15y}$
$2\sqrt{10xy} + 2y\sqrt{15y}$

29. $5\sqrt{3}\left(2\sqrt{8} - 3\sqrt{18}\right)$
$10\sqrt{24} - 15\sqrt{54}$
$10\sqrt{4}\sqrt{6} - 15\sqrt{9}\sqrt{6}$
$20\sqrt{6} - 45\sqrt{6}$
$-25\sqrt{6}$

31. $(\sqrt{3} + 4)(\sqrt{3} - 7)$
$\sqrt{9} - 7\sqrt{3} + 4\sqrt{3} - 28$
$3 - 3\sqrt{3} - 28$
$-25 - 3\sqrt{3}$

33. $(\sqrt{5} - 6)(\sqrt{5} - 3)$
$\sqrt{25} - 3\sqrt{5} - 6\sqrt{5} + 18$
$5 - 9\sqrt{5} + 18$
$23 - 9\sqrt{5}$

35. $(3\sqrt{5} - 2\sqrt{3})(2\sqrt{7} + \sqrt{2})$
$6\sqrt{35} + 3\sqrt{10} - 4\sqrt{21} - 2\sqrt{6}$

37. $(2\sqrt{6} + 3\sqrt{5})(\sqrt{8} - 3\sqrt{12})$
$2\sqrt{48} - 6\sqrt{72} + 3\sqrt{40} - 9\sqrt{60}$
$2\sqrt{16}\sqrt{3} - 6\sqrt{36}\sqrt{2} + 3\sqrt{4}\sqrt{10} - 9\sqrt{4}\sqrt{15}$
$8\sqrt{3} - 36\sqrt{2} + 6\sqrt{10} - 18\sqrt{15}$

39. $(2\sqrt{6} + 5\sqrt{5})(3\sqrt{6} - \sqrt{5})$
$6\sqrt{36} - 2\sqrt{30} + 15\sqrt{30} - 5\sqrt{25}$
$36 + 13\sqrt{30} - 25$
$11 + 13\sqrt{30}$

41. $(3\sqrt{2} - 5\sqrt{3})(6\sqrt{2} - 7\sqrt{3})$
$18\sqrt{4} - 21\sqrt{6} - 30\sqrt{6} + 35\sqrt{9}$
$36 - 51\sqrt{6} + 105$
$141 - 51\sqrt{6}$

43. $(\sqrt{6} + 4)(\sqrt{6} - 4)$
$\sqrt{36} - 4\sqrt{6} + 4\sqrt{6} - 16$
$6 - 16$
-10

45. $(\sqrt{2} + \sqrt{10})(\sqrt{2} - \sqrt{10})$
$\sqrt{4} - \sqrt{20} + \sqrt{20} - \sqrt{100}$
$2 - 10$
-8

47. $(\sqrt{2x} + \sqrt{3y})(\sqrt{2x} - \sqrt{3y})$
$\sqrt{4x^2} - \sqrt{6xy} + \sqrt{6xy} - \sqrt{9y^2}$
$2x - 3y$

49. $2\sqrt[3]{3}\left(5\sqrt[3]{4} + \sqrt[3]{6}\right)$
$10\sqrt[3]{12} + 2\sqrt[3]{18}$

51. $3\sqrt[3]{4}\left(2\sqrt[3]{2} - 6\sqrt[3]{4}\right)$
$6\sqrt[3]{8} - 18\sqrt[3]{16}$
$6(2) - 18\sqrt[3]{8}\sqrt[3]{2}$
$12 - 36\sqrt[3]{2}$

53. $\dfrac{2}{\sqrt{7}+1} =$

$\dfrac{2}{(\sqrt{7}+1)} \bullet \dfrac{(\sqrt{7}-1)}{(\sqrt{7}-1)} =$

$\dfrac{2(\sqrt{7}-1)}{\sqrt{49}-1} = \dfrac{2(\sqrt{7}-1)}{7-1} =$

$\dfrac{2(\sqrt{7}-1)}{6} = \dfrac{\sqrt{7}-1}{3}$

55. $\dfrac{3}{\sqrt{2}-5} = \dfrac{3}{(\sqrt{2}-5)} \bullet \dfrac{(\sqrt{2}+5)}{(\sqrt{2}+5)} =$

$\dfrac{3(\sqrt{2}+5)}{\sqrt{4}-25} = \dfrac{3(\sqrt{2}+5)}{2-25} =$

$\dfrac{3(\sqrt{2}+5)}{-23} = \dfrac{3\sqrt{2}+15}{-23} = \dfrac{-3\sqrt{2}-15}{23}$

57. $\frac{1}{\sqrt{2}+\sqrt{7}} = \frac{1}{(\sqrt{2}+\sqrt{7})} \bullet \frac{(\sqrt{2}-\sqrt{7})}{(\sqrt{2}-\sqrt{7})} =$

$\frac{\sqrt{2}-\sqrt{7}}{\sqrt{4}-\sqrt{49}} = \frac{\sqrt{2}-\sqrt{7}}{2-7} = \frac{\sqrt{2}-\sqrt{7}}{-5} =$

$\frac{-\sqrt{2}+\sqrt{7}}{5} = \frac{\sqrt{7}-\sqrt{2}}{5}$

59. $\frac{\sqrt{2}}{\sqrt{10}-\sqrt{3}} = \frac{\sqrt{2}}{(\sqrt{10}-\sqrt{3})} \bullet \frac{(\sqrt{10}+\sqrt{3})}{(\sqrt{10}+\sqrt{3})} =$

$\frac{\sqrt{2}(\sqrt{10}+\sqrt{3})}{\sqrt{100}-\sqrt{9}} = \frac{\sqrt{20}+\sqrt{6}}{10-3} =$

$\frac{\sqrt{4}\sqrt{5}+\sqrt{6}}{7} = \frac{2\sqrt{5}+\sqrt{6}}{7}$

61. $\frac{\sqrt{3}}{2\sqrt{5}+4} = \frac{\sqrt{3}}{(2\sqrt{5}+4)} \bullet \frac{(2\sqrt{5}-4)}{(2\sqrt{5}-4)} =$

$\frac{\sqrt{3}(2\sqrt{5}-4)}{4\sqrt{25}-16} = \frac{2\sqrt{15}-4\sqrt{3}}{20-16} =$

$\frac{2\sqrt{15}-4\sqrt{3}}{4} = \frac{2(\sqrt{15}-2\sqrt{3})}{4} =$

$\frac{\sqrt{15}-2\sqrt{3}}{2}$

63. $\frac{6}{3\sqrt{7}-2\sqrt{6}} =$

$\frac{6}{(3\sqrt{7}-2\sqrt{6})} \bullet \frac{(3\sqrt{7}+2\sqrt{6})}{(3\sqrt{7}+2\sqrt{6})} =$

$\frac{6(3\sqrt{7}+2\sqrt{6})}{9\sqrt{49}-4\sqrt{36}} = \frac{18\sqrt{7}+12\sqrt{6}}{9(7)-4(6)} =$

$\frac{18\sqrt{7}+12\sqrt{6}}{63-24} = \frac{18\sqrt{7}+12\sqrt{6}}{39} =$

$\frac{3(6\sqrt{7}+4\sqrt{6})}{39} = \frac{6\sqrt{7}+4\sqrt{6}}{13}$

65. $\frac{\sqrt{6}}{3\sqrt{2}+2\sqrt{3}} =$

$\frac{\sqrt{6}}{(3\sqrt{2}+2\sqrt{3})} \bullet \frac{(3\sqrt{2}-2\sqrt{3})}{(3\sqrt{2}-2\sqrt{3})} =$

$\frac{\sqrt{6}(3\sqrt{2}-2\sqrt{3})}{9\sqrt{4}-4\sqrt{9}} = \frac{3\sqrt{12}-2\sqrt{18}}{9(2)-4(3)} =$

$\frac{3\sqrt{4}\sqrt{3}-2\sqrt{9}\sqrt{2}}{18-12} = \frac{6\sqrt{3}-6\sqrt{2}}{6} =$

$\frac{6(\sqrt{3}-\sqrt{2})}{6} = \sqrt{3}-\sqrt{2}$

67. $\frac{2}{\sqrt{x}+4} =$

$\frac{2}{\sqrt{x}+4} \bullet \frac{(\sqrt{x}-4)}{(\sqrt{x}-4)} =$

$\frac{2(\sqrt{x}-4)}{\sqrt{x^2}-16} = \frac{2\sqrt{x}-8}{x-16}$

69. $\frac{\sqrt{x}}{\sqrt{x}-5} =$

$\frac{\sqrt{x}}{(\sqrt{x}-5)} \bullet \frac{(\sqrt{x}+5)}{(\sqrt{x}+5)} =$

$\frac{\sqrt{x}(\sqrt{x}+5)}{\sqrt{x^2}-25} = \frac{\sqrt{x^2}+5\sqrt{x}}{x-25} =$

$\frac{x+5\sqrt{x}}{x-25}$

71. $\frac{\sqrt{x}-2}{\sqrt{x}+6} =$

$\frac{(\sqrt{x}-2)}{(\sqrt{x}+6)} \bullet \frac{(\sqrt{x}-6)}{(\sqrt{x}-6)} =$

$\frac{\sqrt{x^2}-6\sqrt{x}-2\sqrt{x}+12}{\sqrt{x^2}-36} =$

$\frac{x-8\sqrt{x}+12}{x-36}$

73. $\dfrac{\sqrt{x}}{\sqrt{x}+2\sqrt{y}} =$

$\dfrac{\sqrt{x}}{(\sqrt{x}+2\sqrt{y})} \bullet \dfrac{(\sqrt{x}-2\sqrt{y})}{(\sqrt{x}-2\sqrt{y})} =$

$\dfrac{\sqrt{x^2}-2\sqrt{xy}}{\sqrt{x^2}-4\sqrt{y^2}} = \dfrac{x-2\sqrt{xy}}{x-4y}$

75. $\dfrac{3\sqrt{y}}{2\sqrt{x}-3\sqrt{y}} =$

$\dfrac{3\sqrt{y}}{(2\sqrt{x}-3\sqrt{y})} \bullet \dfrac{(2\sqrt{x}+3\sqrt{y})}{(2\sqrt{x}+3\sqrt{y})} =$

$\dfrac{6\sqrt{xy}+9\sqrt{y^2}}{4\sqrt{x^2}-9\sqrt{y^2}} = \dfrac{6\sqrt{xy}+9y}{4x-9y}$

PROBLEM SET 10.5 Radical Equations

1. $\sqrt{5x} = 10$
$(\sqrt{5x})^2 = 10^2$
$5x = 100$
$x = 20$
Check
$\sqrt{5(20)} \stackrel{?}{=} 10$
$\sqrt{100} \stackrel{?}{=} 10$
$10 = 10$
The solution set is $\{20\}$.

3. $\sqrt{2x} + 4 = 0$
$\sqrt{2x} = -4$
$(\sqrt{2x})^2 = (-4)^2$
$2x = 16$
$x = 8$
Check
$\sqrt{2(8)} + 4 \stackrel{?}{=} 0$
$\sqrt{16} + 4 \stackrel{?}{=} 0$
$4 + 4 \stackrel{?}{=} 0$
$8 \neq 0$
The solution set is $\emptyset$.

5. $2\sqrt{n} = 5$
$\sqrt{n} = \dfrac{5}{2}$
$(\sqrt{n})^2 = \left(\dfrac{5}{2}\right)^2$
$n = \dfrac{25}{4}$

Check
$2\sqrt{\dfrac{25}{4}} \stackrel{?}{=} 5$
$2\left(\dfrac{5}{2}\right) \stackrel{?}{=} 5$
$5 = 5$
The solution set is $\left\{\dfrac{25}{4}\right\}$.

7. $3\sqrt{n} - 2 = 0$
$3\sqrt{n} = 2$
$\sqrt{n} = \dfrac{2}{3}$
$(\sqrt{n})^2 = \left(\dfrac{2}{3}\right)^2$
$n = \dfrac{4}{9}$
Check
$3\sqrt{\dfrac{4}{9}} - 2 \stackrel{?}{=} 0$
$3\left(\dfrac{2}{3}\right) - 2 \stackrel{?}{=} 0$
$2 - 2 \stackrel{?}{=} 0$
$0 = 0$
The solution set is $\left\{\dfrac{4}{9}\right\}$.

9. $\sqrt{3y+1} = 4$
$(\sqrt{3y+1})^2 = 4^2$
$3y + 1 = 16$
$3y = 15$
$y = 5$
Check
$\sqrt{3(5)+1} \stackrel{?}{=} 4$
$\sqrt{16} \stackrel{?}{=} 4$
$4 = 4$
The solution set is $\{5\}$.

11. $\sqrt{4y-3}-6=0$
$\sqrt{4y-3}=6$
$(\sqrt{4y-3})^2=6^2$
$4y-3=36$
$4y=39$
$y=\frac{39}{4}$
Check
$\sqrt{4\left(\frac{39}{4}\right)-3}-6\stackrel{?}{=}0$
$\sqrt{39-3}-6\stackrel{?}{=}0$
$\sqrt{36}-6\stackrel{?}{=}0$
$6-6\stackrel{?}{=}0$
$0=0$
The solution set is $\left\{\frac{39}{4}\right\}$.

13. $\sqrt{2x-5}=-1$
$(\sqrt{2x-5})^2=(-1)^2$
$2x-5=1$
$2x=6$
$x=3$
Check
$\sqrt{2(3)-5}\stackrel{?}{=}-1$
$\sqrt{1}\stackrel{?}{=}-1$
$1\neq-1$
The solution set is $\emptyset$.

15. $\sqrt{5x+2}=\sqrt{6x+1}$
$(\sqrt{5x+2})^2=(\sqrt{6x+1})^2$
$5x+2=6x+1$
$2=x+1$
$1=x$
Check
$\sqrt{5(1)+2}\stackrel{?}{=}\sqrt{6(1)+1}$
$\sqrt{7}=\sqrt{7}$
The solution set is $\{1\}$.

17. $\sqrt{3x+1}=\sqrt{7x-5}$
$(\sqrt{3x+1})^2=(\sqrt{7x-5})^2$
$3x+1=7x-5$
$3x=7x-6$
$-4x=-6$
$x=\frac{-6}{-4}=\frac{3}{2}$
Check
$\sqrt{3\left(\frac{3}{2}\right)+1}\stackrel{?}{=}\sqrt{7\left(\frac{3}{2}\right)-\frac{10}{2}}$
$\sqrt{\frac{9}{2}+1}\stackrel{?}{=}\sqrt{\frac{21}{2}-5}$
$\sqrt{\frac{11}{2}}=\sqrt{\frac{11}{2}}$
The solution set is $\left\{\frac{3}{2}\right\}$.

19. $\sqrt{3x-2}-\sqrt{x+4}=0$
$\sqrt{3x-2}=\sqrt{x+4}$
$(\sqrt{3x-2})^2=(\sqrt{x+4})^2$
$3x-2=x+4$
$3x=x+6$
$2x=6$
$x=3$
Check
$\sqrt{3(3)-2}-\sqrt{3+4}\stackrel{?}{=}0$
$\sqrt{7}-\sqrt{7}\stackrel{?}{=}0$
$0=0$
The solution set is $\{3\}$.

21. $5\sqrt{t-1}=6$
$\sqrt{t-1}=\frac{6}{5}$
$(\sqrt{t-1})^2=\left(\frac{6}{5}\right)^2$
$t-1=\frac{36}{25}$
$t=\frac{36}{25}+1=\frac{36}{25}+\frac{25}{25}=\frac{61}{25}$
Check
$5\sqrt{\frac{61}{25}-1}\stackrel{?}{=}6$
$5\sqrt{\frac{36}{25}}\stackrel{?}{=}6$
$5\left(\frac{6}{5}\right)\stackrel{?}{=}6$
$6=6$
The solution set is $\left\{\frac{61}{25}\right\}$.

23. $\sqrt{x^2+7}=4$
$(\sqrt{x^2+7})^2=4^2$
$x^2+7=16$
$x^2-9=0$
$(x+3)(x-3)=0$
$x+3=0$ or $x-3=0$
$x=-3$ or $x=3$
Checking $x=-3$
$\sqrt{(-3)^2+7}\overset{?}{=}4$
$\sqrt{16}\overset{?}{=}4$
$4=4$
Checking $x=3$
$\sqrt{3^2+7}\overset{?}{=}4$
$\sqrt{16}\overset{?}{=}4$
$4=4$
The solution set is $\{-3, 3\}$.

25. $\sqrt{x^2+13x+37}=1$
$(\sqrt{x^2+13x+37})^2=1^2$
$x^2+13x+37=1$
$x^2+13x+36=0$
$(x+4)(x+9)=0$
$x+4=0$ or $x+9=0$
$x=-4$ or $x=-9$
Checking $x=-4$
$\sqrt{(-4)^2+13(-4)+37}\overset{?}{=}1$
$\sqrt{16-52+37}\overset{?}{=}1$
$\sqrt{1}\overset{?}{=}1$
$1=1$
Checking $x=-9$
$\sqrt{(-9)^2+13(-9)+37}\overset{?}{=}1$
$\sqrt{81-117+37}\overset{?}{=}1$
$\sqrt{1}\overset{?}{=}1$
$1=1$
The solution set is $\{-9, -4\}$.

27. $\sqrt{x^2-x+1}=x+1$
$(\sqrt{x^2-x+1})^2=(x+1)^2$
$x^2-x+1=x^2+2x+1$
$-x+1=2x+1$
$-x=2x$
$-3x=0$
$x=0$
Check
$\sqrt{0^2-0+1}\overset{?}{=}0+1$
$\sqrt{1}\overset{?}{=}1$
$1=1$
The solution set is $\{0\}$.

29. $\sqrt{x^2+3x+7}=x+2$
$(\sqrt{x^2+3x+7})^2=(x+2)^2$
$x^2+3x+7=x^2+4x+4$
$3x+7=4x+4$
$3x+3=4x$
$3=x$
Check
$\sqrt{3^2+3(3)+7}\overset{?}{=}3+2$
$\sqrt{9+9+7}\overset{?}{=}5$
$\sqrt{25}\overset{?}{=}5$
$5=5$
The solution set is $\{3\}$.

31. $\sqrt{-4x+17}=x-3$
$(\sqrt{-4x+17})^2=(x-3)^2$
$-4x+17=x^2-6x+9$
$0=x^2-2x-8$
$0=(x-4)(x+2)$
$x-4=0$ or $x+2=0$
$x=4$ or $x=-2$
Checking $x=4$
$\sqrt{-4(4)+17}\overset{?}{=}4-3$
$\sqrt{1}\overset{?}{=}1$
$1=1$
Checking $x=-2$
$\sqrt{-4(-2)+17}\overset{?}{=}-2-3$
$\sqrt{25}\overset{?}{=}-5$
$5\neq-5$
The solution set is $\{4\}$.

33. $\sqrt{n+4}=n+4$
$(\sqrt{n+4})^2=(n+4)^2$
$n+4=n^2+8n+16$
$0=n^2+7n+12$
$0=(n+4)(n+3)$
$n+4=0$ or $n+3=0$
$n=-4$ or $n=-3$
Checking $n=-4$
$\sqrt{-4+4}\overset{?}{=}-4+4$
$\sqrt{0}\overset{?}{=}0$
$0=0$

Checking $n = -3$
$\sqrt{-3+4} \stackrel{?}{=} -3+4$
$\sqrt{1} \stackrel{?}{=} 1$
$1 = 1$
The solution set is $\{-4, -3\}$.

35. $\sqrt{3y} = y - 6$
$(\sqrt{3y})^2 = (y-6)^2$
$3y = y^2 - 12y + 36$
$0 = y^2 - 15y + 36$
$0 = (y-12)(y-3)$
$y - 12 = 0$ or $y - 3 = 0$
$y = 12$ or $y = 3$
Checking $y = 12$
$\sqrt{3(12)} \stackrel{?}{=} 12 - 6$
$\sqrt{36} \stackrel{?}{=} 6$
$6 = 6$
Checking $y = 3$
$\sqrt{3(3)} \stackrel{?}{=} 3 - 6$
$\sqrt{9} \stackrel{?}{=} -3$
$3 \neq -3$
The solution set is $\{12\}$.

37. $4\sqrt{x} + 5 = x$
$4\sqrt{x} = x - 5$
$(4\sqrt{x})^2 = (x-5)^2$
$16x = x^2 - 10x + 25$
$0 = x^2 - 26x + 25$
$0 = (x-25)(x-1)$
$x - 25 = 0$ or $x - 1 = 0$
$x = 25$ or $x = 1$
Checking $x = 25$
$4\sqrt{25} + 5 \stackrel{?}{=} 25$
$4(5) + 5 \stackrel{?}{=} 25$
$25 = 25$
Checking $x = 1$
$4\sqrt{1} + 5 \stackrel{?}{=} 1$
$4 + 5 \stackrel{?}{=} 1$
$9 \neq 1$
The solution set is $\{25\}$.

39. $\sqrt[3]{x-2} = 3$
$\left(\sqrt[3]{x-2}\right)^3 = 3^3$
$x - 2 = 27$
$x = 29$
Check
$\sqrt[3]{29-2} \stackrel{?}{=} 3$
$\sqrt[3]{27} \stackrel{?}{=} 3$
$3 = 3$
The solution set is $\{29\}$.

41. $\sqrt[3]{2x+3} = -3$
$\left(\sqrt[3]{2x+3}\right)^3 = (-3)^3$
$2x + 3 = -27$
$2x = -30$
$x = -15$
Check
$\sqrt[3]{2(-15)+3} \stackrel{?}{=} -3$
$\sqrt[3]{-27} \stackrel{?}{=} -3$
$-3 = -3$
The solution set is $\{-15\}$.

43. $\sqrt[3]{2x+5} = \sqrt[3]{4-x}$
$\left(\sqrt[3]{2x+5}\right)^3 = \left(\sqrt[3]{4-x}\right)^3$
$2x + 5 = 4 - x$
$2x = -1 - x$
$3x = -1$
$x = -\frac{1}{3}$
Check
$\sqrt[3]{2\left(-\frac{1}{3}\right) + 5} \stackrel{?}{=} \sqrt[3]{4 - \left(-\frac{1}{3}\right)}$
$\sqrt[3]{-\frac{2}{3} + 5} \stackrel{?}{=} \sqrt[3]{4 + \frac{1}{3}}$
$\sqrt[3]{4\frac{1}{3}} = \sqrt[3]{4\frac{1}{3}}$
The solution set is $\left\{-\frac{1}{3}\right\}$.

45. $\sqrt{x+19} - \sqrt{x+28} = -1$
$\sqrt{x+19} = \sqrt{x+28} - 1$
$(\sqrt{x+19})^2 = (\sqrt{x+28} - 1)^2$
$x + 19 = (\sqrt{x+28})^2 + 2(-1)\sqrt{x+28} + 1$
$x + 19 = x + 28 - 2\sqrt{x+28} + 1$
$x + 19 = x + 29 - 2\sqrt{x+28}$
$-10 = -2\sqrt{x+28}$
$5 = \sqrt{x+28}$
$(5)^2 = (\sqrt{x+28})^2$

$25 = x + 28$
$-3 = x$
Check
$\sqrt{-3+19} - \sqrt{-3+28} \stackrel{?}{=} -1$
$\sqrt{16} - \sqrt{25} \stackrel{?}{=} -1$
$4 - 5 \stackrel{?}{=} -1$
$-1 = -1$
The solution set is $\{-3\}$.

47. $\sqrt{3x+1} + \sqrt{2x+4} = 3$
$\sqrt{3x+1} = 3 - \sqrt{2x+4}$
$(\sqrt{3x+1})^2 = (3 - \sqrt{2x+4})^2$
$3x + 1 = 9 + 2(3)(-\sqrt{2x+4}) + (-\sqrt{2x+4})^2$
$3x + 1 = 9 - 6\sqrt{2x+4} + 2x + 4$
$3x + 1 = 2x + 13 - 6\sqrt{2x+4}$
$x - 12 = -6\sqrt{2x+4}$
$(x-12)^2 = (-6\sqrt{2x+4})^2$
$x^2 - 24x + 144 = 36(2x+4)$
$x^2 - 24x + 144 = 72x + 144$
$x^2 - 96x = 0$
$x(x-96) = 0$
$x = 0$ or $x - 96 = 0$
$x = 0$ or $x = 96$
Checking $x = 0$
$\sqrt{3(0)+1} + \sqrt{2(0)+4} \stackrel{?}{=} 3$
$\sqrt{1} + \sqrt{4} \stackrel{?}{=} 3$
$1 + 2 \stackrel{?}{=} 3$
$3 = 3$
Checking $x = 96$
$\sqrt{3(96)+1} + \sqrt{2(96)+4} \stackrel{?}{=} 3$
$\sqrt{289} + \sqrt{196} \stackrel{?}{=} 3$
$17 + 14 \stackrel{?}{=} 3$
$31 \neq 3$
The solution set is $\{0\}$.

49. $\sqrt{n-4} + \sqrt{n+4} = 2\sqrt{n-1}$
$(\sqrt{n-4} + \sqrt{n+4})^2 = (2\sqrt{n-1})^2$
$(\sqrt{n\text{-}4})^2 + 2\sqrt{n\text{-}4}\sqrt{n\text{+}4} + (\sqrt{n\text{+}4})^2 = 4(n\text{-}1)$
$n - 4 + 2\sqrt{(n-4)(n+4)} + n + 4 = 4n - 4$
$2n + 2\sqrt{n^2-16} = 4n - 4$
$2\sqrt{n^2-16} = 2n - 4$
$2\sqrt{n^2-16} = 2(n-2)$
$\sqrt{n^2-16} = n - 2$
$(\sqrt{n^2-16})^2 = (n-2)^2$
$n^2 - 16 = n^2 - 4n + 4$
$-16 = -4n + 4$
$-20 = -4n$
$5 = n$
Check
$\sqrt{5-4} + \sqrt{5+4} \stackrel{?}{=} 2\sqrt{5-1}$
$\sqrt{1} + \sqrt{9} \stackrel{?}{=} 2\sqrt{4}$
$1 + 3 \stackrel{?}{=} 2(2)$
$4 = 4$
The solution set is $\{5\}$.

51. $\sqrt{t+3} - \sqrt{t-2} = \sqrt{7-t}$
$(\sqrt{t+3} - \sqrt{t-2})^2 = (\sqrt{7-t})^2$
$(\sqrt{t\text{+}3})^2\text{-}2\sqrt{t\text{+}3}\sqrt{t\text{-}2} + (\text{-}\sqrt{t\text{-}2})^2 = 7 - t$
$t + 3 - 2\sqrt{(t+3)(t-2)} + t - 2 = 7 - t$
$2t + 1 - 2\sqrt{t^2+t-6} = 7 - t$
$-2\sqrt{t^2+t-6} = -3t + 6$
$(-2\sqrt{t^2+t-6})^2 = (-3t+6)^2$
$4(t^2+t-6) = 9t^2 - 36t + 36$
$4t^2 + 4t - 24 = 9t^2 - 36t + 36$
$0 = 5t^2 - 40t + 60$
$0 = 5(t^2 - 8t + 12)$
$0 = 5(t-6)(t-2)$
$t - 6 = 0$ or $t - 2 = 0$
$t = 6$ or $t = 2$
Checking $t = 6$
$\sqrt{6+3} - \sqrt{6-2} \stackrel{?}{=} \sqrt{7-6}$
$\sqrt{9} - \sqrt{4} \stackrel{?}{=} \sqrt{1}$
$3 - 2 \stackrel{?}{=} 1$
$1 = 1$
Checking $t = 2$
$\sqrt{2+3} - \sqrt{2-2} \stackrel{?}{=} \sqrt{7-2}$
$\sqrt{5} - \sqrt{0} \stackrel{?}{=} \sqrt{5}$
$\sqrt{5} = \sqrt{5}$
The solution set is $\{2, 6\}$.

53. $\sqrt{30Df} = S$
$\sqrt{30D(0.95)} = S$
$\sqrt{28.5D} = S$
$(\sqrt{28.5D})^2 = S^2$
$28.5D = S^2$
$D = \dfrac{S^2}{28.5}$
$D = \dfrac{40^2}{28.5} = 56$ feet

$$D = \frac{55^2}{28.5} = 106 \text{ feet}$$
$$D = \frac{65^2}{28.5} = 148 \text{ feet}$$

55. $L = \dfrac{8T^2}{\pi^2}$

$$L = \frac{8(2)^2}{\pi^2} = 3.2 \text{ feet}$$
$$L = \frac{8(2.5)^2}{\pi^2} = 5.1 \text{ feet}$$
$$L = \frac{8(3)^2}{\pi^2} = 7.3 \text{ feet}$$

PROBLEM SET 10.6 Merging Exponents and Roots

1. $81^{\frac{1}{2}} = \sqrt{81} = 9$

3. $27^{\frac{1}{3}} = \sqrt[3]{27} = 3$

5. $(-8)^{\frac{1}{3}} = \sqrt[3]{-8} = -2$

7. $-25^{\frac{1}{2}} = -\sqrt{25} = -5$

9. $36^{-\frac{1}{2}} = \dfrac{1}{36^{\frac{1}{2}}} = \dfrac{1}{\sqrt{36}} = \dfrac{1}{6}$

11. $\left(\dfrac{1}{27}\right)^{-\frac{1}{3}} = \dfrac{1^{-\frac{1}{3}}}{27^{-\frac{1}{3}}} = \dfrac{27^{\frac{1}{3}}}{1^{\frac{1}{3}}} = \dfrac{\sqrt[3]{27}}{\sqrt[3]{1}} = \dfrac{3}{1} = 3$

13. $4^{\frac{3}{2}} = \left(\sqrt{4}\right)^3 = 2^3 = 8$

15. $27^{\frac{4}{3}} = \left(\sqrt[3]{27}\right)^4 = 3^4 = 81$

17. $(-1)^{\frac{7}{3}} = \left(\sqrt[3]{-1}\right)^7 = (-1)^7 = -1$

19. $-4^{\frac{5}{2}} = -(\sqrt{4})^5 = -(2)^5 = -32$

21. $\left(\dfrac{27}{8}\right)^{\frac{4}{3}} = \left(\sqrt[3]{\dfrac{27}{8}}\right)^4 = \left(\dfrac{3}{2}\right)^4 = \dfrac{3^4}{2^4} = \dfrac{81}{16}$

23. $\left(\dfrac{1}{8}\right)^{-\frac{2}{3}} = \dfrac{1^{-\frac{2}{3}}}{8^{-\frac{2}{3}}} = \dfrac{8^{\frac{2}{3}}}{1^{\frac{2}{3}}}$

$\dfrac{\left(\sqrt[3]{8}\right)^2}{\left(\sqrt[3]{1}\right)^2} = \dfrac{(2)^2}{(1)^2} = \dfrac{4}{1} = 4$

25. $64^{-\frac{7}{6}} = \dfrac{1}{64^{\frac{7}{6}}} = \dfrac{1}{\left(\sqrt[6]{64}\right)^7} = \dfrac{1}{2^7} = \dfrac{1}{128}$

27. $-25^{\frac{3}{2}} = -(\sqrt{25})^3 = -(5)^3 = -125$

29. $125^{\frac{4}{3}} = \left(\sqrt[3]{125}\right)^4 = 5^4 = 625$

31. $x^{\frac{4}{3}} = \sqrt[3]{x^4}$

33. $3x^{\frac{1}{2}} = 3\sqrt{x}$

35. $(2y)^{\frac{1}{3}} = \sqrt[3]{2y}$

37. $(2x - 3y)^{\frac{1}{2}} = \sqrt{2x - 3y}$

39. $(2a - 3b)^{\frac{2}{3}} = \sqrt[3]{(2a - 3b)^2}$

41. $x^{\frac{2}{3}}y^{\frac{1}{3}} = \sqrt[3]{x^2y}$

43. $-3x^{\frac{1}{5}}y^{\frac{2}{5}} = -3\sqrt[5]{xy^2}$

45. $\sqrt{5y} = (5y)^{\frac{1}{2}} = 5^{\frac{1}{2}}y^{\frac{1}{2}}$

47. $3\sqrt{y} = 3y^{\frac{1}{2}}$

49. $\sqrt[3]{xy^2} = (xy^2)^{\frac{1}{3}} = x^{\frac{1}{3}}y^{\frac{2}{3}}$

51. $\sqrt[4]{a^2b^3} = (a^2b^3)^{\frac{1}{4}} = a^{\frac{2}{4}}b^{\frac{3}{4}} = a^{\frac{1}{2}}b^{\frac{3}{4}}$

53. $\sqrt[5]{(2x - y)^3} = (2x - y)^{\frac{3}{5}}$

55. $5x\sqrt{y} = 5xy^{\frac{1}{2}}$

57. $-\sqrt[3]{x+y} = -(x+y)^{\frac{1}{3}}$

59. $(2x^{\frac{2}{5}})(6x^{\frac{1}{4}})$

$12x^{\frac{2}{5}+\frac{1}{4}}$

$12x^{\frac{8}{20}+\frac{5}{20}}$

$12x^{\frac{13}{20}}$

61. $(y^{\frac{2}{3}})(y^{-\frac{1}{4}})$

$y^{\frac{2}{3}-\frac{1}{4}} = y^{\frac{8}{12}-\frac{3}{12}} = y^{\frac{5}{12}}$

63. $(x^{\frac{2}{5}})(4x^{-\frac{1}{2}})$

$4x^{\frac{2}{5}-\frac{1}{2}} = 4x^{\frac{4}{10}-\frac{5}{10}} = 4x^{-\frac{1}{10}} = \frac{4}{x^{\frac{1}{10}}}$

65. $(4x^{\frac{1}{2}}y)^2$

$4^2x^{\frac{1}{2}(2)}y^2$

$16xy^2$

67. $(8x^6y^3)^{\frac{1}{3}}$

$8^{\frac{1}{3}}x^{6(\frac{1}{3})}y^{3(\frac{1}{3})}$

$2x^2y$

69. $\frac{24x^{\frac{3}{5}}}{6x^{\frac{1}{3}}}$

$4x^{\frac{3}{5}-\frac{1}{3}} = 4x^{\frac{9}{15}-\frac{5}{15}} = 4x^{\frac{4}{15}}$

71. $\frac{48b^{\frac{1}{3}}}{12b^{\frac{3}{4}}}$

$4b^{\frac{1}{3}-\frac{3}{4}} = 4b^{\frac{4}{12}-\frac{9}{12}} = 4b^{-\frac{5}{12}} = \frac{4}{b^{\frac{5}{12}}}$

73. $\left(\frac{6x^{\frac{2}{5}}}{7y^{\frac{2}{3}}}\right)^2 = \frac{6^2x^{\frac{2}{5}(2)}}{7^2y^{\frac{2}{3}(2)}} = \frac{36x^{\frac{4}{5}}}{49y^{\frac{4}{3}}}$

75. $\left(\frac{x^2}{y^3}\right)^{-\frac{1}{2}} = \frac{x^{2(-\frac{1}{2})}}{y^{3(-\frac{1}{2})}} = \frac{x^{-1}}{y^{-\frac{3}{2}}} = \frac{y^{\frac{3}{2}}}{x}$

77. $\left(\frac{18x^{\frac{1}{3}}}{9x^{\frac{1}{4}}}\right)^2 = (2x^{\frac{1}{3}-\frac{1}{4}})^2 = (2x^{\frac{4}{12}-\frac{3}{12}})^2$

$(2x^{\frac{1}{12}})^2 = 2^2x^{\frac{1}{12}(2)} = 4x^{\frac{2}{12}} = 4x^{\frac{1}{6}}$

79. $\left(\frac{60a^{\frac{1}{5}}}{15a^{\frac{3}{4}}}\right)^2 = (4a^{\frac{1}{5}-\frac{3}{4}})^2 = (4a^{\frac{4}{20}-\frac{15}{20}})^2$

$(4a^{-\frac{11}{20}})^2 = 4^2a^{-\frac{11}{20}(2)} = 16a^{-\frac{11}{10}} = \frac{16}{a^{\frac{11}{10}}}$

81. $\sqrt[3]{3}\sqrt{3}$

$3^{\frac{1}{3}} \bullet 3^{\frac{1}{2}} = 3^{\frac{1}{3}+\frac{1}{2}} = 3^{\frac{2}{6}+\frac{3}{6}} = 3^{\frac{5}{6}} =$

$\sqrt[6]{3^5} = \sqrt[6]{243}$

83. $\sqrt[4]{6}\sqrt{6}$

$6^{\frac{1}{4}} \bullet 6^{\frac{1}{2}} = 6^{\frac{1}{4}+\frac{1}{2}} = 6^{\frac{1}{4}+\frac{2}{4}} = 6^{\frac{3}{4}} =$

$\sqrt[4]{6^3} = \sqrt[4]{216}$

85. $\frac{\sqrt[3]{3}}{\sqrt[4]{3}} = \frac{3^{\frac{1}{3}}}{3^{\frac{1}{4}}} = 3^{\frac{1}{3}-\frac{1}{4}} = 3^{\frac{4}{12}-\frac{3}{12}} =$

$3^{\frac{1}{12}} = \sqrt[12]{3}$

87. $\frac{\sqrt[3]{8}}{\sqrt[4]{4}} = \frac{\sqrt[3]{2^3}}{\sqrt[4]{2^2}} = \frac{2^{\frac{3}{3}}}{2^{\frac{2}{4}}} = \frac{2^1}{2^{\frac{1}{2}}} =$

$2^{1-\frac{1}{2}} = 2^{\frac{1}{2}} = \sqrt{2}$

89. $\frac{\sqrt[4]{27}}{\sqrt{3}} = \frac{\sqrt[4]{3^3}}{\sqrt{3}} = \frac{3^{\frac{3}{4}}}{3^{\frac{1}{2}}} =$

$3^{\frac{3}{4}-\frac{1}{2}} = 3^{\frac{3}{4}-\frac{2}{4}} = 3^{\frac{1}{4}} = \sqrt[4]{3}$

Calculator Activities

93a) $\sqrt[3]{1728} = 12$
b) $\sqrt[3]{5832} = 18$
c) $\sqrt[4]{2401} = 7$
d) $\sqrt[4]{65536} = 16$
e) $\sqrt[5]{161051} = 11$
f) $\sqrt[5]{6436343} = 23$

95a) $16^{\frac{5}{2}} = 1024$
b) $25^{\frac{7}{2}} = 78125$
c) $16^{\frac{9}{4}} = 512$
d) $27^{\frac{5}{3}} = 243$
e) $343^{\frac{2}{3}} = 49$
f) $512^{\frac{4}{3}} = 4096$

CHAPTER 10 | Review Problem Set

1. $4^{-3} = \frac{1}{4^3} = \frac{1}{64}$

2. $\left(\frac{2}{3}\right)^{-2} = \left(\frac{3}{2}\right)^2 = \frac{3^2}{2^2} = \frac{9}{4}$

3. $(3^2 \bullet 3^{-3})^{-1} =$
$(3^{-1})^{-1} = 3^{-1(-1)} = 3^1 = 3$

4. $\sqrt[3]{-8} = -2$

5. $\sqrt[4]{\frac{16}{81}} = \frac{2}{3}$

6. $4^{\frac{5}{2}} = (\sqrt{4})^5 = 2^5 = 32$

7. $(-1)^{\frac{2}{3}} = \left(\sqrt[3]{-1}\right)^2 = (-1)^2 = 1$

8. $\left(\frac{8}{27}\right)^{\frac{2}{3}} = \left(\sqrt[3]{\frac{8}{27}}\right)^2 = \left(\frac{2}{3}\right)^2 = \frac{4}{9}$

9. $-16^{\frac{3}{2}} = -(\sqrt{16})^3 = -(4)^3 = -64$

10. $\frac{2^3}{2^{-2}} = 2^{3-(-2)} = 2^5 = 32$

11. $(4^{-2} \bullet 4^2)^{-1} = (4^0)^{-1} = 4^0 = 1$

12. $\left(\frac{3^{-1}}{3^2}\right)^{-1} = (3^{-1-2})^{-1} = (3^{-3})^{-1} =$
$3^3 = 27$

13. $\sqrt{54} = \sqrt{9}\sqrt{6} = 3\sqrt{6}$

14. $\sqrt{48x^3y}$
$\sqrt{16x^2}\sqrt{3xy}$
$4x\sqrt{3xy}$

15. $\frac{4\sqrt{3}}{\sqrt{6}} = 4\sqrt{\frac{3}{6}} = 4\sqrt{\frac{1}{2}} = (4)\frac{\sqrt{1}}{\sqrt{2}} =$
$\frac{4}{\sqrt{2}} \bullet \frac{\sqrt{2}}{\sqrt{2}} = \frac{4\sqrt{2}}{\sqrt{4}} = \frac{4\sqrt{2}}{2} = 2\sqrt{2}$

16. $\sqrt{\frac{5}{12x^3}} = \frac{\sqrt{5}}{\sqrt{12x^3}} = \frac{\sqrt{5}}{\sqrt{4x^2}\sqrt{3x}} =$
$\frac{\sqrt{5}}{2x\sqrt{3x}} \bullet \frac{\sqrt{3x}}{\sqrt{3x}} = \frac{\sqrt{15x}}{2x\sqrt{9x^2}} =$
$\frac{\sqrt{15x}}{2x(3x)} = \frac{\sqrt{15x}}{6x^2}$

17. $\sqrt[3]{56} = \sqrt[3]{8}\sqrt[3]{7} = 2\sqrt[3]{7}$

18. $\frac{\sqrt[3]{2}}{\sqrt[3]{9}} = \frac{\sqrt[3]{2}}{\sqrt[3]{9}} \bullet \frac{\sqrt[3]{3}}{\sqrt[3]{3}} = \frac{\sqrt[3]{6}}{\sqrt[3]{27}} = \frac{\sqrt[3]{6}}{3}$

19. $\sqrt{\frac{9}{5}} = \frac{\sqrt{9}}{\sqrt{5}} = \frac{3}{\sqrt{5}} = \frac{3}{\sqrt{5}} \bullet \frac{\sqrt{5}}{\sqrt{5}} =$
$\frac{3\sqrt{5}}{\sqrt{25}} = \frac{3\sqrt{5}}{5}$

20. $\sqrt{\frac{3x^3}{7}} = \frac{\sqrt{3x^3}}{\sqrt{7}} = \frac{\sqrt{x^2}\sqrt{3x}}{\sqrt{7}} = \frac{x\sqrt{3x}}{\sqrt{7}} =$
$\frac{x\sqrt{3x}}{\sqrt{7}} \bullet \frac{\sqrt{7}}{\sqrt{7}} = \frac{x\sqrt{21x}}{\sqrt{49}} = \frac{x\sqrt{21x}}{7}$

21. $\sqrt[3]{108x^4y^8}$
$\sqrt[3]{27x^3y^6}\sqrt[3]{4xy^2}$
$3xy^2\sqrt[3]{4xy^2}$

22. $\frac{3}{4}\sqrt{150}$
$\frac{3}{4}\sqrt{25}\sqrt{6}$
$\frac{3}{4}(5)\sqrt{6}$
$\frac{15\sqrt{6}}{4}$

23. $\frac{2}{3}\sqrt{45xy^3}$
$\frac{2}{3}\sqrt{9y^2}\sqrt{5xy}$
$\frac{2}{3}(3y)\sqrt{5xy}$
$2y\sqrt{5xy}$

24. $\dfrac{\sqrt{8x^2}}{\sqrt{2x}}$
$\sqrt{\dfrac{8x^2}{2x}}$
$\sqrt{4x}$
$\sqrt{4}\sqrt{x}$
$2\sqrt{x}$

25. $(3\sqrt{8})(4\sqrt{5})$
$12\sqrt{40}$
$12\sqrt{4}\sqrt{10}$
$12(2)\sqrt{10}$
$24\sqrt{10}$

26. $(5\sqrt[3]{2})(6\sqrt[3]{4})$
$30\sqrt[3]{8}$
$30(2)$
60

27. $3\sqrt{2}(4\sqrt{6} - 2\sqrt{7})$
$12\sqrt{12} - 6\sqrt{14}$
$12\sqrt{4}\sqrt{3} - 6\sqrt{14}$
$12(2)\sqrt{3} - 6\sqrt{14}$
$24\sqrt{3} - 6\sqrt{14}$

28. $(\sqrt{x} + 3)(\sqrt{x} - 5)$
$\sqrt{x^2} - 5\sqrt{x} + 3\sqrt{x} - 15$
$x - 2\sqrt{x} - 15$

29. $(2\sqrt{5} - \sqrt{3})(2\sqrt{5} + \sqrt{3})$
$4\sqrt{25} + 2\sqrt{15} - 2\sqrt{15} - \sqrt{9}$
$4(5) - 3$
$20 - 3$
17

30. $(3\sqrt{2} + \sqrt{6})(5\sqrt{2} - 3\sqrt{6})$
$15\sqrt{4} - 9\sqrt{12} + 5\sqrt{12} - 3\sqrt{36}$
$15(2) - 4\sqrt{12} - 3(6)$
$30 - 4\sqrt{12} - 18$
$12 - 8\sqrt{3}$

31. $(2\sqrt{a} + \sqrt{b})(3\sqrt{a} - 4\sqrt{b})$
$6\sqrt{a^2} - 8\sqrt{ab} + 3\sqrt{ab} - 4\sqrt{b^2}$
$6a - 5\sqrt{ab} - 4b$

32. $(4\sqrt{8} - \sqrt{2})(\sqrt{8} + 3\sqrt{2})$
$4\sqrt{64} + 12\sqrt{16} - \sqrt{16} - 3\sqrt{4}$
$4(8) + 11\sqrt{16} - 3(2)$
$32 + 44 - 6$
70

33. $\dfrac{4}{\sqrt{7} - 1} = \dfrac{4}{(\sqrt{7} - 1)} \bullet \dfrac{(\sqrt{7} + 1)}{(\sqrt{7} + 1)} =$
$\dfrac{4(\sqrt{7} + 1)}{\sqrt{49} - 1} = \dfrac{4(\sqrt{7} + 1)}{7 - 1} =$
$\dfrac{4(\sqrt{7} + 1)}{6} = \dfrac{2(\sqrt{7} + 1)}{3}$

34. $\dfrac{\sqrt{3}}{\sqrt{8} + \sqrt{5}} =$
$\dfrac{\sqrt{3}}{\sqrt{8} + \sqrt{5}} \bullet \dfrac{(\sqrt{8} - \sqrt{5})}{(\sqrt{8} - \sqrt{5})} =$
$\dfrac{\sqrt{24} - \sqrt{15}}{\sqrt{64} - \sqrt{25}} = \dfrac{\sqrt{4}\sqrt{6} - \sqrt{15}}{8 - 5} =$
$\dfrac{2\sqrt{6} - \sqrt{15}}{3}$

35. $\dfrac{3}{2\sqrt{3} + 3\sqrt{5}} =$
$\dfrac{3}{(2\sqrt{3} + 3\sqrt{5})} \bullet \dfrac{(2\sqrt{3} - 3\sqrt{5})}{(2\sqrt{3} - 3\sqrt{5})} =$
$\dfrac{3(2\sqrt{3} - 3\sqrt{5})}{4\sqrt{9} - 9\sqrt{25}} = \dfrac{3(2\sqrt{3} - 3\sqrt{5})}{4(3) - 9(5)} =$
$\dfrac{3(2\sqrt{3} - 3\sqrt{5})}{12 - 45} = \dfrac{3(2\sqrt{3} - 3\sqrt{5})}{-33} =$
$\dfrac{2\sqrt{3} - 3\sqrt{5}}{-11} = \dfrac{-2\sqrt{3} + 3\sqrt{5}}{11} =$
$\dfrac{3\sqrt{5} - 2\sqrt{3}}{11}$

36. $\dfrac{3\sqrt{2}}{2\sqrt{6}-\sqrt{10}} =$

$\dfrac{3\sqrt{2}}{(2\sqrt{6}-\sqrt{10})} \bullet \dfrac{(2\sqrt{6}+\sqrt{10})}{(2\sqrt{6}+\sqrt{10})} =$

$\dfrac{6\sqrt{12}+3\sqrt{20}}{4\sqrt{36}-\sqrt{100}} = \dfrac{6\sqrt{4}\sqrt{3}+3\sqrt{4}\sqrt{5}}{4(6)-10} =$

$\dfrac{12\sqrt{3}+6\sqrt{5}}{14} = \dfrac{2(6\sqrt{3}+3\sqrt{5})}{14} =$

$\dfrac{6\sqrt{3}+3\sqrt{5}}{7}$

37. $(x^{-3}y^4)^{-2} = x^{-3(-2)}y^{4(-2)} =$

$x^6y^{-8} = \dfrac{x^6}{y^8}$

38. $\left(\dfrac{2a^{-1}}{3b^4}\right)^{-3} = \dfrac{2^{-3}a^3}{3^{-3}b^{-12}} =$

$\dfrac{3^3a^3b^{12}}{2^3} = \dfrac{27a^3b^{12}}{8}$

39. $(4x^{\frac{1}{2}})(5x^{\frac{1}{5}})$

$20x^{\frac{1}{2}+\frac{1}{5}}$

$20x^{\frac{7}{10}}$

40. $\dfrac{42a^{\frac{3}{4}}}{6a^{\frac{1}{3}}} = 7a^{\frac{3}{4}-\frac{1}{3}} = 7a^{\frac{5}{12}}$

41. $\left(\dfrac{x^3}{y^4}\right)^{-\frac{1}{3}} = \dfrac{x^{3(-\frac{1}{3})}}{y^{4(-\frac{1}{3})}} = \dfrac{x^{-1}}{y^{-\frac{4}{3}}} = \dfrac{y^{\frac{4}{3}}}{x}$

42. $\left(\dfrac{6x^{-2}}{2x^4}\right)^{-2} = (3x^{-6})^{-2} = 3^{-2}x^{12} =$

$\dfrac{x^{12}}{3^2} = \dfrac{x^{12}}{9}$

43. $3\sqrt{45} - 2\sqrt{20} - \sqrt{80}$

$3\sqrt{9}\sqrt{5} - 2\sqrt{4}\sqrt{5} - \sqrt{16}\sqrt{5}$

$3(3)\sqrt{5} - 2(2)\sqrt{5} - 4\sqrt{5}$

$9\sqrt{5} - 4\sqrt{5} - 4\sqrt{5}$

$\sqrt{5}$

44. $4\sqrt[3]{24} + 3\sqrt[3]{3} - 2\sqrt[3]{81}$

$4\sqrt[3]{8}\sqrt[3]{3} + 3\sqrt[3]{3} - 2\sqrt[3]{27}\sqrt[3]{3}$

$4(2)\sqrt[3]{3} + 3\sqrt[3]{3} - 2(3)\sqrt[3]{3}$

$8\sqrt[3]{3} + 3\sqrt[3]{3} - 6\sqrt[3]{3}$

$5\sqrt[3]{3}$

45. $3\sqrt{24} - \dfrac{2}{5}\sqrt{54} + \dfrac{1}{4}\sqrt{96}$

$3\sqrt{4}\sqrt{6} - \dfrac{2}{5}\sqrt{9}\sqrt{6} + \dfrac{1}{4}\sqrt{16}\sqrt{6}$

$3(2)\sqrt{6} - \dfrac{2}{5}(3)\sqrt{6} + \dfrac{1}{4}(4)\sqrt{6}$

$6\sqrt{6} - \dfrac{6}{5}\sqrt{6} + \sqrt{6}$

$7\sqrt{6} - \dfrac{6}{5}\sqrt{6}$

$\dfrac{35\sqrt{6}}{5} - \dfrac{6\sqrt{6}}{5}$

$\dfrac{29\sqrt{6}}{5}$

46. $-2\sqrt{12x} + 3\sqrt{27x} - 5\sqrt{48x}$

$-2\sqrt{4}\sqrt{3x} + 3\sqrt{9}\sqrt{3x} - 5\sqrt{16}\sqrt{3x}$

$-2(2)\sqrt{3x} + 3(3)\sqrt{3x} - 5(4)\sqrt{3x}$

$-4\sqrt{3x} + 9\sqrt{3x} - 20\sqrt{3x}$

$-15\sqrt{3x}$

47. $x^{-2} + y^{-1}$

$\dfrac{1}{x^2} + \dfrac{1}{y}$; LCD $= x^2y$

$\dfrac{1}{x^2} \bullet \dfrac{y}{y} + \dfrac{1}{y} \bullet \dfrac{x^2}{x^2}$

$\dfrac{y}{x^2y} + \dfrac{x^2}{x^2y}$

$\dfrac{y+x^2}{x^2y}$

48. $a^{-2} - 2a^{-1}b^{-1}$

$\frac{1}{a^2} - \frac{2}{ab}$; LCD $= a^2b$

$\frac{1}{a^2} \bullet \frac{b}{b} - \frac{2}{ab} \bullet \frac{a}{a}$

$\frac{b - 2a}{a^2b}$

49. $\sqrt{7x - 3} = 4$

$(\sqrt{7x - 3})^2 = 4^2$

$7x - 3 = 16$

$7x = 19$

$x = \frac{19}{7}$

Check :

$\sqrt{7\left(\frac{19}{7}\right) - 3} \stackrel{?}{=} 4$

$\sqrt{19 - 3} \stackrel{?}{=} 4$

$\sqrt{16} \stackrel{?}{=} 4$

$4 = 4$

The solution set is $\left\{\frac{19}{7}\right\}$.

50. $\sqrt{2y + 1} = \sqrt{5y - 11}$

$(\sqrt{2y + 1})^2 = (\sqrt{5y - 11})^2$

$2y + 1 = 5y - 11$

$-3y = -12$

$y = 4$

Check :

$\sqrt{2(4) + 1} \stackrel{?}{=} \sqrt{5(4) - 11}$

$\sqrt{9} = \sqrt{9}$

The solution set is $\{4\}$.

51. $\sqrt{2x} = x - 4$

$(\sqrt{2x})^2 = (x - 4)^2$

$2x = x^2 - 8x + 16$

$0 = x^2 - 10x + 16$

$0 = (x - 8)(x - 2)$

$x - 8 = 0$ or $x - 2 = 0$

$x = 8$ or $x = 2$

Checking $x = 8$

$\sqrt{2(8)} \stackrel{?}{=} 8 - 4$

$\sqrt{16} \stackrel{?}{=} 4$

$4 = 4$

Checking $x = 2$

$\sqrt{2(2)} \stackrel{?}{=} 2 - 4$

$\sqrt{4} \stackrel{?}{=} -2$

$2 \neq -2$

The solution set is $\{8\}$.

52. $\sqrt{n^2 - 4n - 4} = n$

$(\sqrt{n^2 - 4n - 4})^2 = n^2$

$n^2 - 4n - 4 = n^2$

$-4n - 4 = 0$

$-4n = 4$

$n = -1$

Check:

$\sqrt{(-1)^2 - 4(-1) - 4} \stackrel{?}{=} -1$

$\sqrt{1 + 4 - 4} \stackrel{?}{=} -1$

$\sqrt{1} \stackrel{?}{=} -1$

$1 \neq -1$

The solution set is $\emptyset$.

53. $\sqrt[3]{2x - 1} = 3$

$\left(\sqrt[3]{2x - 1}\right)^3 = 3^3$

$2x - 1 = 27$

$2x = 28$

$x = 14$

Checking $x = 14$

$\sqrt[3]{2(14) - 1} \stackrel{?}{=} 3$

$\sqrt[3]{27} \stackrel{?}{=} 3$

$3 = 3$

The solution set is $\{14\}$.

54. $\sqrt{t^2 + 9t - 1} = 3$

$(\sqrt{t^2 + 9t - 1})^2 = 3^2$

$t^2 + 9t - 1 = 9$

$t^2 + 9t - 10 = 0$

$(t + 10)(t - 1) = 0$

$t + 10 = 0$ or $t - 1 = 0$

$t = -10$ or $t = 1$

Checking $t = -10$

$\sqrt{(-10)^2 + 9(-10) - 1} \stackrel{?}{=} 3$

$\sqrt{100 - 90 - 1} \stackrel{?}{=} 3$

$\sqrt{9} \stackrel{?}{=} 3$

$3 = 3$

Checking $t = 1$
$\sqrt{(1)^2 + 9(1) - 1} \stackrel{?}{=} 3$
$\sqrt{1 + 9 - 1} \stackrel{?}{=} 3$
$\sqrt{9} \stackrel{?}{=} 3$
$3 = 3$
The solution set is $\{-10, 1\}$.

55. $\sqrt{x^2 + 3x - 6} = x$
$(\sqrt{x^2 + 3x - 6})^2 = x^2$
$x^2 + 3x - 6 = x^2$
$3x - 6 = 0$
$3x = 6$
$x = 2$
Check :
$\sqrt{2^2 + 3(2) - 6} \stackrel{?}{=} 2$
$\sqrt{4 + 6 - 6} \stackrel{?}{=} 2$
$\sqrt{4} \stackrel{?}{=} 2$
$2 = 2$
The solution set is $\{2\}$.

56. $\sqrt{x + 1} - \sqrt{2x} = -1$
$\sqrt{x + 1} = \sqrt{2x} - 1$
$(\sqrt{x + 1})^2 = (\sqrt{2x} - 1)^2$
$x + 1 = (\sqrt{2x})^2 - 2\sqrt{2x} + 1$
$x + 1 = 2x - 2\sqrt{2x} + 1$
$-x = -2\sqrt{2x}$
$(-x)^2 = (-2\sqrt{2x})^2$
$x^2 = 4(2x)$
$x^2 = 8x$
$x^2 - 8x = 0$
$x(x - 8) = 0$
$x = 0$ or $x - 8 = 0$
$x = 0$ or $x = 8$
Checking $x = 0$
$\sqrt{0 + 1} - \sqrt{2(0)} \stackrel{?}{=} -1$
$\sqrt{1} - \sqrt{0} \stackrel{?}{=} -1$
$1 \neq -1$
Checking $x = 8$
$\sqrt{8 + 1} - \sqrt{2(8)} \stackrel{?}{=} -1$
$\sqrt{9} - \sqrt{16} \stackrel{?}{=} -1$
$3 - 4 \stackrel{?}{=} -1$
$-1 = -1$
The solution set is $\{8\}$.

57. $(0.00002)(0.0003)$
$(2)(10)^{-5}(3)(10)^{-4}$
$(6)(10)^{-9}$
0.000000006

58. $(120{,}000)(300{,}000)$
$(1.2)(10)^5(3)(10)^5$
$(3.6)(10)^{10}$
$36{,}000{,}000{,}000$

59. $(0.000015)(400{,}000)$
$(1.5)(10)^{-5}(4)(10)^5$
$(6.0)(10)^0$
6

60. $\dfrac{0.000045}{0.0003} = \dfrac{(4.5)(10)^{-5}}{(3)(10)^{-4}} =$
$(1.5)(10)^{-1} = 0.15$

61. $\dfrac{(0.00042)(0.0004)}{0.006} =$

$\dfrac{(4.2)(10)^{-4}(4)(10)^{-4}}{(6)(10)^{-3}} =$

$\dfrac{(16.8)(10)^{-8}}{(6)(10)^{-3}} = (2.8)(10)^{-5} =$

0.000028

62. $\sqrt{0.000004} = \sqrt{(4)(10)^{-6}} =$
$[(4)(10)^{-6}]^{\frac{1}{2}} = (4)^{\frac{1}{2}}(10)^{-6(\frac{1}{2})} =$
$(2)(10)^{-3} = 0.002$

63. $\sqrt[3]{0.000000008} = \sqrt[3]{(8)(10)^{-9}} =$
$[(8)(10)^{-9}]^{\frac{1}{3}} = (8)^{\frac{1}{3}}(10)^{-9(\frac{1}{3})} =$
$(2)(10)^{-3} = 0.002$

64. $(4000000)^{\frac{3}{2}} =$
$[(4)(10)^6]^{\frac{3}{2}} = (4)^{\frac{3}{2}}(10)^{6(\frac{3}{2})} =$
$(8)(10)^9 = 8{,}000{,}000{,}000$

CHAPTER 10 | Test

1. $(4)^{-\frac{5}{2}} = \frac{1}{4^{\frac{5}{2}}} = \frac{1}{(4^{\frac{1}{2}})^5} = \frac{1}{(2)^5} = \frac{1}{32}$

2. $-16^{\frac{5}{4}} = -(16^{\frac{1}{4}})^5 = -(2^5) = -32$

3. $\left(\frac{2}{3}\right)^{-4} = \left(\frac{3}{2}\right)^4 = \frac{3^4}{2^4} = \frac{81}{16}$

4. $\left(\frac{2^{-1}}{2^{-2}}\right)^{-2} = (2^1)^{-2} = 2^{-2} = \frac{1}{2^2} = \frac{1}{4}$

5. $\sqrt{63} = \sqrt{9}\sqrt{7} = 3\sqrt{7}$

6. $\sqrt[3]{108}$
$\sqrt[3]{27}\sqrt[3]{4}$
$3\sqrt[3]{4}$

7. $\sqrt{52x^4y^3} = \sqrt{4x^4y^2}\sqrt{13y} = 2x^2y\sqrt{13y}$

8. $\frac{5\sqrt{18}}{3\sqrt{12}} = \frac{5\sqrt{9}\sqrt{2}}{3\sqrt{4}\sqrt{3}} = \frac{15\sqrt{2}}{6\sqrt{3}} =$
$\frac{5\sqrt{2}}{2\sqrt{3}} \bullet \frac{\sqrt{3}}{\sqrt{3}} = \frac{5\sqrt{6}}{2\sqrt{9}} = \frac{5\sqrt{6}}{6}$

9. $\sqrt{\frac{7}{24x^3}} = \frac{\sqrt{7}}{\sqrt{24x^3}} = \frac{\sqrt{7}}{\sqrt{4x^2}\sqrt{6x}} =$
$\frac{\sqrt{7}}{2x\sqrt{6x}} = \frac{\sqrt{7}}{2x\sqrt{6x}} \bullet \frac{\sqrt{6x}}{\sqrt{6x}} =$
$\frac{\sqrt{42x}}{2x\sqrt{36x^2}} = \frac{\sqrt{42x}}{2x(6x)} = \frac{\sqrt{42x}}{12x^2}$

10. $(4\sqrt{6})(3\sqrt{12})$
$12\sqrt{72}$
$12\sqrt{36}\sqrt{2}$
$12(6)\sqrt{2}$
$72\sqrt{2}$

11. $(3\sqrt{2} + \sqrt{3})(\sqrt{2} - 2\sqrt{3})$
$3\sqrt{4} - 6\sqrt{6} + \sqrt{6} - 2\sqrt{9}$
$3(2) - 5\sqrt{6} - 2(3)$
$6 - 5\sqrt{6} - 6$
$-5\sqrt{6}$

12. $2\sqrt{50} - 4\sqrt{18} - 9\sqrt{32}$
$2\sqrt{25}\sqrt{2} - 4\sqrt{9}\sqrt{2} - 9\sqrt{16}\sqrt{2}$
$2(5)\sqrt{2} - 4(3)\sqrt{2} - 9(4)\sqrt{2}$
$10\sqrt{2} - 12\sqrt{2} - 36\sqrt{2}$
$-38\sqrt{2}$

13. $\frac{3\sqrt{2}}{4\sqrt{3} - \sqrt{8}} = \frac{3\sqrt{2}}{4\sqrt{3} - \sqrt{4}\sqrt{2}} =$

$\frac{3\sqrt{2}}{4\sqrt{3} - 2\sqrt{2}} =$

$\frac{3\sqrt{2}}{(4\sqrt{3} - 2\sqrt{2})} \bullet \frac{(4\sqrt{3} + 2\sqrt{2})}{(4\sqrt{3} + 2\sqrt{2})} =$

$\frac{12\sqrt{6} + 6\sqrt{4}}{16\sqrt{9} - 4\sqrt{4}} = \frac{12\sqrt{6} + 6(2)}{16(3) - 4(2)} =$

$\frac{12\sqrt{6} + 12}{48 - 8} = \frac{12\sqrt{6} + 12}{40} =$

$\frac{4(3\sqrt{6} + 3)}{4(10)} = \frac{3\sqrt{6} + 3}{10}$

14. $\left(\frac{2x^{-1}}{3y}\right)^{-2} = \frac{2^{-2}x^2}{3^{-2}y^{-2}} = \frac{3^2x^2y^2}{2^2} = \frac{9x^2y^2}{4}$

15. $\frac{-84a^{\frac{1}{2}}}{7a^{\frac{4}{5}}} = -12a^{\frac{1}{2} - \frac{4}{5}} =$
$-12a^{-\frac{3}{10}} = -\frac{12}{a^{\frac{3}{10}}}$

16. $x^{-1} + y^{-3}$
$\frac{1}{x} + \frac{1}{y^3}$;LCD $= xy^3$
$\frac{1}{x} \bullet \frac{y^3}{y^3} + \frac{1}{y^3} \bullet \frac{x}{x}$
$\frac{y^3}{xy^3} + \frac{x}{xy^3} = \frac{y^3 + x}{xy^3}$

17. $(3x^{-\frac{1}{2}})(-4x^{\frac{3}{4}}) = -12x^{-\frac{1}{2}+\frac{3}{4}} =$
$-12x^{\frac{1}{4}}$

18. $(3\sqrt{5} - 2\sqrt{3})(3\sqrt{5} + 2\sqrt{3})$
$9\sqrt{25} + 6\sqrt{15} - 6\sqrt{15} - 4\sqrt{9}$
$9(5) - 4(3)$
$45 - 12 = 33$

19. $\dfrac{(0.00004)(300)}{0.00002} = \dfrac{(4)(10)^{-5}(3)(10)^{2}}{(2)(10)^{-5}} =$
$\dfrac{(12)(10)^{-3}}{(2)(10)^{-5}} = (6)(10)^{2} = 600$

20. $\sqrt{0.000009} = \sqrt{(9)(10)^{-6}} =$
$[(9)(10)^{-6}]^{\frac{1}{2}} = (9)^{\frac{1}{2}}(10)^{-6(\frac{1}{2})} =$
$(3)(10)^{-3} = 0.003$

21. $\sqrt{3x + 1} = 3$
$(\sqrt{3x + 1})^2 = 3^2$
$3x + 1 = 9$
$3x = 8$
$x = \dfrac{8}{3}$
Check
$\sqrt{3\left(\dfrac{8}{3}\right) + 1} \stackrel{?}{=} 3$
$\sqrt{8 + 1} \stackrel{?}{=} 3$
$\sqrt{9} \stackrel{?}{=} 3$
$3 = 3$
The solution set is $\left\{\dfrac{8}{3}\right\}$.

22. $\sqrt[3]{3x + 2} = 2$
$(\sqrt[3]{3x + 2})^3 = (2)^3$
$3x + 2 = 8$
$3x = 6$
$x = 2$
Check:
$\sqrt[3]{3(2) + 2} \stackrel{?}{=} 2$
$\sqrt[3]{8} \stackrel{?}{=} 2$
$2 = 2$
The solution set is $\{2\}$.

23. $\sqrt{x} = x - 2$
$(\sqrt{x})^2 = (x - 2)^2$
$x = x^2 - 4x + 4$
$0 = x^2 - 5x + 4$
$0 = (x - 4)(x - 1)$
$x - 4 = 0$ or $x - 1 = 0$
$x = 4$ or $x = 1$
Checking $x = 4$
$\sqrt{4} \stackrel{?}{=} 4 - 2$
$2 = 2$
Checking $x = 1$
$\sqrt{1} \stackrel{?}{=} 1 - 2$
$1 \neq -1$
The solution set is $\{4\}$.

24. $\sqrt{5x - 2} = \sqrt{3x + 8}$
$(\sqrt{5x - 2})^2 = (\sqrt{3x + 8})^2$
$5x - 2 = 3x + 8$
$2x = 10$
$x = 5$
Check:
$\sqrt{5(5) - 2} \stackrel{?}{=} \sqrt{3(5) + 8}$
$\sqrt{23} = \sqrt{23}$
The solution set is $\{5\}$.

25. $\sqrt{x^2 - 10x + 28} = 2$
$(\sqrt{x^2 - 10x + 28})^2 = 2^2$
$x^2 - 10x + 28 = 4$
$x^2 - 10x + 24 = 0$
$(x - 6)(x - 4) = 0$
$x - 6 = 0$ or $x - 4 = 0$
$x = 6$ or $x = 4$
Checking $x = 6$
$\sqrt{6^2 - 10(6) + 28} \stackrel{?}{=} 2$
$\sqrt{36 - 60 + 28} \stackrel{?}{=} 2$
$\sqrt{4} \stackrel{?}{=} 2$
$2 = 2$
Checking $x = 4$
$\sqrt{4^2 - 10(4) + 28} \stackrel{?}{=} 2$
$\sqrt{16 - 40 + 28} \stackrel{?}{=} 2$
$\sqrt{4} \stackrel{?}{=} 2$
$2 = 2$
The solution set is $\{4, 6\}$.

CHAPTERS 1-10 | Cumulative Practice Test

1a. $\left(\frac{3}{4}-\frac{1}{3}\right)^{-1}=\left(\frac{9}{12}-\frac{4}{12}\right)^{-1}=$

$\left(\frac{5}{12}\right)^{-1}=\left(\frac{12}{5}\right)^{1}=\frac{12^1}{5^1}=\frac{12}{5}$

b. $(2^{-1}+3^{-2})^{-2}=\left(\frac{1}{2}+\frac{1}{3^2}\right)^{-2}=$

$\left(\frac{1}{2}+\frac{1}{9}\right)^{-2}=\left(\frac{2}{18}+\frac{9}{18}\right)^{-2}=$

$\left(\frac{11}{18}\right)^{-2}=\left(\frac{18}{11}\right)^{2}=\frac{18^2}{11^2}=\frac{324}{121}$

c. $\sqrt[3]{-\frac{1}{8}}=\frac{\sqrt[3]{-1}}{\sqrt[3]{8}}=\frac{-1}{2}=-\frac{1}{2}$

d. $(-27)^{\frac{4}{3}}=\left((-27)^{\frac{1}{3}}\right)^4=$

$\left(\sqrt[3]{-27}\right)^4=(-3)^4=81$

2. $\frac{-64x^{-1}y^3}{16x^3y^{-2}}=-4x^{-1-3}y^{3-(-2)}=$

$-4x^{-4}y^5=-\frac{4y^5}{x^4}$

3. $(-4x^{-2}y^{-3})(3x^2y^{-1})=$

$-12x^{-2+2}y^{-3-1}=-12x^0y^{-4}=$

$-\frac{12(1)}{y^4}=-\frac{12}{y^4}$

4. $(3x-1)(2x^2+6x-4)=$

$6x^3+18x^2-12x-2x^2-6x+4=$

$6x^3+18x^2-2x^2-12x-6x+4=$

$6x^3+16x^2-18x+4$

5.

$$\begin{array}{r|l} & 4x^2-7x-2 \\ \hline x+6 & 4x^3+17x^2-44x-12 \\ & 4x^3+24x^2 \\ \hline & -7x^2-44x \\ & -7x^2-42x \\ \hline & -2x-12 \\ & -2x-12 \\ \hline & 0 \end{array}$$

Q: $4x^2-7x-2$ R: 0

6. $\begin{pmatrix} 5x-3y=-31 \\ 4x+7y=41 \end{pmatrix}$

Multiply equation 1 by 7 and multiply equation 2 by 3. Then add the equations.

$35x-21y=-217$

$12x+21y=123$

$47x=-94$

$x=-2$

$5x-3y=-31$

$5(-2)-3y=-31$

$-10-3y=-31$

$-3y=-21$

$y=7$

The solution set is $\{(-2,7)\}$.

7a. $3\sqrt{56}$

$3\sqrt{4}\sqrt{14}$

$3(2)\sqrt{14}$

$6\sqrt{14}$

b. $2\sqrt[3]{56}$

$2\sqrt[3]{8}\sqrt[3]{7}$

$2(2)\sqrt[3]{7}$

$4\sqrt[3]{7}$

c. $\frac{3\sqrt{2}}{4\sqrt{6}}=\frac{3\sqrt{2}}{4\sqrt{6}}\cdot\frac{\sqrt{6}}{\sqrt{6}}=\frac{3\sqrt{12}}{4\sqrt{36}}=$

$\frac{3\sqrt{4}\sqrt{3}}{4\cdot 6}=\frac{3(2)\sqrt{3}}{24}=\frac{6\sqrt{3}}{24}=\frac{\sqrt{3}}{4}$

d. $\sqrt{\frac{6}{8}}=\sqrt{\frac{3}{4}}=\frac{\sqrt{3}}{\sqrt{4}}=\frac{\sqrt{3}}{2}$

8. $3\sqrt{2}-2\sqrt{8}+9\sqrt{18}-\sqrt{32}=$
$3\sqrt{2}-2\sqrt{4}\sqrt{2}+9\sqrt{9}\sqrt{2}-\sqrt{16}\sqrt{2}=$
$3\sqrt{2}-2(2)\sqrt{2}+9(3)\sqrt{2}-4\sqrt{2}=$
$3\sqrt{2}-4\sqrt{2}+27\sqrt{2}-4\sqrt{2}=$
$22\sqrt{2}=22(1.414)=31.108=$
31.11 to the nearest hundredth

9. Twenty-five percent of what number is 18?
$25\%(n)=18$
$0.25n=18$
$n=\frac{18}{0.25}=\frac{1800}{25}=72$
Twenty-five percent of 72 is 18.

10. $-4(2a-b)+6(b-2a)-(3a+4b)$
$-8a+4b+6b-12a-3a-4b$
$-8a-12a-3a+4b+6b-4b$
$-23a+6b$ for $a=-15$ and $b=14$
$-23(-15)+6(14)=345+84=429$

11. $\frac{56x^{-1}y^{-3}}{8x^{-2}y^{-4}}=7x^{-1-(-2)}y^{-3-(-4)}=$
$7x^1y^1=7xy$ for $x=-\frac{1}{2}$ and $y=\frac{2}{3}$
$7\left(-\frac{1}{2}\right)\left(\frac{2}{3}\right)=-\frac{7}{2}\left(\frac{2}{3}\right)=-\frac{7}{3}$

12 **a.** $52=2\cdot2\cdot13$
b. $80=2\cdot2\cdot2\cdot2\cdot5$
c. $91=7\cdot13$
d. $78=2\cdot3\cdot13$

13. $\frac{\frac{5}{3x}-\frac{2}{y}}{\frac{3}{x}+\frac{5}{3y}}=\frac{\left(\frac{5}{3x}-\frac{2}{y}\right)}{\left(\frac{3}{x}+\frac{5}{3y}\right)}\cdot\frac{3xy}{3xy}=$

$\frac{\left(\frac{5}{3x}\right)3xy-\left(\frac{2}{y}\right)3xy}{\left(\frac{3}{x}\right)3xy+\left(\frac{5}{3y}\right)3xy}=$

$\frac{\frac{15xy}{3x}-\frac{6xy}{y}}{\frac{9xy}{x}+\frac{15xy}{3y}}=\frac{5y-6x}{9y+5x}$

14. $\frac{3x-7}{5}-\frac{2x+1}{4}$; LCD $=20$
$\left(\frac{3x-7}{5}\right)\cdot\frac{4}{4}-\left(\frac{2x+1}{4}\right)\cdot\frac{5}{5}$
$\frac{4(3x-7)}{20}-\frac{5(2x+1)}{20}$
$\frac{4(3x-7)-5(2x+1)}{20}$
$\frac{12x-28-10x-5}{20}$
$\frac{2x-33}{20}$

15. $\frac{5x^2y}{7xy}\div\frac{15y}{14x}$

$\frac{5x^2y}{7xy}\cdot\frac{14x}{15y}$

$\frac{5\cdot14\cdot x\cdot x\cdot x\cdot y}{15\cdot7\cdot x\cdot y\cdot y}$

$\frac{\overset{1}{\cancel{5}}\cdot\overset{2}{\cancel{14}}\cdot\cancel{x}\cdot x\cdot x\cdot\cancel{y}}{\underset{3}{\cancel{15}}\cdot\underset{1}{\cancel{7}}\cdot\cancel{x}\cdot\cancel{y}\cdot y}$

$\frac{2x^2}{3y}$

16. $\left(\frac{x^2+5x}{3x^2+2x-8}\right)\left(\frac{2x^2-8}{x^3+3x^2-10x}\right)$

$\frac{x(x+5)}{(3x-4)(x+2)}\cdot\frac{2(x-2)(x+2)}{x(x+5)(x-2)}$

$\frac{2}{3x-4}$

17. $x(x-4)-3(x-4)=0$
$(x-4)(x-3)=0$
$x-4=0$ or $x-3=0$
$x=4$ or $x=3$
The solution set is $\{3,4\}$.

18. $(x+2)(x-5) = -6$
$x^2 - 3x - 10 = -6$
$x^2 - 3x - 4 = 0$
$(x+1)(x-4) = 0$
$x+1=0$ or $x-4=0$
$x = -1$ or $x = 4$
The solution set is $\{-1, 4\}$.

19. $(3x-5)(2x+7) = 0$
$3x-5=0$ or $2x+7=0$
$3x = 5$ or $2x = -7$
$x = \frac{3}{5}$ or $x = -\frac{7}{2}$
The solution set is $\left\{-\frac{7}{2}, \frac{5}{3}\right\}$.

20. $|2x-5| = -4$ has no solutions because the absolute value of a number cannot be negative. Therefore the solution set is Ø.

21. $\sqrt{5+2x} = 1 + \sqrt{2x}$
$(\sqrt{5+2x})^2 = (1+\sqrt{2x})^2$
$5 + 2x = 1 + 2\sqrt{2x} + (\sqrt{2x})^2$
$5 + 2x = 1 + 2\sqrt{2x} + 2x$
$4 = 2\sqrt{2x}$
$2 = \sqrt{2x}$
$(2)^2 = (\sqrt{2x})^2$
$4 = 2x$
$2 = x$
Check $x = 2$.
$\sqrt{5+2x} = 1 + \sqrt{2x}$
$\sqrt{5+2(2)} = 1 + \sqrt{2(2)}$
$\sqrt{5+4} = 1 + \sqrt{4}$
$\sqrt{9} = 1 + 2$
$3 = 3$
The solution set is $\{2\}$.

22. Let x represent the altitude of the triangle. Then $3x-1$ represents the side perpendicular to the altitude.
Area $= 51$ square inches.
$A = \frac{1}{2}bh$
$51 = \frac{1}{2}\left(3x-1\right)x$
$102 = (3x-1)x$
$102 = 3x^2 - x$
$0 = 3x^2 - x - 102$
$0 = (3x+17)(x-6)$
$3x + 17 = 0$ or $x - 6 = 0$
$3x = -17$ or $x = 6$
$x = -\frac{17}{3}$ or $x = 6$
Discard $x = -\frac{17}{3}$, since length cannot be negative. The length of the altitude is 6 inches and the length of the side is $3(6) - 1 = 17$ inches.

23. Let $x =$ Pedro's present age.
Let $6 + x =$ Brad's present age.
Also $x - 5 =$ Pedro's age 5 years ago, and $6 + x - 5 =$ Brad's age 5 years ago.
$\left(\text{Pedro's age 5 years ago}\right) = \frac{3}{4}\left(\text{Brad's age 5 years ago}\right)$
$x - 5 = \frac{3}{4}\left(6 + x - 5\right)$
$4(x-5) = 4\left[\frac{3}{4}\left(1+x\right)\right]$
$4x - 20 = 3(1+x)$
$4x - 20 = 3 + 3x$
$x - 20 = 3$
$x = 23$
Pedro is 23 years old and Brad is 29 years old.

24. Let c represent the selling price.
Profit is a percent of cost.
Selling price = Cost + Profit

$$97.50 = c + (30\%)(c)$$
$$97.50 = c + 0.30(c)$$
$$97.50 = 1.30c$$
$$75 = c$$

The original cost was \$75.00.

25. Let $x =$ the width before cutouts and
$x + 4 =$ the length before cutouts.

After the cutouts the width $= x - 4$ and the length $= x + 4 - 4 = x$. These two dimensions form the base of the box.

$V = lwh$
$42 = (x)(x - 4)(2)$
$42 = 2x(x - 4)$
$0 = 2x^2 - 8x - 42$
$0 = x^2 - 4x - 21$
$0 = (x - 7)(x + 3)$
$x - 7 = 0$ or $x + 3 = 0$
$x = 7$ or $x = -3$
Discard the root $x = -3$.
The original piece of cardboard has width of 7 inches and length of $7 + 4 = 11$ inches.

Chapter 11 Quadratic Equations and Inequalities

PROBLEM SET 11.1 Complex Numbers

1. False
3. True
5. True
7. True

9. $(6+3i)+(4+5i)$
$(6+4)+(3+5)i$
$10+8i$

11. $(-8+4i)+(2+6i)$
$(-8+2)+(4+6)i$
$-6+10i$

13. $(3+2i)-(5+7i)$
$3+2i-5-7i$
$(3-5)+(2-7)i$
$-2-5i$

15. $(-7+3i)-(5-2i)$
$-7+3i-5+2i$
$(-7-5)+(3+2)i$
$-12+5i$

17. $(-3-10i)+(2-13i)$
$(-3+2)+(-10-13)i$
$-1-23i$

19. $(4-8i)-(8-3i)$
$4-8i-8+3i$
$(4-8)+(-8+3)i$
$-4-5i$

21. $(-1-i)-(-2-4i)$
$-1-i+2+4i$
$(-1+2)+(-1+4)i$
$1+3i$

23. $\left(\frac{3}{2}+\frac{1}{3}i\right)+\left(\frac{1}{6}-\frac{3}{4}i\right)$
$\left(\frac{3}{2}+\frac{1}{6}\right)+\left(\frac{1}{3}-\frac{3}{4}\right)i$
$\frac{5}{3}-\frac{5}{12}i$

25. $\left(-\frac{5}{9}+\frac{3}{5}i\right)-\left(\frac{4}{3}-\frac{1}{6}i\right)$
$-\frac{5}{9}+\frac{3}{5}i-\frac{4}{3}+\frac{1}{6}i$
$\left(-\frac{5}{9}-\frac{4}{3}\right)+\left(\frac{3}{5}+\frac{1}{6}\right)i$
$-\frac{17}{9}+\frac{23}{30}i$

27. $\sqrt{-81}=\sqrt{-1}\sqrt{81}=i\sqrt{81}=9i$

29. $\sqrt{-14}=\sqrt{-1}\sqrt{14}=i\sqrt{14}$

31. $\sqrt{-\frac{16}{25}}=\sqrt{-1}\sqrt{\frac{16}{25}}=i\left(\frac{4}{5}\right)=\frac{4}{5}i$

33. $\sqrt{-18}=\sqrt{-1}\sqrt{18}=i\sqrt{9}\sqrt{2}=$
$i(3)\sqrt{2}=3i\sqrt{2}$

35. $\sqrt{-75}=\sqrt{-1}\sqrt{75}=i\sqrt{25}\sqrt{3}=$
$i(5)\sqrt{3}=5i\sqrt{3}$

37. $3\sqrt{-28}=3\sqrt{-1}\sqrt{28}=3i\sqrt{4}\sqrt{7}=$
$3i(2)\sqrt{7}=6i\sqrt{7}$

39. $-2\sqrt{-80}=-2\sqrt{-1}\sqrt{80}=$
$-2i\sqrt{16}\sqrt{5}=-2i(4)\sqrt{5}=-8i\sqrt{5}$

41. $12\sqrt{-90}=12\sqrt{-1}\sqrt{90}=$
$12i\sqrt{9}\sqrt{10}=12i(3)\sqrt{10}=36i\sqrt{10}$

43. $\sqrt{-4}\sqrt{-16}=(i\sqrt{4})(i\sqrt{16})=$
$(2i)(4i)=8i^2=8(-1)=-8$

45. $\sqrt{-3}\sqrt{-5}=(i\sqrt{3})(i\sqrt{5})=$
$i^2\sqrt{15}=-\sqrt{15}$

47. $\sqrt{-9}\sqrt{-6}=(i\sqrt{9})(i\sqrt{6})=$
$(3i)(i\sqrt{6})=3i^2\sqrt{6}=$
$3(-1)\sqrt{6}=-3\sqrt{6}$

49. $\sqrt{-15}\sqrt{-5} = (i\sqrt{15})(i\sqrt{5}) =$
$i^2\sqrt{75} = -1\sqrt{25}\sqrt{3} = -1(5)\sqrt{3} = -5\sqrt{3}$

51. $\sqrt{-2}\sqrt{-27} = (i\sqrt{2})(i\sqrt{27}) =$
$i^2\sqrt{54} = -1\sqrt{9}\sqrt{6} =$
$-1(3)\sqrt{6} = -3\sqrt{6}$

53. $\sqrt{6}\sqrt{-8} = (\sqrt{6})(i\sqrt{8}) = i\sqrt{48}$
$i\sqrt{16}\sqrt{3} = 4i\sqrt{3}$

55. $\dfrac{\sqrt{-25}}{\sqrt{-4}} = \dfrac{i\sqrt{25}}{i\sqrt{4}} = \dfrac{5i}{2i} = \dfrac{5}{2}$

57. $\dfrac{\sqrt{-56}}{\sqrt{-7}} = \dfrac{i\sqrt{56}}{i\sqrt{7}} = \sqrt{\dfrac{56}{7}} =$
$\sqrt{8} = \sqrt{4}\sqrt{2} = 2\sqrt{2}$

59. $\dfrac{\sqrt{-24}}{\sqrt{6}} = \dfrac{i\sqrt{24}}{\sqrt{6}} = i\sqrt{\dfrac{24}{6}} = i\sqrt{4} = 2i$

61. $(5i)(4i) = 20i^2 = 20(-1) = -20 + 0i$

63. $(7i)(-6i) = -42i^2 = -42(-1) = 42 + 0i$

65. $(3i)(2 - 5i) = 6i - 15i^2 = 6i - 15(-1) =$
$6i + 15 = 15 + 6i$

67. $(-6i)(-2 - 7i) = 12i + 42i^2 =$
$12i + 42(-1) = 12i - 42 = -42 + 12i$

69. $(3 + 2i)(5 + 4i) = 15 + 12i + 10i + 8i^2 =$
$15 + 22i + 8(-1) = 15 + 22i - 8 =$
$7 + 22i$

71. $(6 - 2i)(7 - i) = 42 - 6i - 14i + 2i^2 =$
$42 - 20i + 2(-1) = 42 - 20i - 2 =$
$40 - 20i$

73. $(-3 - 2i)(5 + 6i) =$
$-15 - 18i - 10i - 12i^2 =$
$-15 - 28i - 12(-1) =$
$-15 - 28i + 12 = -3 - 28i$

75. $(9 + 6i)(-1 - i) = -9 - 9i - 6i - 6i^2 =$
$-9 - 15i - 6(-1) = -9 - 15i + 6 =$
$-3 - 15i$

77. $(4 + 5i)^2$
$(4)^2 + 2(4)(5i) + (5i)^2$
$16 + 40i + 25i^2$
$16 + 40i + 25(-1)$
$16 + 40i - 25$
$-9 + 40i$

79. $(-2 - 4i)^2$
$(-2)^2 + 2(-2)(-4i) + (-4i)^2$
$4 + 16i + 16i^2$
$4 + 16i + 16(-1)$
$4 + 16i - 16$
$-12 + 16i$

81. $(6 + 7i)(6 - 7i)$
$36 - 42i + 42i - 49i^2$
$36 - 49(-1) + 0i$
$36 + 49 + 0i$
$85 + 0i$

83. $(-1 + 2i)(-1 - 2i)$
$1 + 2i - 2i - 4i^2$
$1 - 4(-1) + 0i$
$1 + 4 + 0i$
$5 + 0i$

85. $\dfrac{3i}{2 + 4i} = \dfrac{3i}{(2 + 4i)} \bullet \dfrac{(2 - 4i)}{(2 - 4i)} =$
$\dfrac{6i - 12i^2}{4 - 16i^2} = \dfrac{6i - 12(-1)}{4 - 16(-1)} =$
$\dfrac{6i + 12}{4 + 16} = \dfrac{12 + 6i}{20} =$
$\dfrac{12}{20} + \dfrac{6}{20}i = \dfrac{3}{5} + \dfrac{3}{10}i$

87. $\dfrac{-2i}{3 - 5i} = \dfrac{-2i}{(3 - 5i)} \bullet \dfrac{(3 + 5i)}{(3 + 5i)} =$
$\dfrac{-6i - 10i^2}{9 - 25i^2} = \dfrac{-6i - 10(-1)}{9 - 25(-1)} =$
$\dfrac{-6i + 10}{9 + 25} = \dfrac{10 - 6i}{34} =$
$\dfrac{10}{34} - \dfrac{6}{34}i = \dfrac{5}{17} - \dfrac{3}{17}i$

89. $\frac{-2+6i}{3i}=\frac{(-2+6i)}{3i}\bullet\frac{(-i)}{(-i)}=$

$\frac{2i-6i^2}{-3i^2}=\frac{2i-6(-1)}{-3(-1)}=\frac{2i+6}{3}=$

$=\frac{6+2i}{3}=\frac{6}{3}+\frac{2}{3}i=2+\frac{2}{3}i$

91. $\frac{2}{7i}=\frac{2}{7i}\bullet\frac{(-i)}{(-i)}=\frac{-2i}{-7i^2}=$

$\frac{-2i}{-7(-1)}=\frac{-2i}{7}=0-\frac{2}{7}i$

93. $\frac{2+6i}{1+7i}=\frac{(2+6i)}{(1+7i)}\bullet\frac{(1-7i)}{(1-7i)}=$

$\frac{2-14i+6i-42i^2}{1-49i^2}=\frac{2-8i-42(-1)}{1-49(-1)}=$

$\frac{2-8i+42}{1+49}=\frac{44-8i}{50}=\frac{44}{50}-\frac{8}{50}i=$

$\frac{22}{25}-\frac{4}{25}i$

95. $\frac{3+6i}{4-5i}=\frac{(3+6i)}{(4-5i)}\bullet\frac{(4+5i)}{(4+5i)}=$

$\frac{12+15i+24i+30i^2}{16-25i^2}=$

$\frac{12+39i+30(-1)}{16+25}=\frac{12+39i-30}{41}=$

$\frac{-18+39i}{41}=-\frac{18}{41}+\frac{39}{41}i$

97. $\frac{-2+7i}{-1+i}=\frac{(-2+7i)}{(-1+i)}\bullet\frac{(-1-i)}{(-1-i)}=$

$\frac{2+2i-7i-7i^2}{1-i^2}=\frac{2-5i-7(-1)}{1-(-1)}=$

$\frac{2-5i+7}{1+1}=\frac{9-5i}{2}=\frac{9}{2}-\frac{5}{2}i$

99. $\frac{-1-3i}{-2-10i}=\frac{(-1-3i)}{(-2-10i)}\bullet\frac{(-2+10i)}{(-2+10i)}=$

$\frac{2-10i+6i-30i^2}{4-100i^2}=\frac{2-4i-30(-1)}{4-100(-1)}=$

$\frac{2-4i+30}{4+100}=\frac{32-4i}{104}=$

$\frac{32}{104}-\frac{4}{104}i=\frac{4}{13}-\frac{1}{26}i$

PROBLEM SET 11.2 Quadratic Equations

1. $x^2-9x=0$
$x(x-9)=0$
$x=0$ or $x-9=0$
$x=0$ or $x=9$
The solution set is $\{0, 9\}$.

3. $x^2=-3x$
$x^2+3x=0$
$x(x+3)=0$
$x=0$ or $x+3=0$
$x=0$ or $x=-3$
The solution set is $\{-3, 0\}$.

5. $3y^2+12y=0$
$3y(y+4)=0$
$3y=0$ or $y+4=0$
$y=0$ or $y=-4$
The solution set is $\{-4, 0\}$.

7. $5n^2-9n=0$
$n(5n-9)=0$
$n=0$ or $5n-9=0$
$n=0$ or $5n=9$
$n=0$ or $n=\frac{9}{5}$
The solution set is $\left\{0, \frac{9}{5}\right\}$.

9. $x^2+x-30=0$
$(x+6)(x-5)=0$
$x+6=0$ or $x-5=0$
$x=-6$ or $x=5$
The solution set is $\{-6, 5\}$.

11. $x^2-19x+84=0$
$(x-12)(x-7)=0$
$x-12=0$ or $x-7=0$
$x=12$ or $x=7$
The solution set is $\{7, 12\}$.

13. $2x^2 + 19x + 24 = 0$
$(2x + 3)(x + 8) = 0$
$2x + 3 = 0$ or $x + 8 = 0$
$2x = -3$ or $x = -8$
$x = -\frac{3}{2}$ or $x = -8$
The solution set is $\left\{-8, -\frac{3}{2}\right\}$.

15. $15x^2 + 29x - 14 = 0$
$(3x + 7)(5x - 2) = 0$
$3x + 7 = 0$ or $5x - 2 = 0$
$3x = -7$ or $5x = 2$
$x = -\frac{7}{3}$ or $x = \frac{2}{5}$
The solution set is $\left\{-\frac{7}{3}, \frac{2}{5}\right\}$.

17. $25x^2 - 30x + 9 = 0$
$(5x - 3)(5x - 3) = 0$
$(5x - 3)^2 = 0$
$5x - 3 = 0$
$5x = 3$
$x = \frac{3}{5}$
The solution set is $\left\{\frac{3}{5}\right\}$.

19. $6x^2 - 5x - 21 = 0$
$(2x + 3)(3x - 7) = 0$
$2x + 3 = 0$ or $3x - 7 = 0$
$2x = -3$ or $3x = 7$
$x = -\frac{3}{2}$ or $x = \frac{7}{3}$
The solution set is $\left\{-\frac{3}{2}, \frac{7}{3}\right\}$.

21. $3\sqrt{x} = x + 2$
$(3\sqrt{x})^2 = (x + 2)^2$
$9x = x^2 + 4x + 4$
$0 = x^2 - 5x + 4$
$0 = (x - 4)(x - 1)$
$x - 4 = 0$ or $x - 1 = 0$
$x = 4$ or $x = 1$
Checking $x = 4$
$3\sqrt{4} \stackrel{?}{=} 4 + 2$
$3(2) \stackrel{?}{=} 6$
$6 = 6$
Checking $x = 1$
$3\sqrt{1} \stackrel{?}{=} 1 + 2$
$3(1) \stackrel{?}{=} 3$
$3 = 3$
The solution set is $\{1, 4\}$.

23. $\sqrt{2x} = x - 4$
$(\sqrt{2x})^2 = (x - 4)^2$
$2x = x^2 - 8x + 16$
$0 = x^2 - 10x + 16$
$0 = (x - 8)(x - 2)$
$x - 8 = 0$ or $x - 2 = 0$
$x = 8$ or $x = 2$
Checking $x = 8$
$\sqrt{2(8)} \stackrel{?}{=} 8 - 4$
$\sqrt{16} \stackrel{?}{=} 4$
$4 = 4$
Checking $x = 2$
$\sqrt{2(2)} \stackrel{?}{=} 2 - 4$
$\sqrt{4} \stackrel{?}{=} -2$
$2 \neq -2$
The solution set is $\{8\}$.

25. $\sqrt{3x} + 6 = x$
$\sqrt{3x} = x - 6$
$(\sqrt{3x})^2 = (x - 6)^2$
$3x = x^2 - 12x + 36$
$0 = x^2 - 15x + 36$
$0 = (x - 12)(x - 3)$
$x - 12 = 0$ or $x - 3 = 0$
$x = 12$ or $x = 3$
Checking $x = 12$
$\sqrt{3(12)} + 6 \stackrel{?}{=} 12$
$\sqrt{36} + 6 \stackrel{?}{=} 12$
$12 = 12$
Checking $x = 3$
$\sqrt{3(3)} + 6 \stackrel{?}{=} 3$
$\sqrt{9} + 6 \stackrel{?}{=} 3$
$3 + 6 \stackrel{?}{=} 3$
$9 \neq 3$
The solution set is $\{12\}$.

27. $x^2 - 5kx = 0$
$x(x - 5k) = 0$
$x = 0$ or $x - 5k = 0$
$x = 0$ or $x = 5k$
The solution set is $\{0, 5k\}$.

29. $x^2 = 16k^2x$
$x^2 - 16k^2x = 0$
$x(x - 16k^2) = 0$
$x = 0$ or $x - 16k^2 = 0$
$x = 0$ or $x = 16k^2$
The solution set is $\{0, 16k^2\}$.

31. $x^2 - 12kx + 35k^2 = 0$
$(x - 5k)(x - 7k) = 0$
$x - 5k = 0$ or $x - 7k = 0$
$x = 5k$ or $x = 7k$
The solution set is $\{5k, 7k\}$.

33. $2x^2 + 5kx - 3k^2 = 0$
$(2x - k)(x + 3k) = 0$
$2x - k = 0$ or $x + 3k = 0$
$2x = k$ or $x = -3k$
$x = \frac{k}{2}$ or $x = -3k$
The solution set is $\left\{-3k, \frac{k}{2}\right\}$.

35. $x^2 = 1$
$x = \pm\sqrt{1}$
$x = \pm 1$
The solution set is $\{-1, 1\}$.

37. $x^2 = -36$
$x = \pm\sqrt{-36}$
$x = \pm i\sqrt{36}$
$x = \pm 6i$
The solution set is $\{-6i, 6i\}$.

39. $x^2 = 14$
$x = \pm\sqrt{14}$
The solution set is $\{-\sqrt{14}, \sqrt{14}\}$.

41. $n^2 - 28 = 0$
$n^2 = 28$
$n = \pm\sqrt{28}$
$n = \pm 2\sqrt{7}$
The solution set is $\{-2\sqrt{7}, 2\sqrt{7}\}$.

43. $3t^2 = 54$
$t^2 = 18$
$t = \pm\sqrt{18}$
$t = \pm 3\sqrt{2}$
The solution set is $\{-3\sqrt{2}, 3\sqrt{2}\}$.

45. $2t^2 = 7$
$t^2 = \frac{7}{2}$
$t = \pm\sqrt{\frac{7}{2}}$
$t = \pm\frac{\sqrt{7}}{\sqrt{2}} \bullet \frac{\sqrt{2}}{\sqrt{2}}$
$t = \pm\frac{\sqrt{14}}{2}$
The solution set is $\left\{-\frac{\sqrt{14}}{2}, \frac{\sqrt{14}}{2}\right\}$.

47. $15y^2 = 20$
$y^2 = \frac{20}{15}$
$y^2 = \frac{4}{3}$
$y = \pm\sqrt{\frac{4}{3}} = \pm\frac{2}{\sqrt{3}}$
$y = \pm\frac{2}{\sqrt{3}} \bullet \frac{\sqrt{3}}{\sqrt{3}}$
$y = \pm\frac{2\sqrt{3}}{3}$
The solution set is $\left\{-\frac{2\sqrt{3}}{3}, \frac{2\sqrt{3}}{3}\right\}$.

49. $10x^2 + 48 = 0$
$10x^2 = -48$
$x^2 = -\frac{48}{10}$
$x^2 = -\frac{24}{5}$
$x = \pm\sqrt{-\frac{24}{5}}$
$x = \pm\frac{i\sqrt{24}}{\sqrt{5}}$

$x = \pm \frac{i\sqrt{4}\sqrt{6}}{\sqrt{5}}$

$x = \pm \frac{2i\sqrt{6}}{\sqrt{5}} \bullet \frac{\sqrt{5}}{\sqrt{5}}$

$x = \pm \frac{2i\sqrt{30}}{5}$

The solution set is $\left\{ -\frac{2i\sqrt{30}}{5}, \frac{2i\sqrt{30}}{5} \right\}$.

51. $24x^2 = 36$

$x^2 = \frac{36}{24}$

$x^2 = \frac{3}{2}$

$x = \pm \sqrt{\frac{3}{2}}$

$x = \pm \frac{\sqrt{3}}{\sqrt{2}} \bullet \frac{\sqrt{2}}{\sqrt{2}}$

$x = \pm \frac{\sqrt{6}}{2}$

The solution set is $\left\{ -\frac{\sqrt{6}}{2}, \frac{\sqrt{6}}{2} \right\}$.

53. $(x-2)^2 = 9$

$x - 2 = \pm \sqrt{9}$

$x - 2 = \pm 3$

$x = 2 \pm 3$

$x = 2 + 3$ or $x = 2 - 3$

$x = 5$ or $x = -1$

The solution set is $\{-1, 5\}$.

55. $(x+3)^2 = 25$

$x + 3 = \pm \sqrt{25}$

$x + 3 = \pm 5$

$x = -3 \pm 5$

$x = -3 + 5$ or $x = -3 - 5$

$x = 2$ or $x = -8$

The solution set is $\{-8, 2\}$.

57. $(x+6)^2 = -4$

$x + 6 = \pm \sqrt{-4}$

$x + 6 = \pm i\sqrt{4}$

$x + 6 = \pm 2i$

$x = -6 \pm 2i$

The solution set is $\{-6 - 2i, -6 + 2i\}$.

59. $(2x-3)^2 = 1$

$2x - 3 = \pm \sqrt{1}$

$2x - 3 = \pm 1$

$2x = 3 \pm 1$

$x = \frac{3 \pm 1}{2}$

$x = \frac{3+1}{2}$ or $x = \frac{3-1}{2}$

$x = 2$ or $x = 1$

The solution set is $\{1, 2\}$.

61. $(n-4)^2 = 5$

$n - 4 = \pm \sqrt{5}$

$n = 4 \pm \sqrt{5}$

The solution set is $\{4 - \sqrt{5}, 4 + \sqrt{5}\}$.

63. $(t+5)^2 = 12$

$t + 5 = \pm \sqrt{12}$

$t + 5 = \pm \sqrt{4}\sqrt{3}$

$t + 5 = \pm 2\sqrt{3}$

$t = -5 \pm 2\sqrt{3}$

The solution set is $\{-5 - 2\sqrt{3}, -5 + 2\sqrt{3}\}$.

65. $(3y-2)^2 = -27$

$3y - 2 = \pm \sqrt{-27}$

$3y - 2 = \pm i\sqrt{27}$

$3y - 2 = \pm i\sqrt{9}\sqrt{3}$

$3y - 2 = \pm 3i\sqrt{3}$

$3y = 2 \pm 3i\sqrt{3}$

$y = \frac{2 \pm 3i\sqrt{3}}{3}$

The solution set is $\left\{ \frac{2 \pm 3i\sqrt{3}}{3} \right\}$.

67. $3(x+7)^2 + 4 = 79$

$3(x+7)^2 = 75$

$(x+7)^2 = 25$

$x + 7 = \pm \sqrt{25}$

$x + 7 = \pm 5$

$x = -7 \pm 5$

$x = -7 - 5$ or $x = -7 + 5$

$x = -12$ or $x = -2$

The solution set is $\{-12, -2\}$.

69. $2(5x-2)^2+5=25$
$2(5x-2)^2=20$
$(5x-2)^2=10$
$5x-2=\pm\sqrt{10}$
$5x=2\pm\sqrt{10}$
$x=\dfrac{2\pm\sqrt{10}}{5}$
The solution set is $\left\{\dfrac{2\pm\sqrt{10}}{5}\right\}$.

71. $4^2+6^2=c^2$
$16+36=c^2$
$52=c^2$
$\sqrt{52}=c$
$\sqrt{4}\sqrt{13}=c$
$c=2\sqrt{13}$ centimeters

73. $a^2+8^2=12^2$
$a^2+64=144$
$a^2=80$
$a=\sqrt{80}$
$a=\sqrt{16}\sqrt{5}$
$a=4\sqrt{5}$ inches

75. $15^2+b^2=17^2$
$225+b^2=289$
$b^2=64$
$b=\sqrt{64}=8$
$b=8$ yards

77. $6^2+6^2=c^2$
$36+36=c^2$
$72=c^2$
$\sqrt{72}=c$
$\sqrt{36}\sqrt{2}=c$
$6\sqrt{2}=c$
$c=6\sqrt{2}$ inches

79. $a^2+a^2=8^2$
$2a^2=64$
$a^2=32$
$a=\sqrt{32}$
$a=\sqrt{16}\sqrt{2}$
$a=4\sqrt{2}$ meters; $b=4\sqrt{2}$ meters

81. $a=3$ then $c=2a=2(3)=6$
$3^2+b^2=6^2$
$9+b^2=36$
$b^2=27$
$b=\sqrt{27}$
$b=\sqrt{9}\sqrt{3}=3\sqrt{3}$ inches
$c=6$ inches and $b=3\sqrt{3}$ inches.

83. $c=14$ then $a=\dfrac{1}{2}c=\dfrac{1}{2}(14)=7$
$7^2+b^2=14^2$
$49+b^2=196$
$b^2=147$
$b=\sqrt{147}=\sqrt{49}\sqrt{3}=7\sqrt{3}$
$a=7$ centimeters and $b=7\sqrt{3}$ centimeters

85. $b=10;\ a=\dfrac{1}{2}c$
$\left(\dfrac{1}{2}c\right)^2+10^2=c^2$
$\dfrac{1}{4}c^2+100=c^2$
$4\left(\dfrac{1}{4}c^2+100\right)=4(c^2)$
$c^2+400=4c^2$
$400=3c^2$
$\dfrac{400}{3}=c^2$
$\sqrt{\dfrac{400}{3}}=c$
$c=\dfrac{\sqrt{400}}{\sqrt{3}}=\dfrac{20}{\sqrt{3}}\bullet\dfrac{\sqrt{3}}{\sqrt{3}}=\dfrac{20\sqrt{3}}{3}$
$a=\dfrac{1}{2}\left(\dfrac{20\sqrt{3}}{3}\right)=\dfrac{10\sqrt{3}}{3}$
$c=\dfrac{20\sqrt{3}}{3}$ feet and $a=\dfrac{10\sqrt{3}}{3}$ feet.

87. $a^2+16^2=24^2$
$a^2+256=576$
$a^2=320$
$a=\sqrt{320}\approx 17.9$ feet
The ladder is 17.9 feet from the foundation of the house.

89. $16^2 + 34^2 = c^2$
$256 + 1156 = c^2$
$1412 = c^2$
$c = \sqrt{1412} \approx 38$ meters
The diagonal is 38 meters.

91. Let x = length of a side.
$x^2 + x^2 = 75^2$
$2x^2 = 5625$
$x^2 = 2812.5$
$x = \sqrt{2812.5} \approx 53$ meters
The length of a side is 53 meters.

Further Investigations

95. The diagonal of the base:
$8^2 + 6^2 = c^2$
$64 + 36 = c^2$
$100 = c^2$
$\sqrt{100} = c$
$10 = c$
The right triangle to find the diagonal has sides $a = 10$ and $b = 4$.
$c^2 = 10^2 + 4^2$
$c^2 = 100 + 16$
$c^2 = 116$
$c = \sqrt{116}$
$c \approx 10.8$ centimeters

97. $s^2 + s^2 = h^2$
$2s^2 = h^2$
$\sqrt{2s^2} = h$
$\sqrt{s^2}\sqrt{2} = h$
$s\sqrt{2} = h$

PROBLEM SET 11.3 Completing the Square

1a. $x^2 - 4x - 60 = 0$
$(x - 10)(x + 6) = 0$
$x - 10 = 0$ or $x - 6 = 0$
$x = 10$ or $x = -6$

b. $x^2 - 4x - 60 = 0$
$x^2 - 4x = 60$
$x^2 - 4x + 4 = 60 + 4$
$(x - 2)^2 = 64$
$x - 2 = \pm\sqrt{64}$
$x - 2 = \pm 8$
$x = 2 \pm 8$
$x = 2 + 8$ or $x = 2 - 8$
$x = 10$ or $x = -6$
The solution set is $\{-6, 10\}$.

3a. $x^2 - 14x = -40$
$x^2 - 14x + 40 = 0$
$(x - 10)(x - 4) = 0$
$x - 10 = 0$ or $x - 4 = 0$
$x = 10$ or $x = 4$

b. $x^2 - 14x = -40$
$x^2 - 14x + 49 = -40 + 49$
$(x - 7)^2 = 9$
$x - 7 = \pm\sqrt{9}$
$x - 7 = \pm 3$
$x = 7 \pm 3$
$x = 7 + 3$ or $x = 7 - 3$
$x = 10$ or $x = 4$
The solution set is $\{4, 10\}$.

5a. $x^2 - 5x - 50 = 0$
$(x - 10)(x + 5) = 0$
$x - 10 = 0$ or $x + 5 = 0$
$x = 10$ or $x = -5$

b. $x^2 - 5x - 50 = 0$
$x^2 - 5x = 50$
$x^2 - 5x + \frac{25}{4} = 50 + \frac{25}{4}$
$\left(x - \frac{5}{2}\right)^2 = \frac{225}{4}$
$x - \frac{5}{2} = \pm\sqrt{\frac{225}{4}}$
$x - \frac{5}{2} = \pm\frac{15}{2}$
$x = \frac{5}{2} \pm \frac{15}{2}$
$x = \frac{5}{2} + \frac{15}{2}$ or $x = \frac{5}{2} - \frac{15}{2}$

$x = \frac{20}{2}$ or $x = \frac{-10}{2}$
$x = 10$ or $x = -5$
The solution set is $\{-5, 10\}$.

7a. $x(x + 7) = 8$
$x^2 + 7x = 8$
$x^2 + 7x - 8 = 0$
$(x + 8)(x - 1) = 0$
$x + 8 = 0$ or $x - 1 = 0$
$x = -8$ or $x = 1$

b. $x(x + 7) = 8$
$x^2 + 7x = 8$
$x^2 + 7x + \frac{49}{4} = 8 + \frac{49}{4}$
$\left(x + \frac{7}{2}\right)^2 = \frac{81}{4}$
$x + \frac{7}{2} = \pm\sqrt{\frac{81}{4}}$
$x + \frac{7}{2} = \pm\frac{9}{2}$
$x = -\frac{7}{2} + \frac{9}{2}$ or $x = -\frac{7}{2} - \frac{9}{2}$
$x = \frac{2}{2} = 1$ or $x = -\frac{16}{2} = -8$
The solution set is $\{-8, 1\}$.

9a. $2n^2 - n - 15 = 0$
$(2n + 5)(n - 3) = 0$
$2n + 5 = 0$ or $n - 3 = 0$
$2n = -5$ or $n = 3$
$n = -\frac{5}{2}$ or $n = 3$

b. $2n^2 - n - 15 = 0$
$n^2 - \frac{1}{2}n - \frac{15}{2} = 0$
$n^2 - \frac{1}{2}n = \frac{15}{2}$
$n^2 - \frac{1}{2}n + \frac{1}{16} = \frac{15}{2} + \frac{1}{16}$
$\left(n - \frac{1}{4}\right)^2 = \frac{121}{16}$
$n - \frac{1}{4} = \pm\sqrt{\frac{121}{16}}$
$n - \frac{1}{4} = \pm\frac{11}{4}$
$n = \frac{1}{4} \pm \frac{11}{4}$
$n = \frac{1}{4} + \frac{11}{4}$ or $n = \frac{1}{4} - \frac{11}{4}$
$n = \frac{12}{4}$ or $n = -\frac{10}{4}$
$n = 3$ or $n = -\frac{5}{2}$
The solution set is $\left\{-\frac{5}{2}, 3\right\}$.

11a. $3n^2 + 7n - 6 = 0$
$(3n - 2)(n + 3) = 0$
$3n - 2 = 0$ or $n + 3 = 0$
$3n = 2$ or $n = -3$
$n = \frac{2}{3}$ or $n = -3$

b. $3n^2 + 7n - 6 = 0$
$n^2 + \frac{7}{3}n - 2 = 0$
$n^2 + \frac{7}{3}n = 2$
$n^2 + \frac{7}{3}n + \frac{49}{36} = 2 + \frac{49}{36}$
$\left(n + \frac{7}{6}\right)^2 = \frac{121}{36}$
$n + \frac{7}{6} = \pm\sqrt{\frac{121}{36}}$
$n + \frac{7}{6} = \pm\frac{11}{6}$
$n = -\frac{7}{6} \pm \frac{11}{6}$
$n = -\frac{7}{6} + \frac{11}{6}$ or $n = -\frac{7}{6} - \frac{11}{6}$
$n = \frac{4}{6} = \frac{2}{3}$ or $n = -\frac{18}{6} = -3$
The solution set is $\left\{-3, \frac{2}{3}\right\}$.

13a. $n(n + 6) = 160$
$n^2 + 6n = 160$
$n^2 + 6n - 160 = 0$
$(n + 16)(n - 10) = 0$
$n + 16 = 0$ or $n - 10 = 0$
$n = -16$ or $n = 10$

b. $n(n + 6) = 160$
$n^2 + 6n = 160$
$n^2 + 6n + 9 = 160 + 9$
$(n + 3)^2 = 169$
$n + 3 = \pm\sqrt{169}$
$n + 3 = \pm 13$
$n = -3 + 13$ or $n = -3 - 13$
$n = 10$ or $n = -16$
The solution set is $\{-16, 10\}$.

15. $x^2 + 4x - 2 = 0$
$x^2 + 4x = 2$
$x^2 + 4x + 4 = 2 + 4$
$(x + 2)^2 = 6$
$x + 2 = \pm\sqrt{6}$
$x = -2 \pm \sqrt{6}$
The solution set is $\{-2 \pm \sqrt{6}\}$.

17. $x^2 + 6x - 3 = 0$
$x^2 + 6x = 3$
$x^2 + 6x + 9 = 3 + 9$
$(x + 3)^2 = 12$
$x + 3 = \pm\sqrt{12}$
$x + 3 = \pm 2\sqrt{3}$
$x = -3 \pm 2\sqrt{3}$
The solution set is $\{-3 \pm 2\sqrt{3}\}$.

19. $y^2 - 10y = 1$
$y^2 - 10y + 25 = 1 + 25$
$(y - 5)^2 = 26$
$y - 5 = \pm\sqrt{26}$
$y = 5 \pm \sqrt{26}$
The solution set is $\{5 \pm \sqrt{26}\}$.

21. $n^2 - 8n + 17 = 0$
$n^2 - 8n = -17$
$n^2 - 8n + 16 = -17 + 16$
$(n - 4)^2 = -1$
$n - 4 = \pm\sqrt{-1}$
$n - 4 = \pm i$
$n = 4 \pm i$
The solution set is $\{4 \pm i\}$.

23. $n(n + 12) = -9$
$n^2 + 12n = -9$
$n^2 + 12n + 36 = -9 + 36$
$(n + 6)^2 = 27$
$n + 6 = \pm\sqrt{27}$
$n + 6 = \pm 3\sqrt{3}$
$n = -6 \pm 3\sqrt{3}$
The solution set is $\{-6 \pm 3\sqrt{3}\}$.

25. $n^2 + 2n + 6 = 0$
$n^2 + 2n = -6$
$n^2 + 2n + 1 = -6 + 1$
$(n + 1)^2 = -5$
$n + 1 = \pm\sqrt{-5}$
$n + 1 = \pm i\sqrt{5}$
$n = -1 \pm i\sqrt{5}$
The solution set is $\{-1 \pm i\sqrt{5}\}$.

27. $x^2 + 3x - 2 = 0$
$x^2 + 3x = 2$
$x^2 + 3x + \frac{9}{4} = 2 + \frac{9}{4}$
$\left(x + \frac{3}{2}\right)^2 = \frac{17}{4}$
$x + \frac{3}{2} = \pm\sqrt{\frac{17}{4}}$
$x + \frac{3}{2} = \pm\frac{\sqrt{17}}{2}$
$x = -\frac{3}{2} \pm \frac{\sqrt{17}}{2}$
$x = \frac{-3 \pm \sqrt{17}}{2}$
The solution set is $\left\{\frac{-3 \pm \sqrt{17}}{2}\right\}$.

29. $x^2 + 5x + 1 = 0$
$x^2 + 5x = -1$
$x^2 + 5x + \frac{25}{4} = -1 + \frac{25}{4}$
$\left(x + \frac{5}{2}\right)^2 = \frac{21}{4}$
$x + \frac{5}{2} = \pm\sqrt{\frac{21}{4}}$
$x + \frac{5}{2} = \pm\frac{\sqrt{21}}{2}$
$x = -\frac{5}{2} \pm \frac{\sqrt{21}}{2}$
$x = \frac{-5 \pm \sqrt{21}}{2}$
The solution set is $\left\{\frac{-5 \pm \sqrt{21}}{2}\right\}$.

31. $y^2 - 7y + 3 = 0$
$y^2 - 7y = -3$
$y^2 - 7y + \frac{49}{4} = -3 + \frac{49}{4}$
$\left(y - \frac{7}{2}\right)^2 = \frac{37}{4}$
$y - \frac{7}{2} = \pm\sqrt{\frac{37}{4}}$

$$y-\frac{7}{2}=\pm\frac{\sqrt{37}}{2}$$
$$y=\frac{7}{2}\pm\frac{\sqrt{37}}{2}$$
$$y=\frac{7\pm\sqrt{37}}{2}$$
The solution set is $\left\{\frac{7\pm\sqrt{37}}{2}\right\}$.

33. $2x^2+4x-3=0$
$$x^2+2x-\frac{3}{2}=0$$
$$x^2+2x=\frac{3}{2}$$
$$x^2+2x+1=\frac{3}{2}+1$$
$$(x+1)^2=\frac{5}{2}$$
$$x+1=\pm\sqrt{\frac{5}{2}}$$
$$x+1=\pm\frac{\sqrt{5}}{\sqrt{2}}\bullet\frac{\sqrt{2}}{\sqrt{2}}$$
$$x+1=\pm\frac{\sqrt{10}}{2}$$
$$x=-1\pm\frac{\sqrt{10}}{2}$$
$$x=-\frac{2}{2}\pm\frac{\sqrt{10}}{2}$$
$$x=\frac{-2\pm\sqrt{10}}{2}$$
The solution set is $\left\{\frac{-2\pm\sqrt{10}}{2}\right\}$.

35. $3n^2-6n+5=0$
$$n^2-2n+\frac{5}{3}=0$$
$$n^2-2n=-\frac{5}{3}$$
$$n^2-2n+1=-\frac{5}{3}+1$$
$$(n-1)^2=-\frac{2}{3}$$
$$n-1=\pm\sqrt{-\frac{2}{3}}$$
$$n-1=\pm\frac{i\sqrt{2}}{\sqrt{3}}$$
$$n-1=\pm\frac{i\sqrt{2}}{\sqrt{3}}\bullet\frac{\sqrt{3}}{\sqrt{3}}$$
$$n-1=\pm\frac{i\sqrt{6}}{3}$$
$$n=1\pm\frac{i\sqrt{6}}{3}$$
$$n=\frac{3}{3}\pm\frac{i\sqrt{6}}{3}$$
$$n=\frac{3\pm i\sqrt{6}}{3}$$
The solution set is $\left\{\frac{3\pm i\sqrt{6}}{3}\right\}$.

37. $3x^2+5x-1=0$
$$x^2+\frac{5}{3}x-\frac{1}{3}=0$$
$$x^2+\frac{5}{3}x=\frac{1}{3}$$
$$x^2+\frac{5}{3}x+\frac{25}{36}=\frac{1}{3}+\frac{25}{36}$$
$$\left(x+\frac{5}{6}\right)^2=\frac{37}{36}$$
$$x+\frac{5}{6}=\pm\sqrt{\frac{37}{36}}$$
$$x+\frac{5}{6}=\pm\frac{\sqrt{37}}{6}$$
$$x=-\frac{5}{6}\pm\frac{\sqrt{37}}{6}$$
$$x=\frac{-5\pm\sqrt{37}}{6}$$
The solution set is $\left\{\frac{-5\pm\sqrt{37}}{6}\right\}$.

39. $x^2+8x-48=0$
$(x+12)(x-4)=0$
$x+12=0$ or $x-4=0$
$x=-12$ or $x=4$
The solution set is $\{-12, 4\}$.

41. $2n^2 - 8n = -3$

$n^2 - 4n = -\frac{3}{2}$

$n^2 - 4n + 4 = -\frac{3}{2} + 4$

$(n-2)^2 = \frac{5}{2}$

$n - 2 = \pm\sqrt{\frac{5}{2}}$

$n - 2 = \pm\frac{\sqrt{5}}{\sqrt{2}} \bullet \frac{\sqrt{2}}{\sqrt{2}}$

$n - 2 = \pm\frac{\sqrt{10}}{2}$

$n = 2 \pm \frac{\sqrt{10}}{2} = \frac{4}{2} \pm \frac{\sqrt{10}}{2}$

$n = \frac{4 \pm \sqrt{10}}{2}$

The solution set is $\left\{\frac{4 \pm \sqrt{10}}{2}\right\}$.

43. $(3x-1)(2x+9) = 0$

$3x - 1 = 0$ or $2x + 9 = 0$

$3x = 1$ or $2x = -9$

$x = \frac{1}{3}$ or $x = -\frac{9}{2}$

The solution set is $\left\{-\frac{9}{2}, \frac{1}{3}\right\}$.

45. $(x+2)(x-7) = 10$

$x^2 - 5x - 14 = 10$

$x^2 - 5x - 24 = 0$

$(x-8)(x+3) = 0$

$x - 8 = 0$ or $x + 3 = 0$

$x = 8$ or $x = -3$

The solution set is $\{-3, 8\}$.

47. $(x-3)^2 = 12$

$x - 3 = \pm\sqrt{12}$

$x - 3 = \pm 2\sqrt{3}$

$x = 3 \pm 2\sqrt{3}$

The solution set is $\{3 \pm 2\sqrt{3}\}$.

49. $3n^2 - 6n + 4 = 0$

$n^2 - 2n + \frac{4}{3} = 0$

$n^2 - 2n = -\frac{4}{3}$

$n^2 - 2n + 1 = -\frac{4}{3} + 1$

$(n-1)^2 = -\frac{1}{3}$

$n - 1 = \pm\sqrt{-\frac{1}{3}}$

$n - 1 = \pm\frac{i}{\sqrt{3}}$

$n - 1 = \pm\frac{i}{\sqrt{3}} \bullet \frac{\sqrt{3}}{\sqrt{3}}$

$n - 1 = \pm\frac{i\sqrt{3}}{3}$

$n = 1 \pm \frac{i\sqrt{3}}{3}$

$n = \frac{3}{3} \pm \frac{i\sqrt{3}}{3}$

$n = \frac{3 \pm i\sqrt{3}}{3}$

The solution set is $\left\{\frac{3 \pm i\sqrt{3}}{3}\right\}$.

51. $n(n+8) = 240$

$n^2 + 8n = 240$

$n^2 + 8n - 240 = 0$

$(n+20)(n-12) = 0$

$n + 20 = 0$ or $n - 12 = 0$

$n = -20$ or $n = 12$

The solution set is $\{-20, 12\}$.

53. $3x^2 + 29x = -66$

$3x^2 + 29x + 66 = 0$

$(3x+11)(x+6) = 0$

$3x + 11 = 0$ or $x + 6 = 0$

$3x = -11$ or $x = -6$

$x = -\frac{11}{3}$ or $x = -6$

The solution set is $\left\{-6, -\frac{11}{3}\right\}$.

55. $6n^2 + 23n + 21 = 0$

$(3n+7)(2n+3) = 0$

$3n + 7 = 0$ or $2n + 3 = 0$

$3n = -7$ or $2n = -3$

$x = -\frac{7}{3}$ or $x = -\frac{3}{2}$

The solution set is $\left\{-\frac{7}{3}, -\frac{3}{2}\right\}$.

57. $x^2 + 12x = 4$
$x^2 + 12x + 36 = 4 + 36$
$(x+6)^2 = 40$
$x + 6 = \pm\sqrt{40}$
$x + 6 = \pm 2\sqrt{10}$
$x = -6 \pm 2\sqrt{10}$
The solution set is $\{-6 \pm 2\sqrt{10}\}$.

59. $12n^2 - 7n + 1 = 0$
$(4n-1)(3n-1) = 0$
$4n - 1 = 0$ or $3n - 1 = 0$
$4n = 1$ or $3n = 1$
$n = \frac{1}{4}$ or $n = \frac{1}{3}$
The solution set is $\left\{\frac{1}{4}, \frac{1}{3}\right\}$.

61. $ax^2 + bx + c = 0$
$x^2 + \frac{b}{a}x + \frac{c}{a} = 0$
$x^2 + \frac{b}{a}x = -\frac{c}{a}$
$x^2 + \frac{b}{a}x + \frac{b^2}{4a^2} = -\frac{c}{a} + \frac{b^2}{4a^2}$
$\left(x + \frac{b}{2a}\right)^2 = \frac{b^2}{4a^2} - \frac{c}{a}$
$\left(x + \frac{b}{2a}\right)^2 = \frac{b^2}{4a^2} - \frac{c}{a} \bullet \frac{4a}{4a}$
$\left(x + \frac{b}{2a}\right)^2 = \frac{b^2}{4a^2} - \frac{4ac}{4a^2}$
$\left(x + \frac{b}{2a}\right)^2 = \frac{b^2 - 4ac}{4a^2}$
$x + \frac{b}{2a} = \pm\sqrt{\frac{b^2 - 4ac}{4a^2}}$
$x + \frac{b}{2a} = \pm\frac{\sqrt{b^2 - 4ac}}{2a}$
$x = -\frac{b}{2a} \pm \frac{\sqrt{b^2 - 4ac}}{2a}$
$x = \frac{-b \pm \sqrt{b^2 - 4ac}}{2a}$
The solution set is $\left\{\frac{-b \pm \sqrt{b^2 - 4ac}}{2a}\right\}$.

Further Investigations

65. $\frac{x^2}{a^2} + \frac{y^2}{b^2} = 1$ for x
$\frac{x^2}{a^2} = 1 - \frac{y^2}{b^2}$
$\frac{x^2}{a^2} = \frac{b^2 - y^2}{b^2}$
$x^2 = \frac{a^2(b^2 - y^2)}{b^2}$
$x = \sqrt{\frac{a^2(b^2 - y^2)}{b^2}}$
$x = \frac{a\sqrt{b^2 - y^2}}{b}$

67. $A = \pi r^2$ for r
$\frac{A}{\pi} = r^2$
$\sqrt{\frac{A}{\pi}} = r$
$r = \frac{\sqrt{A}}{\sqrt{\pi}} \bullet \frac{\sqrt{\pi}}{\sqrt{\pi}} = \frac{\sqrt{A\pi}}{\sqrt{\pi^2}} = \frac{\sqrt{A\pi}}{\pi}$

69. $x^2 - 5ax + 6a^2 = 0$
$(x - 2a)(x - 3a) = 0$
$x - 2a = 0$ or $x - 3a = 0$
$x = 2a$ or $x = 3a$
The solution set is $\{2a, 3a\}$.

71. $6x^2 + ax - 2a^2 = 0$
$(3x + 2a)(2x - a) = 0$
$3x + 2a = 0$ or $2x - a = 0$
$3x = -2a$ or $2x = a$
$x = -\frac{2a}{3}$ or $x = \frac{a}{2}$
The solution set is $\left\{-\frac{2a}{3}, \frac{a}{2}\right\}$.

73. $9x^2 - 12bx + 4b^2 = 0$
$(3x - 2b)(3x - 2b) = 0$
$(3x - 2b)^2 = 0$
$3x - 2b = 0$
$3x = 2b$
$x = \frac{2b}{3}$
The solution set is $\left\{\frac{2b}{3}\right\}$.

PROBLEM SET 11.4 Quadratic Formula

1. $x^2+4x-21=0$
b^2-4ac
$4^2-4(1)(-21)=16+84=100$
Since $b^2-4ac>0$, there will be two real solutions.
$(x+7)(x-3)=0$
$x+7=0$ or $x-3=0$
$x=-7$ or $x=3$
The solution set is $\{-7, 3\}$.

3. $9x^2-6x+1=0$
b^2-4ac
$(-6)^2-4(9)(1)=36-36=0$
Since $b^2-4ac=0$, there will be one real solution.
$(3x-1)(3x-1)=0$
$(3x-1)^2=0$
$3x-1=0$
$3x=1$
$x=\frac{1}{3}$
The solution set is $\left\{\frac{1}{3}\right\}$.

5. $x^2-7x+13=0$
b^2-4ac
$(-7)^2-4(1)(13)=49-52=-3$
Since $b^2-4ac<0$, there will be two complex solutions.
$x=\frac{-(-7)\pm\sqrt{(-7)^2-4(1)(13)}}{2(1)}$
$x=\frac{7\pm\sqrt{49-52}}{2}$
$x=\frac{7\pm\sqrt{-3}}{2}$
$x=\frac{7\pm i\sqrt{3}}{2}$
The solution set is $\left\{\frac{7\pm i\sqrt{3}}{2}\right\}$.

7. $15x^2+17x-4=0$
b^2-4ac
$(17)^2-4(15)(-4)=289+240=529$
Since $b^2-4ac>0$, there will be two real solutions.
$(3x+4)(5x-1)=0$
$3x+4=0$ or $5x-1=0$
$3x=-4$ or $5x=1$
$x=-\frac{4}{3}$ or $x=\frac{1}{5}$
The solution set is $\left\{-\frac{4}{3}, \frac{1}{5}\right\}$.

9. $3x^2+4x=2$
$3x^2+4x-2=0$
b^2-4ac
$(4)^2-4(3)(-2)=16+24=40$
Since $b^2-4ac>0$, there will be two real solutions.
$x=\frac{-(4)\pm\sqrt{40}}{2(3)}$
$x=\frac{-4\pm2\sqrt{10}}{6}$
$x=\frac{2(-2\pm\sqrt{10})}{6}$
$x=\frac{-2\pm\sqrt{10}}{3}$
The solution set is $\left\{\frac{-2\pm\sqrt{10}}{3}\right\}$.

11. $x^2+2x-1=0$
$x=\frac{-(2)\pm\sqrt{(2)^2-4(1)(-1)}}{2(1)}$
$x=\frac{-2\pm\sqrt{4+4}}{2}$
$x=\frac{-2\pm\sqrt{8}}{2}$
$x=\frac{-2\pm2\sqrt{2}}{2}$
$x=\frac{2(-1\pm\sqrt{2})}{2}$
$x=-1\pm\sqrt{2}$
The solution set is $\{-1\pm\sqrt{2}\}$.

13. $n^2+5n-3=0$
$n=\frac{-(5)\pm\sqrt{(5)^2-4(1)(-3)}}{2(1)}$

$$n = \frac{-5 \pm \sqrt{25+12}}{2}$$
$$n = \frac{-5 \pm \sqrt{37}}{2}$$
The solution set is $\left\{\frac{-5 \pm \sqrt{37}}{2}\right\}$.

15. $a^2 - 8a = 4$
$a^2 - 8a - 4 = 0$
$$a = \frac{-(-8) \pm \sqrt{(-8)^2 - 4(1)(-4)}}{2(1)}$$
$$a = \frac{8 \pm \sqrt{64+16}}{2}$$
$$a = \frac{8 \pm \sqrt{80}}{2}$$
$$a = \frac{8 \pm 4\sqrt{5}}{2}$$
$$a = \frac{2(4 \pm 2\sqrt{5})}{2}$$
$a = 4 \pm 2\sqrt{5}$
The solution set is $\{4 \pm 2\sqrt{5}\}$.

17. $n^2 + 5n + 8 = 0$
$$n = \frac{-(5) \pm \sqrt{(5)^2 - 4(1)(8)}}{2(1)}$$
$$n = \frac{-5 \pm \sqrt{25-32}}{2}$$
$$n = \frac{-5 \pm \sqrt{-7}}{2}$$
$$n = \frac{-5 \pm i\sqrt{7}}{2}$$
The solution set is $\left\{\frac{-5 \pm i\sqrt{7}}{2}\right\}$.

19. $x^2 - 18x + 80 = 0$
$$x = \frac{-(-18) \pm \sqrt{(-18)^2 - 4(1)(80)}}{2(1)}$$
$$x = \frac{18 \pm \sqrt{324-320}}{2}$$
$$x = \frac{18 \pm \sqrt{4}}{2}$$
$$x = \frac{18 \pm 2}{2}$$
$x = \frac{18+2}{2}$ or $x = \frac{18-2}{2}$
$x = 10$ or $x = 8$
The solution set is $\{8, 10\}$.

21. $-y^2 = -9y + 5$
$0 = y^2 - 9y + 5$
$$y = \frac{-(-9) \pm \sqrt{(-9)^2 - 4(1)(5)}}{2(1)}$$
$$y = \frac{9 \pm \sqrt{81-20}}{2}$$
$$y = \frac{9 \pm \sqrt{61}}{2}$$
The solution set is $\left\{\frac{9 \pm \sqrt{61}}{2}\right\}$.

23. $2x^2 + x - 4 = 0$
$$x = \frac{-(1) \pm \sqrt{(1)^2 - 4(2)(-4)}}{2(2)}$$
$$x = \frac{-1 \pm \sqrt{1+32}}{4}$$
$$x = \frac{-1 \pm \sqrt{33}}{4}$$
The solution set is $\left\{\frac{-1 \pm \sqrt{33}}{4}\right\}$.

25. $4x^2 + 2x + 1 = 0$
$$x = \frac{-2 \pm \sqrt{2^2 - 4(4)(1)}}{2(4)}$$
$$x = \frac{-2 \pm \sqrt{4-16}}{8}$$
$$x = \frac{-2 \pm \sqrt{-12}}{8}$$
$$x = \frac{-2 \pm i\sqrt{12}}{8}$$
$$x = \frac{-2 \pm 2i\sqrt{3}}{8}$$
$$x = \frac{2(-1 \pm i\sqrt{3})}{8}$$
$$x = \frac{-1 \pm i\sqrt{3}}{4}$$
The solution set is $\left\{\frac{-1 \pm i\sqrt{3}}{4}\right\}$.

27. $3a^2 - 8a + 2 = 0$

$a = \frac{-(-8) \pm \sqrt{(-8)^2 - 4(3)(2)}}{2(3)}$

$a = \frac{8 \pm \sqrt{64 - 24}}{6}$

$a = \frac{8 \pm \sqrt{40}}{6}$

$a = \frac{8 \pm 2\sqrt{10}}{6}$

$a = \frac{2(4 \pm \sqrt{10})}{6}$

$a = \frac{4 \pm \sqrt{10}}{3}$

The solution set is $\left\{\frac{4 \pm \sqrt{10}}{3}\right\}$.

29. $-2n^2 + 3n + 5 = 0$

$n = \frac{-(3) \pm \sqrt{(3)^2 - 4(-2)(5)}}{2(-2)}$

$n = \frac{-3 \pm \sqrt{9 + 40}}{-4}$

$n = \frac{-3 \pm \sqrt{49}}{-4}$

$n = \frac{-3 \pm 7}{-4}$

$n = \frac{-3 + 7}{-4}$ or $n = \frac{-3 - 7}{-4}$

$n = -1$ or $n = \frac{-10}{-4} = \frac{5}{2}$

The solution set is $\left\{-1, \frac{5}{2}\right\}$.

31. $3x^2 + 19x + 20 = 0$

$x = \frac{-19 \pm \sqrt{19^2 - 4(3)(20)}}{2(3)}$

$x = \frac{-19 \pm \sqrt{361 - 240}}{6}$

$x = \frac{-19 \pm \sqrt{121}}{6}$

$x = \frac{-19 \pm 11}{6}$

$x = \frac{-19 + 11}{6}$ or $x = \frac{-19 - 11}{6}$

$x = \frac{-8}{6} = -\frac{4}{3}$ or $x = \frac{-30}{6} = -5$

The solution set is $\left\{-5, -\frac{4}{3}\right\}$.

33. $36n^2 - 60n + 25 = 0$

$n = \frac{-(-60) \pm \sqrt{(-60)^2 - 4(36)(25)}}{2(36)}$

$n = \frac{60 \pm \sqrt{3600 - 3600}}{72}$

$n = \frac{60}{72} = \frac{5}{6}$

The solution set is $\left\{\frac{5}{6}\right\}$.

35. $4x^2 - 2x = 3$

$4x^2 - 2x - 3 = 0$

$x = \frac{-(-2) \pm \sqrt{(-2)^2 - 4(4)(-3)}}{2(4)}$

$x = \frac{2 \pm \sqrt{4 + 48}}{8}$

$x = \frac{2 \pm \sqrt{52}}{8}$

$x = \frac{2 \pm 2\sqrt{13}}{8}$

$x = \frac{2(1 \pm \sqrt{13})}{8}$

$x = \frac{1 \pm \sqrt{13}}{4}$

The solution set is $\left\{\frac{1 \pm \sqrt{13}}{4}\right\}$.

37. $5x^2 - 13x = 0$

$x = \frac{-(-13) \pm \sqrt{(-13)^2 - 4(5)(0)}}{2(5)}$

$x = \frac{13 \pm \sqrt{169}}{10}$

$x = \frac{13 \pm 13}{10}$

$x = \frac{13 + 13}{10}$ or $x = \frac{13 - 13}{10}$

$x = \frac{26}{10} = \frac{13}{5}$ or $x = 0$

The solution set is $\left\{0, \frac{13}{5}\right\}$.

Problem Set 11.4

39. $3x^2 = 5$

$3x^2 - 5 = 0$

$$x = \frac{-0 \pm \sqrt{0^2 - 4(3)(-5)}}{2(3)}$$

$$x = \frac{\pm\sqrt{60}}{6} = \frac{\pm 2\sqrt{15}}{6} = \pm\frac{\sqrt{15}}{3}$$

The solution set is $\left\{\pm\frac{\sqrt{15}}{3}\right\}$.

41. $6t^2 + t - 3 = 0$

$$t = \frac{-1 \pm \sqrt{1^2 - 4(6)(-3)}}{2(6)}$$

$$t = \frac{-1 \pm \sqrt{1 + 72}}{12}$$

$$t = \frac{-1 \pm \sqrt{73}}{12}$$

The solution set is $\left\{\frac{-1 \pm \sqrt{73}}{12}\right\}$.

43. $n^2 + 32n + 252 = 0$

$$n = \frac{-32 \pm \sqrt{32^2 - 4(1)(252)}}{2(1)}$$

$$n = \frac{-32 \pm \sqrt{1024 - 1008}}{2}$$

$$n = \frac{-32 \pm \sqrt{16}}{2}$$

$$n = \frac{-32 \pm 4}{2}$$

$$n = \frac{-32 + 4}{2} \quad \text{or} \quad n = \frac{-32 - 4}{2}$$

$$n = -14 \quad \text{or} \quad n = -18$$

The solution set is $\{-18, -14\}$.

45. $12x^2 - 73x + 110 = 0$

$$x = \frac{-(-73) \pm \sqrt{(-73)^2 - 4(12)(110)}}{2(12)}$$

$$x = \frac{73 \pm \sqrt{5329 - 5280}}{24}$$

$$x = \frac{73 \pm \sqrt{49}}{24}$$

$$x = \frac{73 \pm 7}{24}$$

$$x = \frac{73 + 7}{24} \quad \text{or} \quad x = \frac{73 - 7}{24}$$

$$x = \frac{80}{24} = \frac{10}{3} \quad \text{or} \quad x = \frac{66}{24} = \frac{11}{4}$$

The solution set is $\left\{\frac{11}{4}, \frac{10}{3}\right\}$.

47. $-2x^2 + 4x - 3 = 0$

$$x = \frac{-4 \pm \sqrt{4^2 - 4(-2)(-3)}}{2(-2)}$$

$$x = \frac{-4 \pm \sqrt{16 - 24}}{-4}$$

$$x = \frac{-4 \pm \sqrt{-8}}{-4}$$

$$x = \frac{-4 \pm i\sqrt{8}}{-4}$$

$$x = \frac{-4 \pm 2i\sqrt{2}}{-4}$$

$$x = \frac{-2(2 \pm i\sqrt{2})}{-4}$$

$$x = \frac{2 \pm i\sqrt{2}}{2}$$

The solution set is $\left\{\frac{2 \pm i\sqrt{2}}{2}\right\}$.

49. $-6x^2 + 2x + 1 = 0$

$$x = \frac{-2 \pm \sqrt{2^2 - 4(-6)(1)}}{2(-6)}$$

$$x = \frac{-2 \pm \sqrt{4 + 24}}{-12}$$

$$x = \frac{-2 \pm \sqrt{28}}{-12}$$

$$x = \frac{-2 \pm 2\sqrt{7}}{-12}$$

$$x = \frac{-2(1 \pm \sqrt{7})}{-12}$$

$$x = \frac{1 \pm \sqrt{7}}{6}$$

The solution set is $\left\{\frac{1 \pm \sqrt{7}}{6}\right\}$.

Further Investigations

55. $x^2 - 16x - 24 = 0$

$x = \dfrac{-(-16) \pm \sqrt{(-16)^2 - 4(1)(-24)}}{2(1)}$

$x = \dfrac{16 \pm \sqrt{256 + 96}}{2}$

$x = \dfrac{16 \pm \sqrt{352}}{2}$

$x = \dfrac{16 + \sqrt{352}}{2}$ or $x = \dfrac{16 - \sqrt{352}}{2}$

$x = 17.381$ or $x = -1.381$

The solution set is $\{-1.381, 17.381\}$.

57. $x^2 + 10x - 46 = 0$

$x = \dfrac{-10 \pm \sqrt{(10)^2 - 4(1)(-46)}}{2(1)}$

$x = \dfrac{-10 \pm \sqrt{100 + 184}}{2}$

$x = \dfrac{-10 \pm \sqrt{284}}{2}$

$x = \dfrac{-10 + \sqrt{284}}{2}$ or $x = \dfrac{-10 - \sqrt{284}}{2}$

$x = 3.426$ or $x = -13.426$

The solution set is $\{-13.426, 3.426\}$.

59. $x^2 + 9x + 3 = 0$

$x = \dfrac{-9 \pm \sqrt{(9)^2 - 4(1)(3)}}{2(1)}$

$x = \dfrac{-9 \pm \sqrt{81 - 12}}{2}$

$x = \dfrac{-9 \pm \sqrt{69}}{2}$

$x = \dfrac{-9 + \sqrt{69}}{2}$ or $x = \dfrac{-9 - \sqrt{69}}{2}$

$x = -0.347$ or $x = -8.653$

The solution set is $\{-8.653, -0.347\}$.

61. $5x^2 - 9x + 1 = 0$

$x = \dfrac{-(-9) \pm \sqrt{(-9)^2 - 4(5)(1)}}{2(5)}$

$x = \dfrac{9 \pm \sqrt{81 - 20}}{10}$

$x = \dfrac{9 \pm \sqrt{61}}{10}$

$x = \dfrac{9 + \sqrt{61}}{10}$ or $x = \dfrac{9 - \sqrt{61}}{10}$

$x = 1.681$ or $x = 0.119$

The solution set is $\{0.119, 1.681\}$.

63. $3x^2 - 12x - 10 = 0$

$x = \dfrac{-(-12) \pm \sqrt{(-12)^2 - 4(3)(-10)}}{2(3)}$

$x = \dfrac{12 \pm \sqrt{144 + 120}}{6}$

$x = \dfrac{12 \pm \sqrt{264}}{6}$

$x = \dfrac{12 + \sqrt{264}}{6}$ or $x = \dfrac{12 - \sqrt{264}}{6}$

$x = 4.708$ or $x = -0.708$

The solution set is $\{-0.708, 4.708\}$.

65. $4x^2 - kx + 1 = 0$

$b^2 - 4ac = 0$

$(-k)^2 - 4(4)(1) = 0$

$k^2 \quad 16 = 0$

$(k + 4)(k - 4) = 0$

$k + 4 = 0$ or $k - 4 = 0$

$k = -4$ or $k = 4$

PROBLEM SET 11.5 More Quadratic Equations and Applications

1. $x^2 - 4x - 6 = 0$

$x^2 - 4x = 6$

$x^2 - 4x + 4 = 6 + 4$

$(x - 2)^2 = 10$

$x - 2 = \pm\sqrt{10}$

$x = 2 \pm \sqrt{10}$

The solution set is $\{2 \pm \sqrt{10}\}$.

3. $3x^2 + 23x - 36 = 0$

$(3x - 4)(x + 9) = 0$

$3x - 4 = 0$ or $x + 9 = 0$
$3x = 4$ or $x = -9$
$x = \frac{4}{3}$ or $x = -9$
The solution set is $\left\{-9, \frac{4}{3}\right\}$.

5. $x^2 - 18x = 9$
$x^2 - 18x + 81 = 9 + 81$
$(x - 9)^2 = 90$
$x - 9 = \pm\sqrt{90}$
$x - 9 = \pm 3\sqrt{10}$
$x = 9 \pm 3\sqrt{10}$
The solution set is $\{9 \pm 3\sqrt{10}\}$.

7. $2x^2 - 3x + 4 = 0$
$$x = \frac{-(-3) \pm \sqrt{(-3)^2 - 4(2)(4)}}{2(2)}$$
$$x = \frac{3 \pm \sqrt{9 - 32}}{4}$$
$$x = \frac{3 \pm \sqrt{-23}}{4}$$
$$x = \frac{3 \pm i\sqrt{23}}{4}$$
The solution set is $\left\{\frac{3 \pm i\sqrt{23}}{4}\right\}$.

9. $135 + 24n + n^2 = 0$
$n^2 + 24n + 135 = 0$
$(n + 15)(n + 9) = 0$
$n + 15 = 0$ or $n + 9 = 0$
$n = -15$ or $n = -9$
The solution set is $\{-15, -9\}$.

11. $(x - 2)(x + 9) = -10$
$x^2 + 7x - 18 = -10$
$x^2 + 7x - 8 = 0$
$(x + 8)(x - 1) = 0$
$x + 8 = 0$ or $x - 1 = 0$
$x = -8$ or $x = 1$
The solution set is $\{-8, 1\}$.

13. $2x^2 - 4x + 7 = 0$
$$x = \frac{-(-4) \pm \sqrt{(-4)^2 - 4(2)(7)}}{2(2)}$$
$$x = \frac{4 \pm \sqrt{16 - 56}}{4}$$
$$x = \frac{4 \pm \sqrt{-40}}{4}$$
$$x = \frac{4 \pm 2i\sqrt{10}}{4}$$
$$x = \frac{2(2 \pm i\sqrt{10})}{4}$$
$$x = \frac{2 \pm i\sqrt{10}}{2}$$
The solution set is $\left\{\frac{2 \pm i\sqrt{10}}{2}\right\}$.

15. $x^2 - 18x + 15 = 0$
$x^2 - 18x = -15$
$x^2 - 18x + 81 = -15 + 81$
$(x - 9)^2 = 66$
$x - 9 = \pm\sqrt{66}$
$x = 9 \pm \sqrt{66}$
The solution set is $\{9 \pm \sqrt{66}\}$.

17. $20y^2 + 17y - 10 = 0$
$(4y + 5)(5y - 2) = 0$
$4y + 5 = 0$ or $5y - 2 = 0$
$4y = -5$ or $5y = 2$
$y = -\frac{5}{4}$ or $y = \frac{2}{5}$
The solution set is $\left\{-\frac{5}{4}, \frac{2}{5}\right\}$.

19. $4t^2 + 4t - 1 = 0$
$$t = \frac{-4 \pm \sqrt{4^2 - 4(4)(-1)}}{2(4)}$$
$$t = \frac{-4 \pm \sqrt{16 + 16}}{8}$$
$$t = \frac{-4 \pm \sqrt{32}}{8}$$
$$t = \frac{-4 \pm 4\sqrt{2}}{8}$$
$$t = \frac{4(-1 \pm \sqrt{2})}{8}$$
$$t = \frac{-1 \pm \sqrt{2}}{2}$$
The solution set is $\left\{\frac{-1 \pm \sqrt{2}}{2}\right\}$.

21. $n + \frac{3}{n} = \frac{19}{4}; n \neq 0$

$4n\left(n + \frac{3}{n}\right) = 4n\left(\frac{19}{4}\right)$

$4n^2 + 12 = 19n$

$4n^2 - 19n + 12 = 0$

$(4n - 3)(n - 4) = 0$

$4n - 3 = 0$ or $n - 4 = 0$

$4n = 3$ or $n = 4$

$n = \frac{3}{4}$

The solution set is $\left\{\frac{3}{4}, 4\right\}$.

23. $\frac{3}{x} + \frac{7}{x-1} = 1; x \neq 0, x \neq 1$

$x(x-1)\left(\frac{3}{x} + \frac{7}{x-1}\right) = x(x-1)(1)$

$x(x-1)\left(\frac{3}{x}\right) + x(x-1)\left(\frac{7}{x-1}\right) = x^2 - x$

$3(x-1) + 7x = x^2 - x$

$3x - 3 + 7x = x^2 - x$

$10x - 3 = x^2 - x$

$0 = x^2 - 11x + 3$

$x = \frac{-(-11) \pm \sqrt{(-11)^2 - 4(1)(3)}}{2(1)}$

$x = \frac{11 \pm \sqrt{121 - 12}}{2}$

$x = \frac{11 \pm \sqrt{109}}{2}$

The solution set is $\left\{\frac{11 \pm \sqrt{109}}{2}\right\}$.

25. $\frac{12}{x-3} + \frac{8}{x} = 14; x \neq 3, x \neq 0$

$x(x-3)\left(\frac{12}{x-3} + \frac{8}{x}\right) = x(x-3)(14)$

$12x + 8(x-3) = 14x^2 - 42x$

$12x + 8x - 24 = 14x^2 - 42x$

$20x - 24 = 14x^2 - 42x$

$0 = 14x^2 - 62x + 24$

$0 = 2(7x^2 - 31x + 12)$

$0 = 2(7x - 3)(x - 4)$

$7x - 3 = 0$ or $x - 4 = 0$

$7x = 3$ or $x = 4$

$x = \frac{3}{7}$ or $x = 4$

The solution set is $\left\{\frac{3}{7}, 4\right\}$.

27. $\frac{3}{x-1} - \frac{2}{x} = \frac{5}{2}; x \neq 0, x \neq 1$

$2x(x-1)\left(\frac{3}{x-1} - \frac{2}{x}\right) = 2x(x-1)\left(\frac{5}{2}\right)$

$2x(x\text{-}1)\left(\frac{3}{x-1}\right) - 2x(x\text{-}1)\left(\frac{2}{x}\right) = 5x(x\text{-}1)$

$6x - 4(x-1) = 5x^2 - 5x$

$6x - 4x + 4 = 5x^2 - 5x$

$0 = 5x^2 - 7x - 4$

$x = \frac{-(-7) \pm \sqrt{(-7)^2 - 4(5)(-4)}}{2(5)}$

$x = \frac{7 \pm \sqrt{49 + 80}}{10}$

$x = \frac{7 \pm \sqrt{129}}{10}$

The solution set is $\left\{\frac{7 \pm \sqrt{129}}{10}\right\}$.

29. $\frac{6}{x} + \frac{40}{x+5} = 7; x \neq 0, x \neq -5$

$x(x+5)\left(\frac{6}{x} + \frac{40}{x+5}\right) = x(x+5)(7)$

$6(x+5) + 40x = 7x(x+5)$

$6x + 30 + 40x = 7x^2 + 35x$

$0 = 7x^2 - 11x - 30$

$0 = (7x + 10)(x - 3)$

$7x + 10 = 0$ or $x - 3 = 0$

$7x = -10$ or $x = 3$

$x = -\frac{10}{7}$ or $x = 3$

The solution set is $\left\{-\frac{10}{7}, 3\right\}$.

31. $\frac{5}{n-3} - \frac{3}{n+3} = 1; n \neq 3, n \neq -3$

$(n\text{-}3)(n\text{+}3)\left(\frac{5}{n\text{-}3} - \frac{3}{n\text{+}3}\right) = (n\text{-}3)(n\text{+}3)(1)$

$5(n+3) - 3(n-3) = n^2 - 9$

$5n + 15 - 3n + 9 = n^2 - 9$

$2n + 24 = n^2 - 9$

$0 = n^2 - 2n - 33$

$n^2 - 2n - 33 = 0$

$n^2 - 2n = 33$

$n^2 - 2n + 1 = 33 + 1$

$(n-1)^2 = 34$
$n - 1 = \pm\sqrt{34}$
$n = 1 \pm \sqrt{34}$
The solution set is $\{1 \pm \sqrt{34}\}$.

33. $x^4 - 18x^2 + 72 = 0$
$(x^2 - 12)(x^2 - 6) = 0$
$x^2 - 12 = 0$ or $x^2 - 6 = 0$
$x^2 = 12$ or $x^2 = 6$
$x = \pm\sqrt{12}$ or $x = \pm\sqrt{6}$
$x = \pm 2\sqrt{3}$ or $x = \pm\sqrt{6}$
The solution set is $\{\pm\sqrt{6}, \pm 2\sqrt{3}\}$.

35. $3x^4 - 35x^2 + 72 = 0$
$(3x^2 - 8)(x^2 - 9) = 0$
$3x^2 - 8 = 0$ or $x^2 - 9 = 0$
$3x^2 = 8$ or $x^2 = 9$
$x^2 = \frac{8}{3}$ or $x = \pm\sqrt{9}$
$x = \pm\sqrt{\frac{8}{3}}$ or $x = \pm 3$
$x = \pm\frac{2\sqrt{2}}{\sqrt{3}}$ or $x = \pm 3$
$x = \pm\frac{2\sqrt{2}}{\sqrt{3}} \bullet \frac{\sqrt{3}}{\sqrt{3}}$ or $x = \pm 3$
$x = \pm\frac{2\sqrt{6}}{3}$ or $x = \pm 3$
The solution set is $\left\{\pm\frac{2\sqrt{6}}{3}, \pm 3\right\}$.

37. $3x^4 + 17x^2 + 20 = 0$
$(3x^2 + 5)(x^2 + 4) = 0$
$3x^2 + 5 = 0$ or $x^2 + 4 = 0$
$3x^2 = -5$ or $x^2 = -4$
$x^2 = -\frac{5}{3}$ or $x = \pm\sqrt{-4}$
$x = \pm\sqrt{-\frac{5}{3}}$ or $x = \pm 2i$
$x = \pm\frac{i\sqrt{5}}{\sqrt{3}} \bullet \frac{\sqrt{3}}{\sqrt{3}}$ or $x = \pm 2i$
$x = \pm\frac{i\sqrt{15}}{3}$ or $x = \pm 2i$
The solution set is $\left\{\pm\frac{i\sqrt{15}}{3}, \pm 2i\right\}$.

39. $6x^2 - 29x^2 + 28 = 0$
$(3x^2 - 4)(2x^2 - 7) = 0$
$3x^2 - 4 = 0$ or $2x^2 - 7 = 0$
$3x^2 = 4$ or $2x^2 = 7$
$x^2 = \frac{4}{3}$ or $x^2 = \frac{7}{2}$
$x = \pm\sqrt{\frac{4}{3}}$ or $x = \pm\sqrt{\frac{7}{2}}$
$x = \pm\frac{2}{\sqrt{3}}$ or $x = \pm\frac{\sqrt{7}}{\sqrt{2}} \bullet \frac{\sqrt{2}}{\sqrt{2}}$
$x = \pm\frac{2}{\sqrt{3}} \bullet \frac{\sqrt{3}}{\sqrt{3}}$ or $x = \pm\frac{\sqrt{14}}{2}$
$x = \pm\frac{2\sqrt{3}}{3}$ or $x = \pm\frac{\sqrt{14}}{2}$
The solution set is $\left\{\pm\frac{2\sqrt{3}}{3}, \pm\frac{\sqrt{14}}{2}\right\}$.

41. Let $x = 1^{\text{st}}$ whole number and
$x + 1 = 2^{\text{nd}}$ whole number.
$x^2 + (x+1)^2 = 145$
$x^2 + x^2 + 2x + 1 = 145$
$2x^2 + 2x - 144 = 0$
$2(x^2 + x - 72) = 0$
$2(x+9)(x-8) = 0$
$x + 9 = 0$ or $x - 8 = 0$
$x = -9$ or $x = 8$
Discard the root $x = -9$.
The numbers are 8 and 9.

43. Let $x = 1^{\text{st}}$ positive integer and
$x - 3 = 2^{\text{nd}}$ positive integer.
$x(x-3) = 108$
$x^2 - 3x = 108$
$x^2 - 3x - 108 = 0$
$(x-12)(x+9) = 0$
$x - 12 = 0$ or $x + 9 = 0$
$x = 12$ or $x = -9$
Discard the root $x = -9$.
The numbers are 12 and 9.

45. Let $x = 1^{\text{st}}$ number and
$10 - x = 2^{\text{nd}}$ number.
$x(10-x) = 22$
$10x - x^2 = 22$
$x^2 - 10x = -22$

$x^2 - 10x + 25 = -22 + 25$
$(x - 5)^2 = 3$
$x - 5 = \pm\sqrt{3}$
$x = 5 \pm \sqrt{3}$

$x_1 = 5 + \sqrt{3}$	$x_1 = 5 - \sqrt{3}$
$x_2 = 10 - (5 + \sqrt{3})$	$x_2 = 10 - (5 - \sqrt{3})$
$x_2 = 10 - 5 - \sqrt{3}$	$x_2 = 10 - 5 + \sqrt{3}$
$x_2 = 5 - \sqrt{3}$	$x_2 = 5 + \sqrt{3}$

The numbers are $5 + \sqrt{3}$ and $5 - \sqrt{3}$.

47.

Numbers	Reciprocal
x	$\frac{1}{x}$
$9 - x$	$\frac{1}{9-x}$

$\frac{1}{x} + \frac{1}{9-x} = \frac{1}{2}$
$2x(9-x)\left(\frac{1}{x} + \frac{1}{9-x}\right) = 2x(9-x)\left(\frac{1}{2}\right)$
$2(9 - x) + 2x = x(9 - x)$
$18 - 2x + 2x = 9x - x^2$
$18 = 9x - x^2$
$x^2 - 9x + 18 = 0$
$(x - 6)(x - 3) = 0$
$x - 6 = 0$ or $x - 3 = 0$
$x = 6$ or $x = 3$
The numbers are 6 and 3.

49. Let $x =$ one leg and
$21 - x =$ other leg.
$x^2 + (21 - x)^2 = 15^2$
$x^2 + 441 - 42x + x^2 = 225$
$2x^2 - 42x + 216 = 0$
$2(x^2 - 21x + 108) = 0$
$2(x - 9)(x - 12) = 0$
$x - 9 = 0$ or $x - 12 = 0$
$x = 9$ or $x = 12$
The legs are 9 inches and 12 inches.

51. Let $x =$ width of walk.

Area of plot and sidewalk
$= (12 + 2x)(20 + 2x)$

Area of plot $= 12(20) = 240$

Area of Sidewalk $=$ Area of Plot and Sidewalk $-$ Area of Plot

$68 = (12 + 2x)(20 + 2x) - 240$
$68 = 240 + 24x + 40x + 4x^2 - 240$
$0 = 4x^2 + 64x - 68$
$0 = 4(x^2 + 16x - 17)$
$0 = 4(x + 17)(x - 1)$
$x + 17 = 0$ or $x - 1 = 0$
$x = -17$ or $x = 1$
Discard the root $x = -17$.
The width of the walk is 1 meter.

53. Let $x =$ the width of the rectangle.
$P = 2l + 2w$
$44 = 2l + 2x$
$44 - 2x = 2l$
Then $22 - x =$ the length.

$x(22 - x) = 112$
$22x - x^2 = 112$
$0 = x^2 - 22x + 112$
$0 = (x - 8)(x - 14)$
$x - 8 = 0$ or $x - 14 = 0$
$x = 8$ or $x = 14$
The dimensions of the rectangle are 8 inches by 14 inches.

55.

	Rate	Time	Distance
Charlotte	$x + 5$	$\frac{250}{x+5}$	250
Lorraine	x	$\frac{180}{x}$	180

$\frac{250}{x+5} = \frac{180}{x} + 1$
$x(x+5)\left(\frac{250}{x+5}\right) = x(x+5)\left(\frac{180}{x} + 1\right)$

$250x = 180(x + 5) + 1x(x + 5)$
$250x = 180x + 900 + x^2 + 5x$
$250x = x^2 + 185x + 900$
$0 = x^2 - 65x + 900$
$0 = (x - 45)(x - 20)$
$x - 45 = 0$ or $x - 20 = 0$
$x = 45$ or $x = 20$
Case 1: Lorraine drives at 45 mph and Charlotte drives at 50 mph.
Case 2: Lorraine drives at 20 mph and Charlotte drives at 25 mph.

Problem Set 11.5

57.

	Rate	Time	Distance
1st Part	x	$\frac{330}{x}$	330
2nd Part	$x+5$	$\frac{240}{x+5}$	240

$$\frac{330}{x}+\frac{240}{x+5}=10$$
$$x(x+5)\left(\frac{330}{x}+\frac{240}{x+5}\right)=10x(x+5)$$
$$330(x+5)+240x=10x^2+50x$$
$$330x+1650+240x=10x^2+50x$$
$$570x+1650=10x^2+50x$$
$$0=10x^2-520x-1650$$
$$0=10(x^2-52x-165)$$
$$0=10(x-55)(x+3)$$
$$x-55=0 \quad \text{or} \quad x+3=0$$
$$x=55 \quad \text{or} \quad x=-3$$
Discard the root $x=-3$.
He drove at 55 mph for the first part.

59.

	Time in hours	Rate
Terry	$x+2$	$\frac{1}{x+2}$
Tom	x	$\frac{1}{x}$

$$\left(\frac{1}{x+2}+\frac{1}{x}\right)3+\left(\frac{1}{x+2}\right)1=1$$
$$\frac{3}{x+2}+\frac{3}{x}+\frac{1}{x+2}=1$$
$$\frac{4}{x+2}+\frac{3}{x}=1$$
$$x(x+2)\left(\frac{4}{x+2}+\frac{3}{x}\right)=x(x+2)(1)$$
$$4x+3(x+2)=x^2+2x$$
$$4x+3x+6=x^2+2x$$
$$0=x^2-5x-6$$
$$0=(x-6)(x+1)$$
$$x-6=0 \quad \text{or} \quad x+1=0$$
$$x=6 \quad \text{or} \quad x=-1$$
Discard the root $x=-1$.
It takes Tom 6 hours and Terry 8 hours.

61.

	rate pay/hour	time in hours	Pay
Anticipated	$\frac{360}{x}$	x	360
Actual	$\frac{360}{x+6}$	$x+6$	360

$$\frac{360}{x}-2=\frac{360}{x+6}$$
$$x(x+6)\left(\frac{360}{x}-2\right)=x(x+6)\left(\frac{360}{x+6}\right)$$
$$360(x+6)-2x(x+6)=360x$$
$$360x+2160-2x^2-12x=360x$$
$$-2x^2-12x+2160=0$$
$$2x^2+12x-2160=0$$
$$2(x^2+6x-1080)=0$$
$$2(x+36)(x-30)=0$$
$$x+36=0 \quad \text{or} \quad x-30=0$$
$$x=-36 \quad \text{or} \quad x=30$$
Discard the root $x=-36$.
He anticipated it would take was 30 hours.

63.

	price/ person	number of persons	price
Group	$\frac{100}{x+2}$	$x+2$	100
Group Less 2	$\frac{100}{x}$	x	100

$$\frac{100}{x+2}+2.50=\frac{100}{x}$$
$$\frac{100}{x+2}+\frac{5}{2}=\frac{100}{x}$$
$$2x(x+2)\left(\frac{100}{x+2}+\frac{5}{2}\right)=2x(x+2)\left(\frac{100}{x}\right)$$
$$2x(100)+5x(x+2)=200(x+2)$$
$$200x+5x^2+10x=200x+400$$
$$5x^2+210x=200x+400$$
$$5x^2+10x-400=0$$
$$5(x^2+2x-80)=0$$
$$5(x+10)(x-8)=0$$
$$x+10=0 \quad \text{or} \quad x-8=0$$
$$x=-10 \quad \text{or} \quad x=8$$
Discard the root $x=-10$.
There were 8 people that actually contributed.

65.

	price/share	number of shares	price
Purchase	$\frac{720}{x}$	x	720
Sell	$\frac{800}{x-20}$	$x-20$	800

$\frac{720}{x}+8=\frac{800}{x-20}$
$x(x-20)\left(\frac{720}{x}+8\right)=x(x-20)\left(\frac{800}{x-20}\right)$
$720(x-20)+8x(x-20)=800x$
$720x-14400+8x^2-160x=800x$
$8x^2+560x-14400=800x$
$8x^2-240x-14400=0$
$8(x^2-30x-1800)=0$
$8(x-60)(x+30)=0$
$x-60=0$ or $x+30=0$
$x=60$ or $x=-30$
Discard the root $x=-30$.
He sold 40 shares at \$20 per share.

67. $S=\frac{n(n+1)}{2}$
$1275=\frac{n(n+1)}{2}$
$2(1275)=2\left[\frac{n(n+1)}{2}\right]$
$2550=n(n+1)$
$2550=n^2+n$
$0=n^2+n-2550$
$0=(n+51)(n-50)$
$n+51=0$ or $n-50=0$
$n=-51$ or $n=50$
Discard the root $n=-51$.
There needs to be 50 consecutive numbers.

69. $A=P(1+r)^t$
$5724.50=5000(1+r)^2$
$1.1449=(1+r)^2$
$(1+r)^2=1.1449$
$1+r=\pm\sqrt{1.1449}$
$r=-1\pm\sqrt{1.1449}$
$r=-1+\sqrt{1.1449}$ or $r=-1-\sqrt{1.1449}$
$r=0.07$ or $r=-2.07$
Discard the root $r=-2.07$. The rate is 7%.

71. $A=P(1+r)^t$
$8988.80=8000(1+r)^2$
$1.1236=(1+r)^2$
$(1+r)^2=1.1236$
$1+r=\pm\sqrt{1.1236}$
$r=\pm\sqrt{1.1236}-1$
$r=0.06$ or $r=-2.06$
Discard $r=-2.06$. The rate is 6%.

Further Investigations

77. $x-9\sqrt{x}+18=0$
Let $y=\sqrt{x}$, then $y^2=x$.
$y^2-9y+18=0$
$(y-6)(y-3)=0$
$y-6=0$ or $y-3=0$
$y=6$ or $y=3$
Substituting $\sqrt{x}$ for y:
$\sqrt{x}=6$ or $\sqrt{x}=3$
$(\sqrt{x})^2=6^2$ or $(\sqrt{x})^2=3^2$
$x=36$ or $x=9$
Checking $x=36$
$36-9\sqrt{36}+18\stackrel{?}{=}0$
$36-54+18\stackrel{?}{=}0$
$0=0$
Checking $x=9$
$9-9\sqrt{9}+18\stackrel{?}{=}0$
$9-27+18\stackrel{?}{=}0$
$0=0$
The solution set is $\{9, 36\}$.

79. $x+\sqrt{x}-2=0$
Let $y=\sqrt{x}$, then $y^2=x$.
$y^2+y-2=0$
$(y+2)(y-1)=0$
$y+2=0$ or $y-1=0$
$y=-2$ or $y=1$
Substituting $\sqrt{x}$ for y:
$\sqrt{x}=-2$ or $\sqrt{x}=1$
$(\sqrt{x})^2=(-2)^2$ or $(\sqrt{x})^2=1^2$
$x=4$ or $x=1$
Checking $x=4$
$4+\sqrt{4}-2\stackrel{?}{=}0$
$4+2-2\stackrel{?}{=}0$
$4\neq 0$

Checking $x = 1$

$1 + \sqrt{1} - 2 \stackrel{?}{=} 0$

$1 + 1 - 2 \stackrel{?}{=} 0$

$0 = 0$

The solution set is $\{1\}$.

81. $6x^{\frac{2}{3}} - 5x^{\frac{1}{3}} - 6 = 0$

Let $y = x^{\frac{1}{3}}$, then $y^2 = x^{\frac{2}{3}}$.

$6y^2 - 5y - 6 = 0$

$(3y + 2)(2y - 3) = 0$

$3y + 2 = 0$ or $2y - 3 = 0$

$3y = -2$ or $2y = 3$

$y = -\frac{2}{3}$ or $y = \frac{3}{2}$

Substituting $x^{\frac{1}{3}}$ for y:

$x^{\frac{1}{3}} = -\frac{2}{3}$ or $x^{\frac{1}{3}} = \frac{3}{2}$

$(x^{\frac{1}{3}})^3 = \left(-\frac{2}{3}\right)^3$ or $(x^{\frac{1}{3}})^3 = \left(\frac{3}{2}\right)^3$

$x = -\frac{8}{27}$ or $x = \frac{27}{8}$

The solution set is $\left\{-\frac{8}{27}, \frac{27}{8}\right\}$.

83. $12x^{-2} - 17x^{-1} - 5 = 0$

Let $y = x^{-1}$, then $y^2 = x^{-2}$.

$12y^2 - 17y - 5 = 0$

$(4y + 1)(3y - 5) = 0$

$4y + 1 = 0$ or $3y - 5 = 0$

$4y = -1$ or $3y = 5$

$y = -\frac{1}{4}$ or $y = \frac{5}{3}$

Substitute x^{-1} for y:

$x^{-1} = -\frac{1}{4}$ or $x^{-1} = \frac{5}{3}$

$\frac{1}{x} = -\frac{1}{4}$ or $\frac{1}{x} = \frac{5}{3}$

$x = -4$ or $x = \frac{3}{5}$

The solution set is $\left\{-4, \frac{3}{5}\right\}$.

PROBLEM SET 11.6 Quadratic and Other Nonlinear Inequalities

1. $(x + 2)(x - 1) > 0$

$(x + 2)(x - 1) = 0$

$x + 2 = 0$ or $x - 1 = 0$

$x = -2$ or $x = 1$

	-2		1
Test Point	-3	0	2
$x + 2$:	negative	positive	positive
$x - 1$:	negative	negative	positive
product:	positive	negative	positive

The solution set is $(-\infty, -2) \cup (1, \infty)$.

-7 -6 -5 -4 -3 -2 -1 0 1 2 3 4 5 6

3. $(x + 1)(x + 4) < 0$

$(x + 1)(x + 4) = 0$

$x + 1 = 0$ or $x + 4 = 0$

$x = -1$ or $x = -4$

	-4		-1
Test Point	-5	-2	0
$x + 1$:	negative	negative	positive
$x + 4$:	negative	positive	positive
product:	positive	negative	positive

The solution set is $(-4, -1)$.

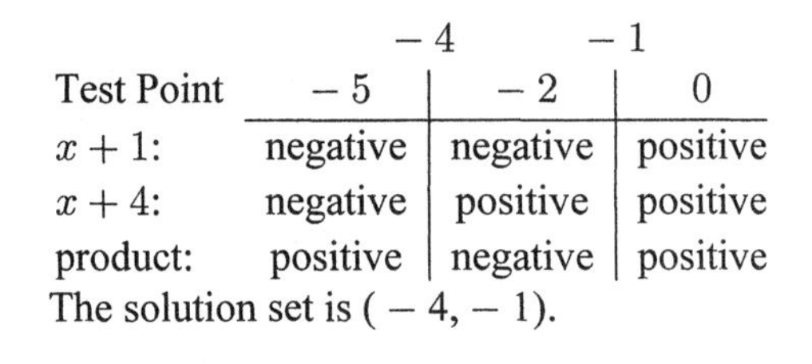

5. $(2x - 1)(3x + 7) \geq 0$

$(2x - 1)(3x + 7) = 0$

$2x - 1 = 0$ or $3x + 7 = 0$

$2x = 1$ or $3x = -7$

$x = \frac{1}{2}$ or $x = -\frac{7}{3}$

Critical points: $-\frac{7}{3}$, $\frac{1}{2}$

Test Point	-3	0	1
$2x-1$:	negative	negative	positive
$3x+7$:	negative	positive	positive
product:	positive	negative	positive

The solution set is $\left(-\infty, -\frac{7}{3}\right] \cup \left[\frac{1}{2}, \infty\right)$.

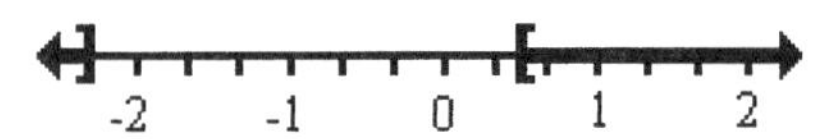

7. $(x+2)(4x-3) \le 0$
$(x+2)(4x-3) = 0$
$x+2=0$ or $4x-3=0$
$x=-2$ or $4x=3$
$x=-2$ or $x=\frac{3}{4}$

Critical points: -2, $\frac{3}{4}$

Test Point	-3	0	1
$x+2$:	negative	positive	positive
$4x-3$:	negative	negative	positive
product:	positive	negative	positive

The solution set is $\left[-2, \frac{3}{4}\right]$.

9. $(x+1)(x-1)(x-3) > 0$
$(x+1)(x-1)(x-3) = 0$
$x+1=0$ or $x-1=0$ or $x-3=0$
$x=-1$ or $x=1$ or $x=3$

Critical points: -1, 1, 3

Test Point	-2	0	2	4
$x+1$:	neg	pos	pos	pos
$x-1$:	neg	neg	pos	pos
$x-3$:	neg	neg	neg	pos
product:	neg	pos	neg	pos

The solution set is $(-1, 1) \cup (3, \infty)$.

11. $x(x+2)(x-4) \le 0$
$x(x+2)(x-4) = 0$
$x=0$ or $x+2=0$ or $x-4=0$
$x=0$ or $x=-2$ or $x=4$

Critical points: -2, 0, 4

Test Point	-3	-1	2	5
x:	neg	neg	pos	pos
$x+2$:	neg	pos	pos	pos
$x-4$:	neg	neg	neg	pos
product:	neg	pos	neg	pos

The solution set is $(-\infty, -2] \cup [0, 4]$.

13. $\frac{x+1}{x-2} > 0;\ x \ne 2$
$x+1=0$ or $x-2=0$
$x=-1$ or $x=2$

Critical points: -1, 2

Test Point	-2	0	3
$x+1$:	negative	positive	positive
$x-2$:	negative	negative	positive
quotient:	positive	negative	positive

The solution set is $(-\infty, -1) \cup (2, \infty)$.

15. $\frac{x-3}{x+2} < 0;\ x \ne -2$
$x-3=0$ or $x+2=0$
$x=3$ or $x=-2$

Critical points: -2, 3

Test Point	-3	0	4
$x-3$:	negative	negative	positive
$x+2$:	negative	positive	positive
quotient:	positive	negative	positive

The solution set is $(-2, 3)$.

Problem Set 11.6

17. $\dfrac{2x-1}{x} \geq 0;\ x \neq 0$

$2x - 1 = 0$ or $x = 0$

$2x = 1$ or $x = 0$

$x = \dfrac{1}{2}$ or $x = 0$

		0	$\frac{1}{2}$
Test Point	-1	$\frac{1}{4}$	1
$2x-1$:	negative	negative	positive
x:	negative	positive	positive
quotient:	positive	negative	positive

The solution set is $(-\infty, 0) \cup \left[\dfrac{1}{2}, \infty\right)$.

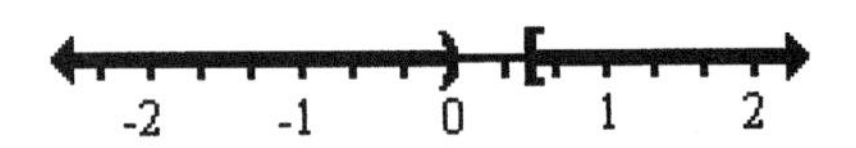

19. $\dfrac{-x+2}{x-1} \leq 0;\ x \neq 1$

$-x + 2 = 0$ or $x - 1 = 0$

$2 = x$ or $x = 1$

		1	2
Test Point	0	$\frac{3}{2}$	3
$-x+2$:	positive	positive	negative
$x-1$:	negative	positive	positive
quotient:	negative	positive	negative

The solution set is $(-\infty, 1) \cup [2, \infty)$.

-7 -6 -5 -4 -3 -2 -1 0 1 2 3 4 5 6

21. $x^2 + 2x - 35 < 0$

$(x+7)(x-5) < 0$

$(x+7)(x-5) = 0$

$x + 7 = 0$ or $x - 5 = 0$

$x = -7$ or $x = 5$

		-7	5
Test Point	-8	0	6
$x+7$:	negative	positive	positive
$x-5$:	negative	negative	positive
product:	positive	negative	positive

The solution sct is $(-7, 5)$.

23. $x^2 - 11x + 28 > 0$

$(x-7)(x-4) > 0$

$(x-7)(x-4) = 0$

$x - 7 = 0$ or $x - 4 = 0$

$x = 7$ or $x = 4$

		4	7
Test Point	0	6	8
$x-7$:	negative	negative	positive
$x-4$:	negative	positive	positive
product:	positive	negative	positive

The solution set is $(-\infty, 4) \cup (7, \infty)$.

25. $3x^2 + 13x - 10 \leq 0$

$(3x-2)(x+5) \leq 0$

$(3x-2)(x+5) = 0$

$3x - 2 = 0$ or $x + 5 = 0$

$3x = 2$ or $x = -5$

$x = \dfrac{2}{3}$ or $x = -5$

		-5	$\frac{2}{3}$
Test Point	-6	0	1
$3x-2$:	negative	negative	positive
$x+5$:	negative	positive	positive
product:	positive	negative	positive

The solution set is $\left[-5, \dfrac{2}{3}\right]$.

27. $8x^2 + 22x + 5 \geq 0$

$(2x+5)(4x+1) \geq 0$

$(2x+5)(4x+1) = 0$

$2x + 5 = 0$ or $4x + 1 = 0$

$2x = -5$ or $4x = -1$

$x = -\dfrac{5}{2}$ or $x = -\dfrac{1}{4}$

		$-\frac{5}{2}$	$-\frac{1}{4}$
Test Point	-3	-1	0
$2x+5$:	negative	positive	positive
$4x+1$:	negative	negative	positive
product:	positive	negative	positive

The solution set is $\left(-\infty, -\dfrac{5}{2}\right] \cup \left[-\dfrac{1}{4}, \infty\right)$.

29. $x(5x - 36) > 32$
$5x^2 - 36x > 32$
$5x^2 - 36x - 32 > 0$
$(5x + 4)(x - 8) > 0$
$(5x + 4)(x - 8) = 0$
$5x + 4 = 0$ or $x - 8 = 0$
$5x = -4$ or $x = 8$
$x = -\frac{4}{5}$ or $x = 8$

$-\frac{4}{5}$ $\quad$ 8

Test Point	-1	0	10
$5x + 4$:	negative	positive	positive
$x - 8$:	negative	negative	positive
product:	positive	negative	positive

The solution set is $\left(-\infty, -\frac{4}{5}\right) \cup (8, \infty)$.

31. $x^2 - 14x + 49 \geq 0$
$(x - 7)(x - 7) \geq 0$
$(x - 7)(x - 7) = 0$
$(x - 7)^2 = 0$
$x - 7 = 0$
$x = 7$

7

Test Point	5	9
$x - 7$:	negative	positive
$x - 7$:	negative	positive
product:	positive	positive

The solution set is $(-\infty, \infty)$.

33. $4x^2 + 20x + 25 \leq 0$
$(2x + 5)(2x + 5) \leq 0$
$(2x + 5)(2x + 5) = 0$
$(2x + 5)^2 = 0$
$2x + 5 = 0$
$2x = -5$
$x = -\frac{5}{2}$

$-\frac{5}{2}$

Test Point	-3	0
$2x + 5$:	negative	positive
$2x + 5$:	negative	positive
product:	positive	positive

The solution set is $\left\{-\frac{5}{2}\right\}$.

35. $(x + 1)(x - 3)^2 > 0$
$(x + 1)(x - 3)(x - 3) > 0$
$(x + 1)(x - 3)^2 = 0$
$x + 1 = 0$ or $x - 3 = 0$
$x = -1$ or $x = 3$

-1 $\quad$ 3

Test Point	-2	0	4
$x + 1$:	negative	positive	positive
$x - 3$:	negative	negative	positive
$x - 3$:	negative	negative	positive
product:	negative	positive	positive

The solution set is $(-1, 3) \cup (3, \infty)$.

37. $\frac{2x}{x + 3} > 4$
$\frac{2x}{x + 3} - 4 > 0$
$\frac{2x}{x + 3} - \frac{4(x + 3)}{x + 3} > 0$
$\frac{2x - 4x - 12}{x + 3} > 0$
$\frac{-2x - 12}{x + 3} > 0; x \neq -3$
$-2x - 12 = 0$ or $x + 3 = 0$
$-2x = 12$ or $x = -3$
$x = -6$ or $x = -3$

-6 $\quad$ -3

Test Point	-7	-5	0
$-2x - 12$:	positive	negative	negative
$x + 3$:	negative	negative	positive
quotient:	negative	positive	negative

The solution set is $(-6, -3)$.

39. $\frac{x - 1}{x - 5} \leq 2$
$\frac{x - 1}{x - 5} - 2 \leq 0$
$\frac{x - 1}{x - 5} - \frac{2(x - 5)}{x - 5} \leq 0$
$\frac{x - 1 - 2(x - 5)}{x - 5} \leq 0$
$\frac{x - 1 - 2x + 10}{x - 5} \leq 0$
$\frac{-x + 9}{x - 5} \leq 0; x \neq 5$

$-x+9=0$ or $x-5=0$
$9=x$ or $x=5$

Test Point	4	6	10
$-x+9$:	positive	positive	negative
$x-5$:	negative	positive	positive
quotient:	negative	positive	negative

(Critical points: 5, 9)

The solution set is $(-\infty, 5) \cup [9, \infty)$.

41. $\frac{x+2}{x-3} > -2$

$\frac{x+2}{x-3} + 2 > 0$

$\frac{x+2}{x-3} + \frac{2(x-3)}{x-3} > 0$

$\frac{x+2+2(x-3)}{x-3} > 0$

$\frac{x+2+2x-6}{x-3} > 0$

$\frac{3x-4}{x-3} > 0; x \neq 3$

$3x-4=0$ or $x-3=0$
$3x=4$ or $x=3$
$x=\frac{4}{3}$ or $x=3$

Test Point	0	2	4
$3x-4$:	negative	positive	positive
$x-3$:	negative	negative	positive
quotient:	positive	negative	positive

(Critical points: $\frac{4}{3}$, 3)

The solution set is $\left(-\infty, \frac{4}{3}\right) \cup (3, \infty)$.

43. $\frac{3x+2}{x+4} \leq 2$

$\frac{3x+2}{x+4} - 2 \leq 0$

$\frac{3x+2}{x+4} - \frac{2(x+4)}{x+4} \leq 0$

$\frac{3x+2-2(x+4)}{x+4} \leq 0$

$\frac{3x+2-2x-8}{x+4} \leq 0$

$\frac{x-6}{x+4} \leq 0; x \neq -4$

$x-6=0$ or $x+4=0$
$x=6$ or $x=-4$

Test Point	-5	0	7
$x-6$:	negative	negative	positive
$x+4$:	negative	positive	positive
quotient:	positive	negative	positive

(Critical points: -4, 6)

The solution set is $(-4, 6]$.

45. $\frac{x+1}{x-2} < 1$

$\frac{x+1}{x-2} - 1 < 0$

$\frac{x+1}{x-2} - \frac{1(x-2)}{x-2} < 0$

$\frac{x+1-1(x-2)}{x-2} < 0$

$\frac{x+1-x+2}{x-2} < 0$

$\frac{3}{x-2} < 0; x \neq 2$

$x-2=0$
$x=2$

Test Point	1	3
3:	positive	positive
$x-2$:	negative	positive
quotient:	negative	positive

(Critical point: 2)

The solution set is $(-\infty, 2)$.

CHAPTER 11 Review Problem Set

1. $(-7+3i)+(9-5i)$
$(-7+9)+(3-5)i$
$2-2i$

2. $(4-10i)-(7-9i)$
$4-10i-7+9i$
$-3-i$

3. $5i(3 - 6i)$
$15i - 30i^2$
$15i - 30(-1)$
$15i + 30$
$30 + 15i$

4. $(5 - 7i)(6 + 8i)$
$30 + 40i - 42i - 56i^2$
$30 - 2\text{i} + 56$
$86 - 2i$

5. $(-2 - 3i)(4 - 8i)$
$-8 + 16i - 12i + 24i^2$
$-8 + 4i - 24$
$-32 + 4i$

6. $(4 - 3i)(4 + 3i)$
$16 + 12i - 12i - 9i^2$
$16 + 9 + 0i$
$25 + 0i$

7. $\frac{4 + 3i}{6 - 2i} = \frac{(4 + 3i)}{(6 - 2i)} \bullet \frac{(6 + 2i)}{(6 + 2i)}$
$\frac{24 + 8i + 18i + 6i^2}{36 - 4i^2} = \frac{24 + 26i - 6}{36 + 4}$
$\frac{18 + 26i}{40} = \frac{18}{40} + \frac{26}{40}i = \frac{9}{20} + \frac{13}{20}i$

8. $\frac{-1 - i}{2 + 5i} =$
$\frac{(-1 - i)}{(-2 + 5i)} \bullet \frac{(-2 - 5i)}{(-2 - 5i)} = \frac{2 + 7i + 5i^2}{4 - 25i^2}$
$\frac{2 + 7i - 5}{4 + 25} = \frac{-3 + 7i}{29} = -\frac{3}{29} + \frac{7}{29}i$

9. $4x^2 - 20x + 25 = 0$
$b^2 - 4ac$
$(-20)^2 - 4(4)(25)$
$400 - 400 = 0$
Since $b^2 - 4ac = 0$, there will be one real solution with multiplicity of two.

10. $5x^2 - 7x + 31 = 0$
$b^2 - 4ac$
$(-7)^2 - 4(5)(31) = -571$
Since $b^2 - 4ac < 0$, there will be two nonreal complex solutions.

11. $7x^2 - 2x - 14 = 0$
$b^2 - 4ac$
$(-2)^2 - 4(7)(-14)$
$4 + 392 = 396$
Since $b^2 - 4ac > 0$, there will be two unequal real solutions.

12. $5x^2 - 2x = 4$
$5x^2 - 2x - 4 = 0$
$b^2 - 4ac$
$(-2)^2 - 4(5)(-4)$
$4 + 80 = 84$
Since $b^2 - 4ac > 0$, there will be two unequal real solutions.

13. $x^2 - 17x = 0$
$x(x - 17) = 0$
$x = 0$ or $x - 17 = 0$
$x = 0$ or $x = 17$
The solution set is $\{0, 17\}$.

14. $(x - 2)^2 = 36$
$x - 2 = \pm 6$
$x = 2 + 6$ or $x = 2 - 6$
$x = 8$ or $x = -4$
The solution set is $\{-4, 8\}$.

15. $(2x - 1)^2 = -64$
$2x - 1 = \pm\sqrt{-64}$
$2x - 1 = \pm 8i$
$2x = 1 \pm 8i$
$x = \frac{1 \pm 8i}{2}$
The solution set is $\left\{\frac{1 \pm 8i}{2}\right\}$.

16. $x^2 - 4x - 21 = 0$
$(x - 7)(x + 3) = 0$
$x - 7 = 0$ or $x + 3 = 0$
$x = 7$ or $x = -3$
The solution set is $\{-3, 7\}$.

17. $x^2 + 2x - 9 = 0$
$x^2 + 2x = 9$
$x^2 + 2x + 1 = 9 + 1$
$(x + 1)^2 = 10$
$x + 1 = \pm\sqrt{10}$

$x = -1 \pm \sqrt{10}$
The solution set is $\{-1 \pm \sqrt{10}\}$.

18. $x^2 - 6x = -34$
$x^2 - 6x + 9 = -34 + 9$
$(x-3)^2 = -25$
$x - 3 = \pm\sqrt{-25}$
$x - 3 = \pm 5i$
$x = 3 \pm 5i$
The solution set is $\{3 \pm 5i\}$.

19. $4\sqrt{x} = x - 5$
$(4\sqrt{x})^2 = (x-5)^2$
$16x = x^2 - 10x + 25$
$0 = x^2 - 26x + 25$
$0 = (x-25)(x-1)$
$x - 25 = 0$ or $x - 1 = 0$
$x = 25$ or $x = 1$
Checking $x = 25$
$4\sqrt{25} \stackrel{?}{=} 25 - 5$
$4(5) \stackrel{?}{=} 20$
$20 = 20$
Checking $x = 1$
$4\sqrt{1} \stackrel{?}{=} 1 - 5$
$4(1) \stackrel{?}{=} -4$
$4 \neq -4$
The solution set is $\{25\}$.

20. $3n^2 + 10n - 8 = 0$
$(3n-2)(n+4) = 0$
$3n - 2 = 0$ or $n + 4 = 0$
$3n = 2$ or $n = -4$
$n = \frac{2}{3}$
The solution set is $\left\{-4, \frac{2}{3}\right\}$.

21. $n^2 - 10n = 200$
$n^2 - 10n - 200 = 0$
$(n-20)(n+10) = 0$
$n - 20 = 0$ or $n + 10 = 0$
$n = 20$ or $n = -10$
The solution set is $\{-10, 20\}$.

22. $3a^2 + a - 5 = 0$
$a = \frac{-1 \pm \sqrt{1^2 - 4(3)(-5)}}{2(3)}$
$a = \frac{-1 \pm \sqrt{1 + 60}}{6}$
$a = \frac{-1 \pm \sqrt{61}}{6}$
The solution set is $\left\{\frac{-1 \pm \sqrt{61}}{6}\right\}$.

23. $x^2 - x + 3 = 0$
$x = \frac{-1(-1) \pm \sqrt{(-1)^2 - 4(1)(3)}}{2(1)}$
$x = \frac{1 \pm \sqrt{1 - 12}}{2}$
$x = \frac{1 \pm \sqrt{-11}}{2}$
$x = \frac{1 \pm i\sqrt{11}}{2}$
The solution set is $\left\{\frac{-1 \pm i\sqrt{11}}{2}\right\}$.

24. $2x^2 - 5x + 6 = 0$
$x = \frac{-(-5) \pm \sqrt{(-5)^2 - 4(2)(6)}}{2(2)}$
$x = \frac{5 \pm \sqrt{25 - 48}}{4}$
$x = \frac{5 \pm \sqrt{-23}}{4}$
$x = \frac{5 \pm i\sqrt{23}}{4}$
The solution set is $\left\{\frac{5 \pm i\sqrt{23}}{4}\right\}$.

25. $2a^2 + 4a - 5 = 0$
$a = \frac{-4 \pm \sqrt{4^2 - 4(2)(-5)}}{2(2)}$
$a = \frac{-4 \pm \sqrt{16 + 40}}{4}$
$a = \frac{-4 \pm \sqrt{56}}{4}$
$a = \frac{-4 \pm 2\sqrt{14}}{4}$
$a = \frac{2(-2 \pm \sqrt{14})}{4}$

$a = \frac{-2 \pm \sqrt{14}}{2}$

The solution set is $\left\{ \frac{-2 \pm \sqrt{14}}{2} \right\}$.

26. $t(t+5) = 36$
$t^2 + 5t = 36$
$t^2 + 5t - 36 = 0$
$(t+9)(t-4) = 0$
$t + 9 = 0$ or $t - 4 = 0$
$t = -9$ or $t = 4$
The solution set is $\{-9, 4\}$.

27. $x^2 + 4x + 9 = 0$
$x^2 + 4x = -9$
$x^2 + 4x + 4 = -9 + 4$
$(x+2)^2 = -5$
$x + 2 = \pm\sqrt{-5}$
$x + 2 = \pm i\sqrt{5}$
$x = -2 \pm i\sqrt{5}$
The solution set is $\{-2 \pm i\sqrt{5}\}$.

28. $(x-4)(x-2) = 80$
$x^2 - 6x + 8 = 80$
$x^2 - 6x - 72 = 0$
$(x-12)(x+6) = 0$
$x - 12 = 0$ or $x + 6 = 0$
$x = 12$ or $x = -6$
The solution set is $\{-6, 12\}$.

29. $\frac{3}{x} + \frac{2}{x+3} = 1; x \neq 0, x \neq -3$
$x(x+3)\left(\frac{3}{x} + \frac{2}{x+3}\right) = x(x+3)(1)$
$3(x+3) + 2x = x(x+3)$
$3x + 9 + 2x = x^2 + 3x$
$0 = x^2 - 2x - 9$
$x^2 - 2x = 9$
$x^2 - 2x + 1 = 9 + 1$
$(x-1)^2 = 10$
$x - 1 = \pm\sqrt{10}$
$x = 1 \pm \sqrt{10}$
The solution set is $\{1 \pm \sqrt{10}\}$.

30. $2x^4 - 23x^2 + 56 = 0$
$(2x^2 - 7)(x^2 - 8) = 0$
$2x^2 - 7 = 0$ or $x^2 - 8 = 0$
$2x^2 = 7$ or $x^2 = 8$
$x^2 = \frac{7}{2}$ or $x = \pm\sqrt{8}$
$x = \pm\sqrt{\frac{7}{2}}$ or $x = \pm 2\sqrt{2}$
$x = \pm\frac{\sqrt{7}}{\sqrt{2}} \bullet \frac{\sqrt{2}}{\sqrt{2}}$
$x = \pm\frac{\sqrt{14}}{2}$
The solution set is $\left\{\pm\frac{\sqrt{14}}{2}, \pm 2\sqrt{2}\right\}$.

31. $\frac{3}{n-2} = \frac{n+5}{4}; n \neq 2$
$3(4) = (n-2)(n+5)$
$12 = n^2 + 3n - 10$
$0 = n^2 + 3n - 22$
$n = \frac{-3 \pm \sqrt{3^2 - 4(1)(-22)}}{2(1)}$
$n = \frac{-3 \pm \sqrt{9 + 88}}{2}$
$n = \frac{-3 \pm \sqrt{97}}{2}$
The solution set is $\left\{\frac{-3 \pm \sqrt{97}}{2}\right\}$.

32. $x^2 + 3x - 10 > 0$
$x^2 + 3x - 10 = 0$
$(x+5)(x-2) = 0$
$x + 5 = 0$ or $x - 2 = 0$
$x = -5$ or $x = 2$

Test Point	-6	0	3
$x+5$:	negative	positive	positive
$x-2$:	negative	negative	positive
product:	positive	negative	positive

(Boundaries: -5 between -6 and 0; 2 between 0 and 3.)

The solution set is $(-\infty, -5) \cup (2, \infty)$.

33. $2x^2 + x - 21 \leq 0$
$(2x+7)(x-3) \leq 0$
$(2x+7)(x-3) = 0$
$2x + 7 = 0$ or $x - 3 = 0$

$2x = -7$ or $x = 3$

$x = -\frac{7}{2}$ or $x = 3$

$-\frac{7}{2}$ ... 3

Test Point	-4	0	4
$2x + 7$:	negative	positive	positive
$x - 3$:	negative	negative	positive
product:	positive	negative	positive

The solution set is $\left[-\frac{7}{2}, 3\right]$.

-7 -6 -5 -4 -3 -2 -1 0 1 2 3 4 5 6

34. $\frac{x-4}{x+6} \geq 0; x \neq -6$

$x - 4 = 0$ $\quad$ $x + 6 = 0$

$x = 4$ $\quad$ $x = -6$

-6 ... 4

Test Point	-7	0	5
$x - 4$:	negative	negative	positive
$x + 6$:	negative	positive	positive
quotient:	positive	negative	positive

The solution set is $(-\infty, -6) \cup [4, \infty)$.

-7 -6 -5 -4 -3 -2 -1 0 1 2 3 4 5 6

35. $\frac{2x-1}{x+1} > 4$

$\frac{2x-1}{x+1} - 4 > 0$

$\frac{2x-1}{x+1} - \frac{4(x+1)}{x+1} > 0$

$\frac{2x-1-4(x+1)}{x+1} > 0$

$\frac{2x-1-4x-4}{x+1} > 0$

$\frac{-2x-5}{x+1} > 0, x \neq -1$

$-2x - 5 = 0$ or $x + 1 = 0$

$-2x = 5$ or $x = -1$

$x = -\frac{5}{2}$

$-\frac{5}{2}$ ... -1

Test Point	-3	-2	0
$-2x - 5$:	positive	negative	negative
$x + 1$:	negative	negative	positive
quotient:	negative	positive	negative

The solution set is $\left(-\frac{5}{2}, -1\right)$.

-7 -6 -5 -4 -3 -2 -1 0 1 2 3 4 5 6

36. Let $x =$ one number and
$6 - x =$ other number.

$x(6 - x) = 2$

$6x - x^2 = 2$

$0 = x^2 - 6x + 2$

$x^2 - 6x = -2$

$x^2 - 6x + 9 = -2 + 9$

$(x - 3)^2 = 7$

$x - 3 = \pm\sqrt{7}$

$x = 3 \pm \sqrt{7}$

The numbers are $3 + \sqrt{7}$ and $3 - \sqrt{7}$.

37.

	price/ share	number of shares	Cost
Purchase	$\frac{250}{x}$	x	250
Sell	$\frac{300}{x-5}$	$x - 5$	300

$\frac{250}{x} + 5 = \frac{300}{x-5}$

$x(x-5)\left(\frac{250}{x} + 5\right) = x(x-5)\left(\frac{300}{x-5}\right)$

$250(x - 5) + 5x(x - 5) = 300x$

$250x - 1250 + 5x^2 - 25x = 300x$

$5x^2 + 225x - 1250 = 300x$

$5x^2 - 75x - 1250 = 0$

$5(x^2 - 15x - 250) = 0$

$5(x - 25)(x + 10) = 0$

$x - 25 = 0$ or $x + 10 = 0$

$x = 25$ or $x = -10$

Discard the root $x = -10$.

She sold 20 shares at $15 per share.

38.

	Rate	Time	Distance
Dave	x	$\frac{270}{x}$	270
Sandy	$x+7$	$\frac{260}{x+7}$	260

$\frac{270}{x} = \frac{260}{x+7} + 1$

$x(x+7)\left(\frac{270}{x}\right) = x(x+7)\left(\frac{260}{x+7} + 1\right)$

$270(x+7) = 260x + x(x+7)$

$270x + 1890 = 260x + x^2 + 7x$

$0 = x^2 - 3x - 1890$

$0 = (x-45)(x+42)$

$x - 45 = 0$ or $x + 42 = 0$

$x = 45$ or $x = -42$

Discard the root $x = -42$.

Dave traveled 45 mp; Sandy traveled 52 mph.

39. Let $s =$ length of a side.

Area of square $= s^2$

Perimeter of square $= 4s$

$s^2 = 2(4s)$

$s^2 = 8s$

$s^2 - 8s = 0$

$s(s-8) = 0$

$s = 0$ or $s - 8 = 0$

$s = 0$ or $s = 8$

Discard the root $s = 0$.

The length of the side of the square is 8 units.

40. Let $x =$ 1st even whole number and

$x + 2 =$ 2nd even whole number.

$x^2 + (x+2)^2 = 164$

$x^2 + x^2 + 4x + 4 = 164$

$2x^2 + 4x - 160 = 0$

$2(x^2 + 2x - 80) = 0$

$2(x+10)(x-8) = 0$

$x + 10 = 0$ or $x - 8 = 0$

$x = -10$ or $x = 8$

Discard the root $x = -10$.

The whole numbers are 8 and 10.

41. Let $x =$ the width.

$P = 2L + 2W$

$38 = 2L + 2x$

$38 - 2x = 2L$

$\frac{38-2x}{2} = L$

$19 - x = L$

Then $19 - x =$ the length.

$A = LW$

$84 = (19 - x)x$

$84 = 19x - x^2$

$0 = -x^2 + 19x - 84$

$0 = x^2 - 19x + 84$

$0 = (x-12)(x-7)$

$x - 12 = 0$ or $x - 7 = 0$

$x = 12$ or $x = 7$

The dimensions of the rectangle are 7 inches by 12 inches.

42.

	Time in Hours	Rate
Billy	$x+2$	$\frac{1}{x+2}$
Reena	x	$\frac{1}{x}$

$\left(\frac{1}{x+2}\right)(2) + \frac{1}{x}(2) + \left(\frac{1}{x+2}\right)1 = 1$

$\frac{2}{x+2} + \frac{2}{x} + \frac{1}{x+2} = 1$

$\frac{3}{x+2} + \frac{2}{x} = 1$

$x(x+2)\left(\frac{3}{x+2} + \frac{2}{x}\right) = x(x+2)(1)$

$3x + 2(x+2) = x(x+2)$

$3x + 2x + 4 = x^2 + 2x$

$0 = x^2 - 3x + 4$

$0 = (x-4)(x+1)$

$x - 4 = 0$ or $x + 1 = 0$

$x = 4$ or $x = -1$

Discard the root $x = -1$.

It would take Reena 4 hours and it would take Billy 6 hours.

43. Let $x =$ the width of the strip.

Area of the 40×60 lot $= 2400$ sq. m.

width of new lot $= x + 40$

length of new lot $= x + 60$

$(x+40)(x+60) = 2400 + 1100$

$x^2 + 100x + 2400 = 3500$

$x^2 + 100x - 1100 = 0$

$(x-10)(x+110) = 0$

$x - 10 = 0$ or $x + 110 = 0$

$x = 10$ or $x = -110$

Discard the root $x = -110$.

The width of the strip should be 10 meters.

CHAPTER 11 Test

1. $(3-4i)(5+6i)$
$15+18i-20i-24i^2$
$15-2i+24$
$39-2i$

2. $\dfrac{2-3i}{3+4i}=\dfrac{(2-3i)}{(3+4i)}\bullet\dfrac{(3-4i)}{(3-4i)}=$
$\dfrac{6-17i+12i^2}{9-16i^2}=\dfrac{6-17i-12}{9+16}=$
$\dfrac{-6-17i}{25}=-\dfrac{6}{25}-\dfrac{17}{25}i$

3. $x^2=7x$
$x^2-7x=0$
$x(x-7)=0$
$x=0$ or $x-7=0$
$x=0$ or $x=7$
The solution set is $\{0, 7\}$.

4. $(x-3)^2=16$
$x-3=\pm\sqrt{16}$
$x-3=\pm 4$
$x=3\pm 4$
$x=3+4$ or $x=3-4$
$x=7$ or $x=-1$
The solution set is $\{-1, 7\}$.

5. $x^2+3x-18=0$
$(x+6)(x-3)=0$
$x+6=0$ or $x-3=0$
$x=-6$ or $x=3$
The solution set is $\{-6, 3\}$.

6. $x^2-2x-1=0$
$x^2-2x=1$
$x^2-2x+1=1+1$
$(x-1)^2=2$
$x-1=\pm\sqrt{2}$
$x=1\pm\sqrt{2}$
The solution set is $\{1\pm\sqrt{2}\}$.

7. $5x^2-2x+1=0$
$x=\dfrac{-(-2)\pm\sqrt{(-2)^2-4(5)(1)}}{2(5)}$
$x=\dfrac{2\pm\sqrt{4-20}}{10}$
$x=\dfrac{2\pm\sqrt{-16}}{10}$
$x=\dfrac{2\pm 4i}{10}$
$x=\dfrac{2(1\pm 2i)}{10}$
$x=\dfrac{1\pm 2i}{5}$
The solution set is $\left\{\dfrac{1\pm 2i}{5}\right\}$.

8. $x^2+30x=-224$
$x^2+30x+224=0$
$(x+16)(x+14)=0$
$x+16=0$ or $x+14=0$
$x=-16$ or $x=-14$
The solution set is $\{-16, -14\}$.

9. $(3x-1)^2+36=0$
$(3x-1)^2=-36$
$3x-1=\pm\sqrt{-36}$
$3x-1=\pm 6i$
$3x=1\pm 6i$
$x=\dfrac{1\pm 6i}{3}$
The solution set is $\left\{\dfrac{1\pm 6i}{3}\right\}$.

10. $(5x-6)(4x+7)=0$
$5x-6=0$ or $4x+7=0$
$5x=6$ or $4x=-7$
$x=\dfrac{6}{5}$ or $x=-\dfrac{7}{4}$
The solution set is $\left\{-\dfrac{7}{4}, \dfrac{6}{5}\right\}$.

11. $(2x+1)(3x-2)=55$
$6x^2-x-2=55$
$6x^2-x-57=0$
$(6x-19)(x+3)=0$
$6x-19=0$ or $x+3=0$
$6x=19$ or $x=-3$

$x = \frac{19}{6}$ or $x = -3$

The solution set is $\left\{-3, \frac{19}{6}\right\}$.

12. $n(3n - 2) = 40$
$3n^2 - 2n = 40$
$3n^2 - 2n - 40 = 0$
$(3n + 10)(n - 4) = 0$
$3n + 10 = 0$ or $n - 4 = 0$
$3n = -10$ or $n = 4$
$n = -\frac{10}{3}$
The solution set is $\left\{-\frac{10}{3}, 4\right\}$.

13. $x^4 + 12x^2 - 64 = 0$
$(x^2 + 16)(x^2 - 4) = 0$
$x^2 + 16 = 0$ or $x^2 - 4 = 0$
$x^2 = -16$ or $x^2 = 4$
$x = \pm\sqrt{-16}$ or $x = \pm\sqrt{4}$
$x = \pm 4i$ or $x = \pm 2$
The solution set is $\{\pm 2, \pm 4i\}$.

14. $\frac{3}{x} + \frac{2}{x+1} = 4; x \neq 0, x \neq -1$
$x(x+1)\left(\frac{3}{x} + \frac{2}{x+1}\right) = x(x+1)(4)$
$3(x+1) + 2x = 4x(x+1)$
$3x + 3 + 2x = 4x^2 + 4x$
$0 = 4x^2 - x - 3$
$0 = (4x + 3)(x - 1)$
$4x + 3 = 0$ or $x - 1 = 0$
$4x = -3$ or $x = 1$
$x = -\frac{3}{4}$
The solution set is $\left\{-\frac{3}{4}, 1\right\}$.

15. $3x^2 - 2x - 3 = 0$
$x = \frac{-(-2) \pm \sqrt{(-2)^2 - 4(3)(-3)}}{2(3)}$
$x = \frac{2 \pm \sqrt{4 + 36}}{6}$
$x = \frac{2 \pm \sqrt{40}}{6}$
$x = \frac{2 \pm 2\sqrt{10}}{6}$
$x = \frac{2(1 \pm \sqrt{10})}{6}$
$x = \frac{1 \pm \sqrt{10}}{3}$
The solution set is $\left\{\frac{1 \pm \sqrt{10}}{3}\right\}$.

16. $4x^2 + 20x + 25 = 0$
$b^2 - 4ac$
$(20)^2 - 4(4)(25)$
$400 - 400 = 0$
Since $b^2 - 4ac = 0$, then there will be two equal real solutions.

17. $4x^2 - 3x = -5$
$4x^2 - 3x + 5 = 0$
$b^2 - 4ac$
$(-3)^2 - 4(4)(5)$
$9 - 80 = -71$
Since $b^2 - 4ac < 0$, there will be two nonreal complex solutions.

18. $x^2 - 3x - 54 \leq 0$
$(x - 9)(x + 6) \leq 0$
$(x - 9)(x + 6) = 0$
$x - 9 = 0$ or $x + 6 = 0$
$x = 9$ or $x = -6$

	-6		9
Test Point	-8	0	10
$x - 9$:	negative	negative	positive
$x + 6$:	negative	positive	positive
product:	positive	negative	positive

The solution set is $[-6, 9]$.

19. $\frac{3x - 1}{x + 2} > 0; x \neq -2$
$3x - 1 = 0$ or $x + 2 = 0$
$3x = 1$ or $x = -2$
$x = \frac{1}{3}$ or $x = -2$

Test Point	-3	0	1
	-2		$\frac{1}{3}$
$3x-1$:	negative	negative	positive
$x+2$:	negative	positive	positive
quotient:	positive	negative	positive

The solution set is $(-\infty, -2) \cup \left(\frac{1}{3}, \infty\right)$.

20. $\dfrac{x-2}{x+6} \geq 3$

$\dfrac{x-2}{x+6} - 3 \geq 0$

$\dfrac{x-2}{x+6} - \dfrac{3(x+6)}{x+6} \geq 0$

$\dfrac{x-2-3(x+6)}{x+6} \geq 0$

$\dfrac{x-2-3x-18}{x+6} \geq 0$

$\dfrac{-2x-20}{x+6} \geq 0; x \neq -6$

$-2x-20=0 \qquad x+6=0$
$-2x=20 \qquad x=-6$
$x=-10$

Test Point	-11	-8	0
	-10		-6
$-2x-20$:	positive	negative	negative
$x+6$:	negative	negative	positive
quotient:	negative	positive	negative

The solution set is $[-10, 6)$.

21.

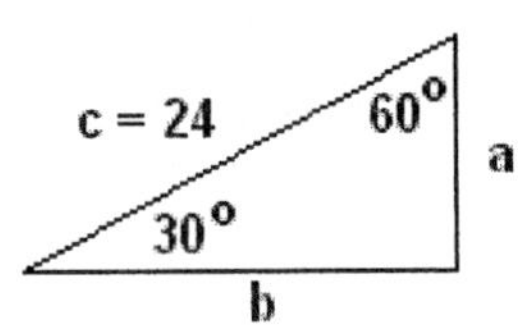

Let a be the side opposite the $30°$ angle, then $a = \frac{1}{2}(24) = 12$.

$a^2 + b^2 = c^2$
$12^2 + b^2 = 24^2$
$144 + b^2 = 576$
$b^2 = 432$
$b = \sqrt{432}$
$b = 20.8$ feet
The ladder reaches 20.8 feet up the building.

22. $a^2 + b^2 = c^2$
$16^2 + 24^2 = c^2$
$256 + 576 = c^2$
$832 = c^2$
$\sqrt{832} = c$
$29 \approx c$
The diagonal is 29 meters.

23.

	price/share	number of shares	Cost
Purchase	$\frac{3000}{x}$	x	3000
Sell	$\frac{3000}{x-50}$	$x-50$	3000

$\dfrac{3000}{x} + 5 = \dfrac{3000}{x-50}$

$x(x-50)\left(\dfrac{3000}{x} + 5\right) = x(x-50)\left(\dfrac{3000}{x-50}\right)$

$3000(x-50) + 5x(x-50) = 3000x$
$3000x - 150{,}000 + 5x^2 - 250x = 3000x$
$5x^2 + 2750x - 150{,}000 = 3000x$
$5x^2 - 250x - 150{,}000 = 0$
$5(x^2 - 50x - 30{,}000) = 0$
$5(x-200)(x+150) = 0$
$x - 200 = 0 \quad$ or $\quad x + 150 = 0$
$x = 200 \quad$ or $\quad x = -150$
Discard the root $x = -150$.
Dana sold $200 - 50 = 150$ shares.

24. Let x = the width.
$P = 2L + 2W$
$41 = 2L + 2x$
$41 - 2x = 2L$
$\dfrac{41-2x}{2} = L$
$20.5 - x = L$
Then $20.5 - x$ = the length.
$A = LW$
$91 = (20.5 - x)x$
$91 = 20.5x - x^2$
$0 = -x^2 + 20.5x - 91$
$0 = x^2 - 20.5x + 91$

$2(0) = 2(x^2 - 20.5x + 91)$
$0 = 2x^2 - 41x + 182$
$0 = (2x - 13)(x - 14)$
$2x - 13 = 0$ or $x - 14 = 0$
$2x = 13$ or $x = 14$
$x = \frac{13}{2} = 6\frac{1}{2}$
The shortest side is $6\frac{1}{2}$ inches.

25. Let $x =$ one number and
$6 - x =$ other number.
$x(6 - x) = 4$
$6x - x^2 = 4$
$-x^2 + 6x = 4$
$x^2 - 6x = -4$
$x^2 - 6x + 9 = -4 + 9$
$(x - 3)^2 = 5$
$x - 3 = \pm\sqrt{5}$
$x = 3 \pm \sqrt{5}$
The larger number is $3 + \sqrt{5}$.

CHAPTERS 1-11 | Cumulative Review

1. $\frac{4a^2b^3}{12a^3b} = \frac{b^2}{3a}$ for $a = 5, b = -8$

$\frac{(-8)^2}{3(5)} = \frac{64}{15}$

2. $\frac{\frac{1}{x} + \frac{1}{y}}{\frac{1}{x} - \frac{1}{y}}$

For $x = 4$ and $y = 7$,

$\frac{\frac{1}{4} + \frac{1}{7}}{\frac{1}{4} - \frac{1}{7}} = \frac{28}{28} \bullet \frac{\left(\frac{1}{4} + \frac{1}{7}\right)}{\left(\frac{1}{4} - \frac{1}{7}\right)}$

$\frac{28\left(\frac{1}{4}\right) + 28\left(\frac{1}{7}\right)}{28\left(\frac{1}{4}\right) - 28\left(\frac{1}{7}\right)} = \frac{7 + 4}{7 - 4} = \frac{11}{3}$

3. $\frac{3}{n} + \frac{5}{2n} - \frac{4}{3n} =$

$\frac{3}{n} \bullet \frac{6}{6} + \frac{5}{2n} \bullet \frac{3}{3} - \frac{4}{3n} \bullet \frac{2}{2} =$

$\frac{18}{6n} + \frac{15}{6n} - \frac{8}{6n} = \frac{25}{6n}$; for $n = 25$

$\frac{25}{6(25)} = \frac{1}{6}$

4. $\frac{4}{x - 1} - \frac{2}{x + 2}$

$\frac{4}{\frac{1}{2} - 1} - \frac{2}{\frac{1}{2} + 2}$, for $x = \frac{1}{2}$

$\frac{4}{-\frac{1}{2}} - \frac{2}{\frac{5}{2}}$

$4\left(\frac{-2}{1}\right) - 2\left(\frac{2}{5}\right)$

$-8 - \frac{4}{5}$

$-8\frac{4}{5} = -\frac{44}{5}$

5. $2\sqrt{2x + y} - 5\sqrt{3x - y}$
For $x = 5$ and $y = 6$,
$2\sqrt{2(5) + 6} - 5\sqrt{3(5) - 6}$
$2\sqrt{16} - 5\sqrt{9}$
$2(4) - 5(3)$
$8 - 15$
-7

6. $(3a^2b)(-2ab)(4ab^3)$
$-24a^{2+1+1}b^{1+1+3}$
$-24a^4b^5$

7. $(x + 3)(2x^2 - x - 4)$
$2x^3 - x^2 - 4x + 6x^2 - 3x - 12$
$2x^3 + 5x^2 - 7x - 12$

8. $\dfrac{6xy^2}{14y} \bullet \dfrac{7x^2y}{8x}$

$\dfrac{6(7)x^3y^3}{(8)(14)xy} = \dfrac{6x^2y^2}{16} = \dfrac{3x^2y^2}{8}$

9. $\dfrac{a^2+6a-40}{a^2-4a} \div \dfrac{2a^2+19a-10}{a^3+a^2}$

$\dfrac{a^2+6a-40}{a^2-4a} \bullet \dfrac{a^3+a^2}{2a^2+19a-10}$

$\dfrac{(a+10)(a-4)}{a(a-4)} \bullet \dfrac{a^2(a+1)}{(2a-1)(a+10)}$

$\dfrac{a(a+1)}{(2a-1)}$

10. $\dfrac{3x+4}{6} - \dfrac{5x-1}{9}$; LCD $= 18$

$\dfrac{(3x+4)}{6} \bullet \dfrac{3}{3} - \dfrac{(5x-1)}{9} \bullet \dfrac{2}{2}$

$\dfrac{3(3x+4)-2(5x-1)}{18}$

$\dfrac{9x+12-10x+2}{18} = \dfrac{-x+14}{18}$

11. $\dfrac{4}{x^2+3x} + \dfrac{5}{x}$

$\dfrac{4}{x(x+3)} + \dfrac{5}{x}$; LCD $= x(x+3)$

$\dfrac{4}{x(x+3)} + \dfrac{5}{x} \bullet \dfrac{(x+3)}{(x+3)}$

$\dfrac{4+5(x+3)}{x(x+3)} = \dfrac{4+5x+15}{x(x+3)}$

$\dfrac{5x+19}{x(x+3)}$

12. $\dfrac{3n^2+n}{n^2+10n+16} \bullet \dfrac{2n^2-8}{3n^3-5n^2-2n}$

$\dfrac{n(3n+1)}{(n+8)(n+2)} \bullet \dfrac{2(n+2)(n-2)}{n(3n+1)(n-2)}$

$\dfrac{2}{n+8}$

13. $\dfrac{3}{5x^2+3x-2} - \dfrac{2}{5x^2-22x+8}$

$\dfrac{3}{(5x-2)(x+1)} - \dfrac{2}{(5x-2)(x-4)}$

LCD $= (5x-2)(x+1)(x-4)$

$\dfrac{3}{(5x-2)(x+1)} \bullet \dfrac{(x-4)}{(x-4)} - \dfrac{2}{(5x-2)(x-4)} \bullet \dfrac{(x+1)}{(x+1)}$

$\dfrac{3(x-4)-2(x+1)}{(5x-2)(x+1)(x-4)}$

$\dfrac{3x-12-2x-2}{(5x-2)(x+1)(x-4)}$

$\dfrac{x-14}{(5x-2)(x+1)(x-4)}$

14.

$$
\begin{array}{r|l}
 & y^2 - 5y + 6 \\
\hline
y-2 & y^3 - 7y^2 + 16y - 12 \\
 & y^3 - 2y^2 \\
\hline
 & -5y^2 + 16y \\
 & -5y^2 + 10y \\
\hline
 & 6y - 12 \\
 & 6y - 12 \\
\hline
 & 0
\end{array}
$$

$y^2 - 5y + 6$

15.

$$
\begin{array}{r|l}
 & x^2 - 3x - 2 \\
\hline
4x-5 & 4x^3 - 17x^2 + 7x + 10 \\
 & 4x^3 - 5x^2 \\
\hline
 & -12x^2 + 7x \\
 & -12x^2 + 15x \\
\hline
 & -8x + 10 \\
 & -8x + 10 \\
\hline
 & 0
\end{array}
$$

$x^2 - 3x - 2$

16. $(3\sqrt{2}+2\sqrt{5})(5\sqrt{2}-\sqrt{5})$

$15\sqrt{4} - 3\sqrt{10} + 10\sqrt{10} - 2\sqrt{25}$

$15(2) + 7\sqrt{10} - 2(5)$

$30 + 7\sqrt{10} - 10$

$20 + 7\sqrt{10}$

17. $(\sqrt{x} - 3\sqrt{y})(2\sqrt{x} + 4\sqrt{y})$
$2\sqrt{x^2} + 4\sqrt{xy} - 6\sqrt{xy} - 12\sqrt{y^2}$
$2x - 2\sqrt{xy} - 12y$

18. $-\sqrt{\frac{9}{64}} = -\frac{3}{8}$

19. $\sqrt[3]{-\frac{8}{27}} = -\frac{2}{3}$

20. $\sqrt[3]{0.008} = \sqrt[3]{(8)(10)^{-3}} =$
$[(8)(10)^{-3}]^{\frac{1}{3}} = (8)^{\frac{1}{3}}(10)^{-3(\frac{1}{3})} =$
$(2)(10)^{-1} = 0.2$

21. $32^{-\frac{1}{5}} = \frac{1}{32^{\frac{1}{5}}} = \frac{1}{2}$

22. $3^0 + 3^{-1} + 3^{-2}$
$1 + \frac{1}{3} + \frac{1}{9}$; LCD=9
$\frac{9}{9} + \frac{3}{9} + \frac{1}{9}$
$\frac{13}{9}$

23. $-9^{\frac{3}{2}} = -(9^{\frac{1}{2}})^3 = -(3)^3 = -27$

24. $\left(\frac{3}{4}\right)^{-2} = \left(\frac{4}{3}\right)^2 = \frac{4^2}{3^2} = \frac{16}{9}$

25. $\frac{1}{\left(\frac{2}{3}\right)^{-3}} = \left(\frac{2}{3}\right)^3 = \frac{2^3}{3^3} = \frac{8}{27}$

26. $3x^4 + 81x$
$3x(x^3 + 27)$
$3x(x + 3)(x^2 - 3x + 9)$

27. $6x^2 + 19x - 20$
$(6x - 5)(x + 4)$

28. $12 + 13x - 14x^2$
$(4 + 7x)(3 - 2x)$

29. $9x^4 + 68x^2 - 32$
$(9x^2 - 4)(x^2 + 8)$
$(3x + 2)(3x - 2)(x^2 + 8)$

30. $2ax - ay - 2bx + by$
$a(2x - y) - b(2x - y)$
$(2x - y)(a - b)$

31. $27x^3 - 8y^3$
$(3x)^3 - (2y)^3$
$(3x - 2y)(9x^2 + 6xy + 4y^2)$

32. $3(x - 2) - 2(3x + 5) = 4(x - 1)$
$3x - 6 - 6x - 10 = 4x - 4$
$-3x - 16 = 4x - 4$
$-7x = 12$
$x = -\frac{12}{7}$
The solution set is $\left\{-\frac{12}{7}\right\}$.

33. $0.06n + 0.08(n + 50) = 25$
$100[0.06n + 0.08(n + 50)] = 100(25)$
$100(0.06n) + 100[0.08(n + 50)] = 2500$
$6n + 8(n + 50) = 2500$
$6n + 8n + 400 = 2500$
$14n = 2100$
$n = 150$
The solution set is $\{150\}$.

34. $4\sqrt{x} + 5 = x$
$4\sqrt{x} = x - 5$
$(4\sqrt{x})^2 = (x - 5)^2$
$16x = x^2 - 10x + 25$
$0 = x^2 - 26x + 25$
$0 = (x - 25)(x - 1)$
$x - 25 = 0$ or $x - 1 = 0$
$x = 25$ or $x = 1$
Checking $x = 25$
$4\sqrt{25} + 5 \stackrel{?}{=} 25$
$4(5) + 5 \stackrel{?}{=} 25$
$25 = 25$
Checking $x = 1$
$4\sqrt{1} + 5 \stackrel{?}{=} 1$
$4(1) + 5 \stackrel{?}{=} 1$
$9 \neq 1$
The solution set is $\{25\}$.

35. $\sqrt[3]{n^2-1} = -1$
$\left(\sqrt[3]{n^2-1}\right)^3 = (-1)^3$
$n^2 - 1 = -1$
$n^2 = 0$
$n = 0$
Check
$\sqrt[3]{0^2-1} \stackrel{?}{=} -1$
$\sqrt[3]{-1} \stackrel{?}{=} -1$
$-1 = -1$
The solution set is $\{0\}$.

36. $6x^2 - 24 = 0$
$6(x^2 - 4) = 0$
$6(x+2)(x-2) = 0$
$x + 2 = 0$ or $x - 2 = 0$
$x = -2$ or $x = 2$
The solution set is $\{-2, 2\}$.

37. $a^2 + 14a + 49 = 0$
$(a+7)(a+7) = 0$
$(a+7)^2 = 0$
$a + 7 = 0$
$a = -7$
The solution set is $\{-7\}$.

38. $3n^2 + 14n - 24 = 0$
$(3n-4)(n+6) = 0$
$3n - 4 = 0$ or $n + 6 = 0$
$3n = 4$ or $n = -6$
$n = \frac{4}{3}$
The solution set is $\left\{-6, \frac{4}{3}\right\}$.

39. $\frac{2}{5x-2} = \frac{4}{6x+1}; x \neq \frac{2}{5}, x \neq -\frac{1}{6}$
$2(6x+1) = 4(5x-2)$
$12x + 2 = 20x - 8$
$-8x = -10$
$x = \frac{-10}{-8} = \frac{5}{4}$
The solution set is $\left\{\frac{5}{4}\right\}$.

40. $\sqrt{2x-1} - \sqrt{x+2} = 0$
$\sqrt{2x-1} = \sqrt{x+2}$
$(\sqrt{2x-1})^2 = (\sqrt{x+2})^2$
$2x - 1 = x + 2$
$x = 3$
Check:
$\sqrt{2(3)-1} - \sqrt{3+2} \stackrel{?}{=} 0$
$\sqrt{5} - \sqrt{5} \stackrel{?}{=} 0$
$0 = 0$
The solution set is $\{3\}$.

41. $5x - 4 = \sqrt{5x-4}$
$(5x-4)^2 = (\sqrt{5x-4})^2$
$25x^2 - 40x + 16 = 5x - 4$
$25x^2 - 45x + 20 = 0$
$5(5x^2 - 9x + 4) = 0$
$5(5x-4)(x-1) = 0$
$5x - 4 = 0$ or $x - 1 = 0$
$5x = 4$ or $x = 1$
$x = \frac{4}{5}$ or $x = 1$
Checking $x = \frac{4}{5}$
$5\left(\frac{4}{5}\right) - 4 \stackrel{?}{=} \sqrt{5\left(\frac{4}{5}\right) - 4}$
$4 - 4 \stackrel{?}{=} \sqrt{4-4}$
$0 = 0$
Checking $x = 1$
$5(1) - 4 \stackrel{?}{=} \sqrt{5(1)-4}$
$5 - 4 \stackrel{?}{=} \sqrt{1}$
$1 = 1$
The solution set is $\left\{\frac{4}{5}, 1\right\}$.

42. $|3x - 1| = 11$
$3x - 1 = 11$ or $3x - 1 = -11$
$3x = 12$ or $3x = -10$
$x = 4$ or $x = -\frac{10}{3}$
The solution set is $\left\{-\frac{10}{3}, 4\right\}$.

43. $(3x-2)(4x-1) = 0$
$3x - 2 = 0$ or $4x - 1 = 0$
$3x = 2$ or $4x = 1$
$x = \frac{2}{3}$ or $x = \frac{1}{4}$
The solution set is $\left\{\frac{1}{4}, \frac{2}{3}\right\}$.

44. $(2x+1)(x-2)=7$
$2x^2-4x+x-2=7$
$2x^2-3x-9=0$
$(2x+3)(x-3)=0$
$2x+3=0$ or $x-3=0$
$2x=-3$ or $x=3$
$x=-\frac{3}{2}$
The solution set is $\left\{-\frac{3}{2}, 3\right\}$.

45. $\frac{5}{6x}-\frac{2}{3}=\frac{7}{10x}$; $x \neq 0$
$30x\left(\frac{5}{6x}-\frac{2}{3}\right)=30x\left(\frac{7}{10x}\right)$
$30x\left(\frac{5}{6x}\right)-30x\left(\frac{2}{3}\right)=3(7)$
$5(5)-10x(2)=21$
$25-20x=21$
$-20x=-4$
$x=\frac{-4}{-20}=\frac{1}{5}$
The solution set is $\left\{\frac{1}{5}\right\}$.

46. $\frac{3}{y+4}+\frac{2y-1}{y^2-16}=\frac{-2}{y-4}$
$\frac{3}{y+4}+\frac{2y-1}{(y+4)(y-4)}=\frac{-2}{y-4}$
LCD $=(y+4)(y-4)$; $y \neq -4, y \neq 4$
$(y+4)(y-4)\left[\frac{3}{y+4}+\frac{2y-1}{(y+4)(y-4)}\right]=(y+4)(y-4)\left(\frac{-2}{y-4}\right)$
$(y+4)(y-4)\left(\frac{3}{y+4}\right)+(y+4)(y-4)\left[\frac{2y-1}{(y+4)(y-4)}\right]=(y+4)(y-4)\left(\frac{-2}{y-4}\right)$
$3(y-4)+2y-1=-2y-8$
$3y-12+2y-1=-2y-8$
$5y-13=-2y-8$
$7y=5$
$y=\frac{5}{7}$
The solution set is $\left\{\frac{5}{7}\right\}$.

47. $6x^4-23x^2-4=0$
$(6x^2+1)(x^2-4)=0$
$(6x^2+1)(x+2)(x-2)=0$
$6x^2+1=0$ or $x+2=0$ or $x-2=0$
$6x^2=-1$ or $x=-2$ or $x=2$
$x^2=-\frac{1}{6}$
$x=\pm\sqrt{-\frac{1}{6}}=\pm\frac{i\sqrt{1}}{\sqrt{6}}=$
$\pm\frac{i}{\sqrt{6}}\cdot\frac{\sqrt{6}}{\sqrt{6}}=\frac{i\sqrt{6}}{\sqrt{36}}=\pm\frac{i\sqrt{6}}{6}$
The solution set is $\left\{\pm\frac{i\sqrt{6}}{6}, -2, 2\right\}$.

48. $3n^3+3n=0$
$3n(n^2+1)=0$
$3n=0$ or $n^2+1=0$
$n=0$ or $n^2=-1$
$n=\pm\sqrt{-1}$
$n=\pm i$
The solution set is $\{0\}$.

49. $n^2-13n-114=0$
$(n+6)(n-19)=0$
$n+6=0$ or $n-19=0$
$n=-6$ or $n=19$
The solution set is $\{-6, 19\}$.

50. $12x^2 + x - 6 = 0$
$(4x + 3)(3x - 2) = 0$
$4x + 3 = 0$ or $3x - 2 = 0$
$4x = -3$ or $3x = 2$
$x = -\frac{3}{4}$ or $x = \frac{2}{3}$
The solution set is $\left\{-\frac{3}{4}, \frac{2}{3}\right\}$.

51. $x^2 - 2x + 26 = 0$
$x = \frac{-(-2) \pm \sqrt{(-2)^2 - 4(1)(26)}}{2(1)}$
$x = \frac{2 \pm \sqrt{4 - 104}}{2}$
$x = \frac{2 \pm \sqrt{-100}}{2}$
$x = \frac{2 \pm 10i}{2} = \frac{2(1 \pm 5i)}{2}$
$x = 1 \pm 5i$
The solution set is $\{1 - 5i, 1 + 5i\}$.

52. $(x + 2)(x - 6) = -15$
$x^2 - 4x - 12 = -15$
$x^2 - 4x + 3 = 0$
$(x - 3)(x - 1) = 0$
$x - 3 = 0$ or $x - 1 = 0$
$x = 3$ or $x = 1$
The solution set is $\{1, 3\}$.

53. $(3x - 1)(x + 4) = 0$
$3x - 1 = 0$ or $x + 4 = 0$
$3x = 1$ or $x = -4$
$x = \frac{1}{3}$ or $x = -4$
The solution set is $\left\{-4, \frac{1}{3}\right\}$.

54. $x^2 + 4x + 20 = 0$
$x = \frac{-4 \pm \sqrt{4^2 - 4(1)(20)}}{2(1)}$
$x = \frac{-4 \pm \sqrt{16 - 80}}{2}$
$x = \frac{-4 \pm \sqrt{-64}}{2}$
$x = \frac{-4 \pm 8i}{2} = \frac{2(-2 \pm 4i)}{2}$
$x = -2 \pm 4i$
The solution set is $\{-2 - 4i, -2 + 4i\}$.

55. $2x^2 - x - 4 = 0$
$x = \frac{-(-1) \pm \sqrt{(-1)^2 - 4(2)(-4)}}{2(2)}$
$x = \frac{1 \pm \sqrt{1 + 32}}{4}$
$x = \frac{1 \pm \sqrt{33}}{4}$
The solution set is $\left\{\frac{1 - \sqrt{33}}{4}, \frac{1 + \sqrt{33}}{4}\right\}$.

56. $6 - 2x \geq 10$
$-2x \geq 4$
$-\frac{1}{2}(-2x) \leq -\frac{1}{2}(4)$
$x \leq -2$
The solution set is $(-\infty, -2]$.

57. $4(2x - 1) < 3(x + 5)$
$8x - 4 < 3x + 15$
$8x < 3x + 19$
$5x < 19$
$x < \frac{19}{5}$
The solution set is $\left(-\infty, \frac{19}{5}\right)$.

58. $\frac{n + 1}{4} + \frac{n - 2}{12} > \frac{1}{6}$
$12\left(\frac{n + 1}{4} + \frac{n - 2}{12}\right) > 12\left(\frac{1}{6}\right)$
$12\left(\frac{n + 1}{4}\right) + 12\left(\frac{n - 2}{12}\right) > 2$
$3(n + 1) + n - 2 > 2$
$3n + 3 + n - 2 > 2$
$4n + 1 > 2$
$4n > 1$
$n > \frac{1}{4}$
The solution set is $\left(\frac{1}{4}, \infty\right)$.

59. $|2x - 1| < 5$
$-5 < 2x - 1 < 5$
$-4 < 2x < 6$
$-2 < x < 3$
The solution set is $(-2, 3)$.

60. $|3x+2| > 11$

$3x+2 < -11$ or $3x+2 > 11$

$3x < -13$ or $3x > 9$

$x < -\frac{13}{3}$ or $x > 3$

The solution set is $\left(-\infty, -\frac{13}{3}\right) \cup (3, \infty)$.

61. $\frac{1}{2}(3x-1) - \frac{2}{3}(x+4) \leq \frac{3}{4}(x-1)$

$12\left[\frac{1}{2}(3x-1) - \frac{2}{3}(x+4)\right] \leq 12\left[\frac{3}{4}(x-1)\right]$

$12\left[\frac{1}{2}(3x-1)\right] - 12\left[\frac{2}{3}(x+4)\right] \leq 9(x-1)$

$6(3x-1) - 8(x+4) \leq 9x-9$

$18x - 6 - 8x - 32 \leq 9x - 9$

$10x - 38 \leq 9x - 9$

$10x \leq 9x + 29$

$x \leq 29$

The solution set is $(-\infty, 29]$.

62. $x^2 - 2x - 8 \leq 0$

$x^2 - 2x - 8 = 0$

$(x-4)(x+2) = 0$

$x - 4 = 0$ or $x + 2 = 0$

$x = 4$ or $x = -2$

$-2 \qquad 4$

Test Point	-3	0	5
$x-4$:	negative	negative	positive
$x+2$:	negative	positive	positive
product:	positive	negative	positive

The solution set is $[-2, 4]$.

63. $3x^2 + 14x - 5 > 0$

$3x^2 + 14x - 5 = 0$

$(3x-1)(x+5) = 0$

$3x - 1 = 0$ or $x + 5 = 0$

$x = \frac{1}{3}$ or $x = -5$

$-5 \qquad \frac{1}{3}$

Test Point	-6	0	1
$3x-1$:	negative	negative	positive
$x+5$:	negative	positive	positive
product:	positive	negative	positive

The solution set is $(-\infty, -5) \cup \left(\frac{1}{3}, \infty\right)$.

64. $\frac{x+2}{x-7} \geq 0;\ x \neq 7$

$x + 2 = 0$ or $x - 7 = 0$

$x = -2$ or $x = 7$

$-2 \qquad 7$

Test Point	-3	0	8
$x+2$:	negative	positive	positive
$x-7$:	negative	negative	positive
quotient:	positive	negative	positive

The solution set is $(-\infty, -2] \cup (7, \infty)$.

65. $\frac{2x-1}{x+3} < 1$

$\frac{2x-1}{x+3} - 1 < 0$

$\frac{2x-1}{x+3} - \frac{x+3}{x+3} < 0$

$\frac{2x-1-(x+3)}{x+3} < 0$

$\frac{2x-1-x-3}{x+3} < 0$

$\frac{x-4}{x+3} < 0;\ x \neq -3$

$x - 4 = 0$ or $x + 3 = 0$

$x = 4$ or $x = -3$

$-3 \qquad 4$

Test Point	-4	0	5
$x-4$:	negative	negative	positive
$x+3$:	negative	positive	positive
quotient:	positive	negative	positive

The solution set is $(-3, 4)$.

66. $\begin{pmatrix} 4x - 9y = 14 \\ x + 5y = -11 \end{pmatrix}$

Multiply equation 2 by -4, and add to equation 1.

$$\begin{aligned} 4x - 9y &= 14 \\ -4x - 20y &= 44 \\ \hline -29y &= 58 \\ y &= -2 \end{aligned}$$

$x + 5y = -11$

$x + 5(-2) = -11$

$x - 10 = -11$

$x = -1$

The solution set is $\{(-1, -2)\}$.

67. $\begin{pmatrix} \frac{1}{2}x - \frac{1}{3}y = 7 \\ \frac{3}{4}x + \frac{2}{3}y = 0 \end{pmatrix}$

Multiply equation 1 by 6 and multiply equation 2 by 12.

$\begin{pmatrix} 3x - 2y = 42 \\ 9x + 8y = 0 \end{pmatrix}$

Multiply equation 1 by 4.
Then add the equations.

$$\begin{aligned} 12x - 8y &= 168 \\ 9x + 8y &= 0 \\ \hline 21x \quad &= 168 \\ x &= 8 \end{aligned}$$

$9x + 8y = 0$
$9(8) + 8y = 0$
$72 + 8y = 0$
$8y = -72$
$y = -9$
The solution set is $\{(8, -9)\}$.

68. $\begin{pmatrix} 7x - 2y = 16 \\ 3x + 5y = 42 \end{pmatrix}$

Multiply equation 1 by 5 and multiply equation 2 by 2. Then add the equations.

$$\begin{aligned} 35x - 10y &= 80 \\ 6x + 10y &= 84 \\ \hline 41x \quad &= 164 \\ x &= 4 \end{aligned}$$

$7x - 2y = 16$
$7(4) - 2y = 16$
$28 - 2y = 16$
$-2y = -12$
$y = 6$
The solution set is $\{(4, 6)\}$.

69. $(-6x^{-3}y^{-4})(5xy^6)$
$-30x^{-3+1}y^{-4+6}$
$-30x^{-2}y^2$
$-\frac{30y^2}{x^2}$

70a. $4\sqrt{54}$
$4\sqrt{9}\sqrt{6}$
$4(3)\sqrt{6}$
$12\sqrt{6}$

b. $2\sqrt[3]{54}$
$2\sqrt[3]{27}\sqrt[3]{2}$
$2(3)\sqrt[3]{2}$
$6\sqrt[3]{2}$

c. $\frac{3\sqrt{6}}{4\sqrt{10}} = \frac{3}{4}\sqrt{\frac{6}{10}} = \frac{3}{4}\sqrt{\frac{3}{5}} =$
$\frac{3\sqrt{3}}{4\sqrt{5}} \cdot \frac{\sqrt{5}}{\sqrt{5}} = \frac{3\sqrt{15}}{4\sqrt{25}} = \frac{3\sqrt{15}}{4(5)} = \frac{3\sqrt{15}}{20}$

d. $\sqrt{24x^3y^5}$
$\sqrt{4x^2y^4}\sqrt{6xy}$
$2xy^2\sqrt{6xy}$

71 **a.** $7652 = (7.652)(10)^3$
b. $0.000026 = (2.6)(10)^{-5}$
c. $1.414 = (1.414)(10)^0$
d. $1000 = (1)(10)^3$

72. Let x = liters of 60% solution.
$(60\%)(x) + (10\%)(14) = (25\%)(x + 14)$
$0.60x + 0.10(14) = 0.25(x + 14)$
$100[0.60x + 0.10(14)] = 100[0.25(x + 14)]$
$60x + 10(14) = 25(x + 14)$
$60x + 140 = 25x + 350$
$35x = 210$
$x = 6$
There needs to be 6 liters of 60% acid solution added.

73. Let x = one part and
$2250 - x$ = other part.
$\frac{x}{2250 - x} = \frac{2}{3}$
$3x = 2(2250 - x)$
$3x = 4500 - 2x$
$5x = 4500$
$x = 900$
It is divided \$900 and \$1350.

74. Dimensions of picture
width $= x$
length $= 2x - 7$
Area $= x(2x - 7) = 2x^2 - 7x$

Dimensions of picture and border
width $= x + 2$
length $= 2x - 5$
Area $= (x + 2)(2x - 5) = 2x^2 - x - 10$
Area of Border $= 62$
$2x^2 - x - 10 - (2x^2 - 7x) = 62$
$2x^2 - x - 10 - 2x^2 + 7x = 62$
$6x - 10 = 62$
$6x = 72$
$x = 12$
The picture is 12 inches by 17 inches.

75.

	Time in hours	Rate
Lolita	x	$\frac{1}{x}$
Doug	10	$\frac{1}{10}$
Together	$3\frac{20}{60} = \frac{10}{3}$	$\frac{1}{\frac{10}{3}} = \frac{3}{10}$

$\frac{1}{x} + \frac{1}{10} = \frac{3}{10}$
$10x\left(\frac{1}{x} + \frac{1}{10}\right) = 10x\left(\frac{3}{10}\right)$
$10x\left(\frac{1}{x}\right) + 10x\left(\frac{1}{10}\right) = 3x$
$10 + x = 3x$
$10 = 2x$
$5 = x$
It would take Lolita 5 hours.

76.

	Price/ golfball	Number of golfballs	Cost
1st	$\frac{14}{x}$	x	14
Re-Order	$\frac{14}{x+1}$	$x + 1$	14

$\frac{14}{x} - 0.25 = \frac{14}{x+1}; x \neq -1, x \neq 0$
$\frac{14}{x} - \frac{1}{4} = \frac{14}{x+1}$
$4x(x+1)\left(\frac{14}{x} - \frac{1}{4}\right) = 4x(x+1)\left(\frac{14}{x+1}\right)$
$4(x+1)(14) - x(x+1) = 4x(14)$
$56x + 56 - x^2 - x = 56x$
$0 = x^2 + x - 56$
$0 = (x + 8)(x - 7)$
$x + 8 = 0$ or $x - 7 = 0$
$x = -8$ or $x = 7$
Discard the root $x = -8$.
She bought 7 golf balls.

77.

	Rate	Time	Distance
First Jogger	$\frac{1}{8}$ mpm	x	$\frac{1}{8}x$
Second Jogger	$\frac{1}{6}$ mpm	x	$\frac{1}{6}x$

$\frac{1}{2} + \frac{1}{8}x = \frac{1}{6}x$
$24\left(\frac{1}{2} + \frac{1}{8}x\right) = 24\left(\frac{1}{6}x\right)$
$24\left(\frac{1}{2}\right) + 24\left(\frac{1}{8}x\right) = 24\left(\frac{1}{6}x\right)$
$12 + 3x = 4x$
$12 = x$
It would take 12 minutes.

78. $P = 1000$ $t = 2$ $A = 1149.90$
$A = P(1 + r)^t$
$1149.90 = 1000(1 + r)^2$
$1.1499 = (1 + r)^2$
$(1 + r)^2 = 1.1499$
$1 + r = \pm\sqrt{1.1499}$
$1 + r = \pm 1.07$
$r = -1 \pm 1.07$
$r = -1 - 1.07$ or $r = -1 + 1.07$
$r = -2.07$ or $r = 0.07$
Discard the negative solution.
r = 0.07 = 7%

79. Let $x =$ the number of rows and
then $2x - 1 =$ the number of chairs per row.
$x(2x - 1) = 120$
$2x^2 - x - 120 = 0$
$(2x + 15)(x - 8) = 0$
$2x + 15 = 0$ or $x - 8 = 0$
$2x = -15$ or $x = 8$
$x = -\frac{15}{2}$
Discard the negative solution.
There will be 8 rows with
$2x - 1 = 2(8) - 1 = 15$ chairs per row.

80. Let x = number of shares purchased. Then $\frac{2800}{x}$ is the price per share when purchased and $\frac{2800}{x} + 6$ is the price per share a month later. Bjorn sold all but 60 shares so the shares sold can be expressed as $x - 60$.

$$\begin{pmatrix}\text{Price per}\\\text{Share}\end{pmatrix} \cdot \begin{pmatrix}\text{Number of}\\\text{Shares}\end{pmatrix} = 2800$$

$$\left(\frac{2800}{x} + 6\right)(x - 60) = 2800$$

$$2800 - \frac{168000}{x} + 6x - 360 = 2800$$

$$-\frac{168000}{x} + 6x - 360 = 0$$

$$x\left(\frac{-168000}{x} + 6x - 360\right) = x(0)$$

$$-168000 + 6x^2 - 360x = 0$$

$$6x^2 - 360x - 168000 = 0$$

$$6(x^2 - 60x - 28000) = 0$$

$$x^2 - 60x - 28000 = 0$$

$$(x - 200)(x + 140) = 0$$

$x - 200 = 0$ or $x + 140 = 0$

$x = 200$ or $x = -140$

Discard the negative solution.

Bjorn sold $x - 60 = 200 - 60 = 140$ shares.

81. $$\begin{pmatrix}\text{measure}\\\text{1st angle}\end{pmatrix} + \begin{pmatrix}\text{measure}\\\text{2nd angle}\end{pmatrix} + \begin{pmatrix}\text{measure}\\\text{3rd angle}\end{pmatrix} = 180$$

$$40 + \begin{pmatrix}\text{measure}\\\text{2nd angle}\end{pmatrix} + \begin{pmatrix}\text{measure}\\\text{3rd angle}\end{pmatrix} = 180$$

$$\begin{pmatrix}\text{measure}\\\text{2nd angle}\end{pmatrix} + \begin{pmatrix}\text{measure}\\\text{3rd angle}\end{pmatrix} = 140$$

Let x represent the measure of the 2nd angle, then $140 - x$ represents the measure of the 3rd angle.

$$\frac{\text{2nd angle}}{\text{3rd angle}} \quad \frac{3}{4} = \frac{x}{140 - x}$$

$$4x = 3(140 - x)$$

$$4x = 420 - 3x$$

$$7x = 420$$

$$x = 60$$

The measures of the remaining angles are 60° and 80°.

82. Let x represent the measure the angle. Then $90 - x$ represents the complement and $180 - x$ represents the supplement.

$$180 - x = 10 + 5(90 - x)$$

$$180 - x = 10 + 450 - 5x$$

$$4x = 280$$

$$x = 70$$

The angle is 70°.

83. Let r represent the other rate.

$$3000(0.07) + 2000r > 380$$

$$210 + 2000r > 380$$

$$2000r > 170$$

$$r > 0.085$$

The other rate must be more than 8.5%.

84. Let x = price of a lemon and y = price of an orange.

$$\begin{pmatrix}3x + 5y = 2.40\\4x + 7y = 3.31\end{pmatrix}$$

Multiply equation 1 by -4 and equation 2 by 3, then add the equations.

$$\begin{aligned}-12x - 20y &= -9.60\\ 12x + 21y &= 9.93\\ \hline y &= 0.33\end{aligned}$$

$$3x + 5y = 2.40$$

$$3x + 5(0.33) = 2.40$$

$$3x + 1.65 = 2.40$$

$$3x = 0.75$$

$$x = 0.25$$

It costs \$0.25 for a lemon and \$0.33 for an orange.

85. Let s represent the selling price. When profit is a percent of selling price,

Selling price = Cost + Profit

$$s = 700 + (30\%)(s)$$

$$s = 700 + 0.3s$$

$$10s = 7000 + 3s$$

$$7s = 7000$$

$$s = 1000$$

Profit based on selling price is $(30\%)(1000) = (0.30)(1000) = \$300$.

When profit is a percent of cost,

Selling price = Cost + Profit

$$s = 700 + (30\%)(700)$$

$$s = 700 + 0.30(700)$$

$$s = 700 + 210 = 910$$

Profit based on cost is \$210.

The difference in profit is $\$300 - \$210 = \$90$.

Chapter 12 Coordinate Geometry: Lines, Parabolas, Circles, Ellipses, and Hyperbolas

PROBLEM SET 12.1 Distance, Slope, and Graphing Techniques

1. $(-2, -1), (7, 11)$

$d = \sqrt{[7-(-2)]^2 + [11-(-1)]^2}$

$d = \sqrt{9^2 + 12^2}$

$d = \sqrt{81 + 144} = \sqrt{225} = 15$

3. $(1, -1), (3, -4)$

$d = \sqrt{(3-1)^2 + [-4-(-1)]^2}$

$d = \sqrt{2^2 + (-3)^2}$

$d = \sqrt{4+9} = \sqrt{13}$

5. $(6, -4), (9, -7)$

$d = \sqrt{(9-6)^2 + [-7-(-4)]^2}$

$d = \sqrt{3^2 + (-3)^2}$

$d = \sqrt{9+9} = \sqrt{18} = \sqrt{9}\sqrt{2} = 3\sqrt{2}$

7. $(-3, 3), (0, -3)$

$d = \sqrt{[0-(-3)]^2 + (-3-3)^2}$

$d = \sqrt{(3)^2 + (-6)^2}$

$d = \sqrt{9+36} = \sqrt{45} = \sqrt{9}\sqrt{5} = 3\sqrt{5}$

9. $(1, -6), (-5, -6)$

$d = \sqrt{(-5-1)^2 + [-6-(-6)]^2}$

$d = \sqrt{(-6)^2 + (0)^2}$

$d = \sqrt{36} = 6$

11. $(1, 7), (4, -2)$

$d = \sqrt{(4-1)^2 + [-2-7]^2}$

$d = \sqrt{3^2 + (-9)^2}$

$d = \sqrt{9+81} = \sqrt{90}$

$d = \sqrt{90} = \sqrt{9}\sqrt{10} = 3\sqrt{10}$

13. $(-3, 1), (5, 7)$, and $(8, 3)$

Distance from $(-3, 1)$ to $(5, 7)$.

$d = \sqrt{[5-(-3)]^2 + (7-1)^2}$

$d = \sqrt{8^2 + 6^2}$

$d = \sqrt{64+36} = \sqrt{100} = 10$

Distance from $(-3, 1)$ to $(8, 3)$.

$d = \sqrt{[8-(-3)]^2 + (3-1)^2}$

$d = \sqrt{11^2 + 2^2}$

$d = \sqrt{121+4} = \sqrt{125}$

$d = \sqrt{125} = \sqrt{25}\sqrt{5} = 5\sqrt{5}$

Distance from $(5, 7)$ to $(8, 3)$.

$d = \sqrt{(8-5)^2 + (3-7)^2}$

$d = \sqrt{3^2 + (-4)^2}$

$d = \sqrt{9+16} = \sqrt{25} = 5$

The potential legs are 10 and 5 and the hypotenuse is $5\sqrt{5}$.

$(10)^2 + (5)^2 \stackrel{?}{=} \left(5\sqrt{5}\right)^2$

$100 + 25 \stackrel{?}{=} (25)(5)$

$125 = 125$

Yes, it is a right triangle.

15. Distance from $(3, 6)$ to $(7, 12)$.

$d = \sqrt{(7-3)^2 + (12-6)^2}$

$d = \sqrt{4^2 + 6^2}$

$d = \sqrt{16+36} = \sqrt{52}$

$d = \sqrt{52} = \sqrt{4}\sqrt{13} = 2\sqrt{13}$

Distance from $(7, 12)$ to $(11, 18)$.

$d = \sqrt{(11-7)^2 + (18-12)^2}$

$d = \sqrt{4^2 + 6^2}$

$d = \sqrt{16+36} = \sqrt{52}$

$d = \sqrt{52} = \sqrt{4}\sqrt{13} = 2\sqrt{13}$

Distance from (11, 18) to (15, 24).

$d = \sqrt{(15-11)^2 + (24-18)^2}$

$d = \sqrt{4^2 + 6^2}$

$d = \sqrt{16+36} = \sqrt{52}$

$d = \sqrt{52} = \sqrt{4}\sqrt{13} = 2\sqrt{13}$

The distances are all equal, so the points do divide the line segment into three segments of equal length.

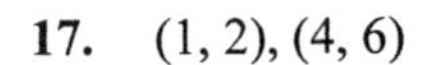

17. (1, 2), (4, 6)

$m = \dfrac{6-2}{4-1} = \dfrac{4}{3}$

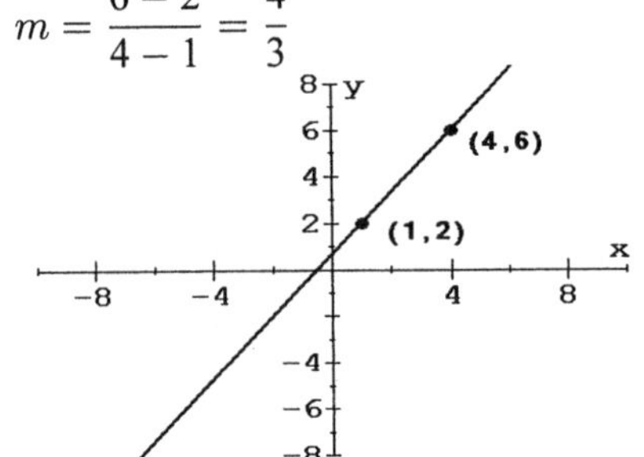

19. (3,4), (3, 8)

$m = \dfrac{8-4}{3-3} = \dfrac{4}{0}$

Slope is undefined.

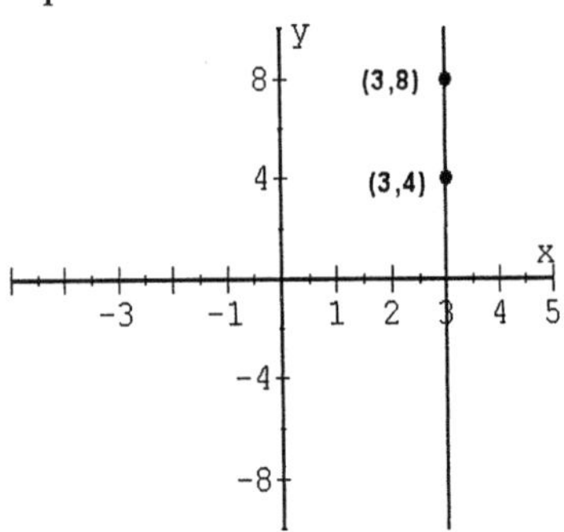

21. (2, 6), (6, − 2)

$m = \dfrac{-2-6}{6-2} = -\dfrac{8}{4} = -2$

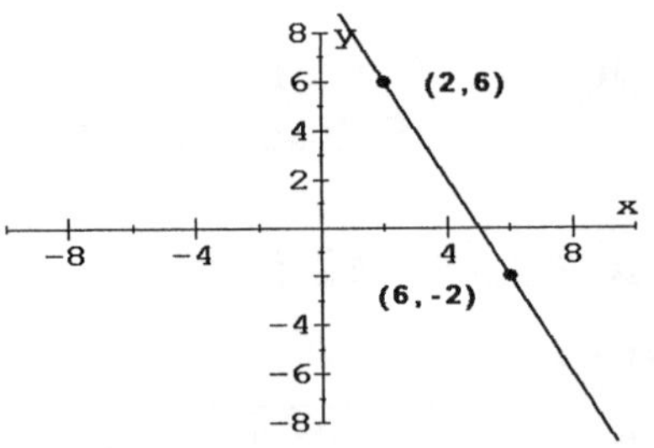

23. (− 6, 1), (− 1, 4)

$m = \dfrac{4-1}{-1-(-6)} = \dfrac{3}{5}$

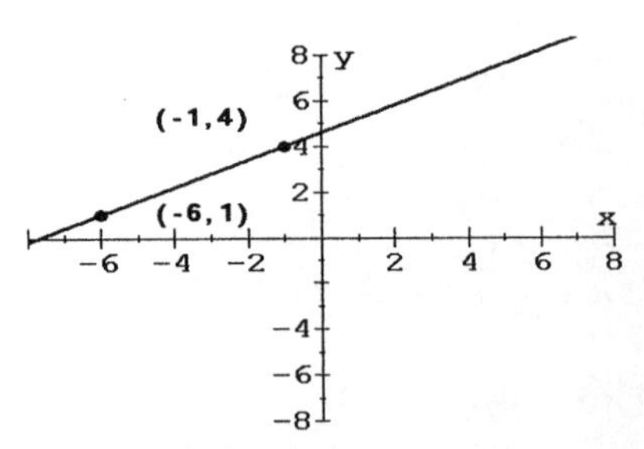

25. (− 2, − 4), (2, − 4)

$m = \dfrac{-4-(-4)}{2-(-2)} = \dfrac{0}{4} = 0$

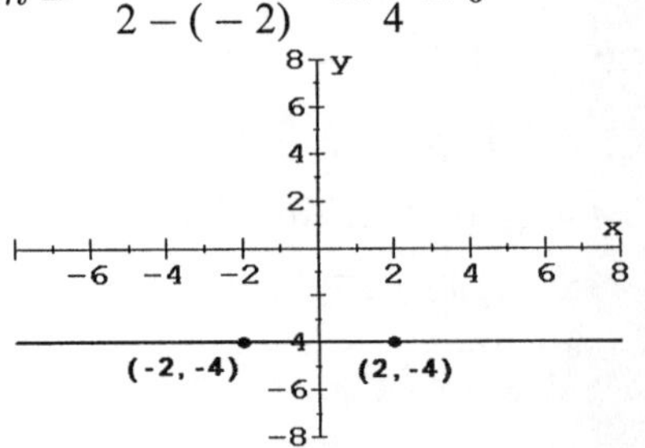

27. (0, − 2), (4, 0)

$m = \dfrac{0-(-2)}{4-0} = \dfrac{2}{4} = \dfrac{1}{2}$

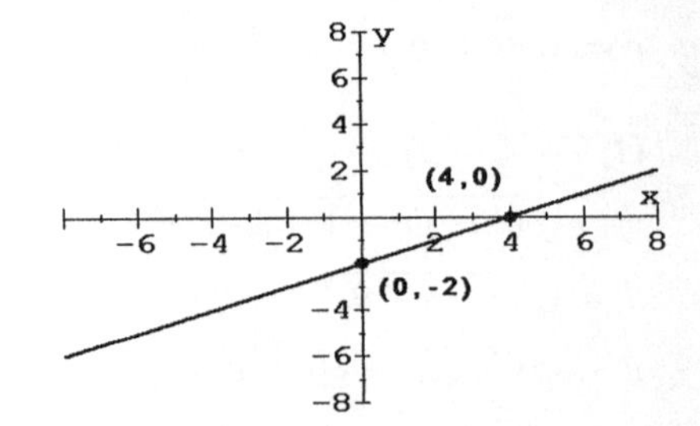

29. $(-2, 4), (x, 6)\ m = \frac{2}{9}$

$\frac{6-4}{x-(-2)} = \frac{2}{9}$

$\frac{2}{x+2} = \frac{2}{9}$

$2(x+2) = 2(9)$

$2x + 4 = 18$

$2x = 14$

$x = 7$

31. $(x, 4), (2, -5)\ m = -\frac{9}{4}$

$\frac{-5-4}{2-x} = -\frac{9}{4}$

$\frac{-9}{2-x} = \frac{-9}{4}$

$-9(4) = -9(2-x)$

$-36 = -18 + 9x$

$-18 = 9x$

$-2 = x$

33.

given point
(-3, 1)
reflection across y-axis
(3, 1)
(-3, -1)
reflection across x-axis
(3, -1)
reflection through origin

35.

reflection through origin
(-7, 2)
reflection across x-axis
(7, 2)
(-7, -2)
reflection across y-axis
(7, -2)
given point

37.

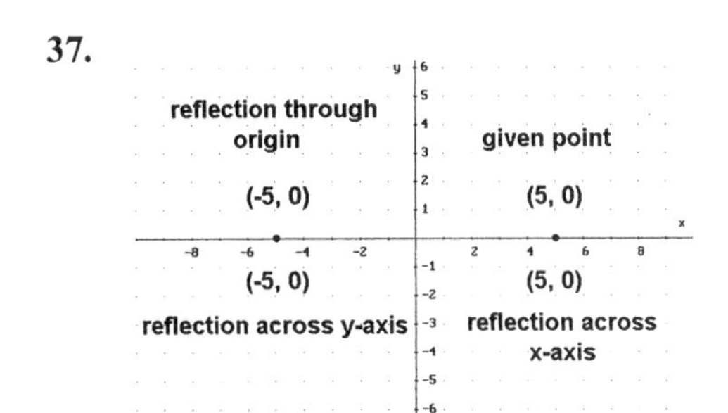

39. $-3x + 2y^2 = -4$

To test symmetry for:

x-axis replace y with $-y$

$-3x + 2(-y)^2 = -4$

$-3x + 2y^2 = -4$

Yes, it is symmetric to the x-axis.

y-axis replace x with $-x$

$-3(-x) + 2y^2 = -4$

$3x + 2y^2 = -4$

No, it is not symmetric to the y-axis.

origin replace x with $-x$ and y with $-y$.

$-3(-x) + 2(-y)^2 = -4$

$3x + 2y^2 = -4$

No, it is not symmetric to the origin.

41. $y = 4x^2 + 13$

To test symmetry for:

x-axis replace y with $-y$

$-y = 4x^2 + 13$

No, it is not symmetric to the x-axis.

y-axis replace x with $-x$

$y = 4(-x)^2 + 13$

$y = 4x^2 + 13$

Yes, it is symmetric to the y-axis.

origin replace x with $-x$ and y with $-y$

$-y = 4(-x)^2 + 13$

$-y = 4x^2 + 13$

No, it is not symmetric to the origin.

43. $2x^2y^2 = 5$

To test symmetry for:

x-axis replace y with $-y$

$2x^2(-y)^2 = 5$

$2x^2y^2 = 5$

Yes, it is symmetric to the x-axis.

y-axis replace x with $-x$
$2(-x)^2y^2 = 5$
$2x^2y^2 = 5$
Yes, it is symmetric to the y-axis.

origin replace x with $-x$
and y with $-y$
$2(-x)^2(-y)^2 = 5$
$2x^2y^2 = 5$
Yes, it is symmetric to the origin.

45. $x^2 - 2x - y^2 = 4$
To test symmetry for:
x-axis replace y with $-y$
$x^2 - 2x - (-y)^2 = 4$
$x^2 - 2x - y^2 = 4$
Yes, it is symmetric to the x-axis.

y-axis replace x with $-x$
$(-x)^2 - 2(-x) - y^2 = 4$
$x^2 + 2x - y^2 = 4$
No, it is not symmetric to the y-axis.

origin replace x with $-x$
and y with $-y$
$(-x^2) - 2(-x) - (-y)^2 = 4$
$x^2 + 2x - y^2 = 4$
No, it is not symmetric to the origin.

47. $y = 2x^2 - 7x - 3$
To test symmetry for:
x-axis replace y with $-y$
$-y = 2x^2 - 7x - 3$
No, it is not symmetric to the x-axis.

y-axis replace x with $-x$
$y = 2(-x)^2 - 7(-x) - 3$
$y = 2x^2 + 7x - 3$
No, it is not symmetric to the y-axis.

origin replace x with $-x$
and y with $-y$
$-y = 2(-x)^2 - 7(-x) - 3$
$-y = 2x^2 + 7x - 3$
No, it is not symmetric to the origin.

49. $y = 2x$
To test symmetry for:
x-axis replace y with $-y$
$-y = 2x$
No, it is not symmetric to the x-axis.

y-axis replace x with $-x$
$y = 2(-x)$
$y = -2x$
No, it is not symmetric to the y-axis.

origin replace x with $-x$
and y with $-y$
$-y = 2(-x)$
$-y = -2x$
$-1(-y) = -1(-2x)$
$y = 2x$
Yes, it is symmetric to the origin.

51. $y = x^4 - x^2 + 2$
To test symmetry for:
x-axis replace y with $-y$
$-y = x^4 - x^2 + 2$
No, it is not symmetric to the x-axis.

y-axis replace x with $-x$
$y = (-x)^4 - (-x)^2 + 2$
$y = x^4 - x^2 + 2$
Yes, it is symmetric to the y-axis.

origin replace x with $-x$
and y with $-y$
$-y = (-x)^4 - (-x)^2 + 2$
$-y = x^4 - x^2 + 2$
No, it is not symmetric to the origin.

53. $y = x - 4$

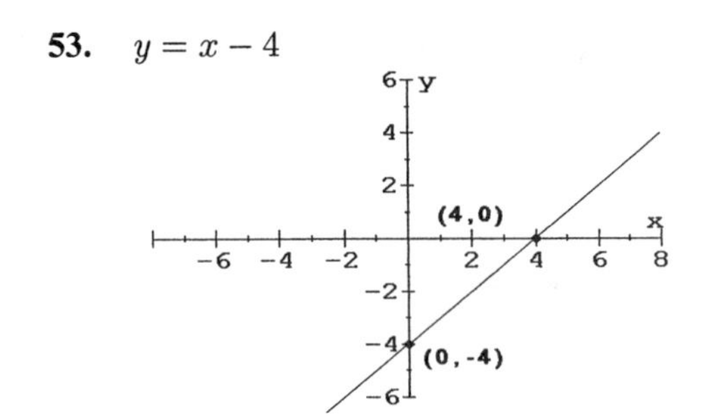

55. 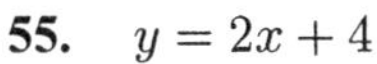 $y = 2x + 4$

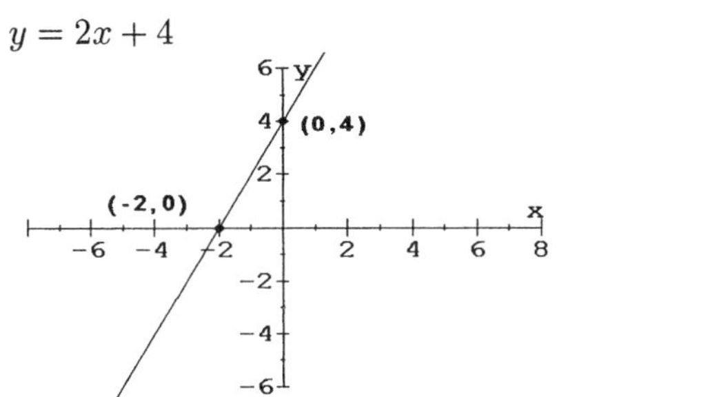

57. $y = -3x - 1$

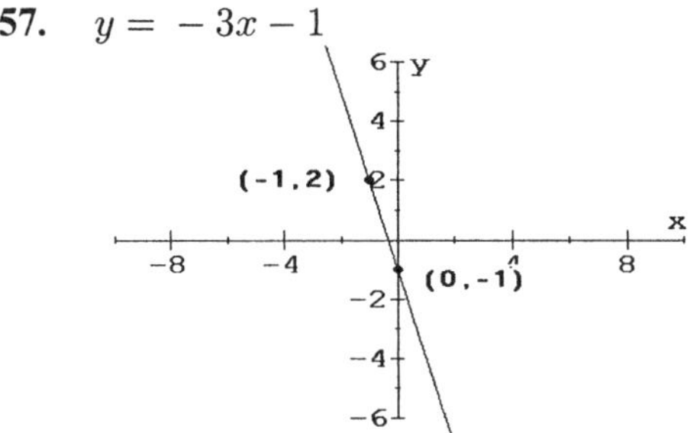

59. $y = x^2 + 2$

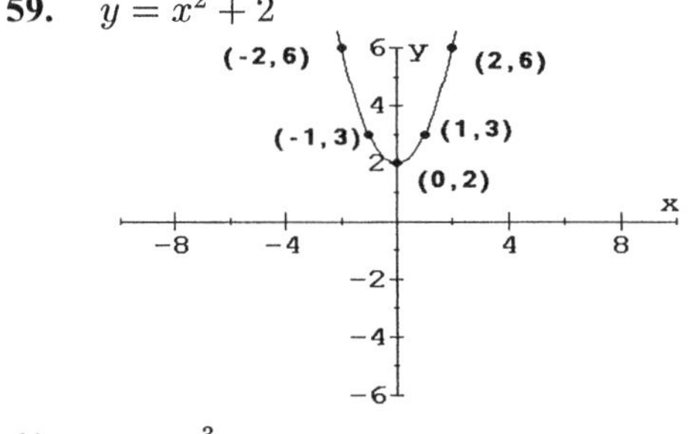

61. $y = x^3$

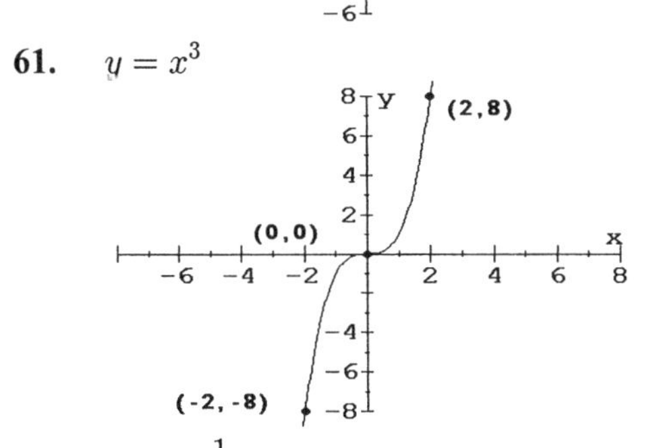

63. $y = \dfrac{-1}{x^2}$

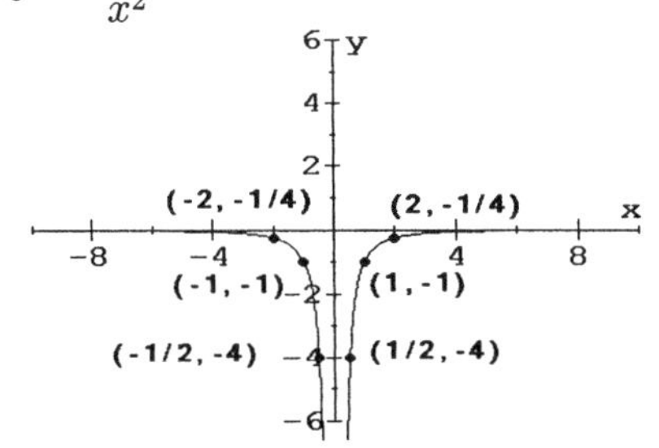

65. $2x - y = 4$

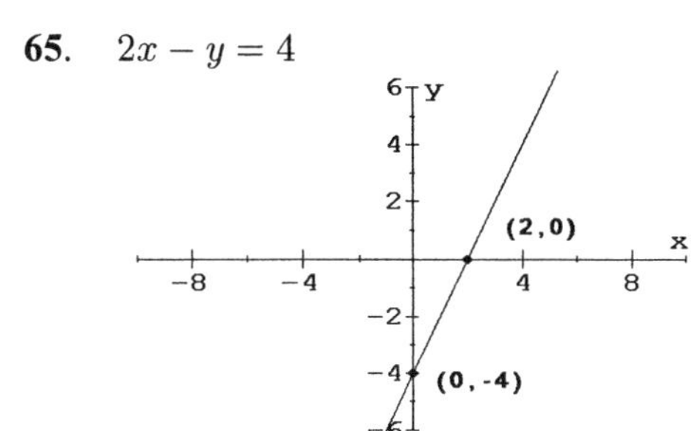

67. $y = -3x^2$

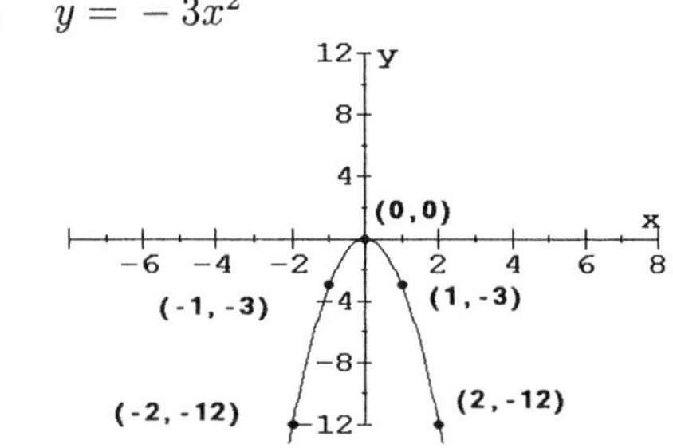

69. $xy = 2$

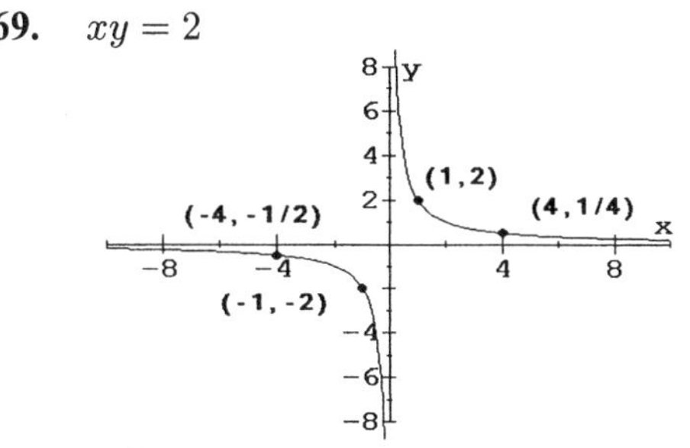

71. $xy^2 = -4$

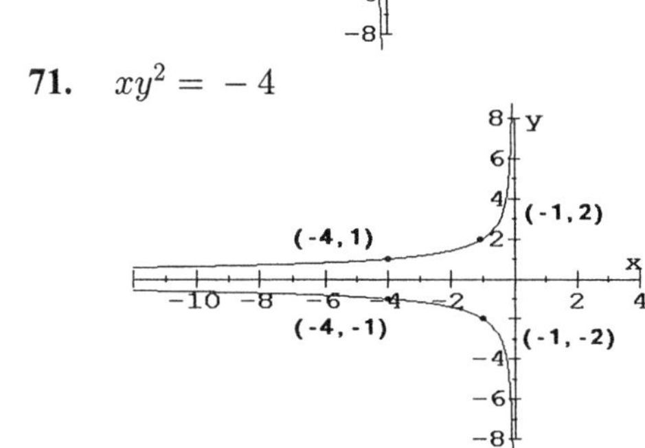

73. $y^2 = x^3$

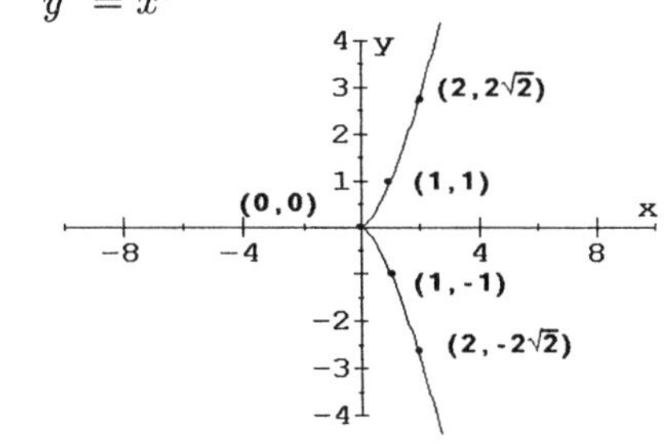

75. $y = \frac{4}{x^2+1}$

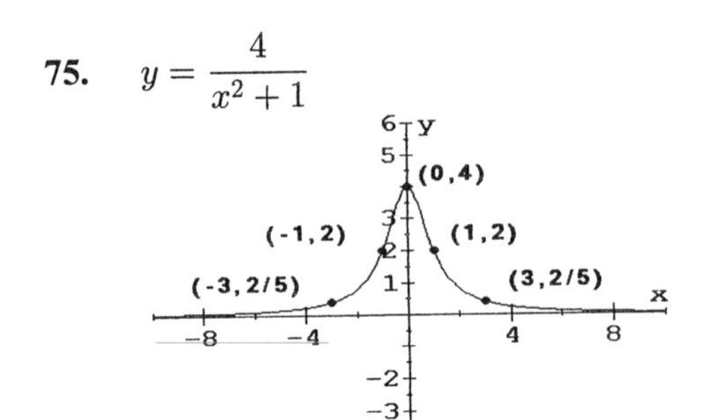

77. $y = x^4$

79. $x = -y^3 + 2$

Further Investigations

85 a. $10 - 1 = 9$ units

$x = 1 + \frac{2}{3}(9) = 1 + 6 = 7$

b. $14 - (-2) = 16$ units

$x = -2 + \frac{3}{4}(16) = -2 + 12 = 10$

c. $7 - (-3) = 10$ units

$x = -3 + \frac{1}{3}(10) = -\frac{9}{3} + \frac{10}{3} = \frac{1}{3}$

d. $6 - (-5) = 11$ units

$x = -5 + \frac{2}{5}(11) = -\frac{25}{5} + \frac{22}{5} = -\frac{3}{5}$

e. $-1 - (-11) = 10$ units

$x = -1 - \frac{3}{5}(10) = -1 - 6 = -7$

f. $3 - (-7) = 10$ units

$x = 3 - \frac{5}{6}(10) = \frac{9}{3} - \frac{25}{3} = -\frac{16}{3}$

PROBLEM SET 12.2 Writing Equations of Lines

1. $m = \frac{1}{2}, (3, 5)$

$y - 5 = \frac{1}{2}(x - 3)$

$2(y - 5) = 2\left[\frac{1}{2}(x - 3)\right]$

$2y - 10 = x - 3$

$-10 = x - 2y - 3$

$-7 = x - 2y$

$x - 2y = -7$

3. $m = 3, (-2, 4)$

$y - 4 = 3[x - (-2)]$

$y - 4 = 3(x + 2)$

$y - 4 = 3x + 6$

$-4 = 3x - y + 6$

$-10 = 3x - y$

$3x - y = -10$

5. $m = -\frac{3}{4}, (-1, -3)$

$y - (-3) = -\frac{3}{4}[x - (-1)]$

$y + 3 = -\frac{3}{4}(x + 1)$

$4(y + 3) = 4\left[-\frac{3}{4}(x + 1)\right]$

$4y + 12 = -3(x + 1)$

$4y + 12 = -3x - 3$

$3x + 4y = -15$

7. $m = \frac{5}{4}, (4, -2)$

$y - (-2) = \frac{5}{4}(x - 4)$

$y + 2 = \frac{5}{4}(x - 4)$

$4(y + 2) = 4\left[\frac{5}{4}(x - 4)\right]$

$4y + 8 = 5(x - 4)$

$4y + 8 = 5x - 20$

$8 = 5x - 4y - 20$

$28 = 5x - 4y$

$5x - 4y = 28$

9. $(2, 1), (6, 5)$

$m = \frac{5 - 1}{6 - 2} = \frac{4}{4} = 1$

$y - 1 = 1(x - 2)$

$y - 1 = x - 2$

$-1 = x - y - 2$

$1 = x - y$

$x - y = 1$

11. $(-2, -3), (2, 7)$

$m = \frac{7 - (-3)}{2 - (-2)} = \frac{10}{4} = \frac{5}{2}$

$y - 7 = \frac{5}{2}(x - 2)$

$2(y - 7) = 2\left[\frac{5}{2}(x - 2)\right]$

$2y - 14 = 5(x - 2)$

$2y - 14 = 5x - 10$

$-14 = 5x - 2y - 10$

$-4 = 5x - 2y$

$5x - 2y = -4$

13. $(-3, 2), (4, 1)$

$m = \frac{1 - 2}{4 - (-3)} = \frac{-1}{7}$

$y - 1 = -\frac{1}{7}(x - 4)$

$7(y - 1) = 7\left[-\frac{1}{7}(x - 4)\right]$

$7y - 7 = -1(x - 4)$

$7y - 7 = -x + 4$

$x + 7y - 7 = 4$

$x + 7y = 11$

15. $(-1, -4), (3, -6)$

$m = \frac{-6 - (-4)}{3 - (-1)} = \frac{-2}{4} = -\frac{1}{2}$

$y - (-6) = -\frac{1}{2}(x - 3)$

$y + 6 = -\frac{1}{2}(x - 3)$

$2(y + 6) = 2\left[-\frac{1}{2}(x - 3)\right]$

$2y + 12 = -1(x - 3)$

$2y + 12 = -x + 3$

$x + 2y + 12 = 3$

$x + 2y = -9$

17. $(0, 0), (5, 7)$

$m = \frac{7 - 0}{5 - 0} = \frac{7}{5}$

$y - 0 = \frac{7}{5}(x - 0)$

$y = \frac{7}{5}x$

$5(y) = 5\left(\frac{7}{5}x\right)$

$5y = 7x$

$0 = 7x - 5y$

$7x - 5y = 0$

19. $m = \frac{3}{7}, b = 4$

$y = \frac{3}{7}x + 4$

21. $m = 2, b = -3$

$y = 2x - 3$

23. $m = -\frac{2}{5}, b = 1$

$y = -\frac{2}{5}x + 1$

25. $m = 0, b = -4$

$y = 0x - 4$

$y = -4$

27. x-intercept of 2 and y-intercept of -4.

$(2, 0), (0, -4)$

$m = \frac{-4-0}{0-2} = \frac{-4}{-2} = 2$
$y = mx + b$
$y = 2x - 4$
$0 = 2x - y - 4$
$4 = 2x - y$
$2x - y = 4$

29. x-intercept of -3 and slope of $-\frac{5}{8}$
$(-3, 0)$, $m = -\frac{5}{8}$
$y - 0 = -\frac{5}{8}[x - (-3)]$
$y = -\frac{5}{8}(x + 3)$
$8(y) = 8\left[-\frac{5}{8}(x+3)\right]$
$8y = -5(x + 3)$
$8y = -5x - 15$
$5x + 8y = -15$

31. $(2, -4)$; parallel to y-axis
$x = 2$
$x + 0y = 2$

33. $(5, 6)$; perpendicular to y-axis
$m = 0$
$y - 6 = 0(x - 5)$
$y - 6 = 0$
$y = 6$
$0x + y = 6$

35. $(1, 3)$; parallel to the line $x + 5y = 9$
Parallel lines have the same slope.
$x + 5y = 9$
$\frac{1}{5}(5y) = \frac{1}{5}(-x + 9)$
$y = -\frac{1}{5}x - \frac{9}{5}$
$m = -\frac{1}{5}$
$y - 3 = -\frac{1}{5}(x - 1)$
$5(y - 3) = 5\left[-\frac{1}{5}(x-1)\right]$
$5y - 15 = -1(x - 1)$
$5y - 15 = -x + 1$
$x + 5y - 15 = 1$
$x + 5y = 16$

37. Contains the origin and is parallel to the line $4x - 7y = 3$
Parallel lines have the same slope.
$4x - 7y = 3$
$-7y = -4x + 3$
$-\frac{1}{7}(-7y) = -\frac{1}{7}(-4x + 3)$
$y = \frac{4}{7}x - \frac{3}{7}$
$m = \frac{4}{7}$
We have the point $(0, 0)$ and $m = \frac{4}{7}$.
$y - 0 = \frac{4}{7}(x - 0)$
$y = \frac{4}{7}x$
$7(y) = 7\left(\frac{4}{7}x\right)$
$7y = 4x$
$0 = 4x - 7y$
$4x - 7y = 0$

39. $(-1, 3)$; perpendicular to the line $2x - y = 4$
Perpendicular lines have slopes that are negative reciprocals.
$2x - y = 4$
$-y = -2x + 4$
$-(-y) = -(-2x + 4)$
$y = 2x + 4$
$m = 2$
The negative reciprocal is $-\frac{1}{2}$.
$y - 3 = -\frac{1}{2}[x - (-1)]$
$y - 3 = -\frac{1}{2}(x + 1)$
$2(y - 3) = 2\left[-\frac{1}{2}(x+1)\right]$
$2y - 6 = -1(x + 1)$
$2y - 6 = -x - 1$
$x + 2y - 6 = -1$
$x + 2y = 5$

41. Contains the origin (0, 0) and is perpendicular to the line $-2x+3y=8$
Perpendicular lines have slopes that are negative reciprocals.

$-2x+3y=8$

$3y=2x+8$

$\frac{1}{3}(3y)=\frac{1}{3}(2x+8)$

$y=\frac{2}{3}x+\frac{8}{3}$

$m=\frac{2}{3}$

The negative reciprocal is $-\frac{3}{2}$.

$y-0=-\frac{3}{2}(x-0)$

$y=-\frac{3}{2}x$

$2(y)=2\left(-\frac{3}{2}x\right)$

$2y=-3x$

$3x+2y=0$

43. $3x+y=7$

$y=-3x+7$

$m=-3$ and $b=7$

45. $3x+2y=9$

$2y=-3x+9$

$\frac{1}{2}(2y)=\frac{1}{2}(-3x+9)$

$y=-\frac{3}{2}x+\frac{9}{2}$

$m=-\frac{3}{2}$ and $b=\frac{9}{2}$

47. $x=5y+12$

$x-12=5y$

$\frac{1}{5}(x-12)=\frac{1}{5}(5y)$

$y=\frac{1}{5}x-\frac{12}{5}$

$m=\frac{1}{5}$ and $b=-\frac{12}{5}$

49. $y=\frac{2}{3}x-4$

51. $y=2x+1$

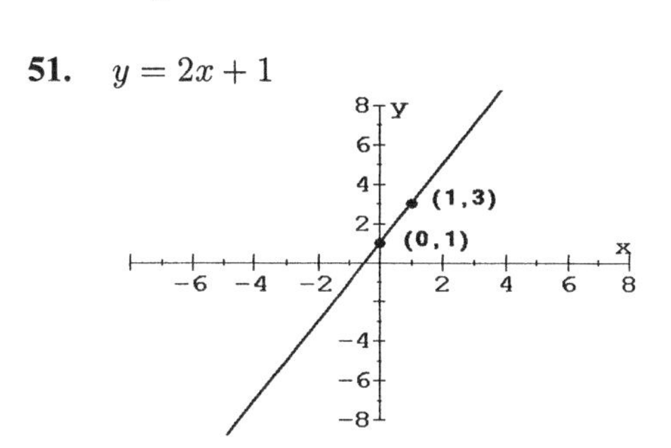

53. $y=-\frac{3}{2}x+4$

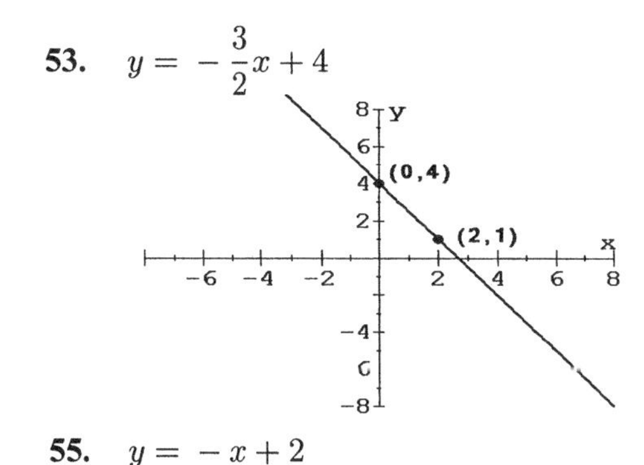

55. $y=-x+2$

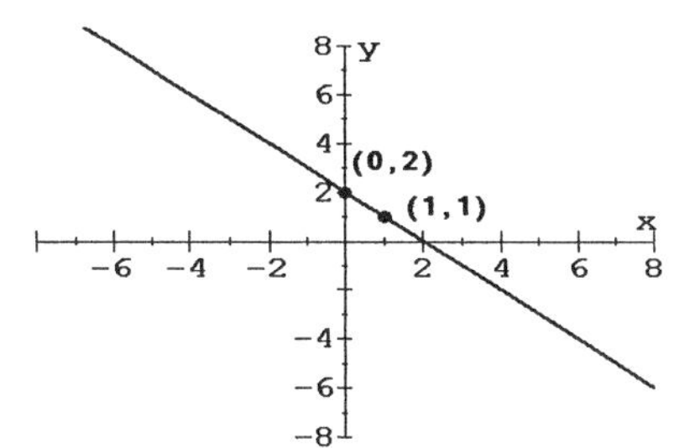

57. Let x = square footage and let y = pounds of fertilizer.
Ordered pairs are: $(5000, 7)$; $(10000, 12)$

$m=\frac{12-7}{10000-5000}=\frac{5}{5000}=\frac{1}{1000}$

$$y - 7 = \frac{1}{1000}\left(x - 5000\right)$$
$$1000(y - 7) = 1000\left[\frac{1}{1000}\left(x - 5000\right)\right]$$
$$1000y - 7000 = x - 5000$$
$$1000y = x + 2000$$
Therefore, the equation is $y = \frac{1}{1000}x + 2.$

59. Let $y =$ temperature in °F
and $x =$ temperature in °C.
Ordered pairs are: (10, 50), (– 5, 23)
$$m = \frac{23 - 50}{-5 - 10} = \frac{-27}{-15} = \frac{9}{5}$$
$$y - 50 = \frac{9}{5}(x - 10)$$
$$5(y - 50) = 5\left[\frac{9}{5}(x - 10)\right]$$
$$5y - 250 = 9(x - 10)$$
$$5y - 250 = 9x - 90$$
$$5y = 9x + 160$$
Therefore, the equation is $y = \frac{9}{5}x + 32.$

65. The slope-intercept form of an equation is $y = mx + b$ where m is the slope and b is the y-intercept.
Write $Ax + By = C$ and $A'x + B'y = C'$ in slope-intercept form.
$$Ax + By = C$$
$$By = -Ax + C$$
$$\frac{1}{B}(By) = \frac{1}{B}(-Ax + C)$$
$$y = -\frac{A}{B}x + \frac{C}{B}$$
Likewise,
$$A'x + B'y = C'$$
$$B'y = -A'x + C'$$
$$\frac{1}{B'}(B'y) = \frac{1}{B'}(-A'x + C')$$
$$y = -\frac{A'}{B'}x + \frac{C'}{B'}$$

a. If $\frac{A}{A'} = \frac{B}{B'}$, then using the cross-product
$$AB' = A'B$$
$$\frac{1}{BB'}(AB') = \frac{1}{BB'}(A'B)$$
$\frac{A}{B} = \frac{A'}{B'}$ and $-\frac{A}{B} = -\frac{A'}{B'}.$
If $-\frac{A}{B} = -\frac{A'}{B'}$, then the slopes of the equations are equal.

If $\frac{B}{B'} \neq \frac{C}{C'}$, then
$$BC' \neq B'C$$
$$\frac{1}{BB'}(BC') \neq \frac{1}{BB'}(B'C)$$
$$\frac{C'}{B'} \neq \frac{C}{B}.$$
If $\frac{C'}{B'} \neq \frac{C}{B}$, then the y-intercepts are not equal.

If slopes are equal and y-intercepts are not equal, then the lines are parallel and not the same line.

b. Perpendicular lines have slopes which are negative reciprocals which means their product is -1.

Multiply the two slopes,
$$\left(-\frac{B}{A}\right)\left(-\frac{B'}{A'}\right) = \frac{BB'}{AA'}$$
If $AA' = -BB'$, then substitute $-BB'$ for AA' into $\frac{BB'}{AA'}$.
$$\frac{BB'}{AA'} = \frac{BB'}{-BB'} = -1$$
Therefore the lines are perpencicular.

PROBLEM SET 12.3 Graphing Parabolas

1. $y = x^2 + 2$

3. $y = x^2 - 1$

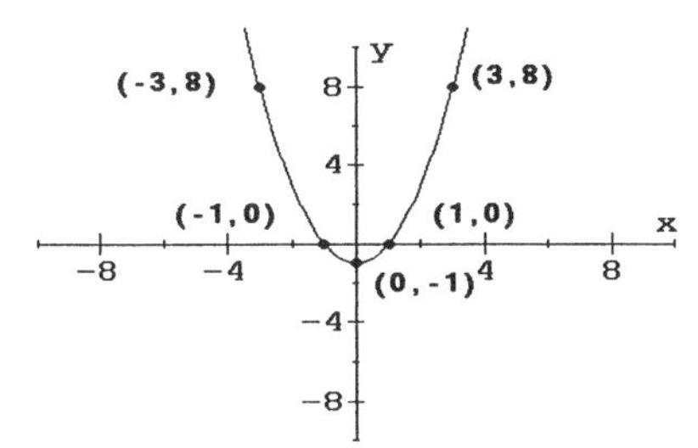

5. $y = 4x^2$

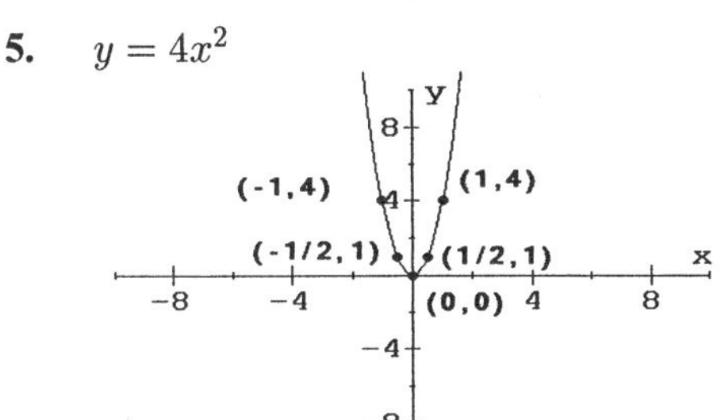

7. $y = -3x^2$

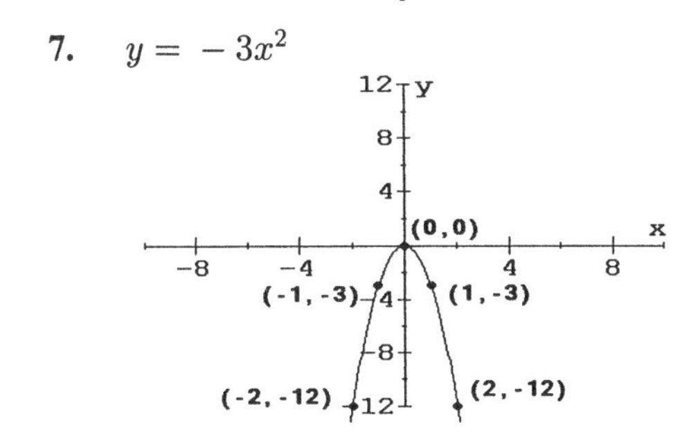

9. $y = \frac{1}{3}x^2$

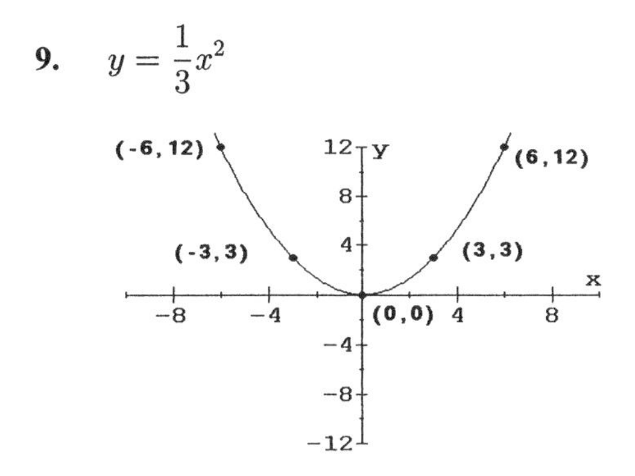

11. $y = -\frac{1}{2}x^2$

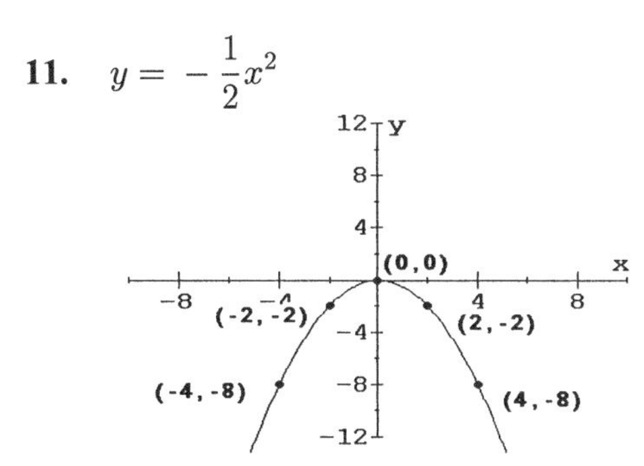

13. $y = (x-1)^2$

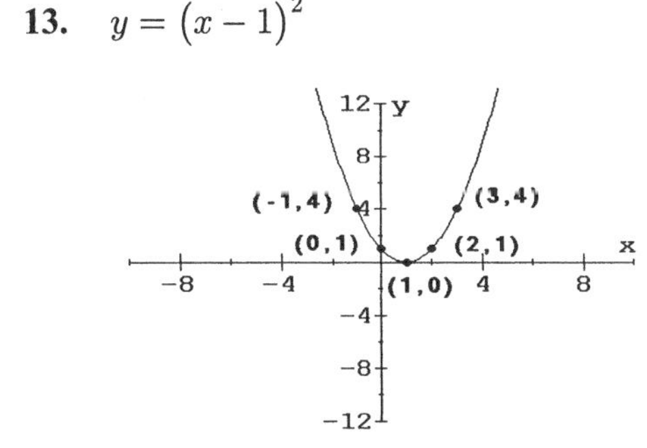

15. $y = (x+4)^2$

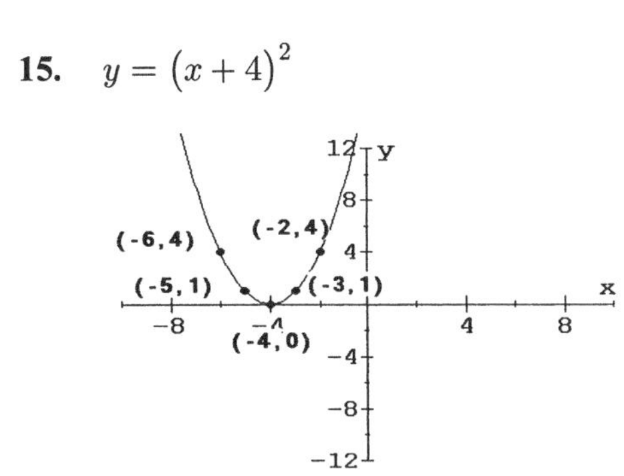

Problem Set 12.3

17. $y = 3x^2 + 2$

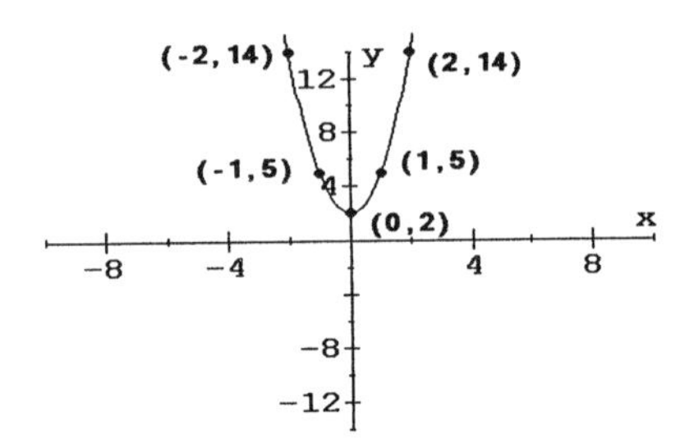

19. $y = -2x^2 - 2$

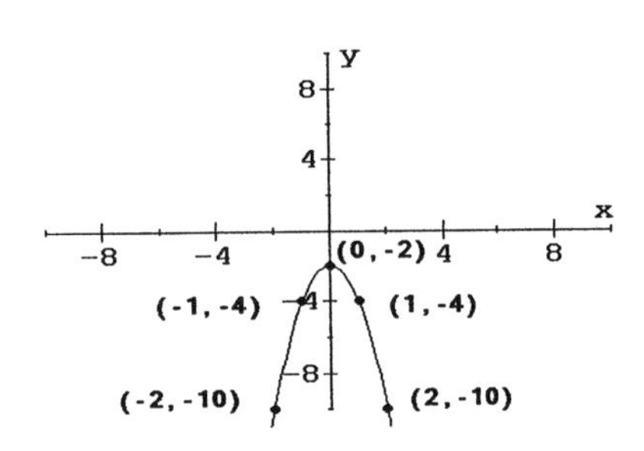

21. $y = (x-1)^2 - 2$

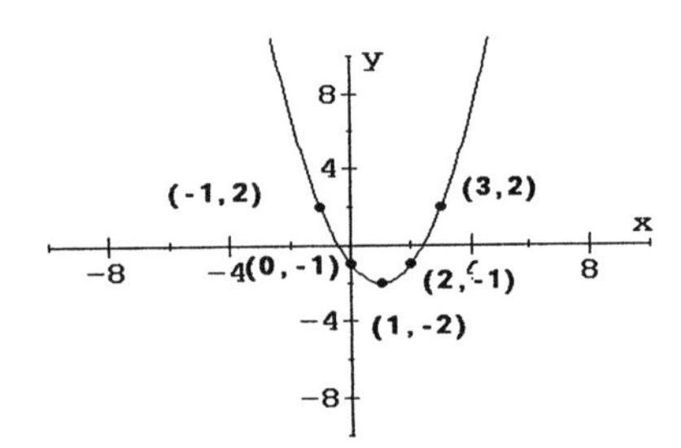

23. $y = (x+2)^2 + 1$

25. $y = 3(x-2)^2 - 4$

27. $y = -(x+4)^2 + 1$

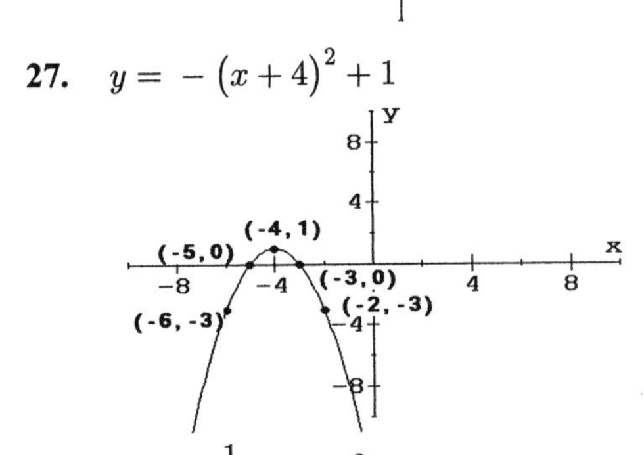

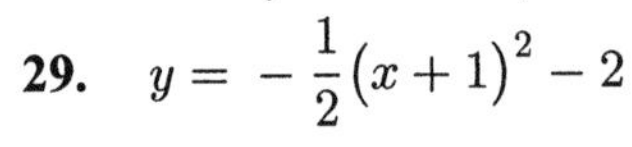

29. $y = -\frac{1}{2}(x+1)^2 - 2$

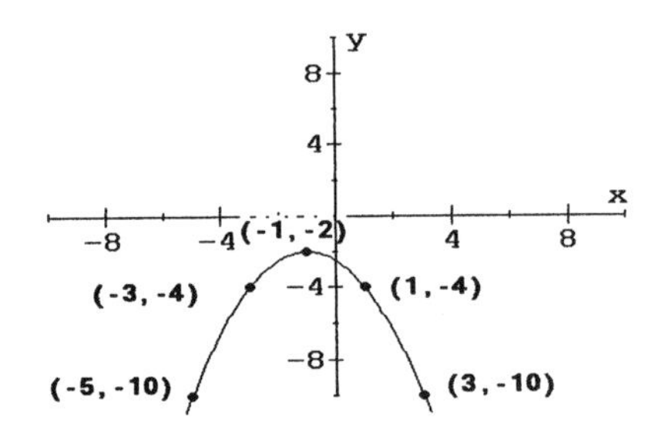

PROBLEM SET 12.4 More Parabolas and Some Circles

1. $y = x^2 - 6x + 13$

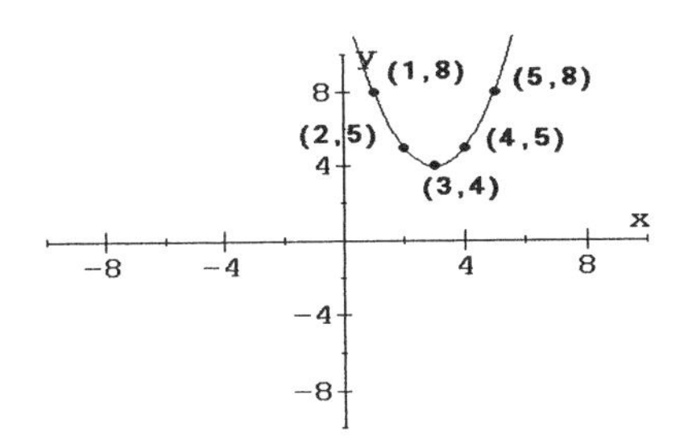

3. $y = x^2 + 2x + 6$

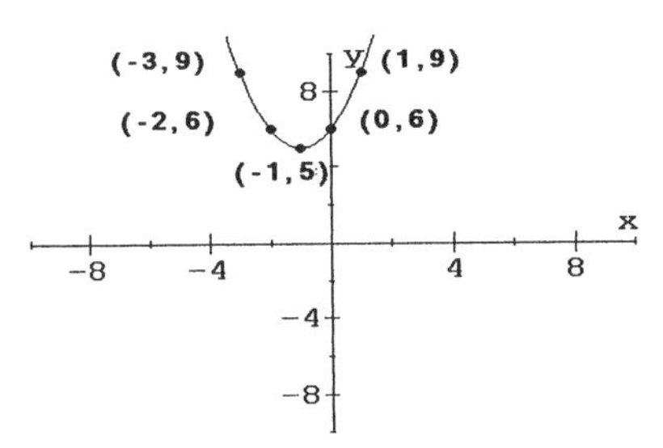

5. $y = x^2 - 5x + 3$

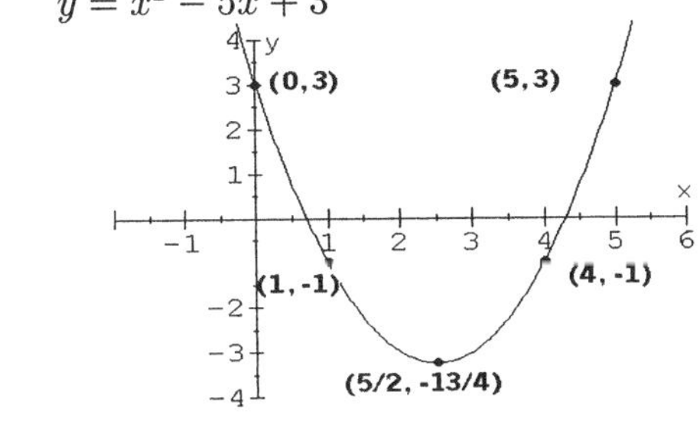

7. $y = x^2 + 7x + 14$

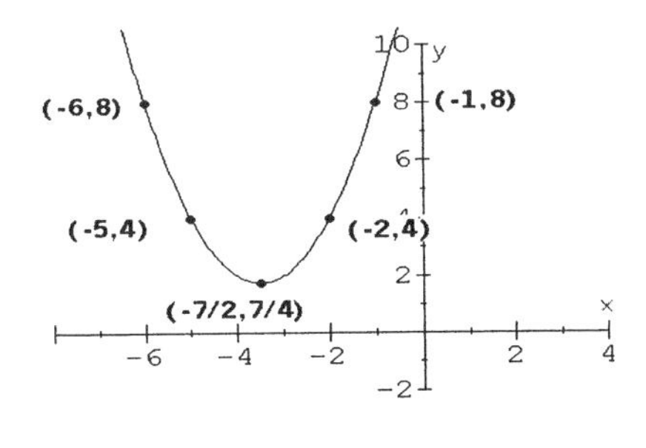

9. $y = 3x^2 - 6x + 5$

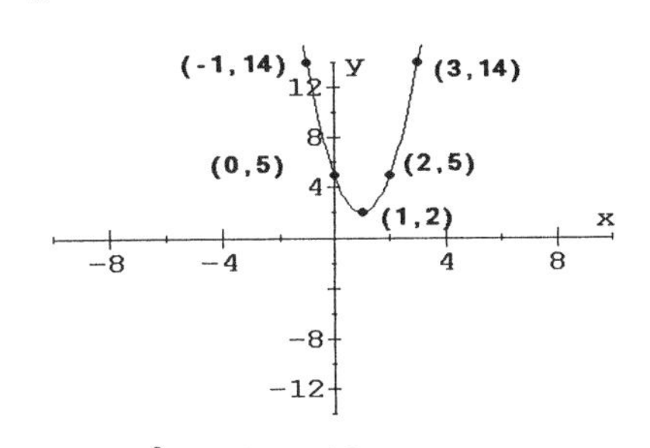

11. $y = 4x^2 - 24x + 32$

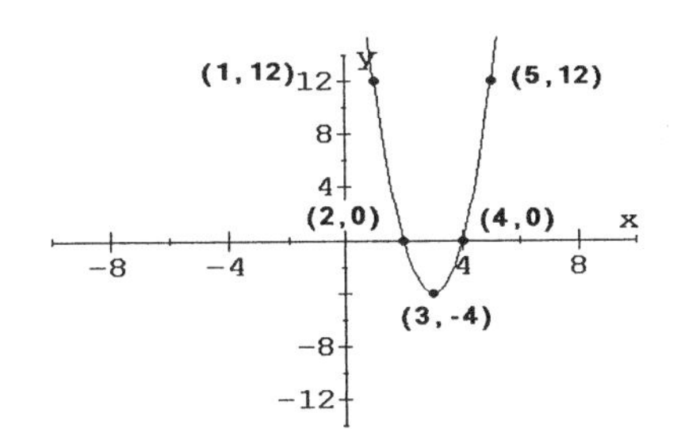

13. $y = -2x^2 - 4x - 5$

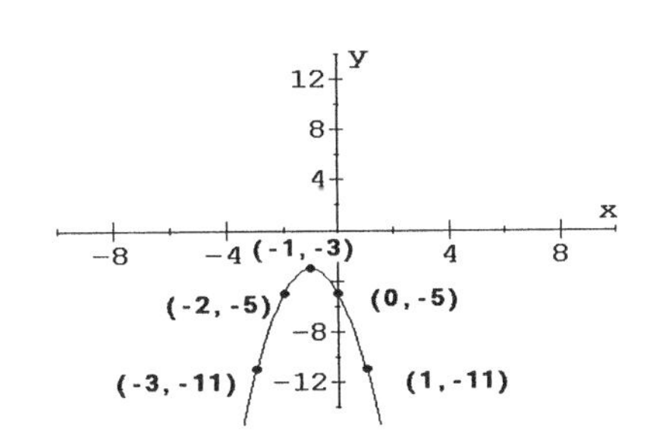

15. $y = -x^2 + 8x - 21$

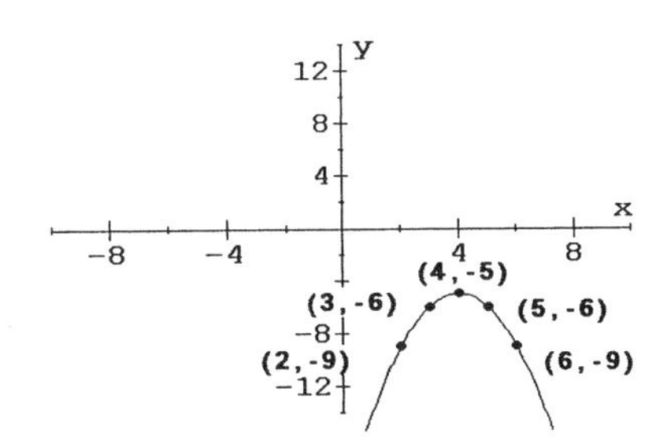

Problem Set 12.4

17. $y = 2x^2 - x + 2$

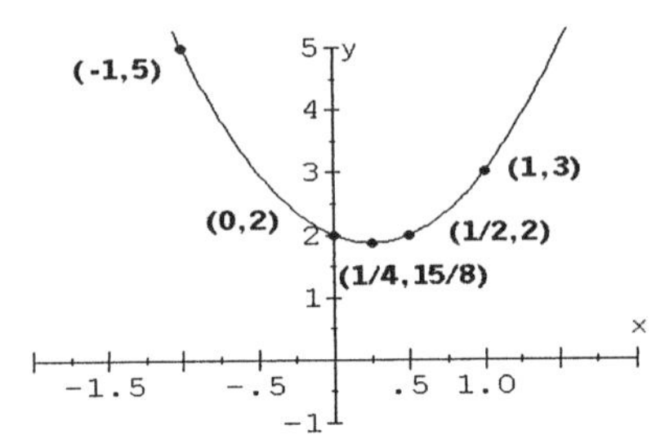

19. $y = 3x^2 + 2x + 1$

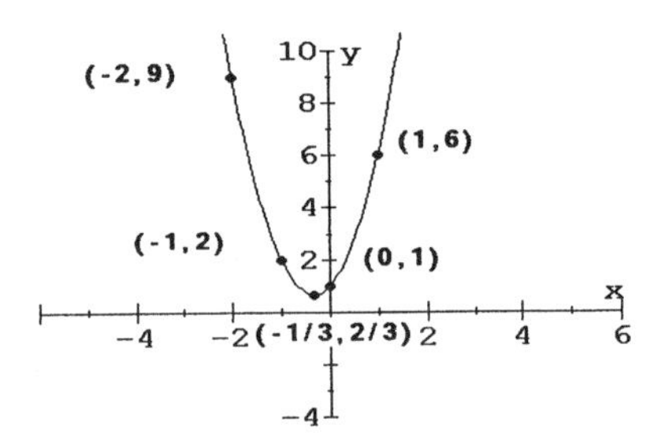

21. $y = -3x^2 - 7x - 2$

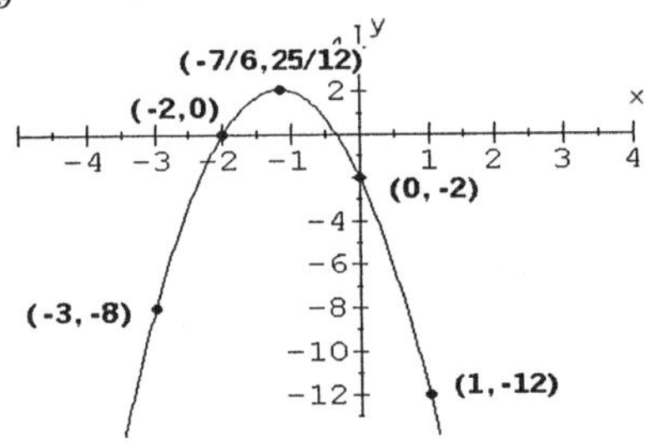

23. $x^2 + y^2 - 2x - 6y - 6 = 0$
$x^2 - 2x + y^2 - 6y = 6$
$x^2 - 2x + 1 + y^2 - 6y + 9 = 6 + 1 + 9$
$(x - 1)^2 + (y - 3)^2 = 16$
center (1, 3), r $= \sqrt{16} = 4$

25. $x^2 + y^2 + 6x + 10y + 18 = 0$
$x^2 + 6x + y^2 + 10y = -18$
$x^2 + 6x + 9 + y^2 + 10y + 25 = -18+9+25$
$(x + 3)^2 + (y + 5)^2 = 16$
center $(-3, -5)$, $r = \sqrt{16} = 4$

27. $x^2 + y^2 = 10$
$(x - 0)^2 + (y - 0)^2 = 10$
center (0, 0), $r = \sqrt{10}$

29. $x^2 + y^2 - 16x + 6y + 71 = 0$
$x^2 - 16x + y^2 + 6y = -71$
$x^2 - 16x + 64 + y^2 + 6y + 9 = -71 + 64 + 9$
$(x - 8)^2 + (y + 3)^2 = 2$
center $(8, -3)$, $r = \sqrt{2}$

31. $x^2 + y^2 + 6x - 8y = 0$
$x^2 + 6x + y^2 - 8y = 0$
$x^2 + 6x + 9 + y^2 - 8y + 16 = 0 + 9 + 16$
$(x + 3)^2 + (y - 4)^2 = 25$
center $(-3, 4)$, $r = \sqrt{25} = 5$

33. $4x^2 + 4y^2 + 4x - 32y + 33 = 0$
$\frac{1}{4}(4x^2 + 4y^2 + 4x - 32y + 33) = \frac{1}{4}(0)$
$x^2 + y^2 + x - 8y + \frac{33}{4} = 0$
$x^2 + x + y^2 - 8y = -\frac{33}{4}$
$x^2 + x + \frac{1}{4} + y^2 - 8y + 16 = -\frac{33}{4} + \frac{1}{4} + 16$
$\left(x + \frac{1}{2}\right)^2 + (y - 4)^2 = 8$
center $\left(-\frac{1}{2}, 4\right)$; $r = \sqrt{8} = 2\sqrt{2}$

35. $x^2 + y^2 = 25$

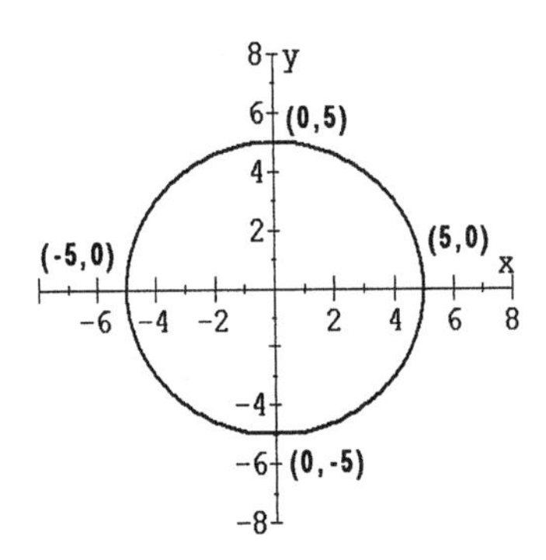

37. $(x-1)^2+(y+2)^2=9$

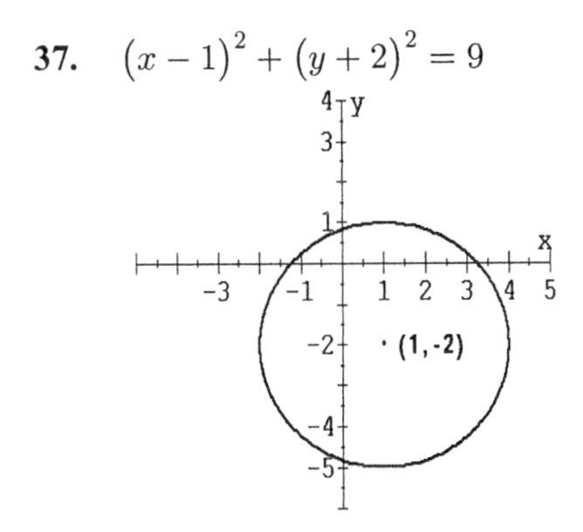

39. $x^2+y^2+6x-2y+6=0$

C(-3,1)

41. $x^2+y^2+4y-5=0$

C(0,-2)

43. $x^2+y^2+4x+4y-8=0$

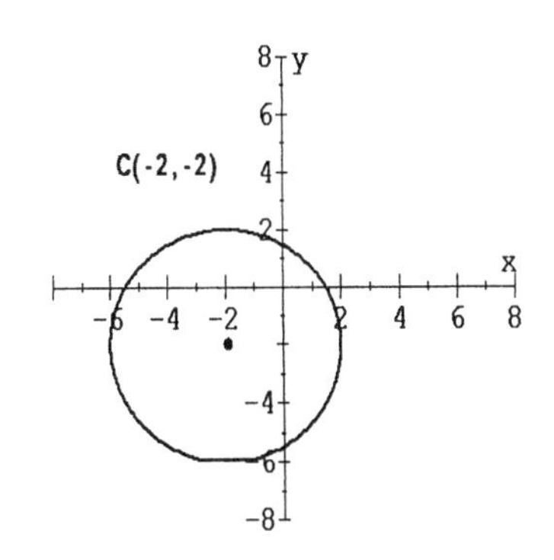

45. center at (3, 5) and $r=5$
$(x-3)^2+(y-5)^2=5^2$
$x^2-6x+9+y^2-10y+25=25$
$x^2+y^2-6x-10y+9=0$

47. center at $(-4, 1)$ and $r=8$
$[x-(-4)]^2+(y-1)^2=8^2$
$(x+4)^2+(y-1)^2=8^2$
$x^2+8x+16+y^2-2y+1=64$
$x^2+y^2+8x-2y-47=0$

49. center at $(-2,-6)$ and $r=3\sqrt{2}$
$[x-(-2)]^2+[y-(-6)]^2=(3\sqrt{2})^2$
$(x+2)^2+(y+6)^2=9(2)$
$x^2+4x+4+y^2+12y+36=18$
$x^2+y^2+4x+12y+22=0$

51. center (0, 0) and $r=2\sqrt{5}$
$(x-0)^2+(y-0)^2=(2\sqrt{5})^2$
$x^2+y^2=4(5)$
$x^2+y^2=20$
$x^2+y^2-20=0$

53. center at $(5,-8)$ and $r=4\sqrt{6}$
$(x-5)^2+[y-(-8)]^2=(4\sqrt{6})^2$
$(x-5)^2+(y+8)^2=16(6)$
$x^2-10x+25+y^2+16y+64=96$
$x^2+y^2-10x+16y-7=0$

55. Passing through (0, 0)
and center at (0, 4)
$r=\sqrt{(0-0)^2+(4-0)^2}$
$r=\sqrt{0+4^2}=\sqrt{4^2}=4$
$(x-0)^2+(y-4)^2=4^2$
$x^2+y^2-8y+16=16$
$x^2+y^2-8y=0$

57. Passing through (0, 0)
and center at $(-4, 3)$
$r=\sqrt{(-4-0)^2+(3-0)^2}$
$r=\sqrt{(-4)^2+3^2}=\sqrt{16+9}=\sqrt{25}=5$
$[x-(-4)]^2+(y-3)^2=5^2$
$(x+4)^2+(y-3)^2=25$
$x^2+8x+16+y^2-6y+9=25$
$x^2+y^2+8x-6y=0$

Further Investigations

63. $h = \frac{D}{-2}, k = \frac{E}{-2}, r = \sqrt{h^2 + k^2 - F}$

a. $x^2 + y^2 - 2x - 8y + 8 = 0$

$h = \frac{-2}{-2} = 1 \qquad k = \frac{-8}{-2} = 4$

$r = \sqrt{1^2 + 4^2 - 8}$

$r = \sqrt{1 + 16 - 8} = \sqrt{9} = 3$

center $(1, 4)$, $r = 3$

b. $x^2 + y^2 + 4x - 14y + 49 = 0$

$h = \frac{4}{-2} = -2 \qquad k = \frac{-14}{-2} = 7$

$r = \sqrt{(-2)^2 + 7^2 - 49}$

$r = \sqrt{4 + 49 - 49} = \sqrt{4} = 2$

center $(-2, 7)$, $r = 2$

c. $x^2 + y^2 + 12x + 8y - 12 = 0$

$h = \frac{12}{-2} = -6 \qquad k = \frac{8}{-2} = -4$

$r = \sqrt{(-6)^2 + (-4)^2 - (-12)}$

$r = \sqrt{36 + 16 + 12} = \sqrt{64} = 8$

center $(-6, -4)$, $r = 8$

d. $x^2 + y^2 - 16x + 20y + 115 = 0$

$h = \frac{-16}{-2} = 8 \qquad k = \frac{20}{-2} = -10$

$r = \sqrt{8^2 + (-10)^2 - (115)}$

$r = \sqrt{64 + 100 - 115} = \sqrt{49} = 7$

center $(8, -10)$, $r = 7$

e. $x^2 + y^2 - 12y - 45 = 0$

$h = \frac{0}{-2} = 0 \qquad k = \frac{-12}{-2} = 6$

$r = \sqrt{0^2 + 6^2 - (-45)}$

$r = \sqrt{36 + 45} = \sqrt{81} = 9$

center $(0, 6)$, $r = 9$

f. $x^2 + y^2 + 14x = 0$

$h = \frac{14}{-2} = -7 \qquad k = \frac{0}{-2} = 0$

$r = \sqrt{(-7)^2 + 0^2 - 0} = \sqrt{49} = 7$

center $(-7, 0)$, $r = 7$

PROBLEM SET 12.5 Graphing Ellipses

1.

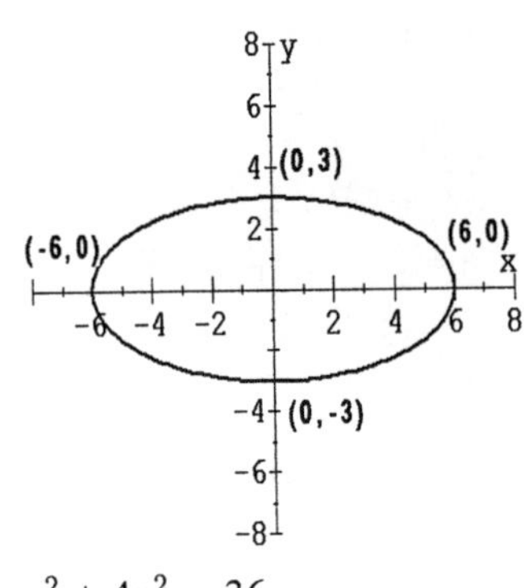

$x^2 + 4y^2 = 36$

Let $x = 0$.

$0^2 + 4y^2 = 36$

$y^2 = 9$

$y = \pm\sqrt{9} = \pm 3$

The endpoints of the minor axis are $(0, 3)$ and $(0, -3)$.

Let $y = 0$.

$x^2 + 4(0)^2 = 36$

$x^2 = 36$

$x = \pm\sqrt{36} = \pm 6$

The endpoints of the major axis are $(6, 0)$ and $(-6, 0)$.

3.

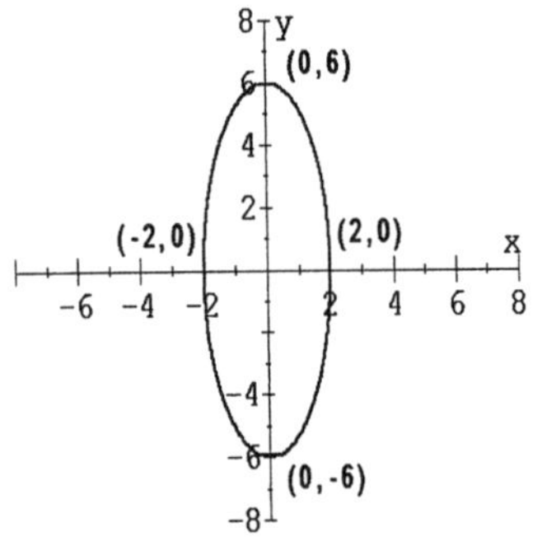

$9x^2 + y^2 = 36$

Let $x = 0$.
$9(0)^2 + y^2 = 36$
$y^2 = 36$
$y = \pm\sqrt{36} = \pm 6$
The endpoints of the major axis are (0, 6) and (0, − 6).
Let $y = 0$.
$9x^2 + 0^2 = 36$
$9x^2 = 36$
$x^2 = 4$
$x = \pm\sqrt{4} = \pm 2$
The endpoints of the minor axis are (2, 0) and (− 2, 0).

5.

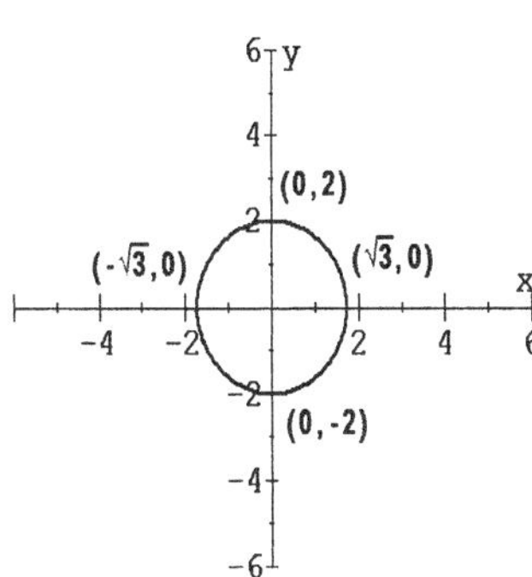

$4x^2 + 3y^2 = 12$
Let $x = 0$.
$4(0)^2 + 3y^2 = 12$
$3y^2 = 12$
$y^2 = 4$
$y = \pm 2$
The endpoints of the major axis are (0, 2) and (0, − 2).
Let $y = 0$.
$4x^2 + 3(0)^2 = 12$
$4x^2 = 12$
$x^2 = 3$
$x = \pm\sqrt{3}$
The endpoints of the minor axis are ($\sqrt{3}$, 0) and ($-\sqrt{3}$, 0).

7.

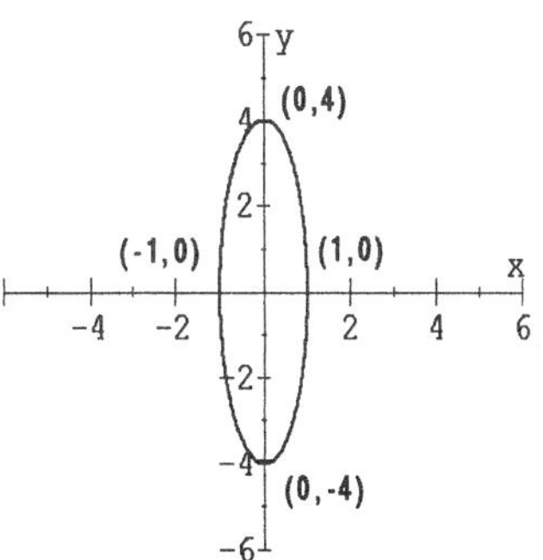

$16x^2 + y^2 = 16$
Let $x = 0$.
$16(0)^2 + y^2 = 16$
$y^2 = 16$
$y = \pm 4$
The endpoints of the major axis are (0, − 4) and (0, 4).
Let $y = 0$.
$16x^2 + 0^2 = 16$
$16x^2 = 16$
$x^2 = 1$
$x = \pm 1$
The endpoints of the minor axis are (− 1, 0) and (1, 0).

9.

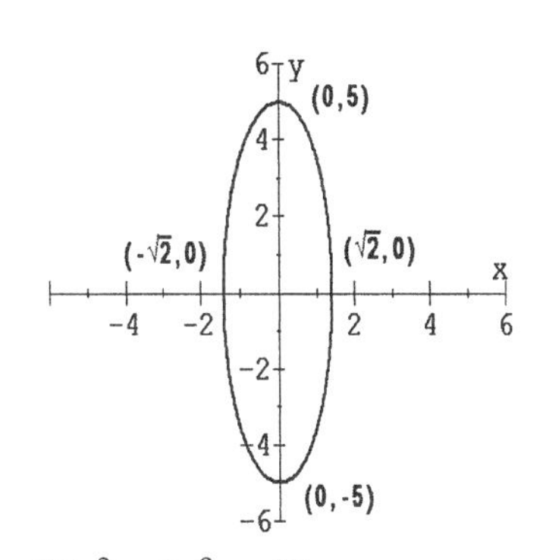

$25x^2 + 2y^2 = 50$
Let $x = 0$.
$25(0)^2 + 2y^2 = 50$
$2y^2 = 50$
$y^2 = 25$
$y = \pm\sqrt{25} = \pm 5$
The endpoints of the major axis are (0, 5) and (0, − 5).

Let $y = 0$.
$25x^2 + 2(0)^2 = 50$
$25x^2 = 50$
$x^2 = 2$
$x = \pm\sqrt{2}$
The endpoints of the minor axis are $(\sqrt{2}, 0)$ and $(-\sqrt{2}, 0)$.

11.

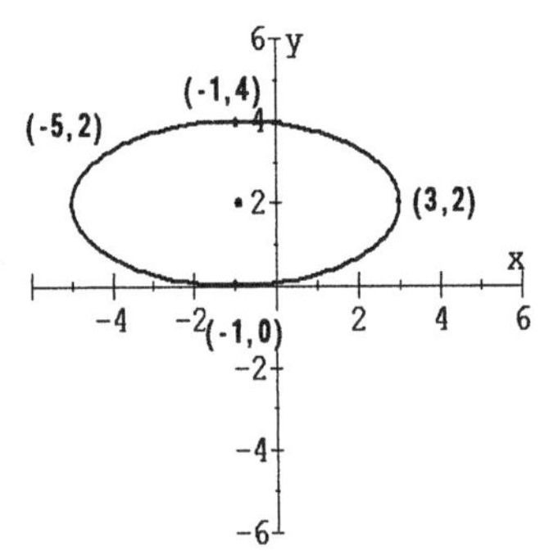

$4x^2 + 8x + 16y^2 - 64y + 4 = 0$
$4(x^2 + 2x) + 16(y^2 - 4y) = -4$
$4(x^2+2x+1) + 16(y^2 - 4y+4) = -4+4+64$
$4(x + 1)^2 + 16(y - 2)^2 = 64$
Let $x = -1$.
$4(-1 + 1)^2 + 16(y - 2)^2 = 64$
$16(y - 2)^2 = 64$
$(y - 2)^2 = 4$
$y - 2 = \pm 2$
$y - 2 = 2$ or $y - 2 = -2$
$y = 4$ or $y = 0$
The endpoints of the minor axis are $(-1, 4)$ and $(-1,0)$.
Let $y = 2$.
$4(x + 1)^2 + 16(2 - 2)^2 = 64$
$4(x + 1)^2 = 64$
$(x + 1)^2 = 16$
$x + 1 = \pm 4$
$x + 1 = 4$ or $x + 1 = -4$
$x = 3$ or $x = -5$
The endpoints of the major axis are $(-5, 2)$ and $(3, 2)$.

13.

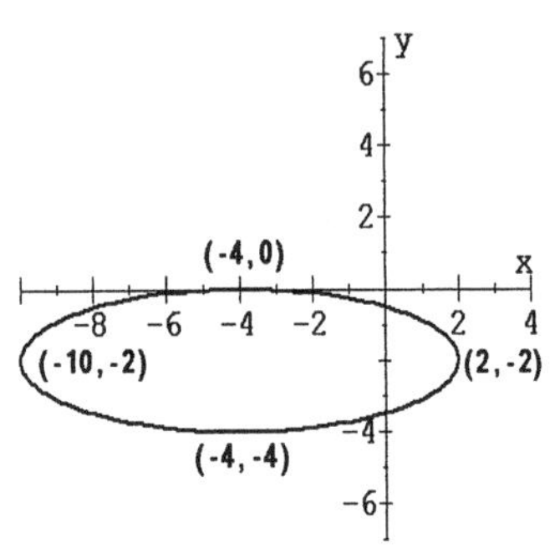

$x^2 + 8x + 9y^2 + 36y + 16 = 0$
$x^2 + 8x + 9(y^2 + 4y) = -16$
$x^2+8x+16 + 9(y^2+4y+4) = -16+16+36$
$(x + 4)^2 + 9(y + 2)^2 = 36$
Let $x = -4$.
$(-4 + 4)^2 + 9(y + 2)^2 = 36$
$9(y + 2)^2 = 36$
$(y + 2)^2 = 4$
$y + 2 = \pm 2$
$y + 2 = 2$ or $y + 2 = -2$
$y = 0$ or $y = -4$
The endpoints of the minor axis are $(-4, 0)$ and $(-4, -4)$.
Let $y = -2$.
$(x + 4)^2 + 9(-2 + 2)^2 = 36$
$(x + 4)^2 = 36$
$x + 4 = \pm 6$
$x + 4 = 6$ or $x + 4 = -6$
$x = 2$ or $x = -10$
The endpoints of the major axis are $(2, -2)$ and $(-10, -2)$.

15.

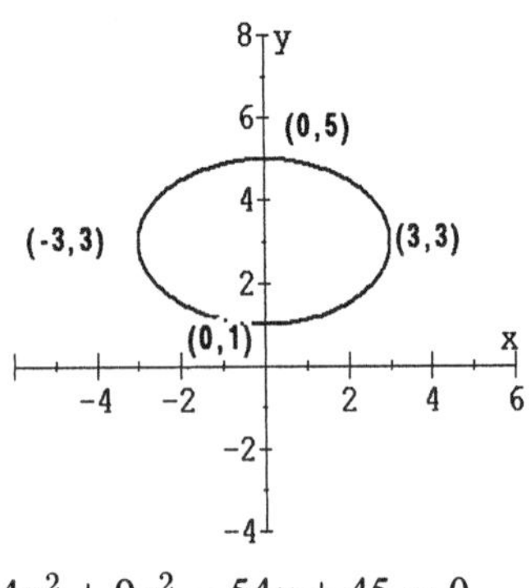

$4x^2 + 9y^2 - 54y + 45 = 0$
$4x^2 + 9(y^2 - 6y) = -45$
$4x^2 + 9(y^2 - 6y + 9) = -45 + 81$
$4x^2 + 9(y - 3)^2 = 36$

Let $x = 0$.
$4(0)^2 + 9(y-3)^2 = 36$
$9(y-3)^2 = 36$
$(y-3)^2 = 4$
$y - 3 = \pm 2$
$y - 3 = 2$ or $y - 3 = -2$
$y = 5$ or $y = 1$
The endpoints of the minor axis are (0, 5) and (0, 1).

Let $y = 3$.
$4x^2 + 9(3-3)^2 = 36$
$4x^2 = 36$
$x^2 = 9$
$x = \pm 3$
The endpoints of the major axis are (3, 3) and (− 3, 3).

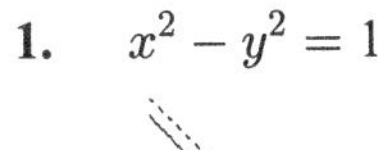

PROBLEM SET 12.6 Graphing Hyperbolas

1. $x^2 - y^2 = 1$

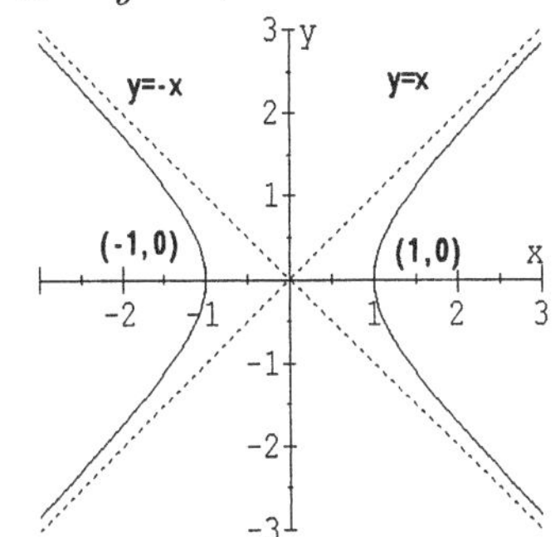

Let $y = 0$.
$x^2 - 0^2 = 1$
$x^2 = 1$
$x = \pm\sqrt{1} = \pm 1$
(1, 0), (− 1, 0)
Let $x = 0$.
$0^2 - y^2 = 1$
$y^2 = -1$
Has no real solutions.
The hyperbola does not intersect the y-axis.
Asymptotes :
$x^2 - y^2 = 0$
$-y^2 = -x^2$
$y^2 = x^2$
$y = \pm\sqrt{x^2}$
$y = \pm x$

3. $y^2 - 4x^2 = 9$

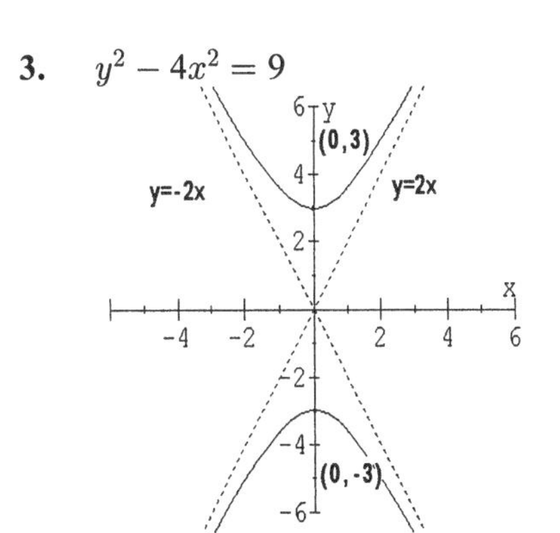

Let $x = 0$.
$y^2 - 4(0)^2 = 9$
$y^2 = 9$
$y = \pm\sqrt{9} = \pm 3$
(0, 3); (0, − 3)
Let $y = 0$.
$0^2 - 4x^2 = 9$
$x^2 = -\frac{9}{4}$
Has no real solutions.
The hyperbola does not intersect the x-axis.
Asymptotes:
$y^2 - 4x^2 = 0$
$y^2 = 4x^2$
$y = \pm\sqrt{4x^2}$
$y = \pm 2x$

Problem Set 12.6

5. $5x^2 - 2y^2 = 20$

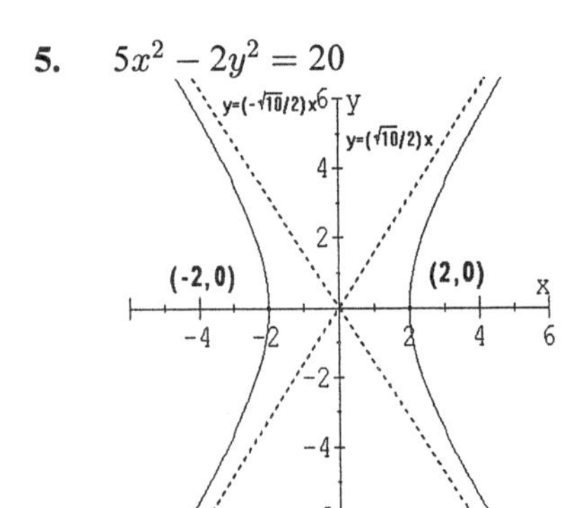

Let $y = 0$.
$5x^2 - 2(0)^2 = 20$
$5x^2 = 20$
$x^2 = 4$
$x = \pm\sqrt{4} = \pm 2$
$(2, 0), (-2, 0)$
Let $x = 0$.
$5(0)^2 - 2y^2 = 20$
$-2y^2 = 20$
$y^2 = -10$
Has no real solutions.
The hyperbola does not intersect the y-axis.
Asymptotes:
$5x^2 - 2y^2 = 0$
$-2y^2 = -5x^2$
$y^2 = \frac{5}{2}x^2$
$y = \pm\sqrt{\frac{5}{2}x^2}$
$y = \pm\frac{\sqrt{5}}{\sqrt{2}} \bullet \frac{\sqrt{2}}{\sqrt{2}}x$
$y = \pm\frac{\sqrt{10}}{2}x$

7. $y^2 - 16x^2 = 4$

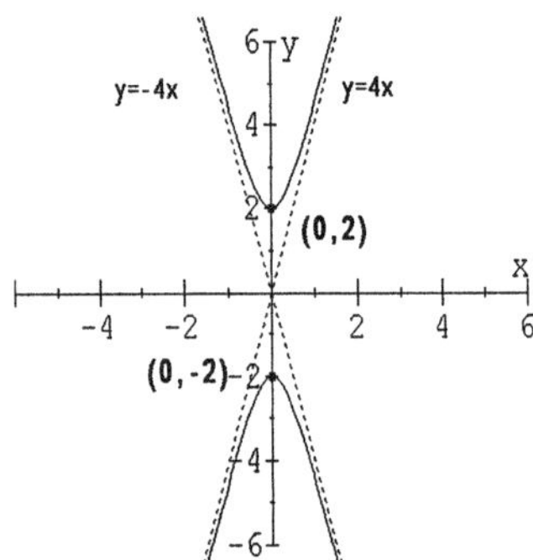

Let $x = 0$.
$y^2 - 16(0)^2 = 4$
$y^2 = 4$
$y = \pm\sqrt{4} = \pm 2$
$(0, 2), (0, -2)$
Let $y = 0$.
$0^2 - 16x^2 = 4$
$x^2 = -\frac{4}{16} = -\frac{1}{4}$
Has no real solutions.
The hyperbola does not cross the x-axis.
Asymptotes:
$y^2 - 16x^2 = 0$
$y^2 = 16x^2$
$y = \pm\sqrt{16x^2}$
$y = \pm 4x$

9. $-4x^2 + y^2 = -4$
$-1(-4x^2 + y^2) = -1(-4)$
$4x^2 - y^2 = 4$

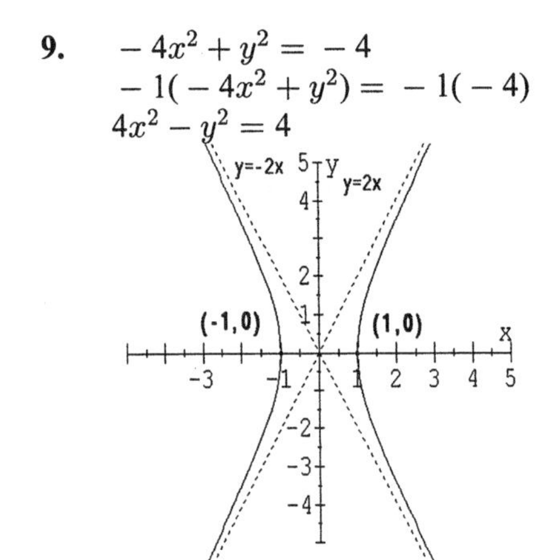

Let $y = 0$.
$4x^2 - 0^2 = 4$
$4x^2 = 4$
$x^2 = 1$
$x = \pm\sqrt{1} = \pm 1$

$(1, 0), (-1, 0)$
Let $x = 0$.
$4(0)^2 - y^2 = 4$
$-y^2 = 4$
$y^2 = -4$
Has no real solutions. The hyperbola does not cross the y-axis.
Asymptotes:
$4x^2 - y^2 = 0$
$-y^2 = -4x^2$
$y^2 = 4x^2$
$y = \pm\sqrt{4x} = \pm 2x$

11. $25y^2 - 3x^2 = 75$

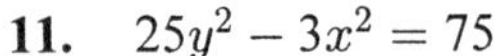

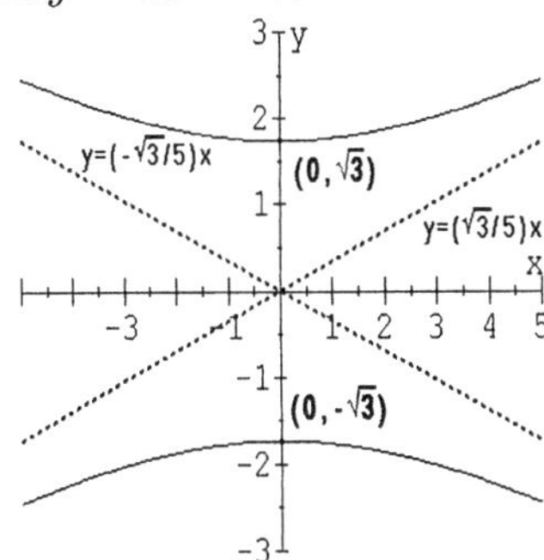

Let $x = 0$.
$25y^2 - 3(0)^2 = 75$
$25y^2 = 75$
$y^2 = 3$
$y = \pm\sqrt{3}$
$(0, \sqrt{3}), (0, -\sqrt{3})$
Let $y = 0$.
$25(0)^2 - 3x^2 = 75$
$-3x^2 = 75$
$x^2 = -25$
Has no real solutions. The hyperbola does not cross the x-axis.
Asymptotes:
$25y^2 - 3x^2 = 0$
$25y^2 = 3x^2$
$y^2 = \frac{3}{25}x^2$
$y = \pm\sqrt{\frac{3}{25}x^2}$
$y = \pm\frac{\sqrt{3}}{5}x$

13. $-4x^2 + 32x + 9y^2 - 18y - 91 = 0$
$-4(x^2 - 8x) + 9(y^2 - 2y) = 91$
$-4(x^2 - 8x+16)+9(y^2 - 2y+1) = 91 - 64+9$
$-4(x - 4)^2 + 9(y - 1)^2 = 36$

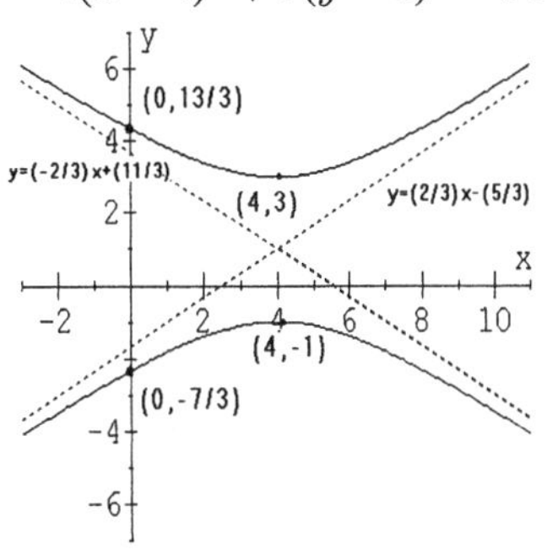

Let $x = 0$.
$-4(0 - 4)^2 + 9(y - 1)^2 = 36$
$-4(-4)^2 + 9(y - 1)^2 = 36$
$-4(16) + 9(y - 1)^2 = 36$
$-64 + 9(y - 1)^2 = 36$
$9(y - 1)^2 = 100$
$(y - 1)^2 = \frac{100}{9}$
$y - 1 = \pm\sqrt{\frac{100}{9}}$
$y - 1 = \pm\frac{10}{3}$
$y = 1 \pm \frac{10}{3}$
$y = \frac{3}{3} \pm \frac{10}{3}$
$y = -\frac{7}{3} \quad y = \frac{13}{3}$
$\left(0, -\frac{7}{3}\right), \left(0, \frac{13}{3}\right)$
Let $x = 4$.
$-4(4 - 4)^2 + 9(y - 1)^2 = 36$
$-4(0)^2 + 9(y - 1)^2 = 36$
$9(y - 1)^2 = 36$
$(y - 1)^2 = 4$
$y - 1 = \pm\sqrt{4} = \pm 2$
$y = 1 \pm 2$
$y = 3 \qquad y = -1$
$\left(4, 3\right), \left(4, -1\right)$
Let $y = 1$.
$-4(x - 4)^2 + 9(1 - 1)^2 = 36$
$-4(x - 4)^2 + 0 = 36$
$-4(x - 4)^2 = 36$

$(x-4)^2 = -9$
Has no real solutions.
The hyperbola does not cross the x-axis.
Asymptotes:
$-4(x-4)^2 + 9(y-1)^2 = 0$
$9(y-1)^2 - 4(x-4)^2 = 0$
$[3(y-1)+2(x-4)][3(y-1)-2(x-4)]=0$
$3(y-1)+2(x-4)=0$ or $3(y-1)-2(x-4)=0$
$3y-3+2x-8=0$ or $3y-3-2x+8=0$
$3y+2x-11=0$ or $3y-2x+5=0$
$3y=-2x+11$ or $3y=2x-5$
$y=-\frac{2}{3}x+\frac{11}{3}$ and $y=\frac{2}{3}x-\frac{5}{3}$

15. 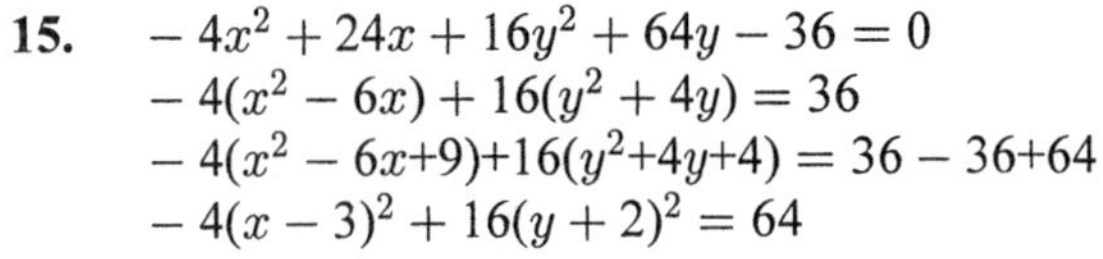
$-4x^2+24x+16y^2+64y-36=0$
$-4(x^2-6x)+16(y^2+4y)=36$
$-4(x^2-6x+9)+16(y^2+4y+4)=36-36+64$
$-4(x-3)^2+16(y+2)^2=64$

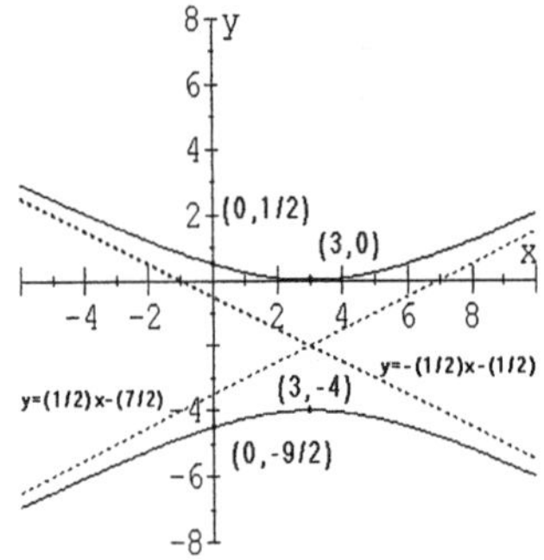

Let $x=3$.
$-4(3-3)^2+16(y+2)^2=64$
$-4(0)+16(y+2)^2=64$
$16(y+2)^2=64$
$(y+2)^2=4$
$y+2=\pm\sqrt{4}=\pm 2$
$y+2=2$ or $y+2=-2$
$y=0$ or $y=-4$
$(3, 0), (3, -4)$
Let $x=0$.
$-4(x-3)^2+16(y+2)^2=64$
$-4(0-3)^2+16(y+2)^2=64$
$-4(9)+16(y+2)^2=64$
$-36+16(y+2)^2=64$
$16(y+2)^2=100$
$(y+2)^2=\frac{100}{16}=\frac{25}{4}$
$y+2=\pm\sqrt{\frac{25}{4}}=\pm\frac{5}{2}$

$y=-2\pm\frac{5}{2}=\frac{-4}{2}\pm\frac{5}{2}$
$y=\frac{1}{2}$ or $y=-\frac{9}{2}$
$\left(0,\frac{1}{2}\right), \left(0,-\frac{9}{2}\right)$
Asymptotes :
$-4(x-3)^2+16(y+2)^2=0$
$16(y+2)^2-4(x-3)^2=0$
$[4(y+2)+2(x-3)][4(y+2)-2(x-3)]=0$
$4(y+2)+2(x-3)=0$ or $4(y+2)-2(x-3)=0$
$4y+8+2x-6=0$ or $4y+8-2x+6=0$
$4y=-2x-2$ or $4y=2x-14$
$y=-\frac{1}{2}x-\frac{1}{2}$ or $y=\frac{1}{2}x-\frac{7}{2}$

17. $4x^2-24x-9y^2=0$
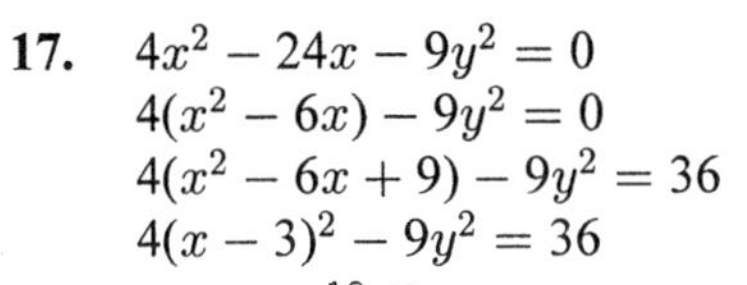
$4(x^2-6x)-9y^2=0$
$4(x^2-6x+9)-9y^2=36$
$4(x-3)^2-9y^2=36$

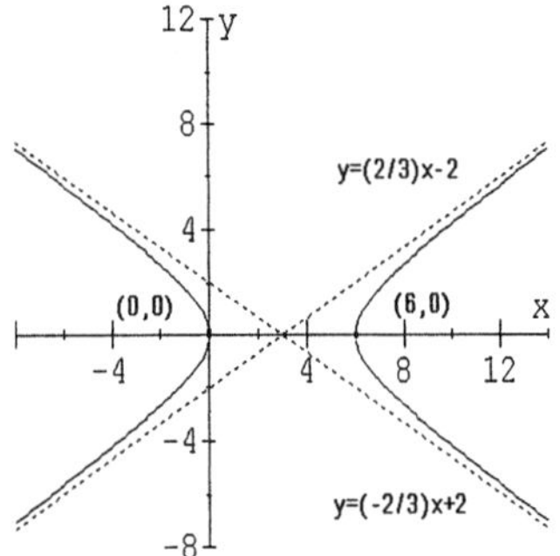

Let $y=0$.
$4(x-3)^2-9(0)^2=36$
$4(x-3)^2=36$
$(x-3)^2=9$
$x-3=\pm\sqrt{9}=\pm 3$
$x-3=3$ or $x-3=-3$
$x=6$ or $x=0$
$(6, 0), (0, 0)$
Asymptotes :
$4(x-3)^2-9y^2=0$
$[2(x-3)+3y][2(x-3)-3y]=0$
$2(x-3)+3y=0$ or $2(x-3)-3y=0$
$2x-6+3y=0$ or $2x-6-3y=0$
$3y=-2x+6$ or $-3y=-2x+6$
$y=-\frac{2}{3}x+2$ or $y=\frac{2}{3}x-2$

19.

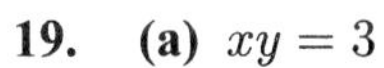

(a) $xy = 3$

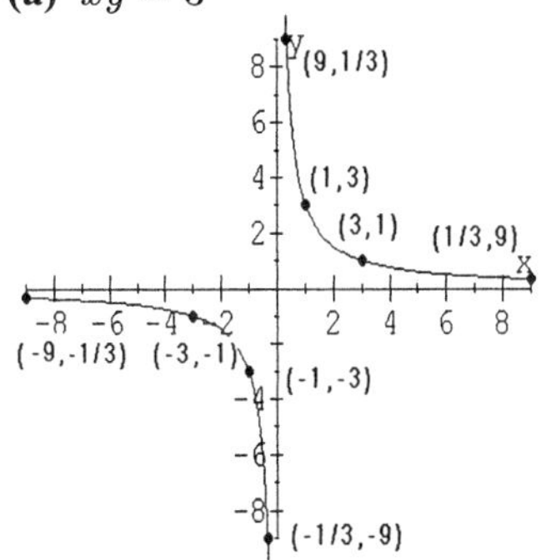

(b) $xy = 5$

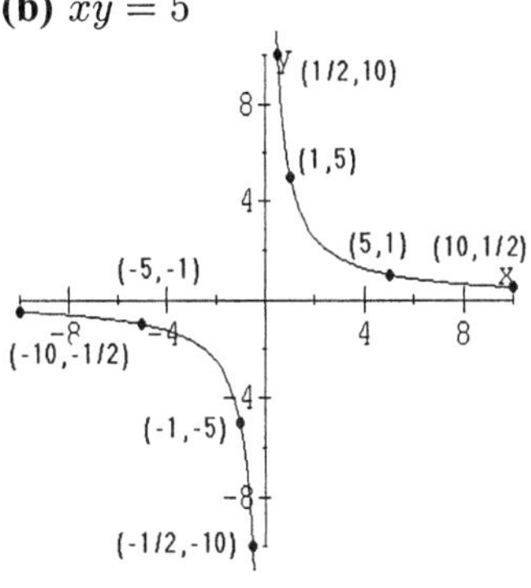

(c) $xy = -2$

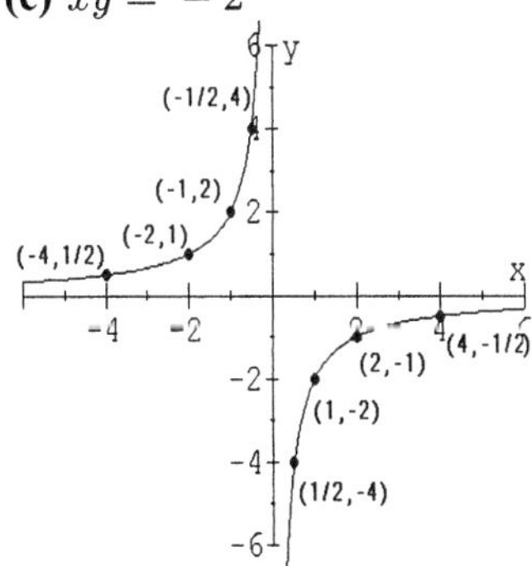

(d) $xy = -4$

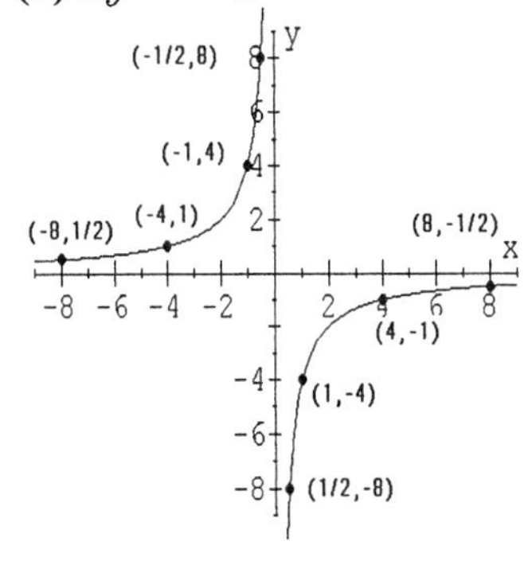

21. **(a)** $x^2 + y^2 = 0$

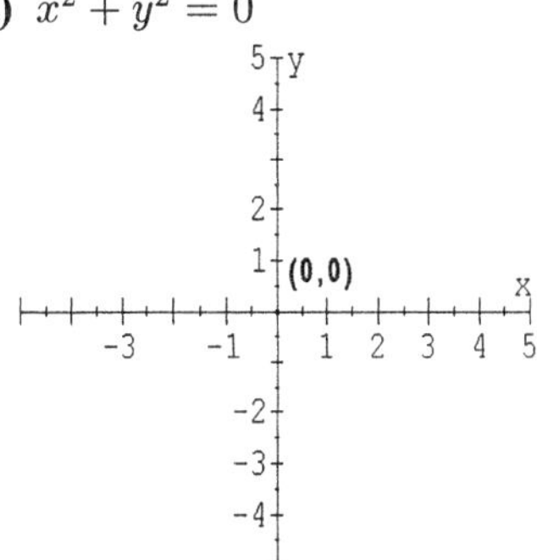

(b) $2x^2 + 3y^2 = 0$

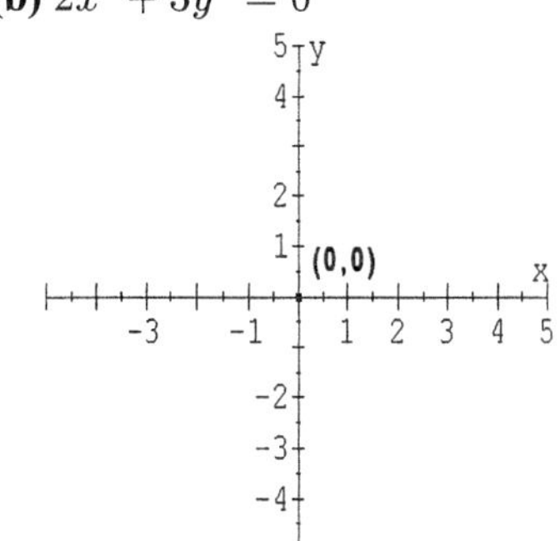

(c) $x^2 - y^2 = 0$

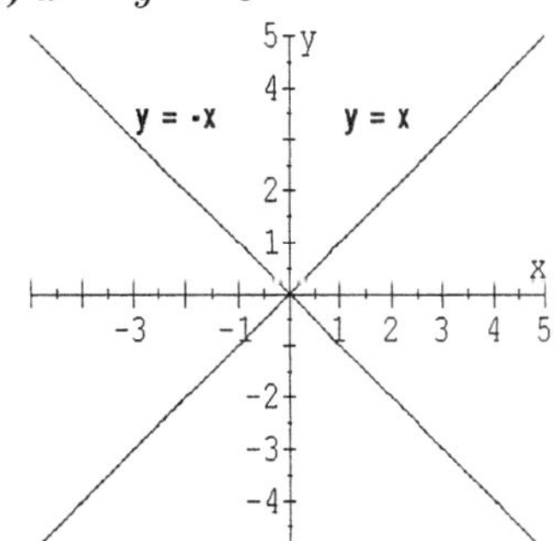

(d) $4y^2 - x^2 = 0$

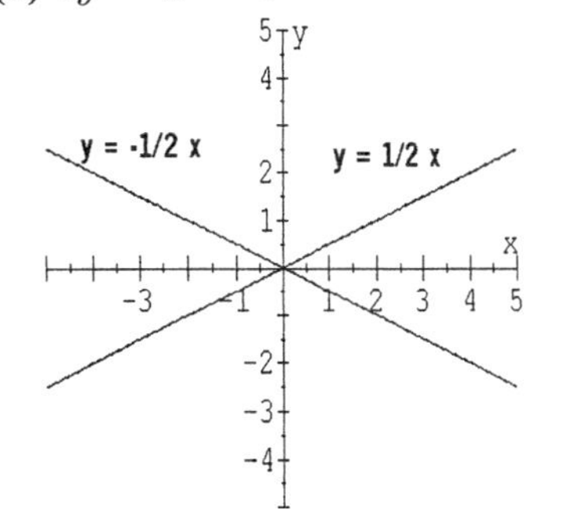

Chapter 12 Review Problem Set

CHAPTER 12 | Review Problem Set

1a. $(3, 4), (-2, -2)$

$m = \dfrac{-2-4}{-2-3} = \dfrac{-6}{-5} = \dfrac{6}{5}$

b. $(-2, 3), (4, -1)$

$m = \dfrac{-1-3}{4-(-2)} = \dfrac{-4}{6} = -\dfrac{2}{3}$

2a. $4x + y = 7$

$y = -4x + 7$

$m = -4$

b. $2x - 7y = 3$

$-7y = -2x + 3$

$-\dfrac{1}{7}(-7y) = -\dfrac{1}{7}(-2x+3)$

$y = \dfrac{2}{7}x - \dfrac{3}{7}$

$m = \dfrac{2}{7}$

3. $(2, 3), (5, -1)$, and $(-4, -5)$

Distance from $(2, 3)$ to $(5, -1)$

$d = \sqrt{(5-2)^2 + (-1-3)^2}$

$d = \sqrt{3^2 + (-4)^2}$

$d = \sqrt{9+16} = \sqrt{25} = 5$

Distance from $(2, 3)$ to $(-4, -5)$

$d = \sqrt{(-5-3)^2 + (-4-2)^2}$

$d = \sqrt{(-8)^2 + (-6)^2}$

$d = \sqrt{64+36} = \sqrt{100} = 10$

Distance from $(5, -1)$ to $(-4, -5)$

$d = \sqrt{(-4-5)^2 + [-5-(-1)]^2}$

$d = \sqrt{(-9)^2 + (-4)^2}$

$d = \sqrt{81+16} = \sqrt{97}$

The lengths of the sides are
5, 10, and $\sqrt{97}$ units.

4. $(-1, 2)$ and $(3, -5)$

$m = \dfrac{-5-2}{3-(-1)} = \dfrac{-7}{4}$

$y - (-5) = -\dfrac{7}{4}(x-3)$

$y + 5 = -\dfrac{7}{4}(x-3)$

$4(y+5) = 4\left[-\dfrac{7}{4}(x-3)\right]$

$4y + 20 = -7(x-3)$

$4y + 20 = -7x + 21$

$7x + 4y + 20 = 21$

$7x + 4y = 1$

5. $m = -\dfrac{3}{7}; b = 4$

$y = -\dfrac{3}{7}x + 4$

$7(y) = 7\left(-\dfrac{3}{7}x + 4\right)$

$7y = -3x + 28$

$3x + 7y = 28$

6. $(-1, -6); m = \dfrac{2}{3}$

$y - (-6) = \dfrac{2}{3}[x - (-1)]$

$y + 6 = \dfrac{2}{3}(x+1)$

$3(y+6) = 3\left[\dfrac{2}{3}(x+1)\right]$

$3y + 18 = 2(x+1)$

$3y + 18 = 2x + 2$

$18 = 2x - 3y + 2$

$16 = 2x - 3y$

$2x - 3y = 16$

7. $(2, 5)$; parallel to the line $x - 2y = 4$
Parallel lines have the same slope.

$x - 2y = 4$

$-2y = -x + 4$

$-\dfrac{1}{2}(-2y) = -\dfrac{1}{2}(-x+4)$

$y = \dfrac{1}{2}x - 2$

$m = \dfrac{1}{2}$

$y - 5 = \dfrac{1}{2}(x-2)$

$2(y-5) = 2\left[\frac{1}{2}(x-2)\right]$
$2y - 10 = x - 2$
$-10 = x - 2y - 2$
$-8 = x - 2y$
$x - 2y = -8$

8. $(-2, -6)$; perpendicular to the line $3x + 2y = 12$
Perpendicular lines have slopes that are negative reciprocals.
$3x + 2y = 12$
$2y = -3x + 12$
$\frac{1}{2}(2y) = \frac{1}{2}(-3x + 12)$
$y = -\frac{3}{2}x + 6$
$m = -\frac{3}{2}$
The negative reciprocal is $\frac{2}{3}$.
$y - (-6) = \frac{2}{3}\left[x - (-2)\right]$
$y + 6 = \frac{2}{3}(x + 2)$
$3(y+6) = 3\left[\frac{2}{3}(x+2)\right]$
$3y + 18 = 2(x + 2)$
$3y + 18 = 2x + 4$
$18 = 2x - 3y + 4$
$14 = 2x - 3y$
$2x - 3y = 14$

9. $x^2 = y + 1$
To test symmetry for:
x-axis replace y with $-y$
$x^2 = (-y) + 1$
$x^2 = -y + 1$
No, it is not symmetric to the x-axis.

y-axis replace x with $-x$
$(-x)^2 = y + 1$
$x^2 = y + 1$
Yes, it is symmetric to the y-axis.

origin replace x with $-x$ and y with $-y$.
$(-x)^2 = (-y) + 1$
$x^2 = -y + 1$
No, it is not symmetric to the origin.

10. $y^2 = x^3$
To test symmetry for:
x-axis replace y with $-y$
$(-y)^2 = x^3$
$y^2 = x^3$
Yes, it is symmetric to the x-axis.

y-axis replace x with $-x$
$y^2 = (-x^3)$
$y^2 = -x^3$
No, it is not symmetric to the y-axis.

origin replace x with $-x$ and y with $-y$.
$(-y)^2 = (-x)^3$
$y^2 = -x^3$
No, it is not symmetric to the origin.

11. $y = -2x$
To test symmetry for:
x-axis replace y with $-y$
$-y = -2x$
$y = 2x$
No, it is not symmetric to the x-axis.

y-axis replace x with $-x$
$y = -2(-x)$
$y = 2x$
No, it is not symmetric to the y-axis.

origin replace x with $-x$ and y with $-y$.
$-y = -2(-x)$
$y = -2x$
Yes, it is symmetric to the origin.

12. $y = \frac{8}{x^2 + 6}$
To test symmetry for:
x-axis replace y with $-y$
$-y = \frac{8}{x^2 + 6}$
$y = -\frac{8}{x^2 + 6}$
No, it is not symmetric to the x-axis.

y-axis replace x with $-x$
$y = \frac{8}{(-x)^2 + 6}$
$y = \frac{8}{x^2 + 6}$
Yes, it is symmetric to the y-axis.

origin replace x with $-x$
and y with $-y$.

$$-y = \frac{8}{(-x)^2+6}$$

$$y = -\frac{8}{x^2+6}$$

No, it is not symmetric to the origin.

13. $x^2 - 2y^2 = -3$

To test symmetry for:

x-axis replace y with $-y$

$x^2 - 2(-y)^2 = -3$

$x^2 - 2y^2 = -3$

Yes, it is symmetric to the x-axis.

y-axis replace x with $-x$

$(-x)^2 - 2y^2 = -3$

$x^2 - 2y^2 = -3$

Yes, it is symmetric to the y-axis.

origin replace x with $-x$
and y with $-y$.

$(-x)^2 - 2(-y)^2 = -3$

$x^2 - 2y^2 = -3$

Yes, it is symmetric to the origin.

14. $2x - y = 6$

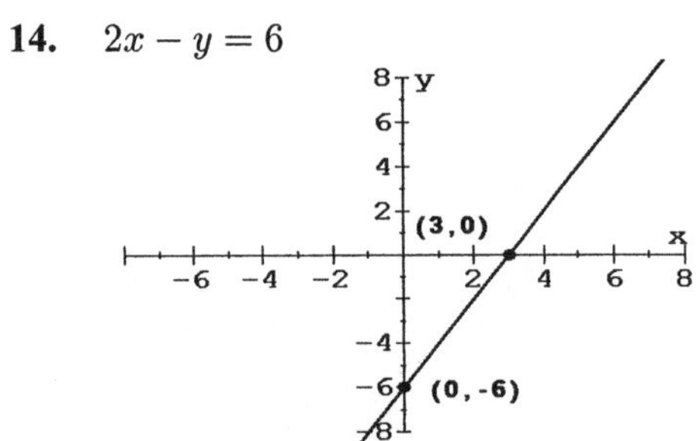

15. $y = -2x^2 - 1$

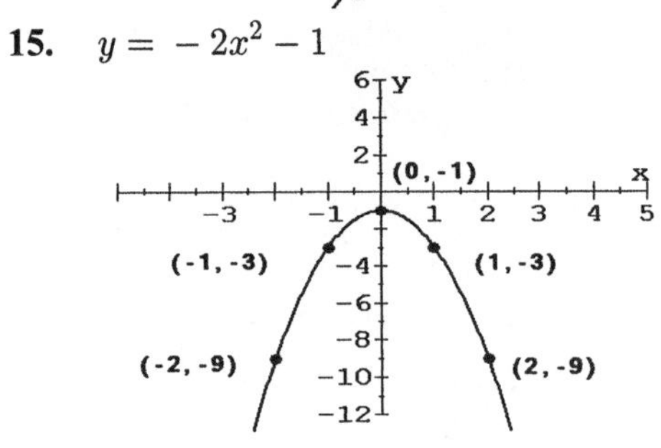

16. $y = -2x - 1$

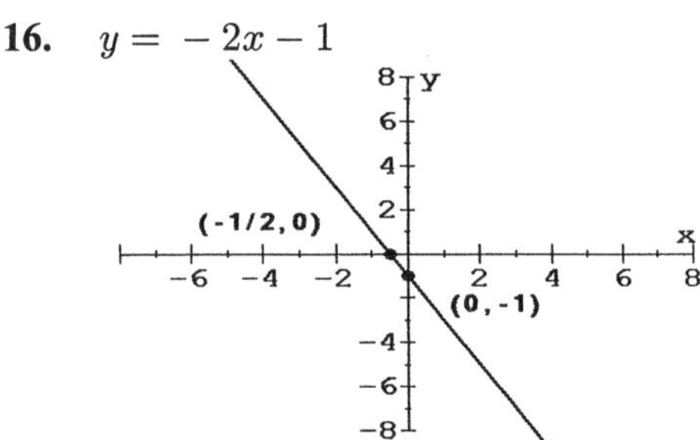

17. $y = -4x$

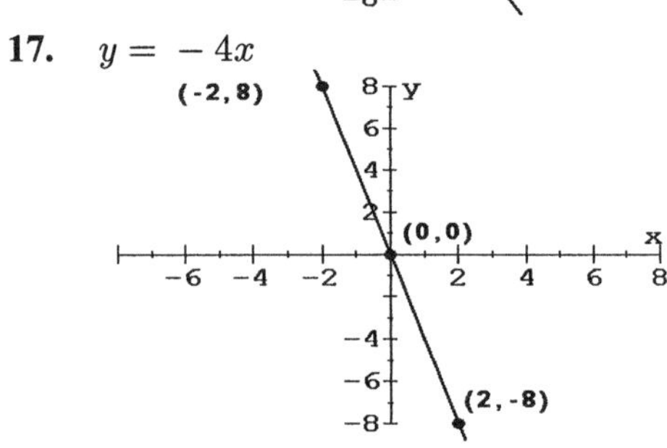

18. $xy^2 = -1$

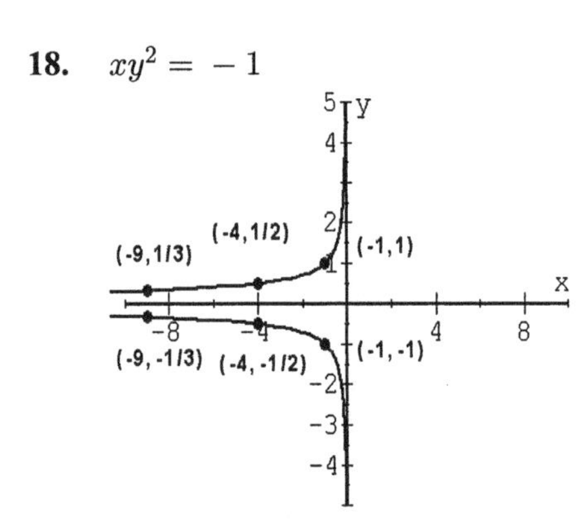

19. $y = \frac{-3}{x^2+1}$

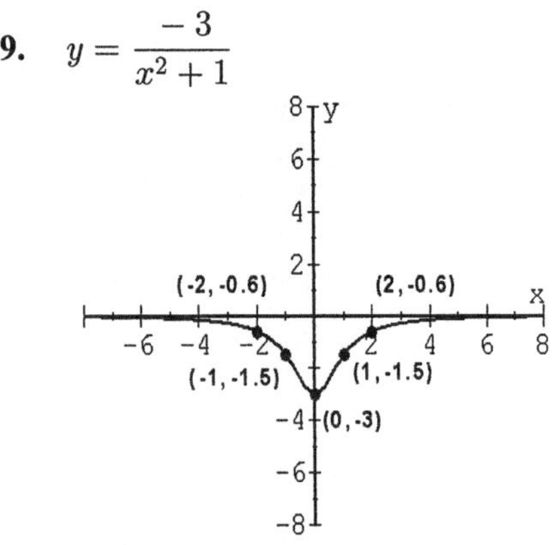

20. 1 mile = 5280 ft.
6% of 1 mile
0.06(5280) = 316.8
It will rise 316.8 feet.

21. $\frac{2}{3} = \frac{x}{12}$
$2(12) = 3x$
$24 = 3x$
$8 = x$
The rise will be 8 inches.

22. $y = x^2 + 6$
$y = (x - 0)^2 + 6$
vertex (0, 6)

23. $y = -x^2 - 8$
$y = -(x - 0)^2 - 8$
vertex (0, – 8)

24. $y = (x + 3)^2 - 1$
$y = [x - (-3)]^2 - 1$
vertex (– 3, – 1)

25. $y = x^2 - 14x + 54$
$y = x^2 - 14x + 49 - 49 + 54$
$y = (x - 7)^2 + 5$
vertex (7, 5)

26. $y = -x^2 + 12x - 44$
$y = -(x^2 - 12x) - 44$
$y = -(x^2 - 12x + 36) + 36 - 44$
$y = -(x - 6)^2 - 8$
vertex (6, – 8)

27. $y = 3x^2 + 24x + 39$
$y = 3(x^2 + 8x) + 39$
$y = 3(x^2 + 8x + 16) - 48 + 39$
$y = 3(x + 4)^2 - 9$
vertex (– 4, – 9)

28. Center (2, – 6) and $r = 5$
$(x - 2)^2 + [y - (-6)]^2 = 5^2$
$(x - 2)^2 + (y + 6)^2 = 25$
$x^2 - 4x + 4 + y^2 + 12y + 36 = 25$
$x^2 + y^2 - 4x + 12y + 15 = 0$

29. Center (– 4, – 8) and $r = 2\sqrt{3}$
$[x - (-4)]^2 + [y - (-8)]^2 = \left(2\sqrt{3}\right)^2$
$(x + 4)^2 + (y + 8)^2 = 4(3)$
$x^2 + 8x + 16 + y^2 + 16y + 64 = 12$
$x^2 + y^2 + 8x + 16y + 68 = 0$

30. Center (0, 5) passes
through (0, 0) so r = 5.
$(x - 0)^2 + (y - 5)^2 = 5^2$
$x^2 + y^2 - 10y + 25 = 25$
$x^2 + y^2 - 10y = 0$

31. $x^2 + 14x + y^2 - 8y + 16 = 0$
$x^2 + 14x + 49 + y^2 - 8y + 16 = 49$
$(x + 7)^2 + (y - 4)^2 = 49$
Center (– 7, 4) with $r = 7$

32. $x^2 + 16x + y^2 + 39 = 0$
$x^2 + 16x + 64 + y^2 = 64 - 39$
$(x + 8)^2 + (y - 0)^2 = 25$
Center (– 8, 0) with $r = 5$

33. $x^2 - 12x + y^2 + 16y = 0$
$x^2 - 12x + 36 + y^2 + 16y + 64 = 36 + 64$
$(x - 6)^2 + (y + 8)^2 = 100$
Center (6, – 8) with $r = 10$

34. $x^2 + y^2 = 24$
$(x - 0)^2 + (y - 0)^2 = 24$
Center (0, 0) with $r = \sqrt{24} = 2\sqrt{6}$

35. $4x^2 + 25y^2 = 100$
Let $y = 0$.
$4x^2 + 25(0)^2 = 100$
$4x^2 = 100$
$x^2 = 25$
$x = \pm 5$
The endpoints of the major axis
are (– 5, 0) and (5, 0).
The length of the major axis
is 5 – (– 5) = 10.
Let $x = 0$.
$4(0)^2 + 25y^2 = 100$
$25y^2 = 100$
$y^2 = 4$
$y = \pm 2$

The endpoints of the minor axis are $(0, -2)$ and $(0, 2)$.
The length of the minor axis is $2 - (-2) = 4$.

36. $2x^2 + 7y^2 = 28$
Let $y = 0$.
$2x^2 + 7(0)^2 = 28$
$2x^2 = 28$
$x^2 = 14$
$x = \pm\sqrt{14}$
The endpoints of the major axis are $(-\sqrt{14}, 0)$ and $(\sqrt{14}, 0)$.
The length of the major axis is $\sqrt{14} - \left(-\sqrt{14}\right) = 2\sqrt{14}$.
Let $x = 0$.
$2(0)^2 + 7y^2 = 28$
$7y^2 = 28$
$y^2 = 4$
$y = \pm 2$
The endpoints of the minor axis are $(0, -2)$ and $(0, 2)$. The length of the minor axis is $2 - (-2) = 4$.

37. $x^2 - 4x + 9y^2 + 54y + 76 = 0$
$x^2 - 4x + 9(y^2 + 6y) = -76$
$x^2 - 4x + 4 + 9(y^2 + 6y + 9) = -76+4+81$
$(x-2)^2 + 9(y+3)^2 = 9$
Let $y = -3$.
$(x-2)^2 + 9(-3+3)^2 = 9$
$(x-2)^2 = 9$
$x - 2 = \pm 3$
$x - 2 = 3$ or $x - 2 = -3$
$x = 5$ or $x = -1$
The endpoints of the major axis are $(-1, -3)$ and $(5, -3)$.
The length of the major axis is $5 - (-1) = 6$.
Let $x = 2$.
$(2-2)^2 + 9(y+3)^2 = 9$
$9(y+3)^2 = 9$
$(y+3)^2 = 1$
$y + 3 = \pm 1$
$y + 3 = 1$ or $y + 3 = -1$
$y = -2$ or $y = -4$
The endpoints of the minor axis are $(2, -2)$ and $(2, -4)$.
The length of the minor axis is $-2 - (-4) = 2$.

38. $9x^2 + 72x + 4y^2 - 8y + 112 = 0$
$9(x^2 + 8x) + 4(y^2 - 2y) = -112$
$9(x^2+8x+16)+4(y^2 - 2y+1) = -112+144+4$
$9(x+4)^2 + 4(y-1)^2 = 36$
Let $x = -4$.
$9(-4+4)^2 + 4(y-1)^2 = 36$
$4(y-1)^2 = 36$
$(y-1)^2 = 9$
$y - 1 = \pm 3$
$y - 1 = -3$ or $y - 1 = 3$
$y = -2$ or $y = 4$
The endpoints of the major axis are $(-4, -2)$ and $(-4, 4)$.
The length of the major axis is $4 - (-2) = 6$.
Let $y = 1$.
$9(x+4)^2 + 4(1-1)^2 = 36$
$9(x+4)^2 = 36$
$(x+4)^2 = 4$
$x + 4 = \pm 2$
$x + 4 = -2$ or $x + 4 = 2$
$x = -6$ or $x = -2$
The endpoints of the minor axis are $(-6, 1)$ and $(-2, 1)$.
The length of the minor axis is $-2 - (-6) = 4$.

39. $4x^2 - 9y^2 = 16$
Replace 16 with 0.
$4x^2 - 9y^2 = 0$
$-9y^2 = -4x^2$
$y^2 = \frac{4}{9}x^2$
$y = \pm\frac{2}{3}x$
Asymptotes are $y = -\frac{2}{3}x$ and $y = \frac{2}{3}x$.

40. $16y^2 - 4x^2 = 17$
Replace 17 with 0.
$16y^2 - 4x^2 = 0$
$16y^2 = 4x^2$
$y^2 = \frac{1}{4}x^2$
$y = \pm\frac{1}{2}x^2$
Asymptotes are $y = -\frac{1}{2}x$ and $y = \frac{1}{2}x$.

41. $25x^2 + 100x - 4y^2 + 24y - 36 = 0$
$25(x^2 + 4x) - 4(y^2 - 6y) = 36$
$25(x^2 + 4x+4) - 4(y^2 - 6y+9) = 36+100 - 36$
$25(x + 2)^2 - 4(y - 3)^2 = 100$

Replace 100 with 0.
$25(x + 2)^2 - 4(y - 3)^2 = 0$
Factoring, using a difference of squares pattern.
$[5(x+2) - 2(y - 3)]\ [5(x+2)+2(y - 3)] = 0$
$5(x+2) - 2(y\text{-}3) = 0$ or $5(x+2) + 2(y\text{-}3) = 0$
$5x+10 - 2y+6 = 0$ or $5x+10+2y - 6 = 0$
$-2y + 5x + 16 = 0$ or $2y + 5x + 4 = 0$
$-2y = -5x - 16$ or $2y = -5x - 4$
$y = \frac{5}{2}x + 8$ or $y = -\frac{5}{2}x - 2$
Asymptotes are
$y = \frac{5}{2}x + 8$ and $y = -\frac{5}{2}x - 2.$

42. $36y^2 - 288y - x^2 + 2x + 539 = 0$
$36(y^2 - 8y) - (x^2 - 2x) = -539$
$36(y^2 - 8y+16) - (x^2 - 2x+1) = -539+576 - 1$
$36(y - 4)^2 - (x - 1)^2 = 36$

Replace 36 with 0.
$36(y - 4)^2 - (x - 1)^2 = 0$
Factoring, using a difference of squares pattern.
$[6(y - 4) - (x - 1)]\ [6(y - 4) + (x - 1)] = 0$
$6(y - 4) - (x - 1)=0$ or $6(y - 4) + (x - 1)=0$
$6y - 24 - x + 1 = 0$ or $6y - 24 + x - 1 = 0$
$6y - x - 23 = 0$ or $6y + x - 25 = 0$
$6y = x + 23$ or $6y = -x + 25$
$y = \frac{1}{6}x + \frac{23}{6}$ or $y = -\frac{1}{6}x + \frac{25}{6}$
Asymptotes are
$y = \frac{1}{6}x + \frac{23}{6}$ and $y = -\frac{1}{6}x + \frac{25}{6}.$

43. $9x^2 + y^2 = 81$

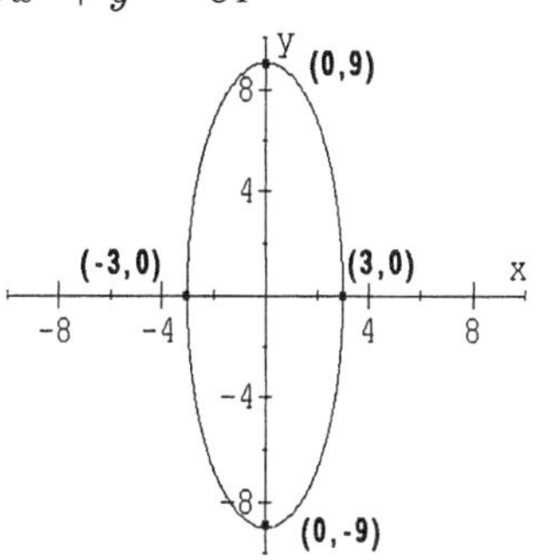

It is an ellipse with center at $(0, 0)$.
Let $x = 0$.
$9(0)^2 + y^2 = 81$
$y^2 = 81$
$y = \pm 9$
The endpoints of the major axis are (0, 9) and (0, − 9).
Let $y = 0$.
$9x^2 + 0^2 = 81$
$9x^2 = 81$
$x^2 = 9$
$x = \pm 3$
The endpoints of the minor axis are (− 3, 0) and (3, 0).

44. $9x^2 - y^2 = 81$
It is a hyperbola.

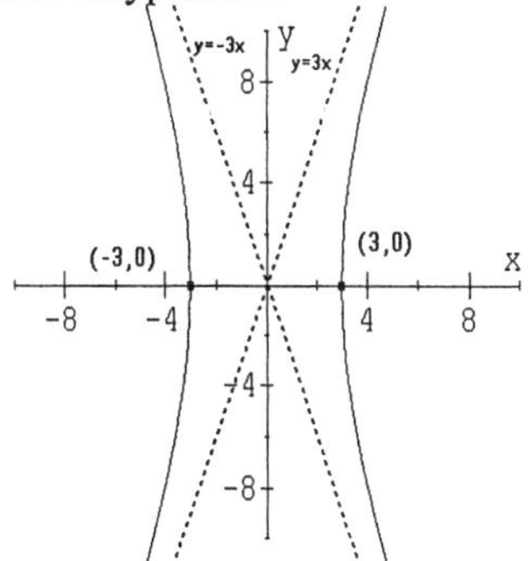

Let $y = 0$.
$9x^2 - 0^2 = 81$
$9x^2 = 81$
$x^2 = 9$
$x = \pm\sqrt{9} = \pm 3$
Vertices are at (− 3, 0) and (3, 0).
To find asymptotes, replace 81 with 0.
$9x^2 - y^2 = 0$
$-y^2 = -9x^2$

$y^2 = 9x^2$
$y = \pm 3x$
Asymptotes are $y = -3x$ and $y = 3x$.

45. $y = -2x^2 + 3$
It is a parabola with vertex at (0, 3) and it opens downward.

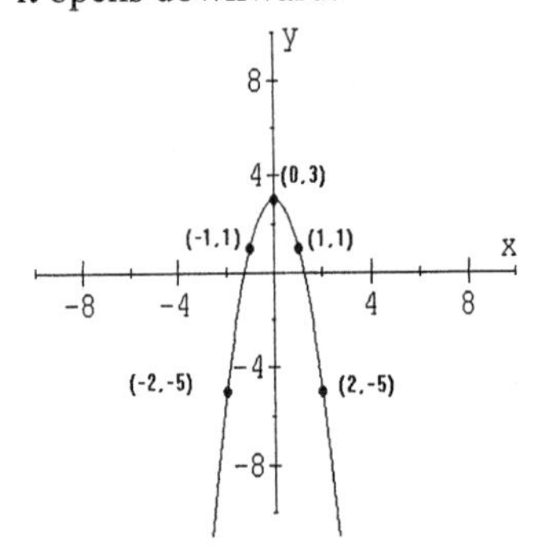

46. $y = 4x^2 - 16x + 19$
$y = 4(x^2 - 4x) + 19$
$y = 4(x^2 - 4x + 4) - 16 + 19$
$y = 4(x - 2)^2 + 3$
It is a parabola with vertex at (2, 3) and opens upward.

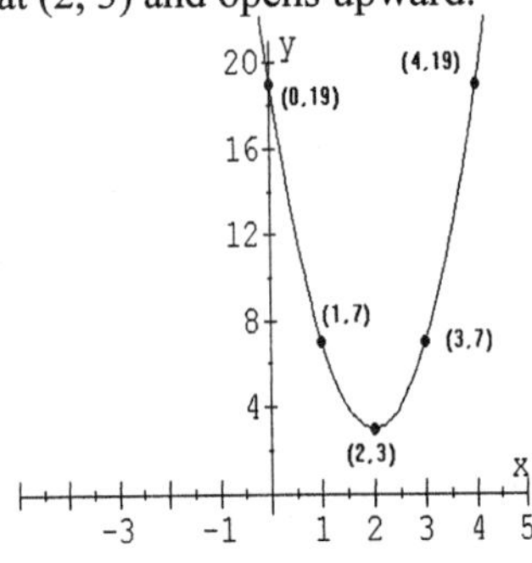

47. $x^2 + 4x + y^2 + 8y + 11 = 0$
$x^2 + 4x + 4 + y^2 + 8y + 16 = 4 + 16 - 11$
$(x + 2)^2 + (y + 4)^2 = 9$
It is a circle with center at $(-2, -4)$ and r = 3.

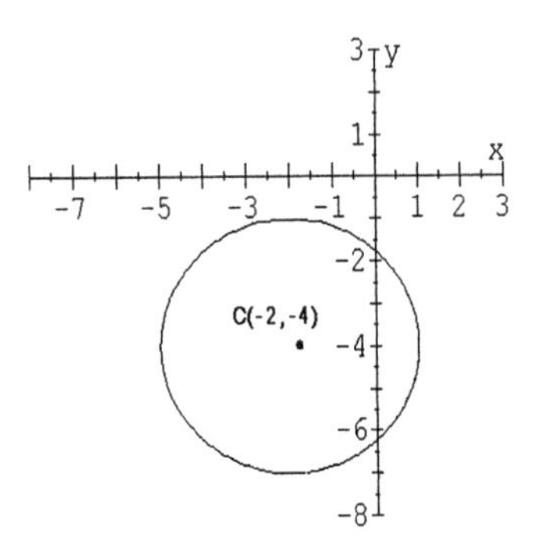

48. $4x^2 - 8x + y^2 + 8y + 4 = 0$
$4(x^2 - 2x) + y^2 + 8y = -4$
$4(x^2 - 2x + 1) + y^2 + 8y + 16 = -4 + 4 + 16$
$4(x - 1)^2 + (y + 4)^2 = 16$
It is an ellipse with center at $(1, -4)$.

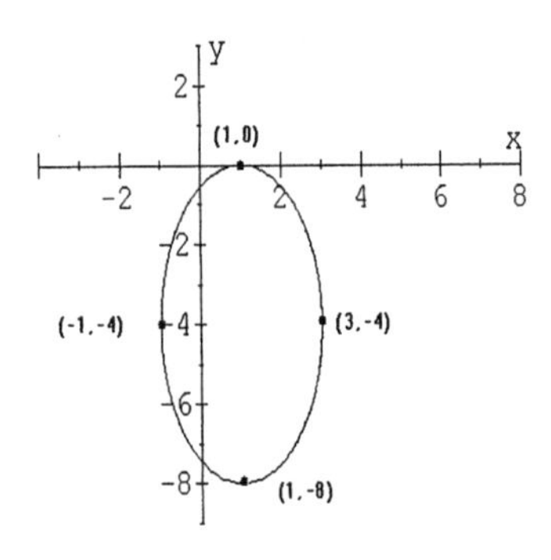

Let $x = 1$.
$4(1 - 1)^2 + (y + 4)^2 = 16$
$(y + 4)^2 = 16$
$y + 4 = \pm 4$
$y + 4 = -4$ or $y + 4 = 4$
$y = -8$ or $y = 0$
The endpoints of the major axis are $(1, -8)$ and $(1, 0)$.
Let $y = -4$.
$4(x - 1)^2 + (-4 + 4)^2 = 16$
$4(x - 1)^2 = 16$
$(x - 1)^2 = 4$
$x - 1 = \pm 2$
$x - 1 = -2$ or $x - 1 = 2$
$x = -1$ or $x = 3$
The endpoints of the minor axis are $(-1, -4)$ and $(3, -4)$.

49. $y^2 + 6y - 4x^2 - 24x - 63 = 0$
$y^2 + 6y - 4(x^2 + 6x) = 63$
$y^2 + 6y + 9 - 4(x^2 + 6x + 9) = 63 + 9 - 36$
$(y + 3)^2 - 4(x + 3)^2 = 36$
It is a hyperbola.

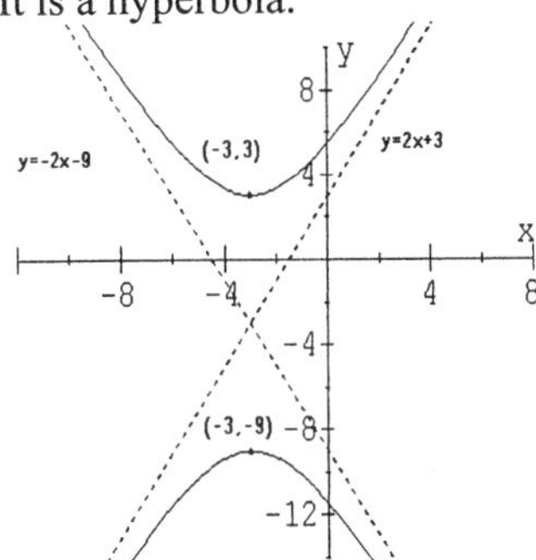

Let $x = -3$
$(y + 3)^2 - 4(-3 + 3)^2 = 36$
$(y + 3)^2 = 36$
$y + 3 = \pm\sqrt{36}$
$y + 3 = \pm 6$
$y = -3 \pm 6$
$y = -9$ or 3
The vertices are at
$(-3, -9)$ and $(-3, 3)$.
To find the asymptotes,
replace 36 with 0.
$(y + 3)^2 - 4(x + 3)^2 = 0$
Factoring, using a difference
of squares pattern.
$[(y + 3) - 2(x + 3)]\,[(y + 3) + 2(x + 3)] = 0$
$(y+3) - 2(x+3) = 0$ or $y+3+2(x+3) = 0$
$y + 3 - 2x - 6 = 0$ or $y + 3 + 2x + 6 = 0$
$y = 2x + 3$ or $y = -2x - 9$
Asymptotes are
$y = 2x + 3$ and $y = -2x - 9$.

50. $y = -2x^2 - 4x - 3$
$y = -2(x^2 + 2x) - 3$
$y = -2(x^2 + 2x + 1) + 2 - 3$
$y = -2(x + 1)^2 - 1$
It is a parabola that opens
downward, with vertex at $(-1, -1)$.

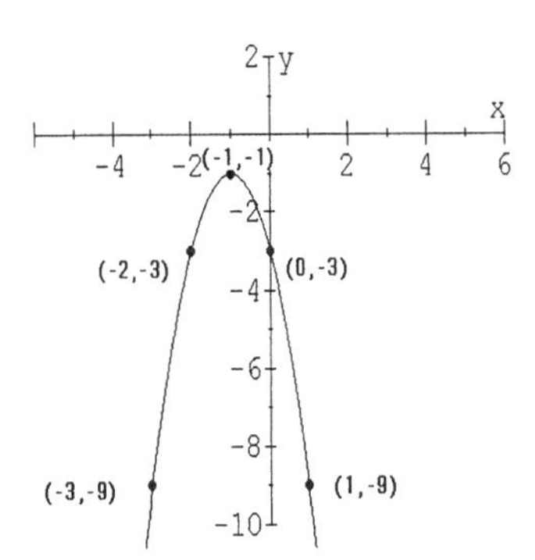

51. $x^2 - y^2 = -9$
$-x^2 + y^2 = 9$
It is a hyperbola.

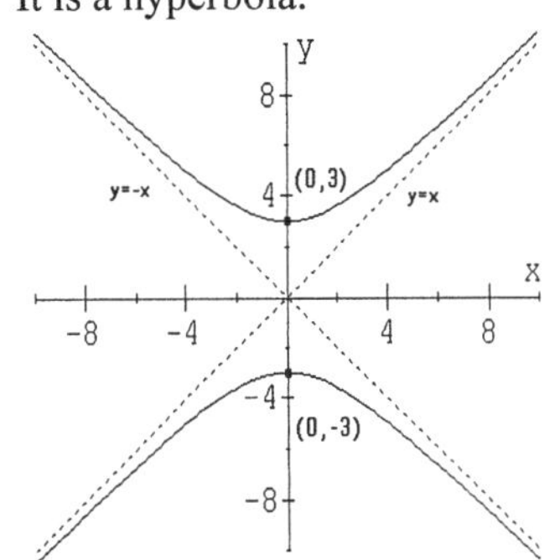

Let $x = 0$.
$-(0)^2 + y^2 = 9$
$y^2 = 9$
$y = \pm 3$
The vertices are at
$(0, -3)$ and $(0, 3)$.
To find the asymptotes, replace 9 with 0.
$-x^2 + y^2 = 0$
$y^2 = x^2$
$y = \pm x$
Asymptotes are $y = x$ and $y = -x$.

52. $4x^2 + 16y^2 + 96y = 0$
$4x^2 + 16(y^2 + 6y) = 0$
$4x^2 + 16(y^2 + 6y + 9) = 144$
$4x^2 + 16(y + 3)^2 = 144$

It is an ellipse with center at $(0, -3)$.

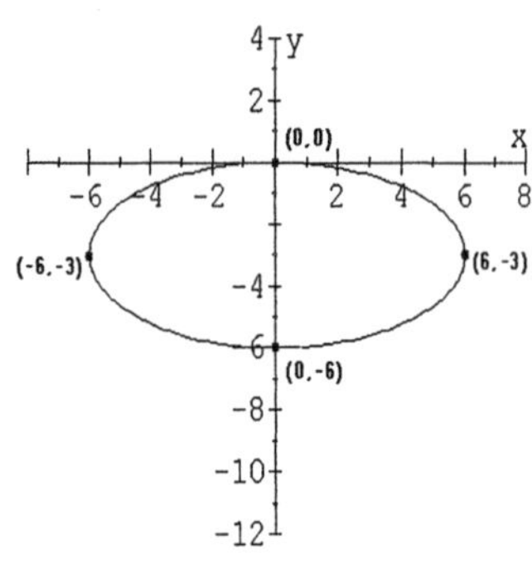

Let $y = -3$.

$4x^2 + 16(-3+3)^2 = 144$

$4x^2 = 144$

$x^2 = 36$

$x = \pm 6$

The endpoints of the major axis are $(-6, -3)$ and $(6, -3)$.

Let $x = 0$.

$4(0)^2 + 16(y+3)^2 = 144$

$16(y+3)^2 = 144$

$(y+3)^2 = 9$

$y + 3 = \pm 3$

$y + 3 = 3$ or $y + 3 = -3$

$y = 0$ or $y = -6$

The endpoints of the minor axis are $(0, 0)$ and $(0, -6)$.

CHAPTER 12 Test

1. $(-2, 4), (3, -2)$

$m = \dfrac{-2-4}{3-(-2)} = \dfrac{-6}{5} = -\dfrac{6}{5}$

2. $3x - 7y = 12$

$-7y = -3x + 12$

$-\frac{1}{7}(-7y) = -\frac{1}{7}(-3x + 12)$

$y = \frac{3}{7}x - \frac{12}{7}$

$m = \frac{3}{7}$

3. $(4, 2)$ and $(-3, -1)$

$d = \sqrt{(-3-4)^2 + (-1-2)^2}$

$d = \sqrt{(-7)^2 + (-3)^2}$

$d = \sqrt{49 + 9} = \sqrt{58}$

4. $m = -\frac{3}{2}, (4, -5)$

$y - (-5) = -\frac{3}{2}(x - 4)$

$y + 5 = -\frac{3}{2}(x - 4)$

$2(y + 5) = 2\left[-\frac{3}{2}(x - 4)\right]$

$2y + 10 = -3(x - 4)$

$2y + 10 = -3x + 12$

$3x + 2y + 10 = 12$

$3x + 2y = 2$

5. $(-4, 2)$ and $(2, 1)$

$m = \dfrac{1-2}{2-(-4)} = \dfrac{-1}{6}$

$y - 1 = -\frac{1}{6}(x - 2)$

$6(y - 1) = 6\left[-\frac{1}{6}(x - 2)\right]$

$6y - 6 = -1(x - 2)$

$6y - 6 = -x + 2$

$6y = -x + 8$

$\frac{1}{6}(6y) = \frac{1}{6}(-x + 8)$

$y = -\frac{1}{6}x + \frac{4}{3}$

6. $(-2, -4)$; parallel to $5x + 2y = 7$

Parallel lines have slopes that are equal.

$5x + 2y = 7$

$2y = -5x + 7$

$\frac{1}{2}(2y) = \frac{1}{2}(-5x + 7)$

$y = -\frac{5}{2}x + \frac{7}{2}$

$m = -\frac{5}{2}$

$y - (-4) = -\frac{5}{2}[x - (-2)]$

$y + 4 = -\frac{5}{2}(x + 2)$

$2(y + 4) = 2\left[-\frac{5}{2}(x + 2)\right]$

$2y + 8 = -5(x + 2)$

$2y + 8 = -5x - 10$
$5x + 2y + 8 = -10$
$5x + 2y = -18$

7. (4, 7); perpendicular to the line $x - 6y = 9$
Perpendicular lines have slopes
that are negative reciprocals.
$x - 6y = 9$
$-6y = -x + 9$
$-\frac{1}{6}(-6y) = -\frac{1}{6}(-x + 9)$
$y = \frac{1}{6}x - \frac{9}{6}$
$m = \frac{1}{6}$
The negative reciprocal is -6.
$y - 7 = -6(x - 4)$
$y - 7 = -6x + 24$
$6x + y - 7 = 24$
$6x + y = 31$

8. $\frac{25}{100} = \frac{120}{x}$
$25x = 12000$
$x = 480$
The horizontal change is 480 feet.

9. $\frac{3}{4} = \frac{32}{x}$
$3x = 4(32)$
$3x = 128$
$x \approx 43$
The run would be 43 centimeters.

10. $y = x^2 - 6x + 9$
$y = (x - 3)^2$
The vertex is (3, 0).

11. $y = 4x^2 + 32x + 62$
$y = 4(x^2 + 8x) + 62$
$y = 4(x^2 + 8x + 16) - 64 + 62$
$y = 4(x + 4)^2 - 2$
The vertex is $(-4, -2)$.

12. Center at (2, 8) and r = 3
$(x - 2)^2 + (y - 8)^2 = 3^2$
$x^2 - 4x + 4 + y^2 - 16y + 64 = 9$
$x^2 + y^2 - 4x - 16y + 59 = 0$

13. $x^2 - 12x + y^2 + 8y + 3 = 0$
$x^2 - 12x + 36 + y^2 + 8y + 16 = 36 + 16 - 3$
$(x - 6)^2 + (y + 4)^2 = 49$
Center at $(6, -4)$ and $r = 7$

14. $9x^2 + 2y^2 = 32$
Let $x = 0$.
$9(0)^2 + 2y^2 = 32$
$2y^2 = 32$
$y^2 = 16$
$y = \pm 4$
The endpoints of the major
axis are $(0, -4)$ and (0, 4).
The length of the major
axis is $4 - (-4) = 8$.

15. $8x^2 - 32x + 5y^2 + 30y + 45 = 0$
$8(x^2 - 4x) + 5(y^2 + 6y) = -45$
$8(x^2 - 4x + 4) + 5(y^2 + 6y + 9) = -45 + 32 + 45$
$8(x - 2)^2 + 5(y + 3)^2 = 32$
Let $y = -3$.
$8(x - 2)^2 + 5(-3 + 3)^2 = 32$
$8(x - 2)^2 = 32$
$(x - 2)^2 = 4$
$x - 2 = \pm 2$
$x - 2 = -2$ or $x - 2 = 2$
$x = 0$ or $x = 4$
The endpoints of the minor axis are
$(0, -3)$ and $(4, -3)$. The length of the
minor axis is $4 - 0 = 4$.

16. $y^2 - 16x^2 = 36$
Replace 36 with 0.
$y^2 - 16x^2 = 0$
$y^2 = 16x^2$
$y = \pm 4x$
The asymptotes are $y = 4x$ and $y = -4x$.

17. $\frac{1}{3}x + \frac{1}{2}y = 2$

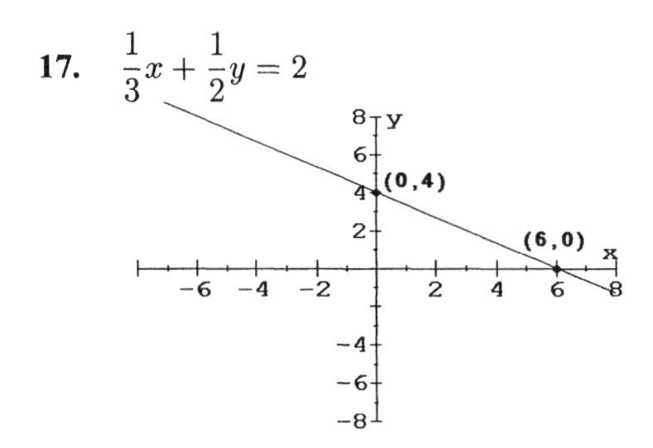

18. 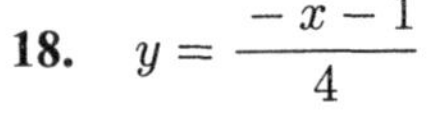

$y = \dfrac{-x-1}{4}$

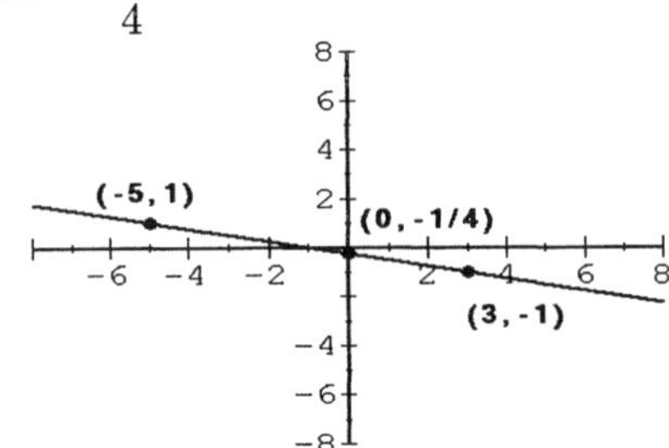

19. $x^2 - 4y^2 = -16$
$-x^2 + 4y^2 = 16$
It is a hyperbola.

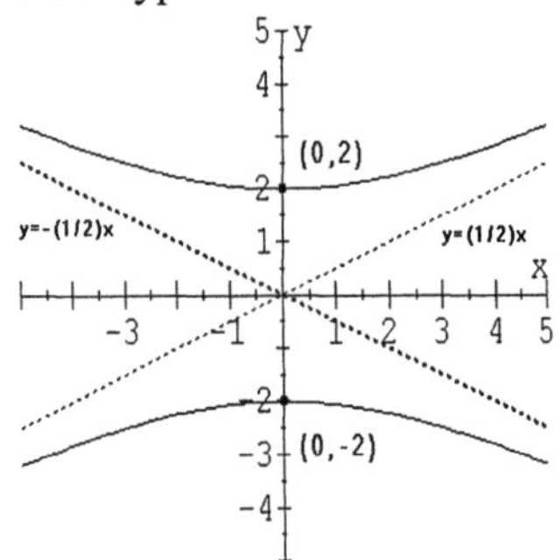

Let $x = 0$.
$-0^2 + 4y^2 = 16$
$4y^2 = 16$
$y^2 = 4$
$y = \pm 2$
The vertices are $(0, 2)$ and $(0, -2)$.
Replace 16 with 0.
$x^2 - 4y^2 = 0$
$-4y^2 = -x^2$
$y^2 = \dfrac{1}{4}x^2$
$y = \pm \dfrac{1}{2}x$
Asymptotes are $y = \dfrac{1}{2}x$ and $y = -\dfrac{1}{2}x$.

20. $y = x^2 + 4x$
$y = x^2 + 4x + 4 - 4$
$y = (x+2)^2 - 4$
It is a parabola that opens
upward with vertex at $(-2, -4)$.

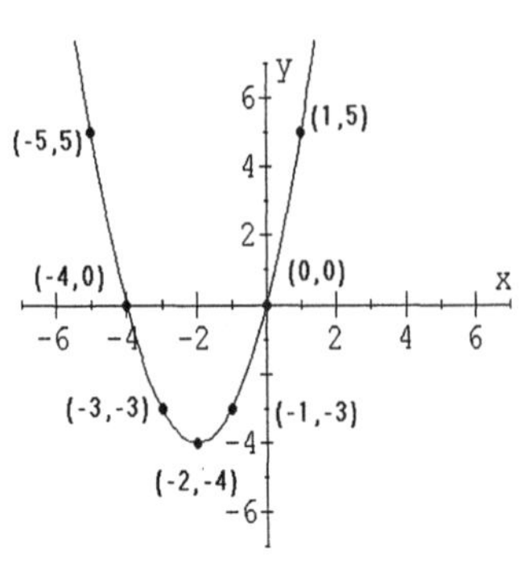

21. $x^2 + 2x + y^2 + 8y + 8 = 0$
$x^2 + 2x + 1 + y^2 + 8y + 16 = -8 + 1 + 16$
$(x+1)^2 + (y+4)^2 = 9$
It is a circle with center at
$(-1, -4)$ and $r = \sqrt{9} = 3$.

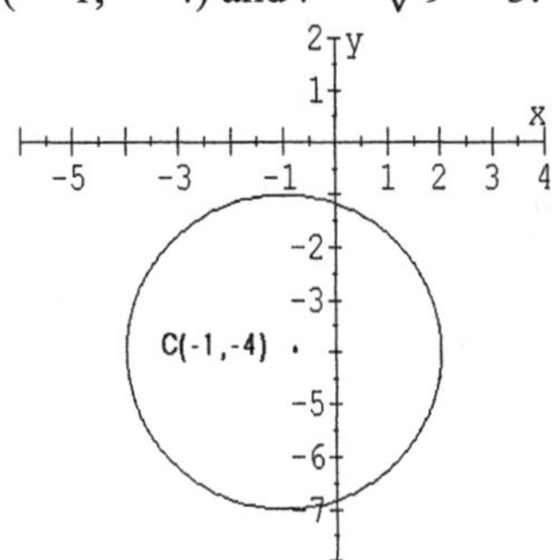

22. $2x^2 + 3y^2 = 12$
It is an ellipse with center at the origin.

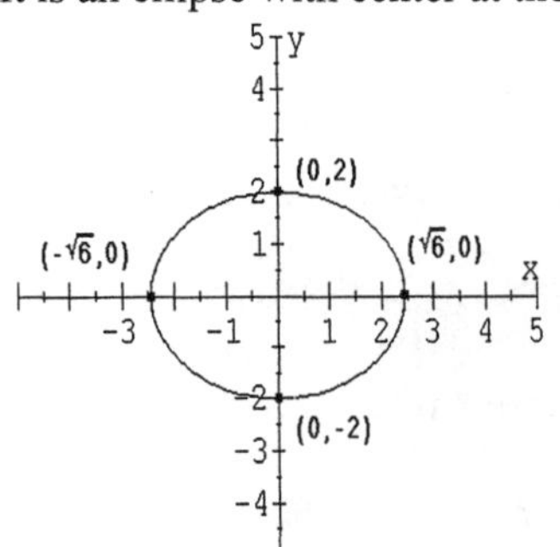

Let $y = 0$.
$2x^2 + 3(0)^2 = 12$
$2x^2 = 12$
$x^2 = 6$
$x = \pm\sqrt{6}$
The endpoints of the major
axis are $\left(-\sqrt{6}, 0\right)$ and $\left(\sqrt{6}, 0\right)$.

Let $x = 0$.
$2(0)^2 + 3y^2 = 12$
$3y^2 = 12$
$y^2 = 4$
$y = \pm 2$
The endpoints of the minor axis are $(0, -2)$ and $(0, 2)$.

23. $y = 2x^2 + 12x + 22$
$y = 2(x^2 + 6x) + 22$
$y = 2(x^2 + 6x + 9) - 18 + 22$
$y = 2(x + 3)^2 + 4$
It is a parabola that opens upwards with the vertex at $(-3, 4)$.

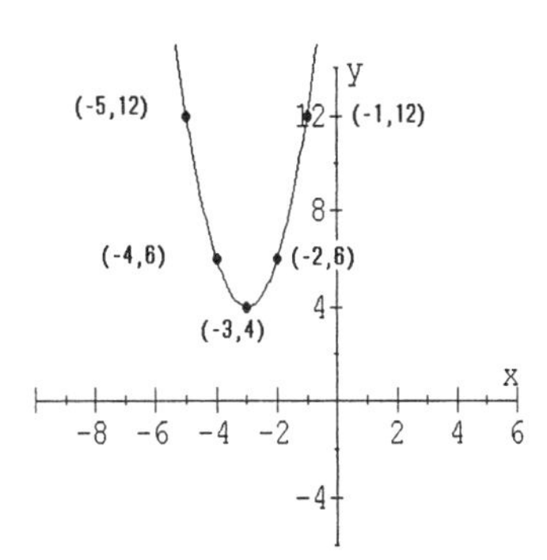

24. $9x^2 - y^2 = 9$
It is a hyperbola.

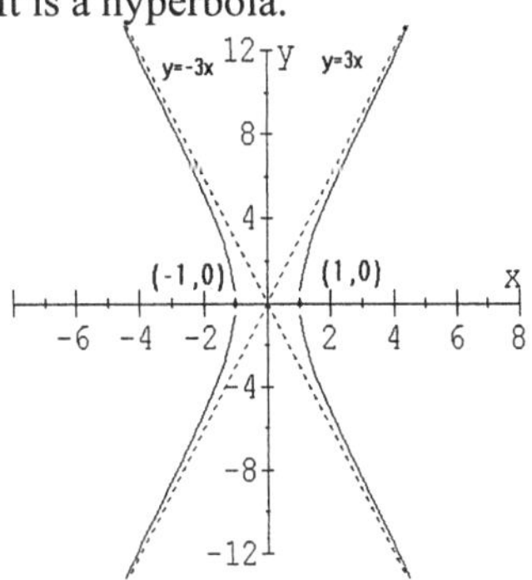

Let $y = 0$.
$9x^2 - (0)^2 = 9$
$9x^2 = 9$
$x^2 = 1$
$x = \pm 1$
The vertices are at $(-1, 0)$ and $(1, 0)$.
To find the asymptotes replace 9 with 0.
$9x^2 - y^2 = 0$
$-y^2 = -9x^2$
$y^2 = 9x^2$
$y = \pm 3x$
Asymptotes are $y = 3x$ and $y = -3x$.

25. $3x^2 - 12x + 5y^2 + 10y - 10 = 0$
$3(x^2 - 4x) + 5(y^2 + 2y) = 10$
$3(x^2 - 4x + 4) + 5(y^2 + 2y + 1) = 10 + 12 + 5$
$3(x - 2)^2 + 5(y + 1)^2 = 27$
It is an ellipse with center at $(2, -1)$.
Let $y = -1$.
$3(x - 2)^2 + 5(-1 + 1)^2 = 27$
$3(x - 2)^2 = 27$
$(x - 2)^2 = 9$
$x - 2 = \pm 3$
$x - 2 = -3$ or $x - 2 = 3$
$x = -1$ or $x = 5$
The endpoints of the major axis are $(-1, -1)$ and $(5, -1)$.

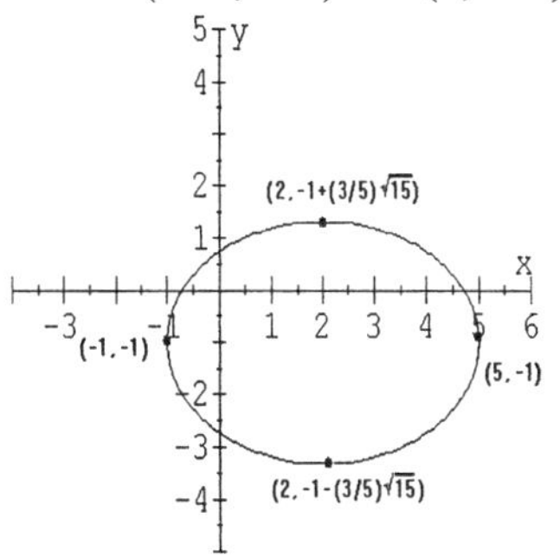

Let $x = 2$.
$3(2 - 2)^2 + 5(y + 1)^2 = 27$
$5(y + 1)^2 = 27$
$$(y + 1)^2 = \frac{27}{5}$$
$$y + 1 = \pm\sqrt{\frac{27}{5}}$$
$$y + 1 = \pm\frac{\sqrt{27}}{\sqrt{5}}$$
$$y + 1 = \pm\frac{3\sqrt{3}}{\sqrt{5}} \bullet \frac{\sqrt{5}}{\sqrt{5}}$$
$$y + 1 = \pm\frac{3\sqrt{15}}{5}$$
$$y = -1 \pm \frac{3\sqrt{15}}{5}$$
The endpoints of the minor axis are $\left(2, -1 - \frac{3\sqrt{15}}{5}\right)$ and $\left(2, -1 + \frac{3\sqrt{15}}{5}\right)$.

CHAPTERS 1-12 | Cumulative Practice Test

1a. $\sqrt[3]{-\frac{8}{27}} = -\frac{2}{3}$

b. $-16^{\frac{5}{4}} = -(16^{\frac{1}{4}})^5 = -(2)^5 = -32$

c. $\left(\frac{2^{-1}}{2^{-2}}\right)^{-2} = \frac{2^2}{2^4} = 2^{2-4} = 2^{-2} = \frac{1}{2^2} = \frac{1}{4}$

d. $\left(\frac{1}{3} - \frac{1}{5}\right)^{-1} = \left(\frac{5}{15} - \frac{3}{15}\right)^{-1} =$

$\left(\frac{2}{15}\right)^{-1} = \left(\frac{15}{2}\right)^{1} = \frac{15^1}{2^1} = \frac{15}{2}$

2a. $\sqrt{96}$
$\sqrt{16}\sqrt{6}$
$4\sqrt{6}$

b. $\sqrt[3]{40}$
$\sqrt[3]{8}\sqrt[3]{5}$
$2\sqrt[3]{5}$

c. $\frac{4\sqrt{2}}{6\sqrt{3}} = \frac{2\sqrt{2}}{3\sqrt{3}} \cdot \frac{\sqrt{3}}{\sqrt{3}} =$

$\frac{2\sqrt{6}}{3\sqrt{9}} = \frac{2\sqrt{6}}{3(3)} = \frac{2\sqrt{6}}{9}$

d. $\frac{2}{\sqrt{7}-\sqrt{2}} = \frac{2}{\sqrt{7}-\sqrt{2}} \cdot \frac{\sqrt{7}+\sqrt{2}}{\sqrt{7}+\sqrt{2}} =$

$\frac{2(\sqrt{7}+\sqrt{2})}{(\sqrt{7}-\sqrt{2})(\sqrt{7}+\sqrt{2})} = \frac{2(\sqrt{7}+\sqrt{2})}{(\sqrt{7})^2-(\sqrt{2})^2} =$

$\frac{2(\sqrt{7}+\sqrt{2})}{7-2} = \frac{2(\sqrt{7}+\sqrt{2})}{5}$

3. $(4xy^2)(-3x^2y^3)(2x)$
$-24x^{1+2+1}y^{2+3}$
$-24x^4y^5$

4. $(2x+1)(3x^2-4x-2)$
$6x^3-8x^2-4x+3x^2-4x-2$
$6x^3-8x^2+3x^2-4x-4x-2$
$6x^3-5x^2-8x-2$

5.

$$
\begin{array}{r|l}
 & x^3+2x^2-x-4 \\
\hline
x+3 & x^4+5x^3+5x^2-7x-12 \\
 & x^4+3x^3 \\
\hline
 & 2x^3+5x^2 \\
 & 2x^3+6x^2 \\
\hline
 & -x^2-7x \\
 & -x^2-3x \\
\hline
 & -4x-12 \\
 & -4x-12 \\
\hline
 & 0
\end{array}
$$

Q: x^3+2x^2-x-4 R: 0

6. $\frac{2x-1}{4} - \frac{3x+2}{3}$; LCD = 12

$\frac{2x-1}{4} \bullet \frac{3}{3} - \frac{3x+2}{3} \bullet \frac{4}{4}$

$\frac{3(2x-1)}{12} - \frac{4(3x+2)}{12}$

$\frac{3(2x-1)-4(3x+2)}{12}$

$\frac{6x-3-12x-8}{12} = \frac{-6x-11}{12}$

7. $(2\sqrt{3}-\sqrt{2})(3\sqrt{3}+2\sqrt{2})$
$6\sqrt{9}-3\sqrt{6}+4\sqrt{6}-2\sqrt{4}$
$6(3)+\sqrt{6}-2(2)$
$18+\sqrt{6}-4$
$14+\sqrt{6}$

8 a. $xy = -6$
To test symmetry for:
x-axis replace y with $-y$
$x(-y) = -6$
$-xy = -6$
$xy = 6$
No, it is not symmetric to the x-axis.

y-axis replace x with $-x$
$(-x)y = -6$
$xy = 6$
No, it is not symmetric to the y-axis.

origin replace x with $-x$ and y with $-y$.
$(-x)(-y) = -6$

$xy = -6$
Yes, it is symmetric to the origin.

b. $x^2 + 3x + y^2 - 4 = 0$
To test symmetry for:
x-axis replace y with $-y$
$x^2 + 3x + (-y)^2 - 4 = 0$
$x^2 + 3x + y^2 - 4 = 0$
Yes, it is symmetric to the x-axis.

y-axis replace x with $-x$
$(-x)^2 + 3(-x) + y^2 - 4 = 0$
$x^2 - 3x + y^2 - 4 = 0$
No, it is not symmetric to the y-axis.

origin replace x with $-x$ and y with $-y$.
$(-x)^2 + 3(-x) + (-y)^2 - 4 = 0$
$x^2 - 3x + y^2 - 4 = 0$
No, it is not symmetric to the origin.

c. $y = x^2 - 14$
To test symmetry for:
x-axis replace y with $-y$
$-y = x^2 - 14$
$y = -x^2 + 14$
No, it is not symmetric to the x-axis.

y-axis replace x with $-x$
$y = (-x)^2 - 14$
$y = x^2 - 14$
Yes, it is symmetric to the y-axis.

origin replace x with $-x$ and y with $-y$.
$-y = (-x)^2 - 14$
$-y = x^2 - 14$
$y = -x^2 + 14$
No, it is not symmetric to the origin.

d. $x^2 - 2y^2 - 3y - 6 = 0$
To test symmetry for:
x-axis replace y with $-y$
$x^2 - 2(-y)^2 - 3(-y) - 6 = 0$
$x^2 - 2y^2 + 3y - 6 = 0$
No, it is not symmetric to the x-axis.

y-axis replace x with $-x$
$(-x)^2 - 2y^2 - 3y - 6 = 0$
$x^2 - 2y^2 - 3y - 6 = 0$
Yes, it is symmetric to the y-axis.

origin replace x with $-x$
and y with $-y$.
$(-x)^2 - 2(-y)^2 - 3(-y) - 6 = 0$
$x^2 - 2y^2 + 3y - 6 = 0$
No, it is not symmetric to the origin.

9. $(3x^2 - 2x - 1) - (3x^2 - 3x + 2)$
$3x^2 - 2x - 1 - 3x^2 + 3x - 2$
$x - 3$ for $x = -18$
$(-18) - 3 = -21$

10. $\dfrac{36x^{-2}y^4}{9x^{-3}y^3} = 4x^{-2-(-3)}y^{4-3} =$
$4x^1y^1 = 4xy$ for $x = 6;\ y = -8$
$4(6)(-8) = -192$

11. $6 = 2 \cdot 3$
$9 = 3 \cdot 3$
$12 = 2 \cdot 2 \cdot 3$
The least common multiple is $2 \cdot 2 \cdot 3 \cdot 3 = 36$.

12. $x - 4y = -6$
$-4y = -x - 6$
$-\frac{1}{4}(-4y) = -\frac{1}{4}(-x - 6)$
$y = \frac{1}{4}x + \frac{3}{2};\ m = \frac{1}{4}$

13. $(2,7),\ (-1,3)$
$m = \dfrac{3-7}{-1-2} = \dfrac{-4}{-3} = \dfrac{4}{3}$
$y - 7 = \frac{4}{3}(x - 2)$
$3(y - 7) = 3\left[\frac{4}{3}(x - 2)\right]$
$3y - 21 = 4(x - 2)$
$3y - 21 = 4x - 8$
$-4x + 3y - 21 = -8$
$-4x + 3y = 13$
$4x - 3y = -13$

14. $(x + 2)(x - 6) = -16$
$x^2 - 4x - 12 = -16$
$x^2 - 4x + 4 = 0$
$(x - 2)(x - 2) = 0$
$x - 2 = 0$ or $x - 2 = 0$
$x = 2$ or $x = 2$
The solution set is $\{2\}$.

15. $\frac{2}{3}(x-1) - \frac{3}{4}(2x+3) = \frac{3}{2}$

$12\left[\frac{2}{3}(x-1) - \frac{3}{4}(2x+3)\right] = 12\left(\frac{3}{2}\right)$

$12\left[\frac{2}{3}(x-1)\right] - 12\left[\frac{3}{4}(2x+3)\right] = 6(3)$

$8(x-1) - 9(2x+3) = 18$

$8x - 8 - 18x - 27 = 18$

$-10x - 35 = 18$

$-10x = 53$

$x = -\frac{53}{10}$

The solution set is $\left\{-\frac{53}{10}\right\}$.

16. $0.06x + 0.07(7000 - x) = 460$

$100[0.06x + 0.07(7000 - x)] = 100(460)$

$6x + 7(7000 - x) = 46000$

$6x + 49000 - 7x = 46000$

$-x + 49000 = 46000$

$-x = -3000$

$x = 3000$

The solution set is $\{3000\}$.

17. $|3x - 5| = 10$ is equivalent to

$3x - 5 = -10$ or $3x - 5 = 10$.

$3x - 5 = -10$ or $3x - 5 = 10$

$3x = -5$ or $3x = 15$

$x = -\frac{5}{3}$ or $x = 5$

The solution set is $\left\{-\frac{5}{3}, 5\right\}$.

18. $3x^2 - x - 1 = 0$

$x = \frac{-(-1) \pm \sqrt{(-1)^2 - 4(3)(-1)}}{2(3)}$

$x = \frac{1 \pm \sqrt{1+12}}{6}$

$x = \frac{1 \pm \sqrt{13}}{6}$

$x = \frac{1 - \sqrt{13}}{6}$ or $x = \frac{1 + \sqrt{13}}{6}$

The solution set is $\left\{\frac{1 \pm \sqrt{13}}{6}\right\}$.

19. $|2x - 3| \le 5$

$-5 \le 2x - 3 \le 5$

$-2 \le 2x \le 8$

$-1 \le x \le 4$

The solution set is $[-1, 4]$.

20. $\frac{x-2}{x-3} > 0; x \ne 3$

$x - 2 = 0$ $\quad$ $x - 3 = 0$

$x = 2$ $\quad$ $x = 3$

Test Point	0	$\frac{5}{2}$	4
$x - 2$:	negative	positive	positive
$x - 3$:	negative	negative	positive
quotient:	positive	negative	positive

(Boundaries: 2 between 0 and $\frac{5}{2}$; 3 between $\frac{5}{2}$ and 4.)

The solution set is $(-\infty, 2) \cup (3, \infty)$.

21. $x^2 - 5x - 6 < 0$

$(x-6)(x+1) < 0$

$(x-6)(x+1) = 0$

$x - 6 = 0$ or $x + 1 = 0$

$x = 6$ or $x = -1$

Test Point	-2	0	7
$x - 6$:	negative	negative	positive
$x + 1$:	negative	positive	positive
product:	positive	negative	positive

(Boundaries: -1 between -2 and 0; 6 between 0 and 7.)

The solution set is $(-1, 6)$.

22. Let x represent the angle, then

$90 - x$ represents the complementary angle.

$x = 6 + \frac{1}{2}(90 - x)$

$2(x) = 2\left[6 + \frac{1}{2}(90 - x)\right]$

$2x = 12 + (90 - x)$

$2x = 102 - x$

$3x = 102$

$x = 34$

One angle measures 34° and its complement measures 56°.

23. Let $x =$ number of quarters
and $y =$ number of dimes.

$$\begin{pmatrix} x + y = 37 \\ 0.25x + 0.10y = 7.45 \end{pmatrix}$$

Solve equation 1 for x:

$x + y = 37$

$x = -y + 37$

Substitute into equation 2:

$0.25x + 0.10y = 7.45$

$0.25(-y + 37) + 0.10y = 7.45$

$100[0.25(-y + 37) + 0.10y] = 100(7.45)$

$25(-y + 37) + 10y = 745$

$-25y + 925 + 10y = 745$

$-15y = -180$

$y = 12$

$x = -y + 37$

$x = -12 + 37$

$x = 25$

There are 25 quarters and 12 dimes.

24. Let t represent Juan's time walking. Then $t - \frac{3}{2}$ represents Cathy's time jogging.
Use Distance $=$ Rate $\cdot$ Time.
The information is recorded in the following table:

	Distance	Rate	Time
Juan	$4t$	4	t
Cathy	$6\left(t - \frac{3}{2}\right)$	6	$t - \frac{3}{2}$

Walking distance $=$ Jogging distance

$4t = 6\left(t - \frac{3}{2}\right)$

$4t = 6t - 9$

$-2t = -9$

$t = \frac{9}{2} = 4\frac{1}{2}$

Juan's time is $4\frac{1}{2}$ hours, so Cathy's time to catch Juan is $\frac{9}{2} - \frac{3}{2} = \frac{6}{2} = 3$ hours.

25. Let $x =$ cost of package of cookies
$y =$ cost of sack of potato chips.

$$\begin{pmatrix} 3x + 2y = 11.35 \\ 2x + 5y = 18.53 \end{pmatrix}$$

Multiply equation 1 by -2 and equation 2 by 3. Then add equations.

$$\begin{aligned} -6x - 4y &= -22.70 \\ 6x + 15y &= 55.59 \\ \hline 11y &= 32.89 \\ y &= 2.99 \end{aligned}$$

Substitute value for y into first equation to find x.

$3x + 2(2.99) = 11.35$

$3x + 5.98 = 11.35$

$3x = 5.37$

$x = 1.79$

A package of cookies costs \$1.79 and a sack of potato chips costs \$2.99.

Chapter 13 Functions

PROBLEM SET 13.1 Relations and Functions

1. D = {1, 2, 3, 4}
R = {5, 8, 11, 14}
It is a function.

3. D = {0, 1}
R = $\{-2\sqrt{6}, -5, 5, 2\sqrt{6}\}$
It is not a function.

5. D = {1, 2, 3, 4, 5}
R = {2, 5, 10, 17, 26}
It is a function.

7. $\{(x, y)|5x - 2y = 6\}$
Domain = {all reals}.
Range = {all reals}.
It is a function.

9. $\{(x, y)|x^2 = y^3\}$
Domain = {all reals}.
To find range solve for y.
$y^3 = x^2$
$y = \sqrt[3]{x^2}$
Since $x^2 \geq 0$, then $\sqrt[3]{x^2} \geq 0$.
Range = $\{y \mid y \geq 0\}$.
It is a function.

11. $f(x) = 7x - 2$
D = {all reals}.

13. $f(x) = \dfrac{1}{x-1}$
$x - 1 = 0$
$x = 1$
D = $\{x|x \neq 1\}$.

15. $g(x) = \dfrac{3x}{4x-3}$
$4x - 3 = 0$
$4x = 3$
$x = \dfrac{3}{4}$
D = $\left\{x|x \neq \dfrac{3}{4}\right\}$.

17. $h(x) = \dfrac{2}{(x+1)(x-4)}$
$x + 1 = 0$ or $x - 4 = 0$
$x = -1$ or $x = 4$
D = $\{x|x \neq -1$ or $x \neq 4\}$.

19. $f(x) = \dfrac{14}{x^2 + 3x - 40}$
$f(x) = \dfrac{14}{(x+8)(x-5)}$
$x + 8 = 0$ or $x - 5 = 0$
$x = -8$ or $x = 5$
D = $\{x|x \neq -8$ or $x \neq 5\}$.

21. $f(x) = \dfrac{-4}{x^2 + 6x}$
$f(x) = \dfrac{-4}{x(x+6)}$
$x = 0$ or $x + 6 = 0$
$x = -6$
D = $\{x|x \neq 0$ or $x \neq -6\}$.

23. $f(t) = \dfrac{4}{t^2 + 9}$
D = {all reals}.

25. $f(t) = \dfrac{3t}{t^2 - 4}$
$f(t) = \dfrac{3t}{(t+2)(t-2)}$
$t + 2 = 0$ or $t - 2 = 0$
$t = -2$ or $t = 2$
D = $\{t|t \neq -2$ or $t \neq 2\}$.

27. $h(x) = \sqrt{x+4}$
$x + 4 \geq 0$
$x \geq -4$
D = $\{x|x \geq -4\}$.

29. $f(s) = \sqrt{4s - 5}$
$4s - 5 \geq 0$
$4s \geq 5$

$s \geq \frac{5}{4}$
$\text{D} = \left\{ s | s \geq \frac{5}{4} \right\}.$

31. $f(x) = \sqrt{x^2 - 16}$
$x^2 - 16 \geq 0$
$(x + 4)(x - 4) \geq 0$
$x + 4 = 0$ or $x - 4 = 0$
$x = -4$ or $x = 4$

Test Point	-5	0	5
$x + 4$:	negative	positive	positive
$x - 4$:	negative	negative	positive
product:	positive	negative	positive

(Boundaries: -4, 4)

$\text{D} = \{x | x \leq -4 \text{ or } x \geq 4\}.$

33. $f(x) = \sqrt{x^2 - 3x - 18}$
$x^2 - 3x - 18 \geq 0$
$(x + 3)(x - 6) \geq 0$
$x - 6 = 0$ or $x + 3 = 0$
$x = 6$ or $x = -3$

Test Point	-4	0	7
$x - 6$:	negative	negative	positive
$x + 3$:	negative	positive	positive
product:	positive	negative	positive

(Boundaries: -3, 6)

$\text{D} = \{x | x \leq -3 \text{ or } x \geq 6\}.$

35. $f(x) = \sqrt{1 - x^2}$
$1 - x^2 \geq 0$
$(1 - x)(1 + x) \geq 0$
$1 - x = 0$ or $1 + x = 0$
$1 = x$ or $x = -1$

Test Point	-2	0	2
$1 - x$:	positive	positive	negative
$1 + x$:	negative	positive	positive
product:	negative	positive	negative

(Boundaries: -1, 1)

$\text{D} = \{x | -1 \leq x \leq 1\}.$

37. $f(x) = 5x - 2$
$f(0) = 5(0) - 2 = -2$
$f(2) = 5(2) - 2 = 10 - 2 = 8$
$f(-1) = 5(-1) - 2 = -5 - 2 = -7$
$f(-4) = 5(-4) - 2 = -20 - 2 = -22$

39. $f(x) = \frac{1}{2}x - \frac{3}{4}$
$f(-2) = \frac{1}{2}(-2) - \frac{3}{4} = -1 - \frac{3}{4} = -\frac{7}{4}$
$f(0) = \frac{1}{2}(0) - \frac{3}{4} = -\frac{3}{4}$
$f\left(\frac{1}{2}\right) = \frac{1}{2}\left(\frac{1}{2}\right) - \frac{3}{4} = \frac{1}{4} - \frac{3}{4} = -\frac{1}{2}$
$f\left(\frac{2}{3}\right) = \frac{1}{2}\left(\frac{2}{3}\right) - \frac{3}{4} = \frac{1}{3} - \frac{3}{4} = -\frac{5}{12}$

41. $g(x) = 2x^2 - 5x - 7$
$g(-1) = 2(-1)^2 - 5(-1) - 7$
$g(-1) = 2 + 5 - 7 = 0$
$g(2) = 2(2)^2 - 5(2) - 7$
$g(2) = 8 - 10 - 7 = -9$
$g(-3) = 2(-3)^2 - 5(-3) - 7$
$g(-3) = 18 + 15 - 7 = 26$
$g(4) = 2(4)^2 - 5(4) - 7$
$g(4) = 32 - 20 - 7 = 5$

43. $h(x) = -2x^2 - x + 4$
$h(-2) = -2(-2)^2 - (-2) + 4$
$h(-2) = -8 + 2 + 4 = -2$
$h(-3) = -2(-3)^2 - (-3) + 4$
$h(-3) = -18 + 3 + 4 = -11$
$h(4) = -2(4)^2 - (4) + 4$
$h(4) = -32 - 4 + 4 = -32$
$h(5) = -2(5)^2 - (5) + 4$
$h(5) = -50 - 5 + 4 = -51$

45. $f(x) = \sqrt{2x + 1}$
$f(3) = \sqrt{2(3) + 1} = \sqrt{7}$
$f(4) = \sqrt{2(4) + 1} = \sqrt{9} = 3$
$f(10) = \sqrt{2(10) + 1} = \sqrt{21}$
$f(12) = \sqrt{2(12) + 1} = \sqrt{25} = 5$

47. $f(x) = \frac{-4}{x + 3}$
$f(1) = \frac{-4}{1 + 3} = \frac{-4}{4} = -1$
$f(-1) = \frac{-4}{-1 + 3} = \frac{-4}{2} = -2$
$f(3) = \frac{-4}{3 + 3} = \frac{-4}{6} = -\frac{2}{3}$
$f(-6) = \frac{-4}{-6 + 3} = \frac{-4}{-3} = \frac{4}{3}$

49. $f(x) = 5x^2 - 2x + 3$ and
$g(x) = -x^2 + 4x - 5$

$f(-2) = 5(-2)^2 - 2(-2) + 3$
$f(-2) = 20 + 4 + 3 = 27$
$f(3) = 5(3)^2 - 2(3) + 3$
$f(3) = 45 - 6 + 3 = 42$
$g(-4) = -(-4)^2 + 4(-4) - 5$
$g(-4) = -16 - 16 - 5 = -37$
$g(6) = -(6)^2 + 4(6) - 5$
$g(6) = -36 + 24 - 5 = -17$

51. $f(x) = 3|x| - 1$ and
$g(x) = -|x| + 1$

$f(-2) = 3|-2| - 1$
$f(-2) = 3(2) - 1 = 5$

$f(3) = 3|3| - 1$
$f(3) = 3(3) - 1 = 8$

$g(-4) = -|-4| + 1$
$g(-4) = -(4) + 1 = -3$

$g(5) = -|5| + 1$
$g(5) = -5 + 1 = -4$

53. $f(x) = -3x + 6$
$f(a) = -3a + 6$
$f(a + h) = -3(a + h) + 6$
$f(a + h) = -3a - 3h + 6$

$$\frac{f(a+h) - f(a)}{h} = \frac{-3a - 3h + 6 - (-3a + 6)}{h}$$
$$= \frac{-3a - 3h + 6 + 3a - 6}{h}$$
$$= \frac{-3h}{h} = -3$$

55. $f(x) = -x^2 - 1$
$f(a) = -a^2 - 1$
$f(a + h) = -(a + h)^2 - 1$
$f(a + h) = -(a^2 + 2ah + h^2) - 1$
$f(a + h) = -a^2 - 2ah - h^2 - 1$

$$\frac{f(a + h) - f(a)}{h} = \frac{-a^2 - 2ah - h^2 - 1 - (-a^2 - 1)}{h}$$
$$= \frac{-a^2 - 2ah - h^2 - 1 + a^2 + 1}{h}$$
$$= \frac{-2ah - h^2}{h}$$
$$= \frac{h(-2a - h)}{h}$$
$$= -2a - h$$

57. $f(x) = 2x^2 - x + 8$
$f(a) = 2a^2 - a + 8$
$f(a + h) = 2(a + h)^2 - (a + h) + 8$
$f(a + h) = 2(a^2 + 2ah + h^2) - a - h + 8$
$f(a + h) = 2a^2 + 4ah + 2h^2 - a - h + 8$

$$\frac{f(a + h) - f(a)}{h} = \frac{2a^2 + 4ah + 2h^2 - a - h + 8 - (2a^2 - a + 8)}{h}$$

$$= \frac{2a^2 + 4ah + 2h^2 - a - h + 8 - 2a^2 + a - 8}{h}$$
$$= \frac{4ah - h + 2h^2}{h}$$
$$= \frac{h(4a - 1 + 2h)}{h}$$
$$= 4a - 1 + 2h$$

59. $f(x) = -4x^2 - 7x - 9$
$f(a) = -4a^2 - 7a - 9$
$f(a+h) = -4(a+h)^2 - 7(a+h) - 9$
$f(a+h) = -4(a^2 + 2ah + h^2) - 7a - 7h - 9$
$f(a+h) = -4a^2 - 8ah - 4h^2 - 7a - 7h - 9$

$$\frac{f(a+h) - f(a)}{h} = \frac{-4a^2 - 8ah - 4h^2 - 7a - 7h - 9 - (-4a^2 - 7a - 9)}{h}$$
$$= \frac{-4a^2 - 8ah - 4h^2 - 7a - 7h - 9 + 4a^2 + 7a + 9}{h}$$
$$= \frac{-8ah - 7h - 4h^2}{h}$$
$$= \frac{h(-8a - 7 - 4h)}{h}$$
$$= -8a - 7 - 4h$$

61. $h(t) = 64t - 16t^2$

$h(1) = 64(1) - 16(1)^2$
$h(1) = 64 - 16 = 48$

$h(2) = 64(2) - 16(2)^2$
$h(2) = 128 - 64 = 64$

$h(3) = 64(3) - 16(3)^2$
$h(3) = 192 - 144 = 48$

$h(4) = 64(4) - 16(4)^2$
$h(4) = 256 - 256 = 0$

63. $C(m) = 50 + 0.32m$

$C(75) = 50 + 0.32(75)$
$C(75) = 50 + 24 = \$74$

$C(150) = 50 + 0.32(150)$
$C(150) = 50 + 48 = \$98$

$C(225) = 50 + 0.32(225)$
$C(225) = 50 + 72 = \$122$

$C(650) = 50 + 0.32(650)$
$C(650) = 50 + 208 = \$258$

65. $I(r) = 2500r$
$I(0.03) = 2500(0.03) = \$75$
$I(0.04) = 2500(0.04) = \$100$
$I(0.135) = 2500(0.05) = \$125$
$I(0.065) = 2500(0.065) = \$162.50$

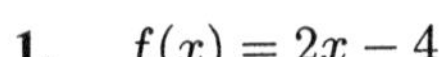

PROBLEM SET 13.2 Functions: Graphs and Applications

1. $f(x) = 2x - 4$

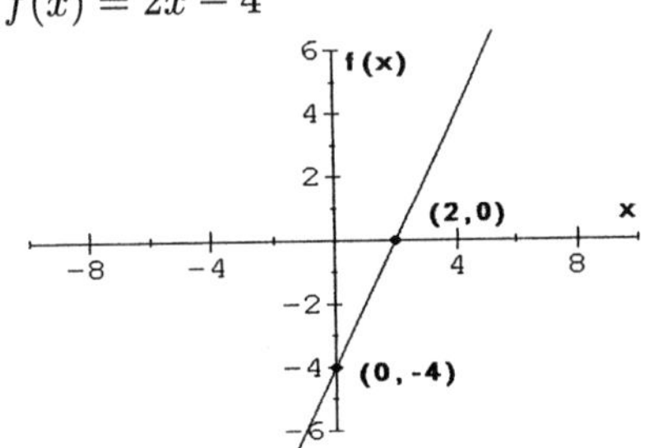

3. $f(x) = -2x^2$

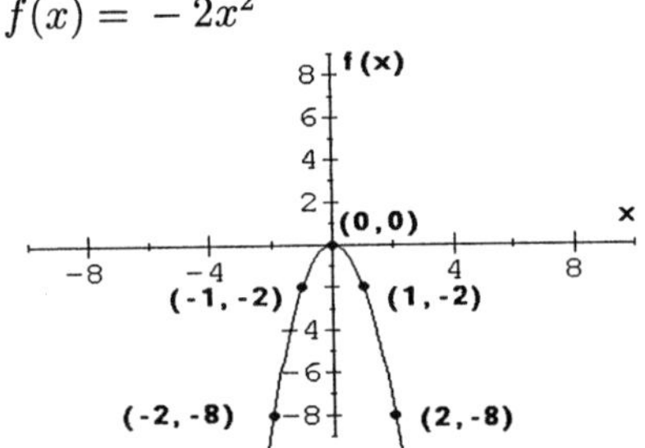

5. $f(x) = -3x$

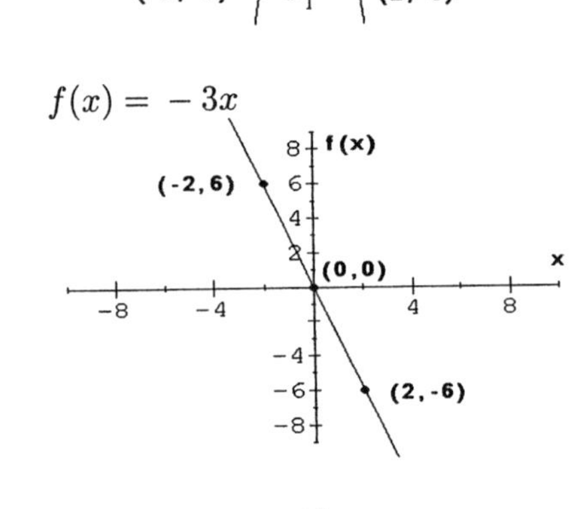

7. $f(x) = -(x+1)^2 - 2$

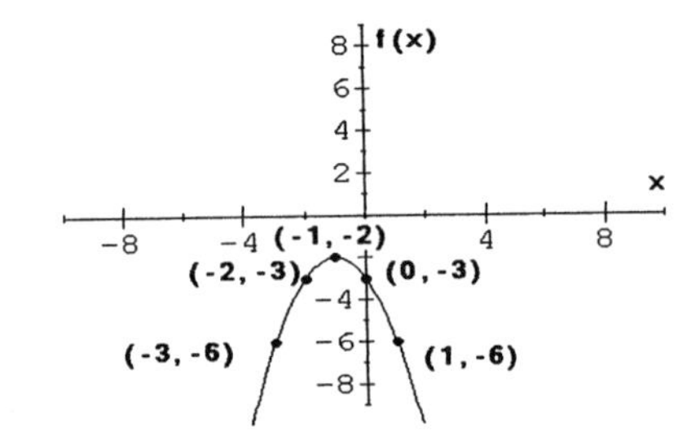

9. $f(x) = -x + 3$

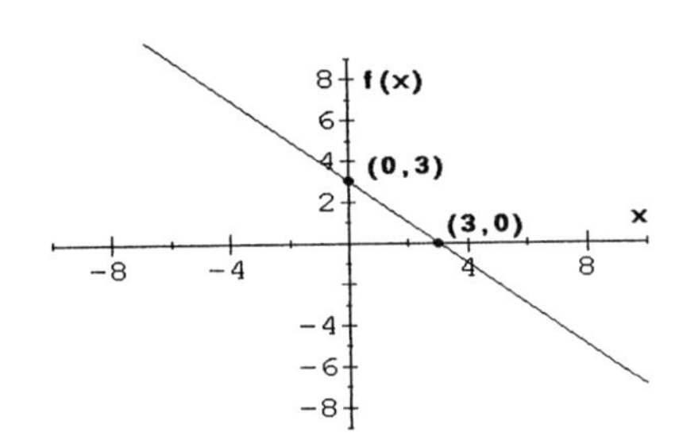

11. $f(x) = x^2 + 2x - 2$

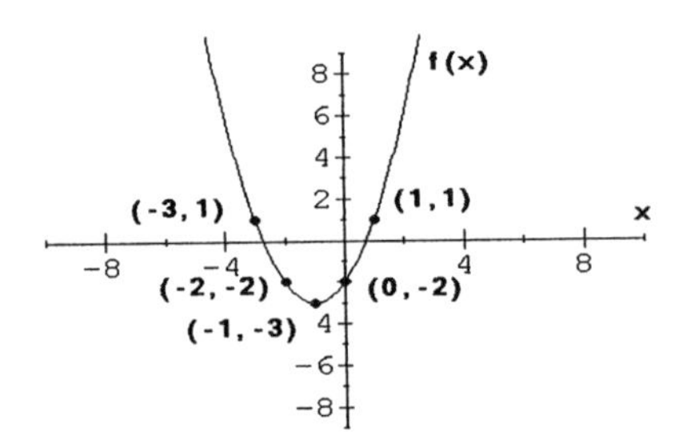

13. $f(x) = -x^2 + 6x - 8$

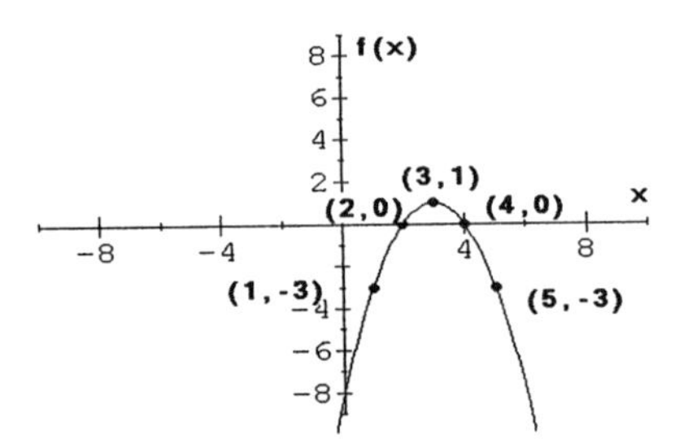

15. $f(x) = -3$

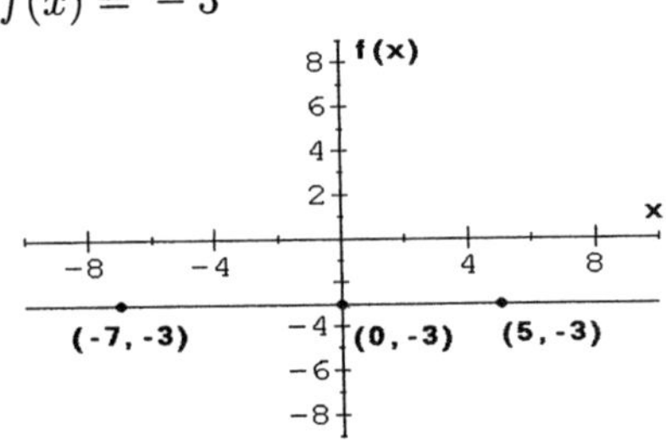

17. $f(x) = 2x^2 - 20x + 52$

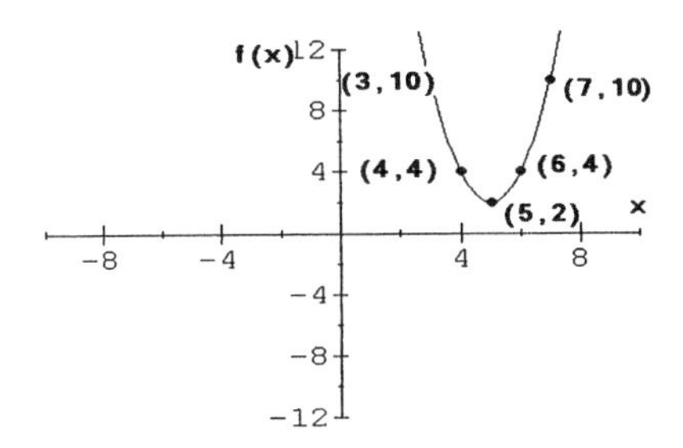

19. $f(x) = -3x^2 + 6x$

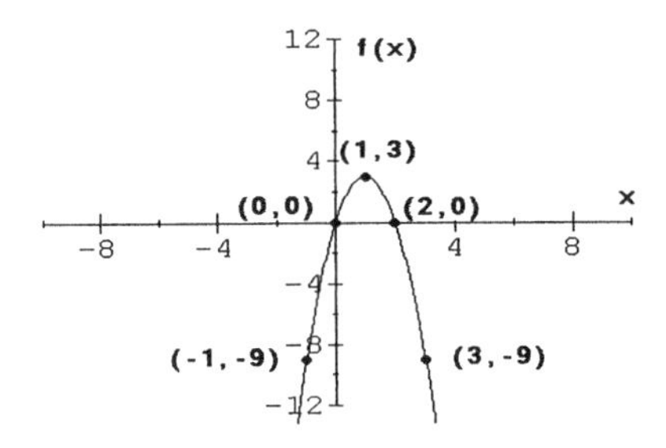

21. $f(x) = x^2 - x + 2$

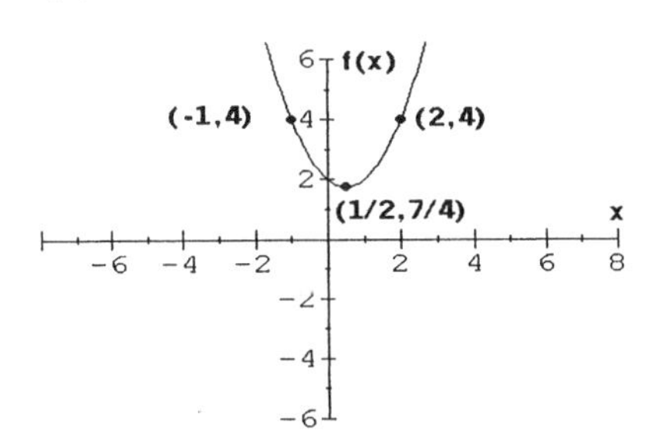

23. $f(x) = 2x^2 + 10x + 11$

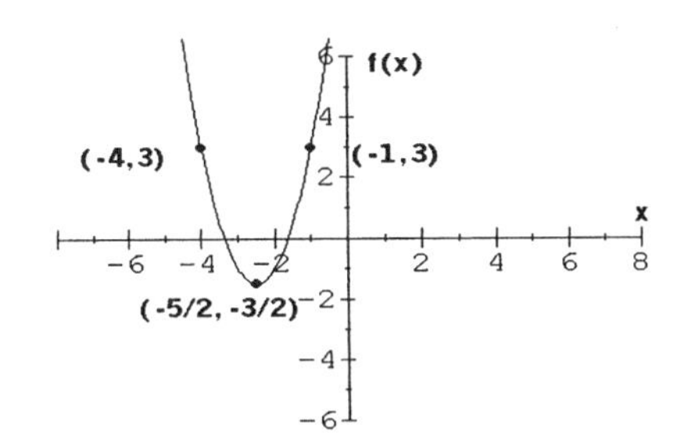

25. $f(x) = -2x^2 - 1$

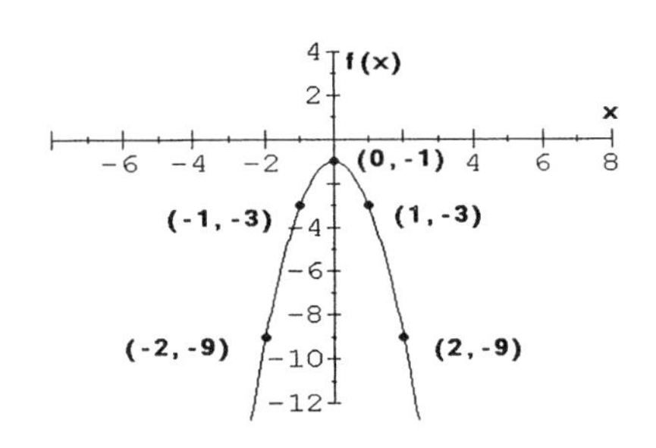

27. $f(x) = -3x^2 + 12x - 7$

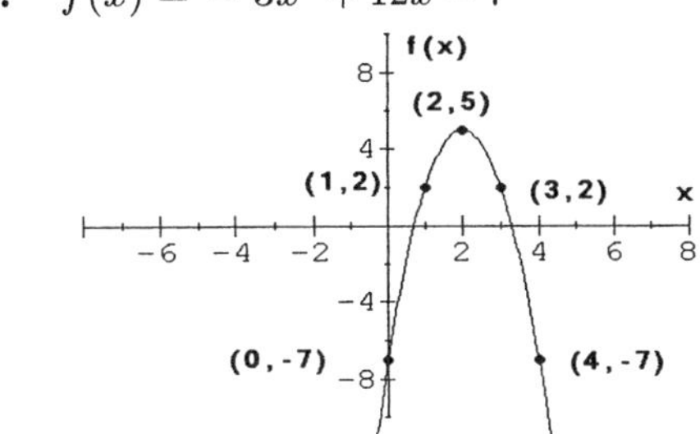

29. $f(x) = -2x^2 + 14x - 25$

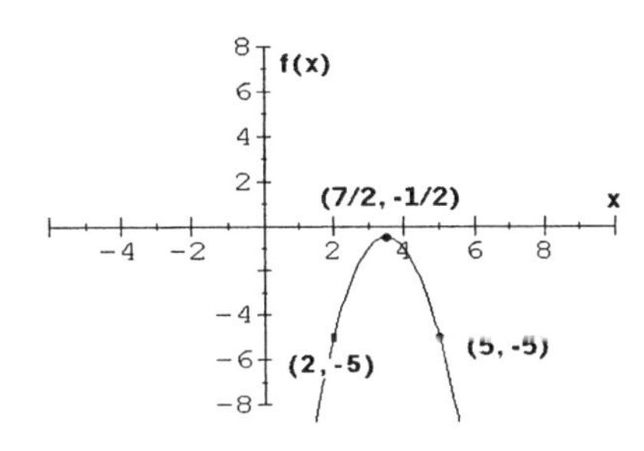

31a. $c(h) = 0.0045h$

$$h = \left(3\ \frac{\text{hours}}{\text{night}}\right)\left(31 \text{ nights}\right) = 93 \text{ hours}$$

$c(93) = 0.0045(93) = 0.42$

The cost is \$0.42.

b.

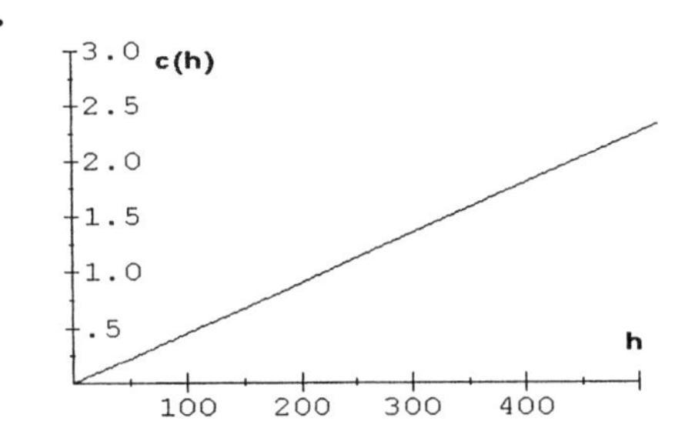

c. $c(225) \approx 1.00$

d. $c(225) = 0.0045(225) = 1.01$
The cost is \$1.01.

33. 150 miles use $f(x) = 26$
The charge would be \$26.

230 miles use
$g(x) = 26 + 0.15(x - 200)$
$g(230) = 26 + 0.15(230 - 200) = 30.5$
The charge would be \$30.50.

360 miles use
$g(x) = 26 + 0.15(x - 200)$
$g(360) = 26 + 0.15(360 - 200) = 50$
The charge would be \$50.

430 miles use
$g(x) = 26 + 0.15(x - 200)$
$g(430) = 26 + 0.15(430 - 200) = 60.50$
The charge would be \$60.50.

35. $s(c) = 1.4c$
$s(1.50) = 1.4(1.50) = \$2.10$
$s(3.25) = 1.4(3.25) = \$4.55$
$s(14.80) = 1.4(14.80) = \$20.72$
$s(21) = 1.4(21) = \$29.40$
$s(24.20) = 1.4(24.20) = \$33.88$

37. Let p represent the marked price.
$c(p) = p - 0.20p = 0.80p$
$c(p) = 0.8p$

$c(9.50) = 0.8(9.50) = \$7.60$
$c(15) = 0.8(15) = \$12.00$
$c(75) = 0.8(75) = \$60.00$
$c(12.50) = 0.8(12.50) = \$10.00$
$c(750) = 0.8(750) = \$600.00$

39. $C(x) = 2x^2 - 320x + 12{,}920$
$C(x) = 2(x^2 - 160x) + 12{,}920$
$C(x) = 2(x^2 - 160x + 6400) - 12{,}800 + 12{,}920$
$C(x) = 2(x - 80)^2 + 120$
The minimum cost is when
80 items are produced.

41. Let x = one number and
$30 - x$ = other number.

$F(x) = x^2 + 10(30 - x)$
$F(x) = x^2 + 300 - 10x$
$F(x) = x^2 - 10x + 25 - 25 + 300$
$F(x) = (x - 5)^2 + 275$
The number are 5 and 25.

43. Let x = width.
Use $P = 2L + 2W$.
$240 = 2L + 2x$
$240 - 2x = 2L$
$\frac{1}{2}\left(240 - 2x\right) = \frac{1}{2}\left(2L\right)$
$120 - x = L$
Use $A = LW$.
$A(x) = x(120 - x)$
$A(x) = 120x - x^2$
$A(x) = -x^2 + 120x$
$A(x) = -(x^2 - 120x)$
$A(x) = -(x^2 - 120x + 3600) + 3600$
$A(x) = -(x - 60)^2 + 3600$
The dimensions should be
60 meters by 60 meters.

45. Let x = number of decreases
$R(x) = \left(15 - \frac{1}{4}x\right)(1000 + 20x)$
$R(x) = 15000 + 300x - 250x - 5x^2$
$R(x) = -5x^2 + 50x + 15000$
$R(x) = -5(x^2 - 10x) + 15000$
$R(x) = -5(x^2 - 10x + 25) + 125 + 15000$
$R(x) = -5(x - 5)^2 + 15125$
There should be 5 rate decreases.
Rate = $15 - 5(0.25) = \$13.75$.
Subscribers = $1000 + 20(5) = 1100$.

PROBLEM SET 13.3 Graphing Made Easy Via Transformations

1. $f(x) = -x^3$

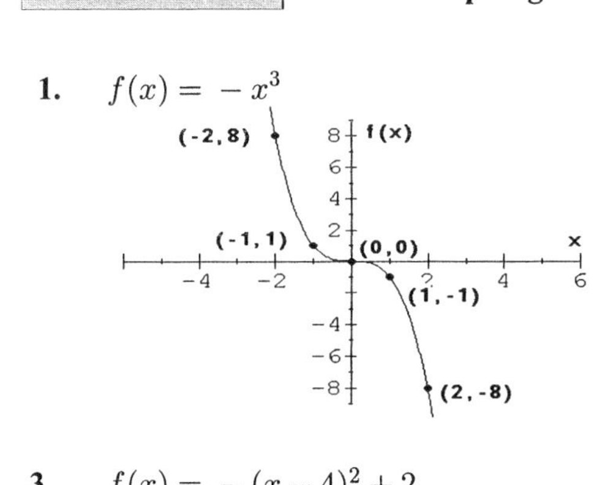

3. $f(x) = -(x-4)^2 + 2$

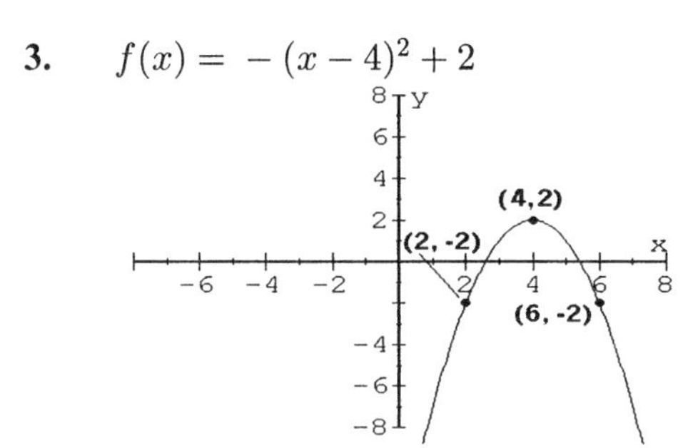

5. $f(x) = \dfrac{1}{x} - 2$

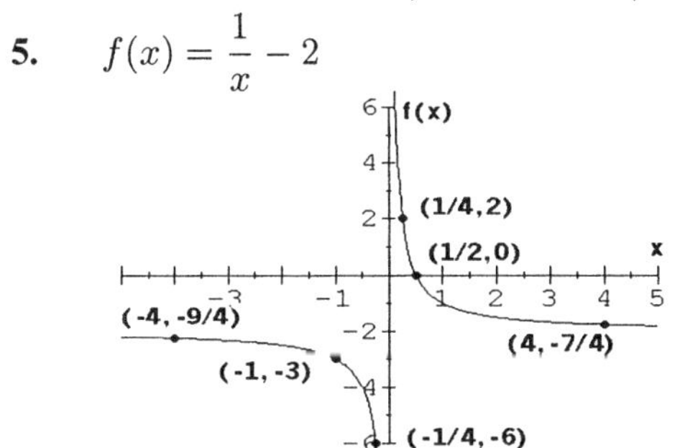

7. $f(x) = \left|x-1\right| + 2$

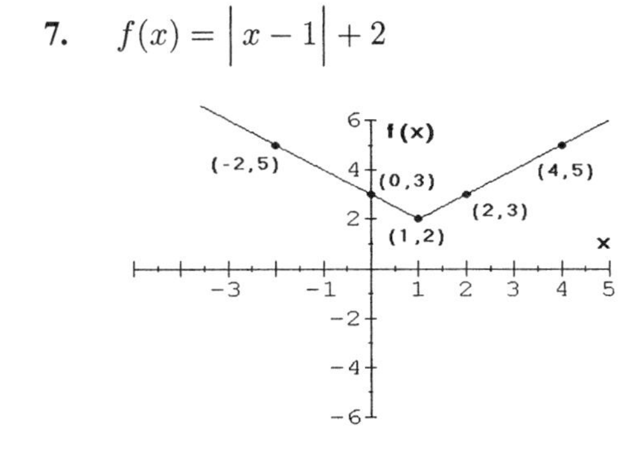

9. $f(x) = \dfrac{1}{2}\left|x\right|$

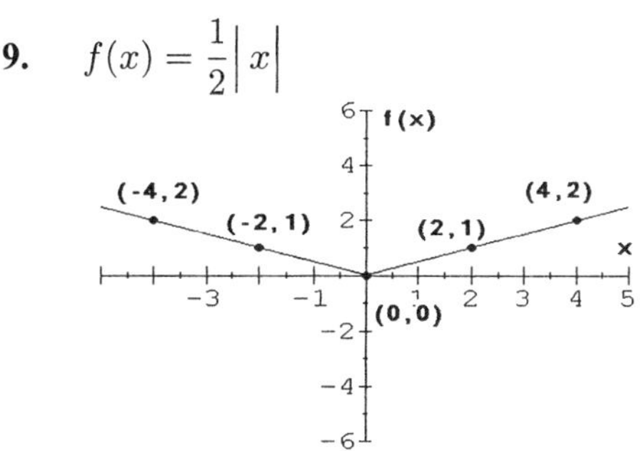

11. $f(x) = -2\sqrt{x}$

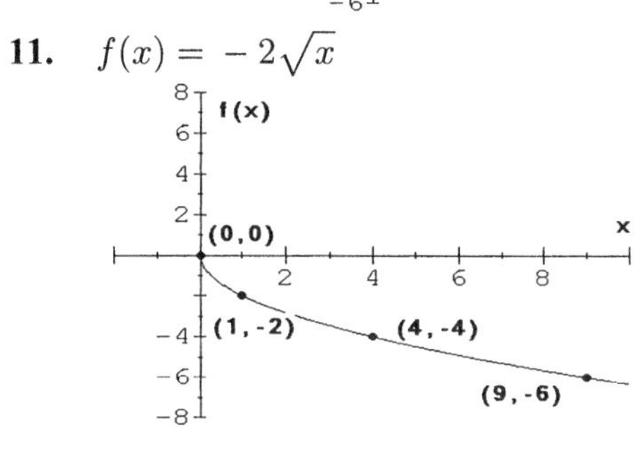

13. $f(x) = \sqrt{x+2} - 3$

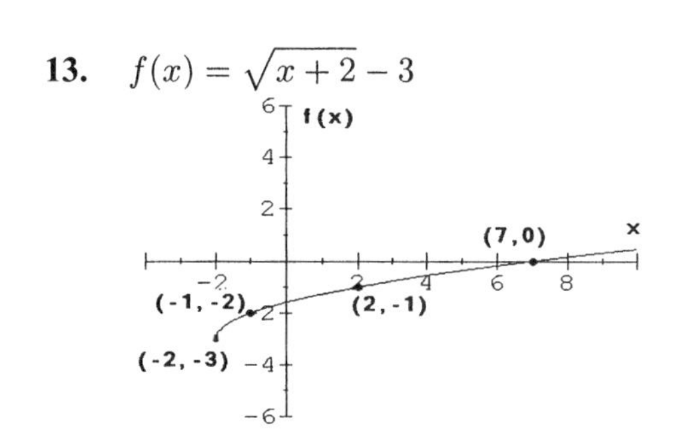

15. $f(x) = \dfrac{2}{x-1} + 3$

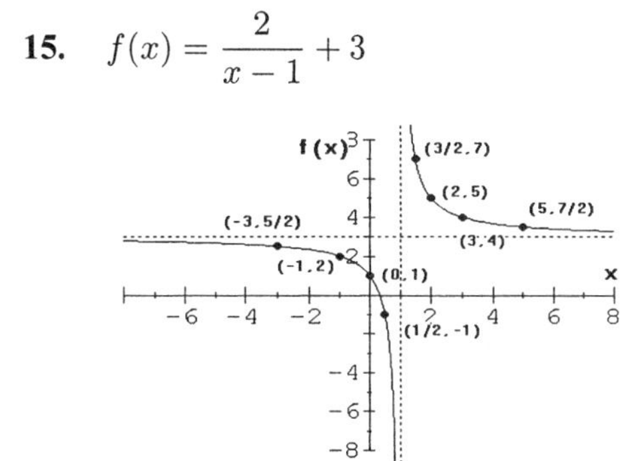

Problem Set 13.3

17. $f(x) = \sqrt{2 - x}$

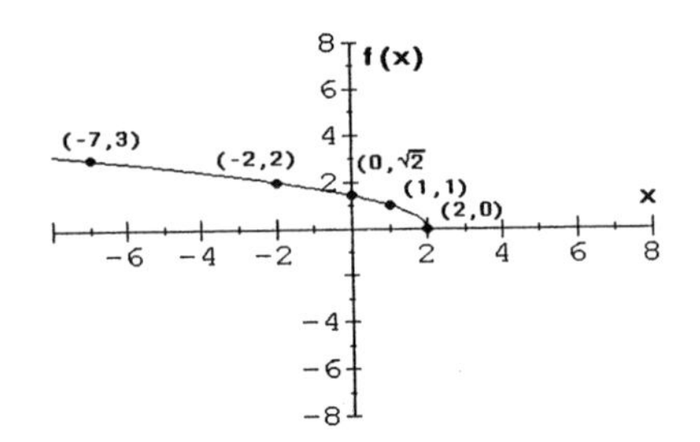

19. $f(x) = -3(x-2)^2 - 1$

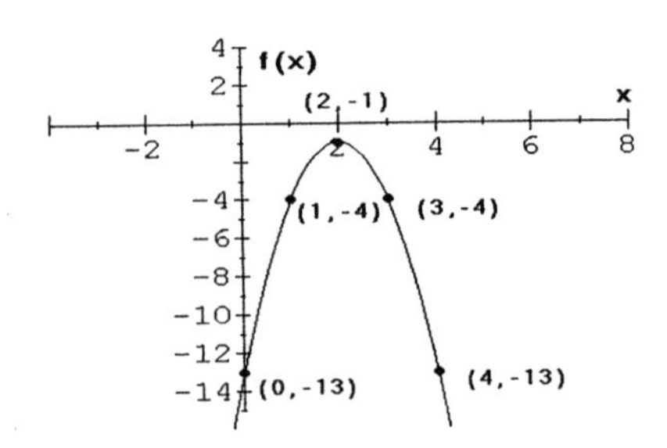

21. $f(x) = 3(x-2)^3 - 1$

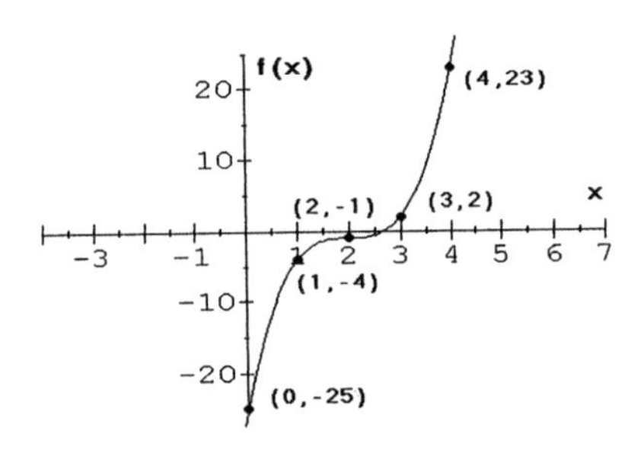

23. $f(x) = 2x^3 + 3$

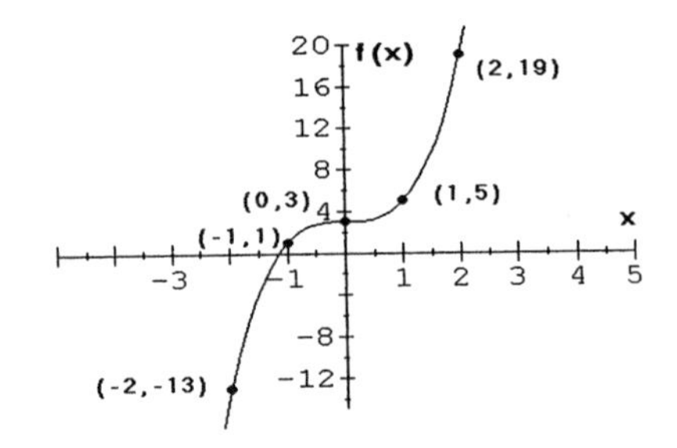

25. $f(x) = -2\sqrt{x+3} + 4$

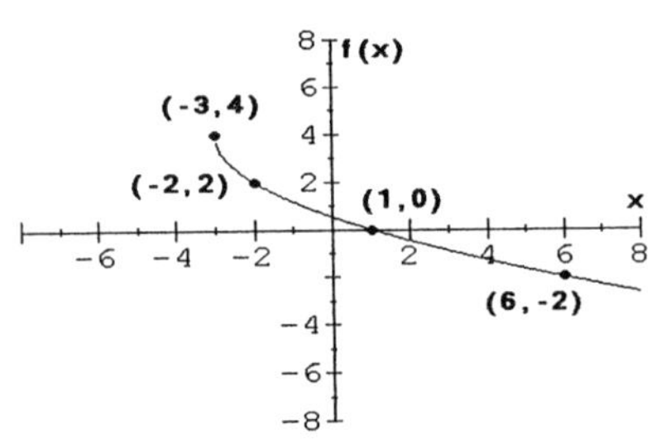

27. $f(x) = \dfrac{-2}{x+2} + 2$

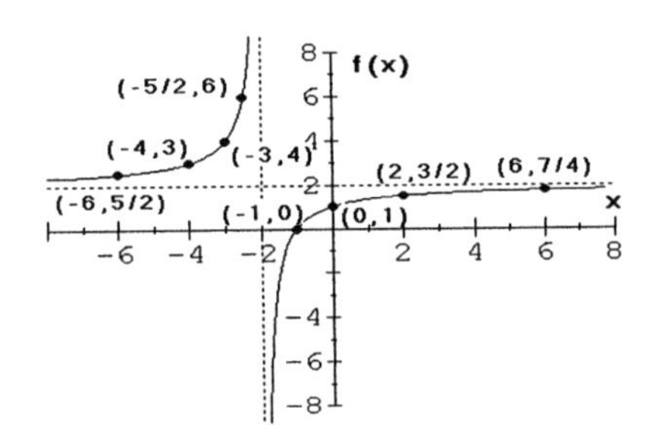

29. $f(x) = \dfrac{x-1}{x}$

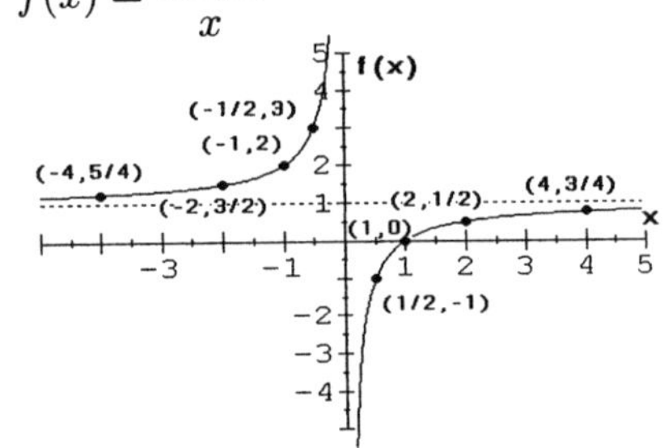

31. $f(x) = -3\left|x+4\right| + 3$

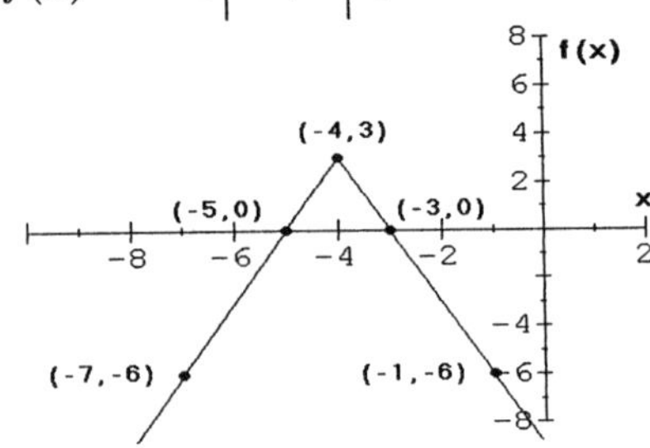

33. $f(x) = 4|x| + 2$

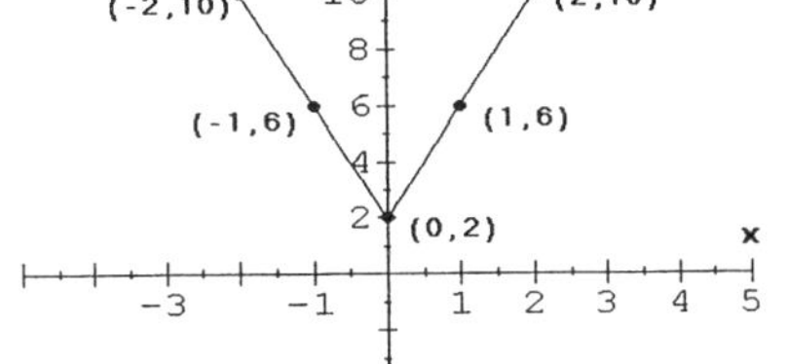

(b) $y = f(x-2)$

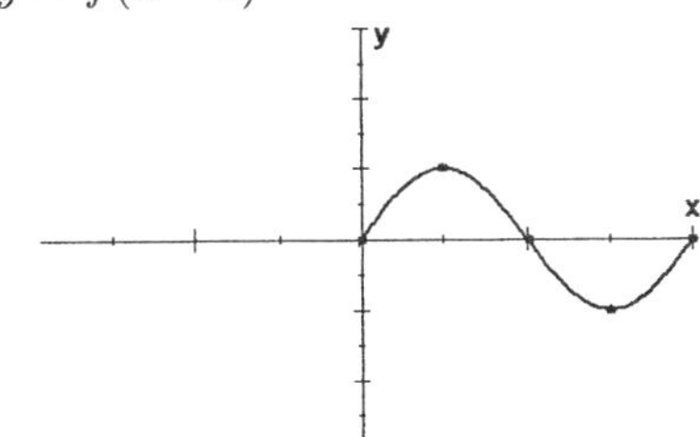

35. $y = f(x)$

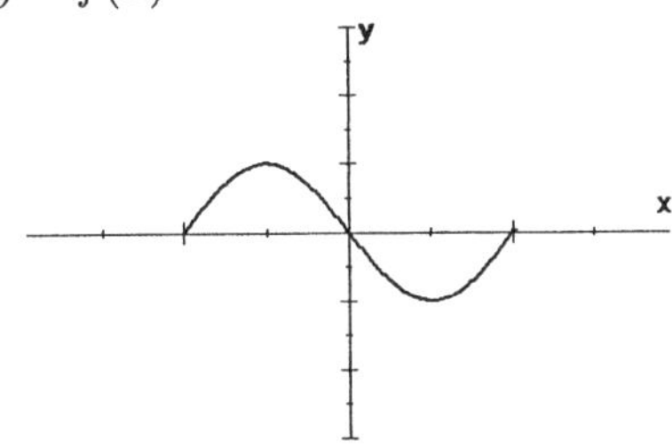

(c) $y = -f(x)$

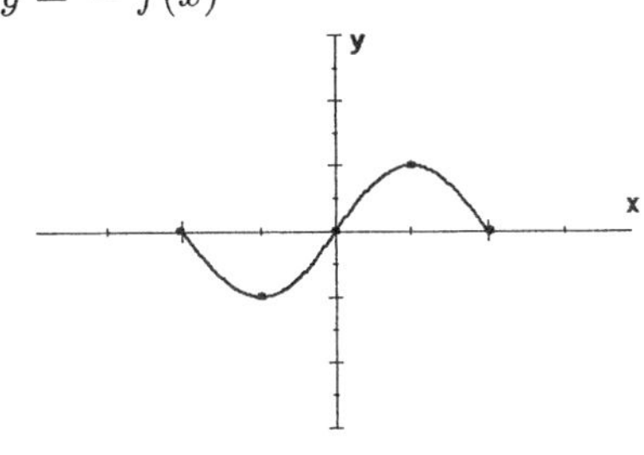

(a) $y = f(x) + 3$

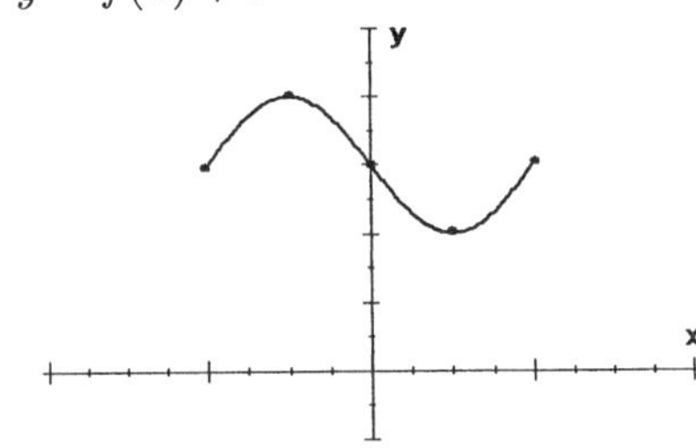

(d) $y = f(x+3) - 4$

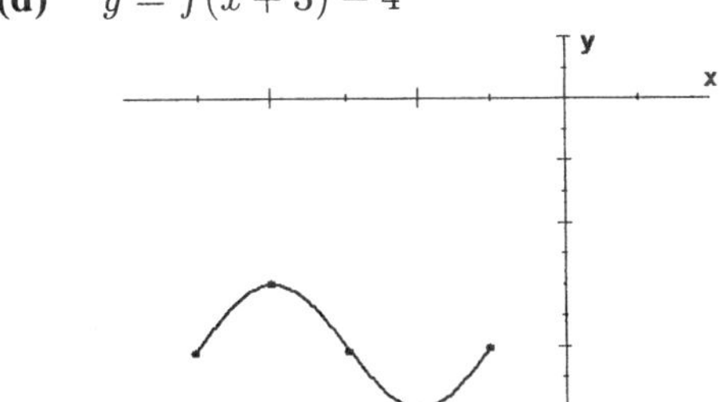

PROBLEM SET 13.4 Composition of Functions

1. $f(x) = 9x - 2$ and $g(x) = -4x + 6$

$(f \circ g)(x) = 9(-4x + 6) - 2$
$(f \circ g)(x) = -36x + 54 - 2$
$(f \circ g)(x) = -36x + 52$

$(f \circ g)(-2) = -36(-2) + 52$
$(f \circ g)(-2) = 72 + 52 = 124$

$(g \circ f)(x) = -4(9x - 2) + 6$
$(g \circ f)(x) = -36x + 8 + 6$
$(g \circ f)(x) = -36x + 14$

$(g \circ f)(4) = -36(4) + 14$
$(g \circ f)(4) = -130$

3. $f(x) = 4x^2 - 1$ and $g(x) = 4x + 5$

$(f \circ g)(x) = 4(4x + 5)^2 - 1$
$(f \circ g)(x) = 4(16x^2 + 40x + 25) - 1$
$(f \circ g)(x) = 64x^2 + 160x + 100 - 1$
$(f \circ g)(x) = 64x^2 + 160x + 99$

$(f \circ g)(1) = 64(1)^2 + 160(1) + 99$
$(f \circ g)(1) = 323$

$(g \circ f)(x) = 4(4x^2 - 1) + 5$
$(g \circ f)(x) = 16x^2 - 4 + 5$
$(g \circ f)(x) = 16x^2 + 1$

$(g \circ f)(4) = 16(4)^2 + 1$
$(g \circ f)(4) = 257$

Problem Set 13.4

5. $f(x) = \frac{1}{x}$ and $g(x) = \frac{2}{x-1}$

$(f \circ g)(x) = \frac{1}{\frac{2}{x-1}} = \frac{x-1}{2}$

$(f \circ g)(2) = \frac{2-1}{2} = \frac{1}{2}$

$(g \circ f)(x) = \frac{2}{\frac{1}{x} - 1}$

$(g \circ f)(x) = \frac{x}{x} \bullet \frac{2}{\left(\frac{1}{x} - 1\right)}$

$(g \circ f)(x) = \frac{2x}{1-x}$

$(g \circ f)(-1) = \frac{2(-1)}{1-(-1)} = \frac{-2}{2} = -1$

7. $f(x) = \frac{1}{x-2}$ and $g(x) = \frac{4}{x-1}$

$(f \circ g)(x) = \frac{1}{\frac{4}{x-1} - 2}$

$(f \circ g)(x) = \frac{(x-1)}{(x-1)} \bullet \frac{1}{\left(\frac{4}{x-1} - 2\right)}$

$(f \circ g)(x) = \frac{x-1}{4-2(x-1)}$

$(f \circ g)(x) = \frac{x-1}{4-2x+2} = \frac{x-1}{-2x+6}$

$(f \circ g)(3) = \frac{3-1}{-2(3)+6} = \frac{2}{0}$

$(f \circ g)(3)$ is undefined.
The domain of f is $\{x|x \neq 2\}$.
Since 2 is not in the domain
of f, then 2 cannot be in
the domain of $g \circ f$.
So $(g \circ f)(2)$ is undefined.

9. $f(x) = \sqrt{3x-2}$ and $g(x) = -x+4$

$(f \circ g)(x) = \sqrt{3(-x+4)-2}$
$(f \circ g)(x) = \sqrt{-3x+12-2}$
$(f \circ g)(x) = \sqrt{-3x+10}$
$(f \circ g)(1) = \sqrt{-3(1)+10} = \sqrt{7}$
$(g \circ f)(x) = -\sqrt{3x-2}+4$
$(g \circ f)(6) = -\sqrt{3(6)-2}+4$
$(g \circ f)(6) = -\sqrt{18-2}+4$
$(g \circ f)(6) = -\sqrt{16}+4 = 0$

11. $f(x) = |4x-5|$ and $g(x) = x^3$

$(f \circ g)(x) = |4(x)^3 - 5|$
$(f \circ g)(x) = |4x^3 - 5|$

$(f \circ g)(-2) = |4(-2)^3 - 5|$
$(f \circ g)(-2) = |-37| = 37$

$(g \circ f)(x) = (|4x-5|)^3$
$(g \circ f)(2) = (|4(2)-5|)^3$
$(g \circ f)(2) = (3)^3 = 27$

13. $f(x) = 3x$ and $g(x) = 5x-1$

$(f \circ g)(x) = 3(5x-1)$
$(f \circ g)(x) = 15x-3$
Domain = {all reals}.

$(g \circ f)(x) = 5(3x)-1$
$(g \circ f)(x) = 15x-1$
Domain = {all reals}.

15. $f(x) = -2x+1$ and $g(x) = 7x+4$

$(f \circ g)(x) = -2(7x+4)+1$
$(f \circ g)(x) = -14x-8+1$
$(f \circ g)(x) = -14x-7$
Domain = {all reals}.

$(g \circ f)(x) = 7(-2x+1)+4$
$(g \circ f)(x) = -14x+7+4$
$(g \circ f)(x) = -14x+11$
Domain = {all reals}.

17. $f(x) = 3x+2$ and $g(x) = x^2+3$

$(f \circ g)(x) = 3(x^2+3)+2$
$(f \circ g)(x) = 3x^2+9+2$
$(f \circ g)(x) = 3x^2+11$
Domain = {all reals}.

$(g \circ f)(x) = (3x+2)^2+3$
$(g \circ f)(x) = 9x^2+12x+4+3$
$(g \circ f)(x) = 9x^2+12x+7$
Domain = {all reals}.

19. $f(x) = 2x^2 - x + 2$ and $g(x) = -x + 3$

$(f \circ g)(x) = 2(-x + 3)^2 - (-x + 3) + 2$
$(f \circ g)(x) = 2(x^2 - 6x + 9) + x - 3 + 2$
$(f \circ g)(x) = 2x^2 - 12x + 18 + x - 1$
$(f \circ g)(x) = 2x^2 - 11x + 17$
Domain = {all reals}.

$(g \circ f)(x) = -(2x^2 - x + 2) + 3$
$(g \circ f)(x) = -2x^2 + x - 2 + 3$
$(g \circ f)(x) = -2x^2 + x + 1$
Domain = {all reals}.

21. $f(x) = \frac{3}{x}$ and $g(x) = 4x - 9$

$(f \circ g)(x) = \frac{3}{4x - 9}$
The domain of f is $\{x | x \neq 0\}$.
Therefore $g(x) \neq 0$.
$4x - 9 = 0$
$4x = 9$
$x = \frac{9}{4}$
The domain of g is {all reals}.
So the domain of $f \circ g$ is $\left\{x | x \neq \frac{9}{4}\right\}$.

$(g \circ f)(x) = 4\left(\frac{3}{x}\right) - 9$
$(g \circ f)(x) = \frac{12}{x} - 9$
$(g \circ f)(x) = \frac{12}{x} - \frac{9x}{x} = \frac{12 - 9x}{x}$
The domain of g is {all reals}.
The domain of f is $\{x | x \neq 0\}$.
So, the domain of $g \circ f$ is $\{x | x \neq 0\}$.

23. $f(x) = \sqrt{x + 1}$ and $g(x) = 5x + 3$

$(f \circ g)(x) = \sqrt{5x + 3 + 1}$
$(f \circ g)(x) = \sqrt{5x + 4}$

Domain of f
$x + 1 \geq 0$
$x \geq -1$
The domain of f is $\{x | x \geq -1\}$.
Therefore $g(x) \geq -1$
$5x + 3 \geq -1$
$5x \geq -4$
$x \geq -\frac{4}{5}$
The domain of g is {all reals}.
So, the domain of $f \circ g$ is
$\left\{x | x \geq -\frac{4}{5}\right\}$.

$(g \circ f)(x) = 5\sqrt{x + 1} + 3$
The domain of g is {all reals}.
The domain of f is $\{x | x \geq -1\}$.
So, the domain of $g \circ f$ is $\{x | x \geq -1\}$.

25. $f(x) = \frac{1}{x}$ and $g(x) = \frac{1}{x - 4}$

$(f \circ g)(x) = \frac{1}{\frac{1}{x - 4}}$
$(f \circ g) = x - 4$
The domain of f is $\{x | x \neq 0\}$.
So, $g(x) \neq 0$, however $g(x)$ will never be zero.
The domain of g is $\{x | x \neq 4\}$.
So, the domain of $f \circ g$ is $\{x | x \neq 4\}$.

$(g \circ f)(x) = \frac{1}{\frac{1}{x} - 4}$
$(g \circ f)(x) = \frac{x}{x} \bullet \frac{1}{\left(\frac{1}{x} - 4\right)}$
$(g \circ f)(x) = \frac{x}{1 - 4x}$
The domain of g is $\{x | x \neq 4\}$.
So, $f(x) \neq 4$.
$\frac{1}{x} = 4$
$x\left(\frac{1}{x}\right) = 4x$
$1 = 4x$
$\frac{1}{4} = x$
The domain of f is $\{x | x \neq 0\}$.
So, the domain of $g \circ f$ is
$\left\{x | x \neq \frac{1}{4}, x \neq 0\right\}$.

27. $f(x) = \sqrt{x}$ and $g(x) = \frac{4}{x}$

$(f \circ g)(x) = \sqrt{\frac{4}{x}}$

$(f \circ g)(x) = \frac{2}{\sqrt{x}} \bullet \frac{\sqrt{x}}{\sqrt{x}}$

$(f \circ g)(x) = \frac{2\sqrt{x}}{x}$

The domain of f is $\{x|x \geq 0\}$.

So, $g(x) \geq 0$.

$\frac{4}{x} \geq 0$

This is a rational inequality.

Test Point	-2	2
	0	
4:	positive	positive
x:	negative	positive
quotient:	negative	positive

$x \geq 0$

The domain of g is $\{x|x \neq 0\}$.

So, the domain of $f \circ g$ is $\{x|x > 0\}$.

$(g \circ f)(x) = \frac{4}{\sqrt{x}}$

$(g \circ f)(x) = \frac{4}{\sqrt{x}} \bullet \frac{\sqrt{x}}{\sqrt{x}}$

$(g \circ f)(x) = \frac{4\sqrt{x}}{x}$

The domain of g is $\{x|x \neq 0\}$.

$f(x) \neq 0$

$\sqrt{x} \neq 0$

$\sqrt{x} = 0$

$(\sqrt{x})^2 = 0^2$

$x = 0$

The domain of f is $\{x|x \geq 0\}$.

So, the domain of $g \circ f$ is $\{x|x > 0\}$.

29. $f(x) = \frac{3}{2x}$ and $g(x) = \frac{1}{x+1}$

$(f \circ g)(x) = \frac{3}{2\left(\frac{1}{x+1}\right)}$

$(f \circ g)(x) = \frac{3}{\frac{2}{x+1}}$

$(f \circ g)(x) = \frac{3(x+1)}{2}$

$(f \circ g)(x) = \frac{3x+3}{2}$

The domain of f is $\{x|x \neq 0\}$.

So, $g(x) \neq 0$. However $g(x)$ will never be zero.

The domain of g is $\{x|x \neq -1\}$.

So, the domain of $f \circ g$ is $\{x|x \neq -1\}$.

$(g \circ f)(x) = \frac{1}{\frac{3}{2x} + 1}$

$(g \circ f)(x) = \frac{2x}{2x} \bullet \frac{1}{\left(\frac{3}{2x} + 1\right)}$

$(g \circ f)(x) = \frac{2x}{3 + 2x}$

The domain of g is $\{x|x \neq -1\}$.

So, $f(x) \neq -1$.

$\frac{3}{2x} = -1$

$3 = -2x$

$-\frac{3}{2} = x$

The domain of f is $\{x|x \neq 0\}$.

So, the domain of $g \circ x$ is

$\left\{x|x \neq -\frac{3}{2} \text{ and } x \neq 0\right\}$.

31. $f(x) = 3x$ and $g(x) = \frac{1}{3}x$

$(f \circ g)(x) = 3\left(\frac{1}{3}x\right) = x$

$(g \circ f)(x) = \frac{1}{3}(3x) = x$

33. $f(x) = 4x + 2$ and $g(x) = \frac{x-2}{4}$

$(f \circ g)(x) = 4\left(\frac{x-2}{4}\right) + 2$

$(f \circ g)(x) = x - 2 + 2$

$(f \circ g)(x) = x$

$(g \circ f)(x) = \frac{4x + 2 - 2}{4}$

$(g \circ f)(x) = \frac{4x}{4}$

$(g \circ f)(x) = x$

35. $f(x) = \frac{1}{2}x + \frac{3}{4}$ and $g(x) = \frac{4x-3}{2}$

$(f \circ g)(x) = \frac{1}{2}\left(\frac{4x-3}{2}\right) + \frac{3}{4}$

$$(f \circ g)(x) = \frac{4x-3}{4} + \frac{3}{4}$$
$$(f \circ g)(x) = \frac{4x-3+3}{4}$$
$$(f \circ g)(x) = \frac{4x}{4}$$
$$(f \circ g)(x) = x$$
$$(g \circ f)(x) = \frac{4\left(\frac{1}{2}x + \frac{3}{4}\right) - 3}{2}$$
$$(g \circ f)(x) = \frac{2x+3-3}{2}$$
$$(g \circ f)(x) = \frac{2x}{2}$$
$$(g \circ f)(x) = x$$

37. $f(x) = -\frac{1}{4}x - \frac{1}{2}$ and $g(x) = -4x - 2$
$$(f \circ g)(x) = -\frac{1}{4}(-4x-2) - \frac{1}{2}$$
$$(f \circ g)(x) = x + \frac{1}{2} - \frac{1}{2}$$
$$(f \circ g)(x) = x$$
$$(g \circ f)(x) = -4\left(-\frac{1}{4}x - \frac{1}{2}\right) - 2$$
$$(g \circ f)(x) = x + 2 - 2$$
$$(g \circ f)(x) = x$$

PROBLEM SET 13.5 Direct and Inverse Variations

1. $y = \frac{k}{x^2}$

3. $C = \frac{kg}{t^3}$

5. $V = kr^3$

7. $S = ke^2$

9. $V = khr^2$

11. $y = kx$
$$8 = k(12)$$
$$\frac{8}{12} = k$$
$$\frac{2}{3} = k$$

13. $y = kx^2$
$$-144 = k(6)^2$$
$$-144 = 36k$$
$$-4 = k$$

15. $V = kBh$
$$96 = k(24)(12)$$
$$96 = 288k$$
$$\frac{96}{288} = k$$
$$\frac{1}{3} = k$$

17. $y = \frac{k}{x}$
$$-4 = \frac{k}{\frac{1}{2}}$$
$$-4 = 2k$$
$$-2 = k$$

19. $r = \frac{k}{t^2}$
$$\frac{1}{8} = \frac{k}{4^2}$$
$$\frac{1}{8} = \frac{k}{16}$$
$$16\left(\frac{1}{8}\right) = 16\left(\frac{k}{16}\right)$$
$$2 = k$$

21. $y = \frac{kx}{z}$
$$45 = \frac{k(18)}{2}$$
$$45 = 9k$$
$$5 = k$$

Problem Set 13.5

23. $y = \frac{kx}{z^2}$

$81 = \frac{k(36)}{(2)^2}$

$81 = \frac{k(36)}{4}$

$81 = 9k$

$9 = k$

25. $y = kx$

$36 = k(48)$

$\frac{36}{48} = k$

$\frac{3}{4} = k$

$y = \frac{3}{4}x$

$y = \frac{3}{4}(12)$

$y = 9$

27. $y = \frac{k}{x}$

$\frac{1}{9} = \frac{k}{12}$

$12\left(\frac{1}{9}\right) = 12\left(\frac{k}{12}\right)$

$\frac{4}{3} = k$

$y = \frac{\frac{4}{3}}{x}$

$y = \frac{4}{3x}$

$y = \frac{4}{3(8)}$

$y = \frac{1}{6}$

29. $A = kbh$

$60 = k(12)(10)$

$60 = 120k$

$\frac{1}{2} = k$

$A = \frac{1}{2}bh$

$A = \frac{1}{2}(16)(14)$

$A = 112$

31. $V = \frac{k}{P}$

$15 = \frac{k}{20}$

$20(15) = 20\left(\frac{k}{20}\right)$

$300 = k$

$V = \frac{300}{P}$

$V = \frac{300}{25}$

$V = 12$ cubic centimeters

33. $V = \frac{kT}{P}$

$48 = \frac{k(320)}{20}$

$48 = 16k$

$3 = k$

$V = \frac{3T}{P}$

$V = \frac{3(280)}{30}$

$V = 28$

35. $P = k\sqrt{l}$

$4 = k\sqrt{12}$

$4 = k(2\sqrt{3})$

$\frac{4}{2\sqrt{3}} = k$

$k = \frac{2}{\sqrt{3}}$

$k = \frac{2}{\sqrt{3}} \bullet \frac{\sqrt{3}}{\sqrt{3}} = \frac{2\sqrt{3}}{3}$

$P = \frac{2\sqrt{3}}{3}\sqrt{l} = \frac{2\sqrt{3l}}{3}$

$P = \frac{2\sqrt{3(3)}}{3} = \frac{2\sqrt{9}}{3} = \frac{2(3)}{3} = 2$ sec.

37. $R = \frac{kl}{d^2}$

$1.5 = \frac{k(200)}{(0.5)^2}$

$1.5 = \frac{k(200)}{0.25}$

$1.5 = 800k$

$\frac{1.5}{800} = k$

$0.001875 = k$

$R = \frac{0.001875l}{d^2}$

$R = \frac{0.001875(400)}{(0.25)^2}$

$R = \frac{0.001875(400)}{0.0625}$

$R = 12$ ohms

39. $I = krt$

a. $385 = k(0.11)(2)$
$385 = 0.22k$
$1750 = k$
$I = 1750rt$
$I = 1750(0.12)(1)$
$I = \$210$

b. $819 = k(0.12)(3)$
$819 = 0.36k$
$2275 = k$
$I = 2275rt$
$I = 2275(0.14)(2)$
$I = \$637$

c. $1960 = k(0.14)(4)$
$1960 = 0.56k$
$3500 = k$
$I = 3500rt$
$I = 3500(0.15)(2)$
$I = \$1050$

41. $V = khr^2$
$549.5 = k(7)(5)^2$
$549.5 = 175k$
$3.14 = k$
$V = 3.14hr^2$
$V = 3.14(14)(9)^2$
$V = 3560.76$ cubic meters

43. $y = \frac{k}{\sqrt{x}}$

$0.08 = \frac{k}{\sqrt{225}}$

$0.08 = \frac{k}{15}$

$15(0.08) = k$

$1.2 = k$

$y = \frac{1.2}{\sqrt{x}}$

$y = \frac{1.2}{\sqrt{625}}$

$y = \frac{1.2}{25} = 0.048$

CHAPTER 13 | Review Problem Set

1. $f = \{(1, 3), (2, 5), (4, 9)\}$
Domain $= \{1, 2, 4\}$.

2. $f(x) = \frac{4}{x - 5}$

$x - 5 = 0$

$x = 5$

Domain $= \{x | x \neq 5\}$.

3. $f(x) = \frac{3}{x^2 + 4x}$

$f(x) = \frac{3}{x(x + 4)}$

$x(x + 4) = 0$

$x = 0$ or $x + 4 = 0$

$x = 0$ or $x = -4$

Domain $= \{x | x \neq 0 \text{ or } x \neq -4\}$.

4. $f(x) = \sqrt{x^2 - 25}$
$x^2 - 25 \geq 0$
$(x + 5)(x - 5) \geq 0$
$x + 5 = 0$ or $x - 5 = 0$
$x = -5$ or $x = 5$

	-5		5	
Test Point	-6	0	6
$x + 5$:	negative	positive	positive
$x - 5$:	negative	negative	positive
product:	positive	negative	positive

Domain $= \{x \mid x \leq -5 \text{ or } x \geq 5\}$.

5. $f(x) = x^2 - 2x - 1$
$f(2) = (2)^2 - 2(2) - 1$
$f(2) = 4 - 4 - 1$
$f(2) = -1$
$f(-3) = (-3)^2 - 2(-3) - 1$
$f(-3) = 9 + 6 - 1$
$f(-3) = 14$
$f(a) = a^2 - 2a - 1$

6. $f(x) = 2x^2 + x - 7$
$f(a) = 2a^2 + a - 7$
$f(a + h) = 2(a + h)^2 + (a + h) - 7$
$f(a + h) = 2(a^2 + 2ah + h^2) + a + h - 7$
$f(a + h) = 2a^2 + 4ah + 2h^2 + a + h - 7$

$$\frac{f(a+h) - f(a)}{h} = \frac{2a^2 + 4ah + 2h^2 + a + h - 7 - (2a^2 + a - 7)}{h}$$

$$\frac{f(a+h) - f(a)}{h} = \frac{2a^2 + 4ah + 2h^2 + a + h - 7 - 2a^2 - a + 7}{h}$$

$$\frac{f(a+h) - f(a)}{h} = \frac{4ah + h + 2h^2}{h}$$

$$\frac{f(a+h) - f(a)}{h} = \frac{h(4a + 1 + 2h)}{h}$$

$$\frac{f(a+h) - f(a)}{h} = 4a + 1 + 2h$$

7. $f(x) = 4$

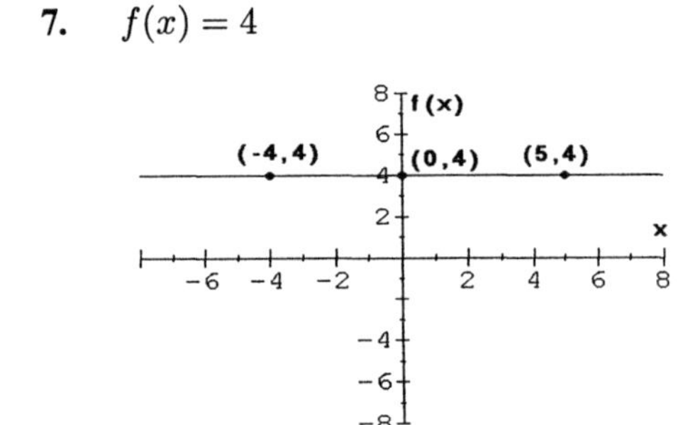

8. $f(x) = -3x + 2$

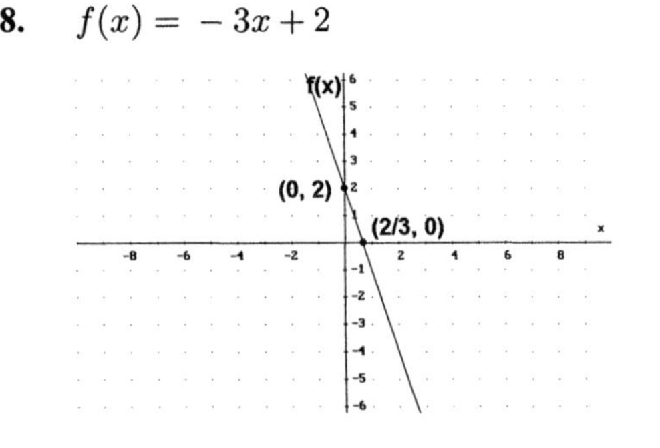

9. $f(x) = x^2 + 2x + 2$

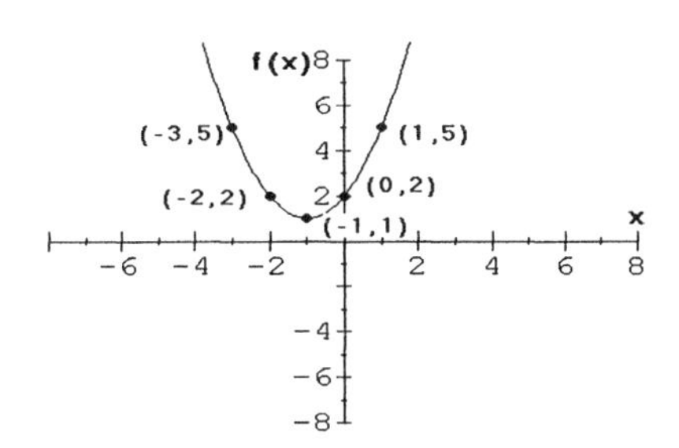

13. $f(x) = \dfrac{1}{x^2}$

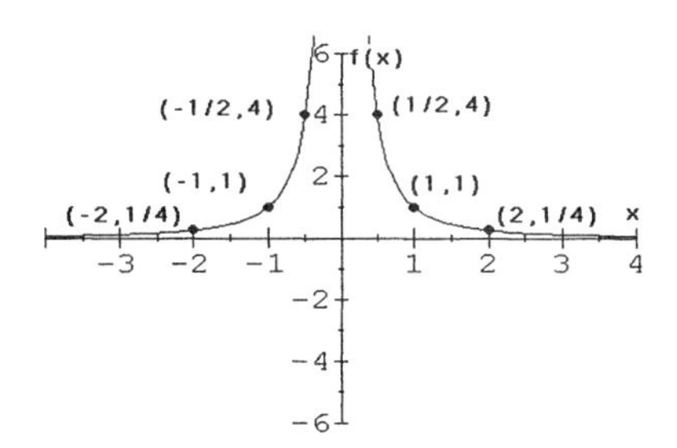

10. $f(x) = \left|x\right| + 4$

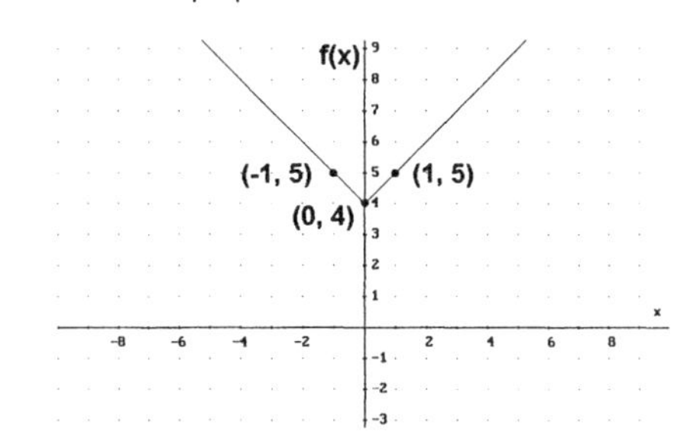

14. $f(x) = -\dfrac{1}{2}x^2$

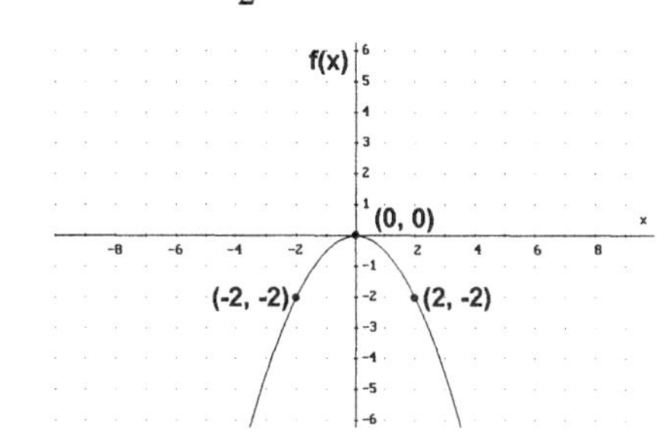

11. $f(x) = -\left|x - 2\right|$

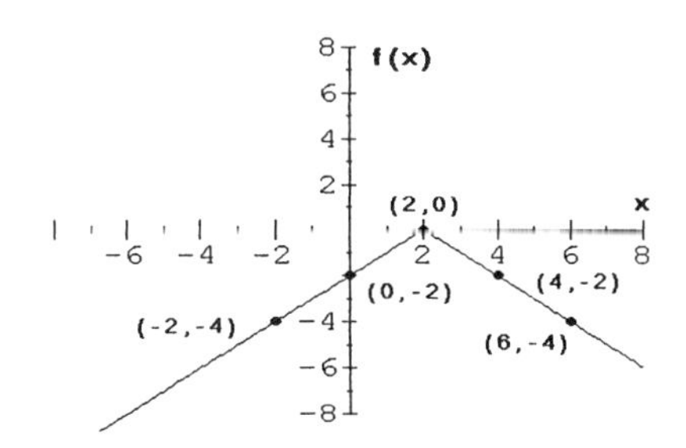

15. $f(x) = -3x^2 + 6x - 2$

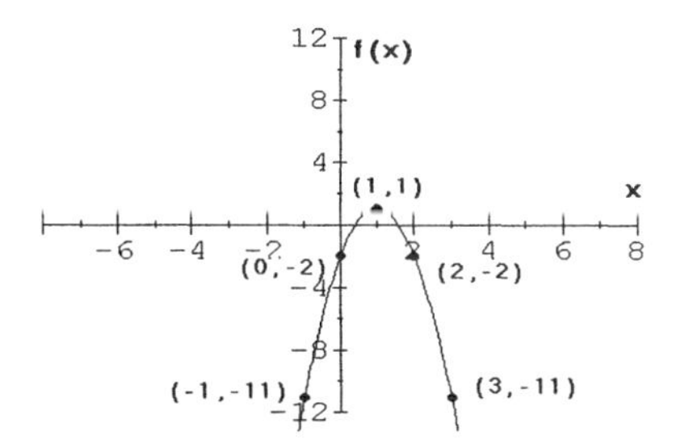

12. $f(x) = \sqrt{x - 2} - 3$

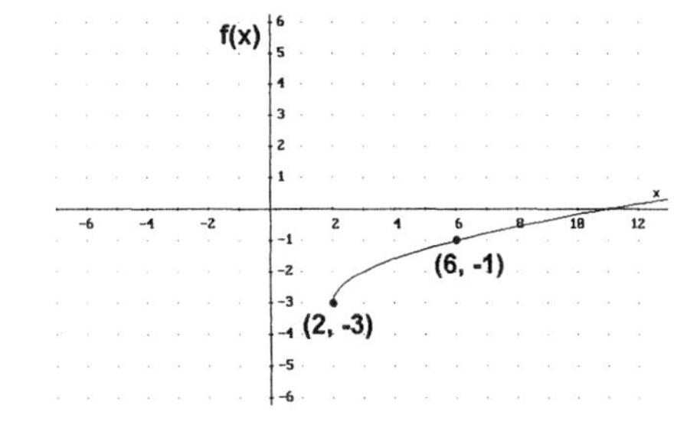

16. $f(x) = -\sqrt{x + 1} - 2$

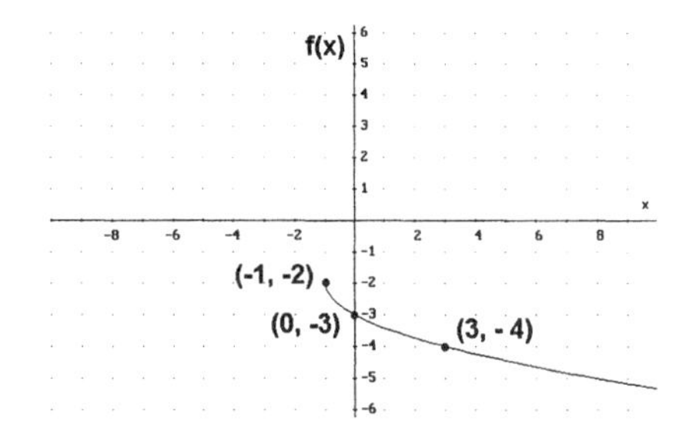

17. **a.** $f(x) = x^2 + 10x - 3$
$f(x) = x^2 + 10x + 25 - 25 - 3$
$f(x) = (x+5)^2 - 28$
vertex: $(-5, -28)$
axis of symmetry: $x = -5$

b. $f(x) = -2x^2 - 14x + 9$
$f(x) = -2(x^2 + 7x) + 9$
$f(x) = -2\left(x^2 + 7x + \frac{49}{4}\right) + \frac{49}{2} + 9$
$f(x) = -2\left(x + \frac{7}{2}\right)^2 + \frac{67}{2}$
vertex: $\left(-\frac{7}{2}, \frac{67}{2}\right)$
axis of symmetry: $x = -\frac{7}{2}$

18. $f(x) = 2x - 3$ and $g(x) = 3x - 4$
$(f \circ g)(x) = 2(3x - 4) - 3$
$(f \circ g)(x) = 6x - 8 - 3$
$(f \circ g)(x) = 6x - 11$
$(g \circ f)(x) = 3(2x - 3) - 4$
$(g \circ f)(x) = 6x - 9 - 4$
$(g \circ f)(x) = 6x - 13$

19. $f(x) = x - 4$ and $g(x) = x^2 - 2x + 3$
$(f \circ g)(x) = x^2 - 2x + 3 - 4$
$(f \circ g)(x) = x^2 - 2x - 1$
$(g \circ f)(x) = (x - 4)^2 - 2(x - 4) + 3$
$(g \circ f)(x) = x^2 - 8x + 16 - 2x + 8 + 3$
$(g \circ f)(x) = x^2 - 10x + 27$

20. $f(x) = x^2 - 5$ and $g(x) = -2x + 5$
$(f \circ g)(x) = (-2x + 5)^2 - 5$
$(f \circ g)(x) = 4x^2 - 20x + 25 - 5$
$(f \circ g)(x) = 4x^2 - 20x + 20$
$(g \circ f)(x) = -2(x^2 - 5) + 5$
$(g \circ f)(x) = -2x^2 + 10 + 5$
$(g \circ f)(x) = -2x^2 + 15$

21. $y = \frac{kx}{z}$
$21 = \frac{k(14)}{6}$
$21 = \frac{7}{3}k$
$\frac{3}{7}(21) = \frac{3}{7}\left(\frac{7}{3}k\right)$
$9 = k$

22. $y = kx\sqrt{z}$
$60 = k(2)\sqrt{9}$
$60 = 6k$
$10 = k$
$y = 10x\sqrt{z}$
$y = 10(3)\sqrt{16}$
$y = 10(3)(4)$
$y = 120$

23. $w = \frac{k}{d^2}$
$200 = \frac{k}{(4000)^2}$
$k = 200(4000)^2$
$k = (3.2)(10)^9$
$w = \frac{(3.2)(10)^9}{d^2}$
$w = \frac{(3.2)(10)^9}{(5000)^2}$
$w = \frac{(3.2)(10)^9}{(2.5)(10)^7}$
$w = (1.28)(10)^2$
$w = 128$ pounds

24. Let x = one number and
$40 - x$ = other number.
$f(x) = x(40 - x)$
$f(x) = 40x - x^2$
$f(x) = -x^2 + 40x$
$f(x) = -(x^2 - 40x)$
$f(x) = -(x^2 - 40x + 400) + 400$
$f(x) = -(x - 20)^2 + 400$
The numbers are 20 and 20.

25. Let x = one number and
$50 - x$ = other number.
$f(x) = x^2 + 6(50 - x)$
$f(x) = x^2 + 300 - 6x$
$f(x) = x^2 - 6x + 300$
$f(x) = x^2 - 6x + 9 - 9 + 300$
$f(x) = (x - 3)^2 + 291$
The numbers are 3 and 47.

26. Let x = number of additional students.
$R(x) = (50 + x)(5 - 0.05x)$
$R(x) = 250 - 2.5x + 5x - 0.05x^2$
$R(x) = -0.05x^2 + 2.5x + 250$
$R(x) = -0.05(x^2 - 50x) + 250$
$R(x) = -0.05(x^2 - 50x+625)+31.25+250$
$R(x) = -0.05(x - 25)^2 + 281.25$
There need to be 25 additional students.

27. $A = ke^2$
$384 = k(8)^2$
$384 = 64k$
$6 = k$
$A = 6e^2$
$A = 6(10)^2$
$A = 600$ square inches

28. $h = \left(4\ \frac{\text{hours}}{\text{night}}\right)(30\text{ nights}) = 120\text{ hours}$
$c(h) = 0.006h$
$c(120) = 0.006(120) = 0.72$
It will cost \$0.72.

29. $c(p) = p - 0.30p = 0.70p$
$c(p) = 0.7p$
$c(65) = 0.7(65) = \$45.50$
$c(48) = 0.7(48) = \$33.60$
$c(15.50) = 0.7(15.50) = \$10.85$

CHAPTER 13 | Test

1. $f(x) = \dfrac{-3}{2x^2 + 7x - 4}$

$f(x) = \dfrac{-3}{(2x - 1)(x + 4)}$

$2x - 1 = 0$ or $x + 4 = 0$
$2x = 1$ or $x = -4$
$x = \frac{1}{2}$ or $x = -4$

Domain $= \left\{x | x \neq \frac{1}{2} \text{ or } x \neq -4\right\}$

2. $f(x) = \sqrt{5 - 3x}$
$5 - 3x \geq 0$
$-3x \geq -5$
$-\frac{1}{3}(-3x) \leq -\frac{1}{3}(-5)$
$x \leq \frac{5}{3}$
Domain $= \left\{x | x \leq \frac{5}{3}\right\}$

3. $f(x) = -\frac{1}{2}x + \frac{1}{3}$
$f(-3) = -\frac{1}{2}(-3) + \frac{1}{3}$
$f(-3) = \frac{3}{2} + \frac{1}{3}$
$f(-3) = \frac{11}{6}$

4. $f(x) = -x^2 - 6x + 3$
$f(-2) = -(-2)^2 - 6(-2) + 3$
$f(-2) = -4 + 12 + 3$
$f(-2) = 11$

5. $f(x) = -2x^2 - 24x - 69$
$f(x) = -2(x^2 + 12x) - 69$
$f(x) = -2(x^2 + 12x + 36) + 72 - 69$
$f(x) = -2(x + 6)^2 + 3$
vertex $(-6, 3)$

6. $f(x) = 3x^2 + 2x - 5$
$f(a) = 3a^2 + 2a - 5$
$f(a + h) = 3(a + h)^2 + 2(a + h) - 5$
$f(a + h) = 3(a^2 + 2ah + h^2) + 2a + 2h - 5$
$f(a + h) = 3a^2 + 6ah + 3h^2 + 2a + 2h - 5$

$$\frac{f(a+h)-f(a)}{h}=\frac{3a^2+6ah+3h^2+2a+2h-5-(3a^2+2a-5)}{h}$$

$$\frac{f(a+h)-f(a)}{h}=\frac{3a^2+6ah+3h^2+2a+2h-5-3a^2-2a+5}{h}$$

$$\frac{f(a+h)-f(a)}{h}=\frac{6ah+2h+3h^2}{h}$$

$$\frac{f(a+h)-f(a)}{h}=\frac{h(6a+2+3h)}{h}$$

$$\frac{f(a+h)-f(a)}{h}=6a+2+3h$$

7. $f(x)=-3x+4$ and $g(x)=7x+2$
$(f\circ g)(x)=-3(7x+2)+4$
$(f\circ g)(x)=-21x-6+4$
$(f\circ g)(x)=-21x-2$

8. $f(x)=2x+5$ and $g(x)=2x^2-x+3$
$(g\circ f)(x)=2(2x+5)^2-(2x+5)+3$
$(g\circ f)(x)=2(4x^2+20x+25)-2x-5+3$
$(g\circ f)(x)=8x^2+40x+50-2x-5+3$
$(g\circ f)(x)=8x^2+38x+48$

9. $f(x)=\dfrac{3}{x-2}$ and $g(x)=\dfrac{2}{x}$

$$(f\circ g)(x)=\frac{3}{\frac{2}{x}-2}$$

$$(f\circ g)(x)=\frac{x}{x}\bullet\frac{3}{\left(\frac{2}{x}-2\right)}$$

$$(f\circ g)(x)=\frac{3x}{2-2x}$$

10. $f(x)=\sqrt{x^2+3x-10}$
$x^2+3x-10\geq 0$
$(x+5)(x-2)\geq 0$
$x+5=0$ or $x-2=0$
$x=-5$ or $x=2$

	-5		2
Test Point	-6	0	3
$x+5$:	negative	positive	positive
$x-2$:	negative	negative	positive
product:	positive	negative	positive

The solution set is $(-\infty,\ -5]\cup[2,\infty)$.
Domain $=\{x \mid x\leq -5 \text{ or } x\geq 2\}$.

11. Let c represent the cost of an item, then $f(c)$ represents the selling price.
Profit is a percent of cost.
Selling price = Cost + Profit
$f(c)=c+(40\%)(c)$
$f(c)=c+0.40(c)$
$f(c)=1.40c$
$f(15)=1.40(15)=\$21$
$f(18)=1.40(18)=\$25.20$
$f(25)=1.40(25)=\$35$

12. Let x = number of additional units sold.
$R(x)=500(50)+x(50-0.10x)$
$R(x)=25000+50x-0.1x^2$
$R(x)=-0.1x^2+50x+25000$
$R(x)=-0.1(x^2-500x)+25000$
$R(x)=-0.1(x^2-500x+62500)$
$+25000-62500$
$R(x)=-0.1(x-250)^2-37500$
There need to be 250 additional units sold, so a total of 750 units will maximize profit.

13. $y = \dfrac{k}{x}$

$\dfrac{1}{2} = \dfrac{k}{-8}$

$-8\left(\dfrac{1}{2}\right) = k$

$-4 = k$

14. $y = kxz$

$18 = k(8)(9)$

$18 = 72k$

$\dfrac{18}{72} = k$

$\dfrac{1}{4} = k$

$y = \dfrac{1}{4}xz$

$y = \dfrac{1}{4}(5)(12)$

$y = 15$

15. Let $x =$ one number and
$60 - x =$ other number.

$f(x) = x^2 + 12(60 - x)$

$f(x) = x^2 + 720 - 12x$

$f(x) = x^2 - 12x + 720$

$f(x) = x^2 - 12x + 36 - 36 + 720$

$f(x) = (x - 6)^2 + 684$

The numbers are 6 and 54.

16. $I = krt$

$140 = k(0.07)(5)$

$140 = 0.35k$

$400 = k$

$I = 400rt$

$I = 400(0.08)(3)$

$I = \$96$

17. The graph of $f(x) = (x - 6)^3 - 4$ is the graph of $f(x) = x^3$ translated 6 units to the right and 4 units downward.

18. The graph of $f(x) = -|x| + 8$ is the graph of $f(x) = |x|$ reflected across the x-$axis$ and translated 8 units upward.

19. The graph of $f(x) = -\sqrt{x + 5} + 7$ is the graph of $f(x) = \sqrt{x}$ reflected across the x-$axis$ and then translated 5 units to the left and 7 units upward.

20. $f(x) = -x - 1$

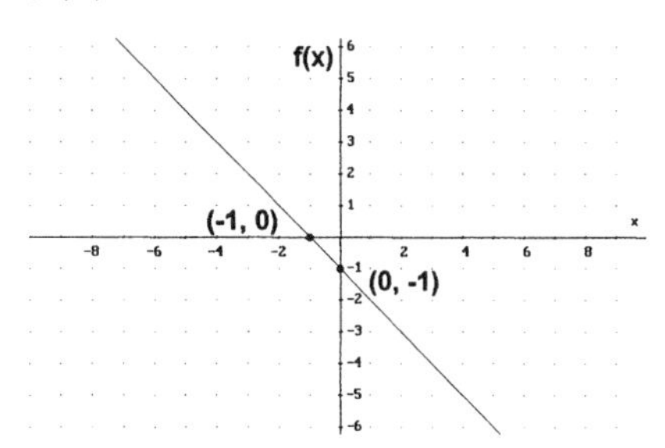

21. $f(x) = -2x^2 - 12x - 14$

22. $f(x) = 2\sqrt{x} - 2$

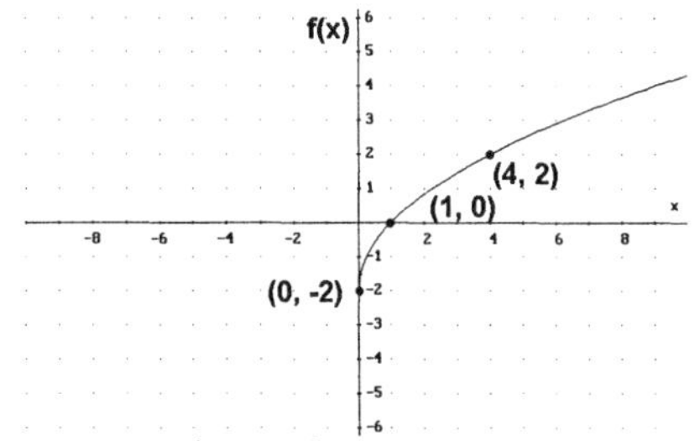

23. $f(x) = 3\left|x - 2\right| - 1$

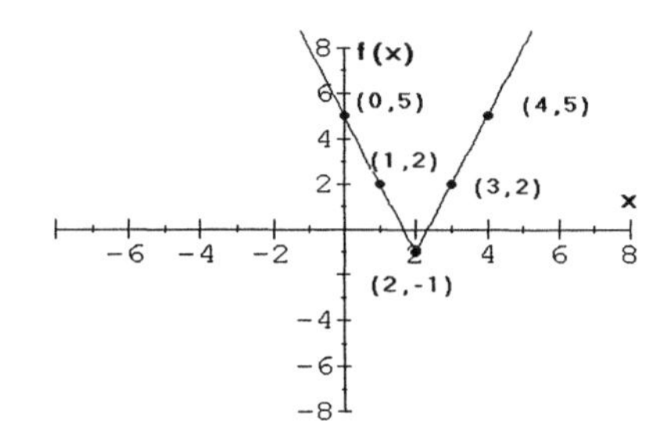

24. $f(x) = -\dfrac{1}{x} + 3$

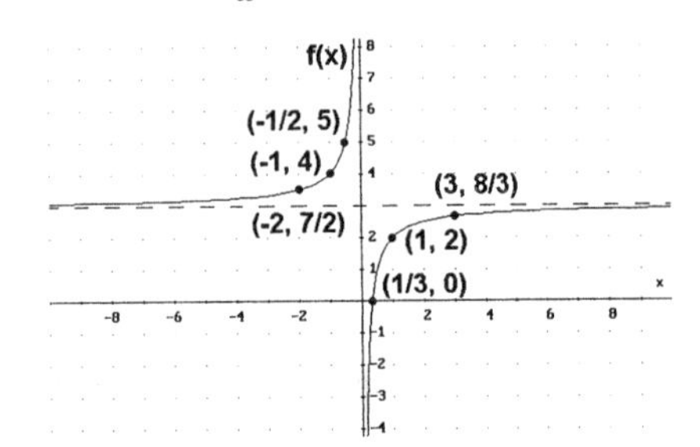

25. $f(x) = \sqrt{-x+2}$

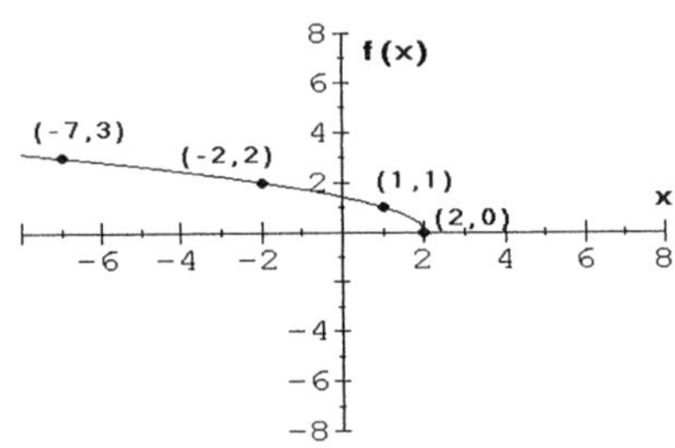

CHAPTERS 1-13 | Cumulative Practice Test

1. $48 = 2 \cdot 2 \cdot 2 \cdot 2 \cdot 3$
$60 = 2 \cdot 2 \cdot 3 \cdot 5$
$84 = 2 \cdot 2 \cdot 3 \cdot 7$

The greatest common factor is $2 \cdot 2 \cdot 3 = 12$.

2a. $2x^2 + y^2 - 3x - 4 = 0$
To test symmetry for:
x-axis replace y with $-y$
$2x^2 + (-y)^2 - 3x - 4 = 0$
$2x^2 + y^2 - 3x - 4 = 0$
Yes, it is symmetric to the x-axis.

y-axis replace x with $-x$
$2(-x)^2 + y^2 - 3(-x) - 4 = 0$
$2x^2 + y^2 + 3x - 4 = 0$
No, it is not symmetric to the y-axis.

origin replace x with $-x$
and y with $-y$.
$2(-x)^2 + (-y)^2 - 3(-x) - 4 = 0$
$2x^2 + y^2 + 3x - 4 = 0$
No, it is not symmetric to the origin.

b. $f(x) = 3x^2 + 4$
To test symmetry for:
x-axis replace $f(x)$ with $-f(x)$
$-f(x) = 3x^2 + 4$
$f(x) = -3x^2 - 4$
No, it is not symmetric to the x-axis.

$f(x)$-axis replace x with $-x$
$f(x) = 3(-x)^2 + 4$
$f(x) = 3x^2 + 4$
Yes, it is symmetric to the $f(x)$-axis.

origin replace x with $-x$
and $f(x)$ with $-f(x)$.
$-f(x) = 3(-x)^2 + 4$
$-f(x) = 3x^2 + 4$
$f(x) = -3x^2 - 4$
No, it is not symmetric to the origin.

c. $xy = 18$
To test symmetry for:
x-axis replace y with $-y$
$x(-y) = 18$
$-xy = 18$
$xy = -18$
No, it is not symmetric to the x-axis.

y-axis replace x with $-x$
$(-x)y = 18$
$-xy = 18$
$xy = -18$
No, it is not symmetric to the y-axis.

origin replace x with $-x$
and y with $-y$.
$(-x)(-y) = 18$
$xy = 18$
Yes, it is symmetric to the origin.

d. $f(x) = x$
To test symmetry for:
x-axis replace y with $-f(x)$
$-f(x) = x$
$f(x) = -x$
No, it is not symmetric to the x-axis.

$f(x)$-axis replace x with $-x$
$f(x) = -x$
No, it is not symmetric to the$f(x)$-axis.
origin replace x with $-x$

and $f(x)$ with $-f(x)$.
$-f(x) = -x$
$f(x) = x$
Yes, it is symmetric to the origin.

3. $f(x) = 2x^2 + 17x - 9$
$0 = 2x^2 + 17x - 9$
$0 = (2x - 1)(x + 9)$
$2x - 1 = 0$ or $x + 9 = 0$
$2x = 1$ or $x = -9$
$x = \frac{1}{2}$ or $x = -9$
The x-intercepts are -9 and $\frac{1}{2}$.

4. $f(x) = 2x^2 + 4x$
$f(x) = 2(x^2 + 2x)$
$f(x) = 2(x^2 + 2x + 1) - 2$
$f(x) = 2(x + 1)^2 - 2$
The vertex is $(-1, -2)$.

5. $\frac{48x^{-3}y^{-2}}{12x^{-4}y^{-3}} = 4x^{-3-(-4)}y^{-2-(-3)} =$
$4x^{-3+4}y^{-2+3} = 4x^1y^1 =$
$4xy$ for $x = -2$ and $y = -4$
$4(-2)(-4) = 32$

6. a. $\sqrt{32x^3y^4} = \sqrt{16x^2y^4}\sqrt{2x} = 4xy^2\sqrt{2x}$

b. $\sqrt[3]{32x^3y^4} = \sqrt[3]{8x^3y^3}\sqrt[3]{4y} = 2xy\sqrt[3]{4y}$

c. $\frac{2\sqrt{6}}{3\sqrt{12}} = \frac{2}{3}\sqrt{\frac{1}{2}} = \frac{2\sqrt{1}}{3\sqrt{2}} = \frac{2}{3\sqrt{2}} =$
$\frac{2}{3\sqrt{2}} \cdot \frac{\sqrt{2}}{\sqrt{2}} = \frac{2\sqrt{2}}{3\sqrt{4}} = \frac{2\sqrt{2}}{3(2)} = \frac{\sqrt{2}}{3}$

d. $\frac{\sqrt{2}}{3\sqrt{2}+\sqrt{3}} = \frac{\sqrt{2}}{3\sqrt{2}+\sqrt{3}} \cdot \frac{3\sqrt{2}-\sqrt{3}}{3\sqrt{2}-\sqrt{3}} =$

$\frac{\sqrt{2}(3\sqrt{2}-\sqrt{3})}{(3\sqrt{2}+\sqrt{3})(3\sqrt{2}-\sqrt{3})} = \frac{3\sqrt{4}-\sqrt{6}}{9\sqrt{4}-\sqrt{9}} =$

$\frac{3(2)-\sqrt{6}}{9(2)-3} = \frac{6-\sqrt{6}}{18-3} = \frac{6-\sqrt{6}}{15}$

7. $\frac{3x^2 + 5x + 2}{3x^2 - x - 2} = \frac{(3x + 2)(x + 1)}{(3x + 2)(x - 1)} =$

$\frac{x+1}{x-1}$ for $x = 21$

$\frac{21+1}{21-1} = \frac{22}{20} = \frac{11}{10}$

8. $(3x + 4)(2x^2 - x - 5)$
$6x^3 - 3x^2 - 15x + 8x^2 - 4x - 20$
$6x^3 - 3x^2 + 8x^2 - 15x - 4x - 20$
$6x^3 + 5x^2 - 19x - 20$

9. $\frac{3}{4x} - \frac{2}{5x} + \frac{7}{10x} =$

$\frac{3}{4x} \cdot \frac{5}{5} - \frac{2}{5x} \cdot \frac{4}{4} + \frac{7}{10x} \cdot \frac{2}{2} =$

$\frac{15}{20x} - \frac{8}{20x} + \frac{14}{20x} = \frac{21}{20x}$

10.

$$\begin{array}{r|l} & 2x^3 - 3x^2 + 4x - 5 \\ \hline x - 5 & 2x^4 - 13x^3 + 19x^2 - 25x + 25 \\ & 2x^4 - 10x^3 \\ \hline & -3x^3 + 19x^2 \\ & -3x^3 + 15x^2 \\ \hline & 4x^2 - 25x \\ & 4x^2 - 20x \\ \hline & -5x + 25 \\ & -5x + 25 \\ \hline & 0 \end{array}$$

Q: $2x^3 - 3x^2 + 4x - 5$ R: 0

11. $\frac{x^3 - 8}{x - 2} = \frac{(x - 2)(x^2 + 2x + 4)}{x - 2}$
$= x^2 + 2x + 4$

12. $(1.414)(10)^{-3} = 0.001414$

13. 41.4 is what percent of 36.
$41.4 = n(36)$
$\frac{41.4}{36} = n$
$1.15 = n$
$n = 115\%$

14. $(6+5i)(-4-2i) =$
$-24-12i-20i-10i^2 =$
$-24-32i-10(-1) =$
$-24-32i+10 =$
$-14-32i$

15 a. $16^{\frac{3}{4}} = (16^{\frac{1}{4}})^3 = (2)^3 = 8$

b. $\left(\frac{3}{2}-\frac{1}{4}\right)^{-1} = \left(\frac{6}{4}-\frac{1}{4}\right)^{-1} =$
$\left(\frac{5}{4}\right)^{-1} = \left(\frac{4}{5}\right)^{1} = \frac{4^1}{5^1} = \frac{4}{5}$

c. $-4^{-\frac{3}{2}} = -\frac{1}{4^{\frac{3}{2}}} = -\frac{1}{(4^{\frac{1}{2}})^3}$
$= -\frac{1}{(4^{\frac{1}{2}})^3} = -\frac{1}{(2)^3} = -\frac{1}{8}$

d. $\sqrt[3]{\frac{64}{27}} = \frac{\sqrt[3]{64}}{\sqrt[3]{27}} = \frac{4}{3}$

16. $-2x-3y=7$
$-3y = 2x+7$
$-\frac{1}{3}(-3y) = -\frac{1}{3}(2x+7)$
$y = -\frac{2}{3}x - \frac{7}{3}$
The slope is $-\frac{2}{3}$.

17. $m = -\frac{3}{4}$; y-intercept $= 5$ which means the line crosses y-axis at $(0,5)$.
Use $y = mx+b$.
$y = -\frac{3}{4}x+5$
Use point-slope form.
$y-5 = -\frac{3}{4}\left(x-0\right)$
$y-5 = -\frac{3}{4}x$
$4(y-5) = 4\left(-\frac{3}{4}x\right)$
$4y-20 = -3x$
$-20 = -3x-4y$
$20 = 3x+4y$

18. $3(2x-1)-(x+2) = 2(-x+3)$
$6x-3-x-2 = -2x+6$
$5x-5 = -2x+6$
$7x-5 = 6$
$7x = 11$
$x = \frac{11}{7}$
The solution set is $\left\{\frac{11}{7}\right\}$.

19. $\frac{3}{2x-1} = \frac{4}{3x-2}$
$3(3x-2) = 4(2x-1)$
$9x-6 = 8x-4$
$x-6 = -4$
$x = 2$
The solution set is $\{2\}$.

20. $3x^2-2x+1=0$
Use the quadratic formula.
$x = \frac{-b\pm\sqrt{b^2-4ac}}{2a}$

$x = \frac{-(-2)\pm\sqrt{(-2)^2-4(3)(1)}}{2(3)}$

$x = \frac{2\pm\sqrt{4-12}}{6} = \frac{2\pm\sqrt{-8}}{6}$

$x = \frac{2\pm\sqrt{-4}\sqrt{2}}{6} = \frac{2\pm 2i\sqrt{2}}{6}$

$x = \frac{2(1\pm i\sqrt{2})}{6} = \frac{1\pm i\sqrt{2}}{3}$

The solution set is $\left\{\frac{1\pm i\sqrt{2}}{3}\right\}$.

21. $\sqrt{x+1}+\sqrt{x-4}=5$
$\sqrt{x-4}=5-\sqrt{x+1}$
$(\sqrt{x-4})^2=(5-\sqrt{x+1})^2$
$x-4=(5-\sqrt{x+1})(5-\sqrt{x+1})$
$x-4=25-5\sqrt{x+1}-5\sqrt{x+1}+(\sqrt{x+1})^2$
$x-4=25-10\sqrt{x+1}+x+1$
$x-4=x+26-10\sqrt{x+1}$
$-30=-10\sqrt{x+1}$
$3=\sqrt{x+1}$
$(3)^2=(\sqrt{x+1})^2$
$9=x+1$
$8=x$
Check $x=8$.
$\sqrt{(8)+1}+\sqrt{(8)-4}\overset{?}{=}5$
$\sqrt{9}+\sqrt{4}\overset{?}{=}5$
$3+2=5$
The solution set is $\{8\}$.

22. Graph $4x^2-y^2=16$.

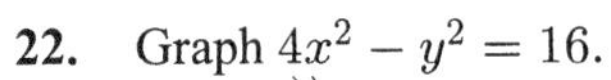

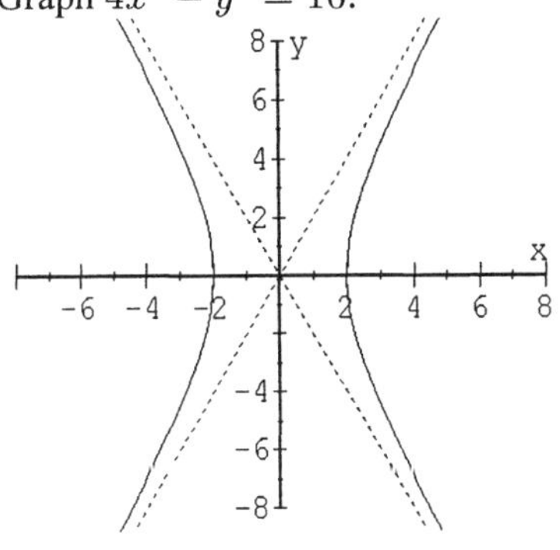

23. Graph $f(x)=-2x^2+8x-5$.

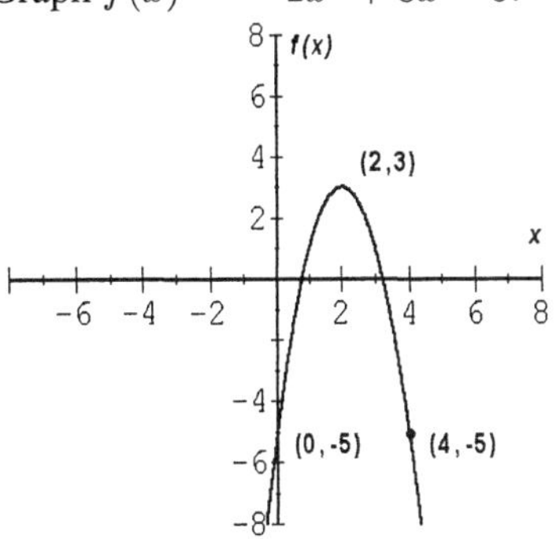

24. Graph $xy^2=-4$.

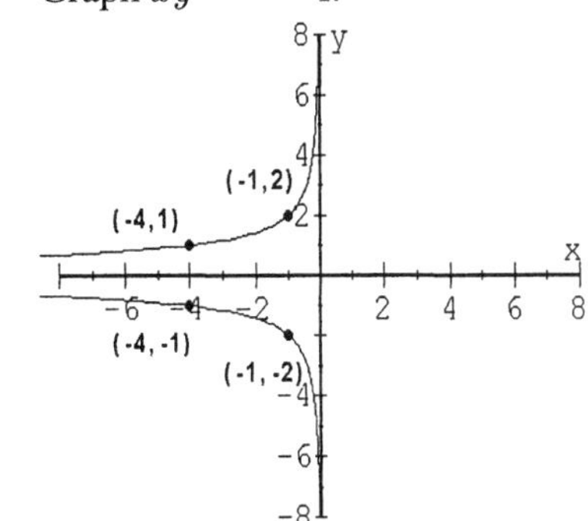

25. Graph $f(x)=-x-3$

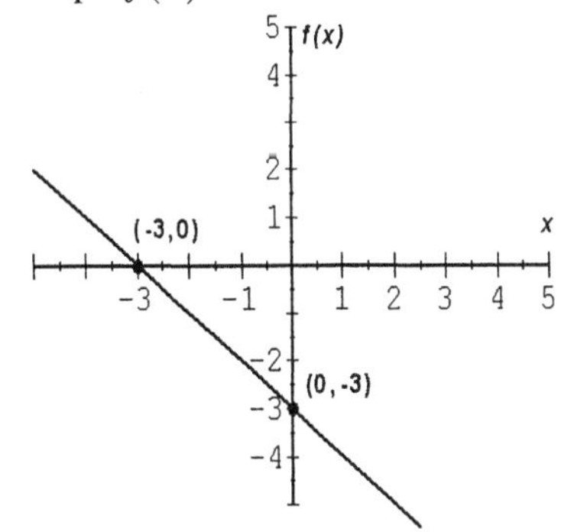

Chapter 14 Exponential and Logarithmic Functions

PROBLEM SET 14.1 Exponents and Exponential Functions

1. $2^x = 64$
$2^x = 2^6$
$x = 6$
The solution set is $\{6\}$.

3. $3^{2x} = 27$
$3^{2x} = 3^3$
$2x = 3$
$x = \frac{3}{2}$
The solution set is $\left\{\frac{3}{2}\right\}$.

5. $\left(\frac{1}{2}\right)^x = \frac{1}{128}$
$\left(\frac{1}{2}\right)^x = \left(\frac{1}{2}\right)^7$
$x = 7$
The solution set is $\{7\}$.

7. $3^{-x} = \frac{1}{243}$
$3^{-x} = \frac{1}{3^5}$
$3^{-x} = 3^{-5}$
$-x = -5$
$x = 5$
The solution set is $\{5\}$.

9. $6^{3x-1} = 36$
$6^{3x-1} = 6^2$
$3x - 1 = 2$
$3x = 3$
$x = 1$
The solution set is $\{1\}$.

11. $5^{x+2} = 125$
$5^{x+2} = 5^3$
$x + 2 = 3$
$x = 1$
The solution set is $\{1\}$.

13. $\left(\frac{3}{4}\right)^n = \frac{64}{27}$
$\left(\frac{3}{4}\right)^n = \left(\frac{4}{3}\right)^3$
$\left(\frac{3}{4}\right)^n = \left(\frac{3}{4}\right)^{-3}$
$n = -3$
The solution set is $\{-3\}$.

15. $16^x = 64$
$(2^4)^x = 2^6$
$2^{4x} = 2^6$
$4x = 6$
$x = \frac{6}{4} = \frac{3}{2}$
The solution set is $\left\{\frac{3}{2}\right\}$.

17. $\left(\frac{1}{2}\right)^{2x} = 64$
$(2^{-1})^{2x} = 2^6$
$2^{-2x} = 2^6$
$-2x = 6$
$x = -3$
The solution set is $\{-3\}$.

19. $6^{2x} + 3 = 39$
$6^{2x} = 36$
$6^{2x} = 6^2$
$2x = 2$
$x = 1$
The solution set is $\{1\}$.

21. $27^{4x} = 9^{x+1}$
$(3^3)^{4x} = (3^2)^{x+1}$
$3^{12x} = 3^{2x+2}$
$12x = 2x + 2$
$10x = 2$
$x = \frac{2}{10} = \frac{1}{5}$

The solution set is $\left\{\frac{1}{5}\right\}$.

23. $9^{4x-2} = \frac{1}{81}$
$9^{4x-2} = \frac{1}{9^2}$
$9^{4x-2} = 9^{-2}$
$4x - 2 = -2$
$4x = 0$
$x = 0$
The solution set is $\{0\}$.

25. $10^x = 0.1$
$10^x = 10^{-1}$
$x = -1$
The solution set is $\{-1\}$.

27. $(2^{x+1})(2^x) = 64$
$2^{2x+1} = 2^6$
$2x + 1 = 6$
$2x = 5$
$x = \frac{5}{2}$
The solution set is $\left\{\frac{5}{2}\right\}$.

29. $(27)(3^x) = 9^x$
$(3^3)(3^x) = (3^2)^x$
$3^{x+3} = 3^{2x}$
$x + 3 = 2x$
$3 = x$
The solution set is $\{3\}$.

31. $(4^x)(16^{3x-1}) = 8$
$(2^2)^x(2^4)^{3x-1} = 2^3$
$(2^{2x})(2^{12x-4}) = 2^3$
$2^{14x-4} = 2^3$
$14x - 4 = 3$
$14x = 7$
$x = \frac{7}{14} = \frac{1}{2}$
The solution set is $\left\{\frac{1}{2}\right\}$.

33. $f(x) = 3^x$

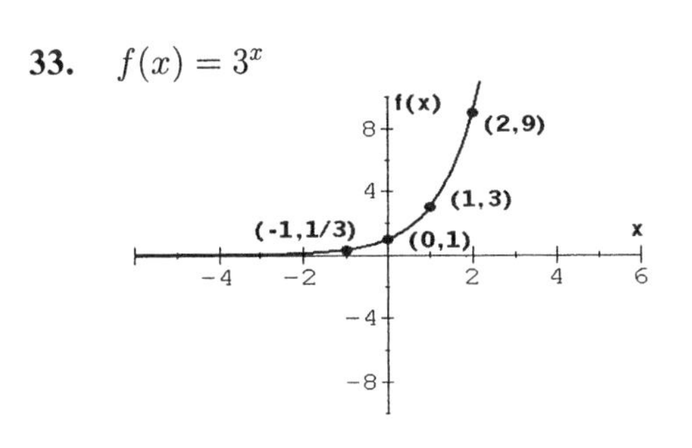

35. $f(x) = \left(\frac{1}{3}\right)^x$

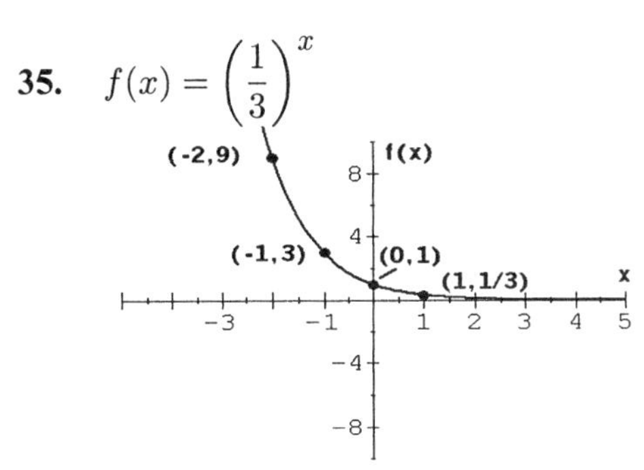

37. $f(x) = \left(\frac{3}{2}\right)^x$

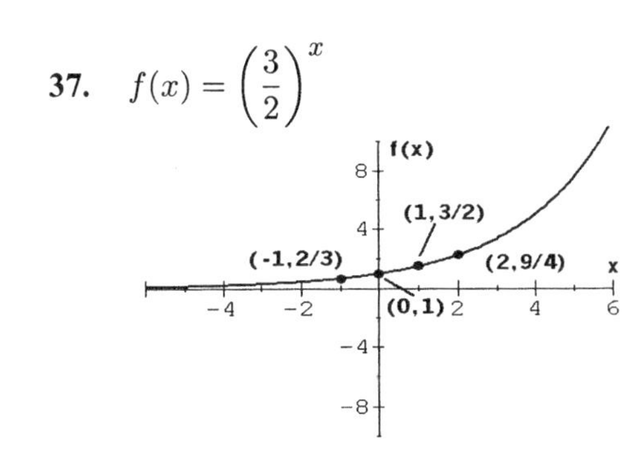

39. $f(x) = 2^x - 3$

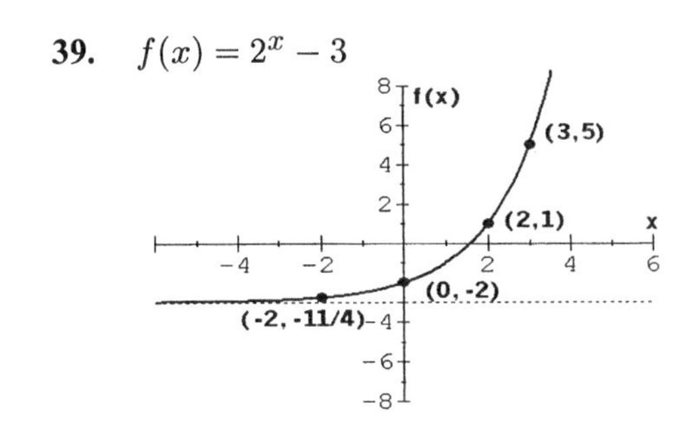

41. $f(x) = 2^{x+2}$

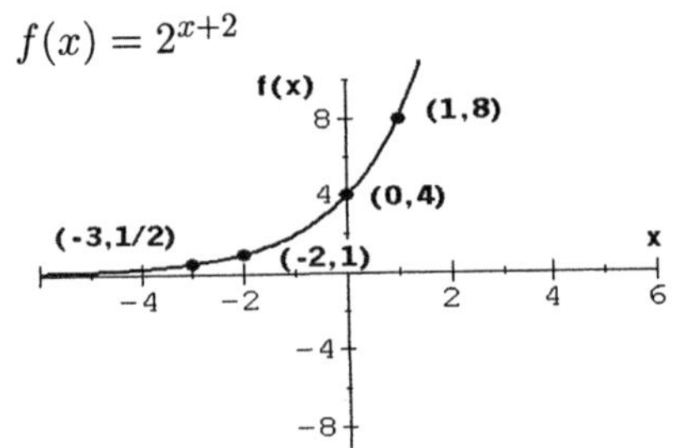

43. $f(x) = -2^{x}$

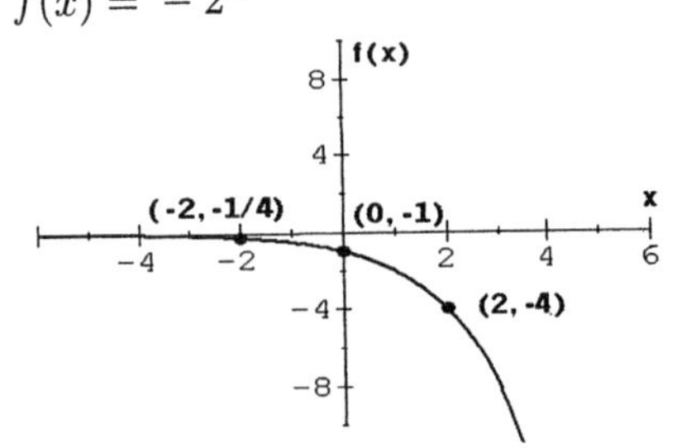

45. $f(x) = 2^{-x-2}$

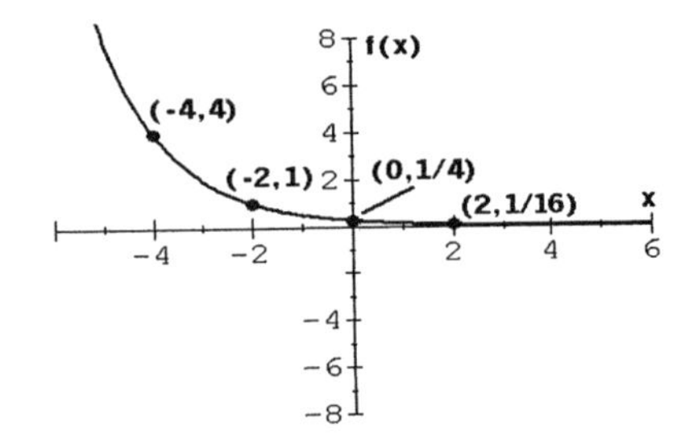

47. $f(x) = 2^{x^2}$

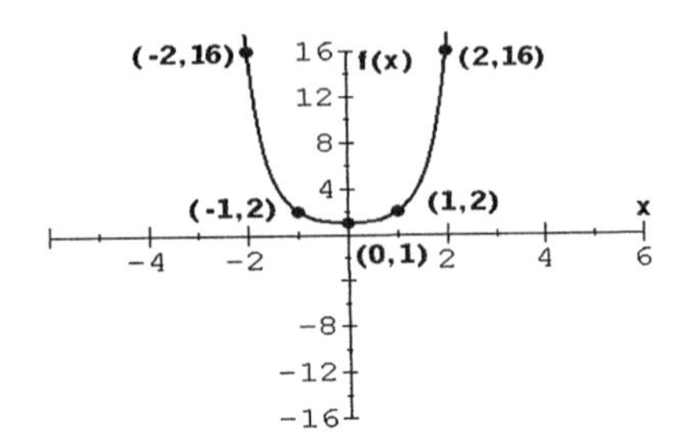

49. $f(x) = 2^{|x|}$

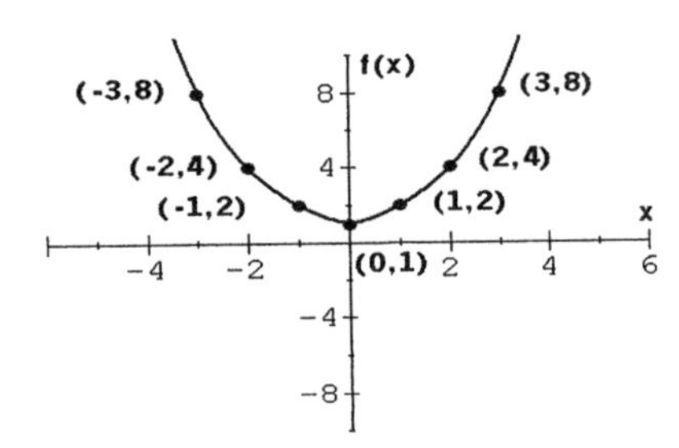

51. $f(x) = 2^{x} - 2^{-x}$

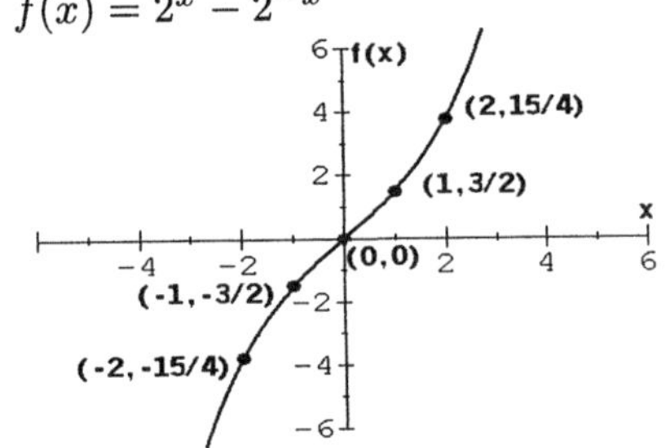

PROBLEM SET 14.2 Applications of Exponential Functions

1. $P = P_0(1.04)^t$

a. $P = 0.89(1.04)^3$
$P = 0.89(1.125)$
$P = \$1.00$

b. $P = 3.43(1.04)^5$
$P = 3.43(1.217)$
$P = \$4.17$

c. $P = 1.99(1.04)^4$
$P = 1.99(1.170)$
$P = \$2.33$

d. $P = 1.05(1.04)^{10}$
$P = 1.05(1.480)$
$P = \$1.55$

e. $P = 18000(1.04)^5$
$P = 18000(1.217)$
$P = \$21{,}900$

f. $P = 120{,}000(1.04)^8$
$P = 120{,}000(1.369)$
$P = \$164{,}228$

g. $P = 500(1.04)^7$
$P = 500(1.316)$
$P = \$658$

3 a. $A = P\left(1 + \frac{r}{n}\right)^{nt}$

$A = 2200\left(1 + \frac{0.025}{1}\right)^{1(6)}$

$A = 2200(1.025)^6$

$A = 2200(1.1597)$

$A = \$2551.33$

b. $A = P\left(1 + \frac{r}{n}\right)^{nt}$

$A = 2200\left(1 + \frac{0.02}{4}\right)^{4(6)}$

$A = 2200(1.005)^{24}$

$A = 2200(1.127)$

$A = \$2479.75$

2.5% compounded annually amounts to more.

5 a. $A = P\left(1 + \frac{r}{n}\right)^{nt}$

$A = 2000\left(1 + \frac{0.035}{4}\right)^{4(5)}$

$A = 2000(1.0875)^{20}$

$A = 2000(1.1903)$

$A = \$2380.68$

b. $A = P\left(1 + \frac{r}{n}\right)^{nt}$

$A = 2000\left(1 + \frac{0.03}{12}\right)^{12(5)}$

$A = 2000(1.0025)^{60}$

$A = 2000(1.1616)$

$A = \$2323.23$

3.5% compounded quarterly amounts to more.

7. $A = Pe^{rt}$

$A = 4000e^{0.03(5)}$

$A = 4000e^{0.15}$

$A = 4000(1.1618)$

$A = \$4647.34$

9. $A = Pe^{rt}$

$A = 1750e^{0.04(3)}$

$A = 1750e^{0.12}$

$A = 1750(1.127497)$

$A = \$1973.12$

11 a. $A = Pe^{rt}$

$A = 10500e^{0.05(4)}$

$A = 10500e^{0.2}$

$A = 10500(1.2214)$

$A = \$12{,}824.73$

b. $A = P\left(1 + \frac{r}{n}\right)^{nt}$

$A = 10500\left(1 + \frac{0.055}{4}\right)^{4(4)}$

$A = 10500(1.01375)^{16}$

$A = 10500(1.24421)$

$A = \$13{,}064.21$

5.5% compounded quarterly amounts to more.

13 a. $A = Pe^{rt}$

$A = 4500e^{0.03(3)}$

$A = 4500e^{0.09}$

$A = 4500(1.09417)$

$A = \$4923.78$

b. $A = P\left(1 + \frac{r}{n}\right)^{nt}$

$A = 4500\left(1 + \frac{0.03}{12}\right)^{12(3)}$

$A = 4500(1.0025)^{36}$

$A = 4500(1.09405)$

$A = \$4923.23$

3% compounded continuously amounts to more.

15. $A = P\left(1 + \frac{r}{n}\right)^{nt}$

$A = 4830\left(1 + \frac{0.109}{12}\right)^{12(3)}$

$A = 4830(1 + 0.00908)^{36}$

$A = 4830(1.00908)^{36}$

$A = 4830(1.384756)$

$A = \$6688.37$

17. $A = P\left(1 + \frac{r}{n}\right)^{nt}$

$A = 3200\left(1 + \frac{0.025}{4}\right)^{4(5)}$

$A = 3200(1 + 0.00625)^{20}$

$A = 3200(1.00625)^{20}$

$A = 3200(1.132708)$

$A = \$3,624.66$

Problem Set 14.2

19. Time is $\frac{3}{52}$. Rate is $0.10(52) = 5.2$.
$A = Pe^{rt}$
$A = 500e^{5.2(\frac{3}{52})}$
$A = 500e^{0.3}$
$A = 500(1.349859) = \$674.93$

21. $A = P\left(1 + \frac{r}{n}\right)^{nt}$
$2700 = 1500\left(1 + \frac{r}{4}\right)^{4(10)}$
$1.8 = \left(1 + \frac{r}{4}\right)^{40}$
$(1.8)^{\frac{1}{40}} = \left[\left(1 + \frac{r}{4}\right)^{40}\right]^{\frac{1}{40}}$
$1.0148 = 1 + \frac{r}{4}$
$0.0148 = \frac{r}{4}$
$0.0592 = r$
$r = 5.9\%$

23. $P(1 + r) = Pe^r$
$1 + r = e^r$
$1 + r = e^{0.0475}$
$1 + r = 1.04865$
$r = 0.04865$
$r = 4.87\%$

25. $A = P\left(1 + \frac{r}{n}\right)^{nt}$
$A = P\left(1 + \frac{0.0525}{4}\right)^4$
$A = P(1.013125)^4$
$A = P(1.0535)$
$A = P\left(1 + \frac{0.053}{2}\right)^2$
$A = P(1.0265)^2$
$A = P(1.0537)$
5.3% compounded semiannually will yield more.

27. $Q = Q_0\left(\frac{1}{2}\right)^{\frac{t}{n}}$
$Q = 400\left(\frac{1}{2}\right)^{\frac{87}{29}}$
$Q = 400\left(\frac{1}{2}\right)^3$
$Q = 400\left(\frac{1}{8}\right)$
$Q = 50$
There will be 50 grams after 87 years.
$Q = 400\left(\frac{1}{2}\right)^{\frac{100}{29}}$
$Q = 400\left(\frac{1}{2}\right)^{3.4483}$
$Q = 400(0.0916)$
$Q = 37$
There will be 37 grams after 100 years.

29. $Q(t) = 1000e^{0.4t}$
$Q(2) = 1000e^{0.4(2)}$
$Q(2) = 1000e^{0.8}$
$Q(2) = 1000(2.2255)$
$Q(2) = 2226$
$Q(3) = 1000e^{0.4(3)}$
$Q(3) = 1000e^{1.2}$
$Q(3) = 1000(3.3201)$
$Q(3) = 3320$
$Q(5) = 1000e^{0.4(5)}$
$Q(5) = 1000e^2$
$Q(5) = 1000(7.3891)$
$Q(5) = 7389$

31. $Q = Q_0e^{0.3t}$
$6640 = Q_0e^{0.3(4)}$
$6640 = Q_0e^{1.2}$
$6640 = Q_0(3.3201)$
$2000 = Q_0$
There were initially 2000 bacteria.

33. $P(a) = 14.7e^{-0.21a}$

a. $P(3.85) = 14.7e^{-0.21(3.85)}$
$P(3.85) = 14.7e^{-0.8085}$
$P(3.85) = 14.7(0.4455)$
$P(3.85) = 6.5$ lbs. per sq. in.

b. $P(1) = 14.7e^{-0.21(1)}$
$P(1) = 14.7e^{-0.21}$
$P(1) = 14.7(0.8106)$
$P(1) = 11.9$ lbs. per sq. in.

c. $1985 \text{ feet} \times \dfrac{1 \text{ mile}}{5280 \text{ feet}} = 0.376 \text{ miles}$

$P(.376) = 14.7e^{-0.21(0.376)}$
$P(.376) = 14.7e^{-0.07896}$
$P(.376) = 14.7(0.9241)$
$P(.376) = 13.6$ lbs. per sq. in.

d. $1090 \text{ feet} \times \dfrac{1 \text{ mile}}{5280 \text{ feet}} = 0.2064$

$P(.2064) = 14.7e^{-0.21(0.2064)}$
$P(.2064) = 14.7e^{-0.0422}$
$P(.2064) = 14.7(0.9576)$
$P(.2064) = 14.1$ lbs. per sq. in.

35. $f(x) = e^x + 1$

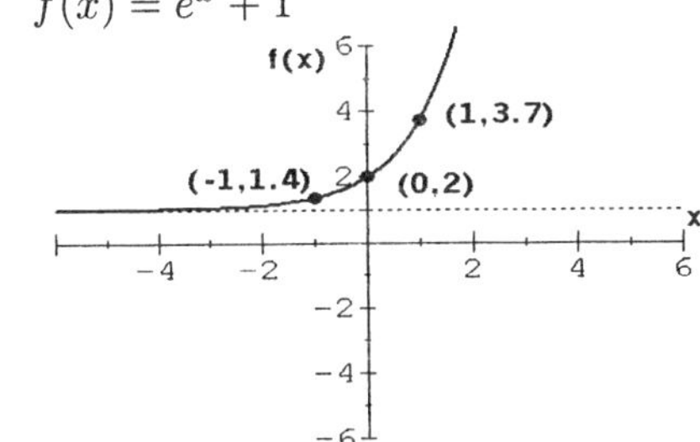

37. $f(x) = 2e^x$

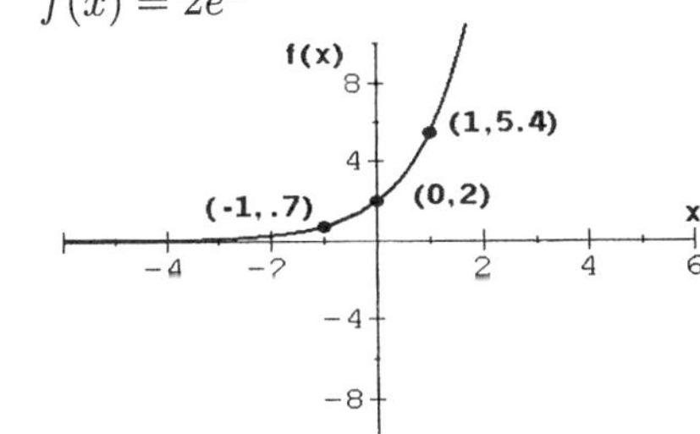

39. $f(x) = e^{2x}$

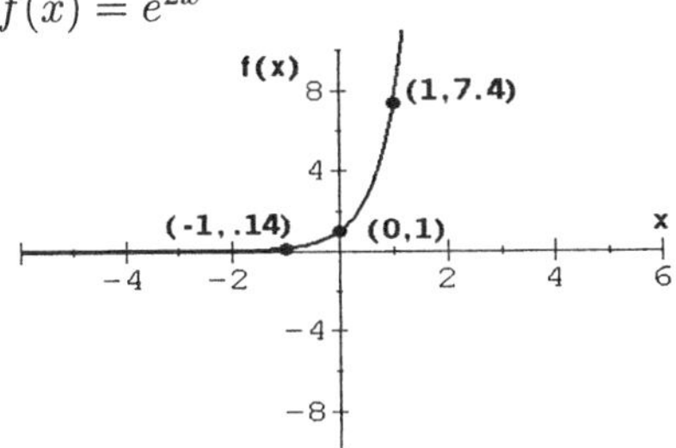

Further Investigations

45. $A = Pe^{rt}$

For 2%

For 5 years
$A = 1000e^{0.02(5)} = 1000e^{0.10}$
$A = 1000(1.1052) = 1105$
For 10 years
$A = 1000e^{0.02(10)} = 1000e^{0.2}$
$A = 1000(1.2214) = 1221$
For 15 years
$A = 1000e^{0.02(15)} = 1000e^{0.3}$
$A = 1000(1.34986) = 1350$
For 20 years
$A = 1000e^{0.02(20)} = 1000e^{0.4}$
$A = 1000(1.4918) = 1492$
For 25 years
$A = 1000e^{0.02(25)} = 1000e^{0.5}$
$A = 1000(1.64872) = 1649$

For 3%

For 5 years
$A = 1000e^{0.03(5)} = 1000e^{0.15}$
$A = 1000(1.1618) = 1162$
For 10 years
$A = 1000e^{0.03(10)} = 1000e^{0.3}$
$A = 1000(1.3499) = 1350$
For 15 years
$A = 1000e^{0.03(15)} = 1000e^{0.45}$
$A = 1000(1.5683) = 1568$
For 20 years
$A = 1000e^{0.03(20)} = 1000e^{0.6}$
$A = 1000(1.8221) = 1822$
For 25 years
$A = 1000e^{0.03(25)} = 1000e^{0.75}$
$A = 1000(2.117) = 2117$

For 4%

For 5 years
$A = 1000e^{0.04(5)} = 1000e^{0.2}$
$A = 1000(1.2214) = 1221$
For 10 years
$A = 1000e^{0.04(10)} = 1000e^{0.4}$
$A = 1000(1.4918) = 1492$
For 15 years
$A = 1000e^{0.04(15)} = 1000e^{0.6}$
$A = 1000(1.8221) = 1822$
For 20 years
$A = 1000e^{0.04(20)} = 1000e^{0.8}$
$A = 1000(2.2255) = 2226$
For 25 years
$A = 1000e^{0.04(25)} = 1000e^{1}$
$A = 1000(2.7183) = 2718$

For 5%

For 5 years

$A = 1000e^{0.05(5)} = 1000e^{0.25}$

$A = 1000(1.284) = 1284$

For 10 years

$A = 1000e^{0.05(10)} = 1000e^{0.5}$

$A = 1000(1.6487) = 1649$

For 15 years

$A = 1000e^{0.05(15)} = 1000e^{0.75}$

$A = 1000(2.117) = 2117$

For 20 years

$A = 1000e^{0.05(20)} = 1000e^{1}$

$A = 1000(2.7183) = 2718$

For 25 years

$A = 1000e^{0.05(25)} = 1000e^{1.25}$

$A = 1000(3.4903) = 3490$

	2%	3%	4%	5%
5 years	\$1105	1162	1221	1284
10 years	1221	1350	1492	1649
15 years	1350	1568	1822	2117
20 years	1492	1822	2226	2718
25 years	1649	2117	2718	3490

47. $A = P\left(1 + \frac{r}{n}\right)^{nt}$

Compounded annually

For 2%

$A = 1000\left(1 + \frac{0.02}{1}\right)^{1(10)} = 1000(1.02)^{10}$

$A = 1000(1.219) = 1219$

For 3%

$A = 1000\left(1 + \frac{0.03}{1}\right)^{1(10)} = 1000(1.03)^{10}$

$A = 1000(1.3439) = 1344$

For 4%

$A = 1000\left(1 + \frac{0.04}{1}\right)^{1(10)} = 1000(1.04)^{10}$

$A = 1000(1.4802) = 1480$

For 5%

$A = 1000\left(1 + \frac{0.05}{1}\right)^{1(10)} = 1000(1.05)^{10}$

$A = 1000(1.6289) = 1629$

Compounded semiannually

For 2%

$A = 1000\left(1 + \frac{0.02}{2}\right)^{2(10)} = 1000(1.01)^{20}$

$A = 1000(1.2202) = 1220$

For 3%

$A = 1000\left(1 + \frac{0.03}{2}\right)^{2(10)} = 1000(1.015)^{20}$

$A = 1000(1.3469) = 1347$

For 4%

$A = 1000\left(1 + \frac{0.04}{2}\right)^{2(10)} = 1000(1.02)^{20}$

$A = 1000(1.4859) = 1486$

For 5%

$A = 1000\left(1 + \frac{0.05}{2}\right)^{2(10)} = 1000(1.025)^{20}$

$A = 1000(1.6386) = 1639$

Compounded quarterly

For 2%

$A = 1000\left(1 + \frac{0.02}{4}\right)^{4(10)} = 1000(1.005)^{40}$

$A = 1000(1.2208) = 1221$

For 3%

$A = 1000\left(1 + \frac{0.03}{4}\right)^{4(10)} = 1000(1.0075)^{40}$

$A = 1000(1.3483) = 1348$

For 4%

$A = 1000\left(1 + \frac{0.04}{4}\right)^{4(10)} = 1000(1.01)^{40}$

$A = 1000(1.4889) = 1489$

For 5%

$A = 1000\left(1 + \frac{0.05}{4}\right)^{4(10)} = 1000(1.0125)^{40}$

$A = 1000(1.6436) = 1644$

Compounded monthly

For 2%

$A = 1000\left(1 + \frac{0.02}{12}\right)^{12(10)} = 1000(1.0017)^{120}$

$A = 1000(1.221) = 1221$

For 3%

$A = 1000\left(1 + \frac{0.03}{12}\right)^{12(10)} = 1000(1.0025)^{120}$

$A = 1000(1.349) = 1349$

For 4%

$A = 1000\left(1 + \frac{0.04}{12}\right)^{12(10)} = 1000(1.0033)^{120}$

$A = 1000(1.4908) = 1491$

For 5%

$A = 1000\left(1 + \frac{0.05}{12}\right)^{12(10)} = 1000(1.0042)^{120}$

$A = 1000(1.647) = 1647$

Compounded continuously

$A = Pe^{rt}$

For 2%

$A = 1000e^{0.02(10)} = 1000e^{0.2}$
$A = 1000(1.221) = 1221$

For 3%

$A = 1000e^{0.03(10)} = 1000e^{0.3}$
$A = 1000(1.3499) = 1350$

For 4%

$A = 1000e^{0.04(10)} = 1000e^{0.4}$
$A = 1000(1.492) = 1492$

For 5%

$A = 1000e^{0.05(10)} = 1000e^{0.5}$
$A = 1000(1.649) = 1649$

Compounded	2%	3%	4%	5%
Annually	$1219	1344	1480	1629
Semiannually	1220	1347	1486	1639
Quarterly	1221	1348	1489	1644
Monthly	1221	1349	1491	1647
Countinuosly	1221	1350	1492	1649

49. $f(x) = \dfrac{e^x + e^{-x}}{2}$

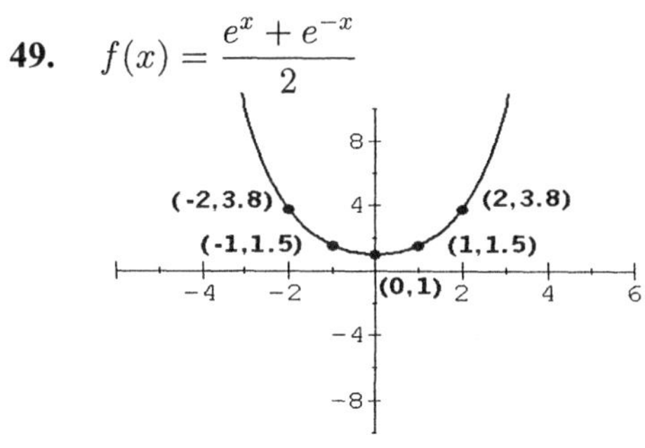

51. $f(x) = \dfrac{e^x - e^{-x}}{2}$

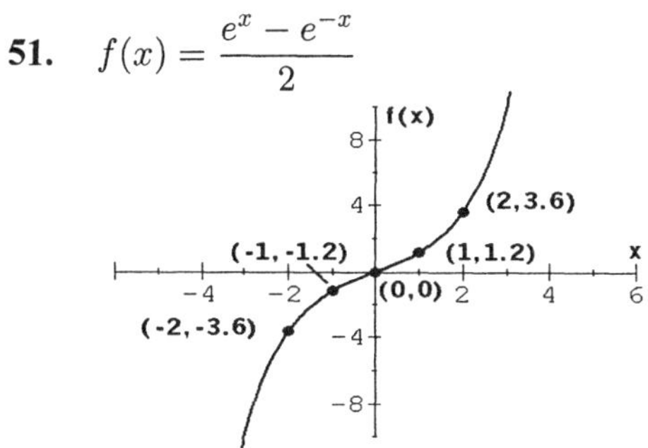

PROBLEM SET 14.3 Inverse Functions

1. Yes, it is one-to-one because no horizontal line intersects the graph in more than one point.

3. No, it is not one-to-one because there is some horizontal line that intersects the graph in more than one point.

5. Yes, it is one-to-one because no horizontal line intersects the graph in more than one point.

7. $f(x) = 5x + 4$
Because the graph is a straight line (slope not equal to 0), it is one-to-one.

9. $f(x) = x^3$
Yes, it is one-to-one because no horizontal line intersects the graph in more than one point.

11. $f(x) = |x| + 1$
No, it is not one-to-one because there is some horizontal line that intersects the graph in more than one point.

13. $f(x) = -x^4$
No, it is not one-to-one because there is some horizontal line that intersects the graph in more than one point.

15. Domain of f : $\{1, 2, 5\}$
Range of f : $\{5, 9, 21\}$
$f^{-1} = \{(5, 1), (9, 2), (21, 5)\}$
Domain of f^{-1} : $\{5, 9, 21\}$
Range of f^{-1} : $\{1, 2, 5\}$

17. Domain of f : $\{0, 2, -1, -2\}$
Range of f : $\{0, 8, -1, -8\}$
$f^{-1} = \{(0, 0), (8, 2), (-1, -1), (-8, -2)\}$
Domain of f^{-1} : $\{0, 8, -1, -8\}$
Range of f^{-1} : $\{0, 2, -1, -2\}$

19. $f(x) = 5x - 9, \quad g(x) = \dfrac{x+9}{5}$
The set of real numbers is the domain and range of both functions.

$$(f \circ g)(x) = 5\left(\frac{x+9}{5}\right) - 9$$
$$= x + 9 - 9 = x$$

$$(g \circ f)(x) = \frac{(5x-9)+9}{5}$$
$$= \frac{5x}{x} = x$$

Since $(f \circ g)(x) = (g \circ f)(x) = x$ they are inverse functions.

21. $f(x) = -\frac{1}{2}x + \frac{5}{6}$, $g(x) = -2x + \frac{5}{3}$

The set of real numbers is the domain and range of both functions.

$$(f \circ g)(x) = -\frac{1}{2}\left(-2x + \frac{5}{3}\right) + \frac{5}{6}$$
$$= x - \frac{5}{6} + \frac{5}{6} = x$$
$$(g \circ f)(x) = -2\left(-\frac{1}{2}x + \frac{5}{6}\right) + \frac{5}{3}$$
$$= x - \frac{5}{3} + \frac{5}{3} = x$$

Since $(f \circ g)(x) = (g \circ f)(x) = x$ they are inverse functions.

23. $f(x) = \frac{1}{x-1}$ for $x > 1$

$g(x) = \frac{x+1}{x}$ for $x > 0$

The domain of g equals the range of f, namely, all real numbers greater than 0.

$$(f \circ g)(x) = \frac{1}{\frac{x+1}{x} - 1} = \frac{1}{\frac{x+1-x}{x}}$$
$$= \frac{1}{\frac{1}{x}} = x$$

The domain of f equals the range of g, namely, all real numbers greater than 1.

$$(g \circ f)(x) = \frac{\frac{1}{x-1} + 1}{\frac{1}{x-1}} = \frac{\frac{1+x-1}{x-1}}{\frac{1}{x-1}}$$
$$= \frac{\frac{x}{x-1}}{\frac{1}{x-1}} = \frac{x}{x-1} \cdot \frac{x-1}{1} = x$$

Since $(f \circ g)(x) = (g \circ f)(x) = x$ they are inverse functions.

25. $f(x) = \sqrt{2x-4}$ for $x \geq 2$

$g(x) = \frac{x^2+4}{2}$ for $x \geq 0$

The domain of g equals the range of f, namely, all real numbers greater than or equal to 0.

$$(f \circ g)(x) = \sqrt{2\left(\frac{x^2+4}{2}\right) - 4}$$
$$= \sqrt{x^2+4-4} = \sqrt{x^2} = x$$

The domain of f equals the range of g, namely, all real numbers greater than or equal to 2.

$$(g \circ f)(x) = \frac{\left(\sqrt{2x-4}\right)^2 + 4}{2}$$
$$= \frac{2x-4+4}{2} = \frac{2x}{2} = x$$

Since $(f \circ g)(x) = (g \circ f)(x) = x$ they are inverse functions.

27. $f(x) = 3x$, $g(x) = -\frac{1}{3}x$

The set of real numbers is the domain and range of both functions.

$$(f \circ g)(x) = 3\left(-\frac{1}{3}x\right) = -x$$
$$(g \circ f)(x) = -\frac{1}{3}(3x) = -x$$

Since $(f \circ g)(x) = (g \circ f)(x) = -x$ they are not inverse functions.

29. $f(x) = x^3$, $g(x) = \sqrt[3]{x}$

The set of real numbers is the domain and range of both functions.

$$(f \circ g)(x) = \left(\sqrt[3]{x}\right)^3 = x$$
$$(g \circ f)(x) = \sqrt[3]{x^3} = x$$

Since $(f \circ g)(x) = (g \circ f)(x) = x$ they are inverse functions.

31. $f(x) = x$, $g(x) = \frac{1}{x}$

$$(f \circ g)(x) = \frac{1}{x}$$

They are not inverse functions.

33. $f(x) = x^2 - 3$ for $x \geq 0$,
$g(x) = \sqrt{x+3}$ for $x \geq -3$
The domain of g equals the range of f, namely, all real numbers greater than or equal to -3.
$(f \circ g)(x) = \left(\sqrt{x+3}\right)^2 - 3 = x+3-3 = x$
The domain of f equals the range of g, namely,all real numbers greater than or equal to 0.
$(g \circ f)(x) = \sqrt{x^2+3-3} = \sqrt{x^2} = x$
Since $(f \circ g)(x) = (g \circ f)(x) = x$ they are inverse functions.

35. $f(x) = \sqrt{x+1}$, $g(x) = x^2 - 1$ for $x \geq 0$
The domain of g equals the range of f, namely, all real numbers greater than or equal to 0.
$(f \circ g)(x) = \sqrt{x^2-1+1} = \sqrt{x^2} = x$
The domain of f equals the range of g, namely,all real numbers greater than or equal to -1.
$(g \circ f)(x) = \left(\sqrt{x+1}\right)^2 - 1 = x+1-1 = x$
Since $(f \circ g)(x) = (g \circ f)(x) = x$ they are inverse functions.

37. $f(x) = x - 4$
$y = x - 4$
To find the inverse function interchange x and y, and solve for y.
$x = y - 4$
$x + 4 = y$
$f^{-1}(x) = x + 4$
The set of real numbers is the domain and range of both functions.
$(f \circ f^{-1})(x) = x + 4 - 4 = x$
$(f^{-1} \circ f)(x) = x - 4 + 4 = x$

39. $f(x) = -3x - 4$
$y = -3x - 4$
To find the inverse function interchange x and y, and solve for y.
$x = -3y - 4$
$x + 4 = -3y$
$\dfrac{x+4}{-3} = y$
$f^{-1}(x) = \dfrac{-x-4}{3}$
The set of real numbers is the domain and range of both functions.
$(f \circ f^{-1})(x) = -3\left(\dfrac{-x-4}{3}\right) - 4$
$= x + 4 - 4 = x$
$(f^{-1} \circ f)(x) = \dfrac{-(-3x-4)-4}{4}$
$= \dfrac{3x}{3} = x$

41. $f(x) = \dfrac{3}{4}x - \dfrac{5}{6}$
$y = \dfrac{3}{4}x - \dfrac{5}{6}$
To find the inverse function interchange x and y, and solve for y.
$x = \dfrac{3}{4}y - \dfrac{5}{6}$
$12x = 9y - 10$
$12x + 10 = 9y$
$y = \dfrac{12x+10}{9}$
$f^{-1}(x) = \dfrac{12x+10}{9} = \dfrac{4}{3}x + \dfrac{10}{9}$
The set of real numbers is the domain and range of both functions.
$(f \circ f^{-1})(x) = \dfrac{3}{4}\left(\dfrac{4}{3}x + \dfrac{10}{9}\right) - \dfrac{5}{6}$
$= x + \dfrac{5}{6} - \dfrac{5}{6} = x$
$(f^{-1} \circ f)(x) = \dfrac{4}{3}\left(\dfrac{3}{4}x - \dfrac{5}{6}\right) + \dfrac{10}{9}$
$= x - \dfrac{10}{9} + \dfrac{10}{9} = x$

43. $f(x) = -\dfrac{2}{3}x$
$y = -\dfrac{2}{3}x$
To find the inverse function interchange x and y, and solve for y.
$x = -\dfrac{2}{3}y$
$-\dfrac{3}{2}x = y$
$f^{-1}(x) = -\dfrac{3}{2}x$

The set of real numbers is the domain and range of both functions.

$(f \circ f^{-1})(x) = -\frac{2}{3}\left(-\frac{3}{2}x\right) = x$

$(f^{-1} \circ f)(x) = -\frac{3}{2}\left(-\frac{2}{3}x\right) = x$

45. $f(x) = \sqrt{x}$ for $x \geq 0$
$y = \sqrt{x}$
To find the inverse function interchange x and y, and solve for y.
$x = \sqrt{y}$
$x^2 = y$
$f^{-1}(x) = x^2$ for $x \geq 0$
The domain and range for both functions is the same namely, all real numbers greater than or equal to 0.

$(f \circ f^{-1})(x) = \sqrt{x^2} = x$

$(f^{-1} \circ f)(x) = \left(\sqrt{x}\right)^2 = x$

47. $f(x) = x^2 + 4$ for $x \geq 0$
$y = x^2 + 4$
To find the inverse function interchange x and y, and solve for y.
$x = y^2 + 4$
$x - 4 = y^2$
$y = \sqrt{x - 4}$
(Note the negative root is discarded.)
$f^{-1}(x) = \sqrt{x-4}$ for $x \geq 4$
The domain of f^{-1} equals the range of f, namely, all real numbers greater than or equal to 4.

$(f \circ f^{-1})(x) = \left(\sqrt{x-4}\right)^2 + 4$
$= x - 4 + 4 = x$

The domain of f equals the range of f^{-1}, namely,all real numbers greater than or equal to 0.

$(f^{-1} \circ f)(x) = \sqrt{x^2 + 4 - 4}$
$= \sqrt{x^2} = x$

49. $f(x) = 1 + \frac{1}{x}$ for $x > 0$

$y = 1 + \frac{1}{x}$ for $x > 0$

To find the inverse function interchange x and y, and solve for y.

$x = 1 + \frac{1}{y}$

$x - 1 = \frac{1}{y}$

$y(x-1) = 1$

$y = \frac{1}{x-1}$

$f^{-1}(x) = \frac{1}{x-1}$ for $x > 1$

The domain of f^{-1} equals the range of f, namely, all real numbers greater than 1.

$(f \circ f^{-1})(x) = 1 + \cfrac{1}{\cfrac{1}{x-1}} = 1 + x - 1 = x$

The domain of f equals the range of f^{-1}, namely,all real numbers greater than 0.

$(f^{-1} \circ f)(x) = \cfrac{1}{1 + \cfrac{1}{x} - 1} = \cfrac{1}{\cfrac{1}{x}} = x$

51. $f(x) = 3x$
$y = 3x$
To find the inverse function interchange x and y, and solve for y.
$x = 3y$

$\frac{1}{3}x = y$

$f^{-1}(x) = \frac{1}{3}x$

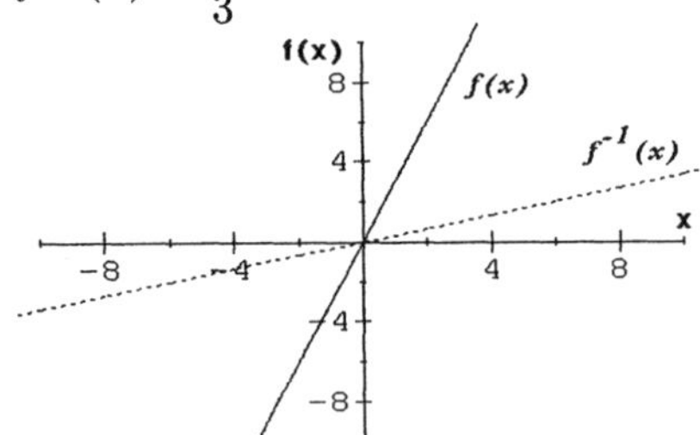

53. $f(x) = 2x + 1$
$y = 2x + 1$
To find the inverse function interchange x and y, and solve for y.
$x = 2y + 1$
$x - 1 = 2y$

$\frac{1}{2}x - \frac{1}{2} = y$

$$f^{-1}(x) = \frac{1}{2}x - \frac{1}{2}$$

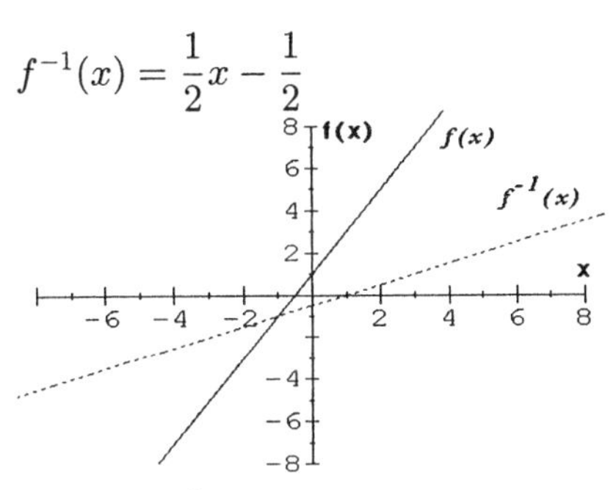

55. $f(x) = \dfrac{2}{x-1}$ for $x > 1$

$y = \dfrac{2}{x-1}$

To find the inverse function interchange x and y, and solve for y.

$x = \dfrac{2}{y-1}$

$(y-1)(x) = 2$

$y - 1 = \dfrac{2}{x}$

$y = \dfrac{2}{x} + 1$

$f^{-1}(x) = \dfrac{2}{x} + 1$ for $x > 0$

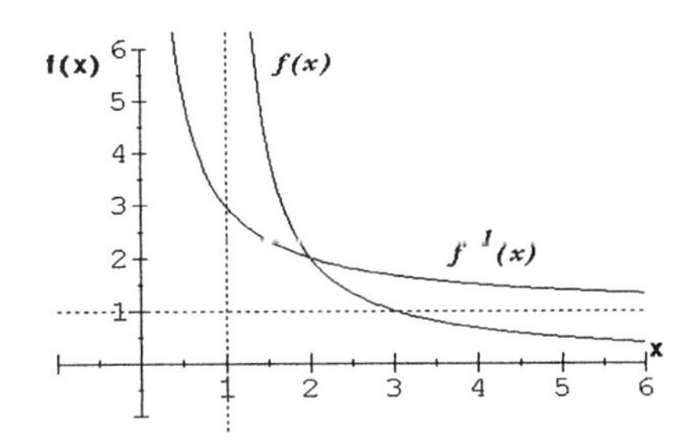

57. $f(x) = x^2 - 4$ for $x \geq 0$

$y = x^2 - 4$

To find the inverse function interchange x and y, and solve for y.

$x = y^2 - 4$

$x + 4 = y^2$

$y = \sqrt{x+4}$

(Note : discard the negative root.)

$f^{-1}(x) = \sqrt{x+4}$ for $x \geq -4$

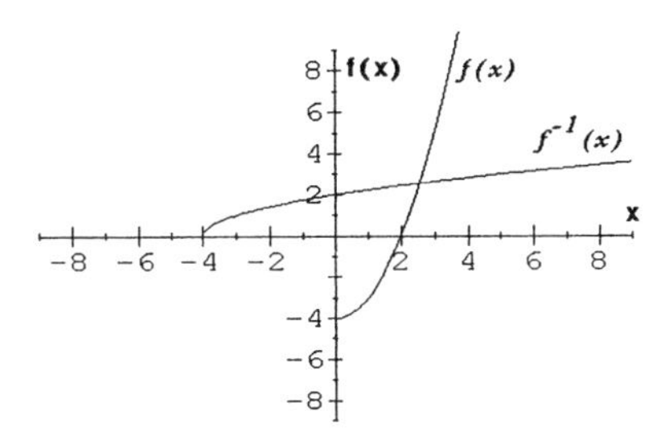

59. $f(x) = x^2 + 1$
Increasing on $[0, \infty)$ and decreasing on $(-\infty, 0]$

61. $f(x) = -3x + 1$
Decreasing on $(-\infty, \infty)$

63. $f(x) = -(x+2)^2 - 1$
Increasing on $(-\infty, -2]$ and decreasing on $[-2, \infty)$

65. $f(x) = -2x^2 - 16x - 35$
Increasing on $(-\infty, -4]$ and decreasing on $[-4, \infty)$

Further Investigations

71a. $f(x) = 3x - 9$

$f(f^{-1}(x)) = 3\left[f^{-1}(x)\right] - 9 = x$

$3\left[f^{-1}(x)\right] = x + 9$

$f^{-1}(x) = \dfrac{x+9}{3}$

b. $f(x) = -2x + 6$

$f(f^{-1}(x)) = -2\left[f^{-1}(x)\right] + 6 = x$

$-2\left[f^{-1}(x)\right] = x - 6$

$f^{-1}(x) = \dfrac{x-6}{-2} = \dfrac{6-x}{2}$

c. $f(x) = -x + 1$

$f(f^{-1}(x)) = -\left[f^{-1}(x)\right] + 1 = x$

$-\left[f^{-1}(x)\right] = x - 1$

$f^{-1}(x) = \dfrac{x-1}{-1} = -x + 1$

$f^{-1}(x) = 1 - x$

d. $f(x) = 2x$

$f(f^{-1}(x)) = 2\left[f^{-1}(x)\right] = x$

$f^{-1}(x) = \dfrac{x}{2}$

e. $f(x) = -5x$

$f(f^{-1}(x)) = -5\left[f^{-1}(x)\right] = x$

$f^{-1}(x) = -\dfrac{x}{5}$

$f^{-1}(x) = -\dfrac{1}{5}x$

f. $f(x) = x^2 + 6 \quad \text{for } x \geq 0$

$f(f^{-1}(x)) = \left[f^{-1}(x)\right]^2 + 6 = x$

$\left[f^{-1}(x)\right]^2 = x - 6$

$f^{-1}(x) = \sqrt{x-6} \quad \text{for } x \geq 6$

PROBLEM SET 14.4 Logarithms

1. $2^7 = 128$
$\log_2 128 = 7$

3. $5^3 = 125$
$\log_5 125 = 3$

5. $10^3 = 1000$
$\log_{10} 1000 = 3$

7. $2^{-2} = \dfrac{1}{4}$
$\log_2 \dfrac{1}{4} = -2$

9. $10^{-1} = 0.1$
$\log_{10} 0.1 = -1$

11. $\log_3 81 = 4$
$3^4 = 81$

13. $\log_4 64 = 3$
$4^3 = 64$

15. $\log_{10} 10,000 = 4$
$10^4 = 10,000$

17. $\log_2 \left(\dfrac{1}{16}\right) = -4$
$2^{-4} = \dfrac{1}{16}$

19. $\log_{10} 0.001 = -3$
$10^{-3} = 0.001$

21. $\log_2 16 = x$
$2^x = 16$
$2^x = 2^4$
$x = 4$

23. $\log_3 81 = x$
$3^x = 81$
$3^x = 3^4$
$x = 4$

25. $\log_6 216 = x$
$6^x = 216$
$6^x = 6^3$
$x = 3$

27. $\log_7 \sqrt{7} = x$
$7^x = \sqrt{7}$
$7^x = 7^{\frac{1}{2}}$
$x = \dfrac{1}{2}$

29. $\log_{10} 1 = x$
$10^x = 1$
$10^x = 10^0$
$x = 0$

31. $\log_{10} 0.1 = x$
$10^x = 0.1$
$10^x = 10^{-1}$
$x = -1$

33. $10^{\log_{10} 5} = x$
$5 = x$

35. $\log_2\left(\frac{1}{32}\right) = x$

$2^x = \frac{1}{32}$

$2^x = \frac{1}{2^5}$

$2^x = 2^{-5}$

$x = -5$

37. $\log_5(\log_2 32) = x$

$\log_5(\log_2 2^5) = x$

$\log_5 5 = x$

$5^x = 5$

$5^x = 5^1$

$x = 1$

39. $\log_{10}(\log_7 7) = x$

$\log_{10} 1 = x$

$10^x = 1$

$10^x = 10^0$

$x = 0$

41. $\log_7 x = 2$

$7^2 = x$

$49 = x$

The solution set is $\{49\}$.

43. $\log_8 x = \frac{4}{3}$

$8^{\frac{4}{3}} = x$

$(8^{\frac{1}{3}})^4 = x$

$(2)^4 = x$

$16 = x$

The solution set is $\{16\}$.

45. $\log_9 x = \frac{3}{2}$

$9^{\frac{3}{2}} = x$

$(9^{\frac{1}{2}})^3 = x$

$3^3 = x$

$27 = x$

The solution set is $\{27\}$.

47. $\log_4 x = -\frac{3}{2}$

$4^{-\frac{3}{2}} = x$

$\frac{1}{4^{\frac{3}{2}}} = x$

$\frac{1}{(4^{\frac{1}{2}})^3} = x$

$\frac{1}{2^3} = x$

$\frac{1}{8} = x$

The solution set is $\left\{\frac{1}{8}\right\}$.

49. $\log_x 2 = \frac{1}{2}$

$x^{\frac{1}{2}} = 2$

$(x^{\frac{1}{2}})^2 = 2^2$

$x = 4$

The solution set is $\{4\}$.

51. $\log_2 35$

$\log_2 (5 \bullet 7)$

$\log_2 5 + \log_2 7$

$2.3219 + 2.8074$

5.1293

53. $\log_2 125$

$\log_2 5^3$

$3 \log_2 5$

$3(2.3219)$

6.9657

55. $\log_2 \sqrt{7}$

$\log_2 7^{\frac{1}{2}}$

$\frac{1}{2}\log_2 7$

$\frac{1}{2}(2.8074)$

1.4037

57. $\log_2 175$

$\log_2 (5^2 \bullet 7)$

$\log_2 5^2 + \log_2 7$

$2\log_2 5 + \log_2 7$

$2(2.3219) + 2.8074$

$4.6438 + 2.8074$

7.4512

Problem Set 14.4

59. $\log_2 80$
$\log_2 (16 \bullet 5)$
$\log_2 16 + \log_2 5$
$\log_2 2^4 + 2.3219$
$4 + 2.3219$
6.3219

61. $\log_8 \left(\frac{5}{11}\right)$
$\log_8 5 - \log_8 11$
$0.7740 - 1.1531$
-0.3791

63. $\log_8 \sqrt{11}$
$\log_8 11^{\frac{1}{2}}$
$\frac{1}{2} \log_8 11$
$\frac{1}{2}(1.1531)$
0.5766

65. $\log_8 88$
$\log_8 (8 \bullet 11)$
$\log_8 8 + \log_8 11$
$1 + 1.1531$
2.1531

67. $\log_8 \left(\frac{25}{11}\right)$
$\log_8 25 - \log_8 11$
$\log_8 5^2 - 1.1531$
$2 \log_8 5 - 1.1531$
$2(0.7740) - 1.1531$
$1.5480 - 1.1531$
0.3949

69. $\log_b xyz$
$\log_b x + \log_b y + \log_b z$

71. $\log_b \left(\frac{y}{z}\right)$
$\log_b y - \log_b z$

73. $\log_b y^3 z^4$
$\log_b y^3 + \log_b z^4$
$3 \log_b y + 4 \log_b z$

75. $\log_b \left(\frac{x^{\frac{1}{2}} y^{\frac{1}{3}}}{z^4}\right)$
$\log_b x^{\frac{1}{2}} + \log_b y^{\frac{1}{3}} - \log_b z^4$
$\frac{1}{2} \log_b x + \frac{1}{3} \log_b y - 4 \log_b z$

77. $\log_b \sqrt[3]{x^2 z}$
$\log_b x^{\frac{2}{3}} z^{\frac{1}{3}}$
$\log_b x^{\frac{2}{3}} + \log_b z^{\frac{1}{3}}$
$\frac{2}{3} \log_b x + \frac{1}{3} \log_b z$

79. $\log_b \left(x \sqrt{\frac{x}{y}}\right)$
$\log_b \frac{x \bullet x^{\frac{1}{2}}}{y^{\frac{1}{2}}}$
$\log_b \frac{x^{\frac{3}{2}}}{y^{\frac{1}{2}}}$
$\log_b x^{\frac{3}{2}} - \log_b y^{\frac{1}{2}}$
$\frac{3}{2} \log_b x - \frac{1}{2} \log_b y$

81. $2 \log_b x - 4 \log_b y$
$\log_b x^2 - \log_b y^4$
$\log_b \left(\frac{x^2}{y^4}\right)$

83. $\log_b x - (\log_b y - \log_b z)$
$\log_b x - \log_b y + \log_b z$
$\log_b x + \log_b z - \log_b y$
$\log_b (xz) - \log_b y$
$\log_b \left(\frac{xz}{y}\right)$

85. $2 \log_b x + 4 \log_b y - 3 \log_b z$
$\log_b x^2 + \log_b y^4 - \log_b z^3$
$\log_b (x^2 y^4) - \log_b z^3$
$\log_b \left(\frac{x^2 y^4}{z^3}\right)$

87. $\frac{1}{2}\log_b x - \log_b x + 4\log_b y$
$\log_b \sqrt{x} - \log_b x + \log_b y^4$
$\log_b \sqrt{x} + \log_b y^4 - \log_b x$
$\log_b (y^4\sqrt{x}) - \log_b x$
$\log_b \left(\frac{y^4\sqrt{x}}{x}\right)$

89. $\log_3 x + \log_3 4 = 2$
$\log_3 4x = 2$
$3^2 = 4x$
$9 = 4x$
$\frac{9}{4} = x$
The solution set is $\left\{\frac{9}{4}\right\}$.

91. $\log_{10} x + \log_{10} (x - 21) = 2$
$\log_{10} x(x - 21) = 2$
$10^2 = x(x - 21)$
$100 = x^2 - 21x$
$x^2 - 21x - 100 = 0$
$(x - 25)(x + 4) = 0$
$x - 25 = 0$ or $x + 4 = 0$
$x = 25$ or $x = -4$
Discard the root $x = -4$.
The solution set is $\{25\}$.

93. $\log_2 x + \log_2 (x - 3) = 2$
$\log_2 x(x - 3) = 2$
$2^2 = x(x - 3)$
$4 = x^2 - 3x$
$x^2 - 3x - 4 = 0$
$(x - 4)(x + 1) = 0$
$x - 4 = 0$ or $x + 1 = 0$
$x = 4$ or $x = -1$
Discard the root $x = -1$.
The solution set is $\{4\}$.

95. $\log_{10} (2x - 1) - \log_{10} (x - 2) = 1$
$\log_{10} \left(\frac{2x-1}{x-2}\right) = 1$
$10^1 = \frac{2x-1}{x-2}; x \neq 2$
$(x - 2)(10) = (x - 2)\left(\frac{2x-1}{x-2}\right)$
$10x - 20 = 2x - 1$
$8x = 19$
$x = \frac{19}{8}$
The solution set is $\left\{\frac{19}{8}\right\}$.

97. $\log_5 (3x - 2) = 1 + \log_5 (x - 4)$
$\log_5 (3x - 2) - \log_5 (x - 4) = 1$
$\log_5 \left(\frac{3x-2}{x-4}\right) = 1$
$5^1 = \frac{3x-2}{x-4}; x \neq 4$
$5(x - 4) = 3x - 2$
$5x - 20 = 3x - 2$
$2x = 18$
$x = 9$
The solution set is $\{9\}$.

99. $\log_8 (x + 7) + \log_8 x = 1$
$\log_8 (x + 7)(x) = 1$
$8^1 = (x + 7)(x)$
$8 = x^2 + 7x$
$x^2 + 7x - 8 = 0$
$(x + 8)(x - 1) = 0$
$x + 8 = 0$ or $x - 1 = 0$
$x = -8$ or $x = 1$
Discard the root $x = -8$.
The solution set is $\{1\}$.

101. $\log_b \left(\frac{r}{s}\right) = \log_b r - \log_b s$
Let $m = \log_b r$ and $n = \log_b s$.
$m = \log_b r$ becomes $b^m = r$.
$n = \log_b s$ becomes $b^n = s$.
The quotient $\frac{r}{s}$ becomes $\frac{r}{s} = \frac{b^m}{b^n} = b^{m-n}$.
Changing $\frac{r}{s} = b^{m-n}$ to logarithmic form
$\log_b \left(\frac{r}{s}\right) = m - n$
$\log_b \left(\frac{r}{s}\right) = \log_b r - \log_b s$

PROBLEM SET 14.5 Logarithmic Functions

1. log 7.24
0.8597

3. log 52.23
1.7179

5. log 3214.1
3.5071

7. log 0.729
− 0.1373

9. log 0.00034
− 3.4685

11. $\log x = 2.6143$
$x = 411.43$

13. $\log x = 4.9547$
$x = 90095$

15. $\log x = 1.9006$
$x = 79.543$

17. $\log x = -1.3148$
$x = 0.048440$

19. $\log x = -2.1928$
$x = 0.0064150$

21. ln 5
1.6094

23. ln 32.6
3.4843

25. ln 430
6.0638

27. ln 0.46
− 0.7765

29. ln 0.0314
− 3.4609

31. $\ln x = 0.4721$
1.6034

33. $\ln x = 1.1425$
$x = 3.1346$

35. $\ln x = 4.6873$
$x = 108.56$

37. $\ln x = -0.7284$
$x = 0.48268$

39. $\ln x = -3.3244$
$x = 0.035994$

41. **(a)** $f(x) = \log x$

x	0.1	0.5	1	2	4	8	10
$\log x$	− 1	− 0.3	0	0.3	0.6	0.9	1

(b) $f(x) = 10^x$

x	− 1	− 0.3	0	0.3	0.6	0.9	1
10^x	0.1	0.5	1.0	2.0	4.0	8.0	10.0

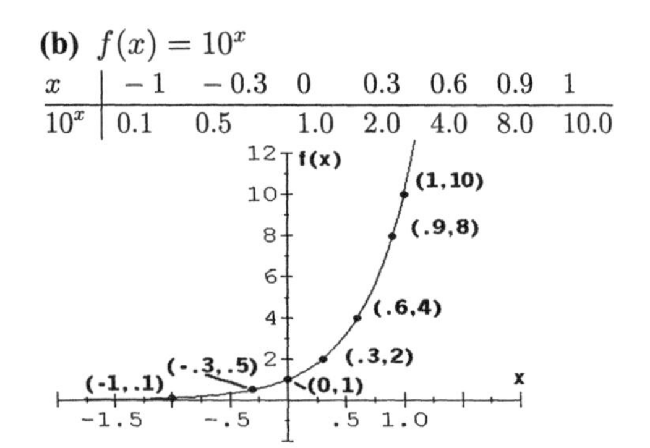

43. $y = \log_{\frac{1}{2}} x$

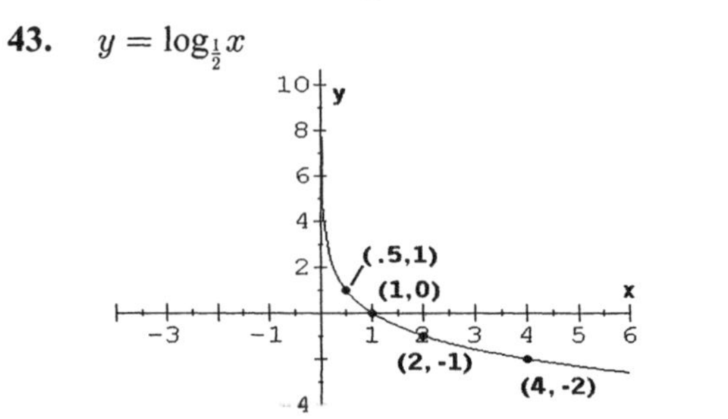

45. $f(x) = \log_3 x \qquad g(x) = 3^x$

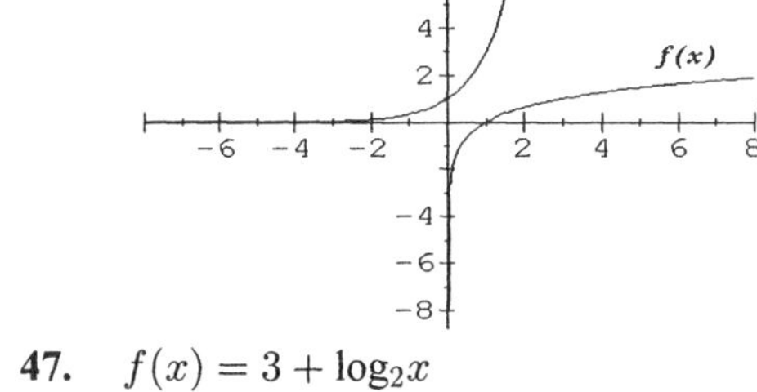

47. $f(x) = 3 + \log_2 x$

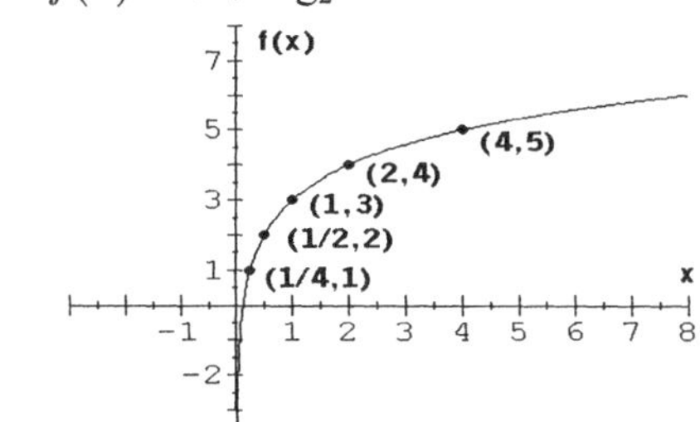

49. $f(x) = \log_2(x+3)$

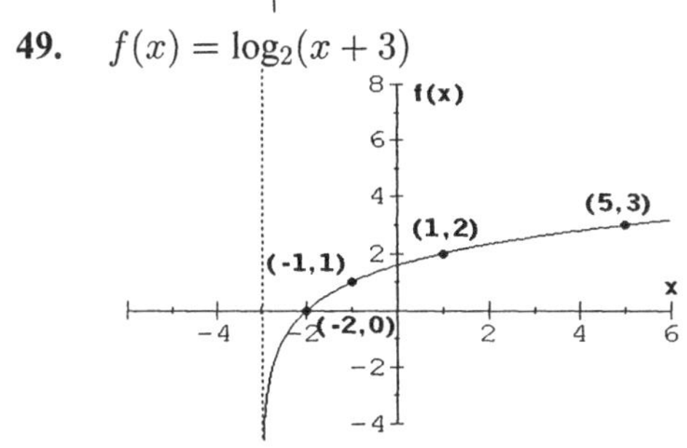

51. $f(x) = \log_2 2x$

53. $f(x) = 2\log_2 x$

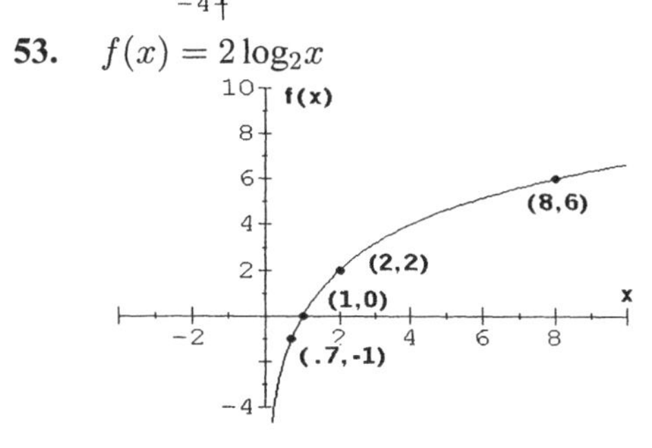

55. $\dfrac{\ln 2}{\ln 7} = \dfrac{0.6931}{1.946} = 0.36$

57. $\dfrac{\ln 5}{2\ln 3} = \dfrac{1.6094}{2(1.0986)} = \dfrac{1.6094}{2.1972} = 0.73$

59. $\dfrac{\ln 2}{0.03} = \dfrac{0.6931}{0.03} = 23.10$

61. $\dfrac{\log 5}{3\log 1.07} = \dfrac{0.6990}{3(0.0294)} = \dfrac{0.6990}{0.0882} = 7.93$

PROBLEM SET 14.6 Exponential Equations, Logarithmic Equations, and Problem Solving

1. $3^x = 13$
$\log 3^x = \log 13$
$x \log 3 = \log 13$
$x(0.4771) = 1.1139$
$x = 2.33$
The solution set is $\{2.33\}$.

3. $4^n = 35$
$\log 4^n = \log 35$
$n \log 4 = \log 35$
$n(0.6021) = 1.5441$
$n = 2.56$
The solution set is $\{2.56\}$.

5. $2^x + 7 = 50$
$2^x = 43$
$\log 2^x = \log 43$
$x \log 2 = \log 43$
$x(0.3010) = 1.6335$
$x = 5.43$
The solution set is $\{5.43\}$.

7. $3^{x-2} = 11$
$\log 3^{x-2} = \log 11$
$(x-2)(\log 3) = \log 11$
$(x-2)(0.4771) = 1.0414$
$0.4771x - 0.9542 = 1.0414$
$0.4771x = 1.9956$
$x = 4.18$
The solution set is $\{4.18\}$.

Problem Set 14.6

9. $5^{3t+1} = 9$
$\log 5^{3t+1} = \log 9$
$(3t + 1)(\log 5) = \log 9$
$(3t + 1)(0.6990) = 0.9542$
$2.097t + 0.6990 = 0.9542$
$2.097t = 0.2552$
$t = 0.12$
The solution set is $\{0.12\}$.

11. $e^x = 27$
$\ln e^x = \ln 27$
$x(\ln e) = \ln 27$
$x = \ln 27$
$x = 3.30$
The solution set is $\{3.30\}$.

13. $e^{x-2} = 13.1$
$\ln e^{x-2} = \ln 13.1$
$x - 2 = 2.57$
$x = 4.57$
The solution set is $\{4.57\}$.

15. $3e^x - 1 = 17$
$3e^x = 18$
$e^x = 6$
$\ln e^x = \ln 6$
$x = 1.79$
The solution set is $\{1.79\}$.

17. $5^{2x+1} = 7^{x+3}$
$\log 5^{2x+1} = \log 7x^{x+3}$
$(2x + 1)(\log 5) = (x + 3)(\log 7)$
$(2x + 1)(0.6990) = (x + 3)(0.8451)$
$1.398x + 0.6990 = 0.8451x + 2.5353$
$0.5529x = 1.8363$
$x = 3.32$
The solution set is $\{3.32\}$.

19. $3^{2x+1} = 2^{3x+2}$
$\log 3^{2x+1} = \log 2^{3x+2}$
$(2x + 1)(\log 3) = (3x + 2)(\log 2)$
$(2x + 1)(0.4771) = (3x + 2)(0.3010)$
$0.9542x + 0.4771 = 0.9030x + 0.6020$
$0.0512x = 0.1249$
$x = 2.44$
The solution set is $\{2.44\}$.

21. $\log x + \log (x + 21) = 2$
$\log x(x + 21) = 2$
$10^2 = x(x + 21)$
$100 = x^2 + 21x$
$0 = x^2 + 21x - 100$
$0 = (x + 25)(x - 4)$
$x + 25 = 0$ or $x - 4 = 0$
$x = -25$ or $x = 4$
Discard the root $x = -25$.
The solution set is $\{4\}$.

23. $\log (3x - 1) = 1 + \log (5x - 2)$
$\log (3x - 1) - \log (5x - 2) = 1$
$\log \dfrac{3x - 1}{5x - 2} = 1$
$10^1 = \dfrac{3x - 1}{5x - 2}$
$10(5x - 2) = 3x - 1$
$50x - 20 = 3x - 1$
$47x = 19$
$x = \dfrac{19}{47}$
The solution set is $\left\{\dfrac{19}{47}\right\}$.

25. $\log (x + 1) = \log 3 - \log (2x - 1)$
$\log (x + 1) + \log (2x - 1) = \log 3$
$\log (x + 1)(2x - 1) = \log 3$
$(x + 1)(2x - 1) = 3$
$2x^2 + x - 1 = 3$
$2x^2 + x - 4 = 0$
$x = \dfrac{-1 \pm \sqrt{1^2 - 4(2)(-4)}}{2(2)}$
$x = \dfrac{-1 \pm \sqrt{33}}{4}$
Discard the root $x = \dfrac{-1 - \sqrt{33}}{4}$.
The solution set is $\left\{\dfrac{-1 + \sqrt{33}}{4}\right\}$.

27. $\log (x + 2) - \log(2x + 1) = \log x$
$\log (x + 2) = \log x + \log (2x + 1)$
$\log (x + 2) = \log x(2x + 1)$
$x + 2 = x(2x + 1)$
$x + 2 = 2x^2 + x$
$2 = 2x^2$
$x^2 = 1$

$x = \pm\sqrt{1} = \pm 1$
Discard the root $x = -1$.
The solution set is $\{1\}$.

29. $\ln(2t+5) = \ln 3 + \ln(t-1)$
$\ln(2t+5) = \ln 3(t-1)$
$2t+5 = 3(t-1)$
$2t+5 = 3t-3$
$8 = t$
The solution set is $\{8\}$.

31. $\log\sqrt{x} = \sqrt{\log x}$
$\log x^{\frac{1}{2}} = (\log x)^{\frac{1}{2}}$
$\frac{1}{2}(\log x) = (\log x)^{\frac{1}{2}}$
$2\left[\frac{1}{2}(\log x)\right] = 2(\log x)^{\frac{1}{2}}$
$\log x = 2(\log x)^{\frac{1}{2}}$
$\log x - 2(\log x)^{\frac{1}{2}} = 0$
let $u = (\log x)^{\frac{1}{2}}$
$u^2 - 2u = 0$
$u(u-2) = 0$
$u = 0$ or $u - 2 = 0$
$u = 0$ or $u = 2$
$(\log x)^{\frac{1}{2}} = 0$ or $(\log x)^{\frac{1}{2}} = 2$
$[(\log x)^{\frac{1}{2}}]^2 = 0^2$ or $[(\log x)^{\frac{1}{2}}]^2 = 2^2$
$\log x = 0$ or $\log x = 4$
$10^0 = x$ or $10^4 = x$
$1 = x$ or $10{,}000 = x$
The solution set is $\{1, 10000\}$.

33. $\log_2 40 = \dfrac{\log 40}{\log 2} = \dfrac{1.6020}{0.3010} = 5.322$

35. $\log_3 16 = \dfrac{\log 16}{\log 3} = \dfrac{1.2041}{0.4771} = 2.524$

37. $\log_4 1.6 = \dfrac{\log 1.6}{\log 4} = \dfrac{0.2041}{0.6021} = 0.339$

39. $\log_5 0.26 = \dfrac{\log 0.26}{\log 5} = \dfrac{-0.5850}{0.6990} = -0.837$

41. $\log_7 500 = \dfrac{\log 500}{\log 7} = \dfrac{2.6990}{0.8451} = 3.194$

43. $A = P\left(1 + \dfrac{r}{n}\right)^{nt}$
$1000 = 750\left(1 + \dfrac{0.12}{4}\right)^{4t}$
$1.3333 = (1.03)^{4t}$
$\log 1.3333 = \log(1.03)^{4t}$
$0.1249 = 4t\log(1.03)$
$0.1249 = 4t(0.0128)$
$0.1249 = 0.0512t$
$2.4 = t$
It will take 2.4 years.

45. $A = Pe^{rt}$
$4000 = 2000e^{0.13t}$
$2 = e^{0.13t}$
$\ln 2 = \ln e^{0.13t}$
$0.6931 = 0.13t$
$5.3 = t$
It will take 5.3 years.

47. $A = Pe^{rt}$
$900 = 500e^{r(10)}$
$1.8 = e^{10r}$
$\ln 1.8 = \ln e^{10r}$
$0.5878 = 10r$
$0.05878 = r$
$5.9\% = r$
The interest rate would need to be 5.9%.

49. $Q = Q_0e^{0.31t}$
$4000 = 400e^{0.34t}$
$10 = e^{0.34t}$
$\ln 10 = \ln e^{0.34t}$
$2.3026 = 0.34t$
$6.8 = t$
It will take 6.8 hours.

51. $P(a) = 14.7e^{-0.21a}$
$11.53 = 14.7e^{-0.21a}$
$0.7844 = e^{-0.21a}$
$\ln 0.7844 = \ln e^{-0.21a}$
$-0.2428 = -0.21a$
$1.156 = a$
$1.156 \text{ miles} \times \dfrac{5280 \text{ feet}}{1 \text{ mile}} = 6104 \text{ feet}$
Cheyenne is approximately 6100 feet above sea level.

53. $Q(t) = Q_0 e^{0.4t}$
$2000 = 500e^{0.4t}$
$4 = e^{0.4t}$
$\ln 4 = \ln e^{0.4t}$
$1.3863 = 0.4t$
$3.5 = t$
It will take 3.5 hours.

55. $R = \log \dfrac{I}{I_0}$
$R = \log \dfrac{5{,}000{,}000 I_0}{I_0}$
$R = \log 5{,}000{,}000$
$R = 6.7$

57. For $R = 7.3$
$I = (10^{7.3}) I_0$
For $R = 6.4$
$I = (10^{6.4}) I_0$
$\dfrac{(10^{7.3}) I_0}{(10^{6.4}) I_0} = 10^{7.3-6.4} = 10^{0.9}$
$10^{0.9} = 7.9$
It is approximately 8 times more intense.

Further Investigations

63. Let $x = \log_a r$
$a^x = r$
$\log_b a^x = \log_b r$
$x \log_b a = \log_b r$
$x = \dfrac{\log_b r}{\log_b a}$
a, b, and r are positive numbers with $a \neq 1$ and $b \neq 1$.

65. $\dfrac{5^x - 5^{-x}}{2} = 3$
$5^x - 5^{-x} = 6$
$5^x - \dfrac{1}{5^x} = 6$
$5^x\left(5^x - \dfrac{1}{5^x}\right) = 5^x(6)$
$5^{2x} - 1 = 6(5^x)$
$(5^x)^2 - 6(5^x) - 1 = 0$
Let $u = 5^x$.
$u^2 - 6u - 1 = 0$
$u = \dfrac{-(-6) \pm \sqrt{(-6)^2 - 4(1)(-1)}}{2(1)}$
$u = \dfrac{6 \pm \sqrt{40}}{2}$
$u = 6.162$ or $u = -0.1623$
Substituting $u = 5^x$
$5^x = 6.162$ or $5^x = -0.1623$ no solution
$\log 5^x = \log 6.162$
$x(\log 5) = \log 6.162$
$x(0.6990) = 0.7897$
$x = 1.13$
The solution set is $\{1.13\}$.

67. $y = \dfrac{e^x - e^{-x}}{2}$
$2y = e^x - e^{-x}$
$e^x(2y) = e^x(e^x - e^{-x})$
$2y(e^x) = (e^x)^2 - 1$
$0 = (e^x)^2 - 2y(e^x) - 1 = 0$
Let $u = e^x$.
$u^2 - 2y(u) - 1 = 0$
$u = \dfrac{-(-2y) \pm \sqrt{(-2y)^2 - 4(1)(-1)}}{2(1)}$
$u = \dfrac{2y \pm \sqrt{4y^2 + 4}}{2}$
$u = \dfrac{2y \pm \sqrt{4(y^2 + 1)}}{2}$
$u = \dfrac{2y \pm 2\sqrt{y^2 + 1}}{2}$
$u = y + \sqrt{y^2 + 1}$
Substituting $u = e^x$
$e^x = y + \sqrt{y^2 + 1}$
$\ln e^x = \ln(y + \sqrt{y^2 + 1})$
$x = \ln(y + \sqrt{y^2 + 1})$

CHAPTER 14 | Review Problem Set

1. $8^{\frac{5}{3}}$
$(8^{\frac{1}{3}})^5 = 2^5 = 32$

2. $-25^{\frac{3}{2}}$
$-(25^{\frac{1}{2}})^3$
$-(5)^3$
-125

3. $(-27)^{\frac{4}{3}}$
$[(-27)^{\frac{1}{3}}]^4 = (-3)^4 = 81$

4. $\log_6 216 = x$
$6^x = 216$
$6^x = 6^3$
$x = 3$

5. $\log_7\left(\frac{1}{49}\right) = x$
$7^x = \frac{1}{49}$
$7^x = \frac{1}{7^2}$
$7^x = 7^{-2}$
$x = -2$

6. $\log_2 \sqrt[3]{2} = x$
$2^x = \sqrt[3]{2}$
$2^x = 2^{\frac{1}{3}}$
$x = \frac{1}{3}$

7. $\log_2\left(\frac{\sqrt[4]{32}}{2}\right) = x$
$2^x = \frac{\sqrt[4]{32}}{2}$
$2^x = \frac{\sqrt[4]{2^5}}{2}$
$2^x = \frac{2^{\frac{5}{4}}}{2^1}$
$2^x = 2^{\frac{5}{4}-1}$
$2^x = 2^{\frac{1}{4}}$
$x = \frac{1}{4}$

8. $\log_{10} 0.00001 = x$
$10^x = 0.00001$
$10^x = 10^{-5}$
$x = -5$

9. $\ln e = x$
$e^x = e$
$x = 1$

10. $7^{\log_7 12} = 12$

11. $\log_{10} 2 + \log_{10} x = 1$
$\log_{10} 2x = 1$
$10^1 = 2x$
$5 = x$
The solution set is $\{5\}$.

12. $\log_3 x = -2$
$3^{-2} = x$
$\frac{1}{9} = x$
The solution set is $\left\{\frac{1}{9}\right\}$.

13. $4^x = 128$
$(2^2)^x = 2^7$
$2^{2x} = 2^7$
$2x = 7$
$x = \frac{7}{2}$
The solution set is $\left\{\frac{7}{2}\right\}$.

14. $3^t = 42$
$\log 3^t = \log 42$
$t(\log 3) = 1.62325$
$t(0.47712) = 1.62325$
$t = 3.40$
The solution set is $\{3.40\}$.

15. $\log_2 x = 3$
$2^3 = x$
$8 = x$
The solution set is $\{8\}$.

16. $\left(\frac{1}{27}\right)^{3x} = 3^{2x-1}$
$(3^{-3})^{3x} = 3^{2x-1}$
$3^{-9x} = 3^{2x-1}$
$-9x = 2x - 1$
$-11x = -1$
$x = \frac{-1}{-11} = \frac{1}{11}$
The solution set is $\left\{\frac{1}{11}\right\}$.

17. $2e^x = 14$
$e^x = 7$
$\ln e^x = \ln 7$
$x = 1.95$
The solution set is $\{1.95\}$.

18. $2^{2x+1} = 3^{x+1}$
$\log 2^{2x+1} = \log 3^{x+1}$
$(2x+1)\log 2 = (x+1)\log 3$
$(2x+1)(0.3010) = (x+1)(0.4771)$
$0.6020x + 0.3010 = 0.4771x + 0.4771$
$0.1249x = 0.1761$
$x = 1.41$
The solution set is $\{1.41\}$.

19. $\ln (x+4) - \ln (x+2) = \ln x$
$\ln (x+4) = \ln x + \ln (x+2)$
$\ln (x+4) = \ln x(x+2)$
$x + 4 = x(x+2)$
$x + 4 = x^2 + 2x$
$0 = x^2 + x - 4$
$x = \dfrac{-1 \pm \sqrt{1^2 - 4(1)(-4)}}{2(1)}$
$x = \dfrac{-1 \pm \sqrt{17}}{2}$
$x = \dfrac{-1 + \sqrt{17}}{2}$ or $x = \dfrac{-1 - \sqrt{17}}{2}$
$x = 1.56$ or $x = -2.56$
Discard the root $x = -2.56$.
The solution set is $\{1.56\}$.

20. $\log x + \log (x-15) = 2$
$\log x(x-15) = 2$
$10^2 = x(x-15)$
$100 = x^2 - 15x$
$0 = x^2 - 15x - 100$
$0 = (x-20)(x+5)$
$x - 20 = 0$ or $x + 5 = 0$
$x = 20$ or $x = -5$
Discard the root $x = -5$.
The solution set is $\{20\}$.

21. $\log (\log x) = 2$
$10^2 = \log x$
$100 = \log x$
$10^{100} = x$
The solution set is $\{10^{100}\}$.

22. $\log(7x-4) - \log(x-1) = 1$
$\log \dfrac{(7x-4)}{(x-1)} = 1$
$10^1 = \dfrac{7x-4}{x-1}$
$10(x-1) = 7x - 4$
$10x - 10 = 7x - 4$
$3x = 6$
$x = 2$
The solution set is $\{2\}$.

23. $\ln (2t-1) = \ln 4 + \ln (t-3)$
$\ln (2t-1) = \ln 4(t-3)$
$2t - 1 = 4(t-3)$
$2t - 1 = 4t - 12$
$11 = 2t$
$\dfrac{11}{2} = t$
The solution set is $\left\{\dfrac{11}{2}\right\}$.

24. $64^{2t+1} = 8^{-t+2}$
$(2^6)^{2t+1} = (2^3)^{-t+2}$
$2^{12t+6} = 2^{-3t+6}$
$12t + 6 = -3t + 6$
$15t = 0$
$t = 0$
The solution set is $\{0\}$.

25. $\log \left(\dfrac{7}{3}\right)$
$\log 7 - \log 3$
$0.8451 - 0.4771$
0.3680

26. $\log 21$
$\log 7(3)$
$\log 7 + \log 3$
$0.8451 + 0.4771$
1.3222

27. $\log 27$
$\log 3^3$
$3 \log 3$
$3(0.4771)$
1.4313

28. $\log 7^{\frac{2}{3}}$

$\frac{2}{3}\log 7$

$\frac{2}{3}(0.8451)$

0.5634

29a. $\log_b\left(\frac{x}{y^2}\right)$

$\log_b x - \log_b y^2$

$\log_b x - 2\log_b y$

b. $\log_b \sqrt[4]{xy^2}$

$\log_b (x^{\frac{1}{4}}y^{\frac{2}{4}})$

$\log_b x^{\frac{1}{4}} + \log_b y^{\frac{2}{4}}$

$\log_b x^{\frac{1}{4}} + \log_b y^{\frac{1}{2}}$

$\frac{1}{4}\log_b x + \frac{1}{2}\log_b y$

c. $\log_b\left(\frac{\sqrt{x}}{y^3}\right)$

$\log_b \sqrt{x} - \log_b y^3$

$\log_b x^{\frac{1}{2}} - \log_b y^3$

$\frac{1}{2}\log_b x - 3\log_b y$

30a) $3\log_b x + 2\log_b y$

$\log_b x^3 + \log_b y^2$

$\log_b x^3y^2$

b) $\frac{1}{2}\log_b y - 4\log_b x$

$\log_b \sqrt{y} - \log_b x^4$

$\log_b \frac{\sqrt{y}}{x^4}$

c) $\frac{1}{2}(\log_b x + \log_b y) - 2\log_b z$

$\frac{1}{2}(\log_b xy) - \log_b z^2$

$\log_b \sqrt{xy} - \log_b z^2$

$\log_b \frac{\sqrt{xy}}{z^2}$

31. $\log_2 3 = \frac{\log 3}{\log 2} = \frac{0.47712}{0.30103} = 1.58$

32. $\log_3 2 = \frac{\log 2}{\log 3} = \frac{0.30103}{0.47712} = 0.63$

33. $\log_4 191 = \frac{\log 191}{\log 4} = \frac{2.28103}{0.60206} = 3.79$

34. $\log_{20} 0.23 = \frac{\log .23}{\log 2} =$

$-\frac{0.63827}{0.30103} = -2.12$

35. $f(x) = \left(\frac{3}{4}\right)^x$

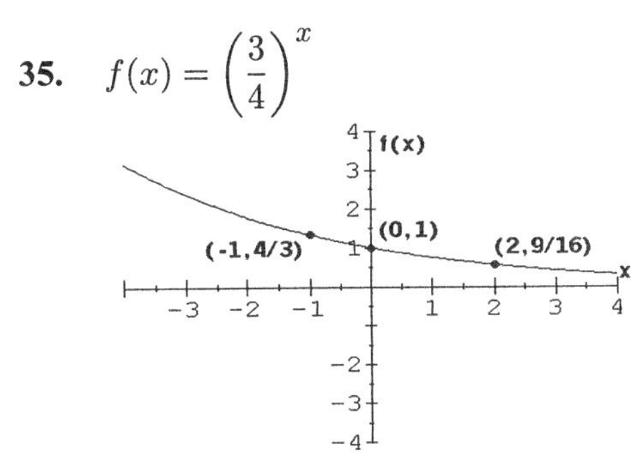

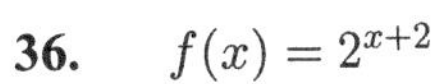

36. $f(x) = 2^{x+2}$

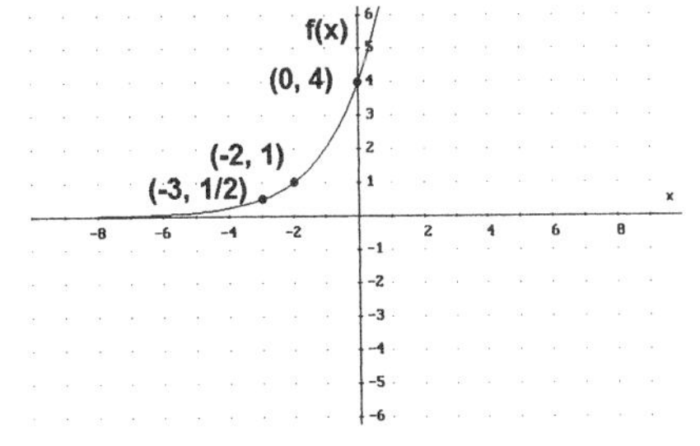

37. $f(x) = e^{x-1}$

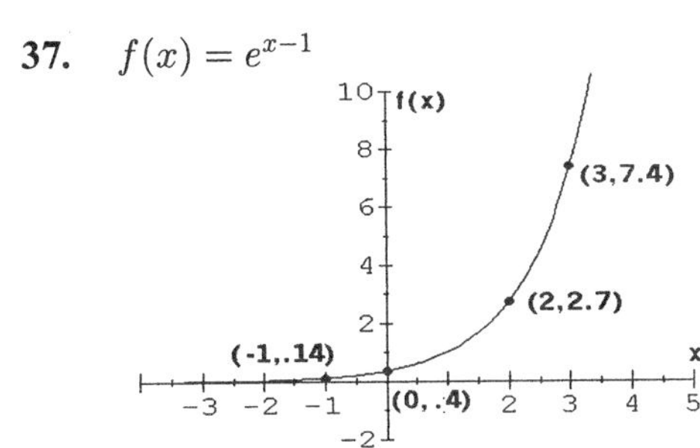

38. 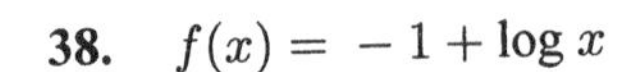
$f(x) = -1 + \log x$

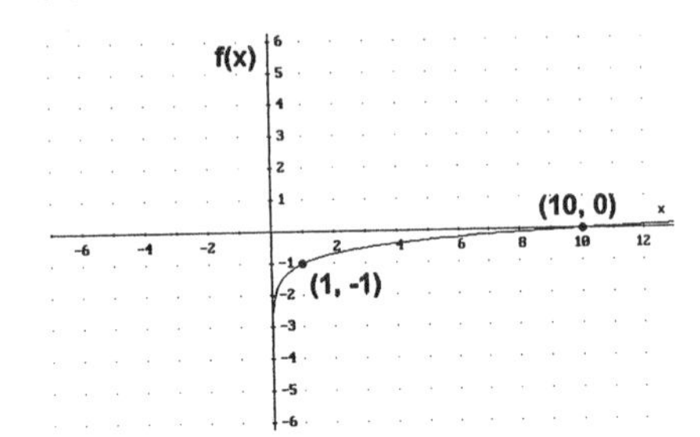

39. $f(x) = 3^x - 3^{-x}$

40. $f(x) = e^{-x^2/2}$

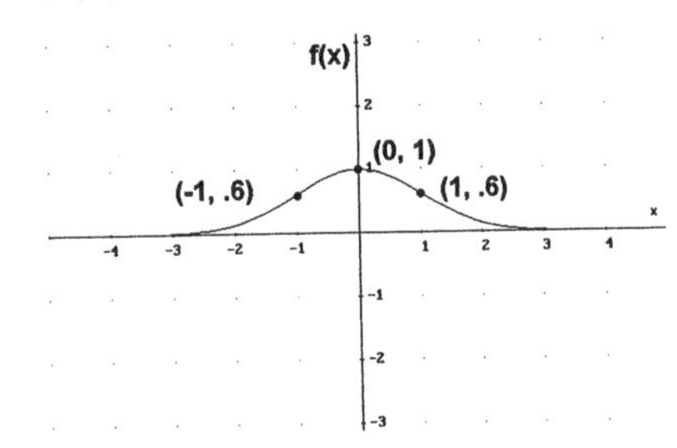

41. $f(x) = \log_2(x-3)$

42. $f(x) = 3\log_3 x$

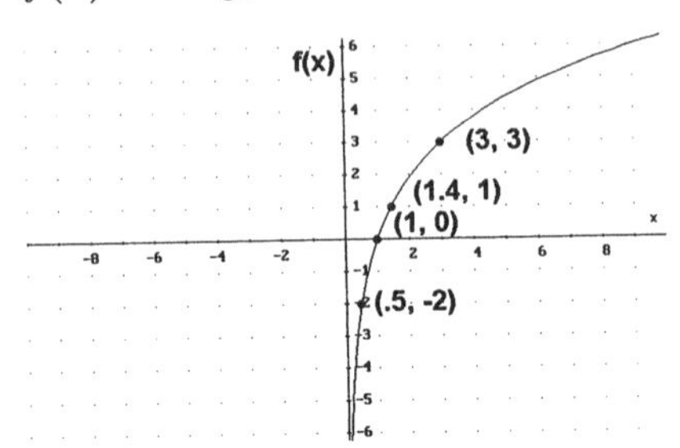

43. $A = P\left(1 + \dfrac{r}{n}\right)^{nt}$

$A = 7250\left(1 + \dfrac{0.07}{4}\right)^{4(10)}$
$A = 7250(1 + 0.0175)^{40}$
$A = 7250(1.0175)^{40}$
$A = 7250(2.0016)$
$A = \$14,511.58$

44. $A = P\left(1 + \dfrac{r}{n}\right)^{nt}$

$A = 12500\left(1 + \dfrac{0.05}{12}\right)^{12(15)}$
$A = 12500(1.004167)^{180}$
$A = 12500(2.1137)$
$A = \$26,421.30$

45. $A = P\left(1 + \dfrac{r}{n}\right)^{nt}$

$A = 2500\left(1 + \dfrac{0.045}{2}\right)^{2(20)}$
$A = 2500(1 + 0.0225)^{40}$
$A = 2500(1.0225)^{40}$
$A = 2500(2.43519)$
$A = \$6,087.97$

46. $f(x) = 7x - 1$ and $g(x) = \dfrac{x+1}{7}$

The set of real numbers is the domain and range of both functions.

$(f \circ g)(x) = 7\left(\dfrac{x+1}{7}\right) - 1$
$= x + 1 - 1 = x$

$(g \circ f)(x) = \dfrac{7x - 1 + 1}{7} = \dfrac{7x}{7} = x$

Yes, they are inverse functions.

47. $f(x) = -\frac{2}{3}x$ and $g(x) = \frac{3}{2}x$

$(f \circ g)(x) = -\frac{2}{3}\left(\frac{3}{2}x\right) = -x$

They are not inverses.

48. $f(x) = x^2 - 6$ for $x \geq 0$

$g(x) = \sqrt{x+6}$ for $x \geq -6$

The domain of g equals the range of f, namely, all real numbers greater than or equal to -6.

$(f \circ g)(x) = \left(\sqrt{x+6}\right)^2 - 6$
$= x + 6 - 6 = x$

The domain of f equals the range of g, namely, all real numbers greater than or equal to 0.

$(g \circ f)(x) = \sqrt{x^2 - 6 + 6}$
$= \sqrt{x^2} = x$

Yes, they are inverse functions.

49. $f(x) = 2 - x^2$ for $x \geq 0$

$g(x) = \sqrt{2-x}$ for $x \leq 2$

The domain of g equals the range of f, namely, all real numbers less than or equal to 2.

$(f \circ g)(x) = 2 - \left(\sqrt{2-x}\right)^2$
$= 2 - (2 - x) = x$

The domain of f equals the range of g, namely, all real numbers greater than or equal to 0.

$(g \circ f)(x) = \sqrt{2 - (2 - x^2)}$
$= \sqrt{2 - 2 + x^2}$
$= \sqrt{x^2} = x$

Yes, the functions are inverses.

50. $f(x) = 4x + 5$

$y = 4x + 5$

Interchange x and y and solve for y.

$x = 4y + 5$

$x - 5 = 4y$

$y = \frac{x-5}{4}$

$f^{-1}(x) = \frac{x-5}{4}$

The domain of f^{-1} equals the range of f, namely, all real numbers.

$(f \circ f^{-1})(x) = 4\left(\frac{x-5}{4}\right) + 5$
$= x - 5 + 5 = x$

The domain of f equals the range of f^{-1}, namely, all real numbers.

$(f^{-1} \circ f)(x) = \frac{4x + 5 - 5}{4}$
$= \frac{4x}{4} = x$

51. $f(x) = -3x - 7$

$y = -3x - 7$

To find the inverse interchange x and y, and solve for y.

$x = -3y - 7$

$x + 7 = -3y$

$-\frac{1}{3}(x + 7) = -\frac{1}{3}(-3y)$

$\frac{x+7}{-3} = y$

$y = \frac{-x-7}{3}$

$f^{-1}(x) = \frac{-x-7}{3}$

The domain of f^{-1} equals the range of f, namely, all real numbers.

$(f \circ f^{-1})(x) = -3\left(\frac{-x-7}{3}\right) - 7$
$= x + 7 - 7 = x$

The domain of f equals the range of f^{-1}, namely, all real numbers.

$(f^{-1} \circ f)(x) = \frac{-(-3x-7)-7}{3}$
$= \frac{3x + 7 - 7}{3} = x$

52. $f(x) = \frac{5}{6}x - \frac{1}{3}$

$y = \frac{5}{6}x - \frac{1}{3}$

Interchange x and y and solve for y.

$x = \frac{5}{6}y - \frac{1}{3}$

$6x = 5y - 2$

$6x + 2 = 5y$

$\frac{6x+2}{5} = y$
$f^{-1}(x) = \frac{6x+2}{5}$
The domain of f^{-1} equals the range of f, namely, all real numbers.

$$(f \circ f^{-1})(x) = \frac{5}{6}\left(\frac{6x+2}{5}\right) - \frac{1}{3} = \frac{6x+2}{6} - \frac{1}{3} = x + \frac{1}{3} - \frac{1}{3} = x$$

The domain of f equals the range of f^{-1}, namely, all real numbers.

$$(f^{-1} \circ f)(x) = \frac{6\left(\frac{5}{6}x - \frac{1}{3}\right) + 2}{5} = \frac{5x - 2 + 2}{5} = \frac{5x}{5} = x$$

53. $f(x) = -2 - x^2$ for $x \geq 0$
Range: $f(x) \leq -2$
$y = -2 - x^2$
To find the inverse interchange x and y and then solve for y.
$x = -2 - y^2$
$x + 2 = -y^2$
$-x - 2 = y^2$
Since the domain of the inverse function is $x \leq -2$ and the range if $f^{-1}(x) \geq 0$, we want only the positive root of y^2.
$y = \sqrt{-x-2}$
$f^{-1}(x) = \sqrt{-x-2}$
The domain of f^{-1} equals the range of f, namely, all real numbers less than or equal to -2.

$$(f \circ f^{-1})(x) = -2 - (\sqrt{-x-2})^2 = -2 - (-x-2) = -2 + x + 2 = x$$

The domain of f equals the range of f^{-1}, namely, all real numbers greater than or equal to 0.

$$(f^{-1} \circ f)(x) = \sqrt{-(-2-x^2) - 2} = \sqrt{2 + x^2 - 2} = \sqrt{x^2} = x$$

54. $-2x^2 + 16x - 35 = 0$
To find vertex, use $x = -\frac{b}{2a}$.
$x = -\frac{(16)}{2(-2)} = \frac{16}{4} = 4$
Critical value $x = 4$
$f(0) = -35$
$f(1) = -21$
Since $0 < 1$ and $f(0) < f(1)$, $f(x)$ is increasing on $(-\infty, 4]$.
$f(5) = -5$
$f(6) = -11$
Since $5 < 6$ and $f(5) > f(6)$, $f(x)$ is decreasing on $[4, \infty)$.

55. $f(x) = 2\sqrt{x-3}$
$x - 3 \geq 0$
$x \geq 3$
Domain is $[3, \infty)$.
$f(3) = 0$
$f(4) = 2$
Since $3 < 4$ and $f(0) < f(2)$, $f(x)$ is increasing on $[3, \infty)$.

56. $A = P\left(1 + \frac{r}{n}\right)^{nt}$
$200 = 100\left(1 + \frac{0.08}{1}\right)^{1t}$
$2 = (1.08)^t$
$\log 2 = \log(1.08)^t$
$0.3010 = t \log(1.08)$
$0.3010 = t(0.0334)$
$9.006 = t$
It will take approximately 9 years.

57. $A = P\left(1 + \frac{r}{n}\right)^{nt}$
$3500 = 1000\left(1 + \frac{0.055}{4}\right)^{4t}$
$3.5 = (1 + 0.01375)^{4t}$
$3.5 = (1.01375)^{4t}$
$\log 3.5 = \log (1.01375)^{4t}$
$0.544068 = 4t \log(1.01375)$
$0.544068 = 4t(0.005931)$
$0.544068 = t(0.23723)$
$22.9 = t$
It will take 22.9 years.

58. $A = Pe^{rt}$
$1000 = 500e^{r(8)}$
$2 = e^{8r}$
$\ln 2 = \ln e^{8r}$
$0.6931 = 8r$
$0.087 = r$
$8.7\% = r$
The rate will be approximately 8.7%.

59. $P(t) = P_0e^{0.02t}$
$P(10) = 50000e^{0.02(10)}$
$P(10) = 50000e^{0.2}$
$P(10) = 50000(1.2214)$
$P(10) = 61{,}070$
$P(15) = 50000e^{0.02(15)}$
$P(15) = 50000e^{0.3}$
$P(15) = 50000(1.34986)$
$P(15) = 67{,}493$
$P(20) = 50000e^{0.02(20)}$
$P(20) = 50000e^{0.4}$
$P(20) = 50000(1.49182)$
$P(20) = 74{,}591$

60. $Q = Qe^{0.29t}$
$2000 = 500e^{0.29t}$
$4 = e^{0.29t}$
$\ln 4 = \ln e^{0.29t}$
$1.3863 = 0.29t$
$4.8 = t$
It will take approximately 4.8 hours.

61. $Q = Q_0\left(\frac{1}{2}\right)^{\frac{t}{n}}$
$$Q = 750\left(\frac{1}{2}\right)^{\frac{100}{40}}$$
$Q = 750(0.5)^{2.5}$
$Q = 750(00.1767767)$
$Q = 133$
There will be 133 grams left after 100 days.

62. $$R = \log\frac{I}{I_0}$$
$$R = \log\frac{125{,}000{,}000I_0}{I_0}$$
$R = \log 125{,}000{,}000$
$R = 8.1$
The Richter number is 8.1.

CHAPTER 14 Test

1. $\log_3\sqrt{3} = x$
$3^x = \sqrt{3}$
$3^x = 3^{\frac{1}{2}}$
$$x = \frac{1}{2}$$

2. $\log_2(\log_2 4)$
$\log_2 2 = x$
$2^x = 2^1$
$x = 1$

3. $-2 + \ln e^3$
$-2 + 3 = 1$

4. $\log_2(0.5) = x$
$2^x = 0.5$
$$2^x = \frac{1}{2}$$
$2^x = 2^{-1}$
$x = -1$

5. $$4^x = \frac{1}{64}$$
$$4^x = \frac{1}{4^3}$$
$4^x = 4^{-3}$
$x = -3$
The solution set is $\{-3\}$.

6. $$9^x = \frac{1}{27}$$
$$(3^2)^x = \frac{1}{3^3}$$
$3^{2x} = 3^{-3}$
$2x = -3$
$$x = -\frac{3}{2}$$
The solution set is $\left\{-\frac{3}{2}\right\}$.

7. $2^{3x-1} = 128$
$2^{3x-1} = 2^7$
$3x - 1 = 7$
$3x = 8$
$x = \frac{8}{3}$
The solution set is $\left\{\frac{8}{3}\right\}$.

8. $\log_9 x = \frac{5}{2}$
$9^{\frac{5}{2}} = x$
$(9^{\frac{1}{2}})^5 = x$
$3^5 = x$
$243 = x$
The solution set is $\{243\}$.

9. $\log x + \log(x + 48) = 2$
$\log x(x + 48) = 2$
$10^2 = x(x + 48)$
$100 = x^2 + 48x$
$0 = x^2 + 48x - 100$
$0 = (x + 50)(x - 2)$
$x + 50 = 0$ or $x - 2 = 0$
$x = -50$ or $x = 2$
Discard the root $x = -50$.
The solution set is $\{2\}$.

10. $\ln x = \ln 2 + \ln(3x - 1)$
$\ln x = \ln 2(3x - 1)$
$e^{\ln x} = e^{\ln 2(3x-1)}$
$x = 2(3x - 1)$
$x = 6x - 2$
$-5x = -2$
$x = \frac{-2}{-5} = \frac{2}{5}$
The solution set is $\left\{\frac{2}{5}\right\}$.

11. $\log_3 100$
$\log_3 (5^2 \bullet 4)$
$\log_3 5^2 + \log_3 4$
$2\log_3 5 + \log_3 4$
$2(1.4650) + 1.2619$
$2.9300 + 1.2619$
4.1919

12. $\log_3\left(\frac{5}{4}\right)$
$\log_3\left(\frac{1}{4}\right)(5)$
$\log_3\left(\frac{1}{4}\right) + \log_3 5$
$\log_3 4^{-1} + 1.4650$
$-1\log_3 4 + 1.4650$
$-1.2619 + 1.4650$
0.2031

13. $\log_3 \sqrt{5}$
$\log_3 5^{\frac{1}{2}}$
$\frac{1}{2}\log_3 5$
$\frac{1}{2}(1.4650)$
0.7325

14. $f(x) = -3x - 6$
$y = -3x - 6$
Interchange x and y and solve for y.
$x = -3y - 6$
$x + 6 = -3y$
$\frac{x + 6}{-3} = y$
$y = \frac{-x - 6}{3}$
$f^{-1}(x) = \frac{-x - 6}{3}$

15. $e^x = 176$
$\ln e^x = \ln 176$
$x = 5.17$
The solution set is $\{5.17\}$.

16. $2^{x-2} = 314$
$\log 2^{x-2} = \log 314$
$(x - 2)\log 2 = \log 314$
$x - 2 = \frac{\log 314}{\log 2}$
$x - 2 = \frac{2.4969}{0.3010}$
$x - 2 = 8.29$
$x = 10.29$
The solution set is $\{10.29\}$.

17. $\log_5 632 = \dfrac{\log 632}{\log 5} = \dfrac{2.800717}{0.698970} = 4.0069$

18. $f(x) = \dfrac{2}{3}x - \dfrac{3}{5}$

$y = \dfrac{2}{3}x - \dfrac{3}{5}$

Interchange x and y and solve for y.

$x = \dfrac{2}{3}y - \dfrac{3}{5}$

$x + \dfrac{3}{5} = \dfrac{2}{3}y$

$\dfrac{3}{2}\left(x + \dfrac{3}{5}\right) = \dfrac{3}{2}\left(\dfrac{2}{3}y\right)$

$\dfrac{3}{2}x + \dfrac{9}{10} = y$

$f^{-1}(x) = \dfrac{3}{2}x + \dfrac{9}{10}$

19. $A = P\left(1 + \dfrac{r}{n}\right)^{nt}$

$A = 3500\left(1 + \dfrac{0.075}{4}\right)^{4(8)}$

$A = 3500(1.01875)^{32}$

$A = 3500(1.812024)$

$A = \$6342.08$

20. $A = P\left(1 + \dfrac{r}{n}\right)^{nt}$

$12500 = 5000\left(1 + \dfrac{0.04}{1}\right)^{1t}$

$2.5 = (1.04)^t$

$\log 2.5 = \log 1.04^t$

$\log 2.5 = t \log 1.04$

$t = \dfrac{\log 2.5}{\log 1.04}$

$t = \dfrac{0.39794}{0.01703}$

$t = 23.4$

It will take 23.4 years.

21. $Q(t) = Q_0 e^{0.23t}$

$2400 = 400e^{0.23t}$

$6 = e^{0.23t}$

$\ln 6 = \ln e^{0.23t}$

$1.79176 = 0.23t$

$7.8 = t$

It will take 7.8 hours.

22. $Q = Q_0\left(\dfrac{1}{2}\right)^{\frac{r}{n}}$

$Q = 7500\left(\dfrac{1}{2}\right)^{\frac{32}{50}}$

$Q = 7500\left(\dfrac{1}{2}\right)^{0.64}$

$Q = 7500(0.64171)$

$Q = 4813$ grams

23. $f(x) = e^x - 2$

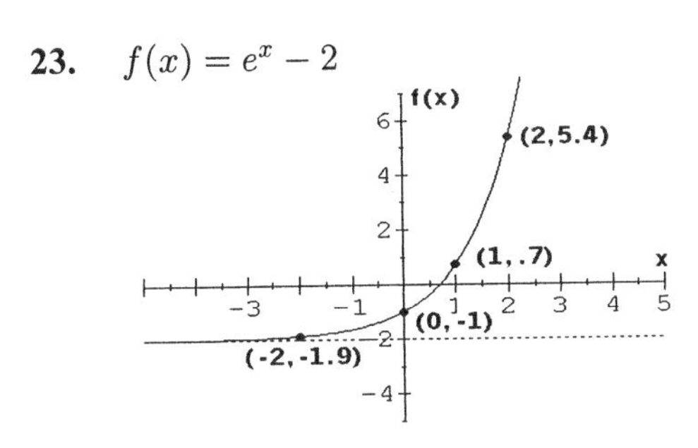

24. $f(x) = -3^{-x}$

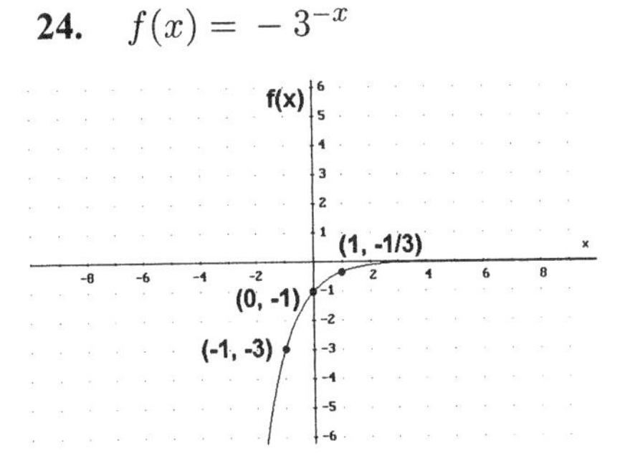

25. $f(x) = \log_2(x - 2)$

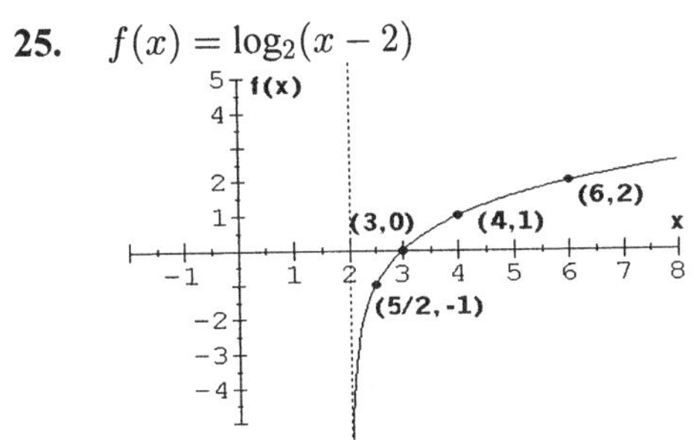

CHAPTERS 1-14 Cumulative Review

1. $-5(x-1)-3(2x+4)+3(3x-1);$
$-5x+5-6x-12+9x-3$
$-2x-10$ for $x=-2$
$-2(-2)-10$
$4-10$
-6

2. $\dfrac{14a^3b^2}{7a^2b}$ for $a=-1, b=4$
$2ab$
$2(-1)(4)$
-8

3. $\dfrac{2}{n}-\dfrac{3}{2n}+\dfrac{5}{3n};$ for $n=4$
$\dfrac{2}{4}-\dfrac{3}{2(4)}+\dfrac{5}{3(4)}$
$\dfrac{1}{2}-\dfrac{3}{8}+\dfrac{5}{12};$ LCD $=24$
$\dfrac{1}{2}\bullet\dfrac{12}{12}-\dfrac{3}{8}\bullet\dfrac{3}{3}+\dfrac{5}{12}\bullet\dfrac{2}{2}$
$\dfrac{12}{24}-\dfrac{9}{24}+\dfrac{10}{24}$
$\dfrac{13}{24}$

4. $4\sqrt{2x-y}+5\sqrt{3x+y}$ for $x=16, y=16$
$4\sqrt{2(16)-16}+5\sqrt{3(16)+16}$
$4\sqrt{16}+5\sqrt{64}$
$4(4)+5(8)$
$16+40=56$

5. $\dfrac{3}{x-2}-\dfrac{5}{x+3};$ for $x=3$
$\dfrac{3}{3-2}-\dfrac{5}{3+3}$
$\dfrac{3}{1}-\dfrac{5}{6};$ LCD $=6$
$\dfrac{18}{6}-\dfrac{5}{6}$
$\dfrac{13}{6}$

6. $(-5\sqrt{6})(3\sqrt{12})$
$-15\sqrt{72}$
$-15\sqrt{36}\sqrt{2}$
$-15(6)\sqrt{2}$
$-90\sqrt{2}$

7. $(2\sqrt{x}-3)(\sqrt{x}+4)$
$2\sqrt{x^2}+8\sqrt{x}-3\sqrt{x}-12$
$2x+5\sqrt{x}-12$

8. $(3\sqrt{2}-\sqrt{6})(\sqrt{2}+4\sqrt{6})$
$3\sqrt{4}+12\sqrt{12}-\sqrt{12}-4\sqrt{36}$
$3(2)+11\sqrt{12}-4(6)$
$6+11(2\sqrt{3})-24$
$-18+22\sqrt{3}$

9. $(2x-1)(x^2+6x-4)$
$2x^3+12x^2-8x-x^2-6x+4$
$2x^3+11x^2-14x+4$

10. $\dfrac{x^2-x}{x+5}\bullet\dfrac{x^2+5x+4}{x^4-x^2}$
$\dfrac{x(x-1)}{x+5}\bullet\dfrac{(x+4)(x+1)}{x^2(x+1)(x-1)}$
$\dfrac{x+4}{x(x+5)}$

11. $\dfrac{16x^2y}{24xy^3}\div\dfrac{9xy}{8x^2y^2}$
$\dfrac{16x^2y}{24xy^3}\bullet\dfrac{8x^2y^2}{9xy}$
$\dfrac{16(8)x^4y^3}{24(9)x^2y^4}$
$\dfrac{16x^2}{27y}$

12. $\frac{x+3}{10}+\frac{2x+1}{15}-\frac{x-2}{18}$; LCD = 90

$\frac{9}{9}\bullet\frac{(x+3)}{10}+\frac{6}{6}\bullet\frac{(2x+1)}{15}-\frac{5}{5}\bullet\frac{(x-2)}{18}$

$\frac{9(x+3)+6(2x+1)-5(x-2)}{90}$

$\frac{9x+27+12x+6-5x+10}{90}$

$\frac{16x+43}{90}$

13. $\frac{7}{12ab}-\frac{11}{15a^2}$; LCD = $60a^2b$

$\frac{7}{12ab}\bullet\frac{5a}{5a}-\frac{11}{15a^2}\bullet\frac{4b}{4b}$

$\frac{35a}{60a^2b}-\frac{44b}{60a^2b}$

$\frac{35a-44b}{60a^2b}$

14. $\frac{8}{x^2-4x}+\frac{2}{x}$

$\frac{8}{x(x-4)}+\frac{2}{x}$; LCD = $x(x-4)$

$\frac{8}{x(x-4)}+\frac{2}{x}\bullet\frac{(x-4)}{(x-4)}$

$\frac{8+2(x-4)}{x(x-4)}$

$\frac{8+2x-8}{x(x-4)}$

$\frac{2x}{x(x-4)}$

$\frac{2}{x-4}$

15.

$$\begin{array}{r|l} & 2x^2-x-4 \\ \hline 4x-1 & 8x^3-6x^2-15x+4 \\ & 8x^3-2x^2 \\ \hline & -4x^2-15x \\ & -4x^2+\ \ x \\ \hline & -16x+4 \\ & -16x+4 \\ \hline & 0 \end{array}$$

$2x^2-x-4$

16. $\frac{\frac{5}{x^2}-\frac{3}{x}}{\frac{1}{y}+\frac{2}{y^2}}$

$\frac{x^2y^2}{x^2y^2}\bullet\frac{\left(\frac{5}{x^2}-\frac{3}{x}\right)}{\left(\frac{1}{y}+\frac{2}{y^2}\right)}$

$\frac{x^2y^2\left(\frac{5}{x^2}\right)-x^2y^2\left(\frac{3}{x}\right)}{x^2y^2\left(\frac{1}{y}\right)+x^2y^2\left(\frac{2}{y^2}\right)}$

$\frac{5y^2-3xy^2}{x^2y+2x^2}$

17. $\frac{\frac{2}{x}-3}{\frac{3}{y}+4}$

$\frac{xy}{xy}\bullet\frac{\frac{2}{x}-3}{\frac{3}{y}+4}$

$\frac{xy\left(\frac{2}{x}-3\right)}{xy\left(\frac{3}{y}+4\right)}$

$\frac{xy\left(\frac{2}{x}\right)-xy\left(3\right)}{xy\left(\frac{3}{y}\right)+xy\left(4\right)}$

$$\frac{2y-3xy}{3x+4xy}$$

18. $$\frac{2-\frac{1}{n-2}}{3+\frac{4}{n+3}}$$

$$\frac{(n-2)(n+3)}{(n-2)(n+3)}\bullet\frac{\left(2-\frac{1}{n-2}\right)}{\left(3+\frac{4}{n+3}\right)}$$

$$\frac{2(n-2)(n+3)-(n+3)}{3(n-2)(n+3)+4(n-2)}$$

$$\frac{(n+3)[2(n-2)-1]}{(n-2)[3(n+3)+4]}$$

$$\frac{(n+3)(2n-4-1)}{(n-2)(3n+9+4)}$$

$$\frac{(n+3)(2n-5)}{(n-2)(3n+13)}$$

19. $$\frac{3a}{2-\frac{1}{a}}-1$$

$$\frac{a}{a}\bullet\frac{3a}{\left(2-\frac{1}{a}\right)}-1$$

$$\frac{3a^2}{2a-1}-1$$

$$\frac{3a^2}{2a-1}-\frac{1(2a-1)}{2a-1}$$

$$\frac{3a^2-1(2a-1)}{2a-1}$$

$$\frac{3a^2-2a+1}{2a-1}$$

20. $20x^2+7x-6$
$(5x-2)(4x+3)$

21. $16x^3+54$
$2(8x^3+27)$
$2(2x+3)(4x^2-6x+9)$

22. $4x^4-25x^2+36$
$(4x^2-9)(x^2-4)$
$(2x+3)(2x-3)(x+2)(x-2)$

23. $12x^3-52x^2-40x$
$4x(3x^2-13x-10)$
$4x(3x+2)(x-5)$

24. $xy-6x+3y-18$
$x(y-6)+3(y-6)$
$(y-6)(x+3)$

25. $10+9x-9x^2$
$(5-3x)(2+3x)$

26. $\left(\frac{2}{3}\right)^{-4}=\left(\frac{3}{2}\right)^{4}=\frac{3^4}{2^4}=\frac{81}{16}$

27. $$\frac{3}{\left(\frac{4}{3}\right)^{-1}}$$

$$3\left(\frac{4}{3}\right)^{1}=4$$

28. $\sqrt[3]{\frac{-27}{64}}=-\frac{3}{4}$

29. $-\sqrt{0.09}=-0.3$

30. $(27)^{-\frac{4}{3}}=\frac{1}{27^{\frac{4}{3}}}=\frac{1}{(27^{\frac{1}{3}})^4}=\frac{1}{3^4}=\frac{1}{81}$

31. $4^0+4^{-1}+4^{-2}$

$$1+\frac{1}{4}+\frac{1}{4^2}$$

$$1+\frac{1}{4}+\frac{1}{16}$$

$$\frac{16}{16}+\frac{4}{16}+\frac{1}{16}$$

$$\frac{21}{16}$$

32. $\left(\dfrac{3^{-1}}{2^{-3}}\right)^{-2} = \dfrac{3^2}{2^6} = \dfrac{9}{64}$

33. $(2^{-3} - 3^{-2})^{-1}$

$\left(\dfrac{1}{2^3} - \dfrac{1}{3^2}\right)^{-1}$

$\left(\dfrac{1}{8} - \dfrac{1}{9}\right)^{-1}$

$\left(\dfrac{9}{72} - \dfrac{8}{72}\right)^{-1}$

$\left(\dfrac{1}{72}\right)^{-1}$

72

34. $\log_2 64 = x$

$2^x = 64$

$2^x = 2^6$

$x = 6$

35. $\log_3 \left(\dfrac{1}{9}\right) = x$

$3^x = \dfrac{1}{9}$

$3^x = \dfrac{1}{3^2}$

$3^x = 3^{-2}$

$x = -2$

36. $(-3x^{-1}y^2)(4x^{-2}y^{-3})$

$-12x^{-3}y^{-1}$

$\dfrac{-12}{x^3y}$

37. $\dfrac{48x^{-4}y^2}{6xy}$

$8x^{-4-1}y^{2-1}$

$8x^{-5}y^1$

$\dfrac{8y}{x^5}$

38. $\left(\dfrac{27a^{-4}b^{-3}}{-3a^{-1}b^{-4}}\right)^{-1}$

$(-9a^{-3}b^1)^{-1}$

$(-9)^{-1}a^3b^{-1}$

$-\dfrac{a^3}{9b}$

39. $\sqrt{80}$

$\sqrt{16}\sqrt{5}$

$4\sqrt{5}$

40. $-2\sqrt{54}$

$-2\sqrt{9}\sqrt{6}$

$-2(3)\sqrt{6}$

$-6\sqrt{6}$

41. $\sqrt{\dfrac{75}{81}} = \dfrac{\sqrt{75}}{\sqrt{81}} = \dfrac{\sqrt{25}\sqrt{3}}{9} = \dfrac{5\sqrt{3}}{9}$

42. $\dfrac{4\sqrt{6}}{3\sqrt{8}} = \dfrac{4}{3}\sqrt{\dfrac{6}{8}} = \dfrac{4}{3}\sqrt{\dfrac{3}{4}}$

$\dfrac{4\sqrt{3}}{3\sqrt{4}} = \dfrac{4\sqrt{3}}{3(2)} = \dfrac{2\sqrt{3}}{3}$

43. $\sqrt[3]{56}$

$\sqrt[3]{8}\sqrt[3]{7}$

$2\sqrt[3]{7}$

44. $\dfrac{\sqrt[3]{3}}{\sqrt[3]{4}} = \dfrac{\sqrt[3]{3}}{\sqrt[3]{4}} \bullet \dfrac{\sqrt[3]{2}}{\sqrt[3]{2}} = \dfrac{\sqrt[3]{6}}{\sqrt[3]{8}} = \dfrac{\sqrt[3]{6}}{2}$

45. $4\sqrt{52x^3y^2}$

$4\sqrt{4x^2y^2}\sqrt{13x}$

$4(2xy)\sqrt{13x}$

$8xy\sqrt{13x}$

46. $\sqrt{\dfrac{2x}{3y}} = \dfrac{\sqrt{2x}}{\sqrt{3y}} = \dfrac{\sqrt{2x}}{\sqrt{3y}} \bullet \dfrac{\sqrt{3y}}{\sqrt{3y}} = \dfrac{\sqrt{6xy}}{3y}$

47. $-3\sqrt{24} + 6\sqrt{54} - \sqrt{6}$

$-3\sqrt{4}\sqrt{6} + 6\sqrt{9}\sqrt{6} - \sqrt{6}$

$-3(2)\sqrt{6} + 6(3)\sqrt{6} - \sqrt{6}$

$-6\sqrt{6}+18\sqrt{6}-\sqrt{6}$
$11\sqrt{6}$

48. $\frac{\sqrt{8}}{3}-\frac{3\sqrt{18}}{4}-\frac{5\sqrt{50}}{2}$
$\frac{2\sqrt{2}}{3}-\frac{9\sqrt{2}}{4}-\frac{25\sqrt{2}}{2}$; LCD = 12
$\frac{4}{4}\bullet\frac{(2\sqrt{2})}{3}-\frac{3}{3}\bullet\frac{(9\sqrt{2})}{4}-\frac{6}{6}\bullet\frac{(25\sqrt{2})}{2}$
$\frac{8\sqrt{2}}{12}-\frac{27\sqrt{2}}{12}-\frac{150\sqrt{2}}{12}$
$\frac{-169\sqrt{2}}{12}$

49. $8\sqrt[3]{3}-6\sqrt[3]{24}-4\sqrt[3]{81}$
$8\sqrt[3]{3}-6\sqrt[3]{8}\sqrt[3]{3}-4\sqrt[3]{27}\sqrt[3]{3}$
$8\sqrt[3]{3}-6(2)\sqrt[3]{3}-4(3)\sqrt[3]{3}$
$8\sqrt[3]{3}-12\sqrt[3]{3}-12\sqrt[3]{3}$
$-16\sqrt[3]{3}$

50. $\frac{\sqrt{3}}{\sqrt{6}-2\sqrt{2}}$

$\frac{\sqrt{3}}{(\sqrt{6}-2\sqrt{2})}\bullet\frac{(\sqrt{6}+2\sqrt{2})}{(\sqrt{6}+2\sqrt{2})}$

$\frac{\sqrt{18}+2\sqrt{6}}{\sqrt{36}-4\sqrt{4}}=\frac{3\sqrt{2}+2\sqrt{6}}{6-4(2)}$

$\frac{3\sqrt{2}+2\sqrt{6}}{6-8}=\frac{3\sqrt{2}+2\sqrt{6}}{-2}$

$\frac{-3\sqrt{2}-2\sqrt{6}}{2}$

51. $\frac{3\sqrt{5}-\sqrt{3}}{2\sqrt{3}+\sqrt{7}}$

$\frac{(3\sqrt{5}-\sqrt{3})}{(2\sqrt{3}+\sqrt{7})}\bullet\frac{(2\sqrt{3}-\sqrt{7})}{(2\sqrt{3}-\sqrt{7})}$

$\frac{6\sqrt{15}-3\sqrt{35}-2\sqrt{9}+\sqrt{21}}{4\sqrt{9}-\sqrt{49}}$

$\frac{6\sqrt{15}-3\sqrt{35}-2(3)+\sqrt{21}}{4(3)-7}$

$\frac{6\sqrt{15}-3\sqrt{35}-6+\sqrt{21}}{5}$

52. $\frac{(0.00016)(300)(0.028)}{0.064}$
$\frac{(1.6)(10)^{-4}(3)(10)^{2}(2.8)(10)^{-2}}{(6.4)(10)^{-2}}$
$\frac{(1.6)(3)(2.8)}{6.4}\bullet\frac{(10)^{-4}(10)^{2}(10)^{-2}}{(10)^{-2}}$
$(2.1)(10)^{-4+2-2-(-2)}$
$(2.1)(10)^{-2}$
0.021

53. $\frac{0.00072}{0.0000024}$
$\frac{(7.2)(10)^{-4}}{(2.4)(10)^{-6}}=3(10)^{-4-(-6)}$
$3(10)^{2}=300$

54. $\sqrt{0.00000009}$
$\sqrt{(9)(10)^{-8}}=\sqrt{9}\sqrt{(10)^{-8}}$
$3(10)^{-4}=0.0003$

55. $(5-2i)(4+6i)$
$20+30i-8i-12i^{2}$
$20+22i+12$
$32+22i$

56. $(-3-i)(5-2i)$
$-15+6i-5i+2i^{2}$
$-15+i-2$
$-17+i$

57. $\frac{5}{4i}=\frac{5}{4i}\bullet\frac{(-i)}{(-i)}=\frac{-5i}{-4i^{2}}=\frac{-5i}{4}$
$0-\frac{5}{4}i$

58. $\dfrac{-1+6i}{7-2i}$

$\dfrac{(-1+6i)}{(7-2i)} \bullet \dfrac{(7+2i)}{(7+2i)}$

$\dfrac{-7-2i+42i+12i^2}{49-4i^2}$

$\dfrac{-7+40i-12}{49+4}$

$\dfrac{-19+40i}{53} = -\dfrac{19}{53} + \dfrac{40}{53}i$

59. (2, − 3) and (− 1, 7)

$m = \dfrac{7-(-3)}{-1-2} = \dfrac{10}{-3} = -\dfrac{10}{3}$

60. $4x - 7y = 9$

$-7y = -4x + 9$

$-\dfrac{1}{7}(-7y) = -\dfrac{1}{7}(-4x+9)$

$y = \dfrac{4}{7}x - \dfrac{9}{7}$

$m = \dfrac{4}{7}$

61. (4, 5) and (− 2, 1)

$d = \sqrt{(-2-4)^2 + (1-5)^2}$

$d = \sqrt{(-6)^2 + (-4)^2}$

$d = \sqrt{36+16} = \sqrt{52}$

$d = \sqrt{4}\sqrt{13} = 2\sqrt{13}$

62. (3, − 1) and (7, 4)

$m = \dfrac{4-(-1)}{7-3} = \dfrac{5}{4}$

$y - 4 = \dfrac{5}{4}(x-7)$

$4(y-4) = 4\left[\dfrac{5}{4}(x-7)\right]$

$4y - 16 = 5(x-7)$

$4y - 16 = 5x - 35$

$-16 = 5x - 4y - 35$

$19 = 5x - 4y$

$5x - 4y = 19$

63. $3x - 4y = 6$; (− 3, − 2)

$-4y = -3x + 6$

$-\dfrac{1}{4}(-4y) = -\dfrac{1}{4}(-3x+6)$

$y = \dfrac{3}{4}x - \dfrac{3}{2}$

Perpendicular lines have slopes that are negative reciprocals.

$m = -\dfrac{4}{3}$; (− 3, − 2)

$y - (-2) = -\dfrac{4}{3}[x - (-3)]$

$y + 2 = -\dfrac{4}{3}(x+3)$

$3(y+2) = 3\left[-\dfrac{4}{3}(x+3)\right]$

$3y + 6 = -4(x+3)$

$3y + 6 = -4x - 12$

$4x + 3y + 6 = -12$

$4x + 3y = -18$

64. $x^2 + 4x + y^2 - 12y + 31 = 0$

$x^2 + 4x + 4 + y^2 - 12y + 36 = -31 + 4 + 36$

$(x+2)^2 + (y-6)^2 = 9$

center (− 2, 6); $r = \sqrt{9} = 3$

65. $y = x^2 + 10x + 21$

$y = x^2 + 10x + 25 - 25 + 21$

$y = (x+5)^2 - 4$

vertex : (− 5, − 4)

66. $x^2 + 4y^2 = 16$

If $y = 0$,

$x^2 + 4(0)^2 = 16$

$x^2 = 16$

$x = \pm\sqrt{16} = \pm 4$

(4, 0), (− 4, 0)

The length of the major axis = 8.

67. $f(x) = -2x - 4$

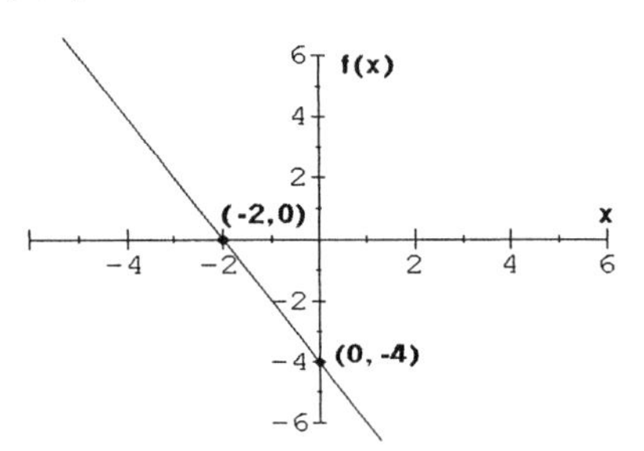

68. $f(x) = -2x^2 - 2$

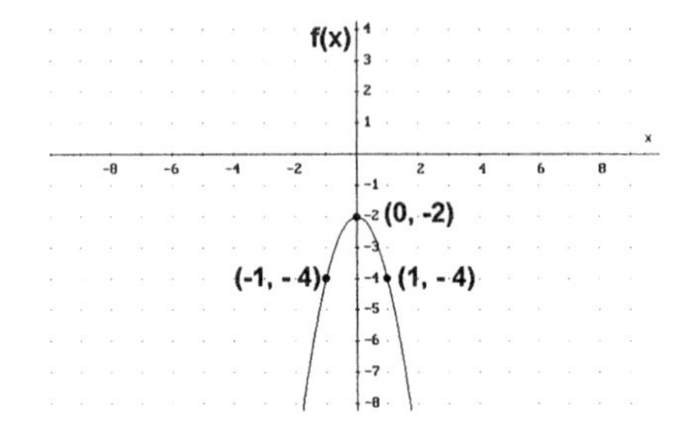

69. $f(x) = x^2 - 2x - 2$

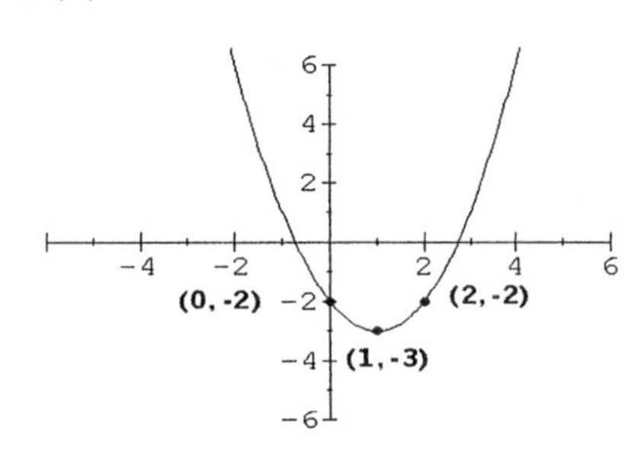

70. $f(x) = \sqrt{x+1} + 2$

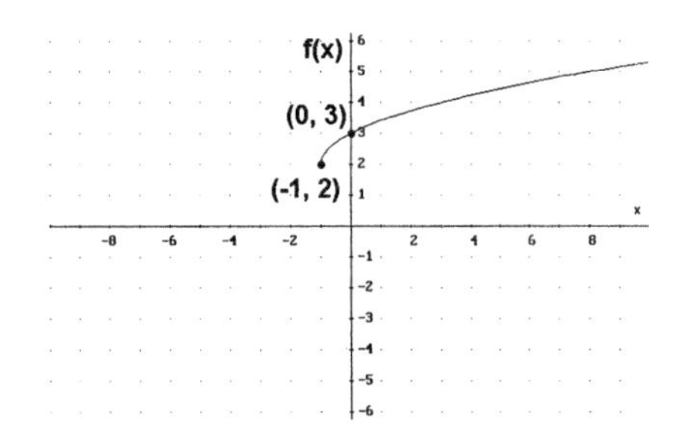

71. $f(x) = 2x^2 + 8x + 9$

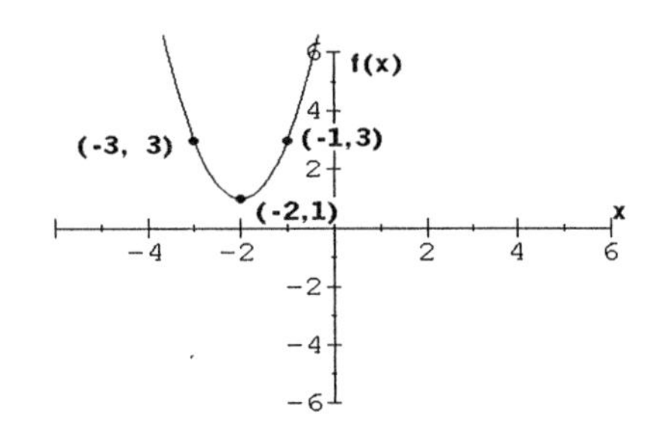

72. $f(x) = -|x-2| + 1$

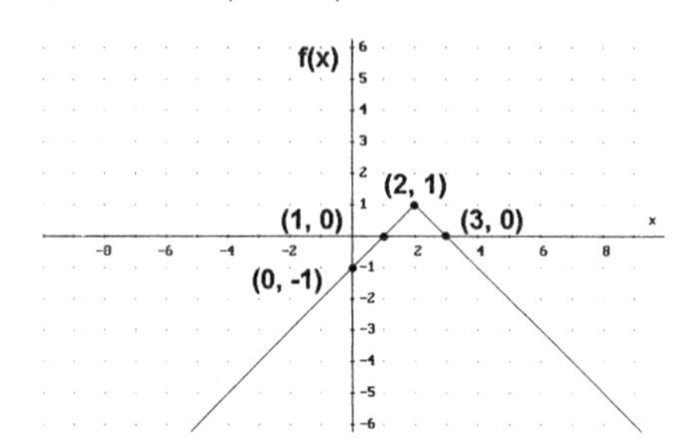

73. $f(x) = 2^x + 2$

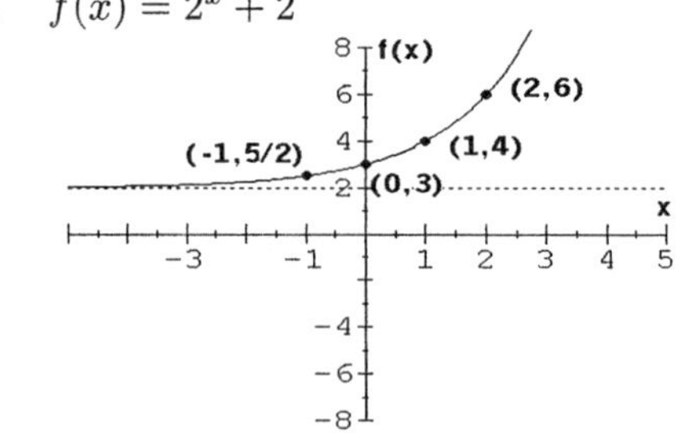

74. $f(x) = \log_2(x-2)$

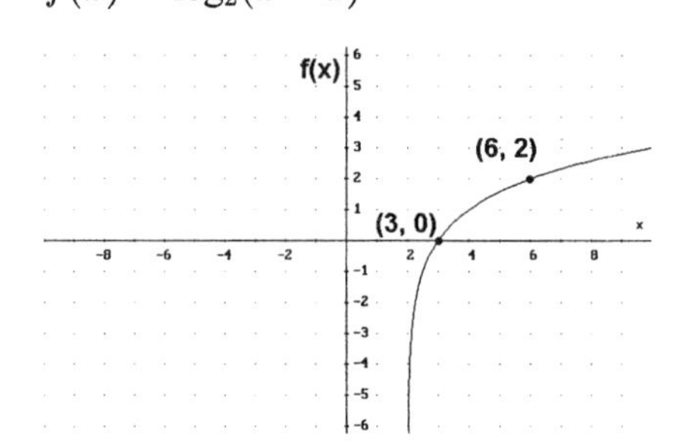

75. $f(x) = -x(x+1)(x-2)$

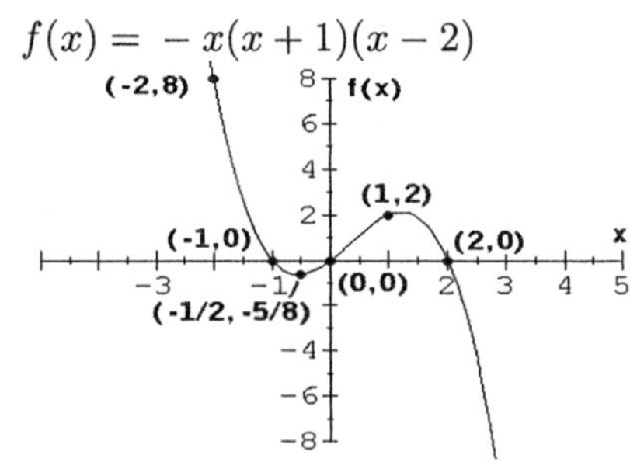

76. $f(x) = \dfrac{-x}{x+2}$

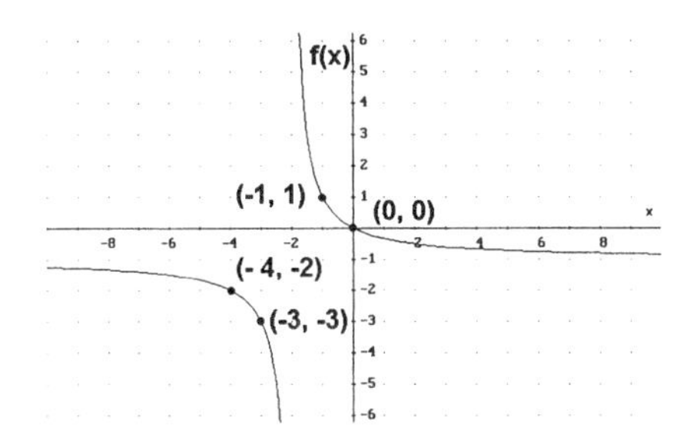

77. $f(x) = x - 3; \qquad g(x) = 2x^2 - x - 1$

$(g \circ f)(x) = g(f(x)) = g(x-3)$
$= 2(x-3)^2 - (x-3) - 1$
$= 2(x^2 - 6x + 9) - x + 3 - 1$
$= 2x^2 - 12x + 18 - x + 2$
$= 2x^2 - 13x + 20$

$(f \circ g)(x) = f(g(x)) = f(2x^2 - x - 1)$
$= (2x^2 - x - 1) - 3$
$= 2x^2 - x - 4$

78. $f(x) = 3x - 7$

$y = 3x - 7$

Interchange x and y, then solve for y.

$x = 3y - 7$

$x + 7 = 3y$

$\frac{1}{3}(x+7) = \frac{1}{3}(3y)$

$\frac{x+7}{3} = y$

$f^{-1}(x) = \frac{x+7}{3}$

79. $f(x) = -\frac{1}{2}x + \frac{2}{3}$

$y = -\frac{1}{2}x + \frac{2}{3}$

Interchange x and y, then solve for y.

$x = -\frac{1}{2}y + \frac{2}{3}$

$6(x) = 6\left(-\frac{1}{2}y + \frac{2}{3}\right)$

$6x = -3y + 4$

$3y + 6x = 4$

$\frac{1}{3}\left(3y + 6x\right) = \left(\frac{1}{3}\right)4$

$y + 2x = \frac{4}{3}$

$y = -2x + \frac{4}{3}$

$f^{-1}(x) = -2x + \frac{4}{3}$

80. $y = kx$

$2 = k\left(-\frac{2}{3}\right)$

$2 = -\frac{2}{3}k$

$-\frac{3}{2}(2) = -\frac{3}{2}\left(-\frac{2}{3}k\right)$

$-3 = k$

81. $y = \frac{k}{x^2}$

$4 = \frac{k}{(3)^2}$

$4 = \frac{k}{9}$

$36 = k$

$y = \frac{36}{(6)^2}$

$y = 1$

82. $V = \frac{k}{P}$

$15 = \frac{k}{20}$

$300 = k$

$V = \frac{300}{P}$

$V = \frac{300}{25} = 12$ cubic centimeters

83. $3(2x-1) - 2(5x+1) = 4(3x+4)$

$6x - 3 - 10x - 2 = 12x + 16$

$-4x - 5 = 12x + 16$

$-16x = 21$

$x = -\frac{21}{16}$

The solution set is $\left\{-\frac{21}{16}\right\}$.

84. $n+\frac{3n-1}{9}-4=\frac{3n+1}{3}$

$9\left(n+\frac{3n-1}{9}-4\right)=9\left(\frac{3n+1}{3}\right)$

$9n+3n-1-36=3(3n+1)$

$12n-37=9n+3$

$3n=40$

$n=\frac{40}{3}$

The solution set is $\left\{\frac{40}{3}\right\}$.

85. $0.92+0.9(x-0.3)=2x-5.95$

$100[0.92+0.9(x-0.3)]=100(2x-5.95)$

$92+90(x-0.3)=200x-595$

$92+90x-27=200x-595$

$65+90x=200x-595$

$-110x=-660$

$x=6$

The solution set is $\{6\}$.

86. $|4x-1|=11$

$4x-1=11$ or $4x-1=-11$

$4x=12$ or $4x=-10$

$x=3$ or $x=-\frac{10}{4}=-\frac{5}{2}$

The solution set is $\left\{-\frac{5}{2},3\right\}$.

87. $3x^2=7x$

$3x^2-7x=0$

$x(3x-7)=0$

$x=0$ or $3x-7=0$

$x=0$ or $3x=7$

$x=0$ or $x=\frac{7}{3}$

The solution set is $\left\{0,\frac{7}{3}\right\}$.

88. $x^3-36x=0$

$x(x^2-36)=0$

$x(x+6)(x-6)=0$

$x=0$ or $x+6=0$ or $x-6=0$

$x=0$ or $x=-6$ or $x=6$

The solution set is $\{-6,0,6\}$.

89. $30x^2+13x-10=0$

$(6x+5)(5x-2)=0$

$6x+5=0$ or $5x-2=0$

$6x=-5$ or $5x=2$

$x=-\frac{5}{6}$ or $x=\frac{2}{5}$

The solution set is $\left\{-\frac{5}{6},\frac{2}{5}\right\}$.

90. $8x^3+12x^2-36x=0$

$4x(2x^2+3x-9)=0$

$4x(2x-3)(x+3)=0$

$4x=0$ or $2x-3=0$ or $x+3=0$

$x=0$ or $2x=3$ or $x=-3$

$x=0$ or $x=\frac{3}{2}$ or $x=-3$

The solution set is $\left\{-3,0,\frac{3}{2}\right\}$.

91. $x^4+8x^2-9=0$

$(x^2+9)(x^2-1)=0$

$x^2+9=0$ or $x^2-1=0$

$x^2=-9$ or $x^2=1$

$x=\pm\sqrt{-9}$ or $x=\pm\sqrt{1}$

$x=\pm 3i$ or $x=\pm 1$

The solution set is $\{\pm 3i,\ \pm 1\}$.

92. $(n+4)(n-6)=11$

$n^2-2n-24=11$

$n^2-2n-35=0$

$(n-7)(n+2)=0$

$n-7=0$ or $n+5=0$

$n=7$ or $n=-5$

The solution set is $\{-5,7\}$.

93. $2-\frac{3x}{x-4}=\frac{14}{x+7};\ x\neq 4,\ x\neq -7$

$(x\text{-}4)(x\text{+}7)\left(2-\frac{3x}{x\text{-}4}\right)=(x\text{-}4)(x\text{+}7)\left(\frac{14}{x\text{+}7}\right)$

$2(x-4)(x+7)-3x(x+7)=14(x-4)$

$2(x^2+3x-28)-3x^2-21x=14x-56$

$2x^2+6x-56-3x^2-21x=14x-56$

$-x^2-15x-56=14x-56$

$0=x^2+29x$

$0=x(x+29)$

$x=0$ or $x+29=0$

$x=0$ or $x=-29$

The solution set is $\{-29,0\}$.

94. $\frac{2n}{6n^2+7n-3} - \frac{n-3}{3n^2+11n-4} = \frac{5}{2n^2+11n+12}$

$\frac{2n}{(3n-1)(2n+3)} - \frac{n-3}{(3n-1)(n+4)} = \frac{5}{(2n+3)(n+4)}$

$(3n-1)(2n+3)(n+4)\left[\frac{2n}{(3n-1)(2n+3)} - \frac{n-3}{(3n-1)(n+4)}\right] = (3n-1)(2n+3)(n+4)\left[\frac{5}{(2n+3)(n+4)}\right]$

$2n(n+4) - (2n+3)(n-3) = 5(3n-1)$

$2n^2 + 8n - (2n^2 - 3n - 9) = 15n - 5$

$2n^2 + 8n - 2n^2 + 3n + 9 = 15n - 5$

$11n + 9 = 15n - 5$

$-4n = -14$

$n = \frac{-14}{-4} = \frac{7}{2}$

The solution set is $\left\{\frac{7}{2}\right\}$.

95. $\sqrt{3y} - y = -6$

$\sqrt{3y} = y - 6$

$(\sqrt{3y})^2 = (y-6)^2$

$3y = y^2 - 12y + 36$

$0 = y^2 - 15y + 36$

$0 = (y-12)(y-3)$

$y - 12 = 0$ or $y - 3 = 0$

$y = 12$ or $y = 3$

Checking $y = 12$

$\sqrt{3(12)} - 12 \stackrel{?}{=} -6$

$\sqrt{36} - 12 \stackrel{?}{=} -6$

$6 - 12 \stackrel{?}{=} -6$

$-6 = -6$

Checking $y = 3$

$\sqrt{3(3)} - 3 \stackrel{?}{=} -6$

$\sqrt{9} - 3 \stackrel{?}{=} -6$

$3 - 3 \stackrel{?}{=} -6$

$0 \neq -6$

The solution set is $\{12\}$.

96. $\sqrt{x+19} - \sqrt{x+28} = -1$

$\sqrt{x+19} = \sqrt{x+28} - 1$

$(\sqrt{x+19})^2 = (\sqrt{x+28} - 1)^2$

$x+19 = (\sqrt{x+28})^2 - 2(\sqrt{x+28})(-1) + (-1)^2$

$x + 19 = x + 28 - 2\sqrt{x+28} + 1$

$-10 = -2\sqrt{x+28}$

$5 = \sqrt{x+28}$

$(5)^2 = (\sqrt{x+28})^2$

$25 = x + 28$

$-3 = x$

Check

$\sqrt{-3+19} - \sqrt{-3+28} \stackrel{?}{=} -1$

$\sqrt{16} - \sqrt{25} \stackrel{?}{=} -1$

$4 - 5 \stackrel{?}{=} -1$

$-1 = -1$

The solution set is $\{-3\}$.

97. $(3x-1)^2 = 45$

$3x - 1 = \pm\sqrt{45}$

$3x - 1 = \pm 3\sqrt{5}$

$3x = 1 \pm 3\sqrt{5}$

$x = \frac{1 \pm 3\sqrt{5}}{3}$

The solution set is $\left\{\frac{1 \pm 3\sqrt{5}}{3}\right\}$.

98. $(2x+5)^2 = -32$

$2x + 5 = \pm\sqrt{-32}$

$2x + 5 = \pm 4i\sqrt{2}$

$2x = -5 \pm 4i\sqrt{2}$

$x = \frac{-5 \pm 4i\sqrt{2}}{2}$

The solution set is $\left\{\frac{-5 \pm 4i\sqrt{2}}{2}\right\}$.

99. $2x^2 - 3x + 4 = 0$

$x = \dfrac{-(-3) \pm \sqrt{(-3)^2 - 4(2)(4)}}{2(2)}$

$x = \dfrac{3 \pm \sqrt{9 - 32}}{4}$

$x = \dfrac{3 \pm \sqrt{-23}}{4}$

$x = \dfrac{3 \pm i\sqrt{23}}{4}$

The solution set is $\left\{\dfrac{3 \pm i\sqrt{23}}{4}\right\}$.

100. $3n^2 - 6n + 2 = 0$

$n = \dfrac{-(-6) \pm \sqrt{(-6)^2 - 4(3)(2)}}{2(3)}$

$n = \dfrac{6 \pm \sqrt{36 - 24}}{6}$

$n = \dfrac{6 \pm \sqrt{12}}{6} = \dfrac{6 \pm 2\sqrt{3}}{6}$

$n = \dfrac{2(3 \pm \sqrt{3})}{6} = \dfrac{3 \pm \sqrt{3}}{3}$

The solution set is $\left\{\dfrac{3 \pm \sqrt{3}}{3}\right\}$.

101. $\dfrac{5}{n-3} - \dfrac{3}{n+3} = 1;\ n \neq 3, n \neq -3$

$(n\text{-}3)(n\text{+}3)\left(\dfrac{5}{n-3} - \dfrac{3}{n+3}\right) = (n\text{-}3)(n\text{+}3)(1)$

$5(n+3) - 3(n-3) = n^2 - 9$

$5n + 15 - 3n + 9 = n^2 - 9$

$2n + 24 = n^2 - 9$

$0 = n^2 - 2n - 33$

$n = \dfrac{-(-2) \pm \sqrt{(-2)^2 - 4(1)(-33)}}{2(1)}$

$n = \dfrac{2 \pm \sqrt{4 + 132}}{2}$

$n = \dfrac{2 \pm \sqrt{136}}{2}$

$n = \dfrac{2 \pm 2\sqrt{34}}{2}$

$n = \dfrac{2(1 \pm \sqrt{34})}{2}$

$n = 1 \pm \sqrt{34}$

The solution set is $\{1 \pm \sqrt{34}\}$.

102. $12x^4 - 19x^2 + 5 = 0$

$(4x^2 - 5)(3x^2 - 1) = 0$

$4x^2 - 5 = 0$ or $3x^2 - 1 = 0$

$4x^2 = 5$ or $3x^2 = 1$

$x^2 = \dfrac{5}{4}$ or $x^2 = \dfrac{1}{3}$

$x = \pm\sqrt{\dfrac{5}{4}}$ or $x = \pm\sqrt{\dfrac{1}{3}}$

$x = \pm\dfrac{\sqrt{5}}{2}$ or $x = \pm\dfrac{1}{\sqrt{3}} \bullet \dfrac{\sqrt{3}}{\sqrt{3}}$

$x = \pm\dfrac{\sqrt{5}}{2}$ or $x = \pm\dfrac{\sqrt{3}}{3}$

The solution set is $\left\{\pm\dfrac{\sqrt{5}}{2}, \pm\dfrac{\sqrt{3}}{3}\right\}$.

103. $2x^2 + 5x + 5 = 0$

$x = \dfrac{-5 \pm \sqrt{(5)^2 - 4(2)(5)}}{2(2)}$

$x = \dfrac{-5 \pm \sqrt{25 - 40}}{4}$

$x = \dfrac{-5 \pm \sqrt{-15}}{4}$

$x = \dfrac{-5 \pm i\sqrt{15}}{4}$

The solution set is $\left\{\dfrac{-5 \pm i\sqrt{15}}{4}\right\}$.

104. $(3x - 5)(4x + 1) = 0$

$3x - 5 = 0$ or $4x + 1 = 0$

$3x = 5$ or $4x = -1$

$x = \dfrac{5}{3}$ or $x = -\dfrac{1}{4}$

The solution set is $\left\{-\dfrac{1}{4}, \dfrac{5}{3}\right\}$.

105. $|2x+9| = -4$
Absolute value cannot be a negative number. Therefore, the solution set is Ø.

106. $16^x = 64$
$(2^4)^x = 2^6$
$2^{4x} = 2^6$
$4x = 6$
$x = \frac{6}{4} = \frac{3}{2}$
The solution set is $\left\{\frac{3}{2}\right\}$.

107. $\log_3 x = 4$
$3^4 = x$
$81 = x$
The solution set is $\{81\}$.

108. $\log_{10} x + \log_{10} 25 = 2$
$\log_{10} 25x = 2$
$10^2 = 25x$
$100 = 25x$
$4 = x$
The solution set is $\{4\}$.

109. $\ln(3x-4) - \ln(x+1) = \ln 2$
$\ln \frac{3x-4}{x+1} = \ln 2$
$\frac{3x-4}{x+1} = 2; x \neq -1$
$3x - 4 = 2(x+1)$
$3x - 4 = 2x + 2$
$x - 4 = 2$
$x = 6$
The solution set is $\{6\}$.

110. $27^{4x} = 9^{x+1}$
$(3^3)^{4x} = (3^2)^{x+1}$
$3^{12x} = 3^{2x+2}$
$12x = 2x + 2$
$10x = 2$
$x = \frac{2}{10} = \frac{1}{5}$
The solution set is $\left\{\frac{1}{5}\right\}$.

111. $-5(y-1) + 3 > 3y - 4 - 4y$
$-5y + 5 + 3 > -y - 4$
$-5y + 8 > -y - 4$
$-4y > -12$
$-\frac{1}{4}(-4y) < -\frac{1}{4}(-12)$
$y < 3$
The solution set is $(-\infty, 3)$.

112. $0.06x + 0.08(250 - x) \geq 19$
$100[0.06x + 0.08(250 - x)] \geq 100(19)$
$6x + 8(250 - x) \geq 1900$
$6x + 2000 - 8x \geq 1900$
$-2x \geq -100$
$-\frac{1}{2}(-2x) \leq -\frac{1}{2}(-100)$
$x \leq 50$
The solution set is $(-\infty, 50]$.

113. $|5x - 2| > 13$
$5x - 2 < -13$ or $5x - 2 > 13$
$5x < -11$ or $5x > 15$
$x < -\frac{11}{5}$ or $x > 3$
The solution set is
$\left(-\infty, -\frac{11}{5}\right) \cup (3, \infty)$.

114. $|6x + 2| < 8$
$-8 < 6x + 2 < 8$
$-10 < 6x < 6$
$-\frac{10}{6} < x < \frac{6}{6}$
$-\frac{5}{3} < x < 1$
The solution set is $\left(-\frac{5}{3}, 1\right)$.

115. $\frac{x-2}{5} - \frac{3x-1}{4} \leq \frac{3}{10}$

$20\left(\frac{x-2}{5} - \frac{3x-1}{4}\right) \leq 20\left(\frac{3}{10}\right)$

$20\left(\frac{x-2}{5}\right) - 20\left(\frac{3x-1}{4}\right) \leq 6$
$4(x-2) - 5(3x-1) \leq 6$
$4x - 8 - 15x + 5 \leq 6$

$-11x - 3 \le 6$
$-11x \le 9$
$x \ge -\frac{9}{11}$
The solution set is $\left[-\frac{9}{11}, \infty\right)$.

116. $(x-2)(x+4) \le 0$
$(x-2)(x+4) = 0$
$x - 2 = 0$ or $x + 4 = 0$
$x = 2$ or $x = -4$
$-4 \quad 2$

Test Point	-5	0	3
$x-2$:	negative	negative	positive
$x+4$:	negative	positive	positive
product:	positive	negative	positive

The solution set is $[-4, 2]$.

117. $(3x-1)(x-4) > 0$
$(3x-1)(x-4) = 0$
$3x - 1 = 0$ or $x - 4 = 0$
$3x = 1$ or $x = 4$
$x = \frac{1}{3}$
$\frac{1}{3} \quad 4$

Test Point	0	1	5
$3x-1$:	negative	positive	positive
$x-4$:	negative	negative	positive
product:	positive	negative	positive

The solution set is
$\left(-\infty, \frac{1}{3}\right) \cup (4, \infty)$.

118. $x(x+5) < 24$
$x^2 + 5x < 24$
$x^2 + 5x - 24 < 0$
$x^2 + 5x - 24 = 0$
$(x+8)(x-3) = 0$
$x + 8 = 0$ or $x - 3 = 0$
$x = -8$ or $x = 3$
$-8 \quad 3$

Test Point	-9	0	4
$x+8$:	negative	positive	positive
$x-3$:	negative	negative	positive
product:	positive	negative	positive

The solution set is $(-8, 3)$.

119. $\frac{x-3}{x-7} \ge 0$
$x - 3 = 0$ or $x - 7 = 0$
$x = 3$ or $x = 7$
$3 \quad 7$

Test Point	2	5	8
$x-3$:	negative	positive	positive
$x-7$:	negative	negative	positive
quotient:	positive	negative	negative

The solution set is $(-\infty, 3] \cup (7, \infty)$.

120. $\frac{2x}{x+3} > 4; x \ne -3$

$\frac{2x}{x+3} - 4 > 0$

$\frac{2x}{x+3} - \frac{4(x+3)}{(x+3)} > 0$

$\frac{2x - 4(x+3)}{x+3} > 0$

$\frac{2x - 4x - 12}{x+3} > 0$

$\frac{-2x - 12}{x+3} > 0$

$-2x - 12 = 0$ or $x + 3 = 0$
$-2x = 12$ or $x = -3$
$x = -6$ or $x = -3$
$-6 \quad -3$

Test Point	-7	-5	0
$-2x-12$:	positive	negative	negative
$x+3$:	negative	negative	positive
quotient:	negative	positive	negative

The solution set is $(-6, -3)$.

Chapter 15 Systems of Equations: Matrices and Determinants

PROBLEM SET 15.1 Systems of Two Linear Equations: A Brief Review

1. $\begin{pmatrix} x+y=16 \\ y=x+2 \end{pmatrix}$
Substitute $x+2$ for y in the 1st equation.
$x+y=16$
$x+(x+2)=16$
$2x+2=16$
$2x=14$
$x=7$
$y=x+2$
$y=7+2$
$y=9$
The solution set is $\{(7, 9)\}$.

3. $\begin{pmatrix} x=3y-25 \\ 4x+5y=19 \end{pmatrix}$
Substitute $3y-25$ for x in the 2nd equation.
$4x+5y=19$
$4(3y-25)+5y=19$
$12y-100+5y=19$
$17y=119$
$y=7$
$x=3y-25$
$x=3(7)-25$
$x=-4$
The solution set is $\{(-4, 7)\}$.

5. $\begin{pmatrix} y=\frac{2}{3}x-1 \\ 5x-7y=9 \end{pmatrix}$
Substitute $\frac{2}{3}x-1$ for y in the 2nd equation.
$5x-7y=9$
$5x-7\left(\frac{2}{3}x-1\right)=9$
$5x-\frac{14}{3}x+7=9$
$15x-14x+21=27$
$x+21=27$
$x=6$
$y=\frac{2}{3}x-1$
$y=\frac{2}{3}(6)-1$
$y=3$
The solution set is $\{(6, 3)\}$.

7. $\begin{pmatrix} a=4b+13 \\ 3a+6b=-33 \end{pmatrix}$
Substitute $4b+13$ for a in the 2nd equation.
$3a+6b=-33$
$3(4b+13)+6b=-33$
$12b+39+6b=-33$
$18b=-72$
$b=-4$
$a=4b+13$
$a=4(-4)+13$
$a=-3$
Therefore, $a=-3$ and $b=-4$.

9. $\begin{pmatrix} 2x-3y=4 \\ y=\frac{2}{3}x-\frac{4}{3} \end{pmatrix}$
Substitute $\frac{2}{3}x-\frac{4}{3}$ for y in the 1st equation.
$2x-3y=4$
$2x-3\left(\frac{2}{3}x-\frac{4}{3}\right)=4$
$2x-2x+4=4$
$4=4$
$0=0$
The system is dependent. Let $x=k$, then $y=\frac{2}{3}k-\frac{4}{3}$.
The solution set is $\left\{\left(k, \frac{2}{3}k-\frac{4}{3}\right)\right\}$ where k is any real number.

11. $\begin{pmatrix} u=t-2 \\ t+u=12 \end{pmatrix}$
Substitute $t-2$ for u in the 2nd equation.
$t+u=12$
$t+t-2=12$
$2t=14$

$t = 7$
$u = t - 2$
$u = 7 - 2$
$u = 5$
Therefore, $u = 5$ and $t = 7$.

13. $\begin{pmatrix} 4x + 3y = -7 \\ 3x - 2y = 16 \end{pmatrix}$
$4x + 3y = -7$
$3y = -4x - 7$
$y = -\frac{4}{3}x - \frac{7}{3}$
Substitute into the 2nd equation.
$3x - 2y = 16$
$3x - 2\left(-\frac{4}{3}x - \frac{7}{3}\right) = 16$
$3x + \frac{8}{3}x + \frac{14}{3} = 16$
$9x + 8x + 14 = 48$
$17x = 34$
$x = 2$
$y = -\frac{4}{3}x - \frac{7}{3}$
$y = -\frac{4}{3}(2) - \frac{7}{3}$
$y = -5$
The solution set is $\{(2, -5)\}$.

15. $\begin{pmatrix} 5x - y = 4 \\ y = 5x + 9 \end{pmatrix}$
Substitute $5x + 9$ for y in the 1st equation.
$5x - y = 4$
$5x - (5x + 9) = 4$
$5x - 5x - 9 = 4$
$-9 \neq 4$
The system is inconsistent.
The solution set is $\emptyset$.

17. $\begin{pmatrix} 4x - 5y = 3 \\ 8x + 15y = -24 \end{pmatrix}$
$4x - 5y = 3$
$4x = 5y + 3$
$x = \frac{5}{4}y + \frac{3}{4}$
Substitute into the 2nd equation.
$8x + 15y = -24$
$8\left(\frac{5}{4}y + \frac{3}{4}\right) + 15y = -24$
$10y + 6 + 15y = -24$
$25y = -30$
$y = -\frac{6}{5}$
$x = \frac{5}{4}y + \frac{3}{4}$
$x = \frac{5}{4}\left(-\frac{6}{5}\right) + \frac{3}{4}$
$x = -\frac{3}{4}$
The solution set is $\left\{\left(-\frac{3}{4}, -\frac{6}{5}\right)\right\}$.

19. $\begin{pmatrix} 3x + 2y = 1 \\ 5x - 2y = 3 \end{pmatrix}$
$3x + 2y = 1$
$5x - 2y = 3$
$8x = 4$
$x = \frac{1}{2}$
$3x + 2y = 1$
$3\left(\frac{1}{2}\right) + 2y = 1$
$\frac{3}{2} + 2y = 1$
$2\left(\frac{3}{2}\right) + 2(2y) = 2(1)$
$3 + 4y = 2$
$4y = -1$
$y = -\frac{1}{4}$
The solution set is $\left\{\left(\frac{1}{2}, -\frac{1}{4}\right)\right\}$.

21. $\begin{pmatrix} x - 3y = -22 \\ 2x + 7y = 60 \end{pmatrix}$
Multiply equation 1 by -2 and then add the result to equation 2.
$-2x + 6y = 44$
$2x + 7y = 60$
$13y = 104$
$y = 8$
$x - 3y = -22$
$x - 3(8) = -22$
$x - 24 = -22$
$x = 2$
The solution sct is $\{(2, 8)\}$.

23. $\begin{pmatrix} 4x - 5y = 21 \\ 3x + 7y = -38 \end{pmatrix}$

Multiply equation 1 by -3,
multiply equation 2 by 4,
and then add the resulting equations.

$$\begin{array}{r} -12x + 15y = -63 \\ 12x + 28y = -152 \\ \hline 43y = -215 \\ y = -5 \end{array}$$

$4x - 5y = 21$
$4x - 5(-5) = 21$
$4x = -4$
$x = -1$

The solution set is $\{(-1, -5)\}$.

25. $\begin{pmatrix} 5x - 2y = 19 \\ 5x - 2y = 7 \end{pmatrix}$

Multiply equation 1 by -1 and then
add the result to equation 2.

$$\begin{array}{r} -5x + 2y = -19 \\ 5x - 2y = \quad 7 \\ \hline 0 \neq -12 \end{array}$$

The system is inconsistent.
The solution set is $\emptyset$.

27. $\begin{pmatrix} 5a + 6b = 8 \\ 2a - 15b = 9 \end{pmatrix}$

Multiply equation 1 by 5,
multiply equation 2 by 2,
and then add the resulting equations.

$$\begin{array}{r} 25a + 30b = 40 \\ 4a - 30b = 18 \\ \hline 29a \qquad = 58 \\ a = \frac{58}{29} = 2 \end{array}$$

$5a + 6b = 8$
$5(2) + 6b = 8$
$10 + 6b = 8$
$6b = -2$
$b = -\frac{2}{6} = -\frac{1}{3}$

Therefore, $a = 2$ and $b = -\frac{1}{3}$.

29. $\begin{pmatrix} \frac{2}{3}s + \frac{1}{4}t = -1 \\ \frac{1}{2}s - \frac{1}{3}t = -7 \end{pmatrix}$

Multiply equation 1 by 12 and
multiply equation 2 by 6.

$\begin{pmatrix} 8s + 3t = -12 \\ 3s - 2t = -42 \end{pmatrix}$

Multiply equation 1 by 2,
multiply equation 2 by 3,
and then add the resulting equations.

$$\begin{array}{r} 16s + 6t = -24 \\ 9s - 6t = -126 \\ \hline 25s \qquad = -150 \\ s = -6 \end{array}$$

$3s - 2t = -42$
$3(-6) - 2t = -42$
$-18 - 2t = -42$
$-2t = -24$
$t = 12$

Therefore, $s = -6$ and $t = 12$.

31. $\begin{pmatrix} \frac{x}{2} - \frac{2y}{5} = \frac{-23}{60} \\ \frac{2x}{3} + \frac{y}{4} = \frac{-1}{4} \end{pmatrix}$

Multiply equation 1 by 60 and
multiply equation 2 by 12.

$\begin{pmatrix} 30x - 24y = -23 \\ 8x + 3y = -3 \end{pmatrix}$

Multiply equation 2 by 8 and then
add the result to equation 1.

$$\begin{array}{r} 30x - 24y = -23 \\ 64x + 24y = -24 \\ \hline 94x \qquad = -47 \\ x = -\frac{47}{94} = -\frac{1}{2} \end{array}$$

$8x + 3y = -3$
$8\left(-\frac{1}{2}\right) + 3y = -3$
$-4 + 3y = -3$
$3y = 1$
$y = \frac{1}{3}$

The solution set is $\left\{\left(-\frac{1}{2}, \frac{1}{3}\right)\right\}$.

33. $\begin{pmatrix} \frac{2x}{3} + \frac{y}{2} = \frac{1}{6} \\ 4x + 6y = -1 \end{pmatrix}$

Multiply equation 1 by 6,
multiply equation 2 by -1,
and then add the resulting equations.

$$\begin{array}{r} 4x + 3y = 1 \\ -4x - 6y = 1 \\ \hline -3y = 2 \end{array}$$

$y = -\frac{2}{3}$

$4x + 6y = -1$

$4x + 6\left(-\frac{2}{3}\right) = -1$

$4x - 4 = -1$

$4x = 3$

$x = \frac{3}{4}$

The solution set is $\left\{\left(\frac{3}{4}, -\frac{2}{3}\right)\right\}$.

35. $\begin{pmatrix} 5x - y = -22 \\ 2x + 3y = -2 \end{pmatrix}$

Multiply equation 1 by 3 and then
add the result to equation 2.

$$\begin{array}{r} 15x - 3y = -66 \\ 2x + 3y = -\ \ 2 \\ \hline 17x \qquad = -68 \end{array}$$

$x = -4$

$5x - y = -22$

$5(-4) - y = -22$

$-20 - y = -22$

$-y = -2$

$y = 2$

The solution set is $\{(-4, 2)\}$.

37. $\begin{pmatrix} x = 3y + 1 \\ x = -2y + 3 \end{pmatrix}$

Substitute $3y + 1$ for x in the 2nd equation.

$x = -2y + 3$

$3y + 1 = -2y + 3$

$5y = 2$

$y = \frac{2}{5}$

$x = -2y + 3$

$x = -2\left(\frac{2}{5}\right) + 3$

$x = -\frac{4}{5} + \frac{15}{5}$

$x = \frac{11}{5}$

The solution set is $\left\{\left(\frac{11}{5}, \frac{2}{5}\right)\right\}$.

39. $\begin{pmatrix} 3x - 5y = 9 \\ 6x - 10y = -1 \end{pmatrix}$

Multiply equation 1 by -2 and then
add the result to equation 2.

$$\begin{array}{r} -6x + 10y = -18 \\ 6x - 10y = -\ \ 1 \\ \hline 0 \neq -19 \end{array}$$

The system is inconsistent.
The solution set is $\emptyset$.

41. $\begin{pmatrix} \frac{1}{2}x - \frac{2}{3}y = 22 \\ \frac{1}{2}x + \frac{1}{4}y = 0 \end{pmatrix}$

Multiply equation 1 by 6 and
multiply equation 2 by 4.

$\begin{pmatrix} 3x - 4y = 132 \\ 2x + y = 0 \end{pmatrix}$

Multiply equation 2 by 4 and then
add the result to equation 1.

$$\begin{array}{r} 3x - 4y = 132 \\ 8x + 4y = 0 \\ \hline 11x \qquad = 132 \end{array}$$

$x = 12$

$2x + y = 0$

$2(12) + y = 0$

$y = -24$

The solution set is $\{(12, -24)\}$.

43. $\begin{pmatrix} t = 2u + 2 \\ 9u - 9t = -45 \end{pmatrix}$

Substitue $2u + 2$ for t in the 2nd equation.

$9u - 9t = -45$

$9u - 9(2u + 2) = -45$

$9u - 18u - 18 = -45$

$-9u = -27$

$u = 3$

$t = 2u + 2$

$t = 2(3) + 2 = 8$

Therefore, $t = 8$ and $u = 3$.

45. $\begin{pmatrix} x+y=1000 \\ 0.12x+0.14y=136 \end{pmatrix}$
Multiply equation 2 by 100.
$\begin{pmatrix} x+y=1000 \\ 12x+14y=13600 \end{pmatrix}$
Multiply equation 1 by -12 and then add the result to equation 2.
$-12x-12y=-12000$
$12x+14y=13600$
$2y=1600$
$y=800$
$x+y=1000$
$x+800=1000$
$x=200$
The solution set is $\{(200, 800)\}$.

47. $\begin{pmatrix} y=2x \\ 0.09x+0.12y=132 \end{pmatrix}$
Substitute $2x$ for y in the 2nd equation.
$0.09x+0.12y=132$
$0.09x+0.12(2x)=132$
$0.09x+0.24x=132$
$0.33x=132$
$x=400$
$y=2x$
$y=2(400)$
$y=800$
The solution set is $\{(400, 800)\}$.

49. $\begin{pmatrix} x+y=10.5 \\ 0.5x+0.8y=7.35 \end{pmatrix}$
Multiply equation 1 by 10 and multiply equation 2 by 100.
$\begin{pmatrix} 10x+10y=105 \\ 50x+80y=735 \end{pmatrix}$
Multiply equation 1 by -5 and then add the result to equation 2.
$-50x-50y=-525$
$50x+80y=735$
$30y=210$
$y=7$
$10x+10y=105$
$10x+10(7)=105$
$10x+70=105$
$10x=35$
$x=3.5$
The solution set is $\{(3.5, 7)\}$.

51. Let $x=$ one number
and $y=$ the other number.
$\begin{pmatrix} x+y=53 \\ x-y=19 \end{pmatrix}$
$x+y=53$
$x-y=19$
$2x=72$
$x=36$
$x+y=53$
$36+y=53$
$y=17$
The numbers are 17 and 36.

53. Let x represent the smaller angle and y represent the larger angle.
$\begin{pmatrix} y=15+4x \\ x+y=90 \end{pmatrix}$
Substitute $15+4x$ for y in the 2nd equation.
$x+y=90$
$x+15+4x=90$
$5x=75$
$x=15$
$y=15+4x$
$y=15+4(15)$
$y=15+60$
$y=75$
The angles are 15° and 75°.

55. Let $t=$ tens digit
and $u=$ units digit.
$\begin{pmatrix} t=3u+1 \\ u+t=9 \end{pmatrix}$
Substitute $3u+1$ for t in the 2nd equation.
$u+t=9$
$u+3u+1=9$
$4u+1=9$
$4u=8$
$u=2$
$t=3u+1$
$t=3(2)+1$
$t=6+1$
$t=7$
The number is 72.

57. Let u represent the units digit and t represent the tens digit.

value of original number $= 10t + u$
value of number with digits
reversed $= 10u + t$

Sum of the digits is 7.
$t + u = 7$
Solve the system.
$$\begin{pmatrix} t + u = 7 \\ 10u + t = 9 + 10t + u \end{pmatrix}$$
Simplify the 2nd equation.
$$\begin{pmatrix} t + u = 7 \\ -9t + 9u = 9 \end{pmatrix}$$
Divide the 2nd equation by 9 and then add the result to equation 1.
$$\begin{aligned} t + u &= 7 \\ -t + u &= 1 \\ \hline 2u &= 8 \\ u &= 4 \end{aligned}$$
$t + u = 7$
$t + 4 = 7$
$t = 3$
The original number is 34.

59. Let $x =$ number of sedans rented and $y =$ number of convertibles rented.
$$\begin{pmatrix} x + y = 32 \\ 35x + 48y = 1276 \end{pmatrix}$$
Multiply equation 1 by -35 and then add the result to equation 2.
$$\begin{aligned} -35x - 35y &= -1120 \\ 35x + 48y &= 1276 \\ \hline 13y &= 156 \\ y &= 12 \end{aligned}$$
There were 12 convertibles rented.

61. Let s represent the number of student tickets and n the number of parent tickets.
$$\begin{pmatrix} 6s + 10n = 3360 \\ s + n = 420 \end{pmatrix}$$
Multiply equation 2 by -6 and then add the result to equation 1.
$$\begin{aligned} 6s + 10n &= 3360 \\ -6s - 6n &= -2520 \\ \hline 4n &= 840 \\ n &= 210 \end{aligned}$$
$s + 210 = 420$
$s = 210$
210 student tickets and
210 parent tickets were sold.

63. Let $x =$ money invested at 8% and $y =$ money invested at 6%.
$$\begin{pmatrix} x = 3y \\ 0.08x + 0.06y = 1500 \end{pmatrix}$$
Substitute $3y$ for x in the 2nd equation.
$0.08(3y) + 0.06y = 1500$
$0.24y + 0.06y = 1500$
$0.30y = 1500$
$y = 5000$
$x = 3y$
$x = 3(5000)$
$x = 15000$
Melinda invested $\$15,000$ at 8% and $\$5,000$ invested at 6%.

65. Let x represent the rate of the kayak in still water and y the rate of the current.

	Rate	Time	Distance
upstream	$x - y$	4	$4(x - y)$
downstream	$2x + y$	1	$1(x + y)$

$$\begin{pmatrix} 4(x - y) = 20 \\ 1(2x + y) = 19 \end{pmatrix}$$
Divide the 1st equation by 4 and solve equation 2 for y. Substitute into equation 1.
$x - y = 5$
$y = -2x + 19$
$x - (-2x + 19) = 5$
$x + 2x - 19 = 5$
$3x = 24$
$x = 8$
$8 - y = 5$
$-y = -3$
$y = 3$
The rate of the current is 3 mph.

67. Let $x =$ price of 1 gallon of green latex paint and $y =$ price of 1 gallon of primer.
$$\begin{pmatrix} 4x + 2y = 116 \\ 3x + y = 80 \end{pmatrix}$$
Multiply the 1st equation by $-\frac{1}{2}$ and then add the result to equation 2.

$$\begin{aligned} -2x - y &= -58 \\ 3x + y &= 80 \\ \hline x \quad &= 22 \end{aligned}$$

The price of a gallon of green latex paint is $22.00.

69. Let f represent the number of five-dollar bills and t the number of ten-dollar bills.

$$\begin{pmatrix} f = 12 + t \\ 5f + 10t = 330 \end{pmatrix}$$

Substitute $12 + t$ for f in the 2nd equation.

$5f + 10t = 330$
$5(12 + t) + 10t = 330$
$60 + 5t + 10t = 330$
$60 + 15t = 330$
$15t = 270$
$t = 18$
$f = 12 + t$
$f = 12 + 18$
$f = 30$

There are 18 ten-dollar bills and 30 five-dollar bills.

Further Investigations

75.
$$\begin{pmatrix} \frac{1}{x} + \frac{2}{y} = \frac{7}{12} \\ \frac{3}{x} - \frac{2}{y} = \frac{5}{12} \end{pmatrix}$$

$$\begin{aligned} \frac{1}{x} + \frac{2}{y} &= \frac{7}{12} \\ \frac{3}{x} - \frac{2}{y} &= \frac{5}{12} \\ \hline \frac{4}{x} \quad &= \frac{12}{12} \end{aligned}$$

$\frac{4}{x} = 1$
$4 = x$

$\frac{1}{x} + \frac{2}{y} = \frac{7}{12}$
$\frac{1}{4} + \frac{2}{y} = \frac{7}{12}$
$\frac{2}{y} = \frac{4}{12}$
$4y = 24$
$y = 6$

The solution set is $\{(4, 6)\}$.

77.
$$\begin{pmatrix} \frac{3}{x} - \frac{2}{y} = \frac{13}{6} \\ \frac{2}{x} + \frac{3}{y} = 0 \end{pmatrix}$$

Multiply equation 1 by 3, multiply equation 2 by 2, and then add the resulting equations.

$$\begin{aligned} \frac{9}{x} - \frac{6}{y} &= \frac{13}{2} \\ \frac{4}{x} + \frac{6}{y} &= 0 \\ \hline \frac{13}{x} \quad &= \frac{13}{2} \end{aligned}$$

$x = 2$

$\frac{2}{x} + \frac{3}{y} = 0$
$\frac{2}{2} + \frac{3}{y} = 0$
$\frac{3}{y} = -1$
$-y = 3$
$y = -3$

The solution set is $\{(2, -3)\}$.

79.
$$\begin{pmatrix} \frac{5}{x} - \frac{2}{y} = 23 \\ \frac{4}{x} + \frac{3}{y} = \frac{23}{2} \end{pmatrix}$$

Multiply equation 1 by 3, multiply equation 2 by 2, and then add the resulting equations.

$$\begin{aligned} \frac{15}{x} - \frac{6}{y} &= 69 \\ \frac{8}{x} + \frac{6}{y} &= 23 \\ \hline \frac{23}{x} \quad &= 92 \end{aligned}$$

$23 = 92x$
$\frac{23}{92} = x$
$x = \frac{1}{4}$

$\frac{5}{x} - \frac{2}{y} = 23$
$\frac{5}{\frac{1}{4}} - \frac{2}{y} = 23$

$20 - \frac{2}{y} = 23$
$-\frac{2}{y} = 3$
$-2 = 3y$
$-\frac{2}{3} = y$
The solution set is $\left\{\left(\frac{1}{4}, -\frac{2}{3}\right)\right\}$.

81. Answers vary.

PROBLEM SET 15.2 Systems of Three Linear Equations in Three Variables

1. $\begin{pmatrix} 2x - 3y + 4z = 10 \\ 5y - 2z = -16 \\ 3z = 9 \end{pmatrix}$

$3z = 9$
$z = 3$
Substitute 3 for z in the 2nd equation.
$5y - 2z = -16$
$5y - 2(3) = -16$
$5y - 6 = -16$
$5y = -10$
$y = -2$
Substitute -2 for y and 3 for z in the 1st equation.
$2x - 3y + 4z = 10$
$2x - 3(-2) + 4(3) = 10$
$2x + 6 + 12 = 10$
$2x = -8$
$x = -4$
The solution set is $\{(-4, -2, 3)\}$.

3. $\begin{pmatrix} x + 2y - 3z = 2 \\ 3y - z = 13 \\ 3y + 5z = 25 \end{pmatrix}$

Multiply the 2nd equation by -1 and then add the result to equation 3.

$$\begin{array}{r} -3y + z = -13 \\ 3y + 5z = 25 \\ \hline 6z = 12 \end{array}$$

$z = 2$
Substitute 2 for z in the 2nd equation.
$3y - z = 13$
$3y - 2 = 13$
$3y = 15$
$y = 5$
Substitute -2 for x and 5 for y in the 1st equation.
$x + 2y - 3z = 2$
$x + 2(5) - 3(2) = 2$
$x + 10 - 6 = 2$
$x + 4 = 2$
$x = -2$
The solution set is $\{(-2, 5, 2)\}$.

5. $\begin{pmatrix} 3x + 2y - 2z = 14 \\ x - 6y = 16 \\ 2x + 5z = -2 \end{pmatrix}$

Multiply the 2nd equation by -2 and then add the result to equation 3.

$$\begin{array}{r} -2x + 12z = -32 \\ 2x + 5z = -2 \\ \hline 17z = -34 \end{array}$$

$z = -2$
Substitute -2 for z in the 2nd equation.
$x - 6z = 16$
$x - 6(-2) = 16$
$x + 12 = 16$
$x = 4$
Substitute 4 for x and -2 for z in the 1st equation.
$3x + 2y - 2z = 14$
$3(4) + 2y - 2(-2) = 14$
$12 + 2y + 4 = 14$
$2y + 16 = 14$
$2y = -2$
$y = -1$
The solution set is $\{(4, -1, -2)\}$.

7. $\begin{pmatrix} x - 2y + 3z = 7 \\ 2x + y + 5z = 17 \\ 3x - 4y - 2z = 1 \end{pmatrix}$

Multiply equation 1 by -2 and add to equation 2 to replace equation 2.
Multiply equation 1 by -3 and add to equation 3 to replace equation 3.

$$\begin{pmatrix} x - 2y + 3z = 7 \\ 5y - z = 3 \\ 2y - 11z = -20 \end{pmatrix}$$

Multiply equation 2 by -11 and add to equation 3 to replace equation 3.

$$\begin{pmatrix} x - 2y + 3z = 7 \\ 5y - z = 3 \\ -53y = -53 \end{pmatrix}$$

$-53y = -53$
$y = 1$

Substitute 1 for y in the 2nd equation.

$5y - z = 3$
$5(1) - z = 3$
$5 - z = 3$
$-z = -2$
$z = 2$

Substitute 1 for y and 2 for z in the 1st equation.

$x - 2y + 3z = 7$
$x - 2(1) + 3(2) = 7$
$x - 2 + 6 = 7$
$x + 4 = 7$
$x = 3$

The solution set is $\{(3, 1, 2)\}$.

9. $$\begin{pmatrix} 2x + y + 2z = 12 \\ 4x + y - z = -6 \\ -y - 5z = 10 \end{pmatrix}$$

Multiply equation 1 by -2 and add to equation 2 to replace equation 2.

$$\begin{pmatrix} 2x + y + 2z = 12 \\ -y - 5z = -30 \\ -y - 5z = 10 \end{pmatrix}$$

Multiply equation 2 by -1 and add to equation 3 to replace equation 3.

$$\begin{pmatrix} 2x + y + 2z = 12 \\ -y - 5z = -30 \\ 0 = 4 \end{pmatrix}$$

The result $0 = 4$ is a false statement. Therefore the solution set is Ø.

11. $$\begin{pmatrix} 2x - y + z = 0 \\ 3x - 2y + 4z = 11 \\ 5x + y - 6z = -32 \end{pmatrix}$$

Add the 1st and 3rd equations.

$$\begin{array}{l} 2x - y + z = 0 \\ \underline{5x + y - 6z = -32} \\ 7x \quad - 5z = -32 \end{array}$$

Multiply the 3rd equation by 2 and then add the result to equation 2.

$$\begin{array}{l} 3x - 2y + 4z = 11 \\ \underline{10x + 2y - 12z = -64} \\ 13x \quad - 8z = -53 \end{array}$$

$$\begin{pmatrix} 7x - 5z = -32 \\ 13x - 8z = -53 \end{pmatrix}$$

Multiply equation 1 by -8, multiply equation 2 by 5, and then add the resulting equations.

$$\begin{array}{l} -56x + 40z = 256 \\ \underline{65x - 40z = -265} \\ 9x \quad = -9 \\ x = -1 \end{array}$$

Substitute -1 for x in the 1st equation.

$7x - 5z = -32$
$7(-1) - 5z = -32$
$-7 - 5z = -32$
$-5z = -25$
$z = 5$

Substitute -1 for x and 5 for z in the 1st equation.

$2x - y + z = 0$
$2(-1) - y + 5 = 0$
$-2 - y + 5 = 0$
$-y + 3 - 0$
$-y = -3$
$y = 3$

The solution set is $\{(-1, 3, 5)\}$.

13. $$\begin{pmatrix} 3x + 2y - z = -11 \\ 2x - 3y + 4z = 11 \\ 5x + y - 2z = -17 \end{pmatrix}$$

Multiply equation 1 by 4 and add to equation 2 to replace equation 2. Multiply equation 1 by -2 and add to equation 3 to replace equation 3.

$$\begin{pmatrix} 3x + 2y - z = 11 \\ 14x + 5y = -33 \\ -x - 3y = 5 \end{pmatrix}$$

Multiply equation 3 by 14 and add to equation 2 to replace equation 2.

$$\begin{pmatrix} 3x + 2y - z = -11 \\ -37y = 37 \\ -x - 3y = 5 \end{pmatrix}$$

$-37y = 37$
$y = -1$
Substitute -1 for y in the 3rd equation.
$-x - 3y = 5$
$-x - 3(-1) = 5$
$-x + 3 = 5$
$-x = 2$
$x = -2$
Substitute -2 for x and
-1 for y in the 1st equation.
$3x + 2y - z = -11$
$3(-2) + 2(-1) - z = -11$
$-6 - 2 - z = -11$
$-8 - z = -11$
$-z = -3$
$z = 3$
The solution set is $\{(-2, -1, 3)\}$.

15. $$\begin{pmatrix} 2x + 3y - 4z = -10 \\ 4x - 5y + 3z = 2 \\ 2y + z = 8 \end{pmatrix}$$

Multiply the 1st equation by -2 and then add the result to equation 2.

$$\begin{array}{r} -4x - 6y + 8z = 20 \\ 4x - 5y + 3z = 2 \\ \hline -11y + 11z = 22 \\ -y + z = 2 \end{array}$$

$$\begin{pmatrix} -y + z = 2 \\ 2y + z = 8 \end{pmatrix}$$

Multiply the 1st equation by -1 and then add the result to equation 2.

$$\begin{array}{rcr} y - z & = & -2 \\ 2y + z & = & 8 \\ \hline 3y & = & 6 \end{array}$$

$y = 2$
Substitute 2 for y in the 1st equation.
$-y + z = 2$
$-2 + z = 2$
$z = 4$
Substitute 2 for y and
4 for z in the 1st equation.
$2x + 3y - 4z = -10$
$2x + 3(2) - 4(4) = -10$
$2x + 6 - 16 = -10$
$2x - 10 = -10$
$2x = 0$
$x = 0$
The solution set is $\{(0, 2, 4)\}$.

17. $$\begin{pmatrix} x + 3y - 2z = -4 \\ 2x + 7y + z = 5 \\ y + 5z = 13 \end{pmatrix}$$

Multiply the 1st equation by -2 and then add the result to equation 2.

$$\begin{array}{r} -2x - 6y + 4z = 8 \\ 2x + 7y + z = 5 \\ \hline y + 5z = 13 \end{array}$$

Multiply the 3rd equation by -1 and then add to the above result.

$$\begin{array}{r} -y - 5z = -13 \\ y + 5z = 13 \\ \hline 0 = 0 \end{array}$$

This is a true statement for all values of x, y and z.
Let $z = k$, therefore
$y + 5z = 13$
$y = -5z + 13$
$y = -5k + 13$
Also, substitute values for y and z into
$x + 3y - 2z = -4$ to find x.
$x + 3(-5k + 13) - 2(k) = -4$
$x - 15k + 39 - 2k = -4$
$x - 17k + 39 = -4$
$x - 17k = -43$
$x = 17k - 43$
The solution set is
$\{(17k - 43, -5k + 13, k) \mid k \text{ is a real number.}\}$

19. $$\begin{pmatrix} 3x + 2y - 2z = 14 \\ 2x - 5y + 3z = 7 \\ 4x - 3y + 7z = 5 \end{pmatrix}$$

Multiply equation 1 by 2,
multiply equation 2 by -3,
and then add the resulting equations.

$$\begin{array}{rcr} 6x + 4y - 4z & = & 28 \\ -6x + 15y - 9z & = & -21 \\ \hline 19y - 13z & = & 7 \end{array}$$

Multiply the 2nd equation by -2 and then add the result to equation 3.

$$\begin{array}{r} -4x+10y-6z=-14 \\ 4x-3y+7z=5 \\ \hline 7y+z=-9 \end{array}$$

$$\begin{pmatrix} 19y-13z=7 \\ 7y+z=-9 \end{pmatrix}$$

Multiply the 2nd equation by 13 and then add the result to equation 1.

$$\begin{array}{r} 19y-13z=7 \\ 91y+13z=-117 \\ \hline 110y=-110 \\ y=-1 \end{array}$$

Substitute -1 for y in the 2nd equation.

$7y+z=-9$
$7(-1)+z=-9$
$-7+z=-9$
$z=-2$

Substitute -1 for y and -2 for z in the 1st equation.

$3x+2y-2z=14$
$3x+2(-1)-2(-2)=14$
$3x-2+4=14$
$3x+2=14$
$3x=12$
$x=4$

The solution set is $\{(4, -1, -2)\}$.

21. $\begin{pmatrix} 2x-3y+4z=-12 \\ 4x+2y-3z=-13 \\ 6x-5y+7z=-31 \end{pmatrix}$

Multiply the 1st equation by -2 and then add the result to equation 2.

$$\begin{array}{r} -4x+6y-8z=24 \\ 4x+2y-3z=-13 \\ \hline 8y-11z=11 \end{array}$$

Multiply the 1st equation by -3 and then add the result to equation 3.

$$\begin{array}{r} -6x+9y-12z=36 \\ 6x-5y+7z=-31 \\ \hline 4y-5z=5 \end{array}$$

$$\begin{pmatrix} 8y-11z=11 \\ 4y-5z=5 \end{pmatrix}$$

Multiply the 2nd equation by -2 and then add the result to equation 1.

$$\begin{array}{r} 8y-11z=11 \\ -8y+10z=-10 \\ \hline -z=1 \\ z=-1 \end{array}$$

Substitute -1 for z in the 2nd equation.

$4y-5z=5$
$4y-5(-1)=5$
$4y+5=5$
$4y=0$
$y=0$

Substitute 0 for y and -1 for z in the 1st equation.

$2x-3y+4z=-12$
$2x-3(0)+4(-1)=-12$
$2x-0-4=-12$
$2x-4=-12$
$2x=-8$
$x=-4$

The solution set is $\{(-4, 0, -1)\}$.

23. $\begin{pmatrix} 5x-3y-6z=22 \\ x-y+z=-3 \\ -3x+7y-5z=23 \end{pmatrix}$

Multiply equation 2 by -5 and add to equation 1 to replace equation 1.
Multiply equation 2 by 3 and add to equation 3 to replace equation 3.

$$\begin{pmatrix} 2y-11z=37 \\ x-y+z=-3 \\ 4y-2z=14 \end{pmatrix}$$

Multiply equation 1 by -2 and add to equation 3 to replace equation 3.

$$\begin{pmatrix} 2y-11z=37 \\ x-y+z=-3 \\ 20z=-60 \end{pmatrix}$$

$20z=-60$
$z=-3$

Substitute -3 for z in the 1st equation.

$2y-11z=37$
$2y-11(-3)=37$
$2y+33=37$
$2y=4$
$y=2$

Substitute 2 for y and -3 for z in the 2nd equation.

$x-y+z=-3$
$x-2+(-3)=-3$

$x - 5 = -3$
$x = 2$
The solution set is $\{(2, 2, -3)\}$.

25. Let $x =$ the 1st number,
$y =$ the 2nd number and
$z =$ the 3rd number.

$$\begin{pmatrix} x + y + z = 20 \\ x + z = 2y + 2 \\ z - x = 3y \end{pmatrix}$$

$$\begin{pmatrix} x + y + z = 20 \\ x - 2y + z = 2 \\ -x - 3y + z = 0 \end{pmatrix}$$

Add equation 1 and equation 3.

$$\begin{array}{r} x + y + z = 20 \\ -x - 3y + z = 0 \\ \hline -2y + 2z = 20 \end{array}$$

$\frac{1}{2}(-2y + 2z) = \frac{1}{2}(20)$
$-y + z = 10$

Add equation 2 and equation 3.

$$\begin{array}{r} x - 2y + z = 2 \\ -x - 3y + z = 0 \\ \hline -5y + 2z = 2 \end{array}$$

$$\begin{pmatrix} -y + z = 10 \\ -5y + 2z = 2 \end{pmatrix}$$

Multiply equation 1 by -2, then add to equation 2.

$$\begin{array}{r} 2y - 2z = -20 \\ -5y + 2z = 2 \\ \hline -3y \quad = -18 \end{array}$$

$y = 6$
$-y + z = 10$
$-6 + z = 10$
$z = 16$
$x + y + z = 20$
$x + 6 + 16 = 20$
$x + 22 = 20$
$x = -2$
The numbers are -2, 6, and 16.

27. Let $x =$ the cost of the helmet,
$y =$ the cost of the jacket and
$z =$ the cost of the gloves.

$x + y + z = 650$
$y = x + 100 \quad \Rightarrow \quad -x + y = 100$
$x + z = y - 50 \quad \Rightarrow \quad x - y + z = -50$

$$\begin{pmatrix} x + y + z = 650 \\ -x + y = 100 \\ x - y + z = -50 \end{pmatrix}$$

Multiply equation (1) by -1 and add the result to equation (3).

$$\begin{array}{r} -x - y - z = -650 \\ x - y + z = -50 \\ \hline -2y \quad = -700 \end{array}$$

$y = 350$
Substitute $y = 350$ into equation (2).
$-x + 350 = 100$
$-x = -250$
$x = 250$
Substitute $x = 250$ and $y = 350$ into equation (1).
$250 + 350 + z = 650$
$600 + z = 650$
$z = 50$
The helmet cost $250,
the jacket cost $350,
and the gloves cost $50.

29. Let $a =$ the measure of $\angle A$
$b =$ the measure of $\angle B$
$c =$ the measure of $\angle C$.

$a = 5b \quad \Rightarrow \quad a - 5b = 0$
$b + c = a - 60 \quad \Rightarrow \quad -a + b + c = -60$
$a + b + c = 180$

$$\begin{pmatrix} a + b + c = 180 \\ -a + b + c = -60 \\ a - 5b = 0 \end{pmatrix}$$

Multiply equation (1) by -1 and add the result to equation (2).

$$\begin{array}{r} -a - b - c = -180 \\ -a + b + c = -60 \\ \hline -2a \quad = -240 \end{array}$$

$a = 120$
Substitute $a = 120$ into equation (3).
$120 - 5b = 0$
$-5b = -120$
$b = 24$
Substitute $a = 120$ and $b = 24$ into equation (1).

$$120 + 24 + c = 180$$
$$144 + c = 180$$
$$c = 36$$

The measure of $\angle A = 120°$,
the measure of $\angle B = 24°$, and
the measure of $\angle C = 36°$.

31. Let $x =$ wages per hour for the plumber,
$y =$ wages per hour for the apprentice,
$z =$ wages per hour for the laborer.

$$x + y + z = 80$$
$$x = y + z + 20 \quad \Rightarrow \quad x - y - z = 20$$
$$x = 5z \quad \Rightarrow \quad x - 5z = 0$$

$$\begin{pmatrix} x + y + z = 80 \\ x - y - z = 20 \\ x - 5z = 0 \end{pmatrix}$$

Add equation (1) and equation (2).

$$\begin{aligned} x + y + z &= 80 \\ x - y - z &= 20 \\ \hline 2x \qquad &= 100 \\ x &= 50 \end{aligned}$$

Substitute $x = 50$ into equation (3).

$$50 - 5z = 0$$
$$-5z = -50$$
$$z = 10$$

Substitute $x = 50$ and $z = 10$ into equation (1).

$$50 + y + 10 = 80$$
$$y + 60 = 80$$
$$y = 20$$

The plumber's wages are $50 per hour, the apprentice's wages are $20 per hour, and the laborer's wages are $10 per hour.

33. Let x represent the number of pounds of almonds, y the number of pounds of pecans, and z the number of pounds of peanuts.

$$\begin{pmatrix} x + y + z = 20 \\ 3.50x + 4y + 2z = (2.70)(20) \\ z = 3y \end{pmatrix}$$

Substitute $3y$ for z in the 1st and 2nd equations.

$$x + y + (3y) = 20$$
$$3.50x + 4y + 2(3y) = 54$$

$$x + 4y = 20$$
$$3.50x + 10y = 54$$

$$-5(x + 4y) = -5(20)$$
$$2(3.50x + 10y) = 2(54)$$

$$\begin{aligned} -5x - 20y &= -100 \\ 7x + 20y &= 108 \\ \hline 2x \qquad &= 8 \\ x &= 4 \end{aligned}$$

$$x + 4y = 20$$
$$4 + 4y = 20$$
$$4y = 16$$
$$y = 4$$
$$x + y + z = 20$$
$$4 + 4 + z = 20$$
$$8 + z = 20$$
$$z = 12$$

There are 4 pounds of almonds, 4 pounds of pecans, and 12 pounds of peanuts.

35. Let $x =$ the number of nickels,
$y =$ the number of dimes, and
$z =$ the number of quarters.

$$\begin{pmatrix} x + y + z = 42 \\ x + y = z - 2 \\ 0.05x + 0.10y + 0.25z = 7.15 \end{pmatrix}$$

$$\begin{pmatrix} x + y + z = 42 \\ x + y - z = -2 \\ 0.05x + 0.10y + 0.25z = 7.15 \end{pmatrix}$$

Multiply equation 1 by -1 and add to equation 2 to replace equation 2.

$$\begin{pmatrix} x + y + z = 42 \\ -2z = -44 \\ 0.05x + 0.10y + 0.25z = 7.15 \end{pmatrix}$$

$$-2z = -44$$
$$z = 22$$

Substitute 22 for z in the 1st and 2nd equations.

$$\begin{pmatrix} x + y + 22 = 42 \\ 0.05x + 0.10y + 0.25(22) = 7.15 \end{pmatrix}$$

$$\begin{pmatrix} x + y = 20 \\ 0.05x + 0.10y = 1.65 \end{pmatrix}$$

Multiply equation 1 by -0.05 and add to equation 2 to replace equation 2.

$$\begin{pmatrix} x + y = 20 \\ 0.05y = 0.65 \end{pmatrix}$$

$$0.05y = 0.65$$
$$y = 13$$

Substitute 13 for y in the 1st equation.
$x + y = 20$
$x + 13 = 20$
$x = 7$
There are 7 nickels, 13 dimes, and 22 quarters.

37. Let x represent the smallest angle, y the middle angle, and z the largest angle.

$$\begin{pmatrix} x + y + z = 180 \\ z = 2x \\ x + z = 2y \end{pmatrix}$$

Substitute $2x$ for z in the 1st and 3rd equations.

$$\begin{pmatrix} x + y + 2x = 180 \\ x + 2x = 2y \end{pmatrix}$$

$$\begin{pmatrix} 3x + y = 180 \\ 3x - 2y = 0 \end{pmatrix}$$

Multiply the 2nd equation by -1 and then add the result to equation 1.

$$\begin{array}{r} 3x + y = 180 \\ -3x + 2y = 0 \\ \hline 3y = 180 \\ y = 60 \end{array}$$

Substitute 60 for y in the 1st equation.
$3x + y = 180$
$3x + 60 = 180$
$3x = 120$
$x = 40$
Substitute 40 for x in the 2nd equation.
$z = 2x$
$z = 2(40)$
$z = 80$
The angles measure 40°, 60°, and 80°.

39. Let $x =$ money invested at 2%, $y =$ money invested at 3%, and $z =$ money invested at 4%.

$$\begin{pmatrix} x + y + z = 30000 \\ 0.02x + 0.03y + 0.04z = 1000 \\ x + y = z \end{pmatrix}$$

Substitute z for $x + y$ in the 1st equation.
$x + y + z = 30000$
$z + z = 30000$
$2z = 30000$
$z = 15000$
Substitute 15000 for z in the 1st and 2nd equations.

$$\begin{pmatrix} x + y + 15000 = 30000 \\ 0.02x + 0.03y + 0.04(15000) = 1000 \end{pmatrix}$$

$$\begin{pmatrix} x + y = 15000 \\ 0.02x + 0.03y = 400 \end{pmatrix}$$

Multiply equation 1 by -0.02 and add to equation 2 to replace equation 2.

$$\begin{pmatrix} x + y = 15000 \\ 0.01y = 100 \end{pmatrix}$$

$0.01y = 100$
$y = 10000$
Substitute 10000 for y in the 1st equation.
$x + y = 15000$
$x + 10000 = 15000$
$x = 5000$
$5,000 is invested at 2%,
$10,000 is invested at 3%, and
$15,000 is invested at 4%.

41. Let x represent the number of Type A birdhouses, y the number of Type B, and z the number of Type C.

$$\begin{pmatrix} 0.1x + 0.2y + 0.1z = 35 \\ 0.4x + 0.4y + 0.3z = 95 \\ 0.2x + 0.1y + 0.3z = 62.5 \end{pmatrix}$$

Multiply each equation by 10.

$$\begin{pmatrix} x + 2y + z = 350 \\ 4x + 4y + 3z = 950 \\ 2x + y + 3z = 625 \end{pmatrix}$$

Multiply the 1st equation by -2 and then add the result to equation 3.

$$\begin{array}{r} -2x - 4y - 2z = -700 \\ 2x + y + 3z = 625 \\ \hline -3y + z = -75 \end{array}$$

Multiply the 1st equation by -4 and then add the result to equation 2.

$$\begin{array}{r} -4z - 8y - 4z = -1400 \\ 4x + 4y + 3z = 950 \\ \hline -4y - z = -450 \end{array}$$

$$\begin{array}{r} -3y + z = -75 \\ -4y - z = -450 \\ \hline -7y = -525 \end{array}$$

$y = 75$
$-3y + z = -75$
$-3(75) + z = -75$
$-225 + z = -75$

$z = 150$
$x + 2y + z = 350$
$x + 2(75) + 150 = 350$
$x + 150 + 150 = 350$
$x + 300 = 350$
$x = 50$

There will be 50 Type A birdhouses, 75 Type B birdhouses and 150 Type C birdhouses.

PROBLEM SET 15.3 A Matrix Approach to Solving Systems

1. Yes

3. Yes

5. No

7. No

9. Yes

11. $\left[\begin{array}{cc|c} 1 & -3 & 14 \\ 3 & 2 & -13 \end{array}\right]$

$-3(\text{row } 1) + (\text{row } 2)$ replace $(\text{row } 2)$

$\left[\begin{array}{cc|c} 1 & -3 & 14 \\ 0 & 11 & -55 \end{array}\right]$

$\frac{1}{11}(\text{row } 2)$

$\left[\begin{array}{cc|c} 1 & -3 & 14 \\ 0 & 1 & -5 \end{array}\right]$

$3(\text{row } 2) + (\text{row } 1)$ replace $(\text{row } 1)$

$\left[\begin{array}{cc|c} 1 & 0 & 1 \\ 0 & 1 & -5 \end{array}\right]$

The solution set is $\{(-1, -5)\}$.

13. $\left[\begin{array}{cc|c} 3 & -4 & 33 \\ 1 & 7 & -39 \end{array}\right]$

Exchange row 1 and row 2.

$\left[\begin{array}{cc|c} 1 & 7 & -39 \\ 3 & -4 & 33 \end{array}\right]$

$-3(\text{row } 1) + (\text{row } 2)$ replace $(\text{row } 2)$

$\left[\begin{array}{cc|c} 1 & 7 & -39 \\ 0 & -25 & 150 \end{array}\right]$

$-\frac{1}{25}(\text{row } 2)$ replace $(\text{row } 2)$

$\left[\begin{array}{cc|c} 1 & 7 & -39 \\ 0 & 1 & -6 \end{array}\right]$

$-7(\text{row } 2) + (\text{row } 1)$ replace $(\text{row } 1)$

$\left[\begin{array}{cc|c} 1 & 0 & 3 \\ 0 & 1 & -6 \end{array}\right]$

The solution set is $\{(3, -6)\}$.

15. $\left[\begin{array}{cc|c} 1 & -6 & -2 \\ 2 & -12 & 5 \end{array}\right]$

$-2(\text{row } 1) + (\text{row } 2)$ replace $(\text{row } 2)$

$\left[\begin{array}{cc|c} 1 & -6 & -2 \\ 0 & 0 & 9 \end{array}\right]$

The system is inconsistent.
The solution set is $\emptyset$.

17. $\left[\begin{array}{cc|c} 3 & -5 & 39 \\ 2 & 7 & -67 \end{array}\right]$

$(\text{row } 1) - (\text{row } 2)$ replace $(\text{row } 1)$

$\left[\begin{array}{cc|c} 1 & -12 & 106 \\ 2 & 7 & -67 \end{array}\right]$

$-2(\text{row } 1) + (\text{row } 2)$ replace $(\text{row } 2)$

$\left[\begin{array}{cc|c} 1 & -12 & 106 \\ 0 & 31 & -279 \end{array}\right]$

$\frac{1}{31}(\text{row } 2)$ replace $(\text{row } 2)$

$\left[\begin{array}{cc|c} 1 & -12 & 106 \\ 0 & 1 & -9 \end{array}\right]$

$12(\text{row } 2) + (\text{row } 1)$ replace $(\text{row } 1)$

$\left[\begin{array}{cc|c} 1 & 0 & -2 \\ 0 & 1 & -9 \end{array}\right]$

The solution set is $\{(-2, -9)\}$.

19. $\left[\begin{array}{ccc|c} 1 & -2 & -3 & -6 \\ 3 & -5 & -1 & 4 \\ 2 & 1 & 2 & 2 \end{array}\right]$

$-3(\text{row } 1) + (\text{row } 2)$ replace $(\text{row } 2)$
$-2(\text{row } 1) + (\text{row } 3)$ replace $(\text{row } 3)$

$\left[\begin{array}{ccc|c} 1 & -2 & -3 & -6 \\ 0 & 1 & 8 & 22 \\ 0 & 5 & 8 & 14 \end{array}\right]$

2(row 2) + (row 1) replace (row 1)
− 5(row 2) + (row 3) replace (row 3)

$$\left[\begin{array}{ccc|c} 1 & 0 & 13 & 38 \\ 0 & 1 & 8 & 22 \\ 0 & 0 & -32 & -96 \end{array}\right]$$

$-\frac{1}{32}$(row 3)

$$\left[\begin{array}{ccc|c} 1 & 0 & 13 & 38 \\ 0 & 1 & 8 & 22 \\ 0 & 0 & 1 & 3 \end{array}\right]$$

− 8(row 3) + (row 2) replace (row 2)
− 13(row 3) + (row 1) replace (row 1)

$$\left[\begin{array}{ccc|c} 1 & 0 & 0 & -1 \\ 0 & 1 & 0 & -2 \\ 0 & 0 & 1 & 3 \end{array}\right]$$

The solution set is $\{(-1, -2, 3)\}$.

21. $$\left[\begin{array}{ccc|c} -2 & -5 & 3 & 11 \\ 1 & 3 & -3 & -12 \\ 3 & -2 & 5 & 31 \end{array}\right]$$

Exchange rows 1 and 2.

$$\left[\begin{array}{ccc|c} 1 & 3 & -3 & -12 \\ -2 & -5 & 3 & 11 \\ 3 & -2 & 5 & 31 \end{array}\right]$$

2(row 1) + (row 2) replace (row 2)
− 3(row 1) + (row 3) replace (row 3)

$$\left[\begin{array}{ccc|c} 1 & 3 & -3 & -12 \\ 0 & 1 & -3 & -13 \\ 0 & -11 & 14 & 67 \end{array}\right]$$

− 3(row 2) + (row 1) replace (row 1)
11(row 2) + (row 3) replace (row 3)

$$\left[\begin{array}{ccc|c} 1 & 0 & 6 & 27 \\ 0 & 1 & -3 & -13 \\ 0 & 0 & -19 & -76 \end{array}\right]$$

$-\frac{1}{19}$(row 3)

$$\left[\begin{array}{ccc|c} 1 & 0 & 6 & 27 \\ 0 & 1 & -3 & -13 \\ 0 & 0 & 1 & 4 \end{array}\right]$$

− 6(row 3) + (row 1) replace (row 1)
3(row 3) + (row 2) replace (row 2)

$$\left[\begin{array}{ccc|c} 1 & 0 & 0 & 3 \\ 0 & 1 & 0 & -1 \\ 0 & 0 & 1 & 4 \end{array}\right]$$

The solution set is $\{(3, -1, 4)\}$.

23. $$\left[\begin{array}{ccc|c} 1 & -3 & -1 & 2 \\ 3 & 1 & -4 & -18 \\ -2 & 5 & 3 & 2 \end{array}\right]$$

− 3(row 1) + (row 2) replace (row 2)
2(row 1) + (row 3) replace (row 3)

$$\left[\begin{array}{ccc|c} 1 & -3 & -1 & 2 \\ 0 & 10 & -1 & -24 \\ 0 & -1 & 1 & 6 \end{array}\right]$$

$\frac{1}{10}$(row 2)

$$\left[\begin{array}{ccc|c} 1 & -3 & -1 & 2 \\ 0 & 1 & -\frac{1}{10} & -\frac{12}{5} \\ 0 & -1 & 1 & 6 \end{array}\right]$$

3(row 2) + (row 1) replace (row 1)
(row 2) + (row 3) replace (row 3)

$$\left[\begin{array}{ccc|c} 1 & 0 & -\frac{13}{10} & -\frac{26}{5} \\ 0 & 1 & -\frac{1}{10} & -\frac{12}{5} \\ 0 & 0 & \frac{9}{10} & \frac{18}{5} \end{array}\right]$$

$\frac{10}{9}$(row 3)

$$\left[\begin{array}{ccc|c} 1 & 0 & -\frac{13}{10} & -\frac{26}{5} \\ 0 & 1 & -\frac{1}{10} & -\frac{12}{5} \\ 0 & 0 & 1 & 4 \end{array}\right]$$

$\frac{1}{10}$(row 3) + (row 2) replace (row 2)
$\frac{13}{10}$(row 3) + (row 1) replace (row 1)

$$\left[\begin{array}{ccc|c} 1 & 0 & 0 & 0 \\ 0 & 1 & 0 & -2 \\ 0 & 0 & 1 & 4 \end{array}\right]$$

The solution set is $\{(0, -2, 4)\}$.

25. $$\left[\begin{array}{ccc|c} 1 & -1 & 2 & 1 \\ -3 & 4 & -1 & 4 \\ -1 & 2 & 3 & 6 \end{array}\right]$$

3(row 1) + (row 2) replace (row 2)
(row 1) + (row 3) replace (row 3)

$$\left[\begin{array}{ccc|c} 1 & -1 & 2 & 1 \\ 0 & 1 & 5 & 7 \\ 0 & 1 & 5 & 7 \end{array}\right]$$

(row 2) − (row 3) replace (row 3)

$$\begin{bmatrix} 1 & -1 & 2 & | & 1 \\ 0 & 1 & 5 & | & 7 \\ 0 & 0 & 0 & | & 0 \end{bmatrix}$$

The row of zeros indicates a dependent system.

$y + 5z = 7$

$y = 7 - 5z$

$x - y + 2z = 1$

$x - (7 - 5z) + 2z = 1$

$x - 7 + 7z = 1$

$x = 8 - 7z$

Let $k = z$. Then $x = 8 - 7k$ and $y = 7 - 5k$.

The solution set is $\{(8 - 7k, 7 - 5k, k)\}$, where k is any real number.

27. $$\begin{bmatrix} -2 & 1 & 5 & | & -5 \\ 3 & 8 & -1 & | & -34 \\ 1 & 2 & 1 & | & -12 \end{bmatrix}$$

Interchange row 1 and row 3.

$$\begin{bmatrix} 1 & 2 & 1 & | & -12 \\ 3 & 8 & -1 & | & -34 \\ -2 & 1 & 5 & | & -5 \end{bmatrix}$$

-3(row 1) + (row 2) replace (row 2)

2(row 1) + (row 3) replace (row 3)

$$\begin{bmatrix} 1 & 2 & 1 & | & -12 \\ 0 & 2 & -4 & | & 2 \\ 0 & 5 & 7 & | & -29 \end{bmatrix}$$

$\frac{1}{2}$(row 2)

$$\begin{bmatrix} 1 & 2 & 1 & | & -12 \\ 0 & 1 & -2 & | & 1 \\ 0 & 5 & 7 & | & -29 \end{bmatrix}$$

-2(row 2) + (row 1) replace (row 1)

-5(row 2) + (row 3) replace (row 3)

$$\begin{bmatrix} 1 & 0 & 5 & | & -14 \\ 0 & 1 & -2 & | & 1 \\ 0 & 0 & 17 & | & -34 \end{bmatrix}$$

$\frac{1}{17}$(row 3)

$$\begin{bmatrix} 1 & 0 & 5 & | & -14 \\ 0 & 1 & -2 & | & 1 \\ 0 & 0 & 1 & | & -2 \end{bmatrix}$$

2(row 3) + (row 2) replace (row 2)

-5(row 3) + (row 1) replace (row 1)

$$\begin{bmatrix} 1 & 0 & 0 & | & -4 \\ 0 & 1 & 0 & | & -3 \\ 0 & 0 & 1 & | & -2 \end{bmatrix}$$

The solution set is $\{(-4, -3, -2)\}$.

29. $$\begin{bmatrix} 2 & 3 & -1 & | & 7 \\ 3 & 4 & 5 & | & -2 \\ 5 & 1 & 3 & | & 13 \end{bmatrix}$$

-1(row 1) + (row 2) replace (row 1)

$$\begin{bmatrix} 1 & 1 & 6 & | & -9 \\ 3 & 4 & 5 & | & -2 \\ 5 & 1 & 3 & | & 13 \end{bmatrix}$$

-3(row 1) + (row 2) replace (row 2)

-5(row 1) + (row 3) replace (row 3)

$$\begin{bmatrix} 1 & 1 & 6 & | & -9 \\ 0 & 1 & -13 & | & 25 \\ 0 & -4 & -27 & | & 58 \end{bmatrix}$$

(row 1) $-$ (row 2) replace (row 1)

4(row 2) + (row 3) replace (row 3)

$$\begin{bmatrix} 1 & 0 & 19 & | & -34 \\ 0 & 1 & -13 & | & 25 \\ 0 & 0 & -79 & | & 158 \end{bmatrix}$$

$-\frac{1}{79}$(row 3)

$$\begin{bmatrix} 1 & 0 & 19 & | & -34 \\ 0 & 1 & -13 & | & 25 \\ 0 & 0 & 1 & | & -2 \end{bmatrix}$$

-19(row 3) + (row 1) replace (row 1)

13(row 3) + (row 2) replace (row 2)

$$\begin{bmatrix} 1 & 0 & 0 & | & 4 \\ 0 & 1 & 0 & | & -1 \\ 0 & 0 & 1 & | & -2 \end{bmatrix}$$

The solution set is $\{(4, -1, -2)\}$.

31. $$\begin{bmatrix} 1 & -3 & -2 & 1 & | & -3 \\ -2 & 7 & 1 & -2 & | & -1 \\ 3 & -7 & -3 & 3 & | & -5 \\ 5 & 1 & 4 & -2 & | & 18 \end{bmatrix}$$

2(row 1) + (row 2) replace (row 2)

-3(row 1) + (row 3) replace (row 3)

-5(row 1) + (row 4) replace (row 4)

$$\begin{bmatrix} 1 & -3 & -2 & 1 & | & -3 \\ 0 & 1 & -3 & 0 & | & -7 \\ 0 & 2 & 3 & 0 & | & 4 \\ 0 & 16 & 14 & -7 & | & 33 \end{bmatrix}$$

3(row 2) + (row 1) replace (row 1)
− 2(row 2) + (row 3) replace (row 3)
− 16(row 2) + (row 4) replace (row 4)

$$\left[\begin{array}{cccc|c} 1 & 0 & -11 & 1 & -24 \\ 0 & 1 & -3 & 0 & -7 \\ 0 & 0 & 9 & 0 & 18 \\ 0 & 0 & 62 & -7 & 145 \end{array}\right]$$

$\frac{1}{9}$(row 3)

$$\left[\begin{array}{cccc|c} 1 & 0 & -11 & 1 & -24 \\ 0 & 1 & -3 & 0 & -7 \\ 0 & 0 & 1 & 0 & 2 \\ 0 & 0 & 62 & -7 & 145 \end{array}\right]$$

11(row 3) + (row 1) replace (row 1)
3(row 3) + (row 2) replace (row 2)
− 62(row 3) + (row 4) replace (row 4)

$$\left[\begin{array}{cccc|c} 1 & 0 & 0 & 1 & -2 \\ 0 & 1 & 0 & 0 & -1 \\ 0 & 0 & 1 & 0 & 2 \\ 0 & 0 & 0 & -7 & 21 \end{array}\right]$$

$-\frac{1}{7}$(row 4)

$$\left[\begin{array}{cccc|c} 1 & 0 & 0 & 1 & -2 \\ 0 & 1 & 0 & 0 & -1 \\ 0 & 0 & 1 & 0 & 2 \\ 0 & 0 & 0 & 1 & -3 \end{array}\right]$$

− (row 4) + (row 1) replace (row 1)

$$\left[\begin{array}{cccc|c} 1 & 0 & 0 & 0 & 1 \\ 0 & 1 & 0 & 0 & -1 \\ 0 & 0 & 1 & 0 & 2 \\ 0 & 0 & 0 & 1 & -3 \end{array}\right]$$

The solution set is $\{(1, -1, 2, -3)\}$.

33.

$$\left[\begin{array}{cccc|c} 1 & 3 & -1 & 2 & -2 \\ 2 & 7 & 2 & -1 & 19 \\ -3 & -8 & 3 & 1 & -7 \\ 4 & 11 & -2 & -3 & 19 \end{array}\right]$$

− 2(row 1) + (row 2) replace (row 2)
3(row 1) + (row 3) replace (row 3)
− 4(row 1) + (row 4) replace (row 4)

$$\left[\begin{array}{cccc|c} 1 & 3 & -1 & 2 & -2 \\ 0 & 1 & 4 & -5 & 23 \\ 0 & 1 & 0 & 7 & -13 \\ 0 & -1 & 2 & -11 & 27 \end{array}\right]$$

− 3(row 2) + (row 1) replace (row 1)
(row 2) − (row 3) replace (row 3)
(row 3) + (row 4) replace (row 4)

$$\left[\begin{array}{cccc|c} 1 & 0 & -13 & 17 & -71 \\ 0 & 1 & 4 & -5 & 23 \\ 0 & 0 & 4 & -12 & 36 \\ 0 & 0 & 2 & -4 & 14 \end{array}\right]$$

− 2(row 4) + (row 3) replace (row 4)
$\frac{1}{4}$(row 3)

$$\left[\begin{array}{cccc|c} 1 & 0 & -13 & 17 & -71 \\ 0 & 1 & 4 & -5 & 23 \\ 0 & 0 & 1 & -3 & 9 \\ 0 & 0 & 0 & -4 & 8 \end{array}\right]$$

$-\frac{1}{4}$(row 4) replace (row 4)

$$\left[\begin{array}{cccc|c} 1 & 0 & -13 & 17 & -71 \\ 0 & 1 & 4 & -5 & 23 \\ 0 & 0 & 1 & -3 & 9 \\ 0 & 0 & 0 & 1 & -2 \end{array}\right]$$

3(row 4) + (row 3) replace (row 3)
5(row 4) + (row 2) replace (row 2)
− 17(row 4) + (row 1) replace (row 1)

$$\left[\begin{array}{cccc|c} 1 & 0 & -13 & 0 & -37 \\ 0 & 1 & 4 & 0 & 13 \\ 0 & 0 & 1 & 0 & 3 \\ 0 & 0 & 0 & 1 & -2 \end{array}\right]$$

− 4(row 3) + (row 2) replace (row 2)
13(row 3) + (row 1) replace (row 1)

$$\left[\begin{array}{cccc|c} 1 & 0 & 0 & 0 & 2 \\ 0 & 1 & 0 & 0 & 1 \\ 0 & 0 & 1 & 0 & 3 \\ 0 & 0 & 0 & 1 & -2 \end{array}\right]$$

The solution set is $\{(2, 1, 3, -2)\}$.

35.

$$\left[\begin{array}{cccc|c} 1 & 0 & 0 & 0 & -2 \\ 0 & 1 & 0 & 0 & 4 \\ 0 & 0 & 1 & 0 & -3 \\ 0 & 0 & 0 & 1 & 0 \end{array}\right]$$

The solution set is $\{(-2, 4, -3, 0)\}$.

37.

$$\left[\begin{array}{cccc|c} 1 & 0 & 0 & 0 & -8 \\ 0 & 1 & 0 & 0 & 5 \\ 0 & 0 & 1 & 0 & -2 \\ 0 & 0 & 0 & 0 & 1 \end{array}\right]$$

The last row represents
$0x_1 + 0x_2 + 0x_3 + 0x_4 = 1$
$0 = 1$
This is false, so the system is inconsistent and the solution is $\emptyset$.

39. $\left[\begin{array}{cccc|c} 1 & 0 & 0 & 3 & 5 \\ 0 & 1 & 0 & 0 & -1 \\ 0 & 0 & 1 & 4 & 2 \\ 0 & 0 & 0 & 0 & 0 \end{array}\right]$

The bottom row at all zeros represents an identity. From the other row we get

$x_3 + 4x_4 = 2$

$x_2 = -1$

$x_1 + 3x_4 = 5$

Let $x_4 = k$.

Then $x_3 + 4k = 2$,

$x_3 = -4k + 2$ and

$x_1 + 3k = 5$

$x_1 = -3k + 5$

The solution set is

$\{(-3k + 5, -1, -4k + 2, k)\}$

where k is any real number.

41. $\left[\begin{array}{cccc|c} 1 & 3 & 0 & 0 & 9 \\ 0 & 0 & 1 & 0 & 2 \\ 0 & 0 & 0 & 1 & -3 \\ 0 & 0 & 0 & 0 & 0 \end{array}\right]$

$x_4 = -3$

$x_3 = 2$

$x_1 + 3x_2 = 9$

$x_1 = 9 - 3x_2$

Let $k = x_2$.

The solution set is

$\{(9 - 3k, k, 2, -3)\}$

where k is any real number.

45. $\left[\begin{array}{ccc|c} 1 & -2 & 3 & 4 \\ 3 & -5 & -1 & 7 \end{array}\right]$

-3(row 1) + (row 2) replace (row 2)

$\left[\begin{array}{ccc|c} 1 & -2 & 3 & 4 \\ 0 & 1 & -10 & -5 \end{array}\right]$

2(row 2) + (row 1) replace (row 1)

$\left[\begin{array}{ccc|c} 1 & 0 & -17 & -6 \\ 0 & 1 & -10 & -5 \end{array}\right]$

$y - 10z = -5$

$y = 10z - 5$

$x - 17z = -6$

$x = 17z - 6$

Let $z = k$.

The solution set is $\{(17k - 6, 10k - 5, k)\}$, where k is a real number.

47. $\left[\begin{array}{ccc|c} 2 & -4 & 3 & 8 \\ 3 & 5 & -1 & 7 \end{array}\right]$

$-\frac{3}{2}$(row 1) + (row 2) replace (row 2)

$\left[\begin{array}{ccc|c} 2 & -4 & 3 & 8 \\ 0 & 11 & -\frac{11}{2} & -5 \end{array}\right]$

$\frac{4}{11}$(row 2) + (row 1) replace (row 1)

$\left[\begin{array}{ccc|c} 2 & 0 & 1 & \frac{68}{11} \\ 0 & 11 & -\frac{11}{2} & -5 \end{array}\right]$

$11y - \frac{11}{2}z = -5$

$11y = \frac{11}{2}z - 5$

$y = \frac{1}{2}z - \frac{5}{11}$

$2x + z = \frac{68}{11}$

$2x = -z + \frac{68}{11}$

$x = -\frac{1}{2}z + \frac{34}{11}$

Let $z = k$.

The solution set is

$\left\{\left(-\frac{1}{2}k + \frac{34}{11}, \frac{1}{2}k - \frac{5}{11}, k\right)\right\}$,

where k is any real number.

49. $\left[\begin{array}{ccc|c} 1 & -2 & 4 & 9 \\ 2 & -4 & 8 & 3 \end{array}\right]$

-2(row 1) + (row 2) replace (row 2)

$\left[\begin{array}{ccc|c} 1 & -26 & 4 & 9 \\ 0 & 0 & 0 & -15 \end{array}\right]$

$0x + 0y + 0z = -15$

$0 \neq -15$

Therefore the solution set is $\emptyset$.

PROBLEM SET 15.4 Determinants

1. $\begin{vmatrix} 4 & 3 \\ 2 & 7 \end{vmatrix} = (4)(7) - (2)(3) = 22$

3. $\begin{vmatrix} -3 & 2 \\ 7 & 5 \end{vmatrix} = -15 - 14 = -29$

5. $\begin{vmatrix} 2 & -3 \\ 8 & -2 \end{vmatrix} = (2)(-2) - (8)(-3) = 20$

7. $\begin{vmatrix} -2 & -3 \\ -1 & -4 \end{vmatrix} = 8 - 3 = 5$

9. $\begin{vmatrix} \frac{1}{2} & \frac{1}{3} \\ -3 & -6 \end{vmatrix}$

$= \left(\frac{1}{2}\right)(-6) - (-3)\left(\frac{1}{3}\right) = -2$

11. $\begin{vmatrix} \frac{1}{2} & \frac{2}{3} \\ \frac{3}{4} & -\frac{1}{3} \end{vmatrix} = -\frac{1}{6} - \frac{1}{2} = -\frac{2}{3}$

13. $\begin{vmatrix} 1 & 2 & -1 \\ 3 & 1 & 2 \\ 2 & 4 & 3 \end{vmatrix}$

-3(row 1) + (row 2) replace (row 2)
-2(row 1) + (row 3) replace (row 3)

$\begin{vmatrix} 1 & 2 & -1 \\ 0 & -5 & 5 \\ 0 & 0 & 5 \end{vmatrix}$

Expand about column 1.

$D = 1(-1)^{1+1}\begin{vmatrix} -5 & 5 \\ 0 & 5 \end{vmatrix}$

$D = 1(-25 - 0) = -25$

15. $\begin{vmatrix} 1 & -4 & 1 \\ 2 & 5 & -1 \\ 3 & 3 & 4 \end{vmatrix}$

-2(row 1) + (row 2) replace (row 2)
-3(row 1) + (row 3) replace (row 3)

$\begin{vmatrix} 1 & -4 & 1 \\ 0 & 13 & -3 \\ 0 & 15 & 1 \end{vmatrix}$

Expand about column 1.

$D = 1(-1)^{1+1}\begin{vmatrix} 13 & -3 \\ 15 & 1 \end{vmatrix}$

$D = 13 - (-45) = 58$

17. $\begin{vmatrix} 6 & 12 & 3 \\ -1 & 5 & 1 \\ -3 & 6 & 2 \end{vmatrix}$

6(row 2) + (row 1) replace (row 1)
-3(row 2) + (row 3) replace (row 3)

$\begin{vmatrix} 0 & 42 & 9 \\ -1 & 5 & 1 \\ 0 & -9 & -1 \end{vmatrix}$

Expand about column 1.

$D = (-1)(-1)^{2+1}\begin{vmatrix} 42 & 9 \\ -9 & -1 \end{vmatrix}$

$D = 1(-42 + 81) = 39$

19. $\begin{vmatrix} 2 & -1 & 3 \\ 0 & 3 & 1 \\ 1 & -2 & -1 \end{vmatrix}$

-2(row 3) + (row 1) replace (row 1)

$\begin{vmatrix} 0 & 3 & 5 \\ 0 & 3 & 1 \\ 1 & -2 & -1 \end{vmatrix}$

Expand about column 1.

$D = 1(-1)^{3+1}\begin{vmatrix} 3 & 5 \\ 3 & 1 \end{vmatrix}$

$D = 3 - 15 = -12$

21. $\begin{vmatrix} -3 & -2 & 1 \\ 5 & 0 & 6 \\ 2 & 1 & -4 \end{vmatrix}$

2(row 3) + (row 1) replace (row 1)

$\begin{vmatrix} 1 & 0 & -7 \\ 5 & 0 & 6 \\ 2 & 1 & -4 \end{vmatrix}$

Expand about column 2.

$D = 1(-1)^{3+2}\begin{vmatrix} 1 & -7 \\ 5 & 6 \end{vmatrix}$

$D = (-1)(6 + 35) = -41$

23. $\begin{vmatrix} 3 & -4 & -2 \\ 5 & -2 & 1 \\ 1 & 0 & 0 \end{vmatrix}$

Expand about row 3.

$D = 1(-1)^{3+1}\begin{vmatrix} -4 & -2 \\ -2 & 1 \end{vmatrix}$

$D = -4 - 4 = -8$

25. $\begin{vmatrix} 24 & -1 & 4 \\ 40 & 2 & 0 \\ -16 & 6 & 0 \end{vmatrix}$

Expand about column 3.

$D = 4(-1)^{1+3}\begin{vmatrix} 40 & 2 \\ -16 & 6 \end{vmatrix}$

$D = 4(240 + 32) = 1088$

27. $\begin{vmatrix} 2 & 3 & -4 \\ 4 & 6 & -1 \\ -6 & 1 & -2 \end{vmatrix}$

-2(row 1) + (row 2) replace (row 2)

$\begin{vmatrix} 2 & 3 & -4 \\ 0 & 0 & 7 \\ -6 & 1 & -2 \end{vmatrix}$

Expand about row 2.

$D = 7(-1)^{2+3}\begin{vmatrix} 2 & 3 \\ -6 & 1 \end{vmatrix}$

$D = -7[2 - (-18)]$

$D = -7(20) = -140$

29. $\begin{vmatrix} 1 & -2 & 3 & 2 \\ 2 & -1 & 0 & 4 \\ -3 & 4 & 0 & -2 \\ -1 & 1 & 1 & 5 \end{vmatrix}$

-3(row 4) + (row 1) replace (row 1)

$\begin{vmatrix} 4 & -5 & 0 & -13 \\ 2 & -1 & 0 & 4 \\ -3 & 4 & 0 & -2 \\ -1 & 1 & 1 & 5 \end{vmatrix}$

Expand about column 3.

$D = 1(-1)^{4+3}\begin{vmatrix} 4 & -5 & -13 \\ 2 & -1 & 4 \\ -3 & 4 & -2 \end{vmatrix}$

-5(row 2) + (row 1) replace (row 1)

4(row 2) + (row 3) replace (row 3)

$D = -1\begin{vmatrix} -6 & 0 & -33 \\ 2 & -1 & 4 \\ 5 & 0 & 14 \end{vmatrix}$

Expand about column 2.

$D = -1\left[(-1)(-1)^{2+2}\begin{vmatrix} -6 & -33 \\ 5 & 14 \end{vmatrix}\right]$

$D = (-1)(-1)(-84 + 165)$

$D = 81$

31. $\begin{vmatrix} 3 & -1 & 2 & 3 \\ 1 & 0 & 2 & 1 \\ 2 & 3 & 0 & 1 \\ 5 & 2 & 4 & -5 \end{vmatrix}$

3(row 1) + (row 3) replace (row 3)

2(row 1) + (row 4) replace (row 4)

$\begin{vmatrix} 3 & -1 & 2 & 3 \\ 1 & 0 & 2 & 1 \\ 11 & 0 & 6 & 10 \\ 11 & 0 & 8 & 1 \end{vmatrix}$

Expand about column 2.

$D = -1(-1)^{1+2}\begin{vmatrix} 1 & 2 & 1 \\ 11 & 6 & 10 \\ 11 & 8 & 1 \end{vmatrix}$

$D = (1)\begin{vmatrix} 1 & 2 & 1 \\ 11 & 6 & 10 \\ 11 & 8 & 1 \end{vmatrix}$

-11(row 1) + (row 2) replace (row 2)

-11(row 1) + (row 3) replace (row 3)

$D = \begin{vmatrix} 1 & 2 & 1 \\ 0 & -16 & -1 \\ 0 & -14 & -10 \end{vmatrix}$

Expand about column 1.

$D = 1(-1)^{1+1}\begin{vmatrix} -16 & -1 \\ -14 & -10 \end{vmatrix}$

$D = \begin{vmatrix} -16 & -1 \\ -14 & -10 \end{vmatrix} = 160 - 14 = 146$

33. Property 15.3

35. Property 15.2

37. Property 15.4

39. Property 15.3

41. Property 15.5

Further Investigations

47. – 49. Each of these properties can be verified 3 × 3 matrices by evaluating the determinants of the general 3 × 3 matrices.

PROBLEM SET 15.5 Cramer's Rule

1. $D=\begin{vmatrix}2 & -1\\3 & 2\end{vmatrix}=4-(-3)=7$

$D_x=\begin{vmatrix}-2 & -1\\11 & 2\end{vmatrix}=-4-(-11)=7$

$D_y=\begin{vmatrix}2 & -2\\3 & 11\end{vmatrix}=22-(-6)=28$

$x=\frac{D_x}{D}=\frac{7}{7}=1$

$y=\frac{D_y}{D}=\frac{28}{7}=4$

The solution set is $\{(1,\ 4)\}$.

3. $D=\begin{vmatrix}5 & 2\\3 & -4\end{vmatrix}=-20-6=-26$

$D_x=\begin{vmatrix}5 & 2\\29 & -4\end{vmatrix}=-20-58=-78$

$D_y=\begin{vmatrix}5 & 5\\3 & 29\end{vmatrix}=145-15=130$

$x=\frac{D_x}{D}=\frac{-78}{-26}=3$

$y=\frac{D_y}{D}=\frac{130}{-26}=-5$

The solution set is $\{(3,\ -5)\}$.

5. $D=\begin{vmatrix}5 & -4\\-1 & 2\end{vmatrix}=10-4=6$

$D_x=\begin{vmatrix}14 & -4\\-4 & 2\end{vmatrix}=28-16=12$

$D_y=\begin{vmatrix}5 & 14\\-1 & -4\end{vmatrix}=-20+14=-6$

$x=\frac{D_x}{D}=\frac{12}{6}=2$

$y=\frac{D_y}{D}=\frac{-6}{6}=-1$

The solution set is $\{(2,\ -1)\}$.

7. $D=\begin{vmatrix}-2 & 1\\6 & -3\end{vmatrix}=6-6=0$

Because the equations are not multiples, the system is inconsistent. The solution set is $\emptyset$.

9. $D=\begin{vmatrix}-4 & 3\\4 & -6\end{vmatrix}=24-12=12$

$D_x=\begin{vmatrix}3 & 3\\-5 & -6\end{vmatrix}=-18+15=-3$

$D_y=\begin{vmatrix}-4 & 3\\4 & -5\end{vmatrix}=20-12=8$

$x=\frac{D_x}{D}=\frac{-3}{12}=-\frac{1}{4}$

$y=\frac{D_y}{D}=\frac{8}{12}=\frac{2}{3}$

The solution set is $\left\{\left(-\frac{1}{4},\frac{2}{3}\right)\right\}$.

11. $D=\begin{vmatrix}9 & -1\\8 & 1\end{vmatrix}=9-(-8)=17$

$D_x=\begin{vmatrix}-2 & -1\\4 & 1\end{vmatrix}=-2-(-4)=2$

$D_y=\begin{vmatrix}9 & -2\\8 & 4\end{vmatrix}=36-(-16)=52$

$x=\frac{D_x}{D}=\frac{2}{17}$

$y=\frac{D_y}{D}=\frac{52}{17}$

The solution set is $\left\{\left(\frac{2}{17},\frac{52}{17}\right)\right\}$.

13. $D=\begin{vmatrix}-\frac{2}{3} & \frac{1}{2}\\ \frac{1}{3} & -\frac{3}{2}\end{vmatrix}=1-\frac{1}{6}=\frac{5}{6}$

$$D_x = \begin{vmatrix} -7 & \frac{1}{2} \\ 6 & -\frac{3}{2} \end{vmatrix} = \frac{21}{2} - 3 = \frac{15}{2}$$

$$D_y = \begin{vmatrix} -\frac{2}{3} & -7 \\ \frac{1}{3} & 6 \end{vmatrix} = -4 + \frac{7}{3} = -\frac{5}{3}$$

$$x = \frac{D_x}{D} = \frac{\frac{15}{2}}{\frac{5}{6}} = 9$$

$$y = \frac{D_y}{D} = \frac{-\frac{5}{3}}{\frac{5}{6}} = -2$$

The solution set is $\{(9, -2)\}$.

15. $\begin{pmatrix} 2x + 7y = -1 \\ x = 2 \end{pmatrix}$

$$D = \begin{vmatrix} 2 & 7 \\ 1 & 0 \end{vmatrix} = 0 - 7 = = -7$$

$$D_x = \begin{vmatrix} -1 & 7 \\ 2 & 0 \end{vmatrix} = 0 - 14 = -14$$

$$D_y = \begin{vmatrix} 2 & -1 \\ 1 & 2 \end{vmatrix} = 4 - (-1) = 5$$

$$x = \frac{D_x}{D} = \frac{-14}{-7} = 2$$

$$y = \frac{D_y}{D} = \frac{5}{-7}$$

The solution set is $\left\{\left(2, -\frac{5}{7}\right)\right\}$.

17. $D = \begin{vmatrix} 1 & -1 & 2 \\ 2 & 3 & -4 \\ -1 & 2 & -1 \end{vmatrix}$

Expand about row 1.

$$D = 1(-1)^{1+1}\begin{vmatrix} 3 & -4 \\ 2 & -1 \end{vmatrix} - 1(-1)^{1+2}\begin{vmatrix} 2 & -4 \\ -1 & -1 \end{vmatrix} + 2(-1)^{1+3}\begin{vmatrix} 2 & 3 \\ -1 & 2 \end{vmatrix}$$

$$D = (-3 + 5) + (-2 - 4) + 2(4 + 3)$$

$$D = 5 - 6 + 14 = 13$$

$$D_x = \begin{vmatrix} -8 & -1 & 2 \\ 18 & 3 & -4 \\ 7 & 2 & -1 \end{vmatrix}$$

Expand about row 1.

$$D_x = -8(-1)^{1+1}\begin{vmatrix} 3 & -4 \\ 2 & -1 \end{vmatrix} - 1(-1)^{1+2}\begin{vmatrix} 18 & -4 \\ 7 & -1 \end{vmatrix} + 2(-1)^{1+3}\begin{vmatrix} 18 & 3 \\ 7 & 2 \end{vmatrix}$$

$$D_x = -8(-3+8)+(-18+28)+2(36 - 21)$$

$$D_x = -8(5) + 10 + 2(15) = 0$$

$$D_y = \begin{vmatrix} 1 & -8 & 2 \\ 2 & 18 & -4 \\ -1 & 7 & -1 \end{vmatrix}$$

Expand about row 1.

$$D_y = 1(-1)^{1+1}\begin{vmatrix} 18 & -4 \\ 7 & -1 \end{vmatrix} - 8(-1)^{1+2}\begin{vmatrix} 2 & -4 \\ -1 & -1 \end{vmatrix} + 2(-1)^{1+3}\begin{vmatrix} 2 & 18 \\ -1 & 7 \end{vmatrix}$$

$$D_y = (-18+28)+8(-2 - 4) + 2(14 + 18)$$

$$D_y = 10 + 8(-6) + 2(32) = 26$$

$$D_z = \begin{vmatrix} 1 & -1 & -8 \\ 2 & 3 & 18 \\ -1 & 2 & 7 \end{vmatrix}$$

Expand about row 1.

$$D_z = 1(-1)^{1+1}\begin{vmatrix} 3 & 18 \\ 2 & 7 \end{vmatrix} - 1(-1)^{1+2}\begin{vmatrix} 2 & 18 \\ -1 & 7 \end{vmatrix} - 8(-1)^{1+3}\begin{vmatrix} 2 & 3 \\ -1 & 2 \end{vmatrix}$$

$$D_z = (21 - 36) + 1(14 + 18) - 8(4 + 3)$$

$$D_z = -15 + 32 - 8(7) = -39$$

$$x = \frac{D_x}{D} = \frac{0}{13} = 0$$

$$y = \frac{D_y}{D} = \frac{26}{13} = 2$$

$$z = \frac{D_z}{D} = \frac{-39}{13} = -3$$

The solution set is $\{(0, 2, -3)\}$.

19. $\begin{pmatrix} 2x - 3y + z = -7 \\ -3x + y - z = -7 \\ x - 2y - 5z = -45 \end{pmatrix}$

$$D = \begin{vmatrix} 2 & -3 & 1 \\ -3 & 1 & -1 \\ 1 & -2 & -5 \end{vmatrix}$$

(row 1) + (row 2) replace (row 2)
5(row 1) + (row 3) replace (row 3)

$$D = \begin{vmatrix} 2 & -3 & 1 \\ -1 & -2 & 0 \\ 11 & -17 & 0 \end{vmatrix}$$

Expand about column 3.

$$D = 1(-1)^{1+3}\begin{vmatrix} -1 & -2 \\ 11 & -17 \end{vmatrix}$$

$$D = 17 - (-22) = 39$$

$$D_x = \begin{vmatrix} -7 & -3 & 1 \\ -7 & 1 & -1 \\ -45 & -2 & -5 \end{vmatrix}$$

(row 1) + (row 2) replace (row 2)
5(row 1) + (row 3) replace (row 3)

$$D_x = \begin{vmatrix} -7 & -3 & 1 \\ -14 & -2 & 0 \\ -80 & -17 & 0 \end{vmatrix}$$

Expand about column 3.

$$D_x = 1(-1)^{1+3}\begin{vmatrix} -14 & -2 \\ -80 & -17 \end{vmatrix}$$

$$D_x = -14(-17) - (-80)(-2) = 78$$

$$D_y = \begin{vmatrix} 2 & -7 & 1 \\ -3 & -7 & -1 \\ 1 & -45 & -5 \end{vmatrix}$$

(row 1) + (row 2) replace (row 2)
5(row 1) + (row 3) replace (row 3)

$$D_y = \begin{vmatrix} 2 & -7 & 1 \\ -1 & -14 & 0 \\ 11 & -80 & 0 \end{vmatrix}$$

Expand about column 3.

$$D_y = 1(-1)^{1+3}\begin{vmatrix} -1 & -14 \\ 11 & -80 \end{vmatrix}$$

$$D_y = -1(-80) - (11)(-14) = 234$$

$$D_z = \begin{vmatrix} 2 & -3 & -7 \\ -3 & 1 & -7 \\ 1 & -2 & -45 \end{vmatrix}$$

3(row 2) + (row 1) replace (row 1)
2(row 2) + (row 3) replace (row 3)

$$D_z = \begin{vmatrix} -7 & 0 & -28 \\ -3 & 1 & -7 \\ -5 & 0 & -59 \end{vmatrix}$$

Expand about column 2.

$$D_z = 1(-1)^{2+2}\begin{vmatrix} -7 & -28 \\ -5 & -59 \end{vmatrix}$$

$$D_z = -7(-59) - (-5)(-28) = 273$$

$$x = \frac{D_x}{D} = \frac{78}{39} = 2$$

$$y = \frac{D_y}{D} = \frac{234}{39} = 6$$

$$z = \frac{D_z}{D} = \frac{273}{39} = 7$$

The solution set is $\{(2, 6, 7)\}$.

21. $D = \begin{vmatrix} 4 & 5 & -2 \\ 7 & -1 & 2 \\ 3 & 1 & 4 \end{vmatrix}$

5(row 2) + (row 1) replace (row 1)
(row 3) + (row 2) replace (row 2)

$$D = \begin{vmatrix} 39 & 0 & 8 \\ 10 & 0 & 6 \\ 3 & 1 & 4 \end{vmatrix}$$

Expand about column 2.

$$D = 1(-1)^{3+2}\begin{vmatrix} 39 & 8 \\ 10 & 6 \end{vmatrix}$$

$$D = -1(234 - 80) = -154$$

$$D_x = \begin{vmatrix} -14 & 5 & -2 \\ 42 & -1 & 2 \\ 28 & 1 & 4 \end{vmatrix}$$

5(row 2) + (row 1) replace (row 1)
(row 3) + (row 2) replace (row 2)

$$D_x = \begin{vmatrix} 196 & 0 & 8 \\ 70 & 0 & 6 \\ 3 & 1 & 4 \end{vmatrix}$$

Expand about column 2.

$$D_x = 1(-1)^{3+2}\begin{vmatrix} 196 & 8 \\ 70 & 6 \end{vmatrix}$$

$$D_x = -1(1176 - 560) = -616$$

$$D_y = \begin{vmatrix} 4 & -14 & -2 \\ 7 & 42 & 2 \\ 3 & 28 & 4 \end{vmatrix}$$

2(row 1) + (row 3) replace (row 3)
(row 1) + (row 2) replace (row 2)

$$D_y = \begin{vmatrix} 4 & -14 & -2 \\ 11 & 28 & 0 \\ 11 & 0 & 0 \end{vmatrix}$$

Expand about column 3.

$$D_y = -2(-1)^{1+3}\begin{vmatrix} 11 & 28 \\ 11 & 0 \end{vmatrix}$$

$$D_y = -2(0 - 308) = 616$$

$$D_z = \begin{vmatrix} 4 & 5 & -14 \\ 7 & -1 & 42 \\ 3 & 1 & 28 \end{vmatrix}$$

5(row 2) + (row 1) replace (row 1)
(row 3) + (row 2) replace (row 2)

$$D_z = \begin{vmatrix} 39 & 0 & 8 \\ 10 & 0 & 6 \\ 3 & 1 & 4 \end{vmatrix}$$

Expand about column 2.

$$D_z = 1(-1)^{3+2}\begin{vmatrix} 39 & 196 \\ 10 & 70 \end{vmatrix}$$

$$D_z = -1(2730 - 1960) = -770$$

$$x = \frac{D_x}{D} = \frac{-616}{-154} = 4$$

$$y = \frac{D_y}{D} = \frac{616}{-154} = -4$$

$$z = \frac{D_z}{D} = \frac{-770}{-154} = 5$$

The solution set is $\{(4, -4, 5)\}$.

23. $$\begin{pmatrix} 2x - y + 3z = -17 \\ 3y + z = 5 \\ x - 2y - z = -3 \end{pmatrix}$$

$$D = \begin{vmatrix} 2 & -1 & 3 \\ 0 & 3 & 1 \\ 1 & -2 & -1 \end{vmatrix}$$

−2(row 3) + (row 1) replace (row 1)

$$D = \begin{vmatrix} 0 & 3 & 5 \\ 0 & 3 & 1 \\ 1 & -2 & -1 \end{vmatrix}$$

Expand about column 1.

$$D = 1(-1)^{3+1}\begin{vmatrix} 3 & 5 \\ 3 & 1 \end{vmatrix}$$

$$D = 3 - 15 = -12$$

$$D_x = \begin{vmatrix} -17 & -1 & 3 \\ 5 & 3 & 1 \\ -3 & -2 & -1 \end{vmatrix}$$

(row 3) + (row 2) replace (row 2)
3(row 3) + (row 1) replace (row 1)

$$D_x = \begin{vmatrix} -26 & -7 & 0 \\ 2 & 1 & 0 \\ -3 & -2 & -1 \end{vmatrix}$$

Expand about column 3.

$$D_x = -1(-1)^{3+3}\begin{vmatrix} -26 & -7 \\ 2 & 1 \end{vmatrix}$$

$$D_x = -[-26 - (-14)] = 12$$

$$D_y = \begin{vmatrix} 2 & -17 & 3 \\ 0 & 5 & 1 \\ 1 & -3 & -1 \end{vmatrix}$$

−2(row 3) + (row 1) replace (row 1)

$$D_y = \begin{vmatrix} 0 & -11 & 5 \\ 0 & 5 & 1 \\ 1 & -3 & -1 \end{vmatrix}$$

Expand about column 1.

$$D_y = (1)(-1)^{3+1}\begin{vmatrix} -11 & 5 \\ 5 & 1 \end{vmatrix}$$

$$D_y = -11 - 25 = -36$$

$$D_z = \begin{vmatrix} 2 & -1 & -17 \\ 0 & 3 & 5 \\ 1 & -2 & -3 \end{vmatrix}$$

−2(row 3) + (row 1) replace (row 1)

$$D_z = \begin{vmatrix} 0 & 3 & -11 \\ 0 & 3 & 5 \\ 1 & -2 & -3 \end{vmatrix}$$

Expand about column 1.

$$D_z = 1(-1)^{3+1}\begin{vmatrix} 3 & -11 \\ 3 & 5 \end{vmatrix}$$

$$D_z = 15 - (-33) = 48$$

$$x = \frac{D_x}{D} = \frac{12}{-12} = -1$$

$$y = \frac{D_y}{D} = \frac{-36}{-12} = 3$$

$$z = \frac{D_z}{D} = \frac{48}{-12} = -4$$

The solution set is $\{(-1, 3, -4)\}$.

25. $$D = \begin{vmatrix} 1 & 3 & -4 \\ 2 & -1 & 1 \\ 4 & 5 & -7 \end{vmatrix}$$

5(row 2) + (row 3) replace (row 3)
3(row 2) + (row 1) replace (row 1)

$$D = \begin{vmatrix} 7 & 0 & -1 \\ 2 & -1 & 1 \\ 14 & 0 & -2 \end{vmatrix}$$

Expand about column 2.

$$D = -1(-1)^{2+2}\begin{vmatrix} 7 & -1 \\ 14 & -2 \end{vmatrix}$$

$$D = -[-14-(-14)] = 0$$

$$D_x = \begin{vmatrix} -1 & 3 & -4 \\ 2 & -1 & 1 \\ 0 & 5 & -7 \end{vmatrix}$$

5(row 2) + (row 3) replace (row 3)
3(row 2) + (row 1) replace (row 1)

$$D_x = \begin{vmatrix} 5 & 0 & -1 \\ 2 & -1 & 1 \\ 10 & 0 & -2 \end{vmatrix}$$

Expand about column 2.

$$D_x = -1(-1)^{2+2}\begin{vmatrix} 5 & -1 \\ 10 & -2 \end{vmatrix}$$

$$D_x = -[-10-(-10)] = 0$$

$$D_y = \begin{vmatrix} 1 & -1 & -4 \\ 2 & 2 & 1 \\ 4 & 0 & -7 \end{vmatrix}$$

2(row 1) + (row 2) replace (row 2)

$$D_y = \begin{vmatrix} 1 & -1 & -4 \\ 4 & 0 & -7 \\ 4 & 0 & -7 \end{vmatrix}$$

Expand about column 2.

$$D_y = -1(-1)^{1+2}\begin{vmatrix} 4 & -7 \\ 4 & -7 \end{vmatrix}$$

$$D_y = -28-(-28) = 0$$

$$D_z = \begin{vmatrix} 1 & 3 & -1 \\ 2 & -1 & 2 \\ 4 & 5 & 0 \end{vmatrix}$$

2(row 1) + (row 2) replace (row 2)

$$D_z = \begin{vmatrix} 1 & 3 & -1 \\ 4 & 5 & 0 \\ 4 & 5 & 0 \end{vmatrix}$$

Expand about column 3.

$$D_z = -1(-1)^{1+3}\begin{vmatrix} 4 & 5 \\ 4 & 5 \end{vmatrix}$$

$$D_z = -(20-20) = 0$$

Equations are dependent. The solution set has infinitely many solutions.

27. $$D = \begin{vmatrix} 3 & -2 & -3 \\ 1 & 2 & 3 \\ -1 & 4 & -6 \end{vmatrix}$$

(row 2) + (row 1) replace (row 1)

$$D = \begin{vmatrix} 4 & 0 & 0 \\ 1 & 2 & 3 \\ -1 & 4 & -6 \end{vmatrix}$$

Expand about row 1.

$$D = 4(-1)^{1+1}\begin{vmatrix} 2 & 3 \\ 4 & -6 \end{vmatrix}$$

$$D = 4[-12-12] = -96$$

$$D_x = \begin{vmatrix} -5 & -2 & -3 \\ -3 & 2 & 3 \\ 8 & 4 & -6 \end{vmatrix}$$

(row 2) + (row 1) replace (row 1)

$$D_x = \begin{vmatrix} -8 & 0 & 0 \\ -3 & 2 & 3 \\ 8 & 4 & -6 \end{vmatrix}$$

Expand about row 1.

$$D_x = -8(-1)^{1+1}\begin{vmatrix} 2 & 3 \\ 4 & -6 \end{vmatrix}$$

$$D_x = -8(-12-12) = 192$$

$$D_y = \begin{vmatrix} 3 & -5 & -3 \\ 1 & -3 & 3 \\ -1 & 8 & -6 \end{vmatrix}$$

3(row 3) + (row 1) replace (row 1)
(row 3) + (row 2) replace (row 2)

$$D_y = \begin{vmatrix} 0 & 19 & -21 \\ 0 & 5 & -3 \\ -1 & 8 & -6 \end{vmatrix}$$

Expand about column 1.

$$D_y = -1(-1)^{3+1}\begin{vmatrix} 19 & -21 \\ 5 & -3 \end{vmatrix}$$

$$D_y = -[-57-(-105)] = -48$$

$$D_z = \begin{vmatrix} 3 & -2 & -5 \\ 1 & 2 & -3 \\ -1 & 4 & 8 \end{vmatrix}$$

3(row 3) + (row 1) replace (row 1)
(row 3) + (row 2) replace (row 2)

$$D_z = \begin{vmatrix} 0 & 10 & 19 \\ 0 & 6 & 5 \\ -1 & 4 & 8 \end{vmatrix}$$

Expand about column 1.

$$D_z = -1(-1)^{3+1}\begin{vmatrix} 10 & 19 \\ 6 & 5 \end{vmatrix}$$

$D_z = -[50 - 114] = 64$

$x = \frac{D_x}{D} = \frac{192}{-96} = -2$

$y = \frac{D_y}{D} = \frac{-48}{-96} = \frac{1}{2}$

$z = \frac{D_z}{D} = \frac{64}{-96} = -\frac{2}{3}$

The solution set is $\left\{\left(-2, \frac{1}{2}, -\frac{2}{3}\right)\right\}$.

29. $D = \begin{vmatrix} 1 & -2 & 3 \\ -2 & 4 & -3 \\ 5 & -6 & 6 \end{vmatrix}$

(row 1) + (row 2) replace (row 2)
-2(row 1) + (row 3) replace (row 3)

$D = \begin{vmatrix} 1 & -2 & 3 \\ -1 & 2 & 0 \\ 3 & -2 & 0 \end{vmatrix}$

Expand about column 3.

$D = 3(-1)^{1+3}\begin{vmatrix} -1 & 2 \\ 3 & -2 \end{vmatrix}$

$D = 3(2 - 6) = -12$

$D_x = \begin{vmatrix} 1 & -2 & 3 \\ -3 & 4 & -3 \\ 10 & -6 & 6 \end{vmatrix}$

(row 1) + (row 2) replace (row 2)
-2(row 1) + (row 3) replace (row 3)

$D_x = \begin{vmatrix} 1 & -2 & 3 \\ -2 & 2 & 0 \\ 8 & -2 & 0 \end{vmatrix}$

Expand about column 3.

$D_x = 3(-1)^{1+3}\begin{vmatrix} -2 & 2 \\ 8 & -2 \end{vmatrix}$

$D_x = 3(4 - 16) = 3(-12) = -36$

$D_y = \begin{vmatrix} 1 & 1 & 3 \\ -2 & -3 & -3 \\ 5 & 10 & 6 \end{vmatrix}$

(row 1) + (row 2) replace (row 2)
-2(row 1) + (row 3) replace (row 3)

$D_y = \begin{vmatrix} 1 & 1 & 3 \\ -1 & -2 & 0 \\ 3 & 8 & 0 \end{vmatrix}$

Expand about column 3.

$D_y = 3(-1)^{1+3}\begin{vmatrix} -1 & -2 \\ 3 & 8 \end{vmatrix}$

$D_y = 3[-8 - (-6)] = 3(-2) = -6$

$D_z = \begin{vmatrix} 1 & -2 & 1 \\ -2 & 4 & -3 \\ 5 & -6 & 10 \end{vmatrix}$

2(row 1) + (row 2) replace (row 2)

$D_z = \begin{vmatrix} 1 & -2 & 1 \\ 0 & 0 & -1 \\ 5 & -6 & 10 \end{vmatrix}$

Expand about row 2.

$D_z = -1(-1)^{2+3}\begin{vmatrix} 1 & -2 \\ 5 & -6 \end{vmatrix}$

$D_z = -6 - (-10) = 4$

$x = \frac{D_x}{D} = \frac{-36}{-12} = 3$

$y = \frac{D_y}{D} = \frac{-6}{-12} = \frac{1}{2}$

$z = \frac{D_z}{D} = \frac{4}{-12} = -\frac{1}{3}$

The solution set is $\left\{\left(3, \frac{1}{2}, -\frac{1}{3}\right)\right\}$.

31. $D = \begin{vmatrix} -1 & -1 & 3 \\ -2 & 1 & 7 \\ 3 & 4 & -5 \end{vmatrix}$

-2(row 1) + (row 2) replace (row 2)
3(row 1) + (row 3) replace (row 3)

$D = \begin{vmatrix} -1 & -1 & 3 \\ 0 & 3 & 1 \\ 0 & 1 & 4 \end{vmatrix}$

Expand about column 1.

$D = -1(-1)^{1+1}\begin{vmatrix} 3 & 1 \\ 1 & 4 \end{vmatrix}$

$D = -(12 - 1) = -11$

$D_x = \begin{vmatrix} -2 & -1 & 3 \\ 14 & 1 & 7 \\ 12 & 4 & -5 \end{vmatrix}$

-2(column 2)+(column 1) replace (column 1)
3(column 2) + (column 3) replace (column 3)

$D_x = \begin{vmatrix} 0 & -1 & 0 \\ 12 & 1 & 10 \\ 4 & 4 & 7 \end{vmatrix}$

Expand about row 1.

$D_x = -1(-1)^{1+2}\begin{vmatrix} 12 & 10 \\ 4 & 7 \end{vmatrix}$

$D_x = 84 - 40 = 44$

$$D_y = \begin{vmatrix} -1 & -2 & 3 \\ -2 & 14 & 7 \\ 3 & 12 & -5 \end{vmatrix}$$

-2(row 1) + (row 2) replace (row 2)
3(row 1) + (row 3) replace (row 3)

$$D_y = \begin{vmatrix} -1 & -2 & 3 \\ 0 & 18 & 1 \\ 0 & 6 & 4 \end{vmatrix}$$

Expand about column 1.

$$D_y = -1(-1)^{1+1}\begin{vmatrix} 18 & 1 \\ 6 & 4 \end{vmatrix}$$

$$D_y = -(72-6) = -66$$

$$D_z = \begin{vmatrix} -1 & -1 & -2 \\ -2 & 1 & 14 \\ 3 & 4 & 12 \end{vmatrix}$$

(row 1) + (row 2) replace (row 2)
4(row 1) + (row 3) replace (row 3)

$$D_z = \begin{vmatrix} -1 & -1 & -2 \\ -3 & 0 & 12 \\ -1 & 0 & 4 \end{vmatrix}$$

Expand about column 2.

$$D_z = -1(-1)^{1+2}\begin{vmatrix} -3 & 12 \\ -1 & 4 \end{vmatrix}$$

$$D_z = -12-(-12) = 0$$

$$x = \frac{D_x}{D} = \frac{44}{-11} = -4$$

$$y = \frac{D_y}{D} = \frac{-66}{-11} = 6$$

$$z = \frac{D_z}{D} = \frac{0}{-11} = 0$$

The solution set is $\{(-4, 6, 0)\}$.

Further Investigations

35. (a) If the constant terms are all zero, then $D_x = 0$, $D_y = 0$, and $D_z = 0$.
If $D \neq 0$ then

$$x = \frac{D_x}{D} = \frac{0}{D} = 0$$

$$y = \frac{D_y}{D} = \frac{0}{D} = 0$$

$$z = \frac{D_z}{D} = \frac{0}{D} = 0$$

Hence the solution is $(0, 0, 0)$.

(b) If $D = 0$ and the constants are all zero, then at least one equation is a multiple or linear combination of the other equations. Therefore the system would be dependent.

37. $D = \begin{vmatrix} 2 & -1 & 1 \\ 3 & 2 & 5 \\ 4 & -7 & 1 \end{vmatrix}$

2(row 1) + (row 2) replace (row 2)
-7(row 1) + (row 3) replace (row 3)

$$D = \begin{vmatrix} 2 & -1 & 1 \\ 7 & 0 & 7 \\ -10 & 0 & -6 \end{vmatrix}$$

Expand about column 1.

$$D = -1(-1)^{1+2}\begin{vmatrix} 7 & 7 \\ -10 & -6 \end{vmatrix}$$

$$D = -42-(-70) = 28$$

Since $D \neq 0$, the solution set is $\{(0, 0, 0)\}$.

39. $D = \begin{vmatrix} 2 & -1 & 2 \\ 1 & 2 & 1 \\ 1 & -3 & 1 \end{vmatrix}$

$-$(column 1)+(column 3) replace (column 3)

$$D = \begin{vmatrix} 2 & -1 & 0 \\ 1 & 2 & 0 \\ 1 & -3 & 0 \end{vmatrix} = 0$$

Since $D = 0$, the system has infinitely many solutions.

PROBLEM SET 15.6 Systems Involving Nonlinear Equations

1. $\begin{pmatrix} x^2 + y^2 = 5 \\ x + 2y = 5 \end{pmatrix}$

$x = -2y + 5$
$(-2y+5)^2 + y^2 = 5$
$4y^2 - 20y + 25 + y^2 = 5$
$5y^2 - 20y + 20 = 0$
$5(y^2 - 4y + 4) = 0$
$y^2 - 4y + 4 = 0$
$(y-2)^2 = 0$
$y = 2$

When $y = 2$
$x = -2y + 5$
$x = -2(2) + 5$
$x = 1$
The solution set is $\{(1,\ 2)\}$.

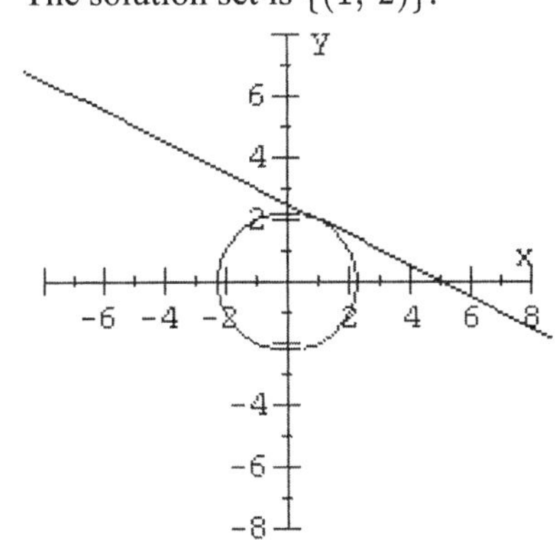

3. $\begin{pmatrix} x^2 + y^2 = 26 \\ x + y = -4 \end{pmatrix}$
$x = -y - 4$
$(-y - 4)^2 + y^2 = 26$
$y^2 + 8y + 16 + y^2 = 26$
$2y^2 + 8y - 10 = 0$
$2(y^2 + 4y - 5) = 0$
$2(y + 5)(y - 1) = 0$
$y + 5 = 0$ or $y - 1 = 0$
$y = -5$ or $y = 1$
When $y = -5$
$x = -y - 4$
$x = -(-5) - 4 = 1$
When $y = 1$
$x = -y + 4$
$x = -1 - 4 = -5$
The solution set is $\{(1,\ -5),\ (-5,\ 1)\}$.

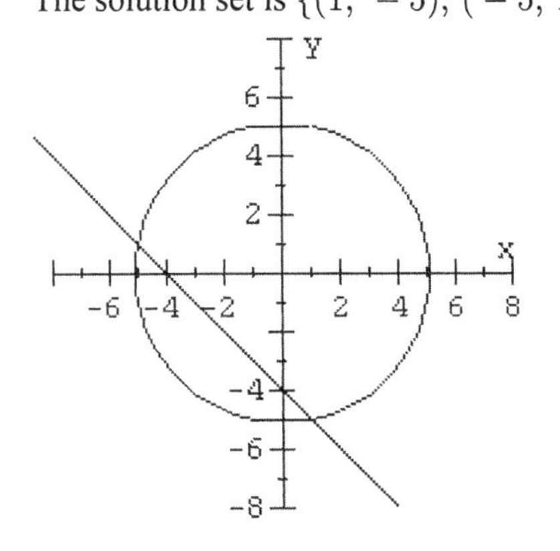

5. $\begin{pmatrix} x^2 + y^2 = 2 \\ x - y = 4 \end{pmatrix}$
$x = y + 4$
$(y + 4)^2 + y^2 = 2$
$y^2 + 8y + 16 + y^2 = 2$
$2y^2 + 8y + 14 = 0$
$2(y^2 + 4y + 7) = 0$
$y^2 + 4y + 7 = 0$

$$y = \frac{-4 \pm \sqrt{4^2 - 4(1)(7)}}{2(1)}$$

$$y = \frac{-4 \pm \sqrt{-12}}{2}$$

$$y = \frac{-4 \pm 2i\sqrt{3}}{2}$$

$$y = -2 \pm i\sqrt{3}$$

When $y = -2 - i\sqrt{3}$
$x = y + 4$
$x = -2 - i\sqrt{3} + 4$
$x = 2 - i\sqrt{3}$
When $y = -2 + i\sqrt{3}$
$x = y + 4$
$x = -2 + i\sqrt{3} + 4$
$x = 2 + i\sqrt{3}$
The solution set is
$\left\{\left(2 + i\sqrt{3},\ -2 + i\sqrt{3}\right),\right.$
$\left.\left(2 - i\sqrt{3},\ -2 - i\sqrt{3}\right)\right\}$.
There are no real solutions.

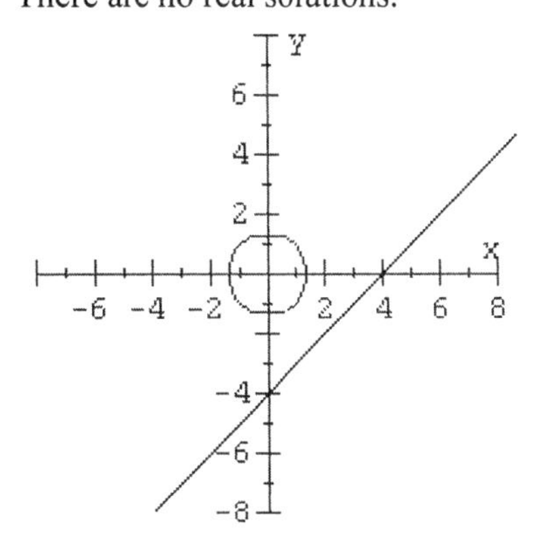

7. $\begin{pmatrix} y = x^2 + 6x + 7 \\ 2x + y = -5 \end{pmatrix}$
$2x + x^2 + 6x + 7 = -5$
$x^2 + 8x + 12 = 0$

Problem Set 15.6

$(x+6)(x+2)=0$
$x+6=0$ or $x+2=0$
$x=-6$ or $x=-2$
When $x=-6$
$y=(-6)^2+6(-6)+7$
$y=7$
When $x=-2$
$y=(-2)^2+6(-2)+7$
$y=-1$
The solution set is $\{(-6,7),(-2,-1)\}$.

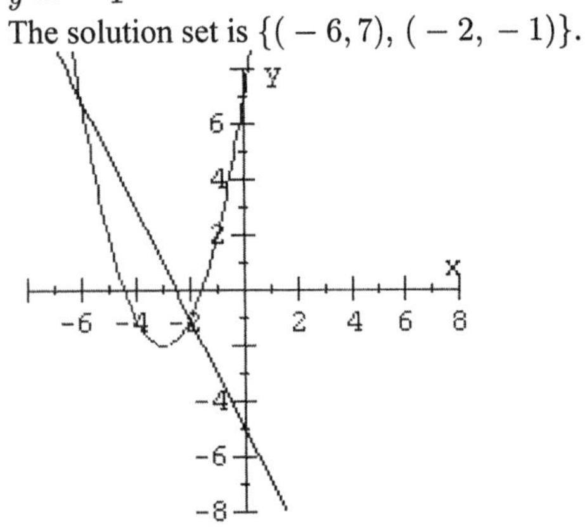

9. $\begin{pmatrix} 2x+y=-2 \\ y=x^2+4x+7 \end{pmatrix}$
$2x+x^2+4x+7=-2$
$x^2+6x+9=0$
$(x+3)^2=0$
$x+3=0$
$x=-3$
When $x=-3$
$2x+y=-2$
$2(-3)+y=-2$
$-6+y=-2$
$y=4$
The solution set is $\{(-3,4)\}$.

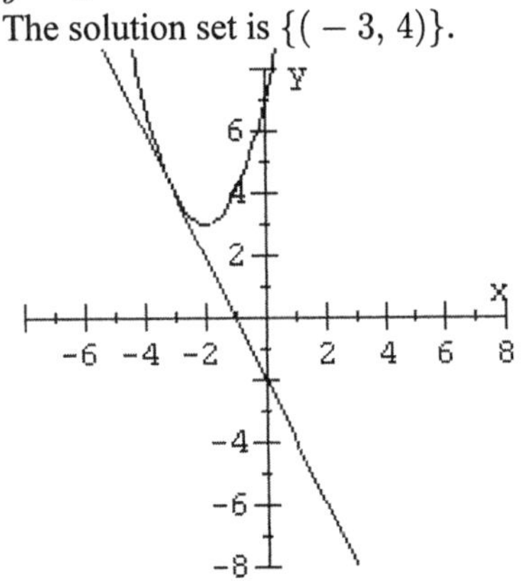

11. $\begin{pmatrix} y=x^2-3 \\ x+y=-4 \end{pmatrix}$
$x+x^2-3=-4$
$x^2+x+1=0$
$$x=\frac{-1\pm\sqrt{1^2-4(1)(1)}}{2(1)}$$
$$x=\frac{-1\pm\sqrt{-3}}{2}$$
$$x=\frac{-1\pm i\sqrt{3}}{2}$$
When $x=\dfrac{-1-i\sqrt{3}}{2}$, $x+y=-4$
$$\frac{-1-i\sqrt{3}}{2}+y=-4$$
$$y=-4-\left(\frac{-1-i\sqrt{3}}{2}\right),$$
$$y=\frac{-8+1+i\sqrt{3}}{2}=\frac{-7+i\sqrt{3}}{2}$$
When $x=\dfrac{-1+i\sqrt{3}}{2}$, $x+y=-4$
$$\frac{-1+i\sqrt{3}}{2}+y=-4$$
$$y=-4-\left(\frac{-1+i\sqrt{3}}{2}\right)$$
$$y=\frac{-8+1-i\sqrt{3}}{2}=\frac{-7-i\sqrt{3}}{2}$$
The solution set is
$$\left\{\left(\frac{-1+i\sqrt{3}}{2},\frac{-7-i\sqrt{3}}{2}\right),\left(\frac{-1-i\sqrt{3}}{2},\frac{-7+i\sqrt{3}}{2}\right)\right\}.$$
There are no real solutions.

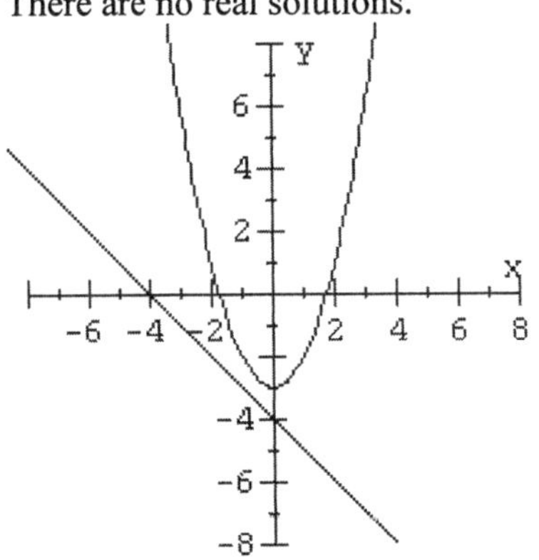

13. $\begin{pmatrix} x^2 + 2y^2 = 9 \\ x - 4y = -9 \end{pmatrix}$

$x = 4y - 9$
$(4y - 9)^2 + 2y^2 = 9$
$16y^2 - 72y + 81 + 2y^2 = 9$
$18y^2 - 72y + 72 = 0$
$18(y^2 - 4y + 4) = 0$
$y^2 - 4y + 4 = 0$
$(y - 2)^2 = 0$
$y - 2 = 0$
$y = 2$
$x = 4y - 9$
$x = 4(2) - 9 = -1$
The solution set is $\{(-1, 2)\}$.

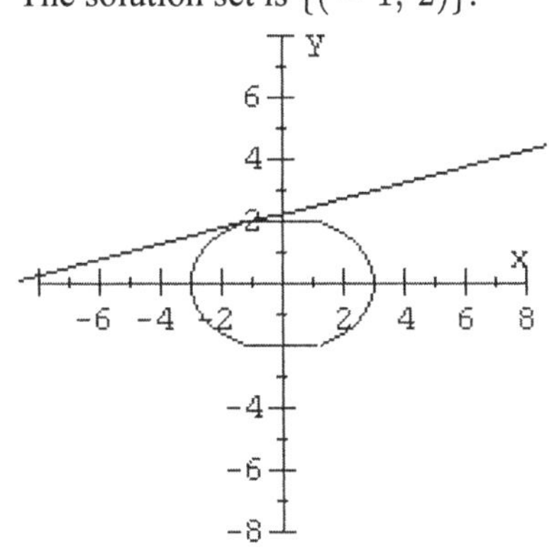

15. $\begin{pmatrix} x + y = -3 \\ x^2 + 2y^2 - 12y - 18 = 0 \end{pmatrix}$

$x = -y - 3$
$(-y - 3)^2 + 2y^2 - 12y - 18 = 0$
$y^2 + 6y + 9 + 2y^2 - 12y - 18 = 0$
$3y^2 - 6y - 9 = 0$
$3(y^2 - 2y - 3) = 0$
$3(y - 3)(y + 1) = 0$
$y - 3 = 0$ or $y + 1 = 0$
$y = 3$ or $y = -1$
When $y = 3$
$x = -3 - 3$
$x = -6$
When $y = -1$
$x = -(-1) - 3$
$x = -2$
The solution set is $\{(-6, 3), (-2, -1)\}$.

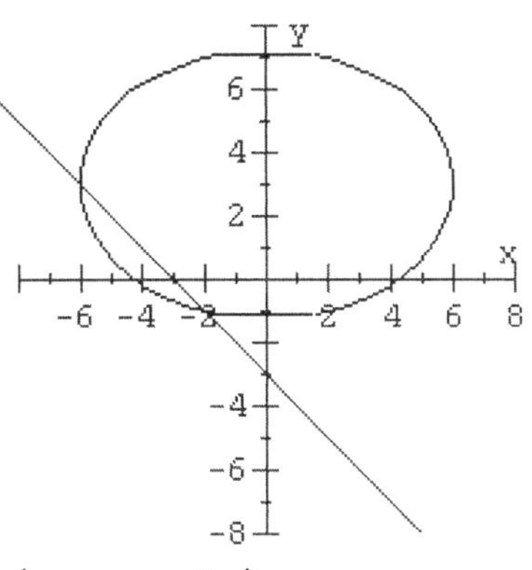

17. $\begin{pmatrix} x - y = 2 \\ x^2 - y^2 = 16 \end{pmatrix}$

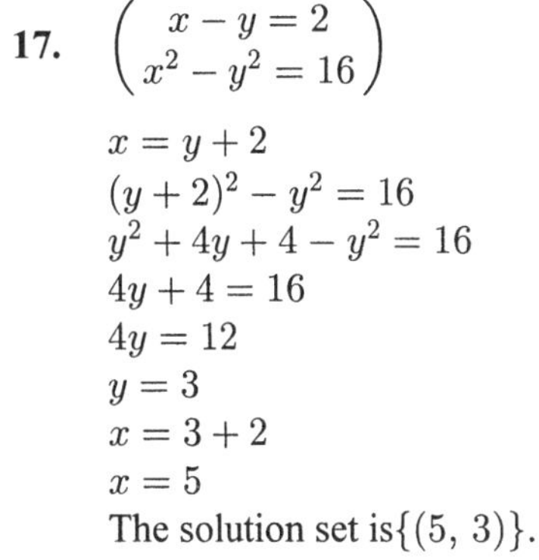

$x = y + 2$
$(y + 2)^2 - y^2 = 16$
$y^2 + 4y + 4 - y^2 = 16$
$4y + 4 = 16$
$4y = 12$
$y = 3$
$x = 3 + 2$
$x = 5$
The solution set is $\{(5, 3)\}$.

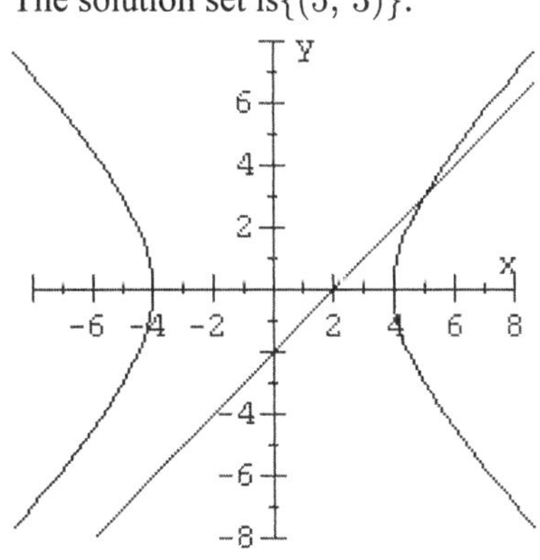

19. $\begin{pmatrix} y = -x^2 + 3 \\ y = x^2 + 1 \end{pmatrix}$

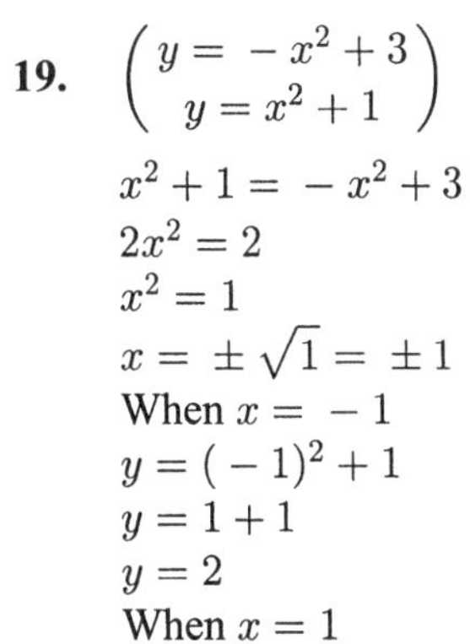

$x^2 + 1 = -x^2 + 3$
$2x^2 = 2$
$x^2 = 1$
$x = \pm\sqrt{1} = \pm 1$
When $x = -1$
$y = (-1)^2 + 1$
$y = 1 + 1$
$y = 2$
When $x = 1$

$y = 1^2 + 1$
$y = 2$
The solution set is $\{(1, 2), (-1, 2)\}$.

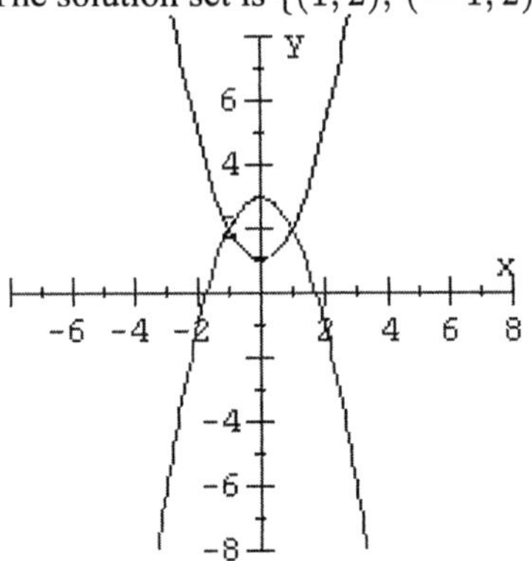

21. $\begin{pmatrix} y = x^2 + 2x - 1 \\ y = x^2 + 4x + 5 \end{pmatrix}$
$x^2 + 2x - 1 = x^2 + 4x + 5$
$2x - 1 = 4x + 5$
$-6 = 2x$
$-3 = x$
$y = x^2 + 2x - 1$
$y = (-3)^2 + 2(-3) - 1$
$y = 2$
The solution set is $\{(-3, 2)\}$.

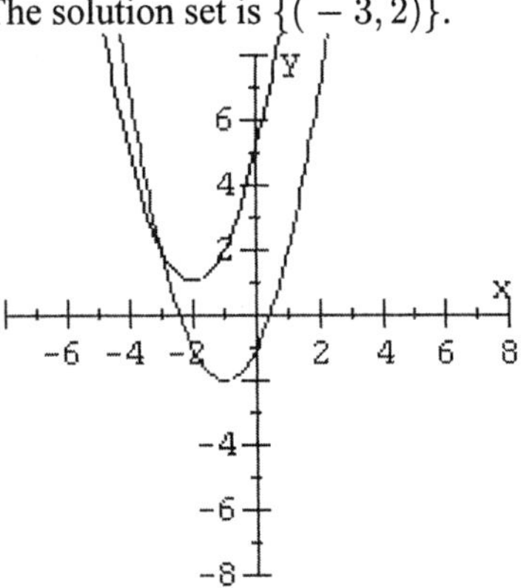

23. $\begin{pmatrix} x^2 - y^2 = 4 \\ x^2 + y^2 = 4 \end{pmatrix}$
Add the equations.
$2x^2 = 8$
$x^2 = 4$
$x = \pm 2$
When $x = -2$
$x^2 + y^2 = 4$
$(-2)^2 + y^2 = 4$
$4 + y^2 = 4$
$y^2 = 0$
$y = 0$
$(-2, 0)$
When $x = 2$
$x^2 + y^2 = 4$
$2^2 + y^2 = 4$
$4 + y^2 = 4$
$y^2 = 0$
$y = 0$
$(2, 0)$
The solution set is $\{(2, 0), (-2, 0)\}$.

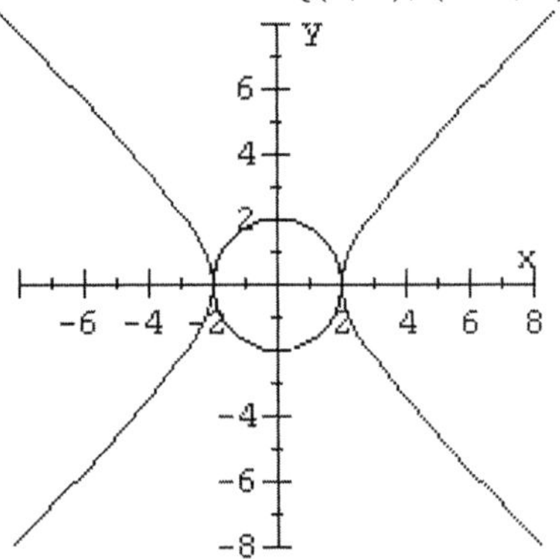

25. $\begin{pmatrix} 8y^2 - 9x^2 = 6 \\ 8x^2 - 3y^2 = 7 \end{pmatrix}$
$3(8y^2 - 9x^2 = 6)$
$8(-3y^2 + 8x^2 = 7)$
Add the equations.

$$\begin{aligned} 24y^2 - 27x^2 &= 18 \\ -24y^2 + 64x^2 &= 56 \\ \hline 37x^2 &= 74 \\ x^2 &= 2 \\ x &= \pm\sqrt{2} \end{aligned}$$

When $x = -\sqrt{2}$
$8x^2 - 3y^2 = 7$
$8(-\sqrt{2})^2 - 3y^2 = 7$
$8(2) - 3y^2 = 7$
$16 - 3y^2 = 7$
$-3y^2 = -9$
$y^2 = 3$
$y = \pm\sqrt{3}$
$(-\sqrt{2}, \sqrt{3})$ and $(-\sqrt{2}, -\sqrt{3})$
When $x = \sqrt{2}$
$8x^2 - 3y^2 = 7$
$8(\sqrt{2})^2 - 3y^2 = 7$
$8(2) - 3y^2 = 7$

$16 - 3y^2 = 7$
$-3y^2 = -9$
$y^2 = 3$
$y = \pm\sqrt{3}$
$(\sqrt{2}, \sqrt{3})$ and $(\sqrt{2}, -\sqrt{3})$
The solution set is
$\left\{\left(\sqrt{2}, \sqrt{3}\right), \left(\sqrt{2}, -\sqrt{3}\right),\right.$
$\left.\left(-\sqrt{2}, \sqrt{3}\right), \left(-\sqrt{2}, -\sqrt{3}\right)\right\}.$

27. $\begin{pmatrix} 2x^2 - 3y^2 = -1 \\ 2x^2 + 3y^2 = 5 \end{pmatrix}$
Add the equations.
$4x^2 = 4$
$x^2 = 1$
$x = \pm 1$
When $x = 1$
$2x^2 + 3y^2 = 5$
$2(1)^2 + 3y^2 = 5$
$2 + 3y^2 = 5$
$3y^2 = 3$
$y^2 = 1$
$y = \pm 1$
When $x = -1$
$2(-1)^2 + 3y^2 = 5$
$2 + 3y^2 = 5$
$3y^3 = 3$
$y^2 = 1$
$y = \pm 1$
The solution set is
$\{(1,1), (1, -1), (-1,1), (-1, -1)\}.$

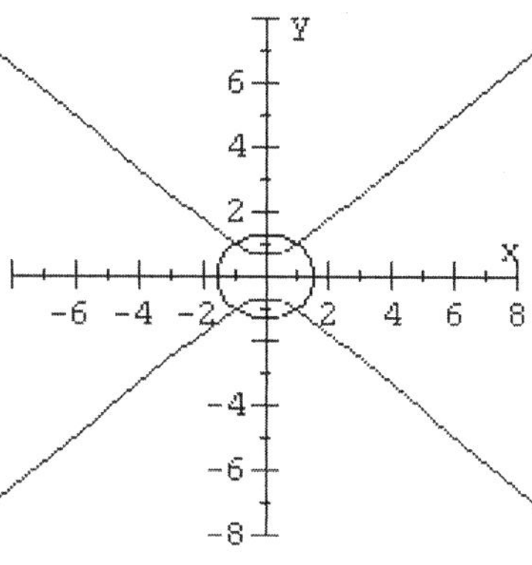

29. $\begin{pmatrix} xy = 3 \\ 2x + 2y = 7 \end{pmatrix}$
$x = \frac{3}{y}$
$2\left(\frac{3}{y}\right) + 2y = 7$
$\frac{6}{y} + 2y = 7$
$6 + 2y^2 = 7y$
$2y^2 - 7y + 6 = 0$
$(2y - 3)(y - 2) = 0$
$2y - 3 = 0$ or $y - 2 = 0$
$2y = 3$ or $y = 2$
$y = \frac{3}{2}$
When $y = \frac{3}{2}$
$x = \frac{3}{y}$
$x = \frac{3}{\frac{3}{2}} = 2$
When $y = 2$
$x = \frac{3}{y}$
$x = \frac{3}{2}$
The solution set is $\left\{\left(2, \frac{3}{2}\right), \left(\frac{3}{2}, 2\right)\right\}.$

Problem Set 15.6

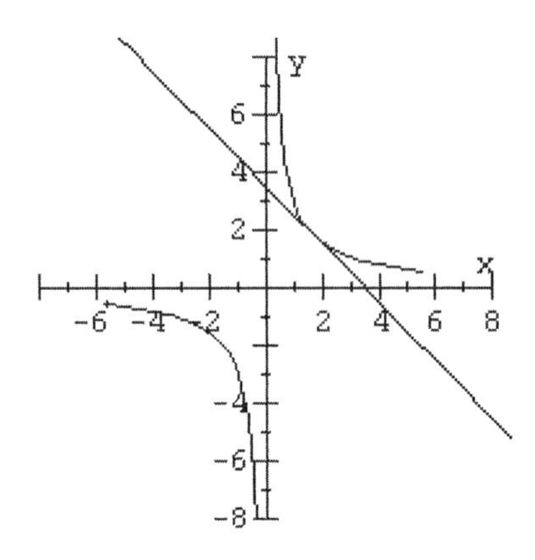

CHAPTER 15 | **Review Problem Set**

1. $\begin{pmatrix} 3x - y = 16 \\ 5x + 7y = -34 \end{pmatrix}$

$3x - y = 16$
$-y = -3x + 16$
$y = 3x - 16$
Substitute $3x - 16$ for y in equation 2.
$5x - 7y = -34$
$5x + 7(3x - 16) = -34$
$5x + 21x - 112 = -34$
$26x = 78$
$x = 3$
$y = 3x - 16$
$y = 3(3) - 16$
$y = 9 - 16$
$y = -7$
The solution set is $\{(3, -7)\}$.

2. $\begin{pmatrix} 6x + 5y = -21 \\ x - 4y = 11 \end{pmatrix}$

$x - 4y = 11$
$x = 4y + 11$
Substitute $4y + 11$ for x in equation 1.
$6x + 5y = -21$
$6(4y + 11) + 5y = -21$
$24y + 66 + 5y = -21$
$29y = -87$
$y = -3$
$x = 4y + 11$
$x = 4(-3) + 11$
$x = -12 + 11$
$x = -1$
The sollution set is $\{(-1, -3)\}$.

3. $\begin{pmatrix} 2x - 3y = 12 \\ 3x + 5y = -20 \end{pmatrix}$

$2x - 3y = 12$
$2x = 3y + 12$
$x = \frac{3}{2}y + 6$
Substitute $\frac{3}{2}y + 6$ for x in equation 2.
$3\left(\frac{3}{2}y + 6\right) + 5y = -20$
$\frac{9}{2}y + 18 + 5y = -20$
$\frac{19}{2}y = -38$
$y = \frac{2}{19}(-38)$
$y = -4$
$x = \frac{3}{2}y + 6$
$x = \frac{3}{2}(-4) + 6$
$x = -6 + 6$
$x = 0$
The solution set is $\{(0, -4)\}$.

4. $\begin{pmatrix} 5x + 8y = 1 \\ 4x + 7y = -2 \end{pmatrix}$

$5x = -8y + 1$

$x = \frac{-8y + 1}{5}$

Substitute $\frac{-8y+1}{5}$ for x in equation 2.

$4\left(\frac{-8y+1}{5}\right)+7y=-2$

$\frac{-32y+4}{5}+7y=-2$

$-32y+4+35y=-10$

$3y=-14$

$y=-\frac{14}{3}$

$x=\frac{-8y+1}{5}$

$x=\frac{1}{5}\left[-8\left(-\frac{14}{3}\right)+1\right]$

$x=\frac{1}{5}\left(\frac{112}{3}+\frac{3}{3}\right)$

$x=\frac{1}{5}\left(\frac{115}{3}\right)$

$x=\frac{23}{3}$

The solution set is $\left\{\left(\frac{23}{3}, -\frac{14}{3}\right)\right\}$.

5. $\begin{pmatrix} 4x-3y=34 \\ 3x+2y=0 \end{pmatrix}$

Multiply equation 1 by 2,
multiply equation 2 by 3,
and then add the resulting equations.

$$\begin{array}{l} 8x-6y=68 \\ \underline{9x+6y=\ \ 0} \\ 17x\qquad=68 \\ \qquad\quad x=4 \end{array}$$

$4x-3y=34$

$4(4)-3y=34$

$16-3y=34$

$-3y=18$

$y=-6$

The solution set is $\{(4, -6)\}$.

6. $\begin{pmatrix} \frac{1}{2}x-\frac{2}{3}y=1 \\ \frac{3}{4}x+\frac{1}{6}y=-1 \end{pmatrix}$

Multiply equation 1 by 6.
Multiply equation 2 by 12.

$\begin{pmatrix} 3x-4y=6 \\ 9x+2y=-12 \end{pmatrix}$

Multiply equation 2 by 2 and then add the result to equation 1.

$$\begin{array}{l} 3x-4y=6 \\ \underline{18x+4y=-24} \\ 21x\qquad=-18 \\ \qquad x=-\frac{18}{21}=-\frac{6}{7} \end{array}$$

$3x-4y=6$

$3\left(-\frac{6}{7}\right)-4y=6$

$-\frac{18}{7}-4y=6$

$-4y=\frac{42}{7}+\frac{18}{7}$

$-4y=\frac{60}{7}$

$y=-\frac{1}{4}\left(\frac{60}{7}\right)$

$y=-\frac{15}{7}$

The solution set is $\left\{\left(-\frac{6}{7}, -\frac{15}{7}\right)\right\}$.

7. $\begin{pmatrix} 2x-y+3z=-19 \\ 3x+2y-4z=21 \\ 5x-4y-z=-8 \end{pmatrix}$

Multiply equation 1 by 2 and add to equation 2 to replace equation 2.
Multiply equation 1 by -4 and add to equation 3 to replace equation 3.

$\begin{pmatrix} 2x-y+3z=-19 \\ 7x+2z=-17 \\ -3x-13z=68 \end{pmatrix}$

Multiply equation 2 by 3 and multiply equation 3 by 7, then add the resulting equations to replace equation 3.

$\begin{pmatrix} 2x-y+3z=-19 \\ 7x+2z=-17 \\ -85z=425 \end{pmatrix}$

$-85z=425$

$z=-5$

$7x+2z=-17$

$7x+2(-5)=-17$

$7x - 10 = -17$
$7x = -7$
$x = -1$
$2x - y + 3z = -19$
$2(-1) - y + 3(-5) = -19$
$-2 - y - 15 = -19$
$-y = -2$
$y = 2$
The solution set is $\{(-1, 2, -5)\}$.

8. $\begin{pmatrix} 3x + 2y - 4z = 4 \\ 5x + 3y - z = 2 \\ 4x - 2y + 3z = 11 \end{pmatrix}$

Multiply equation 2 by -4 and add to equation 1 to replace equation 1.
Multiply equation 2 by 3 and add to equation 3 to replace equation 3.

$\begin{pmatrix} -17x - 10y = -4 \\ 5x + 3y - z = 2 \\ 19x + 7y = 17 \end{pmatrix}$

Multiply equation 1 by 19 and multiply equation 3 by 17, then add the resulting equations to replace equation 3.

$\begin{pmatrix} -17x - 10y = -4 \\ 5x + 3y - z = 2 \\ -71y = 213 \end{pmatrix}$

$-71y = 213$
$y = -3$
$-17x - 10y = -4$
$-17x - 10(-3) = -4$
$-17x + 30 = -4$
$-17x = -34$
$x = 2$
$5x + 3y - z = 2$
$5(2) + 3(-3) - z = 2$
$10 - 9 - z = 2$
$1 - z = 2$
$-z = 1$
$z = -1$
The solution set is $\{(2, -3, -1)\}$.

9. $\begin{pmatrix} x - 3y = 17 \\ -3x + 2y = -23 \end{pmatrix}$

$\left[\begin{array}{cc|c} 1 & -3 & 17 \\ -3 & 2 & -23 \end{array}\right]$

3(row 1) + (row 2) replace (row 2)

$\left[\begin{array}{cc|c} 1 & -3 & 17 \\ 0 & -7 & 28 \end{array}\right]$

$-\frac{1}{7}$(row 2)

$\left[\begin{array}{cc|c} 1 & -3 & 17 \\ 0 & 1 & -4 \end{array}\right]$

3(row 2) + (row 1) replace (row 1)

$\left[\begin{array}{cc|c} 1 & 0 & 5 \\ 0 & 1 & -4 \end{array}\right]$

The solution set is $\{(5, -4)\}$.

10. $\begin{pmatrix} 2x + 3y = 25 \\ 3x - 5y = -29 \end{pmatrix}$

$\left[\begin{array}{cc|c} 2 & 3 & 25 \\ 3 & -5 & -29 \end{array}\right]$

-3(row 1) + 2(row 2) replace (row 2)

$\left[\begin{array}{cc|c} 2 & 3 & 25 \\ 0 & -19 & -133 \end{array}\right]$

$-19y = -133$
$y = 7$
$2x + 3y = 25$
$2x + 3(7) = 25$
$2x + 21 = 25$
$2x = 4$
$x = 2$
The solution set is $\{(2, 7)\}$.

11. $\begin{pmatrix} x - 2y + z = -7 \\ 2x - 3y + 4z = -14 \\ -3x + y - 2z = 10 \end{pmatrix}$

$\left[\begin{array}{ccc|c} 1 & -2 & 1 & -7 \\ 2 & -3 & 4 & -14 \\ -3 & 1 & -2 & 10 \end{array}\right]$

-2(row 1) + (row 2) replace (row 2)
3(row 1) + (row 3) replace (row 3)

$\left[\begin{array}{ccc|c} 1 & -2 & 1 & -7 \\ 0 & 1 & 2 & 0 \\ 0 & -5 & 1 & -11 \end{array}\right]$

2(row 2) + (row 1) replace (row 1)
5(row 2) + (row 3) replace (row 3)

$\left[\begin{array}{ccc|c} 1 & 0 & 5 & -7 \\ 0 & 1 & 2 & 0 \\ 0 & 0 & 11 & -11 \end{array}\right]$

$\frac{1}{11}$(row 3)

$$\left[\begin{array}{ccc|c} 1 & 0 & 5 & -7 \\ 0 & 1 & 2 & 0 \\ 0 & 0 & 1 & -1 \end{array}\right]$$

$-2(\text{row }3) + (\text{row }2)$ replace (row 2)
$-5(\text{row }3) + (\text{row }1)$ replace (row 1)

$$\left[\begin{array}{ccc|c} 1 & 0 & 0 & -2 \\ 0 & 1 & 0 & 2 \\ 0 & 0 & 1 & -1 \end{array}\right]$$

The solution set is $\{(-2, 2, -1)\}$.

12. $\begin{pmatrix} -2x - 7y + z = 9 \\ x + 3y - 4z = -11 \\ 4x + 5y - 3z = -11 \end{pmatrix}$

$$\left[\begin{array}{ccc|c} -2 & -7 & 1 & 9 \\ 1 & 3 & -4 & -11 \\ 4 & 5 & -3 & -11 \end{array}\right]$$

$2(\text{row }2) + (\text{row }1)$ replace (row 1)
$-4(\text{row }2) + (\text{row }3)$ replace (row 3)

$$\left[\begin{array}{ccc|c} 0 & -1 & -7 & -13 \\ 1 & 3 & -4 & -11 \\ 0 & -7 & 13 & 33 \end{array}\right]$$

$-7(\text{row }1) + (\text{row }3)$ replace (row 3)

$$\left[\begin{array}{ccc|c} 0 & -1 & -7 & -13 \\ 1 & 3 & -4 & -11 \\ 0 & 0 & 62 & 124 \end{array}\right]$$

Interchange (row 1) and (row 2).
$\frac{1}{62}(\text{row }3)$ replace (row 3)

$$\left[\begin{array}{ccc|c} 1 & 3 & -4 & -11 \\ 0 & -1 & -7 & -13 \\ 0 & 0 & 1 & 2 \end{array}\right]$$

$3(\text{row }2) + (\text{row }1)$ replace (row 1)
$-1(\text{row }2)$ replace (row 2)

$$\left[\begin{array}{ccc|c} 1 & 0 & -25 & -50 \\ 0 & 1 & 7 & 13 \\ 0 & 0 & 1 & 2 \end{array}\right]$$

$-7(\text{row }3) + (\text{row }2)$ replace (row 2)
$25(\text{row }3) + (\text{row }1)$ replace (row 1)

$$\left[\begin{array}{ccc|c} 1 & 0 & 0 & 0 \\ 0 & 1 & 0 & -1 \\ 0 & 0 & 1 & 2 \end{array}\right]$$

The solution set is $\{(0, -1, 2)\}$.

13. $\begin{pmatrix} 5x + 3y = -18 \\ 4x - 9y = -3 \end{pmatrix}$

$$D = \begin{vmatrix} 5 & 3 \\ 4 & -9 \end{vmatrix} = -45 - 12 = -57$$

$$D_x = \begin{vmatrix} -18 & 3 \\ -3 & -9 \end{vmatrix} = 162 - (-9) = 171$$

$$D_y = \begin{vmatrix} 5 & -18 \\ 4 & -3 \end{vmatrix} = -15 - (-72) = 57$$

$$x = \frac{D_x}{D} = \frac{171}{-57} = -3$$

$$y = \frac{D_y}{D} = \frac{57}{-57} = -1$$

The solution set is $\{(-3, -1)\}$.

14. $\begin{pmatrix} 0.2x + 0.3y = 2.6 \\ 0.5x - 0.1y = 1.4 \end{pmatrix}$

$$D = \begin{vmatrix} 0.2 & 0.3 \\ 0.5 & -0.1 \end{vmatrix}$$

$$D = 0.2(-0.1) - (0.5)(0.3) = -0.17$$

$$D_x = \begin{vmatrix} 2.6 & 0.3 \\ 1.4 & -0.1 \end{vmatrix}$$

$$D_x = 2.6(-0.1) - 1.4(0.3) = -0.68$$

$$D_y = \begin{vmatrix} 0.2 & 2.6 \\ 0.5 & 1.4 \end{vmatrix}$$

$$D_y = 0.2(1.4) - 0.5(2.6) = -1.02$$

$$x = \frac{D_x}{D} = \frac{-0.68}{-0.17} = 4$$

$$y = \frac{D_y}{D} = \frac{-1.02}{-0.17} = 6$$

The solution set is $\{(4, 6)\}$.

15. $\begin{pmatrix} 2x - 3y - 3z = 25 \\ 3x + y + 2z = -5 \\ 5x - 2y - 4z = 32 \end{pmatrix}$

$$D = \begin{vmatrix} 2 & -3 & -3 \\ 3 & 1 & 2 \\ 5 & -2 & -4 \end{vmatrix}$$

$2(\text{row }2) + (\text{row }3)$ replace (row 3)

$$D = \begin{vmatrix} 2 & -3 & -3 \\ 3 & 1 & 2 \\ 11 & 0 & 0 \end{vmatrix}$$

Expand about row 3.

$$D = 11(-1)^{3+1}\begin{vmatrix} -3 & -3 \\ 1 & 2 \end{vmatrix}$$

$$D = 11[-6 - (-3)] = -33$$

$$D_x = \begin{vmatrix} 25 & -3 & -3 \\ -5 & 1 & 2 \\ 32 & -2 & -4 \end{vmatrix}$$

2(row 2) + (row 3) replace (row 3)

$$D_x = \begin{vmatrix} 25 & -3 & -3 \\ -5 & 1 & 2 \\ 22 & 0 & 0 \end{vmatrix}$$

Expand about row 3.

$$D_x = 22(-1)^{3+1}\begin{vmatrix} -3 & -3 \\ 1 & 2 \end{vmatrix}$$

$$D_x = 22[-6-(-3)] = -66$$

$$D_y = \begin{vmatrix} 2 & 25 & -3 \\ 3 & -5 & 2 \\ 5 & 2 & -4 \end{vmatrix}$$

2(row 2) + (row 3) replace (row 3)

$$D_y = \begin{vmatrix} 2 & 25 & -3 \\ 3 & -5 & 2 \\ 11 & 22 & 0 \end{vmatrix}$$

Expand about column 3.

$$D_y = -3(-1)^{1+3}\begin{vmatrix} 3 & -5 \\ 11 & 22 \end{vmatrix}$$

$$+2(-1)^{2+3}\begin{vmatrix} 2 & 25 \\ 11 & 22 \end{vmatrix}$$

$$D_y = -3[66-(-55)] - 2[44-275]$$

$$D_y = -3(121) - 2(-231) = 99$$

$$D_z = \begin{vmatrix} 2 & -3 & 25 \\ 3 & 1 & -5 \\ 5 & -2 & 32 \end{vmatrix}$$

3(row 2) + (row 1) replace (row 1)
2(row 2) + (row 3) replace (row 3)

$$D_z = \begin{vmatrix} 11 & 0 & 10 \\ 3 & 1 & -5 \\ 11 & 0 & 22 \end{vmatrix}$$

Expand about column 2.

$$D_z = 1(-1)^{2+2}\begin{vmatrix} 11 & 10 \\ 11 & 22 \end{vmatrix}$$

$$D_z = 11(22) - 11(10) = 132$$

$$x = \frac{D_x}{D} = \frac{-66}{-33} = 2$$

$$y = \frac{D_y}{D} = \frac{99}{-33} = -3$$

$$z = \frac{D_z}{D} = \frac{132}{-33} = -4$$

The solution set is $\{(2, -3, -4)\}$.

16. $\begin{pmatrix} 3x - y + z = -10 \\ 6x - 2y + 5z = -35 \\ 7x + 3y - 4z = 19 \end{pmatrix}$

$$D = \begin{vmatrix} 3 & -1 & 1 \\ 6 & -2 & 5 \\ 7 & 3 & -4 \end{vmatrix}$$

−2(row 1) + (row 2) replace (row 2)

$$D = \begin{vmatrix} 3 & -1 & 1 \\ 0 & 0 & 3 \\ 7 & 3 & -4 \end{vmatrix}$$

Expand about row 2.

$$D = 3(-1)^{2+3}\begin{vmatrix} 3 & -1 \\ 7 & 3 \end{vmatrix}$$

$$D = -3[9-(-7)] = -48$$

$$D_x = \begin{vmatrix} -10 & -1 & 1 \\ -35 & -2 & 5 \\ 19 & 3 & -4 \end{vmatrix}$$

−2(row 1) + (row 2) replace (row 2)
3(row 1) + (row 3) replace (row 3)

$$D_x = \begin{vmatrix} -10 & -1 & 1 \\ -15 & 0 & 3 \\ -11 & 0 & -1 \end{vmatrix}$$

Expand about column 2.

$$D_x = -1(-1)^{1+2}\begin{vmatrix} -15 & 3 \\ -11 & -1 \end{vmatrix}$$

$$D_x = 15 - (-33) = 48$$

$$D_y = \begin{vmatrix} 3 & -10 & 1 \\ 6 & -35 & 5 \\ 7 & 19 & -4 \end{vmatrix}$$

−5(row 1) + (row 2) replace (row 2)
4(row 1) + (row 3) replace (row 3)

$$D_y = \begin{vmatrix} 3 & -10 & 1 \\ -9 & 15 & 0 \\ 19 & -21 & 0 \end{vmatrix}$$

Expand about column 3.

$$D_y = 1(-1)^{1+3}\begin{vmatrix} -9 & 15 \\ 19 & -21 \end{vmatrix}$$

$$D_y = [-9(-21) - (19)15] = -96$$

$$D_z = \begin{vmatrix} 3 & -1 & -10 \\ 6 & -2 & -35 \\ 7 & 3 & 19 \end{vmatrix}$$

−2(row 1) + (row 2) replace (row 2)

$$D_z = \begin{vmatrix} 3 & -1 & -10 \\ 0 & 0 & -15 \\ 7 & 3 & 19 \end{vmatrix}$$

Expand about row 2.

$$D_z = -15(-1)^{2+3}\begin{vmatrix} 3 & -1 \\ 7 & 3 \end{vmatrix}$$

$D_z = 15[9 - (-7)] = 240$

$x = \dfrac{D_x}{D} = \dfrac{48}{-48} = -1$

$y = \dfrac{D_y}{D} = \dfrac{-96}{-48} = 2$

$z = \dfrac{D_z}{D} = \dfrac{240}{-48} = -5$

The solution set is $\{(-1, 2, -5)\}$.

17. $\begin{pmatrix} 4x + 7y = -15 \\ 3x - 2y = 25 \end{pmatrix}$

Multiply equation 1 by 2,
multiply equation 2 by 7,
and then add the resulting equations.

$$\begin{array}{rl} 8x + 14y = & -30 \\ 21x - 14y = & 175 \\ \hline 29x \qquad = & 145 \end{array}$$

$x = 5$

$4x + 7y = -15$

$4(5) + 7y = -15$

$20 + 7y = -15$

$7y = -35$

$y = -5$

The solution set is $\{(5, -5)\}$.

18. $\begin{pmatrix} \frac{3}{4}x - \frac{1}{2}y = -15 \\ \frac{2}{3}x + \frac{1}{4}y = -5 \end{pmatrix}$

Multiply equation 1 by 4 and
multiply equation 2 by 12.

$\begin{pmatrix} 3x - 2y = -60 \\ 8x + 3y = -60 \end{pmatrix}$

Multiply equation 1 by 3 and multiply equation 2 by 2, then add the resulting equations to replace equation 2.

$\begin{pmatrix} 3x - 2y = -60 \\ 25x = -300 \end{pmatrix}$

$25x = -300$

$x = -12$

$3x - 2y = -60$

$3(-12) - 2y = -60$

$-36 - 2y = -60$

$-2y = -24$

$y = 12$

The solution set is $\{(-12, 12)\}$.

19. $\begin{pmatrix} x + 4y = 3 \\ 3x - 2y = 1 \end{pmatrix}$

$x = -4y + 3$

Substitute $-4y + 3$ for x in equation 2.

$3x - 2y = 1$

$3(-4y + 3) - 2y = 1$

$-12y + 9 - 2y = 1$

$-14y = -8$

$y = \dfrac{-8}{-14} = \dfrac{4}{7}$

$x = -4x + 3$

$x = -4\left(\dfrac{4}{7}\right) + 3$

$x = \dfrac{5}{7}$

The solution set is $\left\{\left(\dfrac{5}{7}, \dfrac{4}{7}\right)\right\}$.

20. $\begin{pmatrix} 7x - 3y = -49 \\ y = \frac{3}{5}x - 1 \end{pmatrix}$

Substitute $\frac{3}{5}x - 1$ for y in equation 1.

$7x - 3y = -47$

$7x - 3\left(\dfrac{3}{5}x - 1\right) = -49$

$7x - \dfrac{9}{5}x + 3 = -49$

$\dfrac{26}{5}x = -52$

$\dfrac{5}{26}\left(\dfrac{26}{5}x\right) = \dfrac{5}{26}(-52)$

$x = -10$

$y = \dfrac{3}{5}x - 1$

$y = \dfrac{3}{5}(-10) - 1$

$y = -6 - 1$

$y = -7$
The solution set is $\{(-10, -7)\}$.

21. $\begin{pmatrix} 2x + 4y = -1 \\ 3x + 6y = 5 \end{pmatrix}$
Multiply equation 1 by -3,
multiply equation 2 by 2,
and then add the resulting equations.

$$\begin{array}{r} -6x - 12y = 3 \\ 6x + 12y = 10 \\ \hline 0 = 13 \end{array}$$

The result, $0 = 13$, is a false statement.
Therefore the solution set is $\emptyset$.

22. $\begin{pmatrix} x - y - z = 4 \\ -3x + 2y + 5z = -21 \\ 5x - 3y - 7z = 30 \end{pmatrix}$
Multiply equation 1 by 3 and add to equation 2 to replace equation 2.
Multiply equation 1 by -5 and add to equation 3 to replace equation 3.

$$\begin{pmatrix} x - y - z = 4 \\ -y + 2z = -9 \\ 2y - 2z = 10 \end{pmatrix}$$

Add equation 2 to equation 3 to replace equation 3.

$$\begin{pmatrix} x - y - z = 4 \\ -y + 2z = -9 \\ y = 1 \end{pmatrix}$$

$y = 1$
$-y + 2z = -9$
$-1 + 2z = -9$
$2z = -8$
$z = -4$
$x - y - z = 4$
$x - 1 - (-4) = 4$
$x + 3 = 4$
$x = 1$
The solution set is $\{(1, 1, -4)\}$.

23. $\begin{pmatrix} 2x - y + z = -7 \\ -5x + 2y - 3z = 17 \\ 3x + y + 7z = -5 \end{pmatrix}$

$$\left[\begin{array}{ccc|c} 2 & -1 & 1 & -7 \\ -5 & 2 & -3 & 17 \\ 3 & 1 & 7 & -5 \end{array}\right]$$

Interchange (row 1) and (row 2).

$$\left[\begin{array}{ccc|c} -5 & 2 & -3 & 17 \\ 2 & -1 & 1 & -7 \\ 3 & 1 & 7 & -5 \end{array}\right]$$

(row 2) + (row 3) replace (row 3)
2(row 2) + (row 1) replace (row 1)

$$\left[\begin{array}{ccc|c} -1 & 0 & -1 & 3 \\ 2 & -1 & 1 & -7 \\ 5 & 0 & 8 & -12 \end{array}\right]$$

2(row 1) + (row 2) replace (row 2)
5(row 1) + (row 3) replace (row 3)

$$\left[\begin{array}{ccc|c} -1 & 0 & -1 & 3 \\ 0 & -1 & -1 & -1 \\ 0 & 0 & 3 & 3 \end{array}\right]$$

$\frac{1}{3}$(row 3) replace (row 3)

$$\left[\begin{array}{ccc|c} -1 & 0 & -1 & 3 \\ 0 & -1 & -1 & -1 \\ 0 & 0 & 1 & 1 \end{array}\right]$$

(row 3) + (row 2) replace (row 2)
(row 3) + (row 1) replace (row 1)

$$\left[\begin{array}{ccc|c} -1 & 0 & 0 & 4 \\ 0 & -1 & 0 & 0 \\ 0 & 0 & 1 & 1 \end{array}\right]$$

– (row 1) replace (row 1)
– (row 2) replace (row 2)

$$\left[\begin{array}{ccc|c} 1 & 0 & 0 & -4 \\ 0 & 1 & 0 & 0 \\ 0 & 0 & 1 & 1 \end{array}\right]$$

The solution set is $\{(-4, 0, 1)\}$.

24. $\begin{pmatrix} 3x - 2y - 5z = 2 \\ -4x + 3y + 11z = 3 \\ 2x - y + z = -1 \end{pmatrix}$
Multiply equation 3 by -2 and add to equation 1 to replace equation 1.
Multiply equation 3 by 3 and add to equation 2 to replace equation 2.

$$\begin{pmatrix} -x - 7z = 4 \\ 2x + 14z = 0 \\ 2x - y + z = -1 \end{pmatrix}$$

Multiply equation 1 by 2 and add to equation 2 to replace equation 2.

$$\begin{pmatrix} -x - 7z = 4 \\ 0 \neq 8 \\ 2x - y + z = -1 \end{pmatrix}$$

Since $0 \neq 8$, the system is inconsistent.
The solution set is $\emptyset$.

25. $\begin{pmatrix} x - y + z = 7 \\ 2x - 3y - z = 2 \\ 3x - 4y = 9 \end{pmatrix}$

Multiply equation 1 by -2 and add to equation 2 to replace equation 2.
Multiply equation 1 by -3 and add to equation 3 to replace equation 3.

$\begin{pmatrix} x - y + z = 7 \\ -y - 3z = -12 \\ -y - 3z = -12 \end{pmatrix}$

Multiply equation 2 by -1 and add to equation 3 to replace equation 3.

$\begin{pmatrix} x - y + z = 7 \\ -y - 3z = -12 \\ 0 = 0 \end{pmatrix}$

Equation 3 is a true statement for all values of x, y, and z.
Let $k = z$.

$-y - 3z = -12$
$-y - 3k = -12$
$-y = 3k - 12$
$y = -3k + 12$
$x - y + z = 7$
$x - (-3k + 12) + k = 7$
$x + 3k - 12 + k = 7$
$x + 4k - 12 = 7$
$x = -4k + 19$

The solution set is
$\{(-4k + 19, -3k + 12, k) \mid k \text{ is a real number}\}$.

26. $\begin{pmatrix} 7x - y + z = -4 \\ -2x + 9y - 3z = -50 \\ x - 5y + 4z = 42 \end{pmatrix}$

$\left[\begin{array}{ccc|c} 7 & -1 & 1 & -4 \\ -2 & 9 & -3 & -50 \\ 1 & -5 & 4 & 42 \end{array}\right]$

Interchange (row 1) and (row 3).

$\left[\begin{array}{ccc|c} 1 & -5 & 4 & 42 \\ -2 & 9 & -3 & -50 \\ 7 & -1 & 1 & -4 \end{array}\right]$

-7(row 1) + (row 3) replace (row 3)
2(row 1) + (row 2) replace (row 2)

$\left[\begin{array}{ccc|c} 1 & -5 & 4 & 42 \\ 0 & -1 & 5 & 34 \\ 0 & 34 & -27 & -298 \end{array}\right]$

34(row 2) + (row 3) replace (row 3)
-5(row 2) + (row 1) replace (row 1)

$\left[\begin{array}{ccc|c} 1 & 0 & -21 & -128 \\ 0 & -1 & 5 & 34 \\ 0 & 0 & 143 & 858 \end{array}\right]$

$\frac{1}{143}$(row 3) replace (row 3)
-1(row 2) replace (row 2)

$\left[\begin{array}{ccc|c} 1 & 0 & -21 & -128 \\ 0 & 1 & -5 & -34 \\ 0 & 0 & 1 & 6 \end{array}\right]$

5(row 3) + (row 2) replace (row 2)
21(row 3) + (row 1) replace (row 1)

$\left[\begin{array}{ccc|c} 1 & 0 & 0 & -2 \\ 0 & 1 & 0 & -4 \\ 0 & 0 & 1 & 6 \end{array}\right]$

The solution set is $\{(-2, -4, 6)\}$.

27. $D = \begin{vmatrix} -2 & 6 \\ 3 & 8 \end{vmatrix} = -16 - 18 = -34$

28. $D = \begin{vmatrix} 5 & -4 \\ 7 & -3 \end{vmatrix} = 5(-3) - 7(-4) = 13$

29. $D = \begin{vmatrix} 2 & 3 & -1 \\ 3 & 4 & -5 \\ 6 & 4 & 2 \end{vmatrix}$

-5(row 1) + (row 2) replace (row 2)
2(row 1) + (row 3) replace (row 3)

$D = \begin{vmatrix} 2 & 3 & -1 \\ -7 & -11 & 0 \\ 10 & 10 & 0 \end{vmatrix}$

Expand about column 3.

$D = -1(-1)^{1+3}\begin{vmatrix} -7 & -11 \\ 10 & 10 \end{vmatrix}$

$D = -[-70 - (-110)] = -40$

30. $D = \begin{vmatrix} 3 & -2 & 4 \\ 1 & 0 & 6 \\ 3 & -3 & 5 \end{vmatrix}$

Expand about row 2.

$D = 1(-1)^{2+1}\begin{vmatrix} -2 & 4 \\ -3 & 5 \end{vmatrix}$

$+ 6(-1)^{2+3}\begin{vmatrix} 3 & -2 \\ 3 & -3 \end{vmatrix}$

$D = -[-10 - (-12)] - 6[-9 - (-6)]$
$D = -2 - 6(-3) = 16$

31. $D = \begin{vmatrix} 5 & 4 & 3 \\ 2 & -7 & 0 \\ 3 & -2 & 0 \end{vmatrix}$

Expand about column 3.

$D = 3(-1)^{1+3}\begin{vmatrix} 2 & -7 \\ 3 & -2 \end{vmatrix}$

$D = 3[-4-(-21)] = 3(17) = 51$

32. $\begin{vmatrix} 5 & -4 & 2 & 1 \\ 3 & 7 & 6 & -2 \\ 2 & 1 & -5 & 0 \\ 3 & -2 & 4 & 0 \end{vmatrix}$

2(row 1) + (row 2) replace (row 2)

$\begin{vmatrix} 5 & -4 & 2 & 1 \\ 13 & -1 & 10 & 0 \\ 2 & 1 & -5 & 0 \\ 3 & -2 & 4 & 0 \end{vmatrix}$

(row 2) + (row 3) replace (row 3)

$-$ 2(row 2) + (row 4) replace (row 4)

$\begin{vmatrix} 5 & -4 & 2 & 1 \\ 13 & -1 & 10 & 0 \\ 15 & 0 & 5 & 0 \\ -23 & 0 & -16 & 0 \end{vmatrix}$

Expand about column 4.

$D = 1(-1)^{1+4}\begin{vmatrix} 13 & -1 & 10 \\ 15 & 0 & 5 \\ -23 & 0 & -16 \end{vmatrix}$

$D = -\begin{vmatrix} 13 & -1 & 10 \\ 15 & 0 & 5 \\ -23 & 0 & -16 \end{vmatrix}$

Expand about column 2.

$D = -(-1)(-1)^{1+2}\begin{vmatrix} 15 & 5 \\ -23 & -16 \end{vmatrix}$

$D = -[-240-(-115)] = 125$

33. $\begin{pmatrix} x^2+y^2 = 17 \\ x-4y = -17 \end{pmatrix}$

$x = 4y - 17$
$x^2 + y^2 = 17$
$(4y-17)^2 + y^2 = 17$
$16y^2 - 136y + 289 + y^2 = 17$
$17y^2 - 136y + 272 = 0$
$17(y^2 - 8y + 16) = 0$
$17(y-4)^2 = 0$
$y - 4 = 0$
$y = 4$
$x = 4(4) - 17$
$x = 16 - 17 = -1$

The solution set is $\{(-1, 4)\}$.

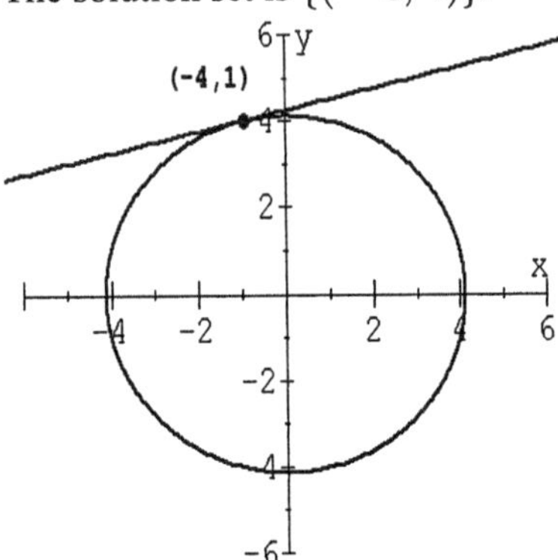

34. $\begin{pmatrix} x^2-y^2 = 8 \\ 3x-y = 8 \end{pmatrix}$

$y = 3x - 8$
$x^2 - (3x-8)^2 = 8$
$x^2 - (9x^2 - 48x + 64) = 8$
$x^2 - 9x^2 + 48x - 64 = 8$
$-8x^2 + 48x - 72 = 0$
$-8(x^2 - 6x + 9) = 0$
$-8(x-3)^2 = 0$
$x - 3 = 0$
$x = 3$
$y = 3(3) - 8$
$y = 9 - 8$
$y = 1$

The solution set is $\{(3, 1)\}$.

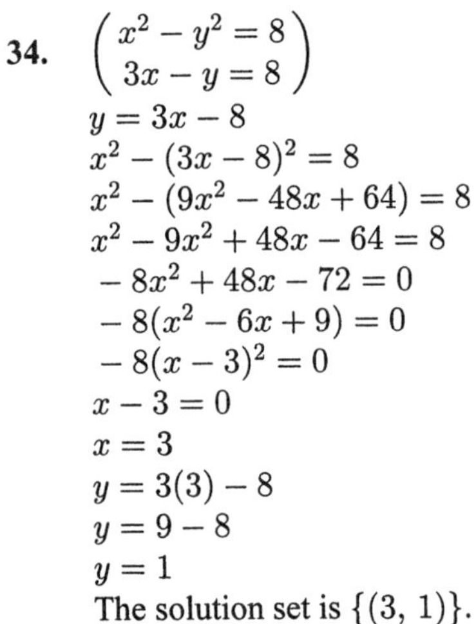
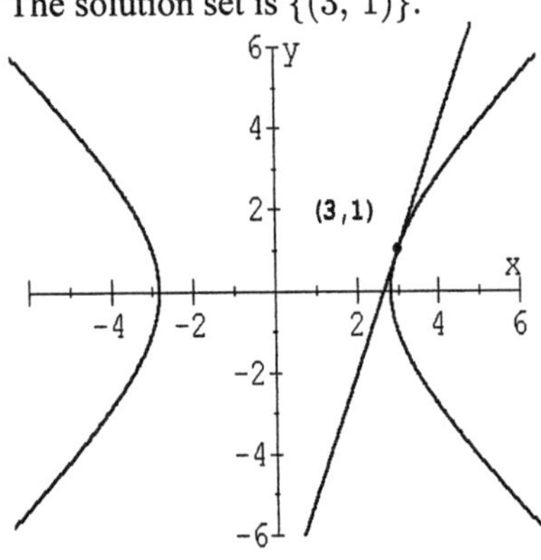

35. $\begin{pmatrix} x-y = 1 \\ y = x^2+4x+1 \end{pmatrix}$

$x - (x^2 + 4x + 1) = 1$
$x - x^2 - 4x - 1 = 1$
$-x^2 - 3x - 2 = 0$
$x^2 + 3x + 2 = 0$
$(x+1)(x+2) = 0$
$x + 1 = 0$ or $x + 2 = 0$
$x = -1$ $\quad x = -2$
When $x = -1$,
$y = x^2 + 4x + 1$
$y = (-1)^2 + 4(-1) + 1$
$y = 1 - 4 + 1$
$y = -2$
When $x = -2$
$y = x^2 + 4x + 1$
$y = (-2)^2 + 4(-2) + 1$
$y = 4 - 8 + 1$
$y = -3$
The solution set is $\{(-1, -2), (-2, -3)\}$.

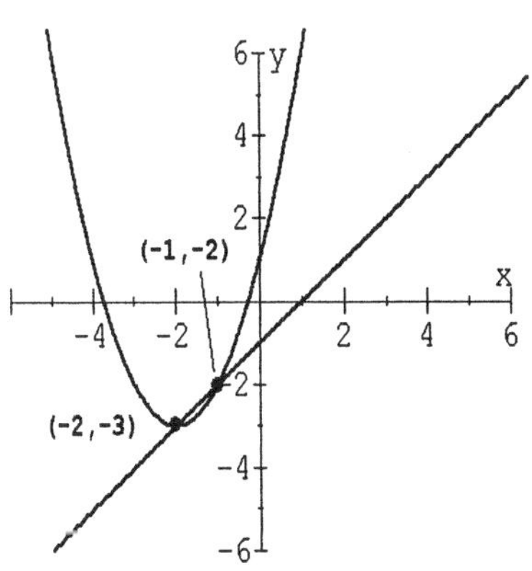

36. $\begin{pmatrix} 4x^2 - y^2 = 16 \\ 9x^2 + 9y^2 = 16 \end{pmatrix}$

Multiply equation 1 by 9 and add to equation 2 to replace equation 2.

$\begin{pmatrix} 4x^2 - y^2 = 16 \\ 45x^2 = 160 \end{pmatrix}$

$45x^2 = 160$

$x^2 = \frac{160}{45} = \frac{32}{9}$

$x = \pm\frac{4\sqrt{2}}{3}$

When $x = \frac{4\sqrt{2}}{3}$, $x^2 = \frac{32}{9}$

$4x^2 - y^2 = 16$
$y^2 = 4x^2 - 16$

$y^2 = 4\left(\frac{32}{9}\right) - 16$

$y^2 = -\frac{16}{9}$

$y = \pm\frac{4i}{3}$

When $x = -\frac{4\sqrt{2}}{3}$, $x^2 = \frac{32}{9}$

Similarly $y = \pm\frac{4i}{3}$

The solution set is

$\left\{\left(\frac{4\sqrt{2}}{3}, \frac{4i}{3}\right), \left(\frac{4\sqrt{2}}{3}, -\frac{4i}{3}\right), \left(-\frac{4\sqrt{2}}{3}, \frac{4i}{3}\right), \left(-\frac{4\sqrt{2}}{3}, -\frac{4i}{3}\right)\right\}$.

There are no real solutions.

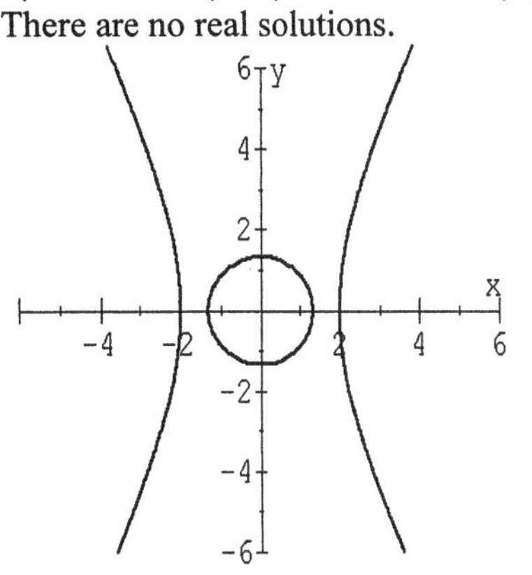

37. $\begin{pmatrix} x^2 + 2y^2 = 8 \\ 2x^2 + 3y^2 = 12 \end{pmatrix}$

$x^2 + 2y^2 = 8$
$x^2 = -2y^2 + 8$

$2x^2 + 3y^2 = 12$
$2(-2y^2 + 8) + 3y^2 = 12$
$-4y^2 + 16 + 3y^2 = 12$
$-y^2 + 16 = 12$
$-y^2 = -4$
$y^2 = 4$
$y = \pm 2$
When $y = -2$
$x^2 = -2y^2 + 8$
$x^2 = -2(-2)^2 + 8$
$x^2 = -2(4) + 8$
$x^2 = -8 + 8$
$x^2 = 0$

$x = 0$
When $y = 2$
$x^2 = -2y^2 + 8$
$x^2 = -2(2)^2 + 8$
$x^2 = -2(4) + 8$
$x^2 = -8 + 8$
$x^2 = 0$
$x = 0$
The solution set is $\{(0, -2), (0, 2)\}$.

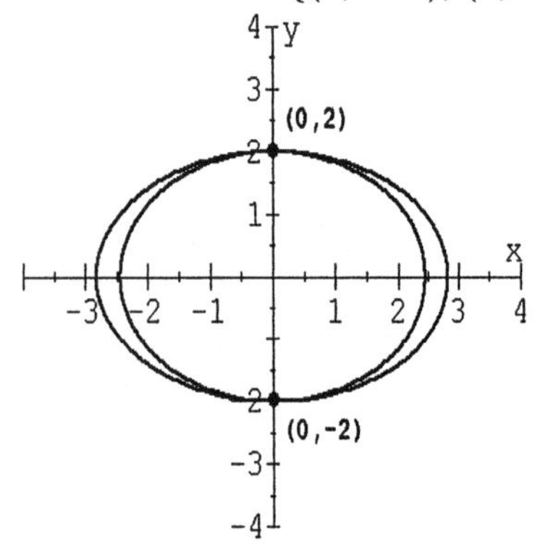

38. $\begin{pmatrix} y^2 - x^2 = 1 \\ 4x^2 + y^2 = 4 \end{pmatrix}$
$y^2 = x^2 + 1$
$4x^2 + x^2 + 1 = 4$
$5x^2 = 3$
$x^2 = \frac{3}{5}$
$x = \pm\sqrt{\frac{3}{5}} = \pm\frac{\sqrt{15}}{5}$
When $x = \frac{\sqrt{15}}{5}$, then $x^2 = \frac{3}{5}$
$y^2 = \frac{3}{5} + 1 = \frac{8}{5}$
$y = \pm\sqrt{\frac{8}{5}} = \pm\frac{2\sqrt{10}}{5}$
When $x = -\frac{\sqrt{15}}{5}$, then $x^2 = \frac{3}{5}$
Same results as above: $y = \pm\frac{2\sqrt{10}}{5}$
The solution set is
$\left\{\left(\frac{\sqrt{15}}{5}, \frac{2\sqrt{10}}{5}\right), \left(\frac{\sqrt{15}}{5}, -\frac{2\sqrt{10}}{5}\right), \left(-\frac{\sqrt{15}}{5}, \frac{2\sqrt{10}}{5}\right), \left(-\frac{\sqrt{15}}{5}, -\frac{2\sqrt{10}}{5}\right)\right\}.$

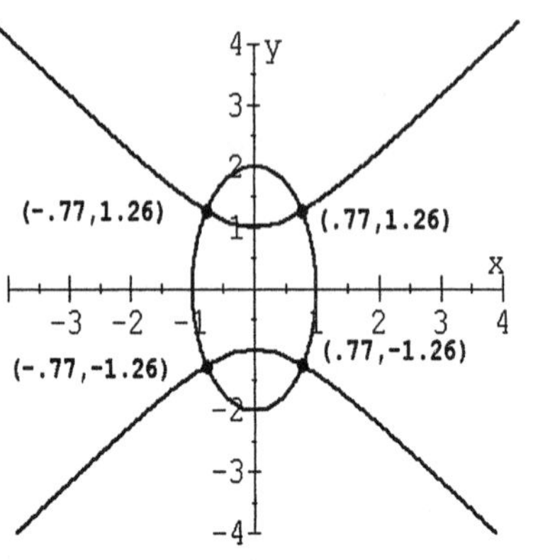

39. Let u = the units digit
and t = the tens digit.
$\begin{pmatrix} u + t = 9 \\ 10u + t = u + 10t - 45 \end{pmatrix}$
$\begin{pmatrix} u + t = 9 \\ 9u - 9t = -45 \end{pmatrix}$
$u + t = 9$
$u = 9 - t$
$9u - 9t = -45$
$9(9 - t) - 9t = -45$
$81 - 9t - 9t = -45$
$81 - 18t = -45$
$-18t = -126$
$t = 7$
$u = 9 - t$
$u = 9 - 7 = 2$
The original number is 72.

40. Let x = money invested at 12%
and y = money invested at 10%.
$\begin{pmatrix} x + y = 2500 \\ 0.12x = 0.10y + 102 \end{pmatrix}$
$\begin{pmatrix} x + y = 2500 \\ 0.12x - 0.10y = 102 \end{pmatrix}$
Multiply equation 1 by 0.10 and add to equation 2 to replace equation 2.
$\begin{pmatrix} x + y = 2500 \\ 0.22x = 352 \end{pmatrix}$
$0.22x = 352$
$x = 1600$
$1600 + y = 2500$
$y = 900$
She invested \$1600 at 12% and \$900 at 10%.

41. Let $n =$ the number of nickels,
$d =$ the number of dimes and
$q =$ the number of quarters.

$$\begin{pmatrix} 0.05n + 0.10d + 0.25q = 17.70 \\ d = 2n - 8 \\ q = n + d + 2 \end{pmatrix}$$

$$\begin{pmatrix} 5n + 10d + 25q = 1770 \\ -2n + d = -8 \\ -n - d + q = 2 \end{pmatrix}$$

Multiply equation 3 by -25 and add to equation 1 to replace equation 1.

$$\begin{pmatrix} 30n + 35d = 1720 \\ -2n + d = -8 \\ -n - d + q = 2 \end{pmatrix}$$

Multiply equation 2 by 15 and add to equation 1 to replace equation 1.

$$\begin{pmatrix} 50d = 1600 \\ -2n + d = -8 \\ -n - d + q = 2 \end{pmatrix}$$

$50d = 1600$
$d = 32$
$-2n + d = -8$
$-2n + 32 = -8$
$-2n = -40$
$n = 20$
$-n - d + q = 2$
$-20 - 32 + q = 2$
$-52 + q = 2$
$q = 54$

There are 20 nickels, 32 dimes, and 54 quarters.

42. Let $a,$ $b,$ and c represent the measures of the angeles respectively.

$$\begin{pmatrix} c = 4a + 10 \\ a + c = 3b \\ a + b + c = 180 \end{pmatrix}$$

$$\begin{pmatrix} -4a + c = 10 \\ a - 3b + c = 0 \\ a + b + c = 180 \end{pmatrix}$$

Multiply equation 3 by 3 and add to equation 2 to replace equation 2.

$$\begin{pmatrix} -4a + c = 10 \\ 4a + 4c = 540 \\ a + b + c = 180 \end{pmatrix}$$

Add equation 1 to equation 2 to replace equation 1.

$$\begin{pmatrix} 5c = 550 \\ 4a + 4c = 540 \\ a + b + c = 180 \end{pmatrix}$$

$5c = 550$
$c = 110$
$4a + 4c = 540$
$4a + 4(110) = 540$
$4a + 440 = 540$
$4a = 100$
$a = 25$
$a + b + c = 180$
$25 + b + 110 = 180$
$b + 135 = 180$
$b = 45$

The measures of the angles are 25°, 45°, and 110°.

CHAPTER 15 | Test

1. III

2. I

3. III

4. II

5. $D = \begin{vmatrix} -2 & 4 \\ -5 & 6 \end{vmatrix} = -12 - (-20) = 8$

6. $D = \begin{vmatrix} \frac{1}{2} & \frac{1}{3} \\ \frac{3}{4} & -\frac{2}{3} \end{vmatrix}$

$D = \frac{1}{2}\left(-\frac{2}{3}\right) - \left(\frac{3}{4}\right)\left(\frac{1}{3}\right)$

$D = -\frac{1}{3} - \frac{1}{4} = -\frac{7}{12}$

7. $D = \begin{vmatrix} -1 & 2 & 1 \\ 3 & 1 & -2 \\ 2 & -1 & 1 \end{vmatrix}$

3(row 1) + (row 2) replace (row 2)
2(row 1) + (row 3) replace (row 3)

$D = \begin{vmatrix} -1 & 2 & 1 \\ 0 & 7 & 1 \\ 0 & 3 & 3 \end{vmatrix}$

Expand about column 1.

$D = -1(-1)^{1+1}\begin{vmatrix} 7 & 1 \\ 3 & 3 \end{vmatrix}$

$D = -(21 - 3) = -18$

8. $D = \begin{vmatrix} 2 & 4 & -5 \\ -4 & 3 & 0 \\ -2 & 6 & 1 \end{vmatrix}$

Expand about column 3.

$D = -5(-1)^{1+3}\begin{vmatrix} -4 & 3 \\ -2 & 6 \end{vmatrix}$

$+1(-1)^{3+3}\begin{vmatrix} 2 & 4 \\ -4 & 3 \end{vmatrix}$

$D = -5[-24 - (-6)] + [6 - (-16)]$
$D = -5(-18) + 22 = 112$

9. The system is dependent, because the equations are multiple. Hence, there are an infinite number of solutions.

10. $\begin{pmatrix} 3x - 2y = -14 \\ 7x + 2y = -6 \end{pmatrix}$

Add equation 1 to equation 2 to replace equation 2.

$\begin{pmatrix} 3x - 2y = -14 \\ 10x = -20 \end{pmatrix}$

$10x = -20$
$x = -2$
$3x - 2y = -14$
$3(-2) - 2y = -14$
$-6 - 2y = -14$
$-2y = -8$
$y = 4$
The solution set is $\{(-2, 4)\}$.

11. $\begin{pmatrix} 4x - 5y = 17 \\ y = -3x + 8 \end{pmatrix}$

Substitute $-3x + 8$ for y in equation 1.

$4x - 5y = 17$
$4x - 5(-3x + 8) = 17$
$4x + 15x - 40 = 17$
$19x = 57$
$x = 3$
$y = -3x + 8$
$y = -3(3) + 8$
$y = -9 + 8 = -1$
The solution set is $\{(3, -1)\}$.

12. $\begin{pmatrix} \frac{3}{4}x - \frac{1}{2}y = -21 \\ \frac{2}{3}x + \frac{1}{6}y = -4 \end{pmatrix}$

Multiply equation 1 by 4 and multiply equation 2 by 6.

$\begin{pmatrix} 3x - 2y = -84 \\ 4x + y = -24 \end{pmatrix}$

Multiply equation 2 by 2 and add to equation 1 to replace equation 1.

$\begin{pmatrix} 11x = -132 \\ 4x + y = -24 \end{pmatrix}$

$11x = -132$
$x = -12$

13. $\begin{pmatrix} 4x - y = 7 \\ 3x + 2y = 2 \end{pmatrix}$

$D = \begin{vmatrix} 4 & -1 \\ 3 & 2 \end{vmatrix} = 8 - (-3) = 11$

$D_y = \begin{vmatrix} 4 & 7 \\ 3 & 2 \end{vmatrix} = 8 - 21 = -13$

$y = \frac{D_y}{D} = -\frac{13}{11}$

14. $\begin{pmatrix} x^2 + y^2 = 16 \\ x^2 - 4y = 8 \end{pmatrix}$

Multiply equation 2 by -1 and add equation 1 to equation 2.

$y^2 + 4y = 8$
$y^2 + 4y - 8 = 0$
This equation is quadratic and has two real solutions.

15. $\left[\begin{array}{ccc|c} 1 & 1 & -4 & 3 \\ 0 & 1 & 4 & 5 \\ 0 & 0 & 3 & 6 \end{array}\right]$

$\frac{1}{3}$(row 3) replace (row 3)

$$\left[\begin{array}{ccc|c} 1 & 1 & -4 & 3 \\ 0 & 1 & 4 & 5 \\ 0 & 0 & 1 & 2 \end{array}\right]$$

-4(row 3) + (row 2) replace (row 2)
4(row 3) + (row 1) replace (row 1)

$$\left[\begin{array}{ccc|c} 1 & 1 & 0 & 11 \\ 0 & 1 & 0 & -3 \\ 0 & 0 & 1 & 2 \end{array}\right]$$

$-$(row 2) + (row 1) replace (row 1)

$$\left[\begin{array}{ccc|c} 1 & 0 & 0 & 14 \\ 0 & 1 & 0 & -3 \\ 0 & 0 & 1 & 2 \end{array}\right]$$

$x = 14$

16. $$\left[\begin{array}{ccc|c} 1 & 2 & -3 & 4 \\ 0 & 1 & 2 & 5 \\ 0 & 0 & 2 & -8 \end{array}\right]$$

$-$(row 3) + (row 2) replace (row 2)

$$\left[\begin{array}{ccc|c} 1 & 2 & -3 & 4 \\ 0 & 1 & 0 & 13 \\ 0 & 0 & 2 & -8 \end{array}\right]$$

$y = 13$

17. $$\begin{pmatrix} x + 3y - z = 5 \\ 2x - y - z = 7 \\ 5x + 8y - 4z = 22 \end{pmatrix}$$

Multiply equation 2 by -4 and add to equation 3 to replace equation 3. Multiply equation 2 by -1 and add to equation 1 to replace equation 1.

$$\begin{pmatrix} -x + 4y = -2 \\ 2x - y - z = 7 \\ -3x + 12y = -6 \end{pmatrix}$$

Multiply equation 1 by -3 and add to equation 3 to replace equation 3.

$$\begin{pmatrix} -x + 4y = -2 \\ 2x - y - z = 7 \\ 0 = 0 \end{pmatrix}$$

Since $0 = 0$ is an identity, the system is dependent. Hence, there are an infinite number of solutions.

18. $$\begin{pmatrix} 3x - y - 2z = 1 \\ 4x + 2y + z = 5 \\ 6x - 2y - 4z = 9 \end{pmatrix}$$

Multiply equation 1 by -2 and add to equation 3 to replace equation 3.

$$\begin{pmatrix} 3x - y - 2z = 1 \\ 4x + 2y + z = 5 \\ 0 = 7 \end{pmatrix}$$

Since $0 \neq 7$, the system is inconsistent. There is no solution.

19. $$\begin{pmatrix} 5x - 3y - 2z = -1 \\ 4y + 7z = 3 \\ 4z = -12 \end{pmatrix}$$

$4z = -12$
$z = -3$
$4y + 7z = 3$
$4y + 7(-3) = 3$
$4y - 21 = 3$
$4y = 24$
$y = 6$
$5x - 3y - 2z = -1$
$5x - 3(6) - 2(-3) = -1$
$5x - 18 + 6 = -1$
$5x - 12 = -1$
$5x = 11$
$x = \frac{11}{5}$

The solution set is $\left\{\left(\frac{11}{5}, 6, -3\right)\right\}$.

20. $$\begin{pmatrix} x - 2y + z = 0 \\ y - 3z = -1 \\ 2y + 5z = -2 \end{pmatrix}$$

Multiply equation 2 by -2 and add to equation 3 to replace equation 3.

$$\begin{pmatrix} x - 2y + z = 0 \\ y - 3z = -1 \\ 11z = 0 \end{pmatrix}$$

$11z = 0$
$z = 0$
$y - 3z = -1$
$y - 3(0) = -1$
$y = -1$
$x - 2y + z = 0$
$x - 2(-1) + 0 = 0$
$x + 2 = 0$
$x = -2$

The solution set is $\{(-2, -1, 0)\}$.

21. $\begin{pmatrix} x - 4y + z = 12 \\ -2x + 3y - z = -11 \\ 5x - 3y + 2z = 17 \end{pmatrix}$

Add equation 1 to equation 2 to replace equation 1.
Multiply equation 2 by 2 and add to equation 3 to replace equation 3.

$\begin{pmatrix} -x - y = 1 \\ -2x + 3y - z = -11 \\ x + 3y = -5 \end{pmatrix}$

Multiply equation 1 by 3 and add to equation 3 to replace equation 3.

$\begin{pmatrix} -x - y = 1 \\ -2x + 3y - z = -11 \\ -2x = -2 \end{pmatrix}$

$-2x = -2$
$x = 1$

22. $\begin{pmatrix} x - 3y + z = -13 \\ 3x + 5y - z = 17 \\ 5x - 2y + 2z = -13 \end{pmatrix}$

Add equation 1 to equation 2 to replace equation 2.
Multiply equation 1 by -2 and add to equation 3 to replace equation 3.

$\begin{pmatrix} x - 3y + z = -13 \\ 4x + 2y = 4 \\ 3x + 4y = 13 \end{pmatrix}$

Multiply equation 2 by -3 and multiply equation 3 by 4, then add the resulting equations to replace equation 3.

$\begin{pmatrix} x - 3y + z = -13 \\ 4x + 2y = 4 \\ 10y = 40 \end{pmatrix}$

$10y = 40$
$y = 4$

23. $\begin{pmatrix} x^2 + 4y^2 = 25 \\ xy = 6 \end{pmatrix}$

Solve $xy = 6$ for x and substitute into equation 1.

$xy = 6$
$x = \frac{6}{y}$
$x^2 + 4y^2 = 25$
$\left(\frac{6}{y}\right)^2 + 4y^2 = 25$
$\frac{36}{y^2} + 4y^2 = 25$
$y^2\left(\frac{36}{y^2} + 4y^2\right) = y^2(25)$
$36 + 4y^4 = 25y^2$
$4y^4 - 25y^2 + 36 = 0$
$(4y^2 - 9)(y^2 - 4) = 0$
$(2y + 3)(2y - 3)(y + 2)(y - 2) = 0$
$2y + 3 = 0$ or $2y - 3 = 0$ or $y + 2 = 0$ or $y - 2 = 0$
$2y = -3$ or $2y = 3$ or $y = -2$ or $y = 2$
$y = -\frac{3}{2}$ or $y = \frac{3}{2}$ or $y = -2$ or $y = 2$

When $y = -\frac{3}{2}$, $x = \frac{6}{y} = \frac{6}{-\frac{3}{2}} = -4$.
When $y = \frac{3}{2}$, $x = \frac{6}{y} = \frac{6}{\frac{3}{2}} = 4$.
When $y = -2$, $x = \frac{6}{-2} = -3$.
When $y = 2$, $x = \frac{6}{2} = 3$.

The solution set is
$\left\{(3, 2), (-3, -2), \left(4, \frac{3}{2}\right), \left(-4, -\frac{3}{2}\right)\right\}$.

24. Let $x =$ the amount of 30% solution and $y =$ amount of 70% solution.

$\begin{pmatrix} x + y = 8 \\ 0.30x + 0.70y = 0.40(8) \end{pmatrix}$

$x = 8 - y$
Substitute $8 - y$ for x in equation 2.
$0.30(8 - y) + 0.70y = 3.2$
$2.40 - 0.30y + 0.70y = 3.2$
$0.40y = 0.8$
$y = 2$
There should be 2 liters of 70% solution.

25. Let $n =$ the number of nickels, $d =$ the number of dimes and $q =$ the number of quarters.

$\begin{pmatrix} 0.05n + 0.10d + 0.25q = 7.25 \\ n + d + q = 43 \\ q = 3n + 1 \end{pmatrix}$

$\begin{pmatrix} 5n + 10d + 25q = 725 \\ n + d + q = 43 \\ -3n + q = 1 \end{pmatrix}$

Multiply equation 2 by -10 and add to equation 1 to replace equation 1.

$$\begin{pmatrix} -5n+15q=295 \\ n+d+q=43 \\ -3n+q=1 \end{pmatrix}$$

Multiply equation 1 by $-\frac{3}{5}$ and add to equation 3 to replace equation 3.

$$\begin{pmatrix} -5n+15q=295 \\ n+d+q=43 \\ -8q=-176 \end{pmatrix}$$

$-8q=-176$
$q=22$
There are 22 quarters.

CHAPTERS 1-15 | Cumulative Review Word Problems

1. Let x represent the cost per hour for labor.
$340=145+3x$
$195=3x$
$65=x$
The cost per hour for labor is \$65.

2. Let $x=$ the first odd integer,
then $x+2=$ the second odd integer,
and $x+4=$ the third odd integer.
Sum of the integers is 57.
$x+x+2+x+4=57$
$3x+6=57$
$3x=51$
$x=17$
The integers are 17, 19 and 21.

3. Let $x=$ the angle, then
$90-x=$ the complement of the angle,
and $180-x=$ the supplement of the angle.
The supplement $=10+5\cdot$ (complement)
$180-x=10+5(90-x)$
$180-x=10+450-5x$
$180-x=460-5x$
$180+4x=460$
$4x=280$
$x=70$
The angle is 70°.

4. Let $x=$ the number of nickels, then
$6+x=$ the number of dimes and
$1+2x=$ the number of quarters
Total number of coins $=63$
$x+6+x+1+2x=63$
$4x+7=63$
$4x=56$
$x=14$
There are 14 nickels, $6+14=20$ dimes and $1+2(14)=29$ quarters.

5. Let $x=$ the smallest angle, then
$10+3x=$ the largest angle, and
$20+x=$ the third angle.
The sum of the measures of the angles of a triangle is 180°.
$x+10+3x+20+x=180$
$5x+30=180$
$5x=150$
$x=30$
The smallest angle is 30°, the largest angle $10+3(30)=100°$, and the third angle is $20+30=50°$.

6. Let $x=$ Kaya's hourly rate, then
$1.5x=$ Kaya's rate for "time and a half"
$40x+6(1.5x)=455.70$
$40x+9x=455.70$
$49x=455.70$
$x=9.30$
Kaya's hourly rate is \$9.30.

7. Let $s=$ the selling price of the DVD player.
$s=c+(\%)(s)$
$s=300+50\%(s)$
$s=300+0.50s$
$0.50s=300$
$s=600$
The selling price is \$600.

8. Let $x=$ amount invested at 8%, then
$300+x=$ amount invested at 9%.
$0.08x+0.09(300+x)=316$
$100[0.08x+0.09(300+x)]=100(316)$

$8x + 9(300 + x) = 31600$
$8x + 2700 + 9x = 31600$
$17x + 2700 = 31600$
$17x = 28900$
$x = 1700$
She invested $1,700 at 8% and $2,000 at 9%.

9. Let $x =$ golf score on 4th day
$\frac{70 + 73 + 76 + x}{4} \leq 72$
$4\left(\frac{70 + 73 + 76 + x}{4}\right) \leq 4(72)$
$70 + 73 + 76 + x \leq 288$
$219 + x \leq 288$
$x \leq 69$
The score is 69 or less.

10. Let $x =$ the number of shares sold, then $x + 10 =$ the number of shares purchased. Use the following formulas.

1) $\left(\text{number of shares sold}\right) \cdot \left(\text{price per share sold}\right) = \left(\text{cost of shares sold}\right)$
$(x) \cdot \left(\text{selling price per share}\right) = 300$
$\text{selling price per share} = \frac{300}{x}$

2) $\left(\text{number of shares purchased}\right) \cdot \left(\text{price per share purchased}\right) = \left(\text{cost of shares purchased}\right)$
$(x + 10) \cdot \left(\text{price per share purchased}\right) = 300$
$\text{price per share purchased} = \frac{300}{x + 10}$

3) $\left(\text{selling price per share}\right) = \left(\text{price per share purchased}\right) + \left(\text{profit per share}\right)$

$\frac{300}{x} = \frac{300}{x + 10} + 5$
$x(x + 10)\left(\frac{300}{x}\right) = x(x + 10)\left[\frac{300}{x + 10} + 5\right]$
$300(x + 10) = 300x + 5x(x + 10)$
$300x + 3000 = 300x + 5x^2 + 50x$
$3000 = 5x^2 + 50x$
$0 = 5x^2 + 50x - 3000$
$0 = x^2 + 10x - 600$
$0 = (x + 30)(x - 20)$
$x + 30 = 0$ or $x - 20 = 0$
$x = -30$ or $x = 20$
Discard $x = -30$.
He originally bought $x + 10 = 20 + 10 = 30$ shares.
The price per share was $\frac{300}{x + 10} = \frac{300}{20 + 10} = \frac{300}{30} = \$10.$

11. Use $P = 2L + 2W$.
$44 = 2L + 2W$
$\frac{1}{2}(44) = \frac{1}{2}(2L + 2W)$
$22 = L + W$
$22 - W = L$

Use $A = LW$.
$112 = (22 - W)W$
$112 = 22W - W^2$
$W^2 - 22W + 112 = 0$
$(W - 8)(W - 14) = 0$
$W - 8 = 0$ or $w - 14 = 0$
$W = 8$ or $W = 14$
If $W = 8$ then $L = 14$ or if W = 14 then $L = 8$.
Dimensions are 8 inches by 14 inches.

12. Let $x =$ the number.
$x^3 = 9x$
$x^3 - 9x = 0$
$x(x^2 - 9) = 0$
$x(x - 3)(x + 3) = 0$
$x = 0$ or $x - 3 = 0$ or $x + 3 = 0$
$x = 0$ or $x = 3$ or $x = -3$
The number is $-3, 0,$ or 3.

13. Let r represent the rate of the south traveling cycle, then $r + 10$ represents the rate of the north traveling cycle. A chart of the information would be as follows:

	Rate	Time	Distance ($d = rt$)
North Cycle	$r + 10$	4.5	$4.5(r + 10)$
South Cycle	r	4.5	$4.5r$

The sum of the distances $= 639$ miles.

$$4.5r + 4.5(r+10) = 639$$
$$4.5r + 4.5r + 45 = 639$$
$$9r = 594$$
$$r = 66$$

The speed of the motorcycles would be 66 miles per hour and 76 miles per hour.

14. Let x represent the amount of pure antifreeze to be added; then x also represents the amount of solution to be drained.

$$\begin{pmatrix}\text{antifreeze in}\\ \text{50\% solution}\end{pmatrix} - \begin{pmatrix}\text{antifreeze to}\\ \text{be drained}\end{pmatrix} + \begin{pmatrix}\text{antifreeze to}\\ \text{be added}\end{pmatrix} = \begin{pmatrix}\text{antifreeze in}\\ \text{final solution}\end{pmatrix}$$
$$(50\%)(10) - (50\%)(x) + (100\%)x = (70\%)(10)$$
$$0.50(10) - 0.50x + x = 0.70(10)$$
$$10[0.50(10) - 0.50x + x] = 10[0.70(10)]$$
$$5(10) - 5x + 10x = 7(10)$$
$$50 + 5x = 70$$
$$5x = 20$$
$$x = 4$$

We must drain 4 quarts of the 50% solution and then replace with 4 quarts of pure antifreeze.

15. Let $t =$ time for investment to double itself with simple interest.
Use $i = prt$.
$750 = 750(0.06)t$
$1 = 0.06t$
$16.67 = t$
The time to double the investment at 6% simple interest is $16\frac{2}{3}$ years.

16. Let $t =$ time for investment to double itself with interest compounded quarterly.
Use $A = P\left(1 + \frac{r}{n}\right)^{4t}$.

$1500 = 750\left(1 + \frac{0.06}{4}\right)^{4t}$
$2 = (1 + 0.015)^{4t}$
$\log 2 = \log(1 + 0.015)^{4t}$
$0.30103 = 4t\log(1.015)$
$0.30103 = 4t(0.00647)$
$0.30103 = t(0.025864)$
$\frac{0.30103}{0.025864} = t$
$11.64 = t$
The investment will double itself in 11.64 years.

17. Let $t =$ time for investment to double itself with interest compounded continuously.
Use $A = Pe^{rt}$.

$1500 = 750e^{0.06t}$
$2 = e^{0.06t}$
$\ln 2 = \ln e^{0.06t}$
$\ln 2 = 0.06t(\ln e)$
$\ln 2 = 0.06t(1)$
$\frac{\ln 2}{0.06} = t$
$11.55 = t$
The investment will double itself in 11.55 years.

18. Let $x =$ the length of the side of the equilateral triangle, then
$x + 2 =$ the length of the side of the square.

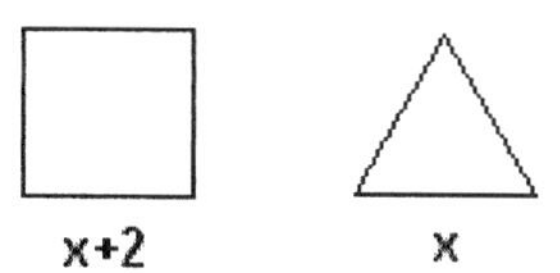

For the square use $P = 4(x + 2)$.
For the triangle use $P = 3x$.

$\begin{pmatrix}\text{Perimeter}\\ \text{of Square}\end{pmatrix} = 2 \cdot \begin{pmatrix}\text{Perimeter}\\ \text{of Triangle}\end{pmatrix} - 4$
$4(x+2) = 2(3x) - 4$
$4x + 8 = 6x - 4$
$-2x = -12$
$x = 6$
The length of the side of the equilateral triangle is 6 centimeters.

19. Let t represent the time Heidi jogs. Then $t - \frac{1}{2}$ represents the time Ed jogs. A chart of the information would be as follows:

	Rate	Time	Distance ($d = rt$)
Heidi	4	t	$4t$
Ed	6	$t - \frac{1}{2}$	$6(t - \frac{1}{2})$

Heidi's distance = Ed's distance
$4t = 6(t - \frac{1}{2})$
$4t = 6t - 3$
$-2t = -3$
$t = \frac{-3}{-2} = \frac{3}{2}$
It will take Ed $\left(t - \frac{1}{2}\right) = \frac{3}{2} - \frac{1}{2} = \frac{2}{2} = 1$ hour to catch Heidi.

20. Let $x =$ the length of the strip to be cut from the sides and ends of the paper.
The reduced dimensions of the paper will be $(8 - 2x)$ by $(14 - 2x)$.
$(8 - 2x)(14 - 2x) = 72$
$2(4 - x)(2)(7 - x) = 72$
$4(4 - x)(7 - x) = 72$
$(4 - x)(7 - x) = 18$
$28 - 11x + x^2 = 18$
$x^2 - 11x - 10 = 0$
$(x - 1)(x - 10) = 0$
$x - 1 = 0$ or $x - 10 = 0$
$x = 1$ or $x = 10$
Discard $x = 10$.
The width of the strip is 1 inch.

21. Let $x =$ the dollar amount for one person, then $2450 - x =$ the dollar amount for the second person.
$\frac{3}{4} = \frac{x}{2450 - x}$
$4x = 3(2450 - x)$
$4x = 7350 - 3x$
$7x = 7350$
$x = 1050$
One person receives $\$1,050$ and the second person receives $\$1,400$.

22. Let r represent the time for the tank to overflow.

	Time	Rate
Sue	t	$\frac{1}{t}$
Dean	2	$\frac{1}{2}$
Together	$1\frac{1}{5} = \frac{6}{5}$	$\frac{5}{6}$

$\frac{1}{t} + \frac{1}{2} = \frac{5}{6}$
$6t\left(\frac{1}{t} + \frac{1}{2}\right) = 6t\left(\frac{5}{6}\right)$
$6 + 3t = 5t$
$6 = 2t$
$3 = t$
It would take Sue 3 hours to complete the task by herself.

23. Let $t =$ tens digit and $u =$ units digit and $10t + u =$ the two-digit number.
$\begin{pmatrix} u = 1 + 2t \\ u + t = 10 \end{pmatrix}$
Substitute $1 + 2t$ for u in the second equation.
$1 + 2t + t = 10$
$1 + 3t = 10$
$3t = 9$
$t = 3$
$u = 1 + 2(3)$
$u = 7$
The two-digit number is $10t + u = 10(3) + 7 = 37$.

24. Let $d =$ the number of days to complete job and $p =$ the number of people to do the job.
$d = \frac{k}{p}$
$8 = \frac{k}{12}$

$96 = k$

$d = \frac{96}{p}$

$d = \frac{96}{20} = \frac{24}{5} = 4\frac{4}{5}$

It would take $4\frac{4}{5}$ days to do the job.

25. Let $c =$ the cost of labor,
$w =$ the number of workers
and $d =$ the number of days of work.
$c = kwd$
$3750 = k(15)(5)$
$3750 = k(75)$
$50 = k$
$c = 50wd$
$c = 50(20)(4)$
$c = 4000$
It will cost \$4, 000.

26. Let r represent the freight train's rate, then $r + 20$ represents the express train's rate.

	Distance	Rate	Time $\left(t = \frac{d}{r}\right)$
Freight	300	r	$\frac{300}{r}$
Express	280	$r + 20$	$\frac{280}{r+20}$

Freight's Time = Express's Time + 2

$$\frac{300}{r} = \frac{280}{r+20} + 2,\ r \neq -20;\ r \neq 0$$
$$r(r+20)\left(\frac{300}{r}\right) = r(r+20)\left(\frac{280}{r+20} + 2\right)$$
$$300(r+20) = 280r + 2r(r+20)$$
$$300r + 6000 = 280r + 2r^2 + 40r$$
$$300r + 6000 = 2r^2 + 320r$$
$$0 = 2r^2 + 20r - 6000$$
$$0 = r^2 + 10r - 3000$$
$$0 = (r+60)(r-50)$$
$$r + 60 = 0 \quad \text{or} \quad r - 50 = 0$$
$$r = -60 \quad \text{or} \quad r = 50$$

The negative solution needs to be discarded. The freight train's rate is 50 miles per hour and the express train's rate is 70 miles per hour.

27. Let $F =$ temperature measured in Fahrenheit. Use $C = \frac{5}{9}\left(F - 32\right)$.
$19 < \frac{5}{9}\left(F - 32\right) < 22$
$9(19) < 9\left[\frac{5}{9}\left(F - 32\right)\right] < 9(22)$
$171 < 5(F - 32) < 198$
$171 < 5F - 160 < 198$
$331 < 5F < 358$
$66.2° < F < 71.6°$
The temperature should be between $66.2°F$ and $71.6°F$.

28. Let $x =$ the number of pork dinners sold and $y =$ the number of rib dinners sold.
Solve the system.
$\begin{pmatrix} x + y = 270 \\ 12x + 15y = 3690 \end{pmatrix}$
Solve for y in the first equation and substitute into the second equation.
$x + y = 270$
$y = 270 - x$
$12x + 15y = 3690$
$12x + 15(270 - x) = 3690$
$12x + 4050 - 15x = 3690$
$-3x = -360$
$x = 120$
There were 120 pork dinners sold and $270 - 120 = 150$ rib dinners sold.

29. Let $x =$ one number and
$y =$ the second number.
Solve the system.
$\begin{pmatrix} x + y = 2 \\ xy = -1 \end{pmatrix}$
Solve for y in the first equation and substitute into the second equation.
$x + y = 2$
$x = 2 - y$
$xy = -1$
$(2 - y)y = -1$
$2y - y^2 = -1$
$-y^2 + 2y + 1 = 0$
$y^2 - 2y - 1 = 0$
Solve using quadratic formula.
$a = 1,\ b = -2,\ c = -1$

$$y=\frac{-(-2)\pm\sqrt{(-2)^2-4(1)(-1)}}{2(1)}$$

$$y=\frac{2\pm\sqrt{4+4}}{2}=\frac{2\pm\sqrt{8}}{2}=$$

$$\frac{2\pm2\sqrt{2}}{2}=\frac{2(1\pm\sqrt{2})}{2}=1\pm\sqrt{2}$$

The numbers are $1+\sqrt{2}$ and $1-\sqrt{2}$.

30. Let x represent Larry's driving rate.

	Rate	Time	Distance
Larry	x	$\frac{156}{x}$	156
Nita	$x+2$	$\frac{108}{x+2}$	108

$$\text{Larry's time}=\text{Nita's time}+1\text{ hour}$$

$$\frac{156}{x}=\frac{108}{x+2}+1$$

$$x(x+2)\left(\frac{156}{x}\right)=x(x+2)\left(\frac{108}{x+2}+1\right)$$

$$156(x+2)=108x+x(x+2)$$

$$156x+312=108x+x^2+2x$$

$$0=x^2-46x-312$$

$$0=(x+6)(x-52)$$

$x+6=0$ or $x-52=0$

$x=-6$ or or $x=52$

Discard the negative value.

Larry's rate is 52 miles per hour and Nita's rate is 54 miles per hour.

31. Let $x=$ the number of seats per row and $x-5=$ the number of rows.

(number of rows)(number of seats/row)=300

$(x-5)x=300$

$x^2-5x=300$

$x^2-5x-300=0$

$(x+15)(x-20)=0$

$x+15=0$ or $x-20=0$

$x=-15$ or $x=20$

Discard $x=-15$.

There are 20 seats per row and 15 rows.

32. Let $r=$ the radius of the circle.

Use $A=\pi r^2$ and $C=2\pi r$.

$$\left(\text{Area of circle}\right)=2\cdot\left(\text{Circumference of circle}\right)$$

$\pi r^2=2(2\pi r)$

$\pi r^2=4\pi r$

$\pi r^2-4\pi r=0$

$\pi r(r-4)=0$

$\pi r=0$ or $r-4=0$

$r=0$ or $r=4$

Discard $r=0$.

The radius is 4 units.

33. Let $x=$ the number of students going on trip.

$$\left(\text{cost of trip}\right)=\left(\text{number of students}\right)\cdot\left(\text{cost per person}\right)$$

$3000=x(\text{cost/person})$

$\frac{3000}{x}=\text{cost/person}$

Use the formula above to generate the following equation.

$$3000=(x+10)\left(\frac{3000}{x}-25\right)$$

$$3000=3000-25x+\frac{30000}{x}-250$$

$$0=-25x+\frac{30000}{x}-250$$

$$x(0)=x\left[-25x+\frac{30000}{x}-250\right]$$

$0=-25x^2+30000-250x$

$0=25x^2+250x-30000$

$0=x^2+10x-1200$

$0=(x+40)(x-30)$

$x+40=0$ or $x-30=0$

$x=-40$ or $x=30$

Discard $x=-40$.

There were 30 students on the trip.

34. Let $x=$ the length of one leg of the right triangle and

$x-2=$ the length of the other leg.

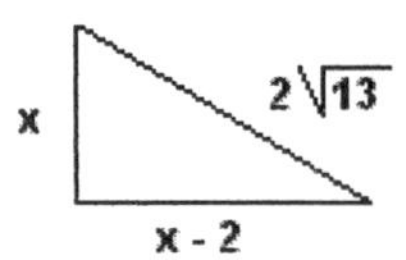

$x^2+(x-2)^2=(2\sqrt{13})^2$

$x^2+x^2-4x+4=4(13)$

$2x^2-4x+4=52$

$2x^2-4x-48=0$

$x^2-2x-24=0$

$(x-6)(x+4)=0$
$x-6=0$ or $x+4=0$
$x=6$ or $x=-4$
Discard $x=-4$.
One leg is 6 yards. The other leg is $x-2=6-2=4$ yards.

35. Let $x=$ the length of each leg.

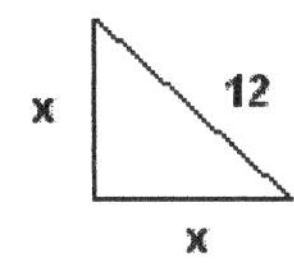

$x^2+x^2=(12)^2$
$2x^2=144$
$x^2=72$
$x=\pm\sqrt{72}$
$x=\pm\sqrt{36}\sqrt{2}$
$x=\pm 6\sqrt{2}$
Discard $-6\sqrt{2}$.
The length of each leg is $6\sqrt{2}$ inches.

36. Let $x=$ the cost of a movie and $y=$ the cost of a game.
$\begin{pmatrix} 3x+2y=20.75 \\ 5x+3y=32.92 \end{pmatrix}$
Multiply the first equation by -3 and multiply the second equation by 2.
Then add the equations.

$$\begin{array}{r} -9x-6y=-62.25 \\ \underline{10x+6y=65.84} \\ x=3.59 \end{array}$$

Substitute value for x into second equation.
$10x+6y=65.84$
$10(3.59)+6y=65.84$
$35.90+6y=65.84$
$6y=29.94$
$y=4.99$
The cost of renting a movie is \$3.59 and the cost of renting a game is \$4.99.

37. Let $x=$ the first whole number,
$x+1=$ the second whole number and
$x+2=$ the third whole number.
$\begin{pmatrix}\text{first}\\\text{whole}\end{pmatrix}+2\cdot\begin{pmatrix}\text{second}\\\text{whole}\end{pmatrix}+3\cdot\begin{pmatrix}\text{third}\\\text{whole}\end{pmatrix}=134$
$x+2(x+1)+3(x+2)=134$
$x+2x+2+3x+6=134$
$6x+8=134$
$6x=126$
$x=21$
The three consecutive whole numbers are 21, 22 and 23.

38. Let $x=$ the number of nickels and
$2x-1=$ the number of dimes.
Value of coins $=\$2.90$.
$0.05x+0.10(2x-1)=2.90$
$100[0.05x+0.10(2x-1)]=100(2.90)$
$5x+10(2x-1)=290$
$5x+20x-10=290$
$25x=300$
$x=12$
There are 12 nickels and $2(12)-1=23$ dimes.

39. Let $x=$ the number.
$14+2x\leq 3x$
$14\leq x$
$x\geq 14$
The solution set is $[14,\infty)$.

40. Let $x=$ the amount of pure acid added.
$100\%(x)+30\%(150)=40\%(x+150)$
$100[1.00x+0.30(150)]=100[0.40(x+150)]$
$100x+30(150)=40(x+150)$
$100x+4500=40x+6000$
$60x=1500$
$x=25$
The amount of pure acid added is 25 milliliters.

Chapter 16 Miscellaneous Topics: Problem Solving

PROBLEM SET 16.1 Arithmetic Sequences

1. $a_n = 3n - 7$
$a_1 = 3(1) - 7 = -4$
$a_2 = 3(2) - 7 = -1$
$a_3 = 3(3) - 7 = 2$
$a_4 = 3(4) - 7 = 5$
$a_5 = 3(5) - 7 = 8$

3. $a_n = -2n + 4$
$a_1 = -2(1) + 4 = 2$
$a_2 = -2(2) + 4 = 0$
$a_3 = -2(3) + 4 = -2$
$a_4 = -2(4) + 4 = -4$
$a_5 = -2(5) + 4 = -6$

5. $a_n = 3n^2 - 1$
$a_1 = 3(1)^2 - 1 = 2$
$a_2 = 3(2)^2 - 1 = 11$
$a_3 = 3(3)^2 - 1 = 26$
$a_4 = 3(4)^2 - 1 = 47$
$a_5 = 3(5)^2 - 1 = 74$

7. $a_n = n(n - 1)$
$a_1 = 1(1 - 1) = 0$
$a_2 = 2(2 - 1) = 2$
$a_3 = 3(3 - 1) = 6$
$a_4 = 4(4 - 1) = 12$
$a_5 = 5(5 - 1) = 20$

9. $a_n = 2^{n+1}$
$a_1 = 2^{1+1} = 4$
$a_2 = 2^{2+1} = 8$
$a_3 = 2^{3+1} = 16$
$a_4 = 2^{4+1} = 32$
$a_5 = 2^{5+1} = 64$

11. $a_n = -5n - 4$
$a_{15} = -5(15) - 4 = -79$
$a_{30} = -5(30) - 4 = -154$

13. $a_n = (-1)^{n+1}$
$a_{25} = (-1)^{25+1} = 1$
$a_{50} = (-1)^{50+1} = -1$

15. $d = 13 - 11 = 2$
$a_n = a_1 + (n - 1)d$
$a_n = 11 + (n - 1)(2)$
$a_n = 11 + 2n - 2$
$a_n = 2n + 9$

17. $d = -1 - 2 = -3$
$a_n = a_1 + (n - 1)d$
$a_n = 2 + (n - 1)(-3)$
$a_n = 2 - 3n + 3$
$a_n = -3n + 5$

19. $d = 2 - \frac{3}{2} = \frac{1}{2}$
$a_n = a_1 + (n - 1)d$
$a_n = \frac{3}{2} + (n - 1)\left(\frac{1}{2}\right)$
$a_n = \frac{3}{2} + \frac{1}{2}n - \frac{1}{2}$
$a_n = \frac{1}{2}n + 1$

21. $d = 6 - 2 = 4$
$a_n = a_1 + (n - 1)d$
$a_n = 2 + (n - 1)4$
$a_n = 2 + 4n - 4$
$a_n = 4n - 2$

23. $d = -6 - (-3) = -3$
$a_n = a_1 + (n - 1)d$
$a_n = -3 + (n - 1)(-3)$
$a_n = -3 - 3n + 3$
$a_n = -3n$

25. $d = 8 - 3 = 5$
$a_n = a_1 + (n - 1)d$
$a_{15} = 3 + (15 - 1)(5)$
$a_{15} = 73$

27. $d = 26 - 15 = 11$
$a_n = a_1 + (n - 1)d$
$a_{30} = 15 + (30 - 1)(11) = 334$

29. $d = \frac{5}{3} - 1 = \frac{2}{3}$
$a_n = a_1(n-1)d$
$a_{52} = 1 + (52-1)\left(\frac{2}{3}\right)$
$a_{52} = 35$

31. $a_n = a_1 + (n-1)d$
$a_6 = 12$
$12 = a_1 + (6-1)d$
$12 = a_1 + 5d$

$a_{10} = 16$
$16 = a_1 + (10-1)d$
$16 = a_1 + 9d$

$$\begin{array}{rl} -12 = & -a_1 - 5d \\ 16 = & a_1 + 9d \\ \hline 4 = & 4d \end{array}$$
$1 = d$
$12 = a_1 + 5d$
$12 = a_1 + 5(1)$
$7 = a_1$

33. $a_n = a_1 + (n-1)d$
$a_3 = 20$
$20 = a_1 + (3-1)d$
$20 = a_1 + 2d$

$a_7 = 32$
$32 = a_1 + (7-1)d$
$32 = a_1 + 6d$

$$\begin{array}{rl} -20 = & -a_1 - 2d \\ 32 = & a_1 + 6d \\ \hline 12 = & 4d \end{array}$$
$3 = d$
$20 = a_1 + 2d$
$20 = a_1 + 2(3)$
$14 = a_1$
$a_{25} = 14 + (25-1)(3)$
$a_{25} = 86$

35. $d = 7 - 5 = 2$
$a_n = a_1 + (n-1)d$
$a_{50} = 5 + (50-1)(2)$
$a_{50} = 103$
$S_n = \frac{n(a_1 + a_n)}{2}$
$S_{50} = \frac{50(5+103)}{2} = 2700$

37. $d = 6 - 2 = 4$
$a_n = a_1 + (n-1)d$
$a_{40} = 2 + (40-1)(4)$
$a_{40} = 158$
$S_n = \frac{n(a_1 + a_n)}{2}$
$S_{40} = \frac{40(2+158)}{2} = 3200$

39. $d = 2 - 5 = -3$
$a_n = a_1 + (n-1)d$
$a_{75} = 5 + (75-1)(-3)$
$a_{75} = -217$
$S_n = \frac{n(a_1 + a_n)}{2}$
$S_{75} = \frac{75(5-217)}{2} = -7950$

41. $d = 1 - \frac{1}{2} = \frac{1}{2}$
$a_n = a_1 + (n-1)d$
$a_{50} = \frac{1}{2} + (50-1)\left(\frac{1}{2}\right)$
$a_{50} = 25$
$S_n = \frac{n(a_1 + a_n)}{2}$
$S_{50} = \frac{50\left(\frac{1}{2} + 25\right)}{2} = 637.5$

43. $d = 5 - 1 = 4$
$a_n = a_1 + (n-1)d$
$197 = 1 + (n-1)(4)$
$197 = 1 + 4n - 4$
$200 = 4n$
$50 = n$
$S_n = \frac{n(a_1 + a_n)}{2}$
$S_{50} = \frac{50(1+197)}{2} = 4950$

45. $d = 8 - 2 = 6$
$a_n = a_1 + (n-1)d$
$146 = 2 + (n-1)6$
$146 = 2 + 6n - 6$
$150 = 6n$

$25 = n$

$S_n = \frac{n(a_{1+} - a_n)}{2}$

$S_{25} = \frac{25(2 + 146)}{2} = 1850$

47. $d = -10 - (-7) = -3$

$a_n = a_1 + (n-1)d$

$-109 = -7 + (n-1)(-3)$

$-109 = -7 - 3n + 3$

$-105 = -3n$

$35 = n$

$S_n = \frac{n(a_1 + a_n)}{2}$

$S_{35} = \frac{35[-7 + (-109)]}{2} = -2030$

49. $d = -3 - (-5) = 2$

$a_n = a_1 + (n-1)d$

$119 = -5 + (n-1)2$

$119 = -5 + 2n - 2$

$126 = 2n$

$63 = n$

$S_n = \frac{n(a_1 + a_n)}{2}$

$S_{63} = \frac{63(-5 + 119)}{2} = 3591$

51. $a_1 = 1,\ d = 2,\ n = 200$

$a_{200} = 1 + (200 - 1)(2)$

$a_{200} = 399$

$S_n = \frac{n(a_1 + a_n)}{2}$

$S_{200} = \frac{200(1 + 399)}{2} = 40,000$

53. $a_1 = 18,\ d = 2,\ a_n = 482$

$a_n = a_1 + (n-1)d$

$482 = 18 + (n-1)2$

$482 = 18 + 2n - 2$

$466 = 2n$

$233 = n$

$S_n = \frac{n(a_1 + a_n)}{2}$

$S_{233} = \frac{233(18 + 482)}{2} = 58,250$

55. $a_n = 5n - 4$

$a_1 = 5(1) - 4 = 1$

$a_{30} = 5(30) - 4 = 146$

$S_n = \frac{n(a_1 + a_n)}{2}$

$S_{30} = \frac{30(1 + 146)}{2} = 2205$

57. $a_n = -4n - 1$

$a_1 = -4(1) - 1 = -5$

$a_{25} = -4(25) - 1 = -101$

$S_n = \frac{n(a_1 + a_n)}{2}$

$S_{25} = \frac{25[-5 + (-101)]}{2} = -1325$

59. $a_n = a_1 + (n-1)d$

$a_1 = 19,500$

$d = 1170$

The salary for 2001 means $n = 22$.

$a_n = 19,500 + (22 - 1)(1170)$

$a_n = 19,500 + 21(1170)$

$a_n = 19,500 + 24,570$

$a_n = \$44,070$

61. $a_n = a_1 + (n-1)d$

$a_1 = 600$

$d = 2150$

$2005 - 2000 = 5,$ means $n = 6$.

$a_n = 600 + (6 - 1)(2150)$

$a_n = 600 + 5(2150)$

$a_n = 600 + 22,500$

$a_n = 11,350$

63. The total number of quarters will be the sum of the series $1 + 2 + 3 + ... + 30$

Use the formula:

$S_n = \frac{n(a_1 + a_n)}{2}$

$n = 30,\ a_1 = 1,\ a_n = 30$

$S_n = \frac{n(a_1 + a_n)}{2} = \frac{30(1 + 30)}{2} =$

$\frac{30(31)}{2} = 15(31) = 465$

At the end of 31 days you would have saved $(465)(0.25) = \$116.25$.

65. $a_n = a_1 + (n-1)d$
$a_1 = 40, \quad d = 4, \quad n = 25$
$a_{25} = 40 + (25-1)(4)$
$a_{25} = 40 + 24(4) = 136$
$S = \dfrac{25(40+136)}{2}$
$S = \dfrac{25(176)}{2}$
$S = 25(88) = 2200$
There are 136 seats in the 25th row and 2200 seats in the auditorium.

67. Interest $= 1500 \times 0.12 \times (1+2+3+...+10)$
$I = 1500 \times 0.12 \times \left[\dfrac{10(1+10)}{2}\right]$
$I = 1500 \times 0.12 \times 55$
$I = 9,900$

Principal $= 1500 \times 10 = 15,000$
Amount $=$ Interest $+$ Principal
$= 9,900 + 15,000 = 24,900$

The value of all investments is $24,900.

PROBLEM SET 16.2 Geometric Sequences

1. $r = \dfrac{6}{3} = 2$
$a_n = a_1 r^{n-1}$
$a_n = 3(2)^{n-1}$

3. $r = \dfrac{9}{3} = 3$
$a_n = a_1 r^{n-1}$
$a_n = 3(3)^{n-1}$
$a_n = 3^n$

5. $r = \dfrac{\frac{1}{8}}{\frac{1}{4}} = \dfrac{1}{2}$
$a_n = a_1 r^{n-1}$
$a_n = \dfrac{1}{4}\left(\dfrac{1}{2}\right)^{n-1}$
$a_n = \left(\dfrac{1}{2}\right)^2\left(\dfrac{1}{2}\right)^{n-1}$
$a_n = \left(\dfrac{1}{2}\right)^{2+n-1}$
$a_n = \left(\dfrac{1}{2}\right)^{n+1}$

7. $r = \dfrac{16}{4} = 4$
$a_n = a_1 r^{n-1}$
$a_n = 4(4)^{n-1}$
$a_n = 4^n$

9. $r = \dfrac{0.3}{1} = 0.3$
$a_n = a_1 r^{n-1}$
$a_n = 1(0.3)^{n-1}$
$a_n = (0.3)^{n-1}$

11. $r = \dfrac{-2}{1} = -2$
$a_n = a_1 r^{n-1}$
$a_n = 1(-2)^{n-1}$
$a_n = (-2)^{n-1}$

13. $r = \dfrac{1}{\frac{1}{2}} = 2$
$a_n = a_1 r^{n-1}$
$a_n = \dfrac{1}{2}(2)^{n-1}$
$a_8 = \dfrac{1}{2}(2)^{8-1} = 2^6 = 64$

15. $r = \dfrac{243}{729} = \dfrac{1}{3}$
$a_n = a_1 r^{n-1}$
$a_9 = 729\left(\dfrac{1}{3}\right)^{9-1}$
$a_9 = \dfrac{1}{9}$

17. $r = \dfrac{-2}{1} = -2$
$a_n = a_1 r^{n-1}$
$a_n = 1(-2)^{n-1}$

Problem Set 16.2

$a_{10} = (-2)^{10-1}$
$a_{10} = (-2)^9$
$a_{10} = -512$

19. $r = \dfrac{\frac{1}{6}}{\frac{1}{2}} = \dfrac{2}{6} = \dfrac{1}{3}$
$a_n = a_1 r^{n-1}$
$a_8 = \dfrac{1}{2}\left(\dfrac{1}{3}\right)^{8-1}$
$a_8 = \dfrac{1}{4374}$

21. $a_5 = \dfrac{32}{3}, \quad r = 2$
$a_n = a_1 r^{n-1}$
$\dfrac{32}{3} = a_1(2)^{5-1}$
$\dfrac{32}{3} = 16a_1$
$\dfrac{2}{3} = a_1$

23. $a_3 = 12$ $\quad a_6 = 96$
$a_3 = a_1 r^{3-1}$ $\quad a_6 = a_1 r^{6-1}$
$12 = a_1 r^2$ $\quad 96 = a_1 r^5$
$a_1 = \dfrac{12}{r^2}$
$96 = \dfrac{12}{r^2}(r^5)$
$96 = 12r^3$
$8 = r^3$
$2 = r$

25. $r = \dfrac{2}{1} = 2$
$S_n = \dfrac{a_1 r^n - a_1}{r-1}$
$S_{10} = \dfrac{1(2)^{10} - 1}{2-1}$
$S_{10} = \dfrac{2^{10} - 1}{1} = 1023$

27. $r = \dfrac{6}{2} = 3$
$S_n = \dfrac{a_1 r^n - a_1}{r-1}$
$S_9 = \dfrac{2(3^9) - 2}{3-1}$
$S_9 = \dfrac{39364}{2} = 19,682$

29. $r = \dfrac{12}{8} = \dfrac{3}{2}$
$S_n = \dfrac{a_1 r^n - a_1}{r-1}$
$S_8 = \dfrac{8\left(\frac{3}{2}\right)^8 - 8}{\frac{3}{2} - 1}$
$S_8 = \dfrac{8\left(\frac{6561}{256}\right) - 8}{\frac{1}{2}} = \dfrac{6305}{16}$
$S_8 = 394\dfrac{1}{16}$

31. $r = \dfrac{8}{-4} = -2$
$S_n = \dfrac{a_1 r^n - a_1}{r-1}$
$S_{10} = \dfrac{-4(-2)^{10} - (-4)}{-2-1}$
$S_{10} = \dfrac{-4092}{-3} = 1364$

33. $r = \dfrac{27}{9} = 3$
$S = 9 + 27 + 81 + ... + 729$
$3S = 27 + 81 + ... + 729 + 2187$
Subtract the 1st equation from the 2nd equation.
$2S = -9 + 2187$
$2S = 2178$
$S = 1089$

35. $r = \dfrac{2}{4} = \dfrac{1}{2}$
$\dfrac{1}{512} = 4\left(\dfrac{1}{2}\right)^{n-1}$
$\dfrac{1}{2048} = \left(\dfrac{1}{2}\right)^{n-1}$

$$\left(\frac{1}{2}\right)^{11} = \left(\frac{1}{2}\right)^{n-1}$$
$$11 = n - 1$$
$$12 = n$$
$$S_{12} = \frac{4\left(\frac{1}{2}\right)^{12} - 4}{\frac{1}{2} - 1}$$
$$S_{12} = \frac{4\left[\left(\frac{1}{2}\right)^{12} - 1\right]}{-\frac{1}{2}}$$
$$S_{12} = -2(4)\left(\frac{1}{4096} - 1\right)$$
$$S_{12} = -8\left(-\frac{4095}{4096}\right) = \frac{32760}{4096}$$
$$S_{12} = 7\frac{511}{512}$$

37. $r = \frac{3}{-1} = -3$
$S = (-1) + 3 + (-9) + ... + (-729)$
$-3S = 3 + (-9) + ... + (-729) + 2187$
Subtract the 1st equation from the 2nd equation.
$-4S = 2187 - (-1)$
$-4S = 2188$
$S = -547$

39. $r = \frac{1}{2}$
$$S_\infty = \frac{a_1}{1-r}$$
$$S_\infty = \frac{2}{1-\frac{1}{2}} = \frac{2}{\frac{1}{2}} = 4$$

41. $r = \frac{\frac{2}{3}}{1} = \frac{2}{3}$
$$S_\infty = \frac{a_1}{1-r}$$
$$S_\infty = \frac{1}{1-\frac{2}{3}} = \frac{1}{\frac{1}{3}} = 3$$

43. $r = \frac{8}{4} = 2$
Because $|r| > 1$, this infinite sequence has no sum.

45. $r = -\frac{3}{9} = -\frac{1}{3}$
$$S_\infty = \frac{a_1}{1-r}$$
$$S_\infty = \frac{9}{1-\left(-\frac{1}{3}\right)} = \frac{9}{\frac{4}{3}}$$
$$S_\infty = 9\left(\frac{3}{4}\right) = \frac{27}{4}$$

47. $r = \frac{\frac{3}{8}}{\frac{1}{2}} = \frac{3}{4}$
$$S_\infty = \frac{a_1}{1-r}$$
$$S_\infty = \frac{\frac{1}{2}}{1-\frac{3}{4}} = \frac{\frac{1}{2}}{\frac{1}{4}} = 2$$

49. $r = -\frac{4}{8} = -\frac{1}{2}$
$$S_\infty = \frac{a_1}{1-r}$$
$$S_\infty = \frac{8}{1-\left(-\frac{1}{2}\right)} = \frac{8}{\frac{3}{2}}$$
$$S_\infty = 8\left(\frac{2}{3}\right) = \frac{16}{3}$$

51. $0.\overline{3} = 0.3 + 0.03 + 0.003 + ...$
$$r = \frac{0.03}{0.3} = 0.1$$
$$S_\infty = \frac{a_1}{1-r}$$
$$S_\infty = \frac{0.3}{1-0.1} = \frac{0.3}{0.9} = \frac{1}{3}$$

53. $0.\overline{26} = 0.26 + 0.0026 + 0.000026 + ...$
$$r = \frac{0.0026}{0.26} = 0.01$$

$$S_\infty = \frac{a_1}{1-r}$$
$$S_\infty = \frac{0.26}{1-0.01} = \frac{0.26}{0.99} = \frac{26}{99}$$

55. $0.\overline{123} = 0.123 + 0.000123 + 0.000000123 + \ldots$
$$r = \frac{0.000123}{0.123} = 0.001$$
$$S_\infty = \frac{a_1}{1-r}$$
$$S_\infty = \frac{0.123}{1-0.001} = \frac{0.123}{0.999} = \frac{123}{999} = \frac{41}{333}$$

57. $0.2\overline{6} = 0.2 + 0.06\overline{6}$
$0.0\overline{6} = 0.06 + 0.006 + 0.0006 + \ldots$
$$r = \frac{0.006}{0.06} = 0.1$$
$$S_\infty = \frac{a_1}{1-r}$$
$$S_\infty = \frac{0.06}{1-0.1} = \frac{0.06}{0.9} = \frac{6}{90} = \frac{1}{15}$$
$$0.2\overline{6} = \frac{2}{10} + \frac{1}{15} = \frac{8}{30} = \frac{4}{15}$$

59. $0.2\overline{14} = 0.2 + 0.0\overline{14}$
$0.0\overline{14} = 0.014 + 0.00014 + 0.0000014 + \ldots$
$$r = \frac{0.00014}{0.014} = \frac{14}{1400} = 0.01$$
$$S_\infty = \frac{a_1}{1-r}$$
$$S_\infty = \frac{0.014}{1-0.01} = \frac{0.014}{0.99} = \frac{14}{990} = \frac{7}{495}$$
$$0.2\overline{14} = 0.2 + 0.0\overline{14} = \frac{2}{10} + \frac{7}{495} =$$
$$\frac{1}{5} + \frac{7}{495} =$$
$$\frac{99}{495} + \frac{7}{495} = \frac{106}{495}$$

61. $2.\overline{3} = 2 + 0.\overline{3}$
$0.\overline{3} = 0.3 + 0.03 + 0.003 + \ldots$
$$r = \frac{0.03}{0.3} = 0.1$$
$$S_\infty = \frac{a_1}{1-r}$$
$$S_\infty = \frac{0.3}{1-0.1} = \frac{0.3}{0.9} = \frac{1}{3}$$
$$2.\overline{3} = 2 + 0.\overline{3} = 2 + \frac{1}{3} = \frac{7}{3}$$

63. Each year's enrollment is 100% + 10% = 110% of the previous year's enrollment. This is a geometric sequence with $a_1 = 5000$ and $r = 110\% = 1.1$. For 1999 let $n = 1$, then for 2003, $n = 5$.
$a_n = a_1 r^{n-1}$
$a_5 = 5000(1.1)^{5-1} = 5000(1.1)^4 = 7320.5$
The predicted enrollment for 2003 is 7320.

65. $a_1 = 16,000, \quad r = \frac{1}{2}, \quad n = 8$
$a_n = a_1 r^{n-1}$
$$a_8 = 16000\left(\frac{1}{2}\right)^{8-1}$$
$a_8 = 125$
After seven days 125 liters remain.

67. If $\frac{1}{3}$ of the water is removed, then $\frac{2}{3}$ of the water remains. This is a geometric sequence witn $a_1 = \frac{2}{3}(5832) = 3888$ and $r = \frac{2}{3}$.
$a_n = a_1 r^{n-1}$
$$a_6 = 3888\left(\frac{2}{3}\right)^{6-1}$$
$$a_6 = 3888\left(\frac{2}{3}\right)^{5}$$
$$a_6 = 3888\left(\frac{32}{243}\right)$$
$a_6 = 512$
At the end of 6 days, 512 gallons remain.

69. The geometric sequence 1, 2, 4, 8, ... represents the savings, in cents, on a day-by-day basis.
$a_n = a_1 r^{n-1}$
$a_{15} = 1(2)^{15-1} = 2^{14}$
$a_{15} = 16384$ cents $= \$163.84$
$$S_n = \frac{a_1 r^n - a_1}{r-1}$$

$S_{15} = \dfrac{1(2)^{15} - 1}{2 - 1} = \dfrac{2^{15} - 1}{1}$

$S_{15} = 32767 = 327.67$

Savings for fifteen days is $327.67.

71. $I = 12,000 \times 0.04 \times (1 + 2 + 3 + ... + 8)$

$I = 12,000 \times 0.04 \times \dfrac{8(8+1)}{2}$

$I = 12,000 \times 0.04 \times 36$

$I = 17,280$

Principal $= 12,000(8) = 96,000$

Amount $=$ Interest $+$ Principal

$= 17,280 + 96,000 = 113,280.$

The total accumulated value of all investments is $113, 280.

73. $a_1 = 1, \; d = 2, \; n = 1,000$

$a_n = a_1 + (n-1)d$

$a_{1000} = 1 + (1000 - 1)(2)$

$a_{1000} = 1999$

$S_n = \dfrac{n(a_1 + a_n)}{2}$

$S_{1000} = \dfrac{1000(1 + 1999)}{2}$

$S_{1000} = 1,000,000$

$1,000,000$ pennies $= \$10,000$

The raffle will take in $10,000.

75. $a_1 = 768, \quad r = \dfrac{1}{2}$

There will be $\dfrac{24}{3} = 8$ time periods, so including the initial term, $n = 9$.

$a_n = a_1 r^{n-1}$

$a_9 = 768\left(\dfrac{1}{2}\right)^{9-1} = \dfrac{768}{256} = 3$

After 24 hours 3 grams remain.

77. For the falling distance,

$a_1 = 100, \quad r = \dfrac{1}{2}, \quad n = 8$

$S_n = \dfrac{a_1 r^n - a_1}{r - 1}$

$S_8 = \dfrac{100\left(\dfrac{1}{2}\right)^8 - 100}{\dfrac{1}{2} - 1}$

$= \dfrac{100\left(\dfrac{1}{256}\right) - 100}{-\dfrac{1}{2}} = \dfrac{6375}{32}$

For the rebound distance,

$a_1 = 50, \; r = \dfrac{1}{2}, \quad n = 7$

$S_n = \dfrac{a_1 r^n - a_1}{r - 1}$

$S_7 = \dfrac{50\left(\dfrac{1}{2}\right)^7 - 50}{\dfrac{1}{2} - 1} = \dfrac{3175}{32}$

Distance traveled $= \dfrac{6375}{32} + \dfrac{3175}{32}$

$= \dfrac{4775}{16} = 298\dfrac{7}{16}$ feet

79. If $\dfrac{1}{3}$ of the air is pumped out on each stroke, then $\dfrac{2}{3}$ of the air remains.

Thus, the geometric sequence $\dfrac{2}{3}, \dfrac{4}{9}, \dfrac{8}{27}, \ldots$ with $r = \dfrac{2}{3}$ represents the air remaining after each stroke.

$a_n = a_1 r^{n-1}$

$a_7 = \dfrac{2}{3}\left(\dfrac{2}{3}\right)^{7-1}$

$a_7 = \dfrac{2}{3}\left(\dfrac{2}{3}\right)^6$

$a_7 = \left(\dfrac{2}{3}\right)^7 = \dfrac{128}{2187}$

$a_7 = 0.059 = 5.9\%$

After seven strokes, 5.9% of air remains.

81. $a_1 = 20, \; r = \dfrac{1}{2}, \; n = 9$

$a_n = a_1 r^{n-1}$

$a_9 = 20\left(\dfrac{1}{2}\right)^{9-1} = \dfrac{5}{64}$

$\dfrac{5}{64}$ of a gallon of water remains.

PROBLEM SET 16.3 Fundamental Principle of Counting

1. $2 \times 10 = 20$

3. $4 \times 3 \times 2 \times 1 = 24$

5. $7 \times 6 \times 4 = 168$

7. $6 \times 2 \times 4 = 48$

9. $2 \times 3 \times 6 = 36$

11. $20 \times 19 \times 18 = 6840$

13. $6 \times 5 \times 4 \times 3 \times 2 \times 1 = 720$

15. $6 \times 5 \times 4 \times 3 \times 2 \times 1 = 720$

17. $\underline{3} \times \underline{3} \times \underline{2} \times \underline{1} \times \underline{2} = 36$

19. $\underline{4} \times \underline{3} \times \underline{1} \times \underline{2} \times \underline{1} = 24$

21. There are three choices of mailboxes for the first letter, three choices of mailboxes for the second letter, etc.
$3 \times 3 \times 3 \times 3 \times 3 = 3^5 = 243$

23. Impossible

25. $6 \times 6 \times 6 = 216$

27. Count the ways of getting 5 or less.

1st die	2nd die
1	1
1	2
1	3
1	4
2	1
2	2
2	3
3	1
3	2
4	1

There are ten ways to get a sum of 5 or less.
There are $6 \times 6 = 36$ possible outcomes.
Therefore there are $36 - 10 = 26$ outcomes of greater than 5.

29. For 3 digit numbers
$2 \times 3 \times 2 = 12$
For 4 digit number
$4 \times 3 \times 2 \times 1 = 24$
$12 + 24 = 36$ numbers

31. $\underline{4} \times \underline{3} \times \underline{3} \times \underline{2} \times \underline{2} \times \underline{1} \times \underline{1} = 144$

33. There are two choices for each of the ten questions.
$2 \times 2 \times 2 \times 2 \times 2 \times 2 \times 2 \times 2 \times 2 \times 2$
$= 2^{10} = 1024$ ways

35. If the first digit is 4
$\underline{1} \times \underline{3} \times \underline{2} \times \underline{1} \times \underline{3} = 18$
If the first digit is 5
$\underline{1} \times \underline{3} \times \underline{2} \times \underline{1} \times \underline{2} = 12$
$18 + 12 = 30$ numbers

37a. $26 \times 26 \times 9 \times 10 \times 10 \times 10 = 6{,}084{,}000$
b. $26 \times 25 \times 9 \times 10 \times 10 \times 10 = 5{,}850{,}000$
c. $26 \times 26 \times 9 \times 9 \times 8 \times 7 = 3{,}066{,}336$
d. $26 \times 25 \times 9 \times 9 \times 8 \times 7 = 2{,}948{,}400$

PROBLEM SET 16.4 Permutations and Combinations

1. $P(5, 3) = 5 \cdot 4 \cdot 3 = 60$

3. $P(6, 4) = 6 \cdot 5 \cdot 4 \cdot 3 = 360$

5. $C(7, 2) = \frac{P(7, 2)}{2!} = \frac{7 \cdot 6}{2 \cdot 1} = 21$

7. $C(10, 5) = \frac{P(10, 5)}{5!} = \frac{10 \cdot 9 \cdot 8 \cdot 7 \cdot 6}{5 \cdot 4 \cdot 3 \cdot 2 \cdot 1} = 252$

9. $C(15, 2) = \frac{P(15, 2)}{2!} = \frac{15 \cdot 14}{2 \cdot 1} = 105$

11. $C(5, 5) = \frac{P(5, 5)}{5!} = \frac{5 \cdot 4 \cdot 3 \cdot 2 \cdot 1}{5 \cdot 4 \cdot 3 \cdot 2 \cdot 1} = 1$

13. $P(4, 4) = 4 \cdot 3 \cdot 2 \cdot 1 = 24$

15. $C(9, 3) = \frac{P(9, 3)}{3!} = \frac{9 \cdot 8 \cdot 7}{3 \cdot 2 \cdot 1} = 84$

17. **a.** $P(8, 3) = 8 \cdot 7 \cdot 6 = 336$
b. $8 \cdot 8 \cdot 8 = 512$

19. $P(4, 4) \times P(5, 5)$
$4 \cdot 3 \cdot 2 \cdot 1 \times 5 \cdot 4 \cdot 3 \cdot 2 \cdot 1 = 2880$

21. $C(7, 4) \times C(8, 4)$

$\frac{P(7, 4)}{4!} \times \frac{P(8, 4)}{4!}$

$\frac{7 \cdot 6 \cdot 5 \cdot 4}{4 \cdot 3 \cdot 2 \cdot 1} \times \frac{8 \cdot 7 \cdot 6 \cdot 5}{4 \cdot 3 \cdot 2 \cdot 1} = 2450$

23. $C(5, 3) = \frac{P(5, 3)}{3!} = \frac{5 \cdot 4 \cdot 3}{3 \cdot 2 \cdot 1} = 10$

25. $C(5, 2) = \frac{P(5, 2)}{2!} = \frac{5 \cdot 4}{2 \cdot 1} = 10$

27. $\frac{7!}{4!3!} = \frac{7 \cdot 6 \cdot 5 \cdot 4 \cdot 3 \cdot 2 \cdot 1}{4 \cdot 3 \cdot 2 \cdot 1 \cdot 3 \cdot 2 \cdot 1} = 35$

29. $\frac{9!}{3!4!2!} = \frac{9 \cdot 8 \cdot 7 \cdot 6 \cdot 5 \cdot 4 \cdot 3 \cdot 2 \cdot 1}{3 \cdot 2 \cdot 1 \cdot 4 \cdot 3 \cdot 2 \cdot 1 \cdot 2 \cdot 1}$
$= 1260$

31. $\frac{7!}{2!1!1!1!1!1!} = \frac{7 \cdot 6 \cdot 5 \cdot 4 \cdot 3 \cdot 2 \cdot 1}{2 \cdot 1} = 2520$

33. $\frac{6!}{4!2!} = \frac{6 \cdot 5 \cdot 4 \cdot 3 \cdot 2 \cdot 1}{4 \cdot 3 \cdot 2 \cdot 1 \cdot 2 \cdot 1} = 15$

35. $C(10, 5) = \frac{P(10, 5)}{5!}$
$= \frac{10 \cdot 9 \cdot 8 \cdot 7 \cdot 6}{5 \cdot 4 \cdot 3 \cdot 2 \cdot 1} = 252$
The order of the two teams is not important so the number of teams is $\frac{252}{2} = 126$.

37. One defective bulb
$C(4, 1) \times C(9, 2)$
$\frac{P(4, 1)}{1!} \times \frac{P(9, 2)}{2!}$
$4 \times \frac{9 \cdot 8}{2 \cdot 1}$
$4 \times 36 = 144$
At least one defective bulb
First find the number of all possible samples and subtract the number of samples that have no defective bulbs.
$C(13, 3) = \frac{P(13, 3)}{3!}$
$= \frac{13 \cdot 12 \cdot 11}{3 \cdot 2 \cdot 1} = 286$
$C(9, 3) = \frac{P(9, 3)}{3!}$
$= \frac{9 \cdot 8 \cdot 7}{3 \cdot 2 \cdot 1} = 84$
$286 - 84 = 202$

39. $C(6, 4) \times C(2, 2)$
$\frac{P(6, 4)}{4!} \times \frac{P(2, 2)}{2!}$

$\frac{6 \cdot 5 \cdot 4 \cdot 3}{4 \cdot 3 \cdot 2 \cdot 1} \times \frac{2 \cdot 1}{2 \cdot 1} = 15$

$C(6,3) \times C(3,3)$

$\frac{P(6,3)}{3!} \times \frac{P(3,3)}{3!}$

$\frac{6 \cdot 5 \cdot 4}{3 \cdot 2 \cdot 1} \times \frac{3 \cdot 2 \cdot 1}{3 \cdot 2 \cdot 1} = 20$

41. Subsets containing A and not B

$C(5,3) = \frac{P(5,3)}{3!} = \frac{5 \cdot 4 \cdot 3}{3 \cdot 2 \cdot 1} = 10$

Subsets containing B and not A

$C(5,3) = \frac{P(5,3)}{3!} = \frac{5 \cdot 4 \cdot 3}{3 \cdot 2 \cdot 1} = 10$

$10 + 10 = 20$

43. 5 points

$C(5,2) = \frac{P(5,2)}{2!} = \frac{5 \cdot 4}{2 \cdot 1} = 10$

6 points

$C(6,2) = \frac{P(6,2)}{2!} = \frac{6 \cdot 5}{2 \cdot 1} = 15$

7 points

$C(7,2) = \frac{P(7,2)}{2!} = \frac{7 \cdot 6}{2 \cdot 1} = 21$

n points

$C(n,2) = \frac{P(n,2)}{2!} = \frac{n(n-1)}{2}$

Further Investigations

47. $\frac{P(6,6)}{6} = \frac{6 \cdot 5 \cdot 4 \cdot 3 \cdot 2 \cdot 1}{6} = 120$

49. $C(n,r) = \frac{P(n,r)}{r!} = \frac{\frac{n!}{(n-r)!}}{r!} =$

$\frac{1}{r!} \cdot \frac{n!}{(n-r)!} = \frac{n!}{r!(n-r)!}$

PROBLEM SET 16.5 Some Basic Probability Ideas

1. $P(E) = \frac{n(E)}{n(S)} = \frac{2}{4} = \frac{1}{2}$

3. $P(E) = \frac{n(E)}{n(S)} = \frac{3}{4}$

5. $P(E) = \frac{n(E)}{n(S)} = \frac{1}{8}$

7. $P(E) = \frac{n(E)}{n(S)} = \frac{7}{8}$

9. $P(E) = \frac{n(E)}{n(S)} = \frac{1}{16}$

11. $P(E) = \frac{n(E)}{n(S)} = \frac{6}{16} = \frac{3}{8}$

13. $P(E) = \frac{n(E)}{n(S)} = \frac{2}{6} = \frac{1}{3}$

15. $P(E) = \frac{n(E)}{n(S)} = \frac{3}{6} = \frac{1}{2}$

17. $P(E) = \frac{n(E)}{n(S)} = \frac{5}{36}$

19. $P(E) = \frac{n(E)}{n(S)} = \frac{6}{36} = \frac{1}{6}$

21. $P(E) = \frac{n(E)}{n(S)} = \frac{11}{36}$

23. $P(E) = \frac{n(E)}{n(S)} = \frac{13}{52} = \frac{1}{4}$

25. $P(E) = \frac{n(E)}{n(S)} = \frac{26}{52} = \frac{1}{2}$

27. $P(E) = \frac{n(E)}{n(S)} = \frac{1}{25}$

29. $P(E) = \frac{n(E)}{n(S)} = \frac{9}{25}$

31. $n(S) = C(5, 2) = 10$
$n(E) = C(4, 1) = 4$
$P(E) = \frac{n(E)}{n(S)} = \frac{4}{10} = \frac{2}{5}$

33. $n(S) = C(5, 2) = 10$
Only one committee with Bill and Carl.
$n(E) = 9$
$P(E) = \frac{n(E)}{n(S)} = \frac{9}{10}$

35. $n(S) = C(8, 5) = 56$
$n(E) = C(6, 3) = 20$
$P(E) = \frac{n(E)}{n(S)} = \frac{20}{56} = \frac{5}{14}$

37. $n(S) = C(8, 5) = 56$
Number of committees
with either Chad or Dominique.
$C(7, 4) + C(7, 4) - C(6, 3)$
$35 + 35 - 20 = 50$
There are 50 committees with Chad and Dominique but now we need to find the number of committees with Chad or Dominique but not both Chad and Dominique.
$n(E) = 50 - C(6, 3) = 50 - 20 = 30$
$P(E) = \frac{n(E)}{n(S)} = \frac{30}{56} = \frac{15}{28}$

39. $n(S) = C(10, 3) = 120$
$n(E) = C(8, 3) = 56$
$P(E) = \frac{n(E)}{n(S)} = \frac{56}{120} = \frac{7}{15}$

41. $n(S) = C(10, 3) = 120$
$n(E) = C(2, 2) \cdot C(8, 1) = 1 \cdot 8 = 8$
$P(E) = \frac{n(E)}{n(S)} = \frac{8}{120} = \frac{1}{15}$

43. $n(S) = P(3, 3) = 6$
$n(E) = 4$
$P(E) = \frac{n(E)}{n(S)} = \frac{4}{6} = \frac{2}{3}$

45. $n(S) = C(5, 4) = 5$
$n(E) = C(4, 4) = 1$
$P(E) = \frac{n(E)}{n(S)} = \frac{1}{5}$

47. $n(S) = P(9, 9) = 362{,}880$
There are two ways, either math books first or history books first.
$n(E) = 2 \times P(4, 4) \times P(5, 5)$
$n(E) = 2 \times 24 \times 120 = 5760$
$P(E) = \frac{n(E)}{n(S)} = \frac{5760}{362{,}880} = \frac{1}{63}$

49. $n(S) = P(4, 4) = 24$
$n(E) = 2 \times P(3, 3)$
$n(E) = 2 \times 6 = 12$
$P(E) = \frac{n(E)}{n(S)} = \frac{12}{24} = \frac{1}{2}$

51. $n(S) = C(11, 4) = 330$
$n(E) = C(6, 2) \times C(5, 2)$
$n(E) = 15 \times 10 = 150$
$P(E) = \frac{n(E)}{n(S)} = \frac{150}{330} = \frac{5}{11}$

53. $n(S) = C(10, 5) = 252$
There are two teams that Ahmed, Bob, and Carl could be on together.
$n(E) = 2 \times C(7, 2)$
$n(E) = 2 \times 21 = 42$
$P(E) = \frac{n(E)}{n(S)} = \frac{42}{252} = \frac{1}{6}$

55. $n(S) = 2^9 = 512$
$n(E) = C(9, 3) = 84$
$P(E) = \frac{n(E)}{n(S)} = \frac{84}{512} = \frac{21}{128}$

57. $n(S) = 2^5 = 32$
Getting 0 heads
$n(E_0) = C(5, 0) = 1$

Getting 1 head
$n(E_1) = C(5, 1) = 5$

Getting 2 heads
$n(E_2) = C(5, 2) = 10$

Getting 3 heads
$n(E_3) = C(5, 3) = 10$

For not getting more than 3 heads
$n(E) = 1 + 5 + 10 + 10 = 26$

$$P(E) = \frac{n(E)}{n(S)} = \frac{26}{32} = \frac{13}{16}$$

59. $n(S) = \dfrac{7!}{2!3!1!1!} = 420$

$n(E) = \dfrac{5!}{3!1!1!} = 20$

$P(E) = \dfrac{20}{420} = \dfrac{1}{21}$

Further Investigations

63. For each suit there are ten cards that could begin the straight flush and there are 4 suits.

$n(E) = 4 \times 10 = 40$

65. $n(E) = 13 \times C(4, 3) \times 12 \times C(4, 2)$
$= 3744$

67. There are ten cards for each of the four suits that could begin the straight flush. So there are 40 cards that could begin the straight flush.

For the second through the fifth cards, the card can be any suit, so there are 4 possibilities for each. This includes the 40 sequences that are straight flushes (of the same suit), so 40 needs to be subtracted.
$n(E) = 40 \times 4 \times 4 \times 4 \times 4 - 40$
$= 10{,}200$

69. $n(E) = \dfrac{13 \times C(4, 2) \times 12 \times C(4, 2)}{2!} \times 44$
$= 123{,}552$

71. P(no pairs)
In poker this means no pairs
and also no straights and no flushes.

From the previous problem
n(one pair) = 1,098,240
n(two pair) = 123,552
n(3 of a kind) = 54,912
n(full house) = 3,744
n(4 of a kind) = 624
n(straight flushes) = 40
n(flushes) = 5108
n(straight) = 10,200

total 1,296,420

n(poker hands) = 2,598,960
n(no pairs) = 2,598,960 − 1,296,420
= 1,302,540

PROBLEM SET 16.6 Binomial Expansions Revisited

1. $(x + y)^8 = x^8 + \binom{8}{1}x^7y + \binom{8}{2}x^6y^2 + \binom{8}{3}x^5y^3 + \binom{8}{4}x^4y^4 + \binom{8}{5}x^3y^5 + \binom{8}{6}x^2y^6 + \binom{8}{7}xy^7 + y^8$

$= x^8 + 8x^7y + 28x^6y^2 + 56x^5y^3 + 70x^4y^4 + 56x^3y^5 + 28x^2y^6 + 8xy^7 + y^8$

3. $(x - y)^6 = x^6 + \binom{6}{1}(x^5)(-y) + \binom{6}{2}(x^4)(-y)^2 + \binom{6}{3}(x^3)(-y)^3 + \binom{6}{4}(x^2)(-y)^4 + \binom{6}{5}(x)(-y)^5 + (-y)^6$

$= x^6 - 6x^5y + 15x^4y^2 - 20x^3y^3 + 15x^2y^4 - 6xy^5 + y^6$

5. $$(a+2b)^4 = a^4 + \binom{4}{1}a^3(2b) + \binom{4}{2}a^2(2b)^2 + \binom{4}{3}a(2b)^3 + (2b)^4$$
$$= a^4 + 8a^3b + 24a^2b^2 + 32ab^3 + 16b^4$$

7. $$(x-3y)^5 = x^5 + \binom{5}{1}x^4(-3y) + \binom{5}{2}x^3(-3y)^2 + \binom{5}{3}x^2(-3y)^3 + \binom{5}{4}x(-3y)^4 + (-3y)^5$$
$$= x^5 - 15x^4y + 90x^3y - 270x^2y^3 + 405xy^4 - 243y^5$$

9. $$(2a-3b)^4 = (2a)^4 + \binom{4}{1}(2a)^3(-3b) + \binom{4}{2}(2a)^2(-3b)^2 + \binom{4}{3}(2a)(-3b)^3 + (-3b)^4$$
$$= 16a^4 - 96a^3b + 216a^2b^2 - 216ab^3 + 81b^4$$

11. $$(x^2+y)^5 = (x^2)^5 + \binom{5}{1}(x^2)^4y + \binom{5}{2}(x^2)^3y^2 + \binom{5}{3}(x^2)^2y^3 + \binom{5}{4}(x^2)y^4 + y^5$$
$$= x^{10} + 5x^8y + 10x^6y^2 + 10x^4y^3 + 5x^2y^4 + y^5$$

13. $$(2x^2-y^2)^4 = (2x^2)^4 + \binom{4}{1}(2x^2)^3(-y^2) + \binom{4}{2}(2x^2)^2(-y^2)^2 + \binom{4}{3}(2x^2)(-y^2)^3 + (-y^2)^4$$
$$= 16x^8 - 32x^6y^2 + 24x^4y^4 - 8x^2y^6 + y^8$$

15. $$(x+3)^6 = x^6 + \binom{6}{1}x^5(3) + \binom{6}{2}x^4(3)^2 + \binom{6}{3}x^3(3)^3 + \binom{6}{4}x^2(3)^4 + \binom{6}{5}x(3)^5 + (3)^6$$
$$= x^6 + 18x^5 + 135x^4 + 540x^3 + 1215x^2 + 1458x + 729$$

17. $$(x-1)^9 = x^9 + \binom{9}{1}x^8(-1) + \binom{9}{2}x^7(-1)^2 + \binom{9}{3}x^6(-1)^3 + \binom{9}{4}x^5(-1)^4 +$$
$$\binom{9}{5}x^4(-1)^5 + \binom{9}{6}x^3(-1)^6 + \binom{9}{7}x^2(-1)^7 + \binom{9}{8}x(-1)^8 + (-1)^9$$
$$= x^9 - 9x^8 + 36x^7 - 84x^6 + 126x^5 - 126x^4 + 84x^3 - 36x^2 + 9x - 1$$

19. $$\left(1+\frac{1}{n}\right)^4 = 1 + \binom{4}{1}\left(\frac{1}{n}\right) + \binom{4}{2}\left(\frac{1}{n}\right)^2 + \binom{4}{3}\left(\frac{1}{n}\right)^3 + \left(\frac{1}{n}\right)^4$$
$$= 1 + \frac{4}{n} + \frac{6}{n^2} + \frac{4}{n^3} + \frac{1}{n^4}$$

Problem Set 16.6

21. $\left(a-\frac{1}{n}\right)^6 = a^6 + \binom{6}{1}a^5\left(-\frac{1}{n}\right) + \binom{6}{2}a^4\left(-\frac{1}{n}\right)^2 + \binom{6}{3}a^3\left(-\frac{1}{n}\right)^3 +$

$\binom{6}{4}a^2\left(-\frac{1}{n}\right)^4 + \binom{6}{5}a\left(-\frac{1}{n}\right)^5 + \left(-\frac{1}{n}\right)^6$

$= a^6 - \frac{6a^5}{n} + \frac{15a^4}{n^2} - \frac{20a^3}{n^3} + \frac{15a^2}{n^4} - \frac{6a}{n^5} + \frac{1}{n^6}$

23. $\left(1+\sqrt{2}\right)^4 = 1 + \binom{4}{1}\sqrt{2} + \binom{4}{2}\left(\sqrt{2}\right)^2 + \binom{4}{3}\left(\sqrt{2}\right)^3 + \left(\sqrt{2}\right)^4$

$= 1 + 4\sqrt{2} + 12 + 8\sqrt{2} + 4$

$= 17 + 12\sqrt{2}$

25. $\left(3-\sqrt{2}\right)^5 = 3^5 + \binom{5}{1}(3)^4\left(-\sqrt{2}\right) + \binom{5}{2}(3)^3\left(-\sqrt{2}\right)^2 +$

$\binom{5}{3}(3)^2\left(-\sqrt{2}\right)^3 + \binom{5}{4}(3)\left(-\sqrt{2}\right)^4 + \left(-\sqrt{2}\right)^5$

$= 243 - 405\sqrt{2} + 540 - 180\sqrt{2} + 60 - 4\sqrt{2}$

$= 843 - 589\sqrt{2}$

27. $(x+y)^{12}$

First four terms

$x^{12} + \binom{12}{1}x^{11}y + \binom{12}{2}x^{10}y^2 + \binom{12}{3}x^9y^3$

$x^{12} + 12x^{11}y + 66x^{10}y^2 + 220x^9y^3$

29. $(x-y)^{20}$

First four terms

$x^{20} + \binom{20}{1}x^{19}(-y) + \binom{20}{2}x^{18}(-y)^2 + \binom{20}{3}x^{17}(-y)^3$

$x^{20} - 20x^{19}y + 190x^{18}y^2 - 1140x^{17}y^3$

31. $(x^2 - 2y^3)^{14}$

First four terms

$(x^2)^{14} + \binom{14}{1}(x^2)^{13}(-2y^3) + \binom{14}{2}(x^2)^{12}(-2y^3)^2 + \binom{14}{3}(x^2)^{11}(-2y^3)^3$

$x^{28} - 28x^{26}y^3 + 364x^{24}y^6 - 2912x^{22}y^9$

33. $\left(a+\frac{1}{n}\right)^9$

First four terms

$a^9+\binom{9}{1}a^8\left(\frac{1}{n}\right)+\binom{9}{2}a^7\left(\frac{1}{n}\right)^2+\binom{9}{3}a^6\left(\frac{1}{n}\right)^3$

$a^9+\frac{9a^8}{n}+\frac{36a^7}{n^2}+\frac{84a^6}{n^3}$

35. $(-x+2y)^{10}$

First four terms

$(-x)^{10}+\binom{10}{1}(-x)^9(2y)+\binom{10}{2}(-x)^8(2y)^2+\binom{10}{3}(-x)^7(2y)^3$

$x^{10}-20x^9y+180x^8y^2-960x^7y^3$

37. $\binom{8}{3}x^5y^3=56x^5y^3$

39. $\binom{9}{4}x^5(-y)^4=126x^5y^4$

41. $\binom{7}{5}(3a)^2b^5=189a^2b^5$

43. $\binom{10}{7}(x^2)^3(y^3)^7=120x^6y^{21}$

45. $\binom{15}{6}(1)^9\left(-\frac{1}{n}\right)^6=\frac{5005}{n^6}$

Further Investigations

51. $(2+i)^6$

$=2^6+\binom{6}{1}(2)^5(i)+\binom{6}{2}(2)^4(i)^2+\binom{6}{3}(2)^3(i)^3+\binom{6}{4}(2)^2(i)^4+\binom{6}{5}(2)(i)^5+i^6$

$=64+192i-240-160i+60+12i-1$

$=-117+44i$

53. $(3-2i)^5$

$=3^5+\binom{5}{1}(3)^4(-2i)+\binom{5}{2}(3)^3(-2i)^2+\binom{5}{3}(3)^2(-2i)^3+\binom{5}{4}(3)(-2i)^4+(-2i)^5$

$=243-810i-1080+720i+240-32i$

$=-597-122i$

CHAPTER 16 Review Problem Set

1. $d=9-3=6$

$a_n=a_1+(n-1)d$

$a_n=3+(n-1)(6)$

$a_n=3+6n-6$

$a_n=6n-3$

2. $r=\frac{1}{\frac{1}{3}}=3$

$a_n=a_1r^{n-1}$

$a_n=\frac{1}{3}(3)^{n-1}$

$a_n = (3^{-1})(3^{n-1})$
$a_n = 3^{n-2}$

3. $r = \dfrac{20}{10} = 2$
$a_n = a_1 r^{n-1}$
$a_n = 10(2)^{n-1}$
$a_n = (5\cdot 2)(2)^{n-1}$
$a_n = 5(2)^n$

4. $d = 2 - 5 = -3$
$a_n = a_1 + (n-1)d$
$a_n = 5 + (n-1)(-3)$
$a_n = 5 - 3n + 3$
$a_n = -3n + 8$

5. $d = -3 - (-5) = 2$
$a_n = a_1 + (n-1)d$
$a_n = -5 + (n-1)(2)$
$a_n = -5 + 2n - 2$
$a_n = 2n - 7$

6. $r = \dfrac{3}{9} = \dfrac{1}{3}$
$a_n = a_1 r^{n-1}$
$a_n = 9\left(\dfrac{1}{3}\right)^{n-1}$
$a_n = 3^2(3^{-1})^{n-1}$
$a_n = 3^2(3)^{-n+1}$
$a_n = 3^{3-n}$

7. $r = \dfrac{2}{-1} = -2$
$a_n = a_1 r^{n-1}$
$a_n = -1(-2)^{n-1}$

8. $d = 15 - 12 = 3$
$a_n = a_1 + (n-1)d$
$a_n = 12 + (n-1)(3)$
$a_n = 12 + 3n - 3$
$a_n = 3n + 9$

9. $d = 1 - \dfrac{2}{3} = \dfrac{1}{3}$
$a_n = a_1 + (n-1)d$
$a_n = \dfrac{2}{3} + (n-1)\left(\dfrac{1}{3}\right)$
$a_n = \dfrac{2}{3} + \dfrac{1}{3}n - \dfrac{1}{3}$
$a_n = \dfrac{1}{3}n + \dfrac{1}{3}$

10. $r = \dfrac{4}{1} = 4$
$a_n = a_1 r^{n-1}$
$a_n = 1(4)^{n-1}$
$a_n = 4^{n-1}$

11. $d = 5 - 1 = 4$
$a_n = a_1 + (n-1)d$
$a_{19} = 1 + (19-1)(4)$
$a_{19} = 73$

12. $d = 2 - (-2) = 4$
$a_n = a_1 + (n-1)d$
$a_{28} = -2 + (28-1)(4)$
$a_{28} = 106$

13. $r = \dfrac{4}{8} = \dfrac{1}{2}$
$a_n = a_1 r^{n-1}$
$a_9 = 8\left(\dfrac{1}{2}\right)^{9-1}$
$a_9 = \dfrac{8}{256} = \dfrac{1}{32}$

14. $r = \dfrac{\frac{81}{16}}{\frac{243}{32}} = \left(\dfrac{81}{16}\right)\left(\dfrac{32}{243}\right) = \dfrac{2}{3}$
$a_n = a_1 r^{n-1}$
$a_8 = \dfrac{243}{32}\left(\dfrac{2}{3}\right)^{8-1}$
$a_8 = \left(\dfrac{3^5}{2^5}\right)\left(\dfrac{2^7}{3^7}\right) = \dfrac{2^2}{3^2} = \dfrac{4}{9}$

15. $d = 4 - 7 = -3$
$a_n = a_1 + (n-1)d$
$a_{34} = 7 + (34-1)(-3)$
$a_{34} = -92$

16. $r = \dfrac{16}{-32} = -\dfrac{1}{2}$

$a_n = a_1 r^{n-1}$

$a_{10} = -32\left(-\dfrac{1}{2}\right)^{10-1}$

$a_{10} = -32\left(-\dfrac{1}{512}\right) = \dfrac{1}{16}$

17. $a_n = a_1 + (n-1)d$

$a_5 = -19 \qquad a_8 = -34$

$-19 = a_1 + 4d \qquad -34 = a_1 + 7d$

$19 = -a_1 - 4d$

$-34 = a_1 + 7d$

$-15 = 3d$

$-5 = d$

18. $a_8 = 37 \qquad a_{13} = 57$

$37 = a_1 + 7d \qquad 57 = a_1 + 12d$

$-37 = -a_1 - 7d$

$57 = a_1 + 12d$

$20 = 5d$

$4 = d$

$37 = a_1 + 7(4)$

$37 = a_1 + 28$

$9 = a_1$

$a_{20} = 9 + (20-1)(4)$

$a_{20} = 85$

19. $a_3 = 5 \qquad a_6 = 135$

$a_n = a_1 r^{n-1}$

$5 = a_1 r^2 \qquad 135 = a_1 r^5$

$a_1 = \dfrac{5}{r^2}$

$135 = \left(\dfrac{5}{r^2}\right) r^5$

$135 = 5r^3$

$27 = r^3$

$3 = r$

$a_1 = \dfrac{5}{(3)^2} = \dfrac{5}{9}$

20. $a_2 = \dfrac{1}{2} \qquad a_6 = 8$

$\dfrac{1}{2} = a_1 r \qquad 8 = a_1 r^5$

$a_1 = \dfrac{1}{2r}$

$8 = \left(\dfrac{1}{2r}\right) r^5$

$16 = r^4$

$r = \pm 2$

$r = 2$ or $r = -2$

21. $r = \dfrac{27}{81} = \dfrac{1}{3} \qquad a_1 = 81$

$S_n = \dfrac{a_1 r^n - a_1}{r - 1}$

$S_9 = \dfrac{81\left(\dfrac{1}{3}\right)^9 - 81}{\dfrac{1}{3} - 1}$

$S_9 = \dfrac{\dfrac{1}{243} - 81}{-\dfrac{2}{3}}$

$S_9 = -\dfrac{3}{2}\left(\dfrac{1}{243} - 81\right)$

$S_9 = -\dfrac{3}{2}\left(\dfrac{-19682}{243}\right) = 121\dfrac{40}{81}$

22. $d = 0 - (-3) = 3$

$a_{70} = -3 + (70-1)(3)$

$a_{70} = 204$

$S_n = \dfrac{n(a_1 + a_n)}{2}$

$S_{70} = \dfrac{70(-3 + 204)}{2} = 7035$

23. $d = 1 - 5 = -4$

$a_{75} = 5 + (75-1)(-4) = -291$

$S_n = \dfrac{n(a_1 + a_n)}{2}$

$S_n = \dfrac{75(5 - 291)}{2} = -10,725$

24. $a_1 = 2^{5-1} = 2^4 = 16$

$a_2 = 2^{5-2} = 2^3 = 8$

$r = \frac{8}{16} = \frac{1}{2}$

$S_n = \frac{a_1 r^n - a_1}{r - 1}$

$S_{10} = \frac{16\left(\frac{1}{2}\right)^{10} - 16}{\frac{1}{2} - 1} = 31\frac{31}{32}$

25. $a_1 = 7(1) + 1 = 8$

$a_{95} = 7(95) + 1 = 666$

$S_{95} = \frac{95(8 + 666)}{2} = 32,015$

26. $d = 7 - 5 = 2$

$137 = 5 + (n - 1)(2)$

$137 = 5 + 2n - 2$

$134 = 2n$

$67 = n$

$S_n = \frac{n(a_1 + a_n)}{2}$

$S_{67} = \frac{67(5 + 137)}{2} = 4757$

27. $r = \frac{16}{64} = \frac{1}{4}$

$a_n = a_1 r^{n-1}$

$\frac{1}{64} = 64\left(\frac{1}{4}\right)^{n-1}$

$\frac{1}{4^3} = 4^3\left(\frac{1}{4}\right)^{n-1}$

$\frac{1}{4^6} = \left(\frac{1}{4}\right)^{n-1}$

$\left(\frac{1}{4}\right)^6 = \left(\frac{1}{4}\right)^{n-1}$

$6 = n - 1$

$7 = n$

$S_n = \frac{a_1 r^n - a_1}{r - 1}$

$S_7 = \frac{64\left(\frac{1}{4}\right)^7 - 64}{\frac{1}{4} - 1}$

$S_7 = \frac{\frac{1}{256} - 64}{-\frac{3}{4}}$

$S_7 = -\frac{4}{3}\left(\frac{1}{256} - 64\right) = 85\frac{21}{64}$

28. $d = 2$

$384 = 8 + (n - 1)(2)$

$384 = 8 + 2n - 2$

$378 = 2n$

$189 = n$

$S_{189} = \frac{189(8 + 384)}{2} = 37,044$

29. $a_1 = 27, \quad d = 3$

$a_n = a_1 + (n - 1)d$

$276 = 27 + (n - 1)(3)$

$276 = 27 + 3n - 3$

$252 = 3n$

$84 = n$

$S_n = \frac{n(a_1 + a_n)}{2}$

$S_{84} = \frac{84(27 + 276)}{2} = 12,726$

30. $r = \frac{16}{64} = \frac{1}{4}$

$S_\infty = \frac{a_1}{1 - r}$

$S_\infty = \frac{64}{1 - \frac{1}{4}} = 85\frac{1}{3}$

31. $0.\overline{36} = 0.36 + 0.0036 + \ldots$

$r = \frac{0.0036}{0.36} = 0.01$

$S_\infty = \frac{a_1}{1 - r}$

$S_\infty = \frac{0.36}{1 - 0.01} = \frac{0.36}{0.99} = \frac{4}{11}$

32. $0.4\overline{5} = 0.4 + (0.05 + 0.005 + \ldots)$

$r = \frac{0.005}{0.05} = 0.1$

$S_\infty = \frac{a_1}{1 - r}$

$$S_{\infty} = \frac{0.05}{1-0.1} = \frac{0.05}{0.90} = \frac{1}{18}$$
$$0.4\overline{5} = \frac{4}{10} + \frac{1}{18} = \frac{41}{90}$$

33. $a_1 = 3750 \quad d = -250 \quad n = 13$
$a_{13} = 3750 + (13-1)(-250)$
$a_{13} = 750$

The savings account will contain \$750.

34. $a_1 = 0.10 \quad d = 0.10 \quad n = 30$

$a_{30} = 0.10 + (30-1)(0.10) = 3.00$

$$S_{30} = \frac{30(0.10+3.00)}{2} = 46.50$$
In April she will save \$46.50.

35. $r = 2,\ a_1 = 0.10,\ n = 15$

$$S_n = \frac{a_1 r^n - a_1}{r-1}$$
$$S_{15} = \frac{0.10(2)^{15} - 0.10}{2-1} = 3,276.70$$

Nancy will save \$3, 276.70.

36. $a_1 = 61,440, \qquad r = \frac{3}{4}$ remaining

$$a_7 = 61,440\left(\frac{3}{4}\right)^{7-1}$$
$a_7 = 10,935$
After six days 10, 935 gallons remain.

37. $P(6, 6) = 6! = 720$

38. $\dfrac{9!}{3!2!1!1!1!1!} = 30,240$

39. $5 \times 6 \times 5 = 150$

40. Sitting Arlene next to Carlos
There are 6 possible seats for Arlene.

The number of orders are
$6 \times 1 \times 1 \times P(5,5) = 720$

Similarily, seating Arlene
next to Carlos gives
$6 \times 1 \times 1 \times P(5,5) = 720.$

So there are $720 + 720 = 1440$
ways to seat Arlene and Carlos
next to each other.

41. $C(6, 3) = \dfrac{P(6, 3)}{3!} = 20$

42. $C(7, 3) \times C(6,2)$
$35 \times 15 = 525$

43. $C(13, 5) = \dfrac{P(13, 5)}{5!} = 1287$

44. Three digit numbers
$2 \times 4 \times 3 = 24$
Four digit numbers
$5 \times 4 \times 3 \times 2 = 120$
Five digit numbers
$5 \times 4 \times 3 \times 2 \times 1 = 120$

$24 + 120 + 120 = 264$

45. Committees with no man
$C(5, 3) = 10$
All possible 3-person committee
$C(9, 3) = 84$
Committee with at least one man

$84 - 10 = 74$

46. If the 1st person is on the
committee and the 2nd person is
not on the committee
$C(6, 3) = 20$
If the 2nd person is on the
committee and the 1st person is
not on the committee
$C(6, 3) = 20$
If neither is on the committee
$C(6, 4) = 15$

$20 + 20 + 15 = 55$

47. Subset with A but not B
$C(6, 3) = 20$
Subset with B but not A
$C(6, 3) = 20$
Subset with A or B but not both
$20 + 20 = 40$

48. $\frac{6!}{4!2!} = 15$

49. $C(3, 2) \times C(4, 1) \times C(5, 1)$
$3 \times 4 \times 5 = 60$

50. $8 \times C(6, 2)$
$8 \times 15 = 120$

51. $n(S) = 2^3 = 8$
$n(E) = C(3, 2) = 3$
$P(E) = \frac{n(E)}{n(S)} = \frac{3}{8}$

52. $n(S) = 2^5 = 32$
$n(E) = C(5, 3) = 10$
$P(E) = \frac{10}{32} = \frac{5}{16}$

53. $n(S) = 6 \times 6 = 36$
$n(E) = 5$
$P(E) = \frac{n(E)}{n(S)} = \frac{5}{36}$

54. $n(S) = 6 \times 6 = 36$
$n(E') = 10$
$n(E) = 36 - 10 = 26$
$P(E) = \frac{26}{36} = \frac{13}{18}$

55. $n(S) = P(5, 5) = 120$
If two people are side by side, there are three seats to fill and then double the number of ways for the order of the two side by side and times four for the different seat positions for the side by side people.
$P(3, 3) \times 2 \times 4 = 6(2)(4) = 48$
Therefore, not side by side is $120 - 48 = 72$
$n(E) = 72$
$P(E) = \frac{n(E)}{n(S)} = \frac{72}{120} = \frac{3}{5}$

56. $n(S) = P(7, 7) = 5040$
$n(E) = 4 \times 3 \times 3 \times 2 \times 2 \times 1 \times 1 = 144$
$P(E) = \frac{n(E)}{n(S)} = \frac{144}{5040} = \frac{1}{35}$

57. $n(S) = 2^6 = 64$
n(no heads) = 1
n(one head) = $C(6, 1) = 6$
n(one or less heads) = $1 + 6 = 7$
n(at least two heads) = $64 - 7 = 57$
$P(E) = \frac{n(E)}{n(S)} = \frac{57}{64}$

58. $P(J,J) = \frac{4}{52} \cdot \frac{3}{51} = \frac{1}{221}$

59. $n(S) = \frac{6!}{3!1!1!1!} = 120$
$n(E) = \frac{5!}{3!1!1!} = 20$
$P(E) = \frac{n(E)}{n(S)} = \frac{20}{120} = \frac{1}{6}$

60. $n(S) = C(7, 3) = 35$
$n(E) = C(6, 3) = 20$
$P(E) = \frac{20}{35} = \frac{4}{7}$

61. $n(S) = C(8, 4) = 70$
Committees with Alice but not Bob
$C(6, 3) = 20$
Committees with Bob but not Alice
$C(6, 3) = 20$
Committees with Alice or Bob but not both
$n(E) = 20 + 20 = 40$
$P(E) = \frac{40}{70} = \frac{4}{7}$

62. $n(S) = C(9, 3) = 84$
$n(E) = C(4, 1) \times C(5, 2) = 40$
$P(E) = \frac{40}{84} = \frac{10}{21}$

63. $n(S) = C(13, 4) = 715$
n(no women) = $C(6, 4) = 15$
n(at least one woman) = $715 - 15 = 700$
$P(E) = \frac{n(E)}{n(S)} = \frac{700}{715} = \frac{140}{143}$

64. $$(x+2y)^5 = x^5 + \binom{5}{1}x^4(2y) + \binom{5}{2}x^3(2y)^2 + \binom{5}{3}x^2(2y)^3 + \binom{5}{4}x(2y)^4 + (2y)^5$$
$$= x^5 + 10x^4y + 40x^3y^2 + 80x^2y^3 + 80xy^4 + 32y^5$$

65. $$(x-y)^8 = x^8 + \binom{8}{1}(x^7)(-y) + \binom{8}{2}(x^6)(-y)^2 + \binom{8}{3}(x^5)(-y)^3 +$$
$$\binom{8}{4}(x^4)(-y)^4 + \binom{8}{5}(x^3)(-y)^5 + \binom{8}{6}(x^2)(-y)^6 + \binom{8}{7}(x)(-y)^7 + (-y)^8$$
$$= x^8 - 8x^7y + 28x^6y^2 - 56x^5y^3 + 70x^4y^4 - 56x^3y^5 + 28x^2y^6 - 8xy^7 + y^8$$

66. $$(a^2-3b^3)^4 = (a^2)^4 + \binom{4}{1}(a^2)^3(-3b^3) + \binom{4}{2}(a^2)^2(-3b^3)^2 + \binom{4}{3}(a^2)(-3b^3)^3 + (-3b^3)^4$$
$$= a^8 - 12a^6b^3 + 54a^4b^6 - 108a^2b^9 + 81b^{12}$$

67. $$\left(x+\frac{1}{n}\right)^6 = x^6 + \binom{6}{1}(x^5)\left(\frac{1}{n}\right) + \binom{6}{2}(x^4)\left(\frac{1}{n}\right)^2 + \binom{6}{3}(x^3)\left(\frac{1}{n}\right)^3$$
$$+ \binom{6}{4}(x^2)\left(\frac{1}{n}\right)^4 + \binom{6}{5}(x)\left(\frac{1}{n}\right)^5 + \left(\frac{1}{n}\right)^6$$
$$= x^6 + \frac{6x^5}{n} + \frac{15x^4}{n^2} + \frac{20x^3}{n^3} + \frac{15x^2}{n^4} + \frac{6x}{n^5} + \frac{1}{n^6}$$

68. $$\left(1-\sqrt{2}\right)^5 = 1^5 + \binom{5}{1}(1)^4\left(-\sqrt{2}\right) + \binom{5}{2}(1)^3\left(-\sqrt{2}\right)^2 + \binom{5}{3}(1)^2\left(-\sqrt{2}\right)^3$$
$$+ \binom{5}{4}(1)\left(-\sqrt{2}\right)^4 + \left(-\sqrt{2}\right)^5$$
$$= 1 - 5\sqrt{2} + 20 - 20\sqrt{2} + 20 - 4\sqrt{2} = 41 - 29\sqrt{2}$$

69. $$(-a+b)^3 = (-a)^3 + \binom{3}{1}(-a)^2b + \binom{3}{2}(-a)b^2 + b^3$$
$$= -a^3 + 3a^2b - 3ab^2 + b^3$$

70. $(x-2y)^{12}$

Fourth term $$= \binom{12}{3}(x)^9(-2y)^3$$
$$= 220(x^9)(-8y^3)$$
$$= -1760x^9y^3$$

71. $$\binom{13}{9}(3a)^4(b^2)^9 = 715(81a^4)(b^{18})$$
$$= 57915a^4b^{18}$$

Chapter 16 Test

CHAPTER 16 | Test

1. $r = \dfrac{\frac{5}{2}}{5} = \dfrac{1}{2}$
$a_n = a_1 r^{n-1}$
$a_n = 5\left(\dfrac{1}{2}\right)^{n-1}$
$a_n = 5(2^{-1})^{n-1}$
$a_n = 5(2)^{1-n}$

2. $d = 16 - 10 = 6$
$a_n = a_1 + (n-1)d$
$a_n = 10 + (n-1)(6)$
$a_n = 6n + 4$

3. $d = 4 - 1 = 3$
$a_n = a_1 + (n-1)d$
$a_{75} = 1 + (75-1)(3)$
$a_{75} = 223$

4. $r = 2 \qquad a_1 = 3$
$S_8 = \dfrac{3(2)^8 - 3}{2-1} = 765$

5. $a_1 = 7(1) - 2 = 5$
$a_{45} = 7(45) - 2 = 313$
$S_{45} = \dfrac{45(5 + 313)}{2} = 7155$

6. $a_1 = 3(2)^1 = 6 \qquad r = 2$
$S_{10} = \dfrac{6(2)^{10} - 6}{2-1} = 6138$

7. $a_1 = 2 \qquad d = 2 \qquad n = 150$
$a_{150} = 2 + (150-1)(2) = 300$
$S_{150} = \dfrac{150(2 + 300)}{2} = 22,650$

8. $r = \dfrac{\frac{3}{2}}{3} = \dfrac{1}{2}$
$S_\infty = \dfrac{3}{1 - \frac{1}{2}} = \dfrac{3}{\frac{1}{2}} = 6$

9. $0.\overline{18} = 0.18 + 0.0018 + \ldots$
$r = \dfrac{0.0018}{0.18} = 0.01$
$S_\infty = \dfrac{0.18}{1 - 0.01} = \dfrac{0.18}{0.99} = \dfrac{2}{11}$

10. $a_1 = 49,152 \qquad r = \dfrac{1}{4}$
$a_8 = 49,152\left(\dfrac{1}{4}\right)^{8-1} = 3$
After seven days, 3 liters remain.

11. $a_1 = 0.10, \quad r = 2, \quad n = 15$
$S_{15} = \dfrac{0.10(2)^{15} - 0.10}{2-1} = 3,276.60$
Therefore, the total saved was $\$3,276.60$.

12. $8, 12, 16, \ldots$
$a_n = a_1 + (n-1)d$
$d = 12 - 8 = 4$
$n = 10, \ a_1 = 8$
$a_{10} = 8 + (10-1)4$
$a_{10} = 8 + (9)4 = 8 + 36 = 44$
$S_n = \dfrac{n(a_1 + a_n)}{2}$
$S_{10} = \dfrac{10(8 + 44)}{2} = \dfrac{10(52)}{2}$
$= 5(52) = 260$
There are 44 members in the last row and 260 members in the band.

13. $6 \times C(10, 2)$
6×45
270

14. If it contains A, then we need to choose 3 elements from the remaining 6 elements.
$C(6, 3) = 20$

15. $C(4,2) \cdot C(4,2) \cdot C(4,1)$
$= 6 \cdot 6 \cdot 4 = 144$

16. There are 4 of the letter S
There are 3 of the letter A
There is 1 letter F
There is 1 letter R
$\dfrac{9!}{4!3!1!1!} = 2520$

17. $C(7,4) \cdot C(5,3)$
$= 35 \cdot 10 = 350$

18. E = sum less than 9
E′ = sum 9 or more
There is 1 way to get a sum of 12
There are 2 ways to get a sum of 11
There are 3 ways to get a sum of 10
There are 4 ways to get a sum of 9
$n(E') = 10$
$P(E') = \dfrac{10}{36} = \dfrac{5}{18}$
$P(E) = 1 - P(E') = 1 - \dfrac{5}{18} = \dfrac{13}{18}$

19. $C(6,3) = 20$
$P(\text{three heads}) = 20\left(\dfrac{1}{2}\right)^6 = \dfrac{5}{16}$

20. $n(s) = P(6, 3) = 120$
E = numbers over 200
E′ = numbers less than 200
$n(E') = 1 \times P(5, 2) = 20$
$P(E') = \dfrac{20}{120} = \dfrac{1}{6}$
$P(E) = 1 - P(E') = 1 - \dfrac{1}{6} = \dfrac{5}{6}$

21. $C(7,4) = 35$ everyone
$C(5, 4) = 5$ excludes Anwar and Bob
$P = \dfrac{5}{35} = \dfrac{1}{7}$

22. $n(S) = C(8, 3) = 56$
E = at least one man
E′ = no men (all women)
$n(E') = C(5, 3) = 10$
$P(E') = \dfrac{10}{56} = \dfrac{5}{28}$
$P(E) = 1 - P(E') = 1 - \dfrac{5}{28} = \dfrac{23}{28}$

23. $\left(2 - \dfrac{1}{n}\right)^6$

$$= 2^6 + \binom{6}{1}(2)^5\left(-\frac{1}{n}\right) + \binom{6}{2}(2)^4\left(-\frac{1}{n}\right)^2 + \binom{6}{3}(2)^3\left(-\frac{1}{n}\right)^3 + \binom{6}{4}(2)^2\left(-\frac{1}{n}\right)^4$$
$$+\binom{6}{5}(2)\left(-\frac{1}{n}\right)^5 + \left(-\frac{1}{n}\right)^6$$
$$= 64 - \frac{192}{n} + \frac{240}{n^2} - \frac{160}{n^3} + \frac{60}{n^4} - \frac{12}{n^5} + \frac{1}{n^6}$$

24. $$(3x + 2y)^5 = (3x)^5 + \binom{5}{1}(3x)^4(2y) + \binom{5}{2}(3x)^3(2y)^2 + \binom{5}{3}(3x)^2(2y)^3 + \binom{5}{4}(3x)(2y)^4 + (2y)^5$$
$$= 243x^5 + 810x^4y + 1080x^3y^2 + 720x^2y^3 + 240xy^4 + 32y^5$$

25. $\left(x - \dfrac{1}{2}\right)^{12}$

$$\text{Ninth term} = \binom{12}{8}(x)^4\left(-\frac{1}{2}\right)^8 = 495(x^4)\left(\frac{1}{256}\right) = \frac{495x^4}{256}$$